AF823070

THE COLLECTED PAPERS OF
WEI-LIANG CHOW

World Scientific Series in 20th Century Mathematics

Published

Vol. 1 The Neumann Compendium
edited by F. Bródy and T. Vámos

Vol. 2 40 Years in Mathematical Physics
by L. D. Faddeev

Vol. 3 Selected Papers of Yu I Manin
by Y. Manin

Vol. 4 A Mathematician and His Mathematical Work
— Selected Papers of S. S. Chern
edited by S. Y. Cheng, P. Li and G. Tian

Vol. 5 Fields Medallists' Lectures
edited by Sir Michael Atiyah and Daniel Iagolnitzer

Vol. 6 Selected Logic Papers
by G. E. Sacks

Vol. 7 Selected Papers of F. W. J. Olver (in 2 Volumes)
edited by Roderick Wong

Vol. 8 The Collected Papers of Wei-Liang Chow
edited by S S Chern and V V Shokurov

World Scientific Series in 20th Century Mathematics Vol. 8

THE COLLECTED PAPERS OF WEI-LIANG CHOW

Edited by

S. S. Chern

Nankai Institute of Mathematics, P. R. China

V. V. Shokurov

Johns Hopkins University, USA

World Scientific
New Jersey • London • Singapore • Hong Kong

Published by

World Scientific Publishing Co. Pte. Ltd.

P O Box 128, Farrer Road, Singapore 912805

USA office: Suite 1B, 1060 Main Street, River Edge, NJ 07661

UK office: 57 Shelton Street, Covent Garden, London WC2H 9HE

British Library Cataloguing-in-Publication Data
A catalogue record for this book is available from the British Library.

THE COLLECTED PAPERS OF WEI-LIANG CHOW

Copyright © 2002 by World Scientific Publishing Co. Pte. Ltd.

All rights reserved. This book, or parts thereof, may not be reproduced in any form or by any means, electronic or mechanical, including photocopying, recording or any information storage and retrieval system now known or to be invented, without written permission from the Publisher.

For photocopying of material in this volume, please pay a copying fee through the Copyright Clearance Center, Inc., 222 Rosewood Drive, Danvers, MA 01923, USA. In this case permission to photocopy is not required from the publisher.

ISBN 981-238-094-9

Printed in Singapore by Mainland Press

WEI-LIANG CHOW
Courtesy of Margot Chow

PREFACE

This volume contains the collection of Chow's papers. It also includes:

- Chow's bibliography of all his publications in mathematics, except for, his unpublished manuscript [Ch 55c] that is posted at `http://mathnt.mat.jhu.edu/mathnew/Chow/`;
- Chow's biography in dates;
- Chern's remarks on Chow, previously published in the Notices of the AMS;
- Lang's comments on Chow's work, a version previously published in Contemporary Mathematics of the AMS; and
- an English translation by W. Huang of [1] (with wan der Waerden), one of the most influential works by Chow.

The main part of this volume is presented by 34 published papers of Chow in mathematics. After some of them we commented.

Acknowledgments: For partial support in preparation of this volume: the Johns Hopkins University, Baltimore, the USA, and the Nankai Institute of Mathematics, Tianjin, People's Republic of China. Picture courtesy of Mrs. Chow. For typesetting and putting all materials in a right form thanks to Carol Young and Mike Smith.

S. S. Chern and V. V. Shokurov*

*The second editor is supported in part by NSF grants DMS 9800807 and DMS 0100991.

Biography

Wei-Liang Chow was born in Shanghai, China, on October 1, 1911. His undergraduate and early graduate work were completed in the United States: at Asbury College in Wilmore, Kentucky, at the University of Kentucky (Lexington), and finally at the University of Chicago, where he received his B.S. degree in 1931 and M.S. degree in 1932. Doctoral studies followed: In 1932, he entered the University of Goettingen in Goettingen, Germany. Then, from 1933 to 1935, he divided his time between the Universities of Leipzig (1933–34) and Hamburg (1934–35), receiving his Ph.D degree in Leipzig in 1936.

Professor Chow's long and distinguished work as a professional mathematician began in China — with professorships at the National Central University in Nanking (1936–37) and the National Tung-Chi University in Shanghai (1946–47) — but ultimately led him to the United States, where he joined the Mathematics faculty of the Johns Hopkins University in Baltimore, Maryland, first as an Associate Professor from 1948 to 1950, then as a full Professor from 1950 until his retirement in 1977.

The late 1940s and early 1950s were a particularly active period for Chow and his pioneering scholarship. He was a Visiting Member of the Institute for Advanced Study in Princeton, NJ (1947–48, then again in 1954–55); an invited speaker at the Conference in Algebraic Geometry and Algebraic Number Theory at the University of Chicago (1949); an official delegate of the Johns Hopkins University to the International Congress of Mathematicians hosted by Harvard University in Cambridge, Massachusetts, and an invited speaker at the Harvard Conference on Algebraic Geometry which immediately followed the Congress (1950); an invited member of the AMS Summer Institute on Lie Groups and Lie Algebra held at Colby College in Waterville, Maine (1953); an invited speaker at Princeton's Conference on Algebraic Geometry in honor of Professor Solomon Lefschetz (1954); and an invited member of the AMS Summer Institute on Algebraic Geometry and Complex Varieties at the University of Colorado in Boulder (1954).

The middle and later portions of Chow's career were similarly productive. In addition to serving as Chairman of the Math Department at Hopkins from 1955 to 1965, he simultaneously enjoyed a long and fruitful relationship with the *American Journal of Mathematics* — a publication whose direction he would help to shape, as Editor-in-Chief, from 1953 to 1977. Lectures at mathematics colloquia of leading universities in the northeastern United States were also a common occurrence during this period, as were a variety of other scientific activities, including the 1964 AMS Summer Institute on Algebraic Geometry and Complex Varieties of the National Academy of Sciences, in Woods Hole, MA, on Cape Cod, which Chow helped to organize.

Professor Chow retired from Hopkins and was named Professor Emeritus in 1977. He did, however, remain active in mathematical circles, serving as a Visiting Professor at Hopkins during the Winter of 1977, as a Visiting Professor at Purdue University in 1978, and as a part-time Visiting Professor at Hopkins from 1979 to 1987. He died in August of 1995 at the age of 83.

CONTENTS

On associated forms and algebraic systems of algebraic manifolds

By

Wei-Liang Chow and B. L. van der Waerden in Leipzig

It is principally important to represent geometrical objects by coordinates. Once this has been done for a specific kind of objects G, then it makes sense to speak of an algebraic manifold or an algebraic system of objects G, and to apply the whole theory of algebraic manifolds (decomposition into irreducible components, notion of dimension, notion of general elements in an irreducible manifold). It is desirable to provide the set of objects G with the structure of an algebraic variety (eventually, after a certain compactification), thus to characterise G by algebraic equations in the coordinates.

The *points* of the projective space S_n are described by $n+1$ homogeneous coordinates, the *subspaces* S_r are described by their Plücker coordinates, and the *hypersurfaces* of degree g are described by the coefficients of their equations. In all these cases the condition described above is satisfied for these objects.

We will obtain a method to describe the *r-dimensional manifolds* M of a fixed degree g in S_n by coordinates: We state the condition when the $r+1$ hyperplanes $u^{(0)}$, $u^{(1)}$, …, $u^{(r)}$ have a common point in M. This condition is a single equation $F(u) = 0$ of degree g in every variable $u^{(0)}$, …, $u^{(r)}$, which is the product of as many factors as M has irreducible components. $F(u)$ is called the *associated form* of the manifold. The cases are included in which some of these irreducible components of M and thus also some factors of the form $F(u)$ are multiply counted. The coefficients of the associated form $F(u)$ are now taken as coordinates of M.

It is easy to see that the manifold M is determined uniquely by the associated form, i.e. by its coordinates. It is not easy to see that *the set of all M of degree g and dimension r in the coordinate space is an algebraic manifold.* The proof, which the first author obtained, is shown in section 1. The notion of algebraic system of manifolds M is discussed in section 2, and is related to the theory of algebraic correspondences.

1 The associated form of a manifold M

Let an r-dimensional irreducible algebraic manifold M in the n-dimensional projective space S_n be defined by the forms f_μ in $K[x] = K[x_0, x_1, \ldots, x_n]$. A general linear subspace S_{n-r} which is defined by r linear forms

$$l_i = \sum_{j=0}^{n} u_j^{(i)} x_j, \qquad (i = 1, \ldots, r),$$

intersects M in a null dimensional manifold which is irreducible with respect to $K(u^{(1)}, \ldots, u^{(r)})$ [1]. These points are denoted by

$$p^{(i)} = (p_0^{(i)}, p_1^{(i)}, \ldots, p_n^{(i)}) \qquad (i = 1, \ldots, g).$$

If we define the g linear forms

$$L_i(u^{(0)}) = \sum_{j=0}^{n} p_j^{(i)} u_j^{(0)},$$

where $u_j^{(0)}$ are indeterminates, then their product

$$G(u^{(0)}) = \prod_{i=1}^{g} L_i(u^{(0)})$$

is a form in $u^{(0)}$ which coincides with its conjugates with respect to $K(u^{(1)}, \ldots, u^{(r)})$. Thus a power of $G(u^{(0)})$ is a form with coefficients in $K(u^{(1)}, \ldots, u^{(r)})$. Multiplying it by a suitable polynomial in $K[u^{(1)}, \ldots, u^{(r)}]$, we obtain a form in $u^{(0)}$ with coefficients in $K[u^{(1)}, \ldots, u^{(r)}]$, which is only divisible by a power product of the $L_i(u^{(0)})$. Since every two such forms have a common factor, which again is such a form, there is up to a factor in K a uniquely determined form $F(u^{(0)})$ having these properties and which is irreducible as a polynomial in $K[u^{(1)}, \ldots, u^{(r)}]$. This form is called the *associated form* of M, and its degree is called *degree* of M[2].

---

[1]For the proof see B. L. v. d. Waerden, ZAG. V, Math. Annalen 110, p. 140.

[2]The number of the different points $p^{(i)}$, which has been called the degree of M, should be called the reduced degree. If K is a perfect field, then $F(u^{(0)}) = G(u^{(0)})$ and the degree equals the reduced degree.

It is clear that two distinct irreducible manifolds cannot have the same associated form, since by factoring the associated form we obtain a general point of the irreducible manifold, and two irreducible manifolds having a common general point must be the same.

The associated form $F(u^{(0)}) = F(u^{(0)}, u^{(1)}, \ldots, u^{(r)})$ of an irreducible manifold which by definition is homogeneous in $u^{(0)}$ is also homogeneous in $u^{(i)}$, with the same degree. The reason is that the u-resultant $D(u)$ of f_μ, l_1, ..., l_r and the additional linear form $l_0 = \sum_{j=0}^{n} u_j^{(0)} x_j$ is also a form in $u^{(0)}$ with coefficients in $K[u^{(1)}, \ldots, u^{(r)}]$ which is only divisible by a power product of the $L_i(u^{(0)})$. This u-resultant $D(u)$ is the greatest common divisor of the system of resultants f_μ, l_0, l_1, ..., l_r. Thus, if two variables are exchanged, then D only changes by a factor in K. Since F is irreducible, D must be a power of F up to an factor in $K[u^{(1)}, \ldots, u^{(r)}]$. But this factor can only be an element of K: If $D(u)$ possessed a factor dependent from $u^{(1)}$ but not from $u^{(0)}$ then $D(u)$ would also possess a factor dependent from $u^{(0)}$ but not from $u^{(1)}$ because of the symmetry. But the latter factor should be a power of F thus F would be independent of $u^{(1)}$ and also from $u^{(2)}$, ..., $u^{(r)}$, which is absurd. So, $D(u)$ is a power of F. In particular, F is homogeneous and of the same degree with respect to all $u^{(i)}$.

An arbitrary pure r-dimensional algebraic manifold M is the union of finitely many r-dimensional irreducible manifolds M_1, M_2, ..., M_h, where any manifold M_i may be taken with multiplicity n_i. The associated form of M_i is denoted by $F_i(u^{(0)}, u^{(1)}, \ldots, u^{(r)})$ with degree g_i. Then we call

$$F(u^{(0)}, u^{(1)}, \ldots, u^{(r)}) = \prod_{i=1}^{h} F_i(u^{(0)}, u^{(1)}, \ldots, u^{(r)})^{n_i}$$

the associated form of M with degree $g = \sum n_i g_i$. It is clear that a manifold is uniquely determined by its associated form, as in the case of irreducible manifolds. This form is homogeneous and of the same degree with respect to all $u^{(i)}$. In the following we call forms which are homogeneous of the same degree g with respect to all $u^{(i)}$, forms of degree g in $K[u^{(0)}, u^{(1)}, \ldots, u^{(r)}]$.

Now the question is, what are the conditions which a form of degree g in $K[u^{(0)}, u^{(1)}, \ldots, u^{(r)}]$ must satisfy, so that it is the associated form of an r-dimensional manifold. The following theorem gives the answer to this question.

Theorem 1 *A form* $F(u^{(0)}, u^{(1)}, \ldots, u^{(r)})$ *of degree* g *in* $K(u^{(0)}, u^{(1)}, \ldots, u^{(r)})$

is the associated form of an r-dimensional algebraic manifold if, and only if, the following conditions are satisfied:

1. *$F(u^{(0)}) = F(u^{(0)}, u^{(1)}, \dots, u^{(r)})$, as a form in $u^{(0)}$, is the product of linear factors in a suitable extension field of $K(u^{(0)}, u^{(1)}, \dots, u^{(r)})$:*

$$F(u^{(0)}) = \prod_{i=1}^{g} L_i(u^{(0)}) = \prod_{i=1}^{g} \Big(\sum_{j=0}^{n} p_j^{(i)} u_j^{(0)} \Big).$$

2. *$L_i(u^{(k)}) = 0$ for all $i = 1, 2, \dots, g$; $k = 1, \dots, r$.*

3. *If $v_j^{(1)}, \dots, v_j^{(r)}$ $(j = 0, 1, \dots, n)$ are elements of any extension field of $K(p^{(i)})$ and if $L_i(v^{(k)}) = 0$ for $k = 1, \dots, r$ (for a fixed i), then $F(u^{(0)}, v^{(1)}, \dots, v^{(r)})$ as a form in $u^{(0)}$ is divisible by $L_i(u^{(0)})$.*

Proof: We may restrict to the case when the form F is irreducible, since the general case may be reduced to this case by writing the form F as the product of its irreducible factors. We have to show that, if the form F satisfies the conditions 1,2,3, then the factors also do, and vice versa. This is clear for condition 1 and 2. For condition 3 we note the following. Let F_1 be an irreducible factor of F,

$$F_1 = \prod_{i=1}^{g'} L_i(u^{(0)}).$$

Let condition 3 be satisfied for F; we show that it is also satisfied for F_1. So, let $v^{(k)}$ be the solutions of the system of linear equations

$$L_i(v^{(k)}) = 0 \qquad (k = 1, \dots, r) \tag{1}$$

for a fixed i $(1 \le i \le g')$. From the theory of linear equations we know that all solutions of this system of equations are obtained by specifying some parameters in the general solution, which contains undetermined linear parameters. A special solution of this system of equations is $v^{(k)} = u^{(k)}$ by 2. We now have to prove that $F_1(u^{(0)}, v^{(1)}, \dots, v^{(r)})$ is divisible by $L_i(u^{(0)})$, if $v^{(k)}$ is the general solution of the aforementioned system of linear equations. By hypothesis $F(u^{(0)}, v^{(1)}, \dots, v^{(r)})$ is divisible by $L_i(u^{(0)})$; thus there is at least one irreducible factor of F which is divisible by $L_i(u^{(0)})$. Let F_2 be this factor. The divisibility is preserved when the undetermined parameters in the general solution $v^{(k)}$ are chosen such that $v^{(k)}$ is the special solution $u^{(k)}$. Hence $F_2(u^{(0)}, u^{(1)}, \ldots, u^{(r)})$ and $F_1(u^{(0)}, u^{(1)}, \ldots, u^{(r)})$ have the common factor $L_i(u^{(0)})$. Since F_1 and F_2 are both irreducible, this implies $F_1 = F_2$ (up to a constant factor). So $F_1(u^{(0)}, v^{(1)}, \ldots, v^{(r)})$ is divisible by $L_i(u^{(0)})$, and condition 3 is proved for F_1.

Now let F be irreducible. The g points $p^{(1)}, \ldots, p^{(g)}$ are then conjugate to each other with respect to $K(u^{(1)}, \ldots, u^{(r)})$. Thus there is an irreducible manifold M which contains all $p^{(i)}$ as general points. By 2, a general $(n-r)$-dimensional linear subspace S_{n-r} with equations

$$\sum_{j=0}^{n} u_j^{(k)} x_j = 0 \qquad (k = 1, \ldots, r)$$

intersects the manifold M in at least g general points. Condition 3 says that S_{n-r} does not contain another general point of M: Let S_{n-r} contain the general point $q = (q_0, q_1, \ldots, q_n)$ of M. By the algebraic equivalence of general points there exists an isomorphism from $K(q)$ to $K(p^{(i)})$. This isomorphism can be extended to an isomorphism from $K(q, u^{(1)}, \ldots, u^{(r)})$ to $K(p^{(i)}, v^{(1)}, \ldots, v^{(r)})$. The equations $\sum u_j^{(k)} q_j = 0$, which say that the point q is in S_{n-r}, are preserved by this isomorphism; so for all $k = 1, 2, \ldots, r$,

$$\sum_{j=0}^{n} v_j^{(k)} p_j^{(i)} = 0 \qquad \text{or} \qquad L_i(v^{(k)}) = 0.$$

By 3 this implies that $F(u^{(0)}, v^{(1)}, \ldots, v^{(r)})$ is divisible by $L_i(u^{(0)})$. Applying the above isomorphism in the inverse direction we obtain that $F(u^{(0)}, u^{(1)}, \ldots, u^{(r)})$ is divisible by $\sum u_j^{(0)} q_j$, i.e. because of 1, that q is one of the points $p^{(i)}$.

Since a general $(n-r)$-dimensional linear subspace intersects M in a finite number of general points, M must be r-dimensional. Now it is clear that F is the associated form of M. By 2, the g points $p^{(1)}, \ldots, p^{(g)}$ are the intersection of M with the general S_{n-r} defined by $u^{(1)}, \ldots, u^{(r)}$. So F is a form in $u^{(0)}$ with coefficients in $K[u^{(1)}, \ldots, u^{(r)}]$, which is divisible only by products of powers of L_i and which is irreducible as a polynomial in $K[u^{(0)}, u^{(1)}, \ldots, u^{(r)}]$, i.e. by definition, the associated form of M.

The fact that 1 and 2 are necessary conditions for an associated form of M follows immediately from the definition of the associated form. Condition 3 is also a necessary condition: It suffices to prove 3 in the case that $v_j^{(k)}$ is the *general* solution of the system of equations (1). In this case the $v_j^{(k)}$ are algebraically independent, since as remarked above, from the general solution $v^{(k)}$ one obtains $u^{(k)}$ by specifying parameters, and even the $u^{(k)}$ are algebraically independent, since they are not determined. Hence there is an isomorphism $K(v^{(1)}, \ldots, v^{(r)}) \cong K(u^{(1)}, \ldots, u^{(r)})$ which takes $v^{(k)}$ to $u^{(k)}$. This isomorphism is extended to an isomorphism $K(v^{(1)}, \ldots, v^{(r)}, p^{(i)}) \cong K(u^{(1)}, \ldots, u^{(r)}, q)$. The point q is contained in the planes $u^{(1)}, \ldots, u^{(r)}$ and in M; thus q is one of the points $p^{(l)}$, and $F(u^{(0)}, u^{(1)}, \ldots, u^{(r)})$ contains the factor $\sum q_j u_j^{(0)}$. Taking the inverse isomorphism it follows that $F(u^{(0)}, v^{(1)}, \ldots, v^{(r)})$ is divisible by $L_i(u^{(0)})$.

We now show that the conditions in the above theorem are expressible by homogeneous algebraic equations between the coefficients a_λ of the form. At first we show that the conditions 1,2,3 are equivalent to algebraic relations between a_λ, $u^{(k)}$ and the coefficients $p_j^{(i)}$ of the linear factors L_i of F. Then we eliminate the $p_j^{(i)}$ from these conditions, and finally we require that the remaining conditions are satisfied identically in the $u_j^{(k)}$.

In order to express condition 1 by homogeneous algebraic equations we put

$$F(u^{(0)}, u^{(1)}, \ldots, u^{(r)}) = \rho \cdot \prod_{i=1}^{g} \Big(\sum_{j=0}^{n} p_j^{(i)} u_j^{(0)} \Big),$$

and compare the coefficients of the power products of the $u_j^{(0)}$ on the left and on the right-hand side:

$$\varphi_k(u^{(1)}, \ldots, u^{(r)}) = \rho \cdot \psi_k(p^{(1)}, \ldots, p^{(g)}) \,,$$

and finally we eliminate the factor ρ, so that the equations become homogeneous:

$$\varphi_k \psi_l - \varphi_l \psi_k = 0. \tag{2}$$

The condition 2 itself has the form of a homogeneous system of equations:

$$\sum_{j=0}^{n} p_j^{(i)} u_j^{(k)} = 0 \qquad (i = 1, \ldots, g; k = 1, \ldots, r). \tag{3}$$

The condition 3 must first be taken to another form. The statement that $F(u^{(0)}, v^{(1)}, \ldots, v^{(r)})$ is divisible by the linear form $L_i(u^{(0)})$ is equivalent to the statement that all zeros of the linear form are also zeros of $F(u^{(0)}, v^{(1)}, \ldots, v^{(r)})$, i.e., that $L_i(v^{(0)}) = 0$ implies $F(v^{(0)}, v^{(1)}, \ldots, v^{(r)}) = 0$. Thus condition 3 is equivalent to the following conditions: If $v^{(0)}, v^{(1)}, \ldots, v^{(r)}$ are elements of an extension field of $K(p^{(i)})$ and $L_i(v^{(k)}) = 0$ for all $k = 0, 1, \ldots, r$, then $F(v^{(0)}, v^{(1)}, \ldots, v^{(r)}) = 0$.

The general solution of the system of linear equations

$$L_i(v) = \sum_{j=0}^{n} p_j^{(i)} v_j = 0$$

is given by

$$v_j = \sum_{l=0}^{n} s_{jl} p_l^{(i)}, \quad s_{jl} = -s_{lj}.$$

So the condition 3 is also equivalent to the following condition: If $s_{jl}^{(k)} = -s_{lj}^{(k)}$ $(k = 0, \ldots, r;\ j, l = 0, \ldots, n)$ are new indeterminates and if we define $v_j^{(k)} = \sum s_{jl}^{(k)} p_l^{(i)}$ $(k = 0, 1, \ldots, r)$, then

$$F(v^{(0)}, v^{(1)}, \ldots, v^{(r)}) = 0. \tag{4}$$

Now, if we choose $v_j^{(k)} = \sum s_{jl}^{(k)} p_l^{(i)}$ in this equation and compare the coefficients of the new indeterminates, then we obtain a system of homogeneous conditions

$$\chi_h(a_\lambda, p_j^{(i)}) = 0. \tag{5}$$

Thus the conditions 1,2,3 are equivalent to the equations (2), (3), (5). We consider these equations as equations which determine $p_j^{(i)}$. The condition for being able to solve these equations is a system of homogeneous equations in a_λ, $u^{(1)}, \ldots, u^{(r)}$. Finally one orders this system by power products of the indeterminates $u_j^{(k)}$ and takes the coefficients of each power product to be zero. Then we obtain a system of homogeneous equations for the a_λ alone. Thus we have proved

Theorem 2 *The necessary and sufficient condition for a form F of degree g in $K[u^{(0)}, u^{(1)}, \ldots, u^{(r)}]$ to be the associated form of a manifold M of degree g and dimension r is a system of homogeneous equations for the coefficients of the form F.*

In the simplest case where the manifold M is a line in the space S_n, the form F is

$$F = \sum\sum p_{ij} u_i^{(0)} u_j^{(1)}.$$

Its coefficients are the Plücker coordinates p_{ij} of the line M. The reformulation of the conditions 1,2,3 as described above is a system of cubic relations between the p_{ij} which is of course equivalent to the well-known linear and quadratic relations

$$p_{ij} = -p_{ji}, \quad p_{ij}p_{kl} + p_{ik}p_{lj} + p_{il}p_{jk} = 0.$$

We have already explained how the associated form is obtained from the equations of a manifold, namely by the construction of the system of resultants from these equations and $r+1$ general linear equations. We now examine how, vice versa, we obtain the equations of the manifold from the associated form.

Theorem 3 *The equations $f_\mu(y) = 0$ of an r-dimensional manifold M are obtained if one substitutes $u_l^{(k)}$ in the associated form $F(u^{(0)}, \ldots, u^{(r)})$ by*

$$v_l^{(k)} = \sum_{j=0}^{n} y_j s_{jl}^{(k)} \qquad (s_{jl}^{(k)} = -s_{lj}^{(k)})$$

and then sets the new form $G(y, s)$ identically zero in the s_{jl}.

Proof: It is sufficient to consider the case that M is irreducible. The fact that $G(y, s) = 0$ for a general point $y = p^{(i)}$ of the manifold is satisfied, is just the reformulation (4) of condition 3. When it holds for some general point, then it holds for every point y.

Conversely, let $G(y, s) = 0$, i.e., $F(v^{(0)}, v^{(1)}, \ldots, v^{(r)}) = 0$ for $v_l^{(k)} = \sum y_j s_{jl}^{(k)}$. We may choose $s_{jl}^{(k)}$ such that $v^{(k)}$ are arbitrary $(r+1)$ hyperplanes joining the point y. We may also choose these hyperplanes such that they intersect the r-dimensional manifold M only in the point y: One chooses the first hyperplane $v^{(0)}$ such that it does not contain an arbitrary point of M, so that its intersection M' with M is $(r-1)$-dimensional; then one chooses the second hyperplane $v^{(1)}$ such that it does not contain a point in each irreducible component of M', thus its intersection with M' is $(r-2)$-dimensional, and so on. Now $F(v^{(0)}, v^{(1)}, \ldots, v^{(r)}) = 0$, and $F(v^{(0)}, v^{(1)}, \ldots, v^{(r)})$ is a divisor of the system of resultants of the equations of M and the equations of the hyperplanes

$v^{(0)}, \ldots, v^{(r)}$; so this system of resultants is zero. Thus there is a common point of M and the hyperplanes $v^{(0)}, \ldots, v^{(r)}$, which is the point y according to the construction of the hyperplanes. Hence y is a point of M.

2 Algebraic systems of manifolds

The associated form of a manifold M is given by the collection of its coefficients a_λ. We consider this collection a_λ as a point in a space $\mathfrak{B}$. Then a manifold M corresponds to its image a in $\mathfrak{B}$ and vice versa. An *algebraic system of manifolds* M is a set of manifolds whose image in $\mathfrak{B}$ is again an algebraic manifold.

By Theorem 2 all manifolds M (of given dimension and degree) form an algebraic system. In the same way one can show that all manifolds M which are contained in a given manifold M_0 in S_n form an algebraic system. For the proof one has to add in the proof of Theorem 2 to the equations (2), (3), (5) the equations which say that the points $p^{(1)}, \ldots, p^{(g)}$ are contained in M_0. If the coordinates of M_0 are known, then one constructs without difficulty the equations for $p^{(1)}, \ldots, p^{(g)}$ by Theorem 3. As in the proof of Theorem 2 one eliminates the coordinates of $p^{(1)}, \ldots, p^{(g)}$ from the whole system of equations, and obtains a system of algebraic relations between the coordinates of M and those of M_0, which say that M is contained in M_0.

Suppose $\mathfrak{S}$ is an algebraic system of manifolds M in S_n. Then there is always an algebraic correspondence between an algebraic manifold $\mathfrak{L}$ and the space S_n, in which to each point x in $\mathfrak{L}$ correspond the points of a manifold $M(x)$ of S_n, such that $M(x)$ runs through the system $\mathfrak{S}$ if x runs through the manifold $\mathfrak{L}$. We choose $\mathfrak{L}$ to be the image of the system $\mathfrak{S}$ in the image space $\mathfrak{B}$; the point x is then the image a of M and the equations of the correspondence are those equations which by Theorem 3 connect a point y of M with the coordinates a_λ.

Eliminating the coordinates a_λ from these equations implies that to every algebraic system of manifolds M belongs a manifold $\mathfrak{T}$ which is covered by the manifolds M. $\mathfrak{T}$ is the image of $\mathfrak{L}$ in the above correspondence.

We now ask if conversely every correspondence $\mathfrak{K}$ between two manifolds $\mathfrak{L}$ and $\mathfrak{T}$ defines an algebraic system of manifolds $\mathfrak{M}$ on the manifold $\mathfrak{T}$, such that to the points of $\mathfrak{L}$ in the correspondence correspond the manifolds M of the system. We may restrict to irreducible correspondences $\mathfrak{K}$. Then $\mathfrak{L}$ and $\mathfrak{T}$ are also irreducible. From the general theory of correspondences, to every point x of $\mathfrak{L}$ corresponds an algebraic manifold $\mathfrak{T}_x$ of $\mathfrak{T}$. However, these manifolds $\mathfrak{T}_x$ do not always form an algebraic system, since there might be manifolds with distinct dimensions, as is well known (see the example at the end). Even if all $\mathfrak{T}_x$ have the same dimension, they might again not form an algebraic system, as it might happen that some $\mathfrak{T}_x$ have a higher degree than others. For example, if one connects all points of a plane curve C^3 with order three, which has a knot point O, with the point O, then one obtains a correspondence in which to a general point of C^3 corresponds a line, but to the point O correspond two lines (the double point tangents). So one has to add some restrictions such that the $\mathfrak{T}_x$ form an algebraic system.

Theorem 4 *If to every point x of $\mathfrak{L}$ in an irreducible correspondence $\mathfrak{K}$ corresponds an r-dimensional manifold $\mathfrak{T}_x$ of points on $\mathfrak{T}$ and if $\mathfrak{L}$ does not contain any multiple points then the image manifolds $\mathfrak{T}_x$ form an algebraic system, provided that its irreducible components are counted with the multiplicities as defined in ZAG. VI, §4.*

In this theorem we suppose the field K to be perfect, as is also supposed in ZAG. VI.

Proof: We take a general point ξ of $\mathfrak{L}$. In order to obtain the associated form of its manifold $\mathfrak{T}_\xi$ one has to determine the intersection points $p^{(1)}, \dots, p^{(g)}$ of $\mathfrak{T}_\xi$ with r general hyperplanes $u^{(1)}, \dots, u^{(r)}$, and take the product

$$F(u^{(0)}) = \rho \cdot \prod_{i=1}^{g} L_i(u^{(0)}) = \rho \cdot \prod_{i=1}^{g} \Big(\sum_{j=0}^{n} p_j^{(i)} u_j^{(0)} \Big). \tag{1}$$

By §1, equation (2), the relations between a_λ, $u^{(k)}$ and $p^{(i)}$ are given by a system of homogeneous relations

$$H(u^{(1)}, \dots, u^{(r)}, a_\lambda, p^{(1)}, \dots, p^{(g)}) = 0, \tag{2}$$

which is equivalent to equation (1).

If we replace the general point ξ by the special point x then the multiplicities of the irreducible parts of $\mathfrak{T}_x$ are found according to ZAG. VI, §4 by intersecting

$\mathfrak{T}_x$ with the general hyperplanes $u^{(1)}, \ldots, u^{(r)}$. One has to see the $p^{(i)}$ change to some intersection points $q^{(i)}$. The multiplicity of a component of $\mathfrak{T}_x$ is the number which says how often an intersection point of this component with the hyperplanes is one of the points $q^{(i)}$. Extending this relation-preserving specialization by a suitable specialization $a_\lambda \to a'_\lambda$, the relations (2), thus also the factorization (1) is preserved. The multiplicities of the components of $\mathfrak{T}_x$ thus equal the multiplicities which their associated forms in the factorization of the associated form $F'(u^{(0)})$ with coefficients a'_λ have. But this means that the specialized a'_λ are exactly the coordinates of the specialized manifold $\mathfrak{T}_x$.

Now, the general pair of points (ξ, a) induces an irreducible correspondence between $\mathfrak{L}$ and the image space $\mathfrak{B}$. The pairs of points (x, a') of this correspondence are exactly those which we obtain from the general pair of points (ξ, a) by the relation-preserving specialization. The points a' which appear in this correspondence form an algebraic manifold $\mathfrak{A}$ in the space $\mathfrak{B}$. So the manifolds $\mathfrak{T}_x$ form an algebraic system in our sense.

If we drop the hypothesis that $\mathfrak{L}$ does not contain any multiple point and that to every point x of $\mathfrak{L}$ corresponds an exactly r-dimensional manifold $\mathfrak{T}_x$, then only the last part of the proof remains true. Thus there exists an irreducible correspondence with general pair of points (ξ, a) in which every special point x of $\mathfrak{L}$ corresponds to one or more, eventually infinitely many points a'. To every point a' there is a manifold $M(a')$ with dimension r, and we can show that the union of these manifolds $M(a')$ is exactly the manifold $\mathfrak{T}_x$. If y is a point of $\mathfrak{T}_x$, then (x, y) is a relation-preserving specialization of the general pair of points (ξ, η) of the correspondence $\mathfrak{K}$, and we extend this correspondence to a specialization $(\xi, \eta, a) \to (x, y, a')$ where a' belongs to the image manifold $\mathfrak{A}$ and where especially those algebraic relations are preserved which say that η is in $M(a)$. Thus y is in the manifold $M(a')$. Conversely every point of $M(a')$ is in $\mathfrak{T}_x$, since the algebraic relations between ξ and a which say that $M(a)$ is on $\mathfrak{T}_\xi$ are preserved by the specialization $\xi \to x$, $a \to a'$.

Thus we have shown:

Theorem 5 *Suppose $\mathfrak{K}$ is an irreducible correspondence between $\mathfrak{L}$ and $\mathfrak{T}$. If ξ is a general point of $\mathfrak{L}$ to which corresponds an r-dimensional manifold $M(a)$ in $\mathfrak{T}$, whose image point is a, then the general pair of points (ξ, a) defines an irreducible correspondence between $\mathfrak{L}$ and an image space $\mathfrak{A}$ in $\mathfrak{B}$, which assigns to every special point x one or more points a'. To these points correspond* *manifolds $M(a')$ whose union is just that manifold $\mathfrak{T}_x$ which corresponds to the point x in the correspondence $\mathfrak{K}$. The manifolds $M(a')$ form an irreducible algebraic system $\mathfrak{S}$.*

If there is some rule which assigns to the general point ξ of a manifold $\mathfrak{L}$ a point η with coordinates from $K(\xi)$, and if η is the general point of a manifold $\mathfrak{M}$, then we will say: *"Suppose ξ runs through the manifold $\mathfrak{L}$, then η runs through the image manifold $\mathfrak{M}$."* This means that (ξ, η) is the general pair of points of an irreducible correspondence between $\mathfrak{L}$ and $\mathfrak{M}$, and this correspondence assigns to each point of $\mathfrak{L}$ at least one point of $\mathfrak{M}$, and vice versa to each point of $\mathfrak{M}$ at least one point of $\mathfrak{L}$.

More generally: If there is some rule which assigns to the general point ξ of a manifold $\mathfrak{L}$ an r-dimensional manifold M_ξ whose equations are in $K(\xi)$, and if M_ξ is the general element of an algebraic system $\mathfrak{S}$ of manifolds, then we will say: *"Suppose ξ runs through the manifold $\mathfrak{L}$, then M_ξ runs through the system $\mathfrak{S}$."* If $\mathfrak{T}$ is the manifold which is covered by $\mathfrak{S}$, then we will also say: *"M_ξ runs through $\mathfrak{T}$."* The first statement means that there is an irreducible correspondence between $\mathfrak{L}$ and $\mathfrak{S}$ which assigns to the general point ξ the manifold M_ξ and to each special point of $\mathfrak{L}$ at least one manifold M of the system $\mathfrak{S}$. If we assign to each manifold M the set of points of M, then we obtain a correspondence between $\mathfrak{S}$ and the manifold $\mathfrak{T}$. These two correspondences are described by the equations

$$f_\mu(x, a) = 0 \quad \text{resp.} \quad g_\nu(a, y) = 0,$$

when a_λ are the coordinates of a manifold M of $\mathfrak{S}$ and where x_k are the coordinates of a point of $\mathfrak{L}$, y_k are the coordinates of a point of $\mathfrak{T}$. Eliminating the a_λ from these equations one obtains a correspondence $\mathfrak{K}$ between $\mathfrak{L}$ and $\mathfrak{T}$, which assigns to a point x all points y of the associated manifolds of the system $\mathfrak{S}$. The relation between this correspondence $\mathfrak{K}$ and the system $\mathfrak{S}$ is described in Theorem 5.

The following example might illustrate Theorem 5. Let g, h be two lines in S_3 which do not intersect, and let $\mathfrak{L}$ be a plane which intersects g and h in the points G and H. There is exactly one line M_ξ joining a general point ξ in $\mathfrak{L}$ which intersects g and h. If ξ runs through the plane, then M_ξ runs through the congruence of lines with Leitstrahlen g, h. That is, M_ξ is the general element of this congruence. To every point x in $\mathfrak{L}$ corresponds exactly one line of the congruence, except for the points G and H, to which corresponds a bundle of lines. If we assign to each point x in $\mathfrak{L}$ all points y which are contained on a line joining x and g, h, then we obtain a correspondence $\mathfrak{K}$ between $\mathfrak{L}$ and S_3, where the points y corresponding to a point x in $\mathfrak{L}$ in general form a line. The points y associated to the points G and H however do not form a line, but a plane which is the union of the congruence of lines passing through G resp. H, according to Theorem 5, which must be so.

(Received on May 24, 1936)

Translated by Wen-ling Huang, Nankai Institute of Mathematics, Tianjin, China.

Die geometrische Theorie der algebraischen Funktionen für beliebige vollkommene Körper*).

Von

Wei-Liang Chow in Nanking (China).

Einleitung.

Die vorliegende Arbeit steckt sich das Ziel, die klassische geometrische Theorie der algebraischen Funktionen einer Veränderlichen für einen beliebigen vollkommenen Körper zu begründen. Dabei stützen wir uns auf die von van der Waerden geschaffenen rein algebraischen Grundlagen der algebraischen Geometrie, deren charakteristisches Merkmal in methodischer Hinsicht die systematische Anwendung der „relationstreuen Spezialisierung" ist[1]). Wir werden im folgenden sehen, daß man bei den Beweisen der grundlegenden Tatsachen unserer Theorie fast ausschließlich mit der Methode der relationstreuen Spezialisierung und der dieser Methode zugrunde liegenden Resultantentheorie auskommen kann. Dadurch erreicht die Theorie eine gewisse methodische Reinheit, die, ganz abgesehen von der daraus folgenden Verallgemeinerungsfähigkeit, auch an sich von Interesse sein dürfte.

Bekanntlich hat F. K. Schmidt[2]) in einer Arbeit über analytische Zahlentheorie die arithmetische Theorie der algebraischen Funktionen einer Veränderlichen für beliebige vollkommene Körper entwickelt. Seine Theorie soll sich übrigens auch auf unvollkommene Körper übertragen lassen[3]), unter der Voraussetzung, daß die Kurve absolut-irreduzibel ist. Ob die hier entwickelte Theorie sich auch auf unvollkommene Körper übertragen läßt, wird hier nicht beantwortet. Der Inhalt von § 1 bis § 3 läßt sich ohne Schwierigkeiten auch für unvollkommene Körper beweisen, aber der Beweis in § 4 für die Existenz eines singularitätenfreien birationalen Bildes läßt sich nicht ohne weiteres übertragen, weil dabei der sich am Schluß des Paragraphen befindende Hilfssatz gebraucht wird. Derselbe Hilfssatz wird am Anfang von § 6 nochmals gebraucht, ließe sich aber hier durch eine andere Betrachtung ersetzen, wenn wir den Existenzsatz

---

*) Diese Arbeit wurde von der Philosophischen Fakultät der Universität Leipzig als Dissertation angenommen.

1) [5], [6], [9].

2) [4].

3) [2].

für den betreffenden Körper schon hätten. In der Tat bildet dieser Existenzsatz den Grundstein, worauf unsere Untersuchungen aufgebaut sind; hat man diesen Satz für unvollkommene Körper bewiesen, so kann man auch die ganze Theorie übertragen. Dabei genügt es schon, wenn man einen Existenzsatz im Kleinen hat, d. h. die Existenz eines birationalen Bildes, das in einem beliebigen vorgegebenen Zweige einfach ist.

In § 1 werden die Grundbegriffe und Tatsachen über algebraische Mannigfaltigkeiten zusammengestellt, in § 2 und § 3 werden die Begriffe der Schnittpunktgruppe und der linearen Schar entwickelt. Der in § 4 gegebene Beweis für die Existenz eines singularitätenfreien birationalen Bildes ist eine Verfeinerung des von Albanese[4]) stammenden. Auf Grund dieses Existenzsatzes wird dann in § 5 der Begriff des Punktes birationalinvariant begründet. Dabei wird auch die Potenzreihenentwicklung begründet und die Äquivalenz der „funktionentheoretischen" mit der „algebraisch-geometrischen" Schnittvielfachheit bewiesen. In § 6 sind wir dem Gedankengang von Severi[5]) eng gefolgt, die Beweise sind jedoch etwas vereinfacht. Den in § 7 gegebenen Beweis der Vollständigkeit der adjungierten Scharen haben wir von Severi[6]) übernommen.

§ 1.

Algebraische Mannigfaltigkeiten. Dimension.

Es sei K ein beliebiger Körper. Eine algebraische Mannigfaltigkeit M ist definiert durch ein System Φ_M von endlichvielen homogenen Polynomen oder Formen aus $K[x_0, x_1, \ldots, x_n] = K[x]$

$$f_i(x_0, x_1, \ldots, x_n).$$

Das System kann auch leer sein; die dadurch definierte Mannigfaltigkeit heißt der projektive Raum S_n. Ein geordnetes System von $n+1$ nicht sämtlich verschwindenden Elementen $\xi = (\xi_0, \xi_1, \ldots, \xi_n)$ aus irgendeinem Oberkörper von K heißt ein Punkt von M, wenn ξ eine gemeinsame Nullstelle aller Formen von Φ_M ist, d. h.

$$f_i(\xi_0, \xi_1, \ldots, \xi_n) = 0.$$

Dabei sind zwei Punkte ξ, ξ' als gleich zu betrachten, wenn es ein Element $\varrho \neq 0$ gibt, so daß $\xi_i = \varrho \xi'_i$ für $i = 0, 1, \ldots, n$. Eine Mannigfaltigkeit M_1 heißt in einer anderen Mannigfaltigkeit M_2 enthalten, wenn jeder Punkt von M_1 zugleich Punkt von M_2 ist. M_1 heißt dann ein Teil von M_2, und zwar ein echter Teil, wenn M_2 nicht zugleich auch in M_1 enthalten ist. Zwei Mannigfaltigkeiten, die ineinander enthalten sind,

[4]) [1].

[5]) [3], S. 144—157.

[6]) [3], S. 149—152.

werden als gleich betrachtet. Alle durch Formensysteme aus $K[x_0, x_1, \ldots, x_n]$ definierten Mannigfaltigkeiten sind offenbar in S_n enthalten.

Der Schnitt von s Mannigfaltigkeiten $M_1, \ldots, M_s$ ist die durch das Vereinigungssystem der Formen von $\Phi_{M_1}, \ldots, \Phi_{M_s}$ definierte Mannigfaltigkeit, die offenbar genau die gemeinsamen Punkte von $M_1, \ldots, M_s$ als ihre Punkte besitzt. Die Summe von $M_1, \ldots, M_s$ ist die durch das System der sämtlichen Produkte von s Formen, je eine aus jedem Φ_{M_i}, definierte Mannigfaltigkeit. Sie hat als ihre Punkte genau die Vereinigungsmenge aller Punkte von $M_1, \ldots, M_s$. Eine Mannigfaltigkeit heißt irreduzibel (in bezug auf K), wenn sie sich nicht als Summe von zwei echten Teilen darstellen läßt, sonst heißt sie reduzibel. Eine reduzible Mannigfaltigkeit läßt sich eindeutig als Summe von endlichvielen irreduziblen darstellen.

Es erweist sich als sehr zweckmäßig, den Mannigfaltigkeitsbegriff in der folgenden Weise zu verschärfen. Die obige Definition werden wir nur für irreduzible Mannigfaltigkeiten gebrauchen. Eine beliebige Mannigfaltigkeit M ist dann definiert als ein System von endlichvielen (nicht notwendig verschiedenen) irreduziblen Mannigfaltigkeiten $M_1, \ldots, M_t$, welche die irreduziblen Bestandteile von M genannt werden[7]). Jede irreduzible Mannigfaltigkeit tritt also in M mit einer gewissen Vielfachheit auf. Kommt eine irreduzible Mannigfaltigkeit in M überhaupt nicht vor, so sagen wir auch, sie tritt in M mit der Vielfachheit Null auf. Eine Mannigfaltigkeit ist also eine Menge von mit Vielfachheiten versehenen irreduziblen Mannigfaltigkeiten. Zwei Mannigfaltigkeiten heißen gleich, wenn sie dieselben irreduziblen Mannigfaltigkeiten mit denselben Vielfachheiten enthalten. Der Begriff der Summe wird in naheliegender Weise übertragen, indem man jeder irreduziblen Mannigfaltigkeit die Summe ihrer Vielfachheiten in den Summanden als ihre Vielfachheit in der Summe zuschreibt. Aber beim Schnittbegriff besteht hier eine Unbestimmtheit, die schon bei den irreduziblen Mannigfaltigkeiten vorkommt. Die obige Definition des Schnittes liefert zwar die irreduziblen Bestandteile, aber nicht die Vielfachheiten. Wir werden uns also vorbehalten, die Vielfachheiten der Schnittbestandteile in jedem Fall geeignet zu definieren[8]).

Ein Punkt $\xi = (\xi_0, \xi_1, \ldots, \xi_n)$ heißt t-dimensional (in bezug auf K), wenn, $\xi_i \neq 0$ vorausgesetzt, $K\left(\frac{\xi_0}{\xi_i}, \ldots, \frac{\xi_n}{\xi_i}\right)$ vom Transzendenzgrad t über K

[7]) Wir unterscheiden hier nicht zwischen einer irreduziblen Mannigfaltigkeit und dem System, das aus dieser irreduziblen Mannigfaltigkeit allein besteht; diese beiden logisch verschiedenen Begriffe werden hier also identifiziert.

[8]) Wir dürfen natürlich immer noch von dem Schnittpunkte zweier beliebiger Mannigfaltigkeiten sprechen, nur bilden diese Schnittpunkte jetzt nicht ohne weiteres eine Mannigfaltigkeit.

ist. Nulldimensionale Punkte werden auch algebraische Punkte genannt. Die Obergrenze r der Dimensionszahlen der Punkte von M heißt die Dimension von M. Wir schreiben dann M_r. Es ist ersichtlich $r \leqq n$, wobei die Gleichheit nur bei dem projektiven Raum S_n erreicht wird. Mannigfaltigkeiten, die keinen Punkt besitzen, geben wir die Dimension -1. Die Summe von endlichvielen r-dimensionalen Mannigfaltigkeiten ist offenbar wieder r-dimensional. Eine Mannigfaltigkeit, die sich als Summe von endlichvielen irreduziblen r-dimensionalen Mannigfaltigkeiten darstellen läßt, heißt rein r-dimensional.

Eine irreduzible nulldimensionale Mannigfaltigkeit besteht aus endlichvielen in bezug auf K konjugierten Punkten. Eine beliebige nulldimensionale Mannigfaltigkeit besteht dann aus endlichvielen mit Vielfachheiten versehenen Punkten. Sie wird auch eine Punktgruppe genannt. Mit jedem Punkt treten also auch alle Konjugierten in einer Punktgruppe auf, und zwar mit derselben Vielfachheit. Umgekehrt ist auch (wenn K vollkommen ist) jede Menge von endlichvielen mit Vielfachheiten versehenen Punkten, in der mit jedem Punkte auch alle in bezug auf K konjugierten mit derselben Vielfachheit auftreten, eine Punktgruppe. Die rein 1-dimensionalen Mannigfaltigkeiten heißen Kurven, die rein $(n-1)$-dimensionalen Hyperflächen. Eine (irreduzible) Hyperfläche läßt sich durch eine einzige (irreduzible) Form darstellen, und umgekehrt definiert eine (irreduzible) Form auch eine (irreduzible) Hyperfläche. Wir werden eine Hyperfläche einfach mit der sie definierenden Form bezeichnen.

Die oben eingeführten Begriffe sind alle in bezug auf den Körper K definiert. Geht man von K zu einem Oberkörper Ω über, so bleiben Mannigfaltigkeiten, Summe, Schnitt in K auch solche in Ω. Die Irreduzibilität ist im allgemeinen nur gegen rein transzendente Erweiterung invariant, die Dimension eines Punktes nur gegen rein algebraische Erweiterung. Die Dimension einer Mannigfaltigkeit ist aber invariant gegenüber beliebiger Erweiterung.

Den nachstehenden Untersuchungen liegt ein vollkommener Körper K zugrunde. Wir werden aber im Laufe unserer Untersuchungen oft Mannigfaltigkeiten heranziehen, die durch Formen mit nicht zu K gehörigen Koeffizienten definiert sind. In solchem Fall ist immer zu verstehen, daß der Körper K dementsprechend durch Adjunktion dieser Koeffizienten zu erweitern ist. Alle weiteren Betrachtungen sind dann in bezug auf diesen erweiterten Körper zu verstehen. Wir werden auch, falls es nicht schon aus dem Zusammenhang klar oder sonst ausdrücklich erwähnt ist, den Bezugskörper oft dadurch andeuten, indem wir uns solcher Ausdrücke wie „K-Mannigfaltigkeit", „K-Hyperfläche", usw. bedienen. In allen anderen Fällen ist immer K der Bezugskörper.

§ 2.

Der Satz von Bezout.

Eine (in bezug auf K) allgemeine Hyperfläche F ist definiert durch eine Form

$$F(u, x) = \Sigma u_{i_1, \ldots, i_m} x_{i_1} \ldots x_{i_m},$$

deren Koeffizienten u neue nicht in K liegende Unbestimmte sind. F ist also eine $(n-1)$-dimensionale $K(u)$-Mannigfaltigkeit und ist offenbar irreduzibel.

M_r sei eine irreduzible r-dimensionale K-Mannigfaltigkeit. Eine allgemeine Hyperfläche $F(u, x)$ schneidet aus M_r eine irreduzible $(r-1)$-dimensionale $K(u)$-Mannigfaltigkeit[9]) aus, die wir als die Schnittmannigfaltigkeit von $F(u, x)$ mit M_r definieren. Daraus folgt durch r-malige Anwendung, daß r (verschiedene) allgemeine Hyperflächen $F_1(u^{(1)}, x), \ldots, F_r(u^{(r)}, x)$ aus M_r eine irreduzible nulldimensionale $K(u^{(1)}, \ldots, u^{(r)})$-Mannigfaltigkeit ausschneiden, die also aus endlichvielen (allgemeinen) Punkten von M_r besteht. Insbesondere hat ein allgemeiner $(n-r)$-dimensionaler linearer Unterraum S_{n-r}, d. h. eine durch r allgemeine Linearformen $\sum_{j=0}^{n} u_j^{(1)} x_j, \ldots, \sum_{j=0}^{n} u_j^{(r)} x_j$ definierte $K(u^{(1)}, \ldots, u^{(r)})$-Mannigfaltigkeit, mit M_r endlichviele Schnittpunkte. Die Anzahl dieser Punkte heißt der Grad von M_r. Der Grad einer beliebigen Mannigfaltigkeit ist gleich der Summe der Grade ihrer irreduziblen Bestandteile, jeder so oft gezählt, wie seine Vielfachheit angibt. Der Grad einer Punktgruppe wird im folgenden auch Ordnung genannt. Man überzeugt sich leicht, daß der Grad einer irreduziblen Hyperfläche gleich dem Grad der definierenden irreduziblen Form ist. Das gilt also insbesondere für eine allgemeine Hyperfläche $F(u, x)$ als eine irreduzible $K(u)$-Mannigfaltigkeit.

$K(u)$ entstehe aus K durch Adjunktion endlichvieler Unbestimmten u_i. Es sei $(\xi^{(1)}, \xi^{(2)}, \ldots, \xi^{(g)})$ eine beliebige irreduzible $K(u)$-Punktgruppe. Ohne Beschränkung der Allgemeinheit können wir annehmen, daß $\xi_0^{(1)}$ und folglich auch alle $\xi_0^{(i)}$ nicht gleich Null sind. Normieren wir die Koordinaten dieser Punkte vorläufig so, daß $\xi_0^{(1)} = \xi_0^{(2)} = \ldots = \xi_0^{(g)} = 1$, so bilden die Linearformen $\sum_{j=0}^{n} \xi_j^{(i)} v_j$ ein vollständiges System von konjugierten Linearformen über $K(u)$. Hat $K(\xi)$ den Exponenten e über $K(u)$, dann ist $\prod_{i=1}^{g} (\sum_{j=0}^{n} \xi_j^{(i)} v_j)^{p^e}$ eine Form in v mit Koeffizienten aus $K(u)$[10]). Wir behaupten: *$\prod_{i=1}^{g} (\sum_{j=0}^{n} \xi_j^{(i)} v_j)^{p^e}$ ist in bezug auf $K(u)$ irreduzibel.*

---

[9]) [9], S. 141.

[10]) p ist die Charakteristik von K, wenn sie nicht Null ist. Im Fall der Charakteristik Null ist $p = 1$ zu setzen.

Beweis: Da $\prod_{i=1}^{g} (\sum_{j=0}^{n} \xi_j^{(i)} v_j)$ invariant über $K(u)$ ist, so gibt es eine ganze Zahl e', so daß $\prod_{i=1}^{g} (\sum_{j=0}^{n} \xi_j^{(i)} v_j)^{p^{e'}}$ eine irreduzible Form in v mit Koeffizienten aus $K(u)$ ist. Dann gibt es offenbar Elemente $\mu_i^{(j)}$ aus $K(u)$ mit $|\mu_i^{(j)}| \neq 0$, so daß $\eta_j^{(i)} = -\Sigma \mu_j^{(k)} \xi_k^{(i)}$ (bei festem j) für alle i verschieden sind. Die $\eta_j^{(i)}$ sind dann Nullstellen des Polynoms $\prod_{i=1}^{g} (v_0 - \eta_j^{(i)})^{p^{e'}}$, also eines Polynoms in $K(u)$ mit g verschiedenen $p^{e'}$-fachen Wurzeln. Das bedeutet, die $\eta_j^{(i)}$ sind höchstens von dem Exponenten e' über $K(u)$. Da $K(\xi) = K(\eta)$ über $K(u)$ den Exponenten e hat, so folgt daraus, daß $e \leqq e'$ ist. Andererseits ist offenbar $e' \leqq e$, mithin ist $e' = e$. Also ist $\prod_{i=1}^{g} (\sum_{j=0}^{n} \xi_j^{(i)} v_j)^{p^e}$ irreduzibel.

Es sei nun $\prod_{i=1}^{g} (\sum_{j=0}^{n} \xi_j^{(i)} v_j)^{p^e} = R(u,v)\, q(u)$, wo $R(u,v)$ ein irreduzibles Element aus $K[u,v]$ ist und $q(u)$ ein Element aus $K(u)$. Normieren wir so, daß $\xi_0^{(i)} = q(u)^{-\frac{1}{gp^e}}$, so fällt in dem Produkt der Faktor $q(u)$ weg, und wir haben

$$R(u,v) = \prod_{i=1}^{g} (\sum_{j=0}^{n} \xi_j^{(i)} v_j)^{p^e}.$$

Die so definierte Normierung der Koordinaten von $\xi^{(i)}$ heißt die normale Normierung und $R(u,v)$ heißt die zugeordnete Form der $K(u)$-Punktgruppe. Die normale Normierung und die zugeordnete Form einer Punktgruppe sind offenbar bis auf einen multiplikativen Faktor aus K eindeutig.

Es sei nun C eine irreduzible K-Kurve. Eine allgemeine Hyperfläche $F(u,x)$ vom Grade m schneidet aus C eine irreduzible $K(u)$-Punktgruppe aus. Die Anzahl der Punkte in dieser Punktgruppe sei mit g_m bezeichnet. g_1 ist also der Grad von C. Wir werden jetzt beweisen: $g_m = m g_1$.

Wir betrachten zwei allgemeine Hyperflächen $F_1(u,x)$ bzw. $F_2(v,x)$ der Grade m bzw. k, deren g_m bzw. g_k Schnittpunkte mit C in normaler Normierung mit $\xi^{(1)}, \xi^{(2)}, \ldots, \xi^{(g_m)}$ bzw. $\eta^{(1)}, \eta^{(2)}, \ldots, \eta^{(g_k)}$ bezeichnet werden mögen. Haben $K(\xi)$ bzw. $K(\eta)$[11]) die Exponenten e_m bzw e_k über $K(u)$ bzw. $K(v)$, dann läßt sich leicht zeigen, daß

$$G_1(u,v) = \prod_{i=1}^{g_k} F_1(u, \eta^{(i)})^{p^{e_k}}$$

[11]) Es soll hier ein für allemal festgesetzt werden, daß $K(\xi)$ immer den Körper bedeuten soll, der aus K durch Adjunktion der Verhältnisse der Koordinaten der Punkte $\xi^{(i)}$, nicht der Koordinaten selbst, entstanden ist.

und

$$G_2(u, v) = \prod_{i=1}^{g_m} F_2(v, \xi^{(i)})^{p^{e_m}}$$

irreduzible Elemente aus $K(u, v)$ sind[12]).

Es ist nun $G_1(u, v) = G_2(u, v)$. Denn F_1, F_2, C haben dann und nur dann einen gemeinsamen Punkt, wenn $F_1(u, x)$ bzw. $F_2(v, x)$ für wenigstens einen der g_k bzw. g_m Schnittpunkte von $F_2(v, x)$ bzw. $F_1(u, x)$ mit C verschwinden, d. h. wenn $G_1(u, v) = 0$ bzw. $G_2(u, v) = 0$. Das bedeutet, $G_1(u, v)$ verschwindet dann und nur dann, wenn $G_2(u,v)$ verschwindet. Da $G_1(u, v)$ und $G_2(u, v)$ beide irreduzibel sind, so folgt daraus nach dem Hilbertschen Nullstellensatz, daß sie gleich sind. Die somit als gleich bewiesenen Formen werden wir jetzt mit $G(u, v)$ bezeichnen.

Alle e sind gleich Null. Denn setzt man $k = m$, so ist $e_k = e_m$, und die u, v werden in $G(u, v)$ nur in p^{e_m}-Potenzen auftreten. Wäre $e_m \neq 0$, so würde $G(u, v)$ reduzibel sein (da K vollkommen ist), was gegen die Definition von $G(u, v)$ ist. Also sind alle e gleich Null, und wir haben

$$G(u, v) = \prod_{i=1}^{g_k} F_1(u, \eta^{(i)}) = \prod_{i=1}^{g_m} F_2(v, \xi^{(i)}).$$

Spezialisieren wir jetzt $F_1(u^*, x) = \prod_{j=1}^{m} L_j(w^{(j)}, x)$, wo die L_j allgemeine Hyperebenen oder Linearformen sind, und bezeichnen wir mit $\Theta^{(j, 1)}, \ldots, \Theta^{(j, g_1)}$ die g_1 Schnittpunkte von L_j mit C, dann haben wir

$$\begin{aligned} G(u^*, v) &= \prod_{i=1}^{g_k} F_1(u^*, \eta^{(i)}) = \prod_{i=1}^{g_k} \prod_{j=1}^{m} L_j(w^{(j)}, \eta^{(i)}) \\ &= \prod_{j=1}^{m} \prod_{i=1}^{g_k} L_j(w^{(j)}, \eta^{(i)}) = \prod_{j=1}^{m} \prod_{i=1}^{g_1} F_2(v, \Theta^{(j, i)}). \end{aligned}$$

Vergleichen wir nun die Grade der rechten und linken Seiten in v, so haben wir $g_m = m g_1$.

Damit ist die Anzahl der Schnittpunkte einer allgemeinen Hyperfläche $F_1(u, x)$ mit der Kurve C bestimmt. Wir wollen nun untersuchen, was mit den Schnittpunkten geschieht, wenn man die allgemeine Hyperfläche spezialisiert, d. h. wenn man die Unbestimmten u durch irgendwelche Elemente aus einem Oberkörper von K ersetzt. Die $m g_1$ Schnittpunkte von $F_1(u, x)$ mit C lassen sich, wenn wir für F_2 eine Linearform $L(v, x) = \Sigma v_j x_j$ nehmen, durch Faktorzerlegung von $G(u, v)$ nach v

[12]) Daß $G_1(u, v)$ bzw. $G_2(u, v)$ in u bzw. v irreduzible Elemente aus $K[u, v]$ sind, ist nach dem Vorangegangenen ziemlich klar. Daß $G_1(u, v)$ bzw. $G_2(u, v)$ auch in v bzw. u irreduzibel sind, sieht man ein, wenn man die Hyperflächen $F_1(u, x)$ bzw. $F_2(v, x)$ in m bzw. k allgemeine Hyperebenen zerfallen läßt.

gewinnen:

$$G(u,v) = \prod_{i=1}^{mg_1} (v_0 \xi_0^{(i)} + \ldots + v_n \xi_n^{(i)}) = \prod_{i=1}^{mg_1} L(v, \xi^{(i)}).$$

Spezialisiert man nun $u \to \mu$, so geht $G(u,v)$ in $G(\mu,v)$ über. Ist identisch $G(\mu,v) = 0$, so ist $\prod_{i=1}^{g_1} F_1(\mu, \eta^{(i)}) = 0$, also etwa $F_1(\mu, \eta^{(i)}) = 0$. Da $\eta^{(i)}$ ein allgemeiner Punkt von C ist, so bedeutet dies, daß $F_1(\mu, x)$ die Kurve C enthält. Nehmen wir also an, daß $F_1(\mu, x)$ die Kurve C nicht enthält und daher mit C nur endlichviele Schnittpunkte hat, so verschwindet $G(\mu, v)$ nicht identisch.

Durch Faktorenzerlegung nach v, [13])

$$G(\mu,v) = \prod_{i=1}^{mg_1} L(v, \alpha^{(i)}) = \prod_{i=1}^{mg_1} (v_0 \alpha_0^{(i)} + \ldots + v_n \alpha_n^{(i)}),$$

erhalten wir mg_1 (nicht notwendig verschiedene) Punkte $\alpha^{(1)}, \ldots, \alpha^{(mg_1)}$, die eine $K(\mu)$-Punktgruppe $P(\mu)$ der Ordnung mg_1 bilden. Da $P(\mu)$ eine relationstreue Spezialisierung von $P(u) = (\xi^{(i)}, \ldots, \xi^{(mg_1)})$ für die Spezialisierung $u \to \mu$ ist, so gehört jeder in $P(\mu)$ enthaltene Punkt zu den Schnittpunkten von $F_1(\mu, x)$ mit C. Umgekehrt ist auch jeder solche Schnittpunkt in $P(\mu)$ enthalten. Denn, sei β ein Schnittpunkt von $F_1(\mu, x)$ mit C, so ist β eine relationstreue Spezialisierung von $\xi^{(1)}$. Ist $F_1(\mu', x)$ eine allgemeine durch β gehende Hyperfläche, dann ist $(u, \xi^{(1)}) \to (\mu', \beta)$ eine relationstreue Spezialisierung. Da nun $(\mu', \beta) \to (\mu, \beta)$ offenbar eine relationstreue Spezialisierung ist, so ist folglich $(u, \xi^{(1)}) \to (\mu, \beta)$ eine relationstreue Spezialisierung. Daraus folgt wegen der Eindeutigkeit der relationstreuen Spezialisierung, daß β in $P(\mu)$ enthalten ist.

Betrachten wir eine zerfallende Form $F = F_1^{r_1} \ldots F_t^{r_t}$ vom Grade m als eine reduzible Hyperfläche, in der die irreduziblen Hyperflächen F_i mit den Vielfachheiten r_i vorkommen, so können wir das bisher in diesem Paragraphen bewiesene in dem folgenden Satz zusammenfassen:

Der Satz von Bezout. *Jede Hyperfläche $F(\mu, x)$ vom Grade m, die eine irreduzible Kurve C vom Grade g nicht enthält, schneidet aus der Kurve C eine $K(\mu)$-Punktgruppe $P(\mu)$ der Ordnung mg aus. Die Punkte von $P(\mu)$ lassen sich durch Faktorenzerlegung einer Form $G(\mu, v)$ nach v erhalten, wobei die Form $G(u, v)$ nur von C und m abhängt. Ist $F(\mu, x)$ eine allgemeine Hyperfläche, so besteht $P(\mu)$ aus mg verschiedenen Punkten.*

---

[13]) Daß $G(\mu, v)$ in Linearfaktoren zerfällt, folgt daraus, daß $G(u, v)$ in Linearfaktoren zerfällt. Denn die Eigenschaft einer Form, in Linearfaktoren zu zerfallen, läßt sich bekanntlich durch das Verschwinden eines Formensystems in den Koeffizienten ausdrücken. Da nun das Formensystem schon für unbestimmte u verschwindet, so muß es natürlich auch für die speziellen Werte μ verschwinden.

§ 3.

Lineare Scharen und rationale Abbildungen.

Ein wichtiger Spezialfall der vorstehenden Begriffsbildungen entsteht, wenn man $u_i \to \sum_j a_j^{(i)} \lambda_j$ spezialisiert, wobei die $a_j^{(i)}$ Elemente aus K und die λ_j nicht in K liegende Unbestimmte sind. Die dadurch aus der allgemeinen Hyperfläche $F_1(u, x)$ entstandene spezialisierte Hyperfläche

$$F_1(\Sigma a \lambda, x) = \lambda_0 F_0(x) + \ldots + \lambda_s F_s(x) = F(\lambda, x)$$

nennen wir die allgemeine Hyperfläche des durch $F_0(x), \ldots, F_s(x)$ erzeugten s-dimensionalen Linearsystems Φ. Wir können annehmen, daß keine Hyperfläche $F(\mu, x)$ des Systems Φ, wo μ Elemente aus irgendeinem Oberkörper von K sind, die Kurve C enthält. Denn, wenn etwa q linear unabhängige Hyperflächen von Φ die Kurve C enthalten, so läßt sich in bekannter Weise ein lineares Teilsystem der Dimension $s - q$ aus Φ herausgreifen, welches dieselben Punktgruppen ausschneidet wie Φ und in welchem keine Hyperfläche die Kurve C enthält. Da es im folgenden nur auf diese Punktgruppen ankommt, so kann man das ganze System durch dieses Teilsystem ersetzen.

Die von $F(\lambda, x)$ auf der Kurve C ausgeschnittene $K(\lambda)$-Punktgruppe $P(\lambda)$ läßt sich nach dem Satz von Bezout durch die Faktorenzerlegung von $G(\Sigma a \lambda, v)$ nach v gewinnen:

$$G(\Sigma a \lambda, v) = \prod_{i=1}^{g_1} (\sum_j \lambda_j F_j(\eta^{(i)})) = \prod_{i=1}^{mg} (\sum_j v_j \alpha_j^{(i)}) = T(\lambda, v).$$

$T(\lambda, v)$ möge nun in einen von λ unabhängigen Teil $T(v)$ und einen von λ abhängigen Teil $T^*(\lambda, v)$ zerfallen. Dementsprechend zerfällt die Punktgruppe $P(\lambda)$ in eine feste Gruppe P und eine von λ abhängige Gruppe $P^*(\lambda)$. Die Punkte $\alpha^{(i)}$ seien so numeriert, daß $T^*(\lambda, v) = \prod_{i=1}^{t} (\sum_j v_j \alpha_j^{(i)})$, also $P^*(\lambda) = (\alpha^{(1)}, \ldots, \alpha^{(t)})$ ist. Die Punkte von P sind die festen Punkte, die Punkte von $P^*(\lambda)$ die allgemeinen Punkte der durch das System Φ auf der Kurve C erzeugten s-dimensionalen linearen K-Schar von Punktgruppen. Bis auf weiteres werden wir immer die feste Gruppe P aus der Schar weglassen. Dann ist $P^*(\lambda)$ die allgemeine Punktgruppe der Schar, und jede spezielle Punktgruppe für die Spezialisierung $\lambda \to \mu$ läßt sich durch Faktorenzerlegung von $T^*(\mu, v)$ nach v erhalten. Die Form $T^*(\lambda, v)$ wird die zugeordnete Form der Schar genannt. Die Ordnung t von $P^*(\lambda)$ heißt auch die Ordnung der Schar. Die Schar selbst bezeichnen wir mit $\mathfrak{G}_t^s = \langle P^*(\lambda) \rangle$. Es ist $s \leqq t$. Denn, wäre $s > t$, so würde es eine Hyperfläche von Φ geben, die durch wenigstens $t + 1$ allgemeine Punkte von C ginge. Das ist aber unmöglich, da die allgemeine Hyperfläche von Φ nur durch t allgemeine Punkte von C geht und keine Hyperfläche von Φ die Kurve C enthält.

Die Form $T^*(\lambda, v)$ *ist irreduzibel* oder, was offenbar damit gleichbedeutend ist, die Form $T(\lambda, v) = T(v)\, T^*(\lambda, v)$ ist in λ irreduzibel (in bezug auf $K(v)$).

Beweis: Es ist $T(\lambda, v) = \prod_{i=1}^{g} (\sum_j \lambda_j F_j(\eta^{(i)}))$. Da die $\eta^{(i)}$ zueinander konjugiert sind, folgt daraus, daß die Form $T(\lambda, v)$, wenn sie in λ reduzibel wäre, eine Potenz einer in λ irreduziblen Form sein müßte. Das bedeutet aber, daß es wenigstens zwei Punkte, etwa $\eta^{(1)}$, $\eta^{(2)}$, gäbe, so daß $\Sigma \lambda_j F_j(\eta^{(1)}) = \Sigma \lambda_j F_j(\eta^{(2)})$. Daraus folgt, daß eine durch $\eta^{(1)}$ gehende Hyperfläche von Φ, z. B. etwa $H = F_2(\eta^{(1)}) F_1(x) - F_1(\eta^{(1)}) F_2(x)$, auch durch $\eta^{(2)}$ gehen müßte. Das ist aber absurd, denn die endlichvielen weiteren Schnittpunkte von H mit C sind schon durch den Punkt $\eta^{(1)}$ bestimmt, und eine allgemeine durch $\eta^{(1)}$ gehende Hyperebene kann offenbar keinen von diesen endlichvielen Punkten unter ihren weiteren Schnittpunkten mit C haben.

Aus der Irreduzibilität von $T^*(\lambda, v)$ folgt, daß $T^*(\lambda, v)$ die zugeordnete Form der irreduziblen $K(\lambda)$-Punktgruppe $P^*(\lambda)$ ist. Damit ist bewiesen:

Der (verschärfte) Satz von Bertini. Die allgemeinen Punkte $\alpha^{(i)} = (\alpha_0^{(i)}, \alpha_1^{(i)}, \ldots, \alpha_n^{(i)})$ *einer linearen Schar* $\langle P^*(\lambda)\rangle$ *auf einer irreduziblen Kurve haben die Vielfachheit* p^e, *wo* e *der Exponent der (zueinander konjugierten) Körper* $K(\alpha^{(i)})$[11] *über* $K(\lambda)$ *ist. Dabei ist* p *gleich der Charakteristik von* K, *falls sie nicht Null ist, gleich Eins, falls sie Null ist.*

Eine rationale Abbildung der Kurve C ist gegeben durch die für die allgemeinen Punkte von C gültigen Formeln:

$$y_0 : y_0 : \ldots : y_t = \varphi_0(x) : \varphi_1(x) : \ldots : \varphi_t(x).$$

Die Bildkurve ist die durch den Punkt $\varphi_i(\xi)$ in S_t definierte Kurve, wobei ξ einen allgemeinen Punkt von C bedeutet. Ohne Beschränkung der Allgemeinheit können wir auch hier annehmen, daß keine Hyperfläche des Linearsystems $\Sigma \lambda_i \varphi_i(x)$ die Kurve C enthält, was damit gleichbedeutend ist, daß die Bildkurve in keinem echten linearen Unterraum von S_t liegt, also zu S_t gehört. Die Abbildung heißt birational, wenn die ξ_i auch von den $\varphi_i(\xi)$ rational abhängig sind, d. h. wenn es Formen $\psi_0(y), \psi_1(y), \ldots, \psi_n(y)$ gibt, so daß

$$\xi_0 : \xi_1 : \ldots : \xi_n = \psi_0(\varphi(\xi)) : \psi_1(\varphi(\xi)) : \ldots : \psi_n(\varphi(\xi)).$$

Satz. *Die rationale Abbildung der Kurve* C

$$y_0 : y_1 : \ldots : y_t = \varphi_0(x) : \varphi_1(x) : \ldots : \varphi_t(x)$$

ist (dann und nur dann) birational, wenn die durch $\Sigma \lambda_i \varphi_i$ *erzeugte lineare Schar* $\langle P^*(\lambda)\rangle$ *auf C einfach ist, d. h. wenn alle durch einen allgemeinen Punkt von C gehenden Punktgruppen von* $\langle P^*(\lambda)\rangle$ *keinen weiteren Punkt gemeinsam haben.*

Beweis: Die Abbildung ist für allgemeine Punkte eineindeutig. Denn, gäbe es zwei allgemeine Punkte ξ, ξ' von C mit demselben Bildpunkt, also

$$\varphi_0(\xi) : \varphi_1(\xi) : \ldots : \varphi_t(\xi) = \varphi_0(\xi') : \varphi_1(\xi') : \ldots : \varphi_t(\xi'),$$

so würden alle durch ξ gehenden Hyperflächen des Systems $\Sigma \lambda_j \varphi_j$ auch durch ξ' gehen. Da ξ' ein allgemeiner Punkt ist, so würde das bedeuten, daß alle durch ξ gehenden Punktgruppen von $\langle P(\lambda)\rangle$ auch durch ξ' gehen müßten, gegen die Voraussetzung, daß $\langle P(\lambda)\rangle$ einfach ist. Also ist der Punkt ξ durch $\psi_i(\xi)$ eindeutig bestimmt. Das bedeutet, wenn $\xi_0 = 1$ normiert ist, daß $\xi_1, \ldots, \xi_n$ über $K(\varphi(\xi))$ keine verschiedenen Konjugierten besitzen. Da $\langle P(\lambda)\rangle$ einfach ist, sind $\xi_1, \ldots, \xi_n$ separabel über $K(\lambda)$. Daraus folgt wegen

$$K(\xi, \lambda_1, \ldots, \lambda_n) \supset K(\varphi(\xi), \lambda_1, \ldots, \lambda_n) \supset K(\lambda_0, \ldots, \lambda_n) = K(\lambda),$$

$$\left(\lambda_0 = \frac{\sum_1^t \lambda_i \varphi_i(\xi)}{\varphi_0(\xi)}\right),$$

daß $K(\xi, \lambda_1, \ldots, \lambda_n)$ über $K(\varphi(\xi), \lambda_1, \ldots, \lambda_n)$, mithin auch $K(\xi)$ über $K(\varphi(\xi))$ separabel ist. Da außerdem $\xi_1, \ldots, \xi_n$ über $K(\varphi(\xi))$ keine verschiedenen Konjugierten besitzen, so ist $K(\xi) = K(\varphi(\xi))$.

Satz (Die Invarianz der linearen Schar). *Die Kurve C sei durch die birationale Abbildung*

$$y_0 : y_1 : \ldots : y_s = \varphi_0(x) : \varphi_1(x) : \ldots : \varphi_s(x),$$
$$x_0 : x_1 : \ldots : x_n = \psi_0(y) : \psi_1(y) : \ldots : \psi_n(y)$$

in die Kurve $\bar{C}$ *in* S_s *abgebildet.* $P^*(\lambda) = (\alpha^{(1)}, \ldots, \alpha^{(t)})$ *sei die allgemeine Punktgruppe der von* $F(\lambda, x)$ *auf C erzeugten linearen Schar,* $\bar{P}^*(\lambda)$ *die allgemeine Punktgruppe der von* $F(\lambda, \psi(y))$ *auf* $\bar{C}$ *erzeugten. Dann ist* $\bar{P}^*(\lambda) = (\varphi(\alpha^{(1)}), \ldots, \varphi(\alpha^{(t)}))$.

Beweis: Sind $\alpha^{(1)}, \ldots, \alpha^{(h)}$ die verschiedenen allgemeinen Schnittpunkte von $F(\lambda, x)$ mit C, so folgt aus der Eineindeutigkeit der birationalen Abbildung für allgemeine Punkte, daß $\varphi(\alpha^{(1)}), \ldots, \varphi(\alpha^{(h)})$ genau die verschiedenen allgemeinen Schnittpunkte von $F(\lambda, \psi(y))$ mit C sind. Also enthält $\bar{P}^*(\lambda)$ die Punkte $\varphi(\alpha^{(1)}), \ldots, \varphi(\alpha^{(h)})$ und keinen anderen. Da nun $K(\alpha^{(i)}) = K(\varphi(\alpha^{(i)}))$ ist, so folgt aus dem (verschärften) Satz von Bertini, daß $\varphi(\alpha^{(i)})$ in $\bar{P}^*(\lambda)$ dieselbe Vielfachheit haben wie $\alpha^{(i)}$ in $\bar{P}^*(\lambda)$. Also ist $\bar{P}^*(\lambda) = (\varphi(\alpha^{(1)}), \ldots, \varphi(\alpha^{(t)}))$.

§ 4.

Die Auflösung der Singularitäten.

Bisher haben wir nur die allgemeinen Punkte der Kurve C betrachtet. Jetzt wenden wir uns den speziellen oder nulldimensionalen Punkten zu. Die Verhältnisse sind hier wesentlich komplizierter, da die speziellen Punkte, im Gegensatz zu den allgemeinen, nicht mehr alle miteinander algebraisch äquivalent sind. Dabei ergibt sich als besonders wichtig für algebraisch-geometrische Untersuchungen der Begriff der Singularitäten, den wir jetzt definieren werden. Ein Punkt $p = (p_0, \ldots, p_n)$ heißt ein k-facher Punkt der Kurve C, wenn eine allgemeine durch p gehende Hyperebene — d. h. eine Hyperebene $\Sigma v_i x_i$, wo die v_i Koordinaten eines allgemeinen Punktes der Hyperebene $\Sigma p_i x_i$ sind — mit C einen k-fachen Schnittpunkt in p hat. Punkte, die nicht einfach sind, heißen die singulären Punkte, deren Gesamtheit die Singularitäten der Kurve C genannt wird. Die allgemeinen Punkte einer irreduziblen Kurve sind offenbar einfach, dagegen zeigen schon einfache Beispiele, daß unter den speziellen Punkten sehr wohl Singularitäten auftreten können.

Es ist eine komplizierte Sache, die Struktur aller möglichen Singularitäten einer irreduziblen Kurve zu untersuchen. Glücklicherweise ist dies für unseren Zweck nicht nötig. Wir werden nämlich zeigen, daß es möglich ist, jede irreduzible Kurve durch eine birationale Abbildung in eine singularitätenfreie Kurve zu transformieren. Dazu seien einige Bemerkungen über rationale Abbildungen vorausgeschickt. Wir haben bisher die rationalen Abbildungen nur für allgemeine Punkte definiert, und zwar durch die Formeln:

$$y_0 : y_1 : \ldots : y_s = \varphi_0(x) : \varphi_1(x) : \ldots : \varphi_s(x).$$

Da jeder Punkt einer irreduziblen Kurve eine relationstreue Spezialisierung eines allgemeinen Punktes ist, so liegt es nahe, die rationale Abbildung dadurch zu ergänzen, daß man einen Punkt δ in S_s als Bildpunkt eines Punktes γ der Kurve C betrachtet, wenn γ, δ eine relationstreue Spezialisierung[14]) von $\xi, \varphi(\xi)$ ist (wobei ξ ein allgemeiner Punkt von C ist). Ein so definierter Bildpunkt δ gehört offenbar zu der durch $\varphi(\xi)$ definierten Bildkurve $\overline{C}$, da δ auch insbesondere eine relationstreue Spezialisierung von $\varphi(\xi)$ sein muß. Jeder Punkt von C besitzt wenigstens einen Bildpunkt, und umgekehrt ist jeder Punkt von $\overline{C}$ auch Bild von wenigstens einem Punkte von C. Das folgt daraus, daß jede partielle relationstreue Spezialisierung sich immer zu einer vollständigen ergänzen läßt[15]).

---

[14]) [5], S. 756.

[15]) [8], S. 131.

Die Abbildung sei nun birational. Wir betrachten die durch das System $\Sigma \lambda_i \varphi_i$ auf der Kurve C erzeugte lineare Schar $\langle P(\lambda)\rangle$. Die Bildpunktgruppe $\overline{P}(\lambda)$ von $P(\lambda)$ ist die von einer allgemeinen Hyperebene $\Sigma \lambda_i y_i$ auf der Kurve $\overline{C}$ ausgeschnittene Punktgruppe. Daraus folgt: Der Grad $\overline{g}$ der Bildkurve ist gleich der Ordnung von $\langle P(\lambda)\rangle$. Ist δ ein k-facher Punkt von $\overline{C}$ und $\Sigma \mu_i y_i$ eine allgemeine durch δ gehende Hyperebene, wo etwa $\mu_0 = -(\delta_1 \lambda_1' + \ldots + \delta_s \lambda_s')$, $\mu_1 = \delta_0 \lambda_1', \ldots, \mu_s = \delta_0 \lambda_s'$, so hat $\overline{P}(\mu)$ außer dem k-fachen Punkte δ noch $\overline{g} - k$ allgemeine Punkte. Dann hat $P(\mu)$ auch $\overline{g} - k$ allgemeine Punkte, die jenen $\overline{g} - k$ Punkten von $\overline{P}(\mu)$ entsprechen[16]), und einige weitere Punkte der gesamten Vielfachheit k, die die in δ abgebildeten Punkte von C sind. Alle in δ abgebildeten Punkte von C sind aber schon dadurch erhalten. Denn, sei γ, δ eine relationstreue Spezialisierung von $\xi, \varphi(\xi) = \overline{\xi}$. Es gelten für $\xi, \varphi(\xi)$ die Relationen: Setzt man $\lambda_0 = -(\overline{\xi}_1 \lambda_1' + \ldots + \overline{\xi}_s \lambda_s')$, $\lambda_1 = \overline{\xi}_0 \lambda_1', \ldots, \lambda_s = \overline{\xi}_0 \lambda_s'$, $v_0 = -(\xi_1 v_1' + \ldots + \xi_n v_n')$, $v_1 = \xi_0 v_1', \ldots, v_n = \xi_0 v_n'$, dann verschwindet $T(\lambda, v)$ (die zugeordnete Form von $\langle P(\lambda)\rangle$) identisch in λ' und v'. Diese Relationen müssen auch für γ, δ gelten, also: Setzt man $\lambda_0 = -(\delta_1 \lambda_1' + \ldots + \delta_s \lambda_s')$, $\lambda_1 = \delta_0 \lambda_1', \ldots, \lambda_s = \delta_0 \lambda_s'$, $v_0 = -(\gamma_1 v_1' + \ldots + \gamma_n v_n')$, $v_1 = \gamma_0 v_1', \ldots, v_n = \gamma_0 v_n'$, dann verschwindet $T(\lambda, v)$ identisch. Das bedeutet, daß $\Sigma v_i \gamma_i$ ein Linearfaktor von $T(\mu, v)$ ist. Folglich ist γ in $P(\mu)$ enthalten. Das Bild von δ (bei der inversen Abbildung) hat also höchstens k Punkte; daraus folgt für $k = 1$, daß bei einer birationalen Abbildung ein einfacher Punkt nur einen Bildpunkt besitzt. Wir haben demzufolge: *Eine birationale Abbildung zwischen zwei singularitätenfreien Kurven ist (für alle Punkte) eineindeutig.*

Läßt man nun die k festen Punkte von $P(\mu)$ weg, so bilden die übrigen $\overline{g} - k$ (allgemeinen) Punkte eine Punktgruppe $Q(\lambda')$, die eine in $\langle P(\lambda)\rangle$ enthaltene lineare $K(\delta)$-Schar $\langle Q(\lambda')\rangle$ der Dimension $s - 1$ und der Ordnung $\overline{g} - k$ definiert. Daraus folgt: Enthält eine lineare Schar $\langle P(\lambda)\rangle$ der Dimension $s \geqq 2$ und der Ordnung g keine Teilschar (in bezug auf einen Oberkörper von K) der Dimension $s - 1$ und der Ordnung $< g - 1$, so ist die durch die entsprechende birationale[17]) Abbildung transformierte Kurve C singularitätenfrei. Durch den Nachweis der Existenz einer solchen Schar werden wir den Beweis des folgenden Satzes erbringen.

---

[16]) Denn $P(\mu)$ bzw. $\overline{P}(\mu)$ entstehen aus $P(\lambda)$ bzw. $\overline{P}(\lambda)$ durch relationstreue Spezialisierung für die Spezialisierung $\lambda \to \mu$. Daraus folgt, daß $(P(\mu), \overline{P}(\mu))$ eine relationstreue Spezialisierung von $(P(\lambda), \overline{P}(\lambda)) = (\alpha^{(1)}, \ldots, \alpha^{(t)}, \varphi(\alpha^{(1)}), \ldots, \varphi(\alpha^{(t)}))$ ist. Daraus folgt, daß jeder Punkt von $P(\mu)$ wenigstens einen Bildpunkt in $\overline{P}(\mu)$ besitzt und umgekehrt.

[17]) Eine solche Schar $\langle P(\lambda)\rangle$ ist offenbar einfach.

Satz. *Jede irreduzible Kurve C läßt sich durch eine birationale Abbildung in eine singularitätenfreie Kurve transformieren.*

Beweis: Wir bemerken zunächst, daß wir uns auf ebene Kurven beschränken können. Denn jede irreduzible Kurve C läßt sich auf eine ebene Kurve birational abbilden. Ist nämlich $\xi = (\xi_0 = 1, \xi_1, \ldots, \xi_n)$ ein allgemeiner Punkt von C, dann ist $K(\xi)$ vom Transzendenzgrad 1 über K, also eine algebraische Erweiterung über $K(y_1)$, wo y_1 ein von K unabhängiges Element aus $K(\xi)$ ist. Da K vollkommen ist, so hat $K(y_1)$ den Grad p über dem Körper $[K(y_1)]^p = K(y_1^p)$, der aus den p-ten Potenzen der Elemente von $K(y_1)$ besteht. Daraus folgt nach Steinitz[18]), daß $K(\xi)$ eine einfache Erweiterung über $K(y_1)$ ist, d. h. $K(\xi) = K(y_1, y_2)$, wo y_2 algebraisch von y_1 abhängt. Setzt man nun $\eta_1 = \eta_0 y_1$, $\eta_2 = \eta_0 y_2$, dann definiert $\eta = (\eta_0, \eta_1, \eta_2)$ eine ebene Kurve, die, wie leicht einzusehen ist, ein birationales Bild von C ist.

C sei also eine ebene Kurve vom Grad m. Das System aller Kurven vom Grade $m-1$ schneidet auf C eine lineare Schar $\mathfrak{G}_{n_1}^{r_1}$ der Ordnung $n_1 = m(m-1)$ und der Dimension $r_1 = \frac{(m-1)(m+2)}{2}$ aus. Enthält $\mathfrak{G}_{n_1}^{r_1}$ keine Teilschar der Dimension $r_1 - 1$ und der Ordnung $\leqq n_1 - 2$, so sind wir schon am Ziel; sonst sei $\mathfrak{G}_{n_2}^{r_2}$ eine Ω_1-Teilschar (wo Ω_1 eine algebraische Erweiterung von K ist) der Dimension $r_2 = r_1 - 1$ und der Ordnung $n_2 \leqq n_1 - 2$. Stimmt $\mathfrak{G}_{n_2}^{r_2}$ mit ihren Konjugierten (relativ zu K) überein, so ist sie eine K-Schar; sonst seien $\mathfrak{G}_{n_2}^{(1)r_2}, \ldots, \mathfrak{G}_{n_2}^{(s)r_2}$ die voneinander verschiedenen konjugierten Scharen. Die Schar $\mathfrak{G}_{n_3}^{r_3} = \mathfrak{G}_{n_2}^{r_2} \frown \mathfrak{G}_{n_2}^{(1)r_2}$ hat dann die Dimension $r_3 = r_2 - 1$ und die Ordnung $n_3 \leqq n_2 - 2$.[19]) Die Schar $\mathfrak{G}_{n_4}^{r_4} = \mathfrak{G}_{n_3}^{r_3} \frown \mathfrak{G}_{n_2}^{(2)r_2}$ hat dann die Dimension $r_4 \geqq r_3 - 1$ und die Ordnung $n_4 \leqq n_3 - 2$ usw. Schließlich hat die Schar

$$\mathfrak{G}_{n_{s+2}}^{r_{s+2}} = \mathfrak{G}_{n_{s+1}}^{r_{s+1}} \frown \mathfrak{G}_{n_2}^{(s)r_2} = \mathfrak{G}_{n_2}^{r_2} \frown \mathfrak{G}_{n_2}^{(1)r_2} \frown \ldots \frown \mathfrak{G}_{n_2}^{(s)r_2}$$

die Dimension $r_{s+2} \geqq r_{s+1} - 1$ und die Ordnung $n_{s+2} \leqq n_{s+1} - 2$. Nun stimmt $\mathfrak{G}_{n_{s+2}}^{r_{s+2}}$ offenbar mit ihren Konjugierten überein und ist daher eine K-Schar. Enthält sie keine Teilschar der Dimension $r_{s+2} - 1$ und der Ordnung $\leqq n_{s+2} - 2$, dann sind wir am Ziel; sonst sei $\mathfrak{G}_{n_{s+3}}^{r_{s+3}}$ eine Ω_2-Teilschar der Dimension $r_{s+3} = r_{s+2} - 1$ und der Ordnung $n_{s+3} \leqq n_{s+2} - 2$. So setzen wir das Verfahren fort. Nach dem i-ten Schritte haben wir dann eine Schar $\mathfrak{G}_{n_i}^{r_i}$ der Dimension $r_i \geqq r_1 - i + 1$ und der Ordnung $n_i \leqq n_1 - 2(i-1)$. Da aber $r_i \leqq n_i$ sein muß, so haben wir $r_1 - i + 1 \leqq n_1 - 2(i-1)$, und daraus

[18]) Journ. f. Math. **137** (1910), S. 244.

[19]) Denn die weggelassenen festen Ω_1-Punktgruppen von $\mathfrak{G}_{n_2}^{r_2}$ und $\mathfrak{G}_{n_2}^{(1)r_2}$ haben keinen gemeinsamen Punkt, sonst würde $\mathfrak{G}_{n_2}^{r_2} = \mathfrak{G}_{n_2}^{(1)r_2}$ sein.

$$i \leqq \frac{(m-1)(m-2)}{2} + 1.$$

Also muß das Verfahren nach höchstens $k = \frac{(m-1)(m-2)}{2} + 1$ Schritten zu Ende kommen. Da $n_i \geqq n_k \geqq 2(m-1) \geqq 2$ (für $i \leqq k$) ist, so liefert der letzte Schritt eine lineare K-Schar mit den erwünschten Eigenschaften.

Bei dem Beweis haben wir den folgenden Hilfssatz benutzt, den wir auch später gebrauchen werden:

Ω sei algebraisch über K. Stimmt eine lineare Ω-Schar $\mathfrak{G}$ mit ihren Konjugierten überein, so ist sie eine lineare K-Schar.

Ist nämlich $\mathfrak{G}$ durch das System

$$\Sigma \lambda_i F_i(x) = \Sigma \lambda_i f_{i0}(x) + \omega \Sigma \lambda_i f_{i1}(x) + \ldots + \omega^s \Sigma \lambda_i f_{is}(x)$$

erzeugt, wo $f_{ij}(x)$ Formen aus $K[x]$ sind, so gehören auch

$$\Sigma \lambda_i f_{i0}(x) + \omega_j \Sigma \lambda_i f_{i1}(x) + \ldots + \omega_j^s \Sigma \lambda_i f_{is}(x),$$

wo $\omega_1, \ldots, \omega_s$ die Konjugierten von ω sind, zu dem System. Da K vollkommen ist, ist $|\omega_j^i| \neq 0$. Daraus folgt, daß die $\Sigma \lambda_i f_{ij}(x)$, daher auch die $f_{ij}(x)$ zu dem System gehören. Das System läßt sich also durch $f_{ij}(x)$ erzeugen, und $\mathfrak{G}$ ist folglich eine lineare K-Schar.

§ 5.

Punktgruppen. Äquivalenz. Vollscharen.

Die Existenz eines singularitätenfreien birationalen Bildes für eine irreduzible Kurve ermöglicht es, eine Geometrie auf der Kurve zu entwickeln. Wir betrachten nämlich alle miteinander birational äquivalenten (d. h. aufeinander birational abbildbaren) Kurven als die verschiedenen projektiven Darstellungen („projektiven Modelle") einer einzigen im abstrakten Sinne zu verstehenden Kurve C. Als Eigenschaften der Kurve C sollen dann nur die birational invarianten gelten. Unter den bisher eingeführten Begriffen ist der einer linearen Schar schon als birational invariant nachgewiesen. Dagegen ist der Begriff eines Punktes, und folglich auch der einer Punktgruppe als einer Menge von endlichvielen mit Vielfachheiten versehenen nulldimensionalen Mannigfaltigkeiten, nicht invariant, da eine birationale Abbildung nicht notwendig für alle Punkte eineindeutig ist[20]). Das gibt Veranlassung, diese Begriffe vom Standpunkt der Geometrie auf der Kurve aus zu modifizieren.

---

[20]) Eine Punktgruppe, betrachtet als eine relationstreue Spezialisierung der allgemeinen Punktgruppe einer bestimmten linearen Schar, ist natürlich birational invariant.

Die Nichtinvarianz des bisher benutzten Punktbegriffes zeigt, daß dieser projektiv invariante Begriff von dem birationalen Standpunkt aus etwas zu allgemein ist. Um die Invarianz wieder herzustellen, hat man also diesen Begriff etwas zu verengen. Das geschieht hier im folgenden in zwei verschiedenen Weisen. Einmal beschränken wir uns auf die singularitätenfreien Kurven und lassen dabei beliebige Punkte zu, ein anderes Mal beschränken wir uns auf die allgemeinen Punkte und lassen dabei beliebige Kurven zu. Den beiden Methoden liegt die im vorigen Paragraphen bewiesene Existenz eines singularitätenfreien birationalen Bildes zugrunde, und aus der Eineindeutigkeit der Zuordnung zwischen den Punkten und den Zweigen einer singularitätenfreien Kurve (die wir bald beweisen werden) geht auch gleichzeitig hervor, daß diese beiden äußerlich verschiedenen Methoden im Grunde genommen äquivalent sind und folglich zu denselben Resultaten führen müssen.

I. Die Gesamtheit aller singularitätenfreien Darstellungen der Kurve C ist offenbar ein invarianter Begriff. Diese Gesamtheit legen wir unseren Untersuchungen zugrunde. Da eine birationale Abbildung zwischen singularitätenfreien Darstellungen für alle Punkte eineindeutig ist, so können wir hier von einem Punkte bzw. einer Punktgruppe der Kurve C sprechen. Ein Punkt bzw. eine Punktgruppe auf C ist also eindeutig bestimmt durch einen Punkt bzw. eine Punktgruppe auf irgendeiner singularitätenfreien Darstellung. Eine Punktgruppe P_n der Ordnung n auf C ist also eine Menge von mit Vielfachheiten versehenen Punkten von C. Mit jedem Punkte treten also auch alle Konjugierten in P_n auf, und zwar mit derselben Vielfachheit. Den schon als invariant nachgewiesenen Begriff einer linearen Schar werden wir von jetzt an prinzipiell nur für singularitätenfreie Darstellungen gebrauchen, jedoch mit der Verallgemeinerung, daß jetzt auch einige feste Punkte, die zusammen eine Punktgruppe bilden, in der Schar zugelassen sind. Das heißt, in der Zerlegung $T(\lambda, v) = T(v)\, T^*(\lambda, v)$ (vgl. § 3) ist $T(v)$ zwar immer noch von λ unabhängig, aber $T^*(\lambda, v)$ kann jetzt auch einige von λ unabhängige Faktoren enthalten. Die aus $T^*(\lambda, v)$ entstandene Punktgruppe $P^*(\lambda)$ wird wieder die allgemeine Punktgruppe der Schar $\langle P^*(\lambda)\rangle$ genannt. Die Ordnung von $P^*(\lambda)$ heißt wieder die Ordnung von $\langle P^*(\lambda)\rangle$. Eine lineare Schar der Dimension r und der Ordnung n werden wir wieder mit $\mathfrak{G}_n^r$ bezeichnen.

Im folgenden werden wir irgendeine singularitätenfreie Darstellung der im abstrakten Sinne zu verstehenden Kurve C auch mit C bezeichnen. Aus dem Zusammenhang wird jedesmal klar sein, welche gemeint ist.

Die Zweckmäßigkeit der Beschränkung auf singularitätenfreie Kurven kommt zur Geltung in dem folgenden einfachen, aber grundlegenden Satz:

Die Gesamtheit aller durch einen Punkt δ *von* C *gehenden Punktgruppen einer linearen Schar* $\mathfrak{G}_n^r$ *auf* C *bildet eine lineare* $K(\delta)$*-Schar der Dimension* $\geqq r-1$, *und zwar genau der Dimension* $r-1$, *wenn* δ *nicht ein fester Punkt von* $\mathfrak{G}_n^r$ *ist.*

Beweis: Ist δ ein fester Punkt der Schar, dann ist nichts mehr zu beweisen. Es sei also δ kein fester Punkt. Wir haben zu beweisen, daß die Bedingung der Teilbarkeit von $T^*(\lambda, v)$ durch $\Sigma\,\delta_i v_i$ sich durch eine einzige lineare Gleichung in λ ausdrücken läßt. Aus

$$T(v)\,T^*(\lambda, v) = T(\lambda, v) = \prod_{i=1}^{g} \left(\sum_j \lambda_j F_j(\eta^{(i)})\right)$$

folgt, daß $T^*(\lambda, v)$ als Form in λ durch die g Linearformen $\sum_j \lambda_j F_j(\eta^{(i)})$ teilbar ist. Läßt man nun $\Sigma\, v_i x_i$ sich in eine allgemeine durch δ gehende Hyperebene $\Sigma\,\mu_i x_i$ spezialisieren, so gehen $\eta^{(1)}, \eta^{(2)}, \ldots, \eta^{(g)}$ in $\Theta^{(1)} = \delta, \Theta^{(2)}, \ldots, \Theta^{(g)}$ über. Daraus folgt, daß $T^*(\lambda, \mu)$ durch $\sum_j \lambda_j F_j(\Theta^{(i)})$ für $i = 2, \ldots, n$ teilbar ist, mithin auch durch

$$\prod_{i=2}^{g} \left(\sum_j \lambda_j F_j(\Theta^{(i)})\right).$$

Folglich ist

$$\frac{T^*(\lambda, \mu)}{\prod_{i=2}^{g} \left(\sum_j \lambda_j F_j(\Theta^{(i)})\right)} = \Sigma\,\lambda_j \varrho_j$$

eine Linearform in λ mit Koeffizienten aus $K(\delta)$. Da δ kein fester Punkt ist, kann $T^*(\lambda, \mu)$, und folglich auch $\Sigma\,\lambda_j \varrho_j$, nicht identisch verschwinden. Also ist das Verschwinden von $T^*(\lambda, \mu)$ für eine von μ unabhängige Wahl der λ gleichbedeutend mit $\Sigma\,\lambda_j \varrho_j = 0$. Da $T^*(\lambda, \mu) = 0$ mit der Teilbarkeit von $T^*(\lambda, v)$ durch $\Sigma\,\delta_i v_i$ gleichbedeutend ist, ist damit der Satz bewiesen.

Es folgt aus diesem Satz: 1. Die Gesamtheit aller durch eine Punktgruppe P_k gehenden Punktgruppen einer linearen Schar $\mathfrak{G}_n^r$ auf C bildet eine lineare Schar der Dimension $\geqq r-k$. Wenn man die feste Punktgruppe P_k wegläßt, so bildet sie eine lineare Schar der Ordnung $n-k$ und der Dimension $\geqq r-k$. Sie heißt dann der *Rest* von $\mathfrak{G}_n^r$ in bezug auf P_k. 2. Schneiden die Hyperflächen $F_0, F_1, \ldots, F_t$ die Kurve C in einer Punktgruppe P, dann schneidet auch $\lambda_0 F_0 + \ldots + \lambda_t F_t$ (für beliebige λ) die Kurve C in P.

Bemerkung. Der obige Satz und seine Folgerungen gelten offenbar auch für eine beliebige (nicht notwendig singularitätenfreie) Kurve D, wenn δ ein einfacher Punkt von D ist bzw. wenn nur einfache Punkte von D in P auftreten [21]). Es läßt sich übrigens genau wie oben der allgemeinere Satz beweisen: *Die Gesamtheit aller durch einen h-fachen Punkt* δ *von* D *gehenden Punktgruppen einer linearen Schar* $\mathfrak{G}_n^r$ *auf* D *bildet höchstens h lineare* K_1*-Scharen (wo* K_1 *ein Oberkörper von* $K(\delta)$ *ist) der Dimension* $\geqq r-1$, *und zwar genau der Dimension* $r-1$, *wenn* δ *nicht ein fester Punkt von* $\mathfrak{G}_n^r$ *ist.* Wenden wir diesen Satz auf die Hyperebenen an, so haben wir die folgende Definition (wo alle Begriffe in bezug auf K_1 zu verstehen sind): Die Gesamtheit aller Hyperebenen durch einen h-fachen Punkt δ ist offenbar ein $(n-1)$-dimensionales Linearsystem. Die Gesamtheit aller Hyperebenen, die mit D in δ einen wenigstens $(h+1)$-fachen Schnittpunkt haben, zerfällt in höchstens h $(n-2)$-dimensionale Linearsysteme $\Phi^{(i)}$. Jedes System $\Phi^{(i)}$ hat eine Gerade als seine Basis. Diese Geraden nennen wir die Tangenten der Kurve D im Punkte δ. Ein einfacher Punkt hat also eine einzige Tangente. Ein h-facher Punkt hat höchstens h Tangenten; hat er genau h verschiedene Tangenten, so heißt er ein gewöhnlicher h-facher Punkt.

II. Die Gesamtheit aller allgemeinen Punkte einer Kurve ist offenbar ein invarianter Begriff. Diese Gesamtheit legen wir unseren Untersuchungen zugrunde. Da eine birationale Abbildung für allgemeine Punkte eineindeutig ist, so können wir hier von einem Punkte bzw. einer Punktgruppe sprechen. Wir werden nun eine besondere Klasse von allgemeinen Punkten definieren, die wir Zweige nennen und die gerade deshalb von Bedeutung sind, weil sie im engsten Zusammenhang mit den speziellen Punkten stehen.

Ω sei die algebraisch-abgeschlossene Hülle von K. Wir betrachten den Körper $\Omega\langle\tau\rangle$, der aus allen Potenzreihen in τ (endlichviele negative Potenzen erlaubt) mit Koeffizienten aus Ω besteht. Dann heißt ein allgemeiner Punkt $\xi(\tau) = (\xi_0(\tau), \xi_1(\tau), \ldots, \xi_n(\tau))$ von C aus $\Omega\langle\tau\rangle$ ein Zweig von C, wenn er der folgenden Bedingung genügt: Es gibt keinen anderen allgemeinen Punkt $\eta(\tau)$ von C aus $\Omega\langle\tau\rangle$ mit der Eigenschaft, daß es ein durch τ^2 teilbares Element $\sigma = a\tau^2 + \ldots$ aus $\Omega\langle\tau\rangle$ gibt, so daß $\xi(\tau) = \eta(\sigma)$. Durch geeignete Normierung der Koordinaten kann man immer erreichen, daß einerseits keine negativen Potenzen von τ in $\xi_i(\tau)$ vorkommen und andererseits auch nicht alle $\xi_i(\tau)$ durch τ teilbar sind. Der Zweig ist dann in reduzierter Form. Der spezielle Punkt $\xi(0)$ heißt der zugehörige spezielle Punkt von $\xi(\tau)$, und $\xi(\tau)$ ein zugehöriger Zweig von $\xi(0)$. Zwei Zweige $\xi^{(1)}$, $\xi^{(2)}$ heißen äquivalent, wenn

[21]) Diese einfachen Punkte von D dürfen natürlich mehrfach in P auftreten.

es ein genau durch τ teilbares Element $\sigma = a\tau + \ldots$ $(a \neq 0)$ aus $\Omega\langle\tau\rangle$ gibt, so daß $\xi^{(1)}(\tau) = \xi^{(2)}(\sigma)$. Äquivalente Zweige gehören offenbar zu demselben Punkte. Es läßt sich in bekannter Weise zeigen, daß die Äquivalenz reflexiv, symmetrisch und transitiv ist, und folglich alle Zweige in Klassen von äquivalenten verteilt sind. Wir werden im folgenden äquivalente Zweige als gleich betrachten; sie sind nur verschiedene Darstellungen desselben Zweiges mit verschiedenen „Ortsuniformisierenden". Man überzeugt sich leicht, daß der Zweig und die Äquivalenz der Zweige birational invariante Begriffe sind. Dagegen hängt der zugehörige Punkt eines Zweiges von den verschiedenen „projektiven Modellen" der Kurve C ab.

Zwischen den speziellen Punkten und den Zweigen einer Kurve bestehen die folgenden Beziehungen:

Satz A. *Es gibt zu jedem (speziellen) k-fachen Punkte* $\alpha = (\alpha_0 = 1, \alpha_1, \ldots, \alpha_n)$ *von* C *genau* r *(inäquivalente) zugehörige Zweige, wenn* α *bei einer birationalen Abbildung von* C *auf eine singularitätenfreie Kurve auf* r *verschiedene Punkte abgebildet ist. Es ist also* $1 \leqq r \leqq k$.

Beweis: Man braucht den Satz nur für den Fall einer singularitätenfreien Kurve zu beweisen, der allgemeine Fall folgt dann unmittelbar daraus, wenn man die gegebene Kurve auf eine singularitätenfreie Kurve birational abbildet. Denn der zugehörige Punkt des Bildzweiges ist ein Bildpunkt des zugehörigen Punktes des Urzweiges, da diese beiden Punkte eine relationstreue Spezialisierung der beiden Zweige bilden. Haben die r Bildpunkte von α r zugehörige Zweige, so sind die Bilder dieser Zweige bei der inversen Abbildung r verschiedene zugehörige Zweige von α.

C sei nun eine singularitätenfreie Kurve. Die Projektion von C auf die (x_0, x_1, x_2)-Ebene sei durch die Kurve $f(x_0, x_1, x_2)$ gegeben. Durch eine geeignete Koordinatentransformation (mit Koeffizienten aus Ω) können wir erreichen, daß 1. die Projektion auf die (x_0, x_1, x_2)-Ebene eine birationale Abbildung ist, und 2. $\frac{\partial f(x_0, x_1, x_2)}{\partial x_2}$ für $x_0 = \alpha_0$, $x_1 = \alpha_1$, $x_2 = \alpha_2$ nicht verschwindet. Die durch die Projektion definierte birationale Abbildung sei:

$$\begin{aligned} x_0 : x_1 : \ldots : x_n \\ = x_0\psi(x_0, x_1, x_2) : x_1\psi(x_0, x_1, x_2) : x_2\psi(x_0, x_1, x_2) \\ : \varphi_3(x_0, x_1, x_2) : \ldots : \varphi_n(x_0, x_1, x_2). \end{aligned}$$

Dann läßt sich in bekannter Weise zeigen, daß

$$\begin{aligned} \xi_0 &= 1, \\ \xi_1 &= \alpha_1 + \tau, \\ \xi_2 &= \alpha_2 + \sum_1^\infty C_i \tau^i = \alpha_2 + \left(-\frac{\partial f}{\partial x_1}\Big/\frac{\partial f}{\partial x_2}\right)_\alpha \tau + \ldots, \\ \xi_i &= \varphi_i(\xi_0, \xi_1, \xi_2)/\psi(\xi_0, \xi_1, \xi_2) \qquad (i = 3, \ldots, n) \end{aligned}$$

ein zu α zugehöriger Zweig von C ist. — Es sei nun $\xi_0^{(1)} \perp (1, \xi_1^{(1)}, \ldots, \xi_n^{(1)})$ irgendein zu α zugehöriger Zweig von C. Setzt man $\sigma = \xi_1^{(1)} - \alpha_1 = a\tau + \ldots$, so folgt aus $f(\xi_0^{(1)}, \xi_1^{(1)}, \xi_2^{(1)}) = f(1, \alpha_1 + \sigma, \xi_2^{(1)}) = 0$ und $\xi_i^{(1)} \psi(\xi_0^{(1)}, \xi_1^{(1)}, \xi_2^{(1)}) = \varphi_i(\xi_0^{(1)}, \xi_1^{(1)}, \xi_2^{(1)})$, daß

$$\xi_0^{(1)} = 1,$$

$$\xi_1^{(1)} = \alpha_1 + \sigma,$$

$$\xi_2^{(1)} = \alpha_2 + \sum_1^{\infty} C_i \sigma^i = \alpha_2 + \left(-\frac{\partial f}{\partial x_1}\Big/\frac{\partial f}{\partial x_2}\right)\sigma + \ldots,$$

$$\xi_i^{(1)} = \varphi_i(\xi_0^{(1)}, \xi_1^{(1)}, \xi_2^{(1)})/\psi(\xi_0^{(1)}, \xi_1^{(1)}, \xi_2^{(1)}) \qquad (i = 3, \ldots, n).$$

Da $\xi^{(1)}$ ein Zweig ist, so ist $a \neq 0$ wegen der Bedingung in der Definition eines Zweiges. Daraus folgt, daß $\xi^{(1)}$ zu ξ äquivalent ist.

Satz B. *$\xi^{(1)}, \xi^{(2)}, \ldots, \xi^{(r)}$ seien die r zugehörigen Zweige von α in reduzierter Form. Eine Hyperfläche $F_1(\mu, x)$ schneidet die Kurve C in α dann und nur dann s-fach, wenn $\prod_{i=1}^{r} F_1(\mu, \xi^{(i)})$ durch τ^s teilbar ist.*

Beweis: C sei durch die birationale Abbildung

$$y_0 : y_1 : \ldots : y_t = \varphi_0(x) : \varphi_1(x) : \ldots : \varphi_t(x)$$

auf die singularitätenfreie Kurve C' in S_t abgebildet. $G(u, v)$ sei die in § 2 für $F_1(u, x)$ und $F_2(v, x)$ definierte Form. Setzt man nun $F_2(v, x) = \Sigma \lambda_i \varphi_i(x)$, so geht $G(u, v)$ in $H(u, \lambda) = H(u) H^*(u, \lambda)$ über, wo $H(u)$ und $H^*(u, \lambda)$ ähnliche Bedeutungen haben wie $T(v)$ und $T^*(\lambda, v)$ in § 3 Dann ist

$$H^*(u, \lambda) = \prod_{i=1}^{g'} F_1(u, \Theta^{(i)}) = \prod_{i=1}^{g} \left(\sum_{j=0}^{t} \lambda_j \omega_j^{(i)}\right),$$

wo $(\Theta^{(1)}, \Theta^{(2)}, \ldots, \Theta^{(g')})$ die allgemeine Punktgruppe von $\Sigma \lambda_i \varphi_i(x)$, also das Bild (auf C) der Schnittpunktgruppe von C' mit der allgemeinen Hyperebene $\Sigma \lambda_i y_i$ ist, und $(\omega^{(1)}, \omega^{(2)}, \ldots, \omega^{(g)})$ das Bild (auf C') der Schnittpunktgruppe von C mit der allgemeinen Hyperfläche $F_1(u, x)$ ist. Die Schnittpunktgruppe von $F_1(\mu, x)$ mit C sei $(\alpha^{(1)}, \alpha^{(2)}, \ldots, \alpha^{(g)})$, wobei $\alpha^{(1)} = \ldots = \alpha^{(s)} = \alpha$ sein mögen. Das Bild dieser Punktgruppe auf C' sei $(\beta^{(1)}, \beta^{(2)}, \ldots, \beta^{(g)})$, wobei $\beta^{(1)} = \ldots = \beta^{(s_1)} = \gamma^{(1)}, \beta^{(s_1+1)} = \ldots = \beta^{(s_1+s_2)} = \gamma^{(2)}, \ldots, \beta^{(s_1+\cdots+s_{r-1}+1)} = \ldots = \beta^{(s_1+\cdots+s_r)} = \gamma^{(r)}$ $\left(\sum_1^r s_i = s\right)$ sein mögen. Die Koordinaten in S_m seien (durch Koordinatentransformation mit Koeffizienten aus Ω) so gewählt, 1. daß die Hyperebenen $\gamma_0^{(1)} y_1 - \gamma_1^{(1)} y_0$, $\gamma_0^{(2)} y_1 - \gamma_1^{(2)} y_0, \ldots, \gamma_0^{(r)} y_1 - \gamma_1^{(r)} y_0$ mit C' nur einfache Schnittpunkte haben, 2. daß für jeden dieser Schnittpunkte die in dem Beweise von Satz A für α aufgestellten Bedingungen gelten, 3. daß $\beta_0^{(i)} \beta_1^{(j)} \neq \beta_1^{(i)} \beta_0^{(j)}$ ist für $\beta^{(i)} \neq \beta^{(j)}$. Dann hat die Hyperebene $\Sigma \Phi_j^{(i)} y_j = \gamma_0^{(i)} y_1 - (\gamma_1^{(i)} + \tau) y_0$

mit C' genau g' verschiedene Zweige von C' als Schnittpunkte. Die Bilder dieser Zweige auf C seien $\xi^{(i,1)} = \xi^{(i)}, \xi^{(i,2)}, \ldots, \xi^{(i,g')}$, wir haben dann

$$H^*(u, \Phi) = \prod_{j=1}^{g'} F_1(u, \xi^{(i,j)}) = \prod_{h=1}^{g} (\Sigma \Phi_j^{(i)} \omega_j^{(h)}).$$

Da $H^*(u, \Phi)$ nicht durch τ teilbar ist, können wir die $\xi^{(i,j)}$ in reduzierter Form nehmen. Spezialisieren wir nun endlich $u \to \mu$, so haben wir

$$H^*(\mu, \Phi) = \prod_{j=1}^{g'} F_1(\mu, \xi^{(i,j)}) = \prod_{h=1}^{g} (\gamma_0^{(i)} \beta_1^{(h)} - \gamma_1^{(i)} \beta_0^{(h)} - \beta_0^{(h)} \tau)$$

oder

$$F_1(\mu, \xi^{(i)}) \prod_{j=2}^{g'} F_1(\mu, \xi^{(i,j)}) = \tau^{s_i} \prod_{\beta^{(h)} \neq \gamma^{(i)}} (\gamma_0^{(i)} \beta_1^{(h)} - \gamma_1^{(i)} \beta_0^{(h)} - \beta_0^{(h)} \tau).$$

Da $\prod\limits_{j=2}^{g'} F_1(\mu, \xi^{(i,j)})$ und $\prod\limits_{\beta^{(h)} \neq \gamma^{(i)}} (\gamma_0^{(i)} \beta_1^{(h)} - \gamma_1^{(i)} \beta_0^{(h)} - \beta_0^{(h)} \tau)$ nicht durch τ teilbar sind, so ist $F_1(\mu, \xi^{(i)})$ durch τ^{s_i} teilbar, mithin ist $\prod\limits_{i=1}^{r} F_1(\mu, \xi^{(i)})$ durch τ^s teilbar.

Nach diesen beiden Sätzen ist es klar, daß alles, was wir bisher für die (speziellen) Punkte bewiesen haben, auch für die Zweige gilt, wenn wir die Schnittvielfachheit einer beliebigen Hyperfläche $F(\mu, x)$ mit C in einem Zweige ξ so definieren: $F(\mu, x)$ schneidet C in ξ s-fach, wenn $F(\mu, \xi)$ durch τ^s teilbar ist. Der Zweigbegriff ist also eine Verfeinerung des Punktbegriffes und ist offenbar birational invariant. Der Satz auf S. 671 bedarf nun kaum eines Beweises und gilt für beliebige Kurven. Da wir im folgenden (außer am Schluß von § 7) ausschließlich mit Zweigen zu tun haben, so werden wir dem üblichen Sprachgebrauch gemäß die Zweige wiederum Punkte nennen.

Definition. *Zwei Punktgruppen P, Q der Ordnung n heißen äquivalent,*

$$P \equiv Q,$$

wenn sie beide in einer linearen Schar der Ordnung n enthalten sind.

Satz. *Aus $P \equiv Q$, $Q \equiv R$ folgt $P \equiv R$.*

Dieser Satz ist in dem folgenden enthalten:

Satz. *Wenn zwei lineare Scharen $\mathfrak{G}_n^r$, $\mathfrak{G}_n^s$ eine gemeinsame Punktgruppe haben, so gibt es eine lineare Schar $\mathfrak{G}_n$, die sie beide enthält.*

Beweis. $\mathfrak{G}_n^r$ bzw. $\mathfrak{G}_n^s$ seien durch $\sum\limits_0^r \lambda_i F_i$ bzw. $\sum\limits_0^s \mu_i G_i$ erzeugt, die weggelassenen festen Punktgruppen seien A bzw. B, die den beiden gemeinsame Punktgruppe P sei (außer A bzw. B) durch F_0 bzw. G_0 ausgeschnitten. Das System

$$\varepsilon F_0 G_0 + \mu_1 G_1 F_0 + \ldots + \mu_s G_s F_0 + \lambda_1 F_1 G_0 + \ldots + \lambda_r F_r G_0$$

schneidet die Kurve in der festen Punktgruppe $A + B + P$. Läßt man diese feste Punktgruppe weg, so erzeugt es eine Schar der Ordnung n, die, wie man leicht einsieht, die beiden Scharen $\mathfrak{G}_n^r$ und $\mathfrak{G}_n^s$ enthält.

Es folgt aus diesem Satz, daß es zu jeder Punktgruppe P_n eine größte lineare Schar der Ordnung n gibt, die F_n enthält. Diese Schar nennen wir die durch P_n bestimmte *Vollschar* $|P_n|$. Ist P_n in keiner linearen Schar enthalten, so sagen wir, die Vollschar $|P_n|$ habe die Dimension 0.

Es gilt: 1. Aus $P \equiv Q$ folgt $P + A \equiv Q + A$, wo A eine beliebige Punktgruppe ist. Denn man kann eine beliebige durch A gehende Hyperfläche dem erzeugenden System der P, Q enthaltenden Schar hinzufügen und die übrigen Schnittpunkte dieser Hyperfläche mit C aus der Schar weglassen. 2. Aus $P \equiv Q$, $A \equiv B$ folgt $P + A \equiv Q + B$. Denn $P + A \equiv Q + A \equiv Q + B$. Die hiernach durch $|P|$, $|A|$ eindeutig bestimmte Vollschar $|P + A|$ nennen wir die Summe von $|P|$ und $|A|$. 3. Der Rest einer Vollschar $|P + A|$ in bezug auf A ist die Vollschar $|P|$. Enthielte nämlich $|P|$ eine nicht in dem Rest liegende Punktgruppe Q, dann würde die Schaar $|P| + A$ eine nicht in $|P + A|$ liegende Punktgruppe $Q + A$ enthalten. Das ist aber absurd, denn $|P + A|$ ist eine Vollschar und muß daher $|P| + A$ enthalten. 4. Ist $A \equiv B$, so ist der Rest von $|P + A|$ in bezug auf B auch gleich $|P|$. Denn aus $P + B \equiv P + A$ folgt $|P + B| = |P + A|$. Wir können also sagen: $|P|$ ist der Rest von $|P + A|$ in bezug auf $|A|$.

§ 6.

Der Riemann-Rochsche Satz.

Bevor wir die folgenden Sätze beweisen, die zu dem Ziel unserer Untersuchungen, dem Riemann-Rochschen Satz, führen, sei hier gleich bemerkt, daß wir uns bei dem Beweise auf den Fall eines algebraisch-abgeschlossenen Körpers beschränken können. Denn es handelt sich hier lediglich um Beziehungen zwischen den Ordnungen und den Dimensionen der Vollscharen auf C. Ist nämlich K nicht schon algebraisch-abgeschlossen, dann gehen wir zu seiner algebraisch-abgeschlossenen Hülle Ω über. Gelten die folgenden Sätze für die Ω-Vollscharen, so gelten sie natürlich insbesondere auch für K-Vollscharen. Es bleibt also nur zu zeigen, daß die durch eine K-Punktgruppe P in bezug auf Ω bestimmte Ω-Vollschar $\mathfrak{G} = |P|$ auch eine K-Vollschar ist. Das ist aber unmittelbar klar. Denn alle Konjugierten von $|P|$ müssen auch die K-Punktgruppe P enthalten und sind daher der Vollschar $\mathfrak{G} = |P|$ gleich. Daraus folgt nach dem Hilfssatz am Schluß von § 4, daß $\mathfrak{G}$ eine K-Vollschar ist. Ebenso ist die kanonische Schar $\mathfrak{G}_{2g-2}^{g-1}$ eine K-Schar, da sie nach Satz 6 die einzige

Schar der Ordnung $2g-2$ und der Dimension $g-1$ ist und daher mit ihren Konjugierten übereinstimmen muß.

Definition. *Das Geschlecht einer irreduziblen Kurve C ist die kleinste Zahl g derart, daß jede Vollschar der Ordnung $> g$ auf C die Dimension $\geqq 1$ hat.*

Daß es eine solche Zahl gibt, sieht man so: C sei durch eine ebene Kurve vom Grade m dargestellt. Die Gesamtheit aller Kurven vom Grade $k = m + \lambda\,(\lambda \geqq 0)$ schneidet auf C eine lineare Schar $\mathfrak{G}^{\varrho}_{km}$ der Ordnung km und der Dimension

$$\varrho = \frac{k(k+3)}{2} - \frac{(k-m)(k-m+3)}{2} - 1 \geqq \binom{m}{2} + \lambda$$

aus. P sei eine beliebige Punktgruppe der Ordnung $\binom{m}{2} + \lambda$, dann enthält $\mathfrak{G}^{\varrho}_{km}$ eine Punktgruppe $P + Q$, wo Q die Ordnung $km - \binom{m}{2} - \lambda$ hat. Der Rest von $\mathfrak{G}^{\varrho}_{km}$ in bezug auf Q ist eine P enthaltende lineare Schar und hat die Dimension

$$r \geqq \varrho - km + \binom{m}{2} + \lambda = m + \lambda - 1 \geqq 1.$$

Folglich hat um so mehr die Vollschar $|P|$ die Dimension $\geqq 1$.

Satz 1. *Eine Vollschar der Ordnung n hat die Dimension $r \geqq n - g$*

Beweis: Der Rest in bezug auf einen nicht festen Punkt ist eine Vollschar der Ordnung $n-1$ und der Dimension $r-1$. Nach r-maliger Anwendung solcher Restbildung haben wir dann eine Vollschar der Ordnung $n-r$ und der Dimension 0. Daraus folgt, daß $n - r \leqq g$ oder $r \geqq n - g$.

Satz 2. *Es gibt für jedes $k \geqq 0$ Vollscharen der Ordnung $n = g + k$ und der Dimension k.*

Beweis: Hätten alle Vollscharen der Ordnung $n = g + k$ die Dimension $\geqq k+1$, so würde die durch eine beliebige P_g und eine Q_k definierte Vollschar $|P_g + Q_k|$ auch die Dimension $\geqq k+1$ haben. Dann würde der Rest von $|P_g + Q_k|$ in bezug auf Q_k, d. h. die Vollschar $|P_g|$, die Dimension $\geqq 1$ haben, gegen die Definition von g.

Definition. *Eine Vollschar $\mathfrak{G}^r_n$ heißt speziell, wenn $r > n - g$ ist.* Alle Vollscharen der Ordnung $n < g$ sind offenbar speziell. Dagegen gibt es nach Satz 2 nicht spezielle Vollscharen für jede Ordnung $n \geqq g$.

Satz 3. *Jede Vollschar der Dimension $r \geqq g$ ist nicht speziell.*

Beweis: Wäre sie speziell, so würde sie die Ordnung $n < r + g$ haben. Der Rest in bezug auf eine nicht spezielle P_g sei $|Q_{n-g}|$. Dann würde der Rest der Schar in bezug auf Q_{n-g}, d. h. die Vollschar $|P_g|$ eine Dimension $\geqq r - n + g \geqq 1$ haben, also wäre P_g doch speziell.

Im Fall $g = 0$ oder $g = 1$ ist jede Vollschar nicht speziell, es gilt also für jede Vollschar die Beziehung $r = n - g$. Wir haben dann nichts mehr zu beweisen. Wir werden also im folgenden annehmen, daß $g > 1$ ist.

Satz 4. (Existenzsatz.) *Es gibt eine lineare Schar* $\mathfrak{G}_{2g-2}^{g-1}$.

Beweis: Wir werden diesen Existenzsatz erst im nächsten Paragraphen beweisen.

Es folgt zunächst: 1. $\mathfrak{G}_{2g-2}^{g-1}$ ist eine Vollschar. Denn jede Vollschar der Dimension $r \geqq g$ ist nicht speziell, muß daher die Ordnung $n \geqq 2g$ haben. 2. $\mathfrak{G}_{2g-2}^{g-1}$ ist speziell, da $g-1 > 2g-2-g = g-2$ ist. 3. Jede in $\mathfrak{G}_{2g-2}^{g-1}$ enthaltene Vollschar ist speziell. Dies gilt in der Tat für jede beliebige spezielle Vollschar $\mathfrak{G}_n^r$. Denn eine Teilvollschar der Ordnung k entsteht dadurch, daß man den Rest in bezug auf eine gewisse P_{n-k} bildet, und hat daher die Dimension $\geqq r-n+k > k-g$, ist also speziell.

Satz 5. (Reduktionssatz.) *Es sei $P+Q$ eine Punktgruppe von $\mathfrak{G}_{2g-2}^{g-1}$ und O ein nicht in Q vorkommender Punkt. Dann ist O ein fester Punkt von $|P+O|$.*

Beweis: $|P+Q+O|$ ist von der Ordnung $2g-1$, mithin nicht speziell, mithin von der Dimension $g-1$. Also ist O ein fester Punkt von $|P+Q+O|$, mithin auch ein fester Punkt von $|P+O|$, da er der Rest von $|P+Q+O|$ in bezug auf die O nicht enthaltende Punktgruppe Q ist.

Satz 6. *Jede spezielle Vollschar ist in $\mathfrak{G}_{2g-2}^{g-1}$ enthalten.*

Beweis: Dies gilt offenbar für spezielle Vollscharen der Dimension $r=0$, also der Ordnung $n \leqq g-1$, da $\mathfrak{G}_{2g-2}^{g-1}$ die Dimension $g-1$ hat. Es sei der Satz für spezielle Vollscharen der Dimension $r-1$ schon bewiesen. Der Rest $\mathfrak{G}_{n-1}^{r-1}$ einer speziellen Vollschar $\mathfrak{G}_n^r$ in bezug auf einen nicht festen Punkt O ist also in $\mathfrak{G}_{2g-2}^{g-1}$ enthalten. Alle durch eine Gruppe von $\mathfrak{G}_{n-1}^{r-1}$ gehenden Gruppen von $\mathfrak{G}_{2g-2}^{g-1}$ müssen aber O enthalten, sonst würde O nach Satz 5 ein fester Punkt von $\mathfrak{G}_n^r$ sein, gegen die Voraussetzung. Das bedeutet aber, daß $\mathfrak{G}_{n-1}^{r-1} + O = \mathfrak{G}^r$ in $\mathfrak{G}_{2g-2}^{g-1}$ enthalten ist.

Nach Satz 6 kann es offenbar nur eine einzige $\mathfrak{G}_{2g-2}^{g-1}$ geben. Diese $\mathfrak{G}_{2g-2}^{g-1}$ nennen wir die *kanonische Schar* von C. Ist der Rest von $\mathfrak{G}_{2g-2}^{g-1}$ in bezug auf eine spezielle Vollschar $\mathfrak{G}_n^r$ von der Dimension $i-1$, so heißt i der *Spezialitätsindex* von $\mathfrak{G}_n^r$. Nicht spezielle Vollscharen sollen den Spezialitätsindex $i=0$ haben.

Satz 7. (Riemann-Rochscher Satz.) *Für eine Vollschar $\mathfrak{G}_n^r$ vom Spezialitätsindex i gilt:*

$$r = n - g + i.$$

Beweis: Der Satz gilt offenbar für $i=0$. Es sei der Satz für $i-1$ schon bewiesen. Der Rest von $\mathfrak{G}_{2g-2}^{g-1}$ in bezug auf eine Vollschar $\mathfrak{G}_n^r$ vom Spezialitätsindex i hat die Dimension $i-1$. O sei ein nicht zu den festen Punkten dieses Restes gehöriger Punkt, dann hat die Vollschar $\mathfrak{G}_n^r + O$ der Ordnung $n+1$ und der Dimension r (Satz 5) den Spezialitätsindex $i-1$. Daraus folgt, daß $r = n+1-g+i-1 = n-g+i$.

§ 7.

Der Beweis des Existenzsatzes.

Wir werden jetzt den im letzten Paragraphen angenommenen Satz beweisen, daß es eine lineare Schar $\mathfrak{G}_{2g-2}^{g-1}$ gibt. Zunächst werden wir zeigen, daß es eine zu dem Zweck besonders geeignete Darstellung der Kurve C gibt: eine ebene Kurve mit nur gewöhnlichen Doppelpunkten (d. h. Doppelpunkten mit zwei verschiedenen Tangenten). Dazu bedienen wir uns der Projektion. Wir nehmen zuerst eine Vollschar $\mathfrak{G}_{r+g}^r$ von der Dimension $r \geqq g+3$, die natürlich einfach ist und daher eine Darstellung der Kurve C in S_r erzeugt. Diese Darstellung $\bar{C}$ besitzt keine Singularitäten (da es keine Schar der Dimension $r-1 \geqq g+2$ und der Ordnung $\leqq r+g-2$ gibt), keine dreifachen Sehnen (da es keine Schar der Dimension $r-2 \geqq g+1$ und der Ordnung $\leqq r+g-3$ gibt), keine Paare in einer Ebene liegender Tangenten, und keine Paare außerhalb $\bar{C}$ schneidender Sehnen (da es keine Schar der Dimension $r-3 \geqq g$ und der Ordnung $\leqq r+g-4$ gibt). Die durch die Gesamtheit aller Sehnen von $\bar{C}$ erzeugte Mannigfaltigkeit N hat den allgemeinen Punkt $\eta = \lambda\xi + \lambda'\xi'$, wo ξ, ξ' zwei voneinander unabhängige allgemeine Punkte von $\bar{C}$ sind. Sie ist daher von der Dimension $\leqq 3$. Die Gesamtheit der Tangenten von $\bar{C}$ bildet eine Mannigfaltigkeit N', die offenbar von der Dimension $\leqq 2$ ist. Nehmen wir nun einen S_{n-3}, der mit N nur endlichviele Schnittpunkte, mit N' und der Kurve $\bar{C}$ aber keinen Schnittpunkt hat, was gewiß möglich ist, und bilden die durch das System aller durch S_{n-3} gehenden Hyperebenen auf $\bar{C}$ erzeugte lineare Schar $\mathfrak{G}^2$, dann definiert $\mathfrak{G}^2$ eine birationale Abbildung, die die erwünschte ebene Kurve liefert. Da N nur endlichviele Schnittpunkte mit S_{n-3} hat und da diese Punkte nicht auf $\bar{C}$ liegen, also nicht Schnittpunkte von zwei Sehnen sind, so ist es klar, daß die durch einen allgemeinen Punkt von $\bar{C}$ gehenden Punktgruppen von $\mathfrak{G}^2$ keinen weiteren Schnittpunkt mit $\bar{C}$ gemeinsam haben können. Also ist die Abbildung birational. Daß die Bildkurve D nur Doppelpunkte besitzt, sieht man so. Den Schnittpunktgruppen von D mit den Geraden des durch einen Punkt δ von D gehenden Büschels entsprechen die Schnittpunktgruppen einer 1-dimensionalen Teilschar $\mathfrak{G}^1$ von $\mathfrak{G}^2$, die von einem durch einen aus S_{n-3} und einem gewissen Punkt von $\bar{C}$ gebildeten S_{n-2} gehenden Büschel von Hyperebenen auf $\bar{C}$ ausgeschnitten wird. Da $\bar{C}$ keine dreifachen Sehnen besitzt, kann ein solches Büschel nicht mehr als zwei feste Schnittpunkte mit $\overline{C}$ haben. Mithin hat $\mathfrak{G}^1$ nicht mehr als zwei feste Punkte, und daraus folgt (§ 4), daß δ höchstens ein Doppelpunkt ist. Ebenso leicht überzeugt man sich, daß, da $\overline{C}$ keine Paare in einer Ebene liegender Tangenten besitzt und da S_{n-3} sich mit N' nicht schneidet, es in jeder der endlichvielen $\mathfrak{G}^1$ (entsprechend den endlichvielen Schnittpunkten von N mit S_{n-3}), die zwei feste Punkte besitzen, zwei verschiedene Punktgruppen gibt, in denen einer der beiden festen Punkte mehrfach auftritt. Das bedeutet, daß die Doppelpunkte von D alle zwei verschiedene Tangenten haben.

Die ebene Kurve D sei vom Grade m und besitze die d Doppelpunkte $\delta^{(1)}, \ldots, \delta^{(d)}$. Die Gesamtheit der durch die $\delta^{(i)}$ gehenden Kurven vom Grade k (der adjungierten Kurven vom Grade k) schneidet D außer in einer festen Punktgruppe, in der jeder $\delta^{(i)}$ zweifach auftritt, noch in einer linearen Schar $\mathfrak{G}_{n_k}^{r_k}$ (der k-ten adjungierten Schar) der Ordnung $n_k = mk - 2d$ und der Dimension

$$r_k = \frac{k(k+3)}{2} - \frac{(k-m)(k-m+3)}{2} - 1 - d$$

$$= n_k - g' \qquad (k \geqq m, d+1),$$

$$r_k = \frac{k(k+3)}{2} - d \qquad (k \leqq m),$$

wo

$$g' = \frac{(m-1)(m-2)}{2} - d$$

ist. Da für hinreichend große k $r_k \leqq n_k - g$ ist, so ist $g' \geqq g > 1$ (unsere Annahme). Für $k = m - 3$ haben wir

$$n_k = 2g' - 2,$$

$$r_k = g' - 1.$$

Wenn wir zeigen können, daß $g' = g$ ist, so haben wir hier schon eine $\mathfrak{G}_{2g-2}$, deren Dimension $\geqq g-1$, also $= g-1$ ist[22]). Im folgenden werden wir die Gleichheit von g und g' dadurch beweisen, daß wir zeigen, daß die oben definierten Scharen $\mathfrak{G}_{n_k}^{r_k}$ für genügend große k Vollscharen sind.

$\mathfrak{G}_{n-m}^{n-m-g}$ sei eine Vollschar der Ordnung $\geqq 2g$, $|P_m|$ sei die von der Gesamtheit aller Geraden erzeugte Schar. Der Rest der nicht speziellen Vollschar $|P_m + \mathfrak{G}_{n-m}^{n-m-g}| = \mathfrak{G}_n^{n-g}$ in bezug auf $|P_m|$ ist die nicht spezielle Vollschar $\mathfrak{G}_{n-m}^{n-m-g}$, aus $|P_m|$ folgen also m unabhängige Bedingungen für $\mathfrak{G}_n^{n-g}$. Nun läßt sich $\mathfrak{G}_n^{n-g}$, wie jede Vollschar überhaupt,

22) Denn eine Schar der Dimension $> g-1$ ist nicht speziell und hat folglich die Ordnung $\geqq 2g$.

als der Rest einer adjungierten Schar $\mathfrak{G}_{n_h}^{r_h}$ in bezug auf eine Punktgruppe K darstellen. Da der Rest von $\mathfrak{G}_{n_h+i}^{r_h+i}$ $(i \geqq 0)$ in bezug auf K die Schar $\mathfrak{G}_n^{n-g}$ enthält, so legt $|P_m|$ natürlich auch diesem Rest m unabhängige Bedingungen auf. Daraus folgt durch e-malige Anwendung, daß $e\,|P_m|$ dem Rest von $\mathfrak{G}_{n_h+e}^{r_h+e}$ in bezug auf K em unabhängige Bedingungen auferlegt. Da $\mathfrak{G}_{n_h+e}^{r_h+e} \subset |\mathfrak{G}_{n_h}^{r_h} + e\,P_m|$ ist, so ist der Rest von $\mathfrak{G}_{n_h+e}^{r_h+e}$ in bezug auf $K + e\,|P_m|$ die Vollschar $\mathfrak{G}_n^{n-g}$, also ist der Rest von $\mathfrak{G}_{n_h+e}^{r_h+e}$ in bezug auf K eine Vollschar $\mathfrak{G}_{n+em}^{n-g+em}$. Wenn wir nun beweisen können, daß K der $\mathfrak{G}_{n_h+e}^{r_h+e}$ für genügend große e genau k (die Ordnung von K) unabhängige Bedingungen auferlegt, dann hat $\mathfrak{G}_{n_h+e}^{r_h+e}$ die Ordnung $n + em + k$ und die Dimension $n + em + k - g$ und ist also eine Vollschar.

Wir haben also zu zeigen, daß K der $\mathfrak{G}_{n_h}^{r_h}$ (für genügend große h) k unabhängige Bedingungen auferlegt. Dazu genügt es zu zeigen, daß es eine adjungierte Kurve vom Grade h gibt, die, außer in der festen Punktgruppe G der Ordnung $2\,d$ (die aus der Adjungiertheit entsteht), die Kurve D genau in einer beliebigen in K enthaltenen Punktgruppe K' und einer mit K punktfremden Gruppe H schneidet. Wir können uns dabei auf den Fall beschränken, wo K' aus einem k'-fach zu zählenden Punkte besteht, der dem Zweige ξ: $\xi_0 = 1$, $\xi_1 = \alpha + \tau$, $\xi_2 = \beta + \sum_1^\infty \beta_i \tau^i$, zugeordnet ist; der allgemeine Fall läßt sich leicht darauf zurückführen. Nun schneidet eine allgemeine k'-fach durch ξ gehende adjungierte Kurve vom Grade h die Kurve D (außer in G) genau k'-fach in ξ und in keinem anderen nicht zu dem zugehörigen Punkte $(1, \alpha, \beta)$ von ξ zugehörigen Zweig von K. Denn dies gilt für die adjungierte Kurve vom Grade h, die aus k' allgemeinen durch $(1, \alpha, \beta)$ gehenden Geraden, d allgemeinen durch die d Doppelpunkte gehenden Geraden, und $h - k' - d$ allgemeinen Geraden besteht. Sie schneidet aber auch (außer in G) den anderen zu $(1, \alpha, \beta)$ gehörigen Zweig (falls $(1, \alpha, \beta)$ ein Doppelpunkt ist) nicht, denn das gilt für die adjungierte Kurve vom Grade h, die aus der Kurve

$$x_2 x_0^{k'-1} - \sum_1^{k'} \beta_i (x_1 - \alpha x_0)^i x_0^{k'-i},$$

$d - 1$ allgemeinen durch die übrigen $d - 1$ Doppelpunkte gehenden Geraden und $h - k' - d + 1$ allgemeinen Geraden besteht. Damit ist alles bewiesen.

Literaturverzeichnis.

Albanese, G.

(1) Trasformazione birazionale di una curva algebrica qualunque in un'altra priva di punti multipli. Atti Accad. naz. Lincei, Rend. (5) 33^1 (1924).

Hasse, H.
(2) Über die Kongruenzzetafunktionen. S.-B. Preuß. Akad. d. Wiss., Phys.-Math. Kl. 1934. XVII.

Severi, F.
(3) Trattato di geometria algebrica. Vol. I, Parte I (1926).

Schmidt, F. K.
(4) Analytische Zahlentheorie in Körpern der Charakteristik p. Math. Zeitschr. **33** (1931).

van der Waerden, B. L.
(5) Der Multiplizitätsbegriff der algebraischen Geometrie. Math. Annalen **97** (1927).
(6) Moderne Algebra; Bd. II (1931).
(7) Zur algebraischen Geometrie. I; Gradbestimmung von Schnittmannigfaltigkeiten einer beliebigen Mannigfaltigkeit mit Hyperflächen. Math. Annalen **108** (1933).
(8) Zur algebraischen Geometrie. V; Ein Kriterium für die Einfachheit von Schnittpunkten. Math. Annalen **110** (1934).
(9) Zur algebraischen Geometrie. VI; Algebraische Korrespondenzen und rationale Abbildungen. Math. Annalen **110** (1934).

(Eingegangen am 26. 2. 1937.)

Einfacher topologischer Beweis des Fundamentalsatzes der Algebra.

Von

Wei-Liang Chow in Shanghai (China).

Deuten wir die komplexen Zahlen x bzw. y als Punkte auf den Kugeloberflächen K_x bzw. K_y, so definiert jede Gleichung $y = x^n + a_1 x^{n-1} + \dots + a_n$ eine (singuläre) 2-Kette Γ auf K_y, deren Urbild K_x ist. Γ ist also ein Zyklus. Gäbe es einen Punkt auf K_y, der nicht von Γ überdeckt wäre, dann würde $\Gamma \sim 0$, wie man leicht einsieht. Andererseits läßt Γ sich durch eine Deformation, die etwa durch

$$y = x^n + (a_1 x^{n-1} + \dots + a_n)\,\lambda,$$
$$\lambda = 1 \rightarrow \lambda = 0,$$

definiert werden möge, in die durch $y = x^n$ definierte 2-Kette Γ' deformieren, die ersichtlich der n-fachen K_y homolog ist. Wir hätten dann $nK_y \sim \Gamma' \sim \Gamma \sim 0$, was offenbar absurd ist. Also überdeckt Γ die ganze K_y, d. h. die Gleichung $y = x^n + a_1 x^{n-1} + \dots + a_n$ ist für jedes y, insbesondere für $y = 0$, lösbar.

(Eingegangen am 9. 7. 1938.)

Über die Multiplizität der Schnittpunkte von Hyperflächen.

Von

Wei-Liang Chow in Shanghai (China).

Wir werden hier einen einfachen rein-algebraischen Beweis für den folgenden Satz bringen:

n Hyperflächen F_i im n-dimensionalen projektiven Raume schneiden sich in einem Punkt ξ, der für die Hyperflächen F_i bzw. s_i-fach ist, im allgemeinen Πs_i-fach; die Schnittmultiplizität wird dann und nur dann größer, wenn die Hyperflächen F_i in ξ eine gemeinsame Tangente besitzen.

Der Beweis, der, dem rein algebraischen Standpunkt gemäß, auf der van der Waerdenschen Multiplizitätstheorie beruht[1]), bildet eine einfache Anwendung von dem Prinzip der Erhaltung der Anzahl. Die zur Anwendung dieses Prinzips nötigen Betrachtungen werden hier durch zwei allgemeinere Sätze A und B über Schnittmultiplizität geliefert. Diese Sätze über Schnittmultiplizität sind Verallgemeinerungen eines Satzes über die Einfachheit der Schnittmultiplizität von Herrn van der Waerden[2]), und sie lassen sich auch mit einer ähnlichen Methode beweisen.

A. Wenn bei einer relationstreuen Spezialisierung der Schnittpunkte der Hyperflächen $G_i(\lambda, x) = 0$ für $\lambda \to \mu$ zwei Schnittpunkte $\xi^{(1)}$ und $\xi^{(2)}$ in einen Schnittpunkt η der spezialisierten Hyperflächen $G_i(\mu, x) = 0$ hineinrücken und wenn der Punkt $\xi^{(1)}$ bzw. s_i-fach für die Hyperflächen $G_i(\lambda, x) = 0$ ist, so haben die Polaren s_i-ter Ordnung von $G_i(\mu, x) = 0$ in bezug auf η eine gemeinsame Erzeugende.

B. Wenn bei einer relationstreuen Spezialisierung der Schnittpunkte der Hyperflächen $G_i(\lambda, x) = 0$ für $\lambda \to \mu$ zwei Schnittpunkte $\xi^{(1)}$ und $\xi^{(2)}$ in einen Schnittpunkt η der spezialisierten Hyperflächen $G_i(\mu, x) = 0$ hineinrücken und wenn die Hyperflächen $G_i(\lambda, x) = 0$ in $\xi^{(1)}$ bzw. s_i-fach sind und dort eine gemeinsame Tangente besitzen, so haben entweder die Polaren s_i-ter Ordnung von $G_i(\mu, x) = 0$ in bezug auf η zwei gemeinsame Erzeugende, oder es gelten gewisse Relationen R, die wir nachher angeben werden.

Beweis von A. Der Beweis lautet fast wörtlich genau so wie bei van der Waerden[2]), nur werden hier statt der Tangentialhyperebenen die Tangentialkegel (die Polaren s_i-ter Ordnung) betrachtet. Wir können $\xi_0^{(1)} = \xi_0^{(2)} = \eta_0 = 1$ annehmen. Die Verbindungslinie von $\xi^{(1)}$ und $\xi^{(2)}$ schneide die Hyperebene $x_0 = 0$ in einem Punkt τ, der bei der relationstreuen Spezialisierung $(\lambda, \xi^{(1)}, \xi^{(2)}) \to (\mu, \eta, \eta)$ in den Punkt ω hineinrücken möge. Wir können $\tau_1 = \omega_1 = 1$ annehmen, dann ist $\xi^{(2)} = \xi^{(1)} + (\xi_1^{(2)} - \xi_1^{(1)})\,\tau$. Entwickeln wir nun $G_i(\lambda, \xi^{(2)}) = G_i(\lambda, \xi^{(1)} + (\xi_1^{(2)} - \xi_1^{(1)})\,\tau)$ nach den Potenzen von $(\xi_1^{(2)} - \xi_1^{(1)})$, so haben wir

$$G_i(\lambda, \xi^{(2)}) = (\xi_1^{(2)} - \xi_1^{(1)})^{s_i} H_{s_i}(\lambda; \xi^{(1)}, \tau) + (\xi_1^{(2)} - \xi_1^{(1)})^{s_i+1} H_{s_i+1}(\lambda; \xi^{(1)}, \tau) + \ldots = 0$$

oder, durch $(\xi_1^{(2)} - \xi_1^{(1)})^{s_i}$ dividiert,

$$H_{s_i}(\lambda; \xi^{(1)}, \tau) + (\xi_1^{(2)} - \xi_1^{(1)})\,[H_{s_i+1}(\lambda; \xi^{(1)}, \tau) + \ldots] = 0.$$

Machen wir diese Gleichungen durch Einführung von $\xi_0^{(1)}, \xi_0^{(2)}, \tau_1$ homogen, machen dann den Übergang $(\lambda, \xi^{(1)}, \xi^{(2)}, \tau) \to (\mu, \eta, \eta, \omega)$ und setzen dann wieder $\eta_0 = \omega_1 = 1$, so bekommen wir

$$H_{s_i}(\mu; \eta, \omega) = 0.$$

Daraus folgt, wie man leicht einsieht, daß die $H_{s_i}(\mu; \eta, x) = 0$ die Verbindungslinie $\overline{\eta\omega}$ von η und ω enthalten.

Beweis von B. Wir können $\eta_0 = \xi_0^{(1)} = \xi_0^{(2)} = 1$ annehmen. Die gemeinsame Tangente von $G_i(\lambda, x) = 0$ schneide die Hyperebene $x_0 = 0$ in einem Punkt σ, der bei der relationstreuen Spezialisierung $(\lambda, \xi^{(1)}) \to (\mu, \eta)$ in ω' hineinrücken möge, wobei wir $\sigma_1 = \omega_1' = 1$ annehmen können. Es gelten dann offenbar $H_{s_i}(\mu; \eta, \omega') = 0$. Die Verbindungslinie von $\xi^{(1)}$ und $\xi^{(2)}$ schneide die Hyperebene $x_0 = 0$ in τ, der bei der relationstreuen Spezialisierung $(\lambda, \xi^{(1)}, \xi^{(2)}) \to (\mu, \eta, \eta)$ in ω hineinrücken möge. Aus A folgt dann $H_{s_i}(\mu; \eta, \omega) = 0$. Ist $\omega \neq \omega'$, so haben die Polaren $H_{s_i}(\mu; \eta, x) = 0$ zwei Erzeugende $\overline{\eta\omega}$, $\overline{\eta\omega'}$ gemeinsam. Es sei nun $\omega = \omega'$. Wir können dann $\tau_1 = \omega_1 = 1$ und folglich $\xi^{(2)} = \xi^{(1)} + (\xi_1^{(2)} - \xi_1^{(1)})\,\tau$ setzen. Daraus folgt wie bei A

$$H_{s_i}(\lambda; \xi^{(1)}, \tau) + (\xi_1^{(2)} - \xi_1^{(1)})\,H_{s_i+1}(\lambda; \xi^{(1)}, \tau) + \ldots = 0. \tag{1}$$

Die Verbindungslinie von σ und τ möge nun die Hyperebene $x_1 = 0$ in einem Punkt ϱ schneiden, der bei der relationstreuen Spezialisierung $(\lambda, \xi^{(1)}, \xi^{(2)}, \tau, \sigma) \to (\mu, \eta, \eta, \omega, \omega)$ in π hineinrücke, wobei wir $\varrho_2 = \pi_2 = 1$ annehmen können. Wir haben dann

$$\tau = \sigma + (\tau_2 - \sigma_2)\,\varrho. \tag{2}$$

Die Gleichung

$$\alpha(\xi_1^{(2)}-\xi_1^{(1)})-\beta(\tau_2-\sigma_2)$$
$$\equiv \alpha\,\tau_1\,\omega_1(\xi_1^{(2)}\xi_0^{(1)}-\xi_1^{(1)}\xi_0^{(2)})-\beta\,\xi_0^{(1)}\xi_0^{(2)}(\tau_2\sigma_1-\tau_1\sigma_2)=0$$

definiert ein Elementenpaar (α,β), das bei der relationstreuen Spezialisierung $(\lambda,\xi^{(1)},\xi^{(2)},\tau,\sigma)\to(\mu,\eta,\eta,\omega,\omega)$ in das Paar A, B hineinrücken möge, wobei A, B nicht beide verschwinden können. Falls $A \neq 0$ ist, dann können wir $A=\alpha=1$ setzen, und wir haben

$$\xi_1^{(2)}-\xi_1^{(1)}=\beta(\tau_2-\sigma_2). \tag{3}$$

Setzen wir nun (2) und (3) in (1) ein und entwickeln die $H(\lambda;\xi^{(1)},\tau)$ nach den Potenzen von $(\tau_2-\sigma_2)$, so haben wir (da $H_{s_i}(\lambda;\xi^{(1)},\sigma)=0$)

$$\begin{aligned}(\tau_2-\sigma_2)\sum_j \varrho_j\,\partial_j H_{s_i}(\lambda;\xi^{(1)},\sigma)&+(\xi_1^{(2)}-\xi_1^{(1)})H_{s_i+1}(\lambda;\xi^{(1)},\sigma)\\ &+(\tau_2-\sigma_2)^2(\ldots)+(\tau_2-\sigma_2)(\xi_1^{(2)}-\xi_1^{(1)})(\ldots)+(\xi_1^{(2)}-\xi_1^{(1)})^2(\ldots)\\ &=(\tau_2-\sigma_2)[\sum_j \varrho_j\,\partial_j H_{s_i}(\lambda;\xi^{(1)},\sigma)+\beta H_{s_i+1}(\lambda;\xi^{(1)},\sigma)]\\ &\qquad+(\tau_2-\sigma_2)^2(\ldots)=0\end{aligned}$$

oder, dividiert durch $(\tau_2-\sigma_2)$,

$$\sum_j \varrho_j\,\partial_j H_{s_i}(\lambda;\xi^{(1)},\sigma)+\beta H_{s_i+1}(\lambda;\xi^{(1)},\sigma)+(\tau_2-\sigma_2)(\ldots)=0. \tag{4}$$

Machen wir nun (4) durch Einführung von $\xi_0^{(1)},\sigma_1,\tau_1,\varrho_2,\alpha$ homogen, machen dann den Übergang $(\lambda,\xi^{(1)},\tau,\sigma,\varrho,\alpha,\beta)\to(\mu,\eta,\omega,\omega,\pi,A,B)$ und setzen dann wieder $\eta_0=\omega_1=\pi_1=1$, so haben wir die Relationen

$$A\sum_{j=2}^{n}\pi_j\,\partial_j H_{s_i}(\mu;\eta,\omega)+B\,H_{s_i+1}(\mu;\eta,\omega)=0, \tag{R}$$

die dadurch, daß wir den Koeffizienten A beibehalten haben, auch für den Fall $A=0$ gelten, wie man sich in ähnlicher Weise wie oben überzeugt.

Beweis des Satzes. Zunächst wenden wir A auf n allgemeine Hyperflächen $F_i(\lambda,x)$ der Ordnung n_i an, die bzw. s_i-fach durch einen festen Punkt η gehen. Für die Spezialisierung $\lambda\to\mu$ sollen diejenigen Hyperflächen $F_i(\mu,x)$ entstehen, die bzw. aus s_i allgemeinen durch η gehenden Hyperebenen und n_i-s_i allgemeinen Hyperebenen bestehen. Da die $F_i(\mu,x)$ offenbar in η einen $\Pi\, s_i$-fachen Schnittpunkt haben und dort keine gemeinsame Tangente besitzen, so müssen sich die Hyperflächen $F_i(\lambda,x)$ auch $\Pi\, s_i$-fach in η schneiden. Damit ist der erste Teil des Satzes schon bewiesen. Daß die Schnittmultiplizität nur dann größer sein kann, wenn eine gemeinsame Tangente vorhanden ist, folgt wieder unmittelbar aus A. Daß dann auch tatsächlich eine größere Schnittmultiplizität auftritt, läßt sich so schließen. Wir wenden B auf n allgemeine Hyperflächen $F_i(\lambda,x)$ der Ordnung n_i an, die bzw. s_i-fach durch einen festen Punkt η, etwa den Punkt $(1,0,\ldots,0)$, gehen und dort eine feste Tangente in der Richtung des Punktes ω, etwa des Punktes $(0, 1, 0, \ldots, 0)$, besitzen. Als spezialisierte Hyperflächen $F_i(\mu, x)$ nehmen wir nun die folgenden: $F_1(\mu, x)$ besteht aus einer allgemeinen quadratischen Hyperfläche $\Sigma a_{st} x_s x_t = 0$, die durch η (aber nicht durch ω) geht und dort eine Tangente in der Richtung ω besitzt (d. h. $\Sigma a_{st} \eta_s \eta_t = \Sigma a_{st} \eta_s \omega_t = 0$, also $a_{00} = a_{01} = a_{10} = 0$, die sonstigen a_{st} sind allgemein), und $s_1 - 1$ allgemein durch η gehenden Hyperebenen und $n_1 - s_1$ allgemeinen Hyperebenen; die anderen $F_i(\mu, x)$ $(i = 2, 3, \ldots, n)$ bestehen bzw. aus einer allgemeinen durch η, ω gehenden Hyperebene $\Sigma b_s^{(i)} x_s = 0$ (d. h. $\Sigma b_s^{(i)} \eta_s = \Sigma b_s^{(i)} \omega_s = 0$, also $b_0^{(i)} = b_1^{(i)} = 0$, die sonstigen $b_s^{(i)}$ sind allgemein), $s_i - 1$ allgemeinen durch η gehenden Hyperebenen und $n_i - s_i$ allgemeinen Hyperebenen[3]).

Die Relationen R lauten dann

$$2(s_i + 1) A \Sigma a_{st} \eta_s \pi_t + B \Sigma a_{st} \omega_s \omega_t = 2(s_i + 1) A \Sigma a_{1t} \pi_t + B a_{11} = 0,$$

$$A \Sigma b_s^{(2)} \pi_s = A (b_2^{(2)} \pi_2 + \ldots + b_n^{(2)} \pi_n) = 0,$$

$$\ldots\ldots\ldots\ldots\ldots\ldots\ldots\ldots\ldots\ldots$$

$$A \Sigma b_s^{(n)} \pi_s = A (b_2^{(n)} \pi_2 + \ldots + b_n^{(n)} \pi_n) = 0.$$

Da die Determinante $| b_s^{(i)} |$ $(i, s = 2, 3, \ldots, n)$ nicht verschwindet (denn die $b_s^{(i)}$ sind ja alle Unbestimmte) und die π_s nicht alle Null sind, so muß $A = 0$ sein, woraus folgt $B a_{11} = 0$, also $B = 0$ (da a_{11} Unbestimmte ist), in Widerspruch zu der Annahme, daß A, B nicht beide Null sind. Also können die Relationen R nicht bestehen, woraus folgt, daß die $F_i(\lambda, x)$ sich in η mit derselben Multiplizität schneiden wie die $F_i(\mu, x)$. Daß die $F_i(\mu, x)$ nun tatsächlich in η einen $(\Pi\, s_i + 1)$-fachen Schnittpunkt haben, läßt sich leicht direkt ausrechnen. Damit ist der Satz vollständig bewiesen.

---

[3]) Der Fall $n_i = s_i$ für alle i, der hier ausgeschlossen ist, läßt sich sehr leicht erledigen. In der Tat sind in diesem Falle alle Hyperflächen $F_i(\lambda, x)$ Kegel mit der Spitze in η. Haben sie eine gemeinsame Tangente in η, so enthalten sie alle diese Gerade, sie haben also dann unendlichviele gemeinsame Schnittpunkte.

(Eingegangen am 9. 7. 1938.)

Über Systeme von linearen partiellen Differentialgleichungen erster Ordnung.

Von

Wei-Liang Chow in Shanghai (China).

C. Carathéodory hat bei seiner Begründung des zweiten Hauptsatzes der Thermodynamik [1]) den folgenden Satz über eine Pfaffsche Gleichung bewiesen: *Wenn eine Pfaffsche Gleichung in jedem Punkte die Eigenschaft hat, daß es in jeder Umgebung von ihm Punkte gibt, die sich nicht durch eine Integralkurve der Gleichung mit ihm verbinden lassen, dann ist die Gleichung vollständig integrierbar.* Dabei ist unter einer Integralkurve einer Pfaffschen Gleichung $\sum_{j=1}^{n} \alpha_j (x_1, \ldots, x_n)\, dx_j = 0$ eine stückweise stetig differenzierbare Kurve [2]) $x_j(t)$ zu verstehen, deren jedes stetig differenzierbare Stück (auch in den Endpunkten) der Gleichung $\sum_{j=1}^{n} \alpha_j (x_1(t), \ldots, x_n(t)) \frac{dx_j}{dt} = 0$ genügt. Diesen Satz, den Carathéodory durch eine geometrische Konstruktion der Integralhyperflächen bewiesen hat, werden wir nun in einer ganz anderen Weise beweisen und gleichzeitig auf Systeme von Pfaffschen Gleichungen verallgemeinern. Unsere Methode besteht darin, daß wir zuerst in bekannter Weise das Pfaffsche System auf ein System von linearen partiellen Differentialgleichungen erster Ordnung zurückführen und dann die von einem Punkte aus durch einen aus den Charakteristiken dieser Differentialgleichungen zusammengesetzten Wege erreichbaren Punkte untersuchen. Dabei stellt es sich heraus, daß dieselben Betrachtungen uns auch die Mittel in die Hand geben, einen neuen Aufbau der Integrationstheorie der Systeme von linearen partiellen Differentialgleichungen erster Ordnung zu gewinnen, der dem üblichen in vielen Hinsichten vorzuziehen ist. Während nämlich die übliche Theorie für ein vollständiges System von den Koeffizienten und den Lösungen nur einmalige stetige Differenzierbarkeit voraussetzt, fordert die Heranziehung der Klammerausdrücke bei den nichtvollständigen Systemen viel mehr. Erstens muß wenigstens genügend oftmalige Differenzierbarkeit von den

---

[1]) Math. Annalen **67** (1909), S. 369.

[2]) Eine Funktion heißt stetig differenzierbar, wenn sie stetige partielle Ableitungen erster Ordnung besitzt. Eine Kurve $x_j(t)$ heißt stetig differenzierbar, wenn die Funktionen $x_j(t)$ so sind. Eine Kurve heißt stückweise stetig differenzierbar, wenn sie stetig und aus endlich vielen stetig differenzierbaren (abgeschlossenen) Kurvenstücken zusammengesetzt ist.

Koeffizienten vorausgesetzt werden, so daß die herangezogenen Klammerausdrücke überhaupt sinnvoll sind; zweitens müssen die Lösungsfunktionen selbst als zweimal stetig differenzierbar angenommen werden, damit die herangezogenen Klammerausdrücke auch diese Funktionen als Lösungen besitzen. Indem wir nun in unserer Theorie statt der Klammerausdrücke eine Art Umformungsausdruck verwenden, können wir auch bei den nichtvollständigen Systemen mit der Voraussetzung der einmaligen stetigen Differenzierbarkeit sowohl bei den Koeffizienten wie bei den Lösungen auskommen. Daß diese Voraussetzung der einmaligen stetigen Differenzierbarkeit selbst im Falle einer einzigen Gleichung nicht ohne weiteres zu unterlassen ist, hat O. Perron durch lehrreiche Beispiele gezeigt [3]).

Es sei $X f = \sum_{j=1}^{n} \beta_j \frac{\partial f}{\partial x_j}$ ein linearer Differentialoperator. Sind die Funktionen $\beta_j(x)$ in der Umgebung eines Punktes $a = (a_1, \ldots, a_n)$ stetig differenzierbar [2]), dann kann bekanntlich aus der Theorie der gewöhnlichen Differentialgleichungen folgendes behauptet werden: Das System von Differentialgleichungen $\frac{d x_j}{d t} = \beta_j(x)$ definiert in der Umgebung von a eine eingliedrige Gruppe (genauer Gruppenkeim) von stetig differenzierbaren Transformationen

$$T(t): \quad x_j' = g_j(x_1, \ldots, x_n, t) = g_j(x, t)$$

mit $g_j(x, 0) = x_j$. Die Differenzierbarkeitsbedingung bedeutet genauer: $g_j(x_1, \ldots, x_n, t)$ sind stetig differenzierbare Funktionen von $x_1, \ldots, x_n$ und t in der Umgebung von $x_1 = a_1, \ldots, x_n = a_n, t = 0$. Der Operator Xf ist also die erzeugende infinitesimale Transformation einer eingliedrigen Gruppe von stetig differenzierbaren Transformationen $T(t)$. Diese Eigenschaft von Xf ist nun invariant gegenüber beliebigen stetig differenzierbaren Transformationen in der Umgebung von a, während die stetige Differenzierbarkeit der Koeffizienten $\beta_j(x)$ dabei verlorengehen kann. In anderen Worten, die stetige Differenzierbarkeit der Transformationsgruppen, aber nicht die der infinitesimalen Transformationen, hat eine geometrische Bedeutung *bei Zugrundelegung der Gruppe aller stetig differenzierbaren Transformationen.* Da wir uns im folgenden auf diesen geometrischen Standpunkt stellen und nur mit geometrisch invarianten Begriffen operieren wollen, so werden wir statt der infinitesimalen Transformation Xf vielmehr die dadurch erzeugte Transformationsgruppe $T(t)$ als Hauptgegenstand unserer Betrachtung stellen, und werden daher von der stetigen Differenzierbarkeit der infinitesimalen Transformationen keinen Gebrauch machen. In der Tat brauchen die infinitesimalen Transformationen der nach unserem später anzugebenden Verfahren konstruierten vollständigen Erweiterung eines gegebenen Systems keineswegs

---

[3]) O. Perron, Math. Zeitschr. **27** (1928), S. 549.

stetig differenzierbar zu sein, auch wenn diese Eigenschaft dem gegebenen System zukommt.

Eine in der Umgebung von a stetig differenzierbare Funktion $\varphi(x_1, \ldots, x_n)$ heißt ein Integral von Xf, wenn sie der Gleichung $X\varphi = \sum \beta_j \frac{\partial \varphi}{\partial x_j} = 0$ genügt. Die Funktion $\varphi(x)$ ist bekanntlich genau dann ein Integral von Xf, wenn sie eine Invariante der durch Xf erzeugten Transformationsgruppe $T(t)$ ist, d. h. $\varphi(x_1, \ldots, x_n) = \varphi(g_1(x, t), \ldots, g_n(x, t))$ für alle $x_1, \ldots, x_n, t$ in der Umgebung von $x_1 = a_1, \ldots, x_n = a_n$, $t = 0$.

Der Punkt a heißt regulär für Xf, wenn nicht alle $\beta_j(a)$ gleich Null sind, sonst heißt er singulär. In einem regulären Punkte hat Xf ein System von $n-1$ Integralen $\varphi_1(x), \ldots, \varphi_{n-1}(x)$ mit der Eigenschaft, daß die Matrix $\left\| \frac{\partial \varphi_i}{\partial x_j} \right\|$ in a den Rang $n-1$ hat. Alle Integrale lassen sich dann in der Form $F(\varphi_1, \ldots, \varphi_{n-1})$ darstellen, wo $F(y_1, \ldots, y_{n-1})$ in der Umgebung von $\varphi_1(a), \ldots, \varphi_{n-1}(a)$ stetig differenzierbar ist. Ein solches System von Integralen heißt ein Hauptsystem. In der Tat, wenn etwa $\beta_1(a) \neq 0$ ist, so kann man die Gleichung $g_1(x, t) = a_1$ nach t lösen, etwa $t = h(x)$. Dann bilden $\varphi_1(x) = g_2(x, h(x)), \ldots, \varphi_{n-1}(x) = g_n(x, h(x))$ ein Hauptsystem von Integralen, und zwar ist $\left| \frac{\partial \varphi_i}{\partial x_j} \right|_a \neq 0$ $(i = 1, \ldots, n-1;\ j = 2, \ldots, n)$. Durch die Transformation $y_1 = h(x)$, $y_2 = \varphi_1(x), \ldots, y_n = \varphi_{n-1}(x)$ wird dann Xf auf die Gestalt $\frac{\partial f}{\partial y_1}$ reduziert, die die Gruppe der Translationen in Richtung der y_1-Achse erzeugt.

Ein lineares Differentialgleichungssystem B_r

$$X_i f = \sum_{j=1}^{n} \beta_{ij}(x) \frac{\partial f}{\partial x_j} = 0 \qquad (i = 1, 2, \ldots, r)$$

hat den Rang r im Punkte a, wenn die Matrix $\|\beta_{ij}(x)\|$ in der Umgebung von a den Rang r hat. Dabei braucht die $\|\beta_{ij}(x)\|$ aber nicht in a selbst den Rang r zu haben. Ein Punkt p heißt regulär für B_r, wenn $\|\beta_{ij}(x)\|$ in p den Rang r hat, anderenfalls heißt der Punkt singulär. Unsere Definition bedeutet also nur, daß es in jeder Umgebung von a reguläre Punkte von B_r gibt. Ein Differentialoperator $X_0 f = \sum \beta_{0j} \frac{\partial f}{\partial x}$ heißt von B_r abhängig, wenn die Matrix $\|\beta_{ij}(x)\|$ $(i = 0, 1, \ldots, r;\ j = 1, \ldots, n)$ in der Umgebung von a den Rang r hat. Zwei Differentialgleichungssysteme B_r heißen äquivalent, wenn die Operatoren $X_i f$ jedes Systems von dem anderen System abhängig sind. Ein gemeinsames Integral von den $X_1 f, \ldots, X_r f$ heißt ein Integral von B_r. $n - r$ Integrale $\varphi_1(x), \ldots, \varphi_{n-r}(x)$ bilden ein Hauptsystem von Integralen von B_r, wenn die Matrix $\left\| \frac{\partial \varphi_i}{\partial x_j} \right\|$ den Rang $n - r$ hat und jedes Integral von B_r sich in der Form $F(\varphi_1, \ldots, \varphi_{n-r})$ darstellen läßt, wobei $F(y_1, \ldots, y_{n-r})$ eine stetig differenzierbare Funktion ist. Man überzeugt sich leicht, daß äquivalente Systeme dieselben Integrale haben.

Die den $X_i f$ entsprechenden stetig differenzierbaren Gruppen von Transformationen seien mit $T_i(t)$ bezeichnet. Wir betrachten nun die Umformung von $T_i(t)$ durch $T_j(\tau)$, also die Transformationen $T_j(\tau)\, T_i(t)\, T_j(-\tau)$, die für jedes feste τ wieder eine eingliedrige Gruppe bilden, die mit der Gruppe $T_i(t)$ ähnlich ist. Da die Transformationen $T_1(t), \ldots, T_r(t)$ alle stetig differenzierbar sind, so sind die Transformationen $T_j(\tau)\, T_i(t)\, T_j(-\tau)$ auch alle stetig differenzierbar. Daraus folgt, daß die Gruppe $T_j(\tau)\, T_i(t)\, T_j(-\tau)$ eine erzeugende infinitesimale Transformation besitzt, die wir mit $T_j(\tau)\, X_i\, T_j(-\tau)$ bezeichnen werden. Ist in der Tat die Transformation $T_j(\tau)$ durch die Gleichungen $x'_j = g_j(x, \tau)$ gegeben, so ist

$$(T_j(\tau)\, X_i\, T_j(-\tau)) f = \sum_{k=1}^{n} \Big(\sum_{l=1}^{n} \beta_{il}(x) \frac{\partial g_k}{\partial x_l} \Big) \frac{\partial f}{\partial x_k},$$

wobei $x_j = g_j(x', -\tau)$ in den Koeffizienten $\sum_{l=1}^{n} \beta_{il}(x) \frac{\partial g_k}{\partial x_l}$ einzusetzen sind. Das System B_r heißt nun vollständig, wenn für alle i, j die Differentialoperatoren $T_j(\tau)\, X_i\, T_j(-\tau)$ für alle genügend kleinen Werte von τ von B_r abhängig sind. Ist das System B_r nicht vollständig, dann kann man es immer durch wiederholte Hinzunahme eines geeigneten „Klammerausdrucks" $T_j(\tau)\, X_i\, T_j(-\tau)$ zu einem vollständigen System B_s vom Range s im Punkte a erweitern. Da ein Integral von B_r gegenüber den Transformationen $T_1(t), \ldots, T_r(t)$ invariant bleibt, so bleibt es auch invariant gegenüber den Transformationen von B_s, woraus folgt, daß es auch ein Integral von B_s ist. Die Systeme B_r und B_s haben also dieselben Integrale. Die Zahl $s - r$ nennen wir den Index von B_r.

Satz A. *Ein vollständiges System B_s besitzt in der Umgebung eines regulären Punktes a ein Hauptsystem von Integralen $\varphi_1(x), \ldots, \varphi_{n-s}(x)$ (und zwar mit $\left| \frac{\partial \varphi_i}{\partial x_j} \right|_{x=a} \neq 0$ $(i = 1, \ldots, n-s;\ j = s+1, \ldots, n)$, wenn $|\beta_{ij}(a)| \neq 0$ $(i, j = 1, \ldots, s)$ ist). Jeder Punkt der durch a gehenden Integralmannigfaltigkeit $\varphi_1(x) = \varphi_1(a), \ldots, \varphi_{n-s}(x) = \varphi_{n-s}(a)$ läßt sich durch einen aus den Bahnkurven von B_s zusammengesetzten Weg mit a verbinden.*

Beweis durch Induktion. Der Satz gilt offenbar für eine Differentialgleichung. Wir nehmen also an, daß der Satz für B_{s-1} schon bewiesen ist, und werden ihn nun für B_s beweisen. Wie wir oben schon erwähnt haben, können wir durch eine Koordinatentransformation eine von den infinitesimalen Transformationen, etwa $X_1 f$, in die Form $X_1 f = \frac{\partial f}{\partial x_1}$ bringen. Die Transformationen $T_1(t)$ sind dann Translationen in Richtung der x_1-Achse. Daraus folgt, daß jedes Integral von B_s eine Funktion von $x_2, \ldots, x_n$, aber nicht von x_1 ist. Da die Matrix $\|\beta_{ij}(a)\|$ den Rang s hat und $\beta_{12}(a) = \beta_{13}(a) = \ldots = \beta_{1n}(a) = 0$ sind, so muß für jedes $i \neq 1$ mindestens ein $\beta_{ij}(a) \neq 0$ mit $j \neq 1$ sein. Das bedeutet, daß jede $X_i f$ $(i \neq 1)$ mindestens ein Integral $\psi_i(x)$ mit $\left(\frac{\partial \psi_i}{\partial x_1}\right)_{x=a} \neq 0$ besitzt. Durch die Transformation

$$U_i\colon\quad (x_1, x_2, \ldots, x_n) \to (\psi_i(x) - \psi_i(a) + a_1, x_2, \ldots, x_n),$$

die jedem Punkte eine Verrückung in Richtung der x_1-Achse erteilt, werden die Transformationen $T_i(t)$ auf die zur x_1-Achse senkrecht stehenden Hyperebenen „projiziert". Die infinitesimale Transformation $U_i X_i U_i^{-1}$ der transformierten Gruppe $T_i'(t) = U_i T_i(t) U_i^{-1}$ sei mit $X_i' f = \sum_{j=1}^{n} \beta_{ij}' \frac{\partial f}{\partial x_j}$ bezeichnet. Es sind dann offenbar $\beta_{21}'(x) = \beta_{31}'(x) = \ldots = \beta_{s1}'(x) = 0$. Der Einfachheit der Bezeichnungen wegen setzen wir auch $X_1' f = \sum \beta_{1j}' \frac{\partial f}{\partial x_j} = X_1 f = \frac{\partial f}{\partial x_1}$. Wir behaupten nun, daß die s infinitesimalen Transformationen $X_1' f$, $X_2' f$, $\ldots$, $X_s' f$ ein System B_s' vom Range s bilden, das mit dem System B_s äquivalent ist. Dazu zeigen wir zunächst, daß die Umformung einer $X_i f$ durch eine Transformation $U\colon (x_1, x_2, \ldots, x_n) \to (\psi(x), x_2, \ldots, x_n)$, die jedem Punkte eine Verrückung in Richtung der x_1-Achse erteilt, eine infinitesimale Transformation $(U X_i U^{-1}) f = \sum_{j=1}^{n} \alpha_{ij} \frac{\partial f}{\partial x_j}$ ergibt, die von dem System B_s abhängig ist. Denn es ist für jeden festen Punkt $c = (c_1, c_2, \ldots, c_n)$

$$\begin{aligned}
\sum_{i=1}^{n} \alpha_{ij}(c) \frac{\partial f}{\partial x_j} &= \Big(\beta_{i1}(c_1', c_2, \ldots, c_n) \frac{\partial \psi(c_1', c_2, \ldots, c_n)}{\partial c_1'} \\
&\quad + \sum_{j=2}^{n} \beta_{ij}(c_1', c_2, \ldots, c_n) \frac{\partial \psi(c_1', c_2, \ldots, c_n)}{\partial c_j}\Big) \frac{\partial f}{\partial x_1} \\
&\quad + \sum_{j=2}^{n} \beta_{ij}(c_1', c_2, \ldots, c_n) \frac{\partial f}{\partial x_j} \\
&= \Big(\beta_{i1}(c_1', c_2, \ldots, c_n) \frac{\partial \psi(c_1', c_2, \ldots, c_n)}{\partial c_1'} - \beta_{i1}(c_1', c_2, \ldots, c_n) \\
&\quad + \sum_{j=2}^{n} \beta_{ij}(c_1', c_2, \ldots, c_n) \frac{\partial \psi(c_1', c_2, \ldots, c_n)}{\partial c_j}\Big) \frac{\partial f}{\partial x_1} \\
&\quad + (T_1(c_1 - c_1') X_i T_1(c_1' - c_1)) f,
\end{aligned}$$

wobei überall $c_1' = \psi^{-1}(c)$ einzusetzen ist. Da das System B_s vollständig ist, so ist $T_1(c_1 - c_1') X_i T(c_1' - c_1)$ und folglich auch $U X_i U^{-1}$ für jeden Punkt c von B_s abhängig, woraus folgt, daß $U X_i U^{-1}$ überhaupt von B_s abhängig ist. Die infinitesimalen Transformationen $X_1' f$, $X_2' f$, $\ldots$, $X_s' f$ sind also alle von dem System B_s abhängig. Daß sie ein System vom Range s bilden, folgt daraus, daß die Matrix

$$\| \beta'_{ij}(a) \| = \left\| \begin{matrix} 1 & 0 & \cdots & 0 \\ 0 & \beta_{22}(a) & \cdots & \beta_{2n}(a) \\ \cdot & \cdot & & \cdot \\ \cdot & \cdot & & \cdot \\ \cdot & \cdot & & \cdot \\ 0 & \beta_{s2}(a) & \cdots & \beta_{sn}(a) \end{matrix} \right\|$$

denselben Rang wie die Matrix

$$\| \beta_{ij}(a) \| = \left\| \begin{matrix} 1 & 0 & \cdots & 0 \\ \beta_{21}(a) & \beta_{22}(a) & \cdots & \beta_{2n}(a) \\ \cdot & \cdot & & \cdot \\ \cdot & \cdot & & \cdot \\ \cdot & \cdot & & \cdot \\ \beta_{s1}(a) & \beta_{s2}(a) & \cdots & \beta_{sn}(a) \end{matrix} \right\|,$$

also den Rang s hat. Daraus folgt, wie man sich leicht überzeugt, daß das System B'_s mit dem System B_s äquivalent ist.

Das System B'_s ist auch vollständig. Denn die Umformung von $X'_i f$ durch $T'_j(\tau)$ ergibt eine infinitesimale Transformation

$$T'_j(\tau)\, X'_i\, T'_j(-\tau) = U_j\, T_j(\tau)\, U_j^{-1}\, U_i\, X_i\, U_i^{-1}\, U_j\, T_j(-\tau)\, U_j^{-1},$$

die aus $X_i f$ durch sukzessive Umformungen durch $U_j^{-1}\,U_i$, $T_j(\tau)$, U_j entsteht. Da jede dieser Umformungen eine von B_s abhängige infinitesimale Transformation in eine solche überführt, so ist $T'_j(\tau)\, X'_i\, T'_j(-\tau)$ von B_s, folglich auch von B'_s, abhängig. Die infinitesimalen Transformationen $X'_2 f, \ldots, X'_s f$ bilden nun auch für sich ein vollständiges System B'_{s-1}. Denn die Transformationen $T'_j(\tau)\, T'_i(t)\, T'_j(-\tau)$, und folglich auch die infinitesimalen Transformationen $T'_j(\tau)\, X'_i\, T'_j(-\tau)$, haben für $i \neq 1$ überhaupt keine Komponente in Richtung der x_1-Achse, d. h. $T'_j(\tau)\, X'_i\, T'_j(-\tau)$ ist schon von $X'_2 f, \ldots, X'_s f$ allein abhängig. Damit ist auch gezeigt, daß die Umformung von B'_{s-1} durch $T'_1(\tau) = T_1(t)$ ein äquivalentes System ergibt. Mit anderen Worten, B'_{s-1} ist bis auf Äquivalenz gegenüber den Translationen in Richtung der x_1-Achse invariant. Da nun x_1 in dem System B'_{s-1} offenbar die Rolle eines Parameters spielt, so können wir x_1 gleich einem festen Wert, etwa $x_1 = a_1$ setzen, und bekommen dann ein System B''_{s-1} in den $n-1$ Veränderlichen $x_2, x_3, \ldots, x_n$. Jedes Integral $\varphi(x_2, \ldots, x_n)$ von B''_{s-1} ist nun wegen der soeben erwähnten Invarianzeigenschaft von B'_{s-1} ein Integral von B'_{s-1} und, da es x_1 nicht enthält, auch ein Integral von B'_s, mithin auch von B_s. Umgekehrt ist jedes Integral von B_s auch ein Integral von B'_s, mithin auch von B'_{s-1}, mithin, da es x_1 nicht enthält, auch von B''_{s-1}.

Da nun B''_{s-1} nach der Induktionsvoraussetzung ein Hauptsystem von $(n-1)-(s-1)=n-s$ Integralen $\varphi_1(x), \ldots, \varphi_{n-s}(x)$ besitzt, und zwar mit $\left|\frac{\partial \varphi_i}{\partial x_j}\right|_{x=a} \neq 0 \ (i=1, \ldots, n-s;\ j=s+1, \ldots, n)$, falls $|\beta''_{ij}(a)| \neq 0$ $(i, j=2, \ldots, s)$ ist, so besitzt B_s auch das Hauptsystem $\varphi_1(x), \ldots, \varphi_{n-s}(x)$, und zwar mit $\left|\frac{\partial \varphi_i}{\partial x_j}\right|_{x=a} \neq 0 \ (i=1, \ldots, n-s;\ j=s+1, \ldots, n)$, falls $|\beta_{ij}(a)| = |\beta'_{ij}(a)| = |\beta''_{kl}(a)| \neq 0 \ (i, j=1, \ldots, s;\ k, l=2, \ldots, s)$ ist. Damit ist der erste Teil des Satzes bewiesen.

Was den zweiten Teil betrifft, ist er nach dem vorangegangenen ziemlich klar. Denn nach der Induktionsvoraussetzung läßt sich jeder Punkt der Integralmannigfaltigkeit $x_1 = a_1,\ \varphi_1(x) = \varphi_1(a), \ldots, \varphi_{n-s}(x) = \varphi_{n-s}(a)$ durch einen aus den Bahnkurven von B''_{s-1} zusammengesetzten Weg mit a verbinden. Daraus folgt, daß jeder Punkt der Mannigfaltigkeit $\varphi_1(x) = \varphi_1(a), \ldots, \varphi_{n-s}(x) = \varphi_{n-s}(a)$ sich durch einen aus den Bahnkurven von B'_s zusammengesetzten Weg mit a verbinden läßt. Da nun die Bahnkurven von $X'_i f$ offenbar aus denen von $X_i f$ und $X_1 f$ zusammengesetzt sind, so läßt sich jeder Punkt der Mannigfaltigkeit $\varphi_1(x) = \varphi_1(a), \ldots, \varphi_{n-s}(x) = \varphi_{n-s}(a)$ auch durch einen aus den Bahnkurven von B_s zusammengesetzten Weg mit a verbinden. Damit ist der Satz vollständig bewiesen.

Aus diesem Satz A folgt dann für ein beliebiges System B_r der folgende

Satz B. *Ein lineares partielles Differentialgleichungssystem B_r hat in jedem Punkt a, der für seine vollständige Erweiterung B_s regulär ist, ein Hauptsystem von $n-s$ Integralen $\varphi_1(x), \ldots, \varphi_{n-s}(x)$. Jeder Punkt der durch a gehenden Integralmannigfaltigkeit $\varphi_1(x) = \varphi_1(a), \ldots, \varphi_{n-s}(x) = \varphi_{n-s}(a)$ läßt sich durch einen aus den Bahnkurven von B_r zusammengesetzten Weg mit a verbinden.*

Zu dem zweiten Teil des Satzes braucht man nur noch zu bemerken, daß die Bahnkurven einer umgeformten Gruppe $T_j(\tau)\, T_i(t)\, T_j(-\tau)$ offenbar durch eine aus den Bahnkurven von $T_i(t)$ und $T_j(t)$ zusammengesetzte Kurve ersetzt werden können. Daß es nötig ist, den Punkt a als einen regulären Punkt nicht nur von B_r, sondern auch von B_s vorauszusetzen, zeigt das folgende Beispiel: Das nicht vollständige System B_2

$$X_1 f = \frac{\partial f}{\partial x_1}, \quad X_2 f = \frac{\partial f}{\partial x_2} + x_1 x_3 \frac{\partial f}{\partial x_3}$$

ist offenbar in dem Punkte $O = (0, 0, 0)$ regulär. Durch Hinzunahme des Klammerausdrucks $(X_1 X_2) f = x_3 \frac{\partial f}{\partial x_3}$ entsteht ein vollständiges System B_3, das aber offenbar nicht mehr in dem Punkte O regulär ist. In der Tat liegen alle durch einen Punkt der Ebene $x_3 = 0$ gehenden Bahnkurven der infinitesimalen Transformation $\alpha X_1 f + \beta X_2 f$, wo $\alpha(x), \beta(x)$ beliebige stetig differenzierbare Funktionen sind, immer in der Ebene $x_3 = 0$, woraus folgt, daß man niemals einen nicht in der Ebene $x_3 = 0$ liegenden Punkt durch einen aus den Bahnkurven von allen solchen infinitesimalen Transformationen $\alpha X_1 f + \beta X_2 f$ zusammengesetzten Weg mit O verbinden kann. Aus dem Satze B folgt dann, daß B_2 nicht in einem im Punkte O regulären vollständigen System B_3 enthalten sein kann.

Ohne Heranziehung des vollständigen Systems B_s kann man auch den zweiten Teil des Satzes B in folgender Weise formulieren:

Satz C. *Ein lineares Differentialgleichungssystem B_r vom Index m im Punkte a hat in jeder Umgebung von a Punkte mit der Eigenschaft, daß die von ihnen aus durch die aus den Bahnkurven von B_r zusammengesetzten Wege erreichbaren Punkte eine $(r + m)$ dimensionale Mannigfaltigkeit bilden.* Oder: *Hat ein System B_r in dem Punkte a die Eigenschaft, daß in jedem Punkte einer Umgebung von a die von ihm aus durch die aus den Bahnkurven von B_r zusammengesetzten Wege erreichbaren Punkte eine Mannigfaltigkeit von höchstens $r + m$ Dimensionen bildet, dann hat das System B_r höchstens den Index m. Gibt es außerdem in jeder Umgebung von a immer Punkte, wobei diese höchste Dimension $r + m$ wirklich auftritt, so hat das System B_r genau den Index m.*

Dies ist die eingangs erwähnte Verallgemeinerung des Carathéodoryschen Satzes. Um sie in eine Aussage über Pfaffsche Systeme zu verwandeln. brauchen wir nur folgendes zu bemerken: Es sei

$$\sum_{j=1}^{n} \alpha_{ij}(x)\, dx_j = 0 \qquad (i = 1, \ldots, r)$$

ein Pfaffsches System A_r vom Range r im Punkte a, wobei die Funktionen $\alpha_{ij}(x)$ in der Umgebung von a stetig differenzierbar sein sollen. Durch die Bedingungen $\sum_{j=1}^{n} \alpha_{ij}\beta_{kj} = 0$ wird bis auf Äquivalenz ein Differentialgleichungssystem B_{n-r} vom Range $n - r$, $\sum_{j=1}^{n} \beta_{ij} \frac{\partial f}{\partial x_j} = 0$ $(i = 1, \ldots, n - r)$ bestimmt. Die Systeme A_r und B_{n-r} heißen dann zueinander assoziiert, sie bestimmen einander bis auf Äquivalenz eindeutig. Die Integralkurven von A_r sind nichts anderes als die aus den Bahnkurven von B_{n-r} zusammengesetzten Kurven. Wir sagen, das System A_r habe den Index m, wenn sein assoziiertes System B_{n-j} den Index m hat. Wir können nun den Satz C in folgender Weise formulieren:

Satz D. *Hat ein Pfaffsches System A_r in dem Punkte a die Eigenschaft, daß in jedem Punkte einer Umgebung von a die von ihm aus durch die Integralkurven von A_r erreichbaren Punkte eine Mannigfaltigkeit von höchstens $n - r + m$ Dimensionen bilden, dann hat das System A_r höchstens den Index m. Gibt es außerdem in jeder Umgebung von a immer Punkte, wobei diese höchste Dimension $n - r + m$ wirklich auftritt, so hat das System genau den Index m.*

Der Satz von Carathéodory entspricht dem Falle $r = 1$, $m = 0$. (Im Falle $m = 0$ ist der letzte Zusatz von Satz D überflüssig, da der Index niemals negativ sein kann.)

(Eingegangen am 19. 11. 1938.)

ON ELECTRIC NETWORKS

By Wei-Liang Chow (周煒良)

Consider an electric network N with n conductors $F_1, F_2, \ldots, F_n$, of resistances $R_1, R_2, \ldots, R_n$ respectively, and $d+1$ junctions $P_1, P_2, \ldots, P_{d+1}$. Let a direction be assigned to each conductor in a quite arbitrary way. To each junction P_i we shall associate a row n-vector, that is a $(1, n)$-matrix $(p_{i1}, p_{i2}, \ldots, p_{in})$, where p_{ij} is equal to $+1$, -1, or 0 according as F_i is directed toward P_i, directed away from P_i, or not connected with P_i. The $d+1$ vectors thus defined shall be denoted again by $P_1, P_2, \ldots, P_{d+1}$. Since the relation $\sum\limits_{i=1}^{d+1} P_i = 0$ evidently holds, these $d+1$ vectors are not linearly independent. But, if we delete any one of them, say P_{d+1}, the rest d vectors will then be linearly independent. This follows from the fact that in any linear relation $\sum\limits_{i=1}^{d+1} C_i P_i = 0$ none of the coefficients C_i can be zero without all the coefficients being zero. For, if $C_{r+1} = \cdots = C_{d+1} = 0$ and $C_i \neq 0\,(i \leq r)$ and F_i be a conductor which connects a junction $P_m\,(m \leq r)$ with a junction $P_t\,(t > r)$,[1] then we have $p_{mj} = \pm 1$ and $p_{ij} = 0\,(i \neq m, i \leq r)$, so that the vector $\sum\limits_{i=1}^{d+1} C_i P_i = \sum\limits_{i=1}^{r} C_i P_i$ will have $\pm C_m \neq 0$ as its j-th element, which is contradictory to our assumption that $\sum\limits_{i=1}^{d+1} C_i P_i = 0$.

We shall call the (d, n)-matrix

$$P \equiv \begin{pmatrix} P_1 \\ \vdots \\ P_d \end{pmatrix} \equiv (p_{ij})$$

the junction matrix on N. Let $X = [x_1, x_2, \ldots, x_n]$ be a column n-vector, where the $x_1, x_2, \ldots, x_n$ are variables. The system of d linear equations in n

(1) At least one such conductor must exist, for we are assuming here that the network N is connected. If N is not connected, the d vectors $P_1, P_2, \ldots, P_d$ will not be linearly independent.

unknowns

$$\sum_{j=1}^{n} p_{ij}x_j = 0\,, \quad (i = 1, 2, \ldots, d)\,,$$

or, in matrix notation,

$$PX = 0$$

has exactly an $(n - d)$-dimensional manifold of solutions. We shall write these solutions again as row n-vectors (for convenience of notation later). It is known that these solutions possess an integral basis, i.e. a set of $s = n - d$ integral solutions

$$Q_1 = (q_{1j})\,, \quad Q_2 = (q_{2j})\,, \quad \ldots\,, \quad Q_s = (q_{sj})$$

such that any integral solution can be expressed as a linear combination of them with integer coefficients. In particular, the solutions of the integral basis may be so chosen that each contains only ± 1 as its non-zero elements. Topologically such solutions correspond to cycles which are sums of oriented simple circuits.[2] We shall consider in the following only such solutions, which shall simply be called circuits. The (s, n)-matrix

$$Q \equiv \begin{pmatrix} Q_1 \\ \vdots \\ Q_s \end{pmatrix} \equiv (q_{ij})$$

will be called a circuit matrix of N. Unlike the junction matrix it depends on the basis chosen.

For the existence of an integral basis of the type mentioned above we shall give here a proof, which will also be useful to us later. We state first a simple lemma, which can be proved very easily by induction (See, for example, Veblen: Analysis Situs, 2nd Edition, p. 174):

[2] For these and other topological notions which, though not necessary for our argument, it will be useful to have in mind, the reader is referred once for all to Veblen: Analysis Situs.

Lemma. If a square matrix is such that each column contains either only one non-zero element, $+1$ or -1, or only two non-zero elements, $+1$ and -1, then the determinant of the matrix, if not zero, is equal to ± 1.

The matrix P must contain at least one set of d linearly independent columns, which we can suppose without loss of generality to be the first d columns. The determinant

$$|p_{ij}| \quad (i, j = 1, 2, \ldots, d)$$

has then according to the above lemma the value ± 1. Write the equation $PX = 0$ in the following form

$$\sum_{j=1}^{d} p_{ij} x_j = - \sum_{j=d+1}^{n} p_{ij} x_j \quad (i = 1, 2, \ldots, d)$$

and solve for $x_1, \ldots, x_d$ in terms of the rest of the variables, we obtain then evidently all integral solutions by letting $x_{d+1}, \ldots, x_n$ run independently through all integral values. The set of s solutions obtained by setting one of the s variables $x_{d+1}, \ldots, x_n$ equal to $+1$ and the rest equal to zero forms then an integral basis of the desired type (again in virtue of the lemma). The integral basis $Q_1, Q_2, \ldots, Q_s$ thus obtained has the further property that the last s columns of the circuit matrix Q form the identity matrix, i.e.

$$q_{i,d+j} = \begin{matrix} 1 & (i = j) \\ 0 & (i \neq j) \end{matrix}$$

This property will be used later. For convenience, we shall say that a matrix has the property (L) if its columns satisfy the conditions stated in the above lemma.

Now let $C_1, C_2, \ldots, C_n$ be the currents, $S_1, S_2, \ldots, S_n$ the e.m.f.'s in the n conductors $F_1, F_2, \ldots, F_n$ respectively. Set

$$R \equiv \begin{pmatrix} R_1 & & O \\ & R_2 & \\ & & \ddots \\ O & & R_n \end{pmatrix}, \quad C \equiv \begin{pmatrix} C_1 \\ \vdots \\ C_n \end{pmatrix}, \quad S = \begin{pmatrix} S_1 \\ \vdots \\ S_n \end{pmatrix},$$

so that R is an (n,n)-matrix with $R_1, R_2, \ldots, R_n$ on the diagonal and zeros elsewhere, and C and S are column vectors. The theory of steady state electric currents gives then the following system of n linear equations

$$\begin{pmatrix} P \\ QR \end{pmatrix} C = \begin{pmatrix} O \\ QS \end{pmatrix} \tag{1}$$

consisting of d junction equation (the first law of Kirchhoff)

$$PC = O \tag{2}$$

and s circuit equations (the second law of Kirchhoff)

$$QRC = QS \tag{3}$$

from which the $n = d + s$ currents $C_1, C_2, \ldots, C_n$ can be determined in terms of R and S,(3) From the theory of linear equations we have the following determinant form of the solutions

$$C_i = \frac{D_i}{D}$$

(3) It is to be noticed that the existence and uniqueness of solution for the system (1) depends on the condition $\begin{vmatrix} P \\ QR \end{vmatrix} \neq 0$, a condition which holds for $R_i > 0$ or more generally for any definite symmetric matrix R. In fact, we have

$$\begin{vmatrix} P \\ QR \end{vmatrix} |P'Q'| = \begin{vmatrix} PP' & O \\ QRQ' & QRP' \end{vmatrix} = |PP'|\,|QRQ'|$$

and hence

$$\begin{vmatrix} P \\ QR \end{vmatrix} = \frac{|PP'|}{\begin{vmatrix} P \\ Q \end{vmatrix}} |QRQ'|$$

Since R is definite and the rows of $\begin{pmatrix} P \\ Q \end{pmatrix}$ are linearly independent by construction, we have $|QRQ'| \neq 0$, $|PP'| \neq 0$, $\begin{vmatrix} P \\ Q \end{vmatrix} \neq 0$, therefore $\begin{vmatrix} P \\ QR \end{vmatrix} \neq 0$. Physically the definiteness of R comes from the consideration that the quadratic form $ZQRQ'Z'$, where $Z = (z_1, \ldots, z_s)$ represents tre cyclic currents, gives the work done (transformed into heat) in the network, and the principle that some work must be done so long as there is some current in the network.

where $D = \begin{vmatrix} P \\ QR \end{vmatrix}$ and the D_i are formed according to the Cramer's Rule. It is obvious that D is a homogeneous polynomial of degree s in the R and the D_i are homogeneous polynomials of degree $s-1$ in the R and of degree 1 in the S. The actual calculations involved in evaluating these determinants according to the usual methods are however very complicated. In an article "On a New Method for the Analysis of Electric Networks"[(4)] Mr. K. T. Wang has shown that in virtue of the special character of the system (1) these calculations can be very much simplified. His method consists of two rules:

(A) The determinant D is found as follows: Consider all the sequences of s integers $(m_1, m_2, \ldots, m_s)$, no two of which are alike, such that the conductors $F_{m_1}, F_{m_2}, \ldots, F_{m_s}$ appear in the circuits $Q_1, Q_2, \ldots, Q_s$ respectively. The same set of s integers may, of course, appear more than once as different sequences. Then the determinant D is up to a sign equal to $\Sigma R_{m_1}, R_{m_2}, \ldots, R_{m_S}$, where the sum extends over all different sets of s integers which appear an odd number times among the above sequences.[(5)]

(B) For the determination of D_i it is assumed for the sake of simplicity that only one conductor, say F_1, contains a battery of unit e.m.f., that is $S = [1, 0, \ldots, 0]$, the determinants in the general case being evidently linear combinations of those in the particular cases. The determinant D_1 can evidently be obtained from D by simply taking all the term of D containing R_1 and suppressing this common factor R_1. In other words, we have $D_1 = \frac{\partial D}{\partial R_1}$. Besides, D_1 stands in the same relation to the rest network after deleting the conductor F_1 (called the minor network) as the determinant D

---

(4) Memoir of the National Research Institute of Engineering. I am sorry I cannot say in which volume of this Memoir this article of Mr. Wang appeared, because I have read it only in the special pamphlet published by Mr. Wang in commemoration of his 60th birthday, where it is only stated that this article is reprinted from the Memoir without giving any further details.

(5) This is a slight modification of Mr. Wang's rule. In fact, the original rule of Mr. Wang, which requires the set of s integers to appear exactly once among the sequences, presupposes a stricter choice of the basis, which can only be realized under certain restrictions on the nature of the network considered. See Note II.

stands to the original network N, so that we can also find D_1 by applying the rule (A) directly to the minor network. One D_1 is obtained the rest of the D_i can then be found by a simple "accounting process" with the help of a diagram of the network. This "accounting process" constitutes Mr. Wang's second rule. We shall not go into the details of this rule here, because we have nothing new to add to it. In Note I at the end of this paper, we shall give a short statement and proof of this rule from our present standpoint.

While for the rule (B) Mr. Wang has given a proof which is quite general, in proof of the rule (A) he has to introduce the restriction that the network be mappable on the sphere. This assumption is quite essential in his proof, where the fact that every conductor, if appears more than once, appears exactly twice in the opposite senses in the circuits of the basis plays a fundamental part in the argument. Such a basis of circuits exists, as will be shown in Note II, if and only if the network in mappable on the sphere. In the following we shall give a very simple new proof of the rule (A) without this restriction. At the same time we shall obtain another rule, dual to the rule (A), which connects the D with the junction matrix P. Calling a set of S integers complementary to a set of d integers if the two sets together give the set of $n = d + s$ integers $1, 2, \ldots, n$, we can state this dual rule as follows:

(A$'$) Consider all the sequences of d different integers $(n_1, n_2, \ldots, n_d)$ such that the conductor $F_{n_1}, F_{n_2}, \ldots, F_{n_d}$ are connected with the junctions $P_1, P_2, \ldots, P_d$ respectively. The determinant D is up to a sign equal to $\Sigma\, R_{m_1} R_{m_2} \cdots R_{m_s}$, where the sum extends over all different sets of s integers which are complementary to sets of d integers which appear an odd number of times among the above sequences.

For proof, we shall start with some simple matrix multiplications. Denote by I_{tt} the identity (t, t)-matrix and by O_{tr} the zero (t, r)-matrix, we have then

$$\begin{pmatrix} O_{dn} & I_{dd} \\ P & O_{dd} \\ Q & O_{sd} \end{pmatrix} \begin{pmatrix} R & P' \\ P & O_{dd} \end{pmatrix} = \begin{pmatrix} P & O_{dd} \\ PR & PP' \\ QR & O_{sd} \end{pmatrix} \tag{4}$$

$$\begin{pmatrix} Q_{sn} & I_{ss} \\ Q & O_{ss} \\ P & O_{ds} \end{pmatrix} \begin{pmatrix} R^{-1} & Q' \\ Q & O_{ss} \end{pmatrix} = \begin{pmatrix} Q & O_{ss} \\ QR^{-1} & QQ' \\ PR^{-1} & Q_{ds} \end{pmatrix} \tag{5}$$

From these we deduce

$$\begin{vmatrix} P \\ Q \end{vmatrix} \begin{vmatrix} R & P' \\ P & O \end{vmatrix} = \pm |PP'| \begin{vmatrix} P \\ QR \end{vmatrix} \tag{6}$$

$$\begin{vmatrix} P \\ Q \end{vmatrix} \begin{vmatrix} R^{-1} & Q' \\ Q & O \end{vmatrix} = \pm |QQ'| \begin{vmatrix} PR^{-1} \\ Q \end{vmatrix} = \pm \frac{|QQ'|}{|R|} \begin{vmatrix} P \\ QR \end{vmatrix} \tag{7}$$

and hence

$$D = \begin{vmatrix} P \\ QR \end{vmatrix} = \pm \frac{\begin{vmatrix} P \\ Q \end{vmatrix}}{|PP'|} \begin{vmatrix} R & P' \\ P & O \end{vmatrix} \tag{8}$$

$$D = \begin{vmatrix} P \\ QR \end{vmatrix} = \pm \frac{\begin{vmatrix} P \\ Q \end{vmatrix} |R|}{|QQ'|} \begin{vmatrix} R^{-1} & Q' \\ Q & O \end{vmatrix} \tag{9}$$

Thus we see that D is up to a constant factor equal to

$$\begin{vmatrix} R & P' \\ P & O \end{vmatrix} \quad \text{or} \quad |R| \begin{vmatrix} R^{-1} & Q' \\ Q & O \end{vmatrix}$$

Observing that R is a diagonal matrix we obtain without difficulty the following:

$$\begin{vmatrix} R & P' \\ P & O \end{vmatrix} = \Sigma R_{m_1} \cdots R_{m_s} |P(n_1, \ldots, n_d)|^2 \tag{10}$$

$$|R| \begin{vmatrix} R^{-1} & Q' \\ Q & O \end{vmatrix} = R_1 \cdots R_n \Sigma R_{n_1}^{-1} \cdots R_{n_d}^{-1} |Q(m_1, \ldots, m_s)|^2$$

$$= \Sigma R_{m_1} \cdots R_{m_s} |Q(m_1, \ldots, m_s)|^2 \tag{11}$$

where $P(n_1, \ldots, n_d)$ is the (s, s)-matrix

$$(p_{ij}), \quad (i = 1, 2, \ldots, d;\ j = n_1, n_2, \ldots, n_d),$$

$Q(m_1, \ldots, m_s)$ is the (s, s)-matrix

$$(q_{ij}), \quad (i = 1, 2, \ldots, s;\ j = m_1, m_2, \ldots, m_s),$$

and the sums extend over all the $\frac{n!}{d!s!}$ partitions of the numbers $1, 2, \ldots, n$ into two complementary sets $(n_1, n_2, \ldots, n_d)$ and $(m_1, m_2, \ldots, m_s)$ of respectively d and s elements each.

Proof of (A′). Since the matrix $P(n_1, \ldots, n_d)$ has evidently the property (L), we have $|P(n_1, \ldots, n_d)|^2 = 1$ if not zero. Hence

$$\begin{vmatrix} R & P' \\ P & O \end{vmatrix} = \Sigma\, R_{m_1} \cdots R_{m_s} \tag{12}$$

where the sum now extends over all sets $(m_1, m_2, \ldots, m_s)$ such that for the complementary set $(n_1, n_2, \ldots, n_d)$ the determinant $|P(n_1, \ldots, n_d)| = \pm 1$. Since each non-zero element of the matrix $P(n_1, \ldots, n_d)$ is either $+1$ or -1, so every non-vanishing term of the determinant $|P(n_1, \ldots, n_d)|$ must also be ± 1. It follows that $|P(n_1, \ldots, n_d)|$ will be an odd or even number according as it contains an odd or even number of non-vanishing terms. But since the only values $|P(n_1, \ldots, n_d)|$ can take are ± 1, 0, so the value of $|P(n_1, \ldots, n_d)|$ will be ± 1 or 0 according as it contains an odd or even number of non-vanishing terms. This is exactly the rule (A′), only somewhat differently expressed.

Proof of (A). What we have said for $\begin{vmatrix} P & P' \\ P & O \end{vmatrix}$ holds evidently also for $\begin{vmatrix} R^{-1} & Q' \\ Q & O \end{vmatrix}$, hence also for $|R| \begin{vmatrix} R^{-1} & Q' \\ Q & O \end{vmatrix}$. Only we have here to prove that the determinant $|Q(m_1, \ldots, m_s)| = \pm 1$ if not zero, for the matrix $Q(m_1, \ldots, m_s)$ will not in general possess the property (L), so that our Lemma cannot be applied. But from the fact $\begin{vmatrix} R & P' \\ P & O \end{vmatrix}$ and $|R| \begin{vmatrix} R^{-1} & Q' \\ Q & O \end{vmatrix}$ are equal up to a constant factor follows that the coefficients of all the terms in $|R| \begin{vmatrix} R^{-1} & Q' \\ Q & O \end{vmatrix}$ are all equal. Hence we have only to prove that there is one determinant $|Q(m_1, \ldots, m_s)| = \pm 1$. This follows however immediately from the last remark in the proof of the existence of the integral basis on

p. 323, since for the integral basis constructed there we have $|Q(d+1, d+2, \ldots, d+s)| = +1$ and since a change of integral basis leaves the determinant $\begin{vmatrix} R^{-1} & Q' \\ Q & O \end{vmatrix}$ evidently unchanged.[6] Hence we have

$$|R| \begin{vmatrix} R^{-1} & Q' \\ Q & O \end{vmatrix} = \Sigma R_{m_1} \cdots R_{m_s} \tag{13}$$

where the summation is as indicated in (A).[7]

So far we have only proved that D is up to a constant factor equal to $\Sigma R_{m_1} \cdots R_{m_s}$. This is really sufficient for our purpose, but for the sake of completeness we shall show that D is only up to a sign equal to $\Sigma R_{m_1} \cdots R_{m_s}$, as stated in (A) and (A′). In fact, we have from (8) (9)

$$\frac{\begin{vmatrix} P \\ Q \end{vmatrix}}{|PP'|} = \pm \frac{\begin{vmatrix} P \\ Q \end{vmatrix}}{|QQ'|}$$

hence

$$\frac{\begin{vmatrix} P \\ Q \end{vmatrix}^2}{|PP'|^2} = \frac{\begin{vmatrix} P \\ Q \end{vmatrix}^2}{|QQ'|^2} = \frac{\begin{vmatrix} P \\ Q \end{vmatrix}^2}{|PP'|\,|QQ'|} = \frac{\begin{vmatrix} P \\ Q \end{vmatrix} |P'\,Q'|}{\begin{vmatrix} PP' & O_{ds} \\ O_{sd} & QQ' \end{vmatrix}} = 1$$

hence

$$\frac{\begin{vmatrix} P \\ Q \end{vmatrix}}{|PP'|} = \pm \frac{\begin{vmatrix} P \\ Q \end{vmatrix}}{|QQ'|} = \pm 1 \tag{14}$$

---

(6) If T is the (s, s)-matrix of determinant ± 1, which effects the change of basis, we have evidently

$$\begin{vmatrix} R^{-1} & Q'T' \\ TQ & O \end{vmatrix} = \begin{vmatrix} I_{nn} & O \\ O & T \end{vmatrix} \begin{vmatrix} R^{-1} & Q' \\ Q & O \end{vmatrix} \begin{vmatrix} I_{nn} & O \\ O & T' \end{vmatrix} = \begin{vmatrix} R^{-1} & Q' \\ Q & O \end{vmatrix}$$

(7) Incidentally we have proved that $|P(n_1, \ldots, n_d)| \neq 0$ is and only if $|Q(m_1, \ldots, m_s)| \neq 0$, where $(n_1, \ldots, n_d)$ and $(m_1, \ldots, m_s)$ are any two complementary sets, a fact which can be easily verified directly and which can also be made the basis of our proof instead of the matrix multiplications.

Therefore

$$D = \pm\Sigma\, R_{m_1} \cdots R_{m_s} \tag{15}$$

where the summation is as indicated in (A) or (A′).

We shall conclude here with two notes, one on the second rule of Mr. Wang and the other on the principle of duality of an electric network, and an example of a simple network not mappable on the sphere.

NOTE I. From our present standpoint the most essential thing in the rule (B) of Mr. Wang is the fact that once the current $C_1 = \frac{D_1}{D}$ is obtained, the rest of the currents $C_2, C_3, \ldots, C_n$ can be found by considering the d junction equations (2) alone. Or, what amounts to the same thing, the determinants $D_2, D_3, \ldots, D_n$ can be found by considering the d equations

$$P\Delta = O \tag{16}$$

where $\Delta = [D_1, D_2, \ldots, D_n]$ is to be regarded as a column vector, once the determinant D_1 is given. (All this, of course, under the assumption stated in (B) that only one conductor, F_1, contains a battery, that is $S = [1, 0, \ldots, 0]$). Here it is essential that D_1 is given as a function of the resistances $R_1, R_2, \ldots, R_n$ (the R_i being regarded as independent variables), not simply as a numerical value in the particular case under consideration. For, in case $s > 1$, the d equation (16) certainly cannot determine the values of the $n-1 > d$ unknowns $D_2, D_3, \ldots, D_n$ uniquely if only the numerical value of D_1 is given. If, however, the D_i are regarded as functions of the resistances, we can then make use of some general properties of these functions (derived, of course, by other considerations than the equations (16)), so that in any particular case it is only necessary to consider the d equations (16).

From the definition of the determinants $D_1, D_2, \ldots, D_n$ we have at once the following properties:

(a) Each D_i is a homogeneous polynomial of degree $s - 1$ in the resistances R.

(b) No term of the determinant D_i can contain the factor R_i.

(c) If $R_{m_1} \cdots R_{m_{s-1}}$ is a term of D_i then $|Q(1, m_1, \ldots, m_{s-1})| \neq 0$. The last property (c) follows from the fact that the column s-vector QS is up to a factor R_1 equal to the first column of Q (under the assumption that $S = [1, 0, \ldots, 0]$). It follows now from (c) and the equation (11) that every term of D_i $(i \geq 2)$ must also appear in D_1. We have therefore, once D_1 is given, only to find out for each term of D_1 in which D_i it appears. This can be done by considering the equations (16) alone if we make use of the property (b). For, let $(n_1, \ldots, n_d)$ be the complementary set of $(1, m_1, \ldots, m_{s-1})$, then only the determinants $D_1, D_{n_1}, \ldots, D_{n_d}$ can contain the term $R_{m_1} \cdots R_{m_{s-1}}$. Setting $D_1 = R_{m_1} \cdots R_{m_{s-1}}, D_{m_1} = \cdots = D_{m_{s-1}} = 0$ in (16), we get a set of d linear equations in d unknowns

$$\sum_{j=1}^{d} p_{in_j} D_{n_j} = -p_{i1} R_{m_1} \cdots R_{m_{s-1}} \tag{17}$$

from which the $D_{n_1}, \ldots, D_{n_d}$ can be determined uniquely. Since the matrix P has the property (L), we see without difficulty that the elements $D_{n_1}, \ldots, D_{n_d}$ of the solution can only be zero or $\pm R_{m_1} \cdots R_{m_{s-1}}$. Repeat this process for each term of D_1 and add together all the terms so obtained for each D_i, we get the desired determinants $D_2, D_s, \ldots, D_n$. The rule (B) of Mr. Wang is simply the formation and solution of the equation (17) translated into topological processes. In fact, the solution of any system of d linear equations in d unknowns

$$\sum_{j=1}^{d} a_{ij} x_j = a_{i0} x_0 \tag{18}$$

when the $(d, d+1)$-matrix (a_{ij}), $(i = 1, \ldots, d,\ j = 0, 1, \ldots, d)$, has the property (L), can be effected by a simple geometrical method with the help of the network of $d+1$ conductors and $d+1$ junctions defined by the junction matrix (a_{ij}) (the last junction is given by the negative sum of the rows of (a_{ij})). This network has evidently $d + 1 - d = 1$ circuit which necessarily passes through the conductor F_0 corresponding to the column $[a_{i0}]$ of $[a_{ij}]$, for we are assuming here that the determinant of the coefficients of the left side of (18) is not equal to zero. If we now let a current x_0 pass through F_0, then the same current with the appropriate sign (according to the orientation of the conductors) must pass through every conductor of this only circuit and the rest conductors will have no current. Since the equations (18) is nothing but the fundamental junction equations for the distribution of current in this particular network (the circuit equation is not needed, since x_0 is given), so we have here a geometrical method for finding the solution of (18). In our present case the network in consideration is the rest network of N after deleting the conductors $F_{m_1}, \ldots, F_{m_{s-1}}$ and the current put through F_1 is equal to $R_{m_1} \cdots R_{m_{s-1}}$.

Since our treatment above is mainly algebraic, it may not be out of place to add a few words here on the geometrical aspect of the subject. It can be easily shown that the determinant D is equal to the sum of all the terms $R_{m_1} R_{m_2} \cdots R_{m_s}$ with the property that the rest network after deleting the conductors $F_{m_1}, \ldots, F_{m_s}$ remains connected but contains no circuit. The same applies, of course, also to the determinant D_1 with respect to the minor network. In other words, to each term of D_1 corresponds a circuit passing through conductor 1, and the determinant D_1 is equal to the sum of all the terms which correspond to such circuits. In general, we shall associate with each circuit of the network a "current element", which is equal to the sum of all terms $R_{m_1} R_{m_2} \cdots R_{m_{s-1}}$ with the property that the rest network after deleting the conductors $F_{m_1}, F_{m_2}, \ldots, F_{m_{s-1}}$ contains only this circuit. Then the current in any conductor F_i caused by a unit e.m.f. in any other conductor F_j is, apart from the denominator D, equal to the sum of the current elements associated with those circuits passing through both conductors, each element of the sum being taken with positive or negative sign according as the two conductors have the same or opposite directions in the circuit. In particular, if the two conductors coincide, the current is, apart from the denominator D, simply the sum of all the current elements associated with circuits passing through this conductor (the signs are now all positive). Incidentally, we have also proved the well known reciprocity theorem that the current caused in the conductor F_i by an e.m.f. in the conductor F_j is equal to the current in F_j caused by the same e.m.f. in F_i. From this we can derive at once the following expression for the current in the conductor 1 in the general case where the e.m.f. vector $S = [S_1, S_2, \ldots, S_n]$ is arbitrary:

$$C_1 = \frac{\sum_{i=1}^{n} S_i D_i}{D}$$

NOTE II. It is possible to stipulate some sort of a duality in an electric network, in which the current C_i and the potential difference V_i of any conductor F_i appear as dual quantities. The fundamental connection between the dual quantities is the Ohm's Law

$$RC = V \tag{19}$$

where V is the column vector $[V_1, \ldots, V_n]$. The dual of the resistance R_i is then the quantity $\frac{1}{R_i}$ known as the conductivity of F_i and (19) can be written in the dual form

$$R^{-1}V = C \tag{20}$$

The fact that the currents are added up at a junction (conductors in parallel) and the p.d.'s are added up in a circuit (conductors in series) show that we should regard the junctions and the circuits as dual system of quantities (the conductors themselves begin self dual). It then follows that the Maxwell's cyclic currents $Z_1, \ldots, Z_s$ (corresponding to the circuits) are the dual of the potentials $Y_1, \ldots, Y_d$ at the junctions, and the total electromotive forces $U_1, \ldots, U_s$ in the circuits correspond to the currents $T_1, \ldots, T_d$ entering (or leaving, if negative,) the network at the junctions. Setting $U = [U_1, \ldots, U_s]$ and $T = [T_1, \ldots, T_d]$, we can write the equations (1) in the more general form

$$\begin{vmatrix} P \\ QR \end{vmatrix} C = \begin{vmatrix} T \\ U \end{vmatrix}. \tag{21}$$

The dual equations are obtained by simply substituting (20) in (21)

$$\begin{vmatrix} PR^{-1} \\ Q \end{vmatrix} V = \begin{vmatrix} T \\ U \end{vmatrix}. \tag{22}$$

The fundamental relation $PQ' = QP' = 0$ is evidently self dual.

If we regard an electric network as defined abstractly by any two integral matrices P, Q (with only ± 1 as their non-zero elements) with the relation $PQ' = 0$ and a diagonal matrix R, we can say that to any electric network defined by P, Q, R corresponds a dual network defined by Q, P, R^{-1}. However, this does not imply that there really exists an electric network in the ordinary sense dual to the given one. For the algebraic duality between P

and Q does not necessarily mean that to the network with the junction matrix P and a suitably chosen circuit matrix Q there exists geometrically a dual network with junction matrix Q and circuit matrix P. This geometrical duality exists, in fact, when and only when the network is mappable on the sphere. To explain this further, it is convenient to augment the junction matrix P by one further row P_{d+1} corresponding to the $(d+1)$-th junction discarded by us at the beginning, so that the augmented junction matrix P will have $d+1$ rows, which satisfy the relation $\sum_{i=1}^{d+1} P_i = 0$. Dually we shall augment the circuit matrix Q also by one further row $Q_{s+1} \equiv -\sum_{i=1}^{s} Q_i$, so that we have here also the relation $\sum_{i=1}^{s+1} Q_i = 0$. One verifies without difficulty that the above mentioned algebraic duality holds also in this extended from.[8] The augmented junction matrix P has now the characteristic property (corresponding to the property (L) of th matrix P) that each column of it contains exactly two non-zero elements $+1$ and -1. This property is in fact a necessary and sufficient condition for P to be the augmented junction matrix of a network. The question is now whether it is possible to choose an augmented basis of circuits such that the corresponding augmented circuit matrix Q will also possess this property. This is evidently possible if the network is mappable on the sphere. For in this case the network will divide the sphere into $n-(d+1)+2 = s+1$ regions (Euler's polyhedron formula), the boundaries of which, when coherently oriented (say all oriented positively in the usual sense), will supply $s+1$ circuits with the desired property. Conversely, if there exist $s+1$ circuits with this property, the network is mappable on the sphere. For, if we take for each circuit Q_i a plane region bounded by a polygon whose sides correspond to the conductors in

(8) Thus we have for the total e.m.f.'s in the circuits $U_1, \dots, U_{s+1}$ the relation $\sum_{i=1}^{d+1} U_i = 0$ and dually for the $T_1, \dots, T_{d+1}$ the relation $\sum_{i=1}^{d+1} T_i = 0$, a condition which is evidently necessary for a steady state current. Also for the cyclic currents $Z_1, \dots, Z_{s+1}$ and for the potentials $Y_1, \dots, Y_{d+1}$ there exists a linear relation, which can be most conveniently taken as $Z_{s+1} = 0$ and $Y_{d+1} = 0$ respectively.

the circuit Q_i with the same cyclic order[9] and then identify every two of the sides of these polygons which correspond to the same conductor in the network, we shall then obtain a surface which is topologically equivalent to the sphere because it is simply connected.

We shall indicate here briefly Mr. Wang's proof of (A) for the special case of a network mappable on the sphere. Consider the following identity already mentioned in[3]:

$$D = \begin{vmatrix} P \\ QR \end{vmatrix} = \frac{|PP'|}{\begin{vmatrix} P \\ Q \end{vmatrix}} |QRQ'| = \pm |QRQ'| \, .$$

The last equality holds in virtue of (14). Consider now the matrix

$$QRQ' \equiv (h_{ij}) = \left(\sum_{k=1}^{n} q_{ik} q_{ik} R_k \right)$$

If the network is mappable on the sphere, we can choose the circuit basis so that the matrix Q has the property (L). Then each element h_{ii} on the diagonal of QRQ' will be equal to the sum of the R_k which appear in the circuit Q_i, each taken with the coefficient $+1$, while the element $h_{ij} (i \neq j)$ off the diagonal will be equal to the sum of the R_k common to both the circuits Q_i and Q_j, each taken with the coefficient -1. From this Mr. Wang then deduces the rule (A) by a direct calculation of the determinant $|QRQ'|$.

Using now the identity

$$\begin{vmatrix} P \\ QR \end{vmatrix} |P'Q'| = \begin{vmatrix} PR^{-1} \\ Q \end{vmatrix} |R| \, |P'Q'| = |R| \begin{vmatrix} PR^{-1}P' & PR^{-1}Q' \\ O & QQ' \end{vmatrix}$$
$$= |R| \, |PR^{-1}P'| \, |QQ'| \, ,$$

and hence

$$D = \begin{vmatrix} P \\ QR \end{vmatrix} = \frac{|QQ'| \, |R|}{|P' \, Q'|} |PR^{-1}P'| = \pm R_1 \cdots R_n \, |PR^{-1}P'| \, ,$$

[9] It can be easily shown that in the present case the circuits are all simple circuits, so that they can be regarded as simple polygons.

we obtain similarly the rule (A′). Here the method holds even for every network without restriction, for the junction matrix P has always the property (L).

EXAMPLE. A simple example of a network not mappable on the sphere is furnished by the following diagram, which is derived from the ordinary Wheatstone bridge by adding a new conductor connecting any two non-adjacent conductors. If we orient the conductors and number the conductors

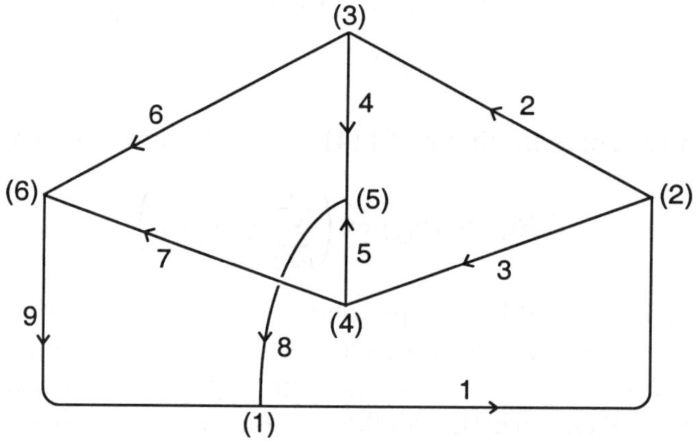

and junctions as in the diagram, we get the following junction matrix, discarding the junction (6),

$$P = \begin{pmatrix} -1 & 0 & 0 & 0 & 0 & 0 & 0 & 1 & 1 \\ 1 & -1 & -1 & 0 & 0 & 0 & 0 & 0 & 0 \\ 0 & 1 & 0 & -1 & 0 & -1 & 0 & 0 & 0 \\ 0 & 0 & 1 & 0 & -1 & 0 & -1 & 0 & 0 \\ 0 & 0 & 0 & 1 & 1 & 0 & 0 & -1 & 0 \end{pmatrix}$$

Taking the circuits 1-2-6-9, 2-3-4-5, 4-5-6-7, 5-8-9-7 as a basis, we get the circuit matrix

$$Q = \begin{pmatrix} 1 & 1 & 0 & 0 & 0 & 1 & 0 & 0 & 1 \\ 0 & -1 & 1 & -1 & 1 & 0 & 0 & 0 & 0 \\ 0 & 0 & 0 & 1 & -1 & -1 & 1 & 0 & 0 \\ 0 & 0 & 0 & 0 & 1 & 0 & -1 & 1 & -1 \end{pmatrix}$$

It is really not necessary to know these matrices P, Q, for we are going to apply Mr. Wang's rules. Instead of first finding D, we shall calculate here directly D_1 by rule (A). Since the circuits 2-3-4-5, 4-5-6-7, 5-8-9-7 form a basis for the minor network, we write down in three columns:

$$\begin{matrix} 2 & 4 & 5 \\ 3 & 5 & 7 \\ 4 & 6 & 8 \\ 5 & 7 & 9 \end{matrix}$$

Among the $4 \times 4 \times 4 = 64$ terms obtained by combining any three elements from the three columns respectively, eighteen are to be discarded for containing the same element more than once, eight are to be discarded for appearing twice: 257, 275, 357, 375, 458, 548, 459, 549; and finally one term appears three times: 457, 547, 475, of which only one should be retained. Thus the determinant D_1 is equal to the sum of the following 36 terms:

$$\begin{aligned} D_1 = {} & 245 \quad 247 \quad 248 \quad 249 \quad 258 \quad 259 \quad 265 \quad 267 \quad 268 \quad 269 \\ & 278 \quad 279 \quad 345 \quad 347 \quad 348 \quad 349 \quad 358 \quad 359 \quad 365 \quad 367 \\ & 368 \quad 369 \quad 378 \quad 379 \quad 457 \quad 465 \quad 467 \quad 468 \quad 469 \quad 478 \\ & 479 \quad 567 \quad 568 \quad 569 \quad 578 \quad 579 \end{aligned}$$

For the determinants $D_2, \ldots, D_9$ we can then apply the rule (B). However, for our particular case we can simplify the calculation by making use of the symmetry of the network with respect to the conductor 1. In fact, the structure of the network and hence also that of the determinants D_i is invariant under the following group G_8 of eight permutations of the conductors (all leaving the conductor 1 invariant):

$$\Theta_1 = I$$

$$\Theta_2 = (23)(45)(67)$$

$$\Theta_3 = (46)(57)(89)$$

$$\Theta_4 = (23)(47)(56)(89) = \Theta_2\Theta_3 = \Theta_3\Theta_2$$

$$\Theta_5 = (28)(39)(56)$$

$$\Theta_6 = (2839)(5764) = \Theta_5\Theta_2$$
$$\Theta_7 = (9382)(4675) = \Theta_5\Theta_3 = \Theta_6^{-1}$$
$$\Theta_8 = (29)(38)(47) = \Theta_5\Theta_4$$

The first four permutations form a subgroup G_4 of order 4, which leaves the orientation of the conductors unchanged, while the last four permutations reverse the orientation of all conductors, including the conductor 1, and hence have also no effect on the structure of the determinants. Now there are only two circuits passing through the conductor 1, which are not equivalent under G_8 or G_4; we can take, for example, the circuits 2-4-8 and 2-4-5-7-9, from which all other circuits passing through conductor 1 can be obtained by applying the permutations of the group G_4 (other permutations of G_8 give the same circuits with only the directions reversed). The current elements associated with these two circuits are

$$A_1 = (2\text{-}4\text{-}8) = 359 \quad 365 \quad 367 \quad 379 \quad 567 \quad 569 \quad 579$$
$$B_1 = (2\text{-}4\text{-}5\text{-}7\text{-}9) = 368$$

which can be found from D_1 by picking out those terms not containing as a factor any of the conductors in the circuit, or also by direct application of the definition of current element to the diagram. The other current elements can then be obtained by applying the permutations of G_4 to the above two elements:

$$A_2 = (3\text{-}5\text{-}8) \quad = 249 \quad 274 \quad 276 \quad 279 \quad 269 \quad 476 \quad 479 \quad 469$$
$$B_2 = (3\text{-}5\text{-}4\text{-}6\text{-}9) = 278$$
$$A_3 = (2\text{-}6\text{-}9) \quad = 378 \quad 347 \quad 345 \quad 348 \quad 358 \quad 745 \quad 748 \quad 758$$
$$B_3 = (2\text{-}6\text{-}7\text{-}5\text{-}8) = 349$$
$$A_4 = (3\text{-}7\text{-}9) \quad = 268 \quad 256 \quad 254 \quad 258 \quad 248 \quad 654 \quad 658 \quad 648$$
$$B_4 = (3\text{-}7\text{-}6\text{-}4\text{-}8) = 259$$

From these eight current elements the determinants D_i can be found at once by the rule stated at the end of Note I. Thus we have

$$D_1 = \sum_{i=1}^{4}(A_i + B_i)$$
$$D_2 = A_1 + A_3 + B_1 + B_3$$

$$D_3 = A_2 + A_4 + B_2 + B_4$$
$$D_4 = A_1 + B_1 - B_2 + B_4$$
$$D_5 = A_2 - B_1 + B_2 + B_3$$
$$D_6 = A_3 + B_2 + B_3 - B_4$$
$$D_7 = A_4 + B_1 - B_3 + B_4$$
$$D_8 = A_1 + A_2 + B_3 + B_4$$
$$D_9 = A_3 + A_4 + B_1 + B_2$$

Finally the determinant D can be obtained from the formula

$$D = 1 \cdot D_1 + 2 \cdot D_2 + 6 \cdot D_6 + 9 \cdot D_9$$

which contains in all $(36 + 18 + 11 + 18) - 2 = 81$ terms (the terms -6259 and 9256 cancel each other, leaving only one term 2569).

In conclusion, we remark that any one of the conductors can be transformed into any other by a permutation of the conductors which leaves the structure of the network invariant. Since, as we have seen above, there are eight permutations which leave any one conductor unchanged and preserve the structure of the network, we conclude that structure of the network is invariant under a group of $9 \times 8 = 72$ permutations. We have seen at the end of Note I that in case of an arbitrary e.m.f. vector the current in the conductor 1 is given by the formula

$$C_1 = \frac{\sum_{i=1}^{9} S_i D_i}{D}$$

Now, in order to find the current in any other conductor, we need only to know a permutation transforming the conductor 1 into this conductor and apply this permutation to the numerator of the above formula (the denominator D is of course invariant under the permutation). However, care must be taken to see that the S_i are given their proper signs after the permutation, for such a permutation may change the orientations of some of the conductors.

(Received May 2, 1939)

Retyped by World Scinetific Publishing Co. Pte. Ltd.

ANNALS OF MATHEMATICS
Vol. 49, No. 3, July, 1948

ON THE ALGEBRAICAL BRAID GROUP

BY WEI-LIANG CHOW

(Received September 22, 1947)

Artin[1] has recently given a new elegant and completely rigorous treatment of his theory of braids. In the same connection Bohnenblust[2] has derived some of the main results, in particular the completeness of the relations of the braid group, by an algebraical analysis of the abstract group defined by the relations (1) (2) below, which he calls the algebraical braid group. In the present paper we shall analyze the algebraical braid group by a different method, which we believe is simpler and more direct. Moreover, we shall be able not only to derive the same results as contained in Bohnenblust's paper, but also determine the center of the braid group.

Let $\mathfrak{B}_n$ be the algebraical braid group generated by the $n - 1$ generators $\sigma_1, \sigma_2, \cdots, \sigma_{n-1}$ with the defining relations:

(1) $$\sigma_i\sigma_k\sigma_i^{-1}\sigma_k^{-1} = 1 \qquad (|i - k| \geqq 2)$$

(2) $$\sigma_i\sigma_{i+1}\sigma_i\sigma_{i+1}^{-1}\sigma_i^{-1}\sigma_{i+1}^{-1} = 1.$$

It can be easily shown that a homomorphism of $\mathfrak{B}_n$ onto the symmetric group of permutation of n numbers $(1, 2, \cdots, n)$ is established by assigning to each generator σ_i the transposition $(i, i + 1)$. Let $\mathfrak{D}_1$ be the subgroup consisting of all elements of $\mathfrak{B}_n$ with the property that the permutations associated with them all leave the number 1 invariant. The index of $\mathfrak{D}_1$ in $\mathfrak{B}_n$ is evidently n, and we can take as a set of representatives of the left cosets to be $M_i = \sigma_i\sigma_{i-1} \cdots \sigma_1$ $(i = 0, 1, \cdots, n - 1)$, where M_0 is taken to be the identity. The application of the well known Reidemeister-Schreier method to $\mathfrak{D}_1$ leads at once to the following generators:

(3) $$\sigma_2, \sigma_3, \cdots, \sigma_{n-1}$$
$$a_i = M_{i-2}^{-1}\sigma_{i-1}^2M_{i-2} \qquad (i = 2, \cdots, n).$$

Since the set M_i satisfies the Schreier condition, a set of defining relations of $\mathfrak{D}_1$ is given by expressing the following relations

(4) $$M_j^{-1}\sigma_i\sigma_k\sigma_i^{-1}\sigma_\kappa^{-1}M_j = 1 \qquad (|i - k| \geqq 2)$$

(5) $$M_j^{-1}\sigma_i\sigma_{i+1}\sigma_i\sigma_{i+1}^{-1}\sigma_i^{-1}\sigma_{i+1}^{-1}M_j = 1$$

[1] E. ARTIN, *Theory of Braids*, Ann. of Math., vol. 48 (1947), pp. 101–126.

[2] F. BOHNENBLUST, *The Algebraical Braid Group*, Ann. of Math., vol. 48 (1947), pp. 127–136.

in terms of the generators (3). This can be done by using the following formulas, derived from relations (1) (2):

$$
(6)\qquad
\begin{aligned}
\sigma_k^{\epsilon} M_i &= M_i \sigma_{k+1}^{\epsilon} && (k \leqq i - 1,\ \epsilon = \pm 1)\\
\sigma_i M_i &= M_{i-1} a_{i+1} &&\\
\sigma_i^{-1} M_i &= M_{i-1} &&\\
\sigma_{i+1} M_i &= M_{i+1} &&\\
\sigma_{i+1}^{-1} M_i &= M_{i+1} a_{i+2}^{-1} &&\\
\sigma_k^{\epsilon} M_i &= M_i \sigma_k^{\epsilon} && (k \geqq i + 2,\ \epsilon = \pm 1).
\end{aligned}
$$

These formulas enable us to express every element of $\mathfrak{B}_n$ in the form $M_i \mathfrak{D}_1$. With the help of (6), the relations (4) (5) can be readily evaluated in terms of the generators (3), and the results are the following sets of relations:[3]

$$
(7)\qquad
\begin{aligned}
\sigma_i \sigma_k \sigma_i^{-1} \sigma_k^{-1} &= 1\\
\sigma_i \sigma_{i+1} \sigma_i \sigma_{i+1}^{-1} \sigma_i^{-1} \sigma_{i+1}^{-1} &= 1
\end{aligned}
$$

$$
(8)\qquad
\begin{aligned}
\sigma_i a_k \sigma_i^{-1} &= a_k && (k \neq i,\ i + 1)\\
\sigma_i a_i \sigma_i^{-1} &= a_{i+1} &&\\
\sigma_i a_{i+1} \sigma_i^{-1} &= a_{i+1}^{-1} a_i a_{i+1}\,. &&
\end{aligned}
$$

We can of course get (8) also by direct verification, but what is essential here is the fact that (7) (8) form a set of *defining* relations of $\mathfrak{D}_1$.

By virtue of relations (8), the generators $a_2, a_3, \cdots, a_n$ generate an invariant subgroup of $\mathfrak{D}_1$, which we shall denote by $\mathfrak{A}_1$. We maintain that $\mathfrak{A}_1$ is a free group with the free generators $a_2, a_3, \cdots, a_n$. For proof it is enough to show the existence of one realization of the abstract group $\mathfrak{D}_1$, in which the elements a_k are free. This realization is given by taking for the generators σ_i and a_k the following automorphisms of a free group $\mathfrak{U}_{n-1}$ of $n - 1$ free generators $u_2, u_3, \cdots, u_n$:

$$
(9)\qquad
\begin{aligned}
\sigma_i:\ & u_i' = u_{i+1}\ ,\quad u_{i+1}' = u_{i+1}^{-1} u_i u_{i+1}\\
& u_j' = u_j \qquad\qquad (j \neq i,\ i + 1)\\
a_k:\ & u_j' = u_k u_j u_k^{-1}.
\end{aligned}
$$

It can be easily verified that these automorphisms satisfy the relations (7) (8), hence the group generated by them is a realization of $\mathfrak{D}_1$. Now the automorphisms a_k evidently generate the group of inner automorphisms of $\mathfrak{U}_{n-1}$, and, since $\mathfrak{U}_{n-1}$ is a free group, it is isomorphic with $\mathfrak{U}_{n-1}$ itself through the correspondence $a_k \leftrightarrow u_k$. As the u_k are free generators, so the elements a_k are free.

The fact that $\mathfrak{A}_1$ is a free group is of fundamental importance for us. One almost immediate consequence is the solution of the "word" problem. The factor group $\mathfrak{D}_1/\mathfrak{A}_1$ is homomorphic to the subgroup of $\mathfrak{D}_1$ generated by the $n - 2$ generators $\sigma_2, \sigma_3, \cdots, \sigma_{n-1}$. Since $\mathfrak{D}_1/\mathfrak{A}_1$ has the defining relations (7),

[3] The immediate results of the calculation are somewhat different from the relations (7) and (8), but can be easily shown to be equivalent to them.

where we substitute the σ_i by the cosets $\mathfrak{A}_1\sigma_i$, and since these relations (7) are also true relations of the elements σ_i themselves, the homomorphism must be an isomorphism. Thus the subgroup of $\mathfrak{D}_1$ generated by the elements $\sigma_2, \sigma_3, \cdots, \sigma_{n-1}$ has also the defining relations (7), and hence it is the algebraical braid group $\mathfrak{B}_{n-1}$ of $n - 2$ generators. In other words, the group $\mathfrak{D}_1$ can be represented as the product $\mathfrak{A}_1\mathfrak{B}_{n-1}$ of the subgroups $\mathfrak{A}_1$ and $\mathfrak{B}_{n-1}$, and the representation of each element of $\mathfrak{D}_1$ as such a product is unique. By induction, we conclude that the group $\mathfrak{B}_n$ can be expressed as a product of the form $M_{i_1}^{(1)}\mathfrak{A}_1 M_{i_2}^{(2)}\mathfrak{A}_2 \cdots M_{i_{n-1}}^{(n-1)}\mathfrak{A}_{n-1}$, or $M_{i_1}^{(1)} \cdots M_{i_{n-1}}^{(n-1)}\mathfrak{A}_1 \cdots \mathfrak{A}_{n-1}$, since each $M^{(k)}$ is commutative with the groups $\mathfrak{A}_j$ $(j < k)$. We have here $M_j^{(k)} \equiv \sigma_j\sigma_{j-1} \cdots \sigma_k$ $(j \geqq k)$, $M_0^{(k)} \equiv 1$, and $\mathfrak{A}_k$ is the group generated by the $n - k$ elements $a_{kj} \equiv \sigma_k^{-1}\sigma_{k+1}^{-1} \cdots \sigma_{j-2}^{-1}\sigma_{j-1}^{2}\sigma_{j-2} \cdots \sigma_{k+1}\sigma_k$ $(j = k + 1, \cdots, n)$.

THEOREM I. *Every element of $\mathfrak{B}_n$ can be expressed in a unique way as a product of the form*

$$B = MA_1 A_2 \cdots A_{n-1} \tag{10}$$

where the A_k are elements of the free groups $\mathfrak{A}_k$ respectively, and the M is an element uniquely determined by the permutation associated with the element.

It is clear that the normal form (10) contains a solution of the "word" problem for the group.

Consider the group $\mathfrak{T}_{n-1}$ of inner automorphisms of $\mathfrak{D}_1$. Since $\mathfrak{A}_1$ is an invariant subgroup of $\mathfrak{D}_1$, the group $\mathfrak{T}_{n-1}$ induces a group of automorphisms of $\mathfrak{A}_1$, which we shall denote by $\mathfrak{S}_{n-1}$. Let the homomorphic mapping $\mathfrak{D}_1 \to \mathfrak{T}_{n-1} \to \mathfrak{S}_{n-1}$ be denoted by Γ. The images of $\sigma_2, \sigma_3, \cdots, \sigma_{n-1}$ under Γ are the automorphisms $a_k \leftrightarrow \sigma_i a_k \sigma_i^{-1}$, given by the formulas (8). They generate a subgroup $\mathfrak{R}_{n-1}$ of $\mathfrak{S}_{n-1}$, which is the geometrical braid group of $(n - 1)$-braids. We shall soon see that the homomorphism $\mathfrak{B}_{n-1} \to \mathfrak{R}_{n-1}$ is an isomorphism, and consequently the completeness of the relations (1) (2)' for the geometrical braid group.

We shall determine the kernel $\mathfrak{K}$ of the homomorphism Γ, which is the centralizer of $\mathfrak{A}_1$ in $\mathfrak{D}_1$, i.e. the group of all elements of $\mathfrak{D}_1$ commutative with all the a_k. We observe first that the permutation associated with any element of $\mathfrak{K}$ must be the identity, for the formulas (8) show that this permutation is also the permutation associated with the corresponding automorphism. Hence every element of $\mathfrak{K}$ must have the normal form $K = A_i A_{i+1} \cdots A_{n-1}$, where A_i is not the identity. Suppose $i > 1$. We have, since K is commutative with a_i,

$$Ka_iK^{-1} = A_i \cdots A_{n-1}a_iA_{n-1}^{-1} \cdots A_i^{-1} = A_ia_iA_i^{-1} = a_i .$$

Transforming the entire equation by M_{i-1}, we obtain

$$(M_{i-1}^{-1}A_iM_{i-1})\,(M_{i-1}^{-1}a_iM_{i-1})\,(M_{i-1}^{-1}A_iM_{i-1})^{-1} = M_{i-1}^{-1}a_iM_{i-1}$$

or

$$(M_{i-1}^{-1}A_iM_{i-1})a_2(M_{i-1}^{-1}A_iM_{i-1})^{-1} = a_2$$

for $M_{i-1}^{-1}a_iM_{i-1} = a_2$ (notice $\sigma_1^{-1}a_2\sigma_1 = a_2$). Now $M_{i-1}^{-1}A_iM_{i-1}$ is an element of $\mathfrak{A}_1$, and, since $\mathfrak{A}_1$ is a free group, we must have

$$M_{i-1}^{-1}A_iM_{i-1} = a_2^m$$

hence

$$A_i = M_{i-1}a_2^mM_{i-1}^{-1} = a_i^m.$$

Since a_i^m is an element of $\mathfrak{A}_1$, this is impossible; hence $i = 1$. This shows that $\mathfrak{B}_{n-1}$ have no element in common with $\mathfrak{K}$ (except the identity), and consequently the homomorphism $\mathfrak{B}_{n-1} \rightarrow \mathfrak{N}_{n-1}$ is an isomorphism.

We set $K = A_1A_2 \cdots A_{n-1} = A_1B_{n-1}$, where A_1 is not the identity. We have the following equations

$$Ka_kK^{-1} = A_1B_{n-1}a_kB_{n-1}^{-1}A_1^{-1} = a_k \qquad (k = 2, \cdots, n)$$

or

$$B_{n-1}a_kB_{n-1}^{-1} = A_1^{-1}a_kA_1 .$$

By multiplying all the equations, we get

$$B_{n-1}(a_2 \cdots a_n)B_{n-1}^{-1} = A_1^{-1}(a_2 \cdots a_n)A_1$$

or, since B_{n-1} is commutative with $(a_2 \cdots a_n)$,

$$(a_2 \cdots a_n) = A_1^{-1}(a_2 \cdots a_n)A_1 .$$

Since $\mathfrak{A}_1$ is a free group, we must have $A_1 = (a_2 \cdots a_n)^m$, and hence

$$K = (a_2 \cdots a_n)^mB_{n-1} .$$

For each value of m there can only be at most one such element K, for the corresponding B_{n-1} must induce the same automorphism in $\mathfrak{N}_{n-1}$ as the element $(a_2 \cdots a_n)^{-1}$, and the mapping $\mathfrak{B}_{n-1} \rightarrow \mathfrak{N}_{n-1}$ is an isomorphism. Now it can be verified by direct calculation that the element $(\sigma_2 \cdots \sigma_{n-1})^{n-1}$ of $\mathfrak{B}_{n-1}$ induces in $\mathfrak{A}_1$ exactly the inner automorphism corresponding to the element $(a_2 \cdots a_n)^{-1}$, i.e. the images of the two elements are the same under the homomorphism Γ. Hence

$$K = (a_2 \cdots a_n)\ (\sigma_2 \cdots \sigma_{n-1})^{n-1} = (\sigma_1\sigma_2 \cdots \sigma_{n-1})^n$$

is an element of $\mathfrak{K}$, and it generates the entire group $\mathfrak{K}$. But $(\sigma_1\sigma_2 \cdots \sigma_{n-1})^n$ is commutative with every element of $\mathfrak{B}_n$, for the same reason that

$$(\sigma_2 \cdots \sigma_{n-1})^{n-1}$$

is commutative with every element of $\mathfrak{B}_{n-1}$; the latter being shown by the fact that it is mapped into the same automorphism under Γ as the element

$$(a_2 \cdots a_n)^{-1},$$

which is commutative with every element of $\mathfrak{B}_{n-1}$, and the fact that the mapping $\mathfrak{B}_{n-1} \rightarrow \mathfrak{R}_{n-1}$ is an isomorphism. Thus $\mathfrak{K}$ is contained in the center of $\mathfrak{B}_n$. On the other hand, the center of $\mathfrak{B}_n$ must be contained in $\mathfrak{D}_1$ (for the corresponding permutation must be the identity for every element of the center), hence in the center of $\mathfrak{D}_1$, hence in $\mathfrak{K}$. So we conclude that $\mathfrak{K}$ is also the center of $\mathfrak{B}_n$ and the center of $\mathfrak{D}_1$.

We sum up our results in the following theorems:

THEOREM II. *The algebraical braid group $\mathfrak{B}_n$ and the geometrical braid group $\mathfrak{R}_n$ are isomorphic. Hence the relations* (1) (2) *form a complete set of defining relations for the geometrical braid group.*

THEOREM III. *The center of $\mathfrak{B}_n$ is the group generated by the element $(\sigma_1\sigma_2 \cdots \sigma_{n-1})^n$. It is also the center of $\mathfrak{D}_1$ and the centralizer of $\mathfrak{A}_1$ in $\mathfrak{D}_1$.*

THE INSTITUTE FOR ADVANCED STUDY

ON COMPACT COMPLEX ANALYTIC VARIETIES.*

By Wei-Liang Chow.

1. Introduction. Let S_n be the projective space of n dimensions over the field k of all complex numbers. For any given choice of the inhomogeneous coordinates $x_1, \cdots, x_n$ in S_n, let $A_n(x)$ be the n-dimensional complex affine space consisting of all the points of S_n which are finite with respect to this coordinate system. A point set U in S_n is said to be analytic in the neighborhood of a point (a) if, choosing the inhomogeneous coordinates so that (a) is a point in $A_n(x)$, there exists a neighborhood R of (a) in $A_n(x)$ such that the intersection $U \cap R$ consists of all the points of R which are common solutions of a finite set of equations,

$$(1) \qquad f_i(x_1, \cdots, x_n) = 0, \qquad (i = 1, \cdots, s),$$

where the $f_i(x_1, \cdots, x_n)$, $i = 1, \cdots, s$, are holomorphic functions in the region R. A point set in S_n is called an analytic variety if it is analytic in the neighborhood of every one of its points. An analytic variety is called compact if it is a compact point set. A point set in S_n is called an algebraic variety if it consists of all the points whose homogeneous coordinates $x_0, x_1, \cdots, x_n$ satisfy a finite set of equations

$$\phi_i(x_0, x_1, \cdots, x_n) = 0, \qquad (i = 1, \cdots, t),$$

where the $\phi_i(x_0, x_1, \cdots, x_n)$, $i = 1, \cdots, t$, are homogeneous polynomials in the $x_0, x_1, \cdots, x_n$. It is easily seen that an algebraic variety is a compact analytic variety. There is a classical conjecture that conversely any compact analytic variety in S_n is also an algebraic variety. So far as we are aware, no proof of this conjecture has ever been offered. It is the purpose of this paper to present a proof of this conjecture.

We begin by recalling some well known properties of an analytic variety.[1] Let R be a neighborhood of the origin of $A_n(x)$ defined by the conditions $|x_i| < e$, $i = 1, \cdots, n$. Let $k\{x_1, \cdots, x_r\}$ be the ring of all power series in the variables $x_1, \cdots, x_r$ which are convergent in the region $|x_i| < e$, $i = 1, \cdots, r$. Consider a set of equations of the following type

* Received April 9, 1949.

[1] For the results of the theory of functions of several complex variables used in this paper, we refer here once for all to Osgood [7], Bochner and Martin [1].

$$(2)\qquad x_{r+1}^{\mu}+B_1x_{r+1}^{\mu-1}+\cdots+B_{\mu-1}x_{r+1}+B_{\mu}=0$$
$$Hx_j=G_j(x_{r+1}),\qquad (j=r+2,\cdots,n),$$

where the $B_1,\cdots,B_\mu$ and H are elements of the ring $k\{x_1,\cdots,x_r\}$ and the $G_j(x_{r+1})$, $j=r+2,\cdots,n$, are polynomials in x_{r+1} with coefficients in $k\{x_1,\cdots,x_r\}$; and the left hand side of the first equation is an irreducible distinguished polynomial in x_{r+1} over $k\{x_1,\cdots,x_r\}$ and H is its discriminant. Let D be the set of all common solutions of the equations (2) in the region R for which $H\neq 0$, as well as those which are limiting points of such solutions. Such a point set D, or any point set in S_n which can be represented as such a set by a suitable choice of the affine space $A_n(x)$ and its coordinates, is called an analytic element in S_n. The number r is called the dimension of the analytic element D, and we shall write D_r to indicate this. The number μ depends in general upon the choice of the affine space and its coordinates; the smallest possible value of this number for a given D_r is called its order. An analytic element is called regular if it has the order one, otherwise it is called singular. The point (a) is called the center of D_r. It is well known from the theory of functions of several complex variables that every point of an analytic variety W has a neighborhood R such that $R\cap W$ consists of a finite number of analytic elements. In other words, an analytical variety is a topological sub-space of S_n which has a system of neighborhoods consisting of analytic elements. An analytic variety is called irreducible if any two analytic elements of it can be obtained from each other by analytic continuations. It is easily seen that all the analytic elements of an irreducible analytic variety W must have the same dimension r; we shall call this number r the dimension of W and we shall write W_r to indicate this. Following a recent practice in algebraic geometry, we shall from now on use the expression "analytic variety" to denote exclusively an irreducible analytic variety; the reducible ones being taken care of later by the more precise concept of an analytic cycle. A point of an analytic variety W_r is called regular if it has a neighborhood consisting of a single regular analytic element; those points of W_r which are not regular are called singular. The set W_r of all singular points of W_r constitutes a finite or enumerably infinite number of analytic varieties of dimensions less than r; we shall call this set $\bar{W}_r$ the singular part of W_r. The set $W_r - W_r$ of all the regular points of W_r is a connected set.

For our present purpose the most important properties of an analytic variety W_r are the following:

(A) *W_r is a topological complex.* This is the triangulation theorem for analytic loci, for the proof of which we shall refer to the literature.[2] In particular, if W_r is compact, it is a finite complex and can be taken as a subcomplex of a simplicial subdivision of the projective space S_n and the singular part $\tilde{W}_r$ can in turn be taken as a subcomplex of W_r.

(B) *W_r can be covered by an enumerable aggregate of regular analytic elements of dimensions* $\leq r$. This can be easily proved by induction, starting with the facts that $W_r - \tilde{W}_r$ can be covered by an enumerable aggregate of analytic elements of dimension r and that $\tilde{W}$ consists of an enumerable aggregate of analytic varieties of dimensions $< r$.

In the following sections we shall show that the main theorem can be proved by using these two properties (A) and (B) alone, without referring directly to the fact that each point of W_r has a neighborhood consisting of a finite number of analytic elements of dimension r. For this reason, we shall give a new (possibly more general) definition of an analytic variety in **2**, essentially by means of these two properties (A) and (B) only. In **2** and **3** we shall study the topological properties of analytic varieties and derive from them the Theorem III, which expresses a special property of a regular analytic element D_r contained in a compact analytic variety of r dimensions. In **4** we then proceed to show (Theorem IV) that this property is really a characteristic property of a regular analytic element D_r which is contained in an algebraic variety of r dimensions. The main theorem (Theorem V) then follows immediately from these two results. In the last section (**5**) we shall indicate some applications of our main theorem.

Finally, we add a few words about the terminology and notations. A topological complex K is a homeomorphic image of an Euclidean complex, and we shall say $2r$-complex if it consists only of simplexes of topological dimension $2r$ and their sides. The boundary of an unoriented complex K (i. e. considered as a chain modulo 2) will be denoted by $\dot{K}$. If the complex K is oriented to become a chain with integral coefficients, then its chain boundary will also be denoted by $\dot{K}$. There is no danger of confusion in this, as the meaning will be clear from the context in each case. The word "dimension" without any further qualification will mean the complex dimension, which is twice the topological dimension; subscripts will be used to denote the former, superscripts the latter.

2. Analytic varieties. An analytic simplex E_r of r dimensions in S_n is a topological $2r$-simplex in S_n, which is a one-to-one analytic image of a topological $2r$-simplex in a complex affine space $A_r(z)$ of r dimensions. More

[2] See Koopman and Brown [4], Lefschetz and Whitehead [6].

specifically this means that, for a suitable choice of the inhomogeneous coordinates $x_1, \cdots, x_n$ in S_n, the points of E_r can be represented parametrically by a set of n equations

$$x_j = f_j(z_1, \cdots, z_r), \qquad j = 1, \cdots, n, \tag{3}$$

where the n functions $f_j(z_1, \cdots, z_r)$, $j = 1, \cdots, n$, are all holomorphic in a region R of the affine space $A_r(z)$ of r complex parameters $z_1, \cdots, z_r$, and the matrix $\| \partial f_j / \partial z_k \|$, $j = 1, \cdots, n$, $k = 1, \cdots, r$, has the maximum rank r at every point of the region R; and that this representation induces a homeomorphism between E_r and a topological $2r$-simplex C^{2r} in the region R. Without loss of generality, we can assume that the coordinate origin $(z) = (0)$ of $A_r(z)$ is contained in the interior of C^{2r}; the corresponding point $(x) = (a)$ in E_r is then called the center of E_r (with respect to the parameters $z_1, \cdots, z_r$). It is obvious that if the analytic simplex E_r is subdivided into a complex, then each simplex of this complex is also an analytic simplex. Since in all our arguments we can always replace any complex by a subdivision in which the simplexes are arbitrarily small, we can therefore assume, without any essential restriction to the concept of an analytic simplex, that for a suitable choice of the coordinates $x_1, \cdots, x_n$, and the parameter space $A_r(z)$, the equations (3) have the form

$$\begin{aligned} &x_1 = z_1, \cdots, x_r = z_r \\ &x_{r+1} = f_{r+1}(z_1, \cdots, z_r), \cdots, x_n = f_n(z_1, \cdots, z_r); \end{aligned} \tag{4}$$

and that the region R has been chosen so small that the functions $f_{r+1}, \cdots, f_n$ can be represented as power series which are convergent in R. Since the equations (4) map the region R into a regular analytic element D_r in S_n, it follows that an analytic simplex E_r is simply a topological $2r$-simplex imbedded in a regular analytic element D_r. On the other hand, the set of interior points $E_r - \dot{E}_r$ of an analytic simplex is itself a regular analytic element. Hence it is rather indifferent whether a point set is covered by an aggregate of analytic simplexes or an aggregate of regular analytic elements; the use of either one instead of the other is only a matter of convenience.

The concept of an analytic complex in S_n is defined by induction as follows. Any topological 0-complex in S_n is an analytic complex K_0 of zero dimension. A topological $2r$-complex (finite or infinite) in S_n is called an analytic complex K_r of r dimensions, if it contains a subcomplex $\bar{K}_r$ which consists of an enumerable aggregate of analytic complexes of dimensions less than r, such that any point of $K_r - (\dot{K}_r + \bar{K}_r)$ has a neighborhood which is a regular analytic element D_r. The points of $K_r - (\dot{K}_r + \bar{K}_r)$ are called regular points of K. The topological complex $\dot{K}_r$ is also called the boundary of K_r. The aggregate of analytic complexes $\tilde{K}_r$ is called the singular part of K_r, though not every point (or even any point) of it need be a singular point in either the topological or analytic sense. An analytic simplex E_r is of course also an analytic complex of r dimensions, the boundary being the boundary $\dot{E}_r$ and the singular part being the empty set. An analytic variety W_r of r dimensions in S_n is a connected analytic complex with the properties that it has no boundary and its singular part $\tilde{W}_r$ consists of an enumerable aggregate of analytic varieties of dimensions less than r. This definition by induction can be completed by the stipulation that an analytic variety W_0 of zero dimension consists of a single point in S_n. An analytic variety is called compact if it is a compact point set; that is, if it is a finite topological complex.

It is clear that the set $W_r - \tilde{W}_r$ of all the regular points of W_r can be covered by an enumerable aggregate of analytic simplexes of r dimensions. From this and the fact that $\tilde{W}_r$ consists of an enumerable aggregate of analytic varieties of lower dimensions, it follows by induction that the analytic variety W_r can be covered by an enumerable aggregate of analytic simplexes of dimensions $\leq r$. However, it is to be noticed that this covering of W_r by an enumerable aggregate of analytic simplexes is not a simplicial subdivision of W_r as a complex in the topological sense, nor is it a covering by neighborhoods of W_r as a topological subspace of S_n. It is simply a representation of W_r as the set-theoretic sum of an enumerable aggregate of analytic simplexes of various dimensions, and this aggregate will be in general infinite even in case of a compact analytic variety W_r.

Let E_r and E_s be two analytic simplexes in S_n contained in the regular analytic elements D_r and D_s respectively. The intersection $D_r \cap D_s$ of the two analytic elements D_r and D_s, if it is not empty, consists of an enumerable aggregate of connected components, each of which is an analytic variety of dimension not less than $r + s - n$. In case such a component has exactly the dimension $r + s - n$, it is called a proper component of the intersection. It is easily seen that if the join of the tangent spaces of D_r and D_s at some one regular point of such a component has the dimension n, then the component has the dimension $r + s - n$ and is therefore proper. In such a case we shall call the component regular. We shall say that two analytic simplexes E_r and E_s are regular with respect to each other if either (1) the intersection $E_r \cap E_s$ is empty, or (2) each point of $E_r \cap E_s$ is contained in a regular component of $D_r \cap D_s$.

It is clear that for $r + s < n$ the first possibility alone can occur, so that in this case two analytic simplexes E_r and E_s are regular with respect to each other if and only if they are disjoint. In case $r+s=n$, the intersection of two analytic simplexes E_r and E_{n-r}, regular with respect to each other, is either empty or consists of a finite number of points at each of which the tangent spaces of D_r and D_{n-r} are transversal to each other. We are mainly interested in the case $r+s=n$, though the following results hold also for the general case.

Let P be the group of all projective transformations in S_n; it is an analytic manifold of $(n+1)^2-1$ dimensions. It is easily seen that if E_r is an analytic simplex in S_n and T is a projective transformation in S_n, then the transformed set TE_r of E_r is also an analytic simplex of r dimensions. It follows then that if K_r is an analytic complex covered by an aggregate of analytic simplexes $\{E\}$, then TK_r is also an analytic complex covered by the aggregate of the transformed simplexes $\{TE\}$. In the proof of the following theorems, we shall make use of the simple fact about everywhere dense sets which is true for every regular separable topological space: The intersection of a finite or enumerable infinite number of everywhere dense open sets is also an everywhere dense set. Since a closed set is nowhere dense if its complement is everywhere dense, it follows that the sum of a finite or enumerably infinite number of nowhere dense closed sets is a set the complement of which is everywhere dense; hence, if this sum is a closed set, then it is also nowhere dense.

THEOREM I. *Let E_r and E_s $(r+s\leq n)$ be two analytic simplexes in S_n and let Ω be the set of all elements T of P such that E_r and TE_s are regular with respect to each other, then Ω is an everywhere dense open set in P.*

Proof. We shall choose the inhomogeneous coordinates $x_1,\cdots,x_n$ in S_n in such a way that both E_r and E_s are in the affine space $A_n(x)$, which is always possible if these simplexes are sufficiently small. Let E_r be given by the one-to-one analytic mapping

$$x_i=f_i(z_1,\cdots,z_r), \qquad i=1,\cdots,n, \tag{5}$$

of a region R' in $A_r(z)$ into $A_n(x)$, so that E_r is the image of a topological $2r$-simplex C^{2r} in R'. Similarly, let E_s be given by the one-to-one analytic mapping

$$x_i=g_i(w_1,\cdots,w_s), \qquad i=1,\cdots,n, \tag{6}$$

of a region R'' in $A_s(w)$ into $A_n(x)$, so that E_s is the image of a topological $2s$-simplex C^{2s} in R''.

We first prove that the set Ω is everywhere dense in P; it is evidently enough to show this for the neighborhood of the identity transformation I. Let $T_{(e)}$ be the affine transformation of $A_n(x)$,

$$x'_i = x_i + e_i, \qquad i = 1, \cdots, n;$$

it is of course also a projective transformation in S_n and is arbitrarily near to the identity transformation I for points (e) sufficiently near to (0). The transformed analytic simplex $T_{(e)}E_s$ is then given by the one-to-one analytic mapping

$$x_i = g_i(w_1, \cdots, w_s) + e_i, \qquad i = 1, \cdots, n,$$

of the topological simplex C^{2s} in the region R''. Consider the analytic mapping

$$(7) \quad u_i = F_i(z, w) = f_i(z_1, \cdots, z_r) - g_i(w_1, \cdots, w_s), \qquad i = 1, \cdots, n,$$

of the region $R = R' \times R''$ in the product space $A_{r+s}(z, w) = A_r(z) \times A_s(w)$ into the space $A_n(u)$. Let C^{2r+2s} be the $2(r+s)$-simplex $C^{2r} \times C^{2s}$ in the region R. It is clear that every common solution of the equations (7) for $(u) = (e)$ in the simplex C^{2r+2s} will correspond to an intersection point of E_r and TE_s and conversely. For $r + s < n$, the image of C^{2r+2s} in $A_n(u)$ under the analytic transformation (7) is a nowhere dense set; hence there exist points (e) in $A_n(u)$ arbitrarily near to the origin such that the equations (7) have no common solution in C^{2r+2s} for $(u) = (e)$. Then the analytic simplexes E_r and $T_{(e)}E_s$ will be disjoint and hence regular with respect to each other. Thus the assertion is proved for this case. For $r + s = n$, we can assume that for all points (e) in a sufficiently small neighborhood of (0) the two analytic simplexes E_r and $T_{(e)}E_s$ have an intersection, for otherwise the assertion is obviously true already. This means that for all points (u) in a sufficiently small neighborhood of the origin in $A_n(u)$ the equations (7) will have a common solution in C^{2r+2s}. It is well known [3] that in this case the Jacobian determinant

$$J(z, w) = | \partial F_i/\partial z_j, \partial F_i/\partial w_k | = | \partial f_i/\partial z_j, -\partial g_i/\partial w_k |,$$

does not vanish identically in the region R. Then the set of all the solutions of the equation $J(z, w) = 0$ in R, if it is not empty, constitutes an enumerable aggregate of analytic varieties of dimension $n - 1$ in R. Let $E^{(i)}$, $i = 1, 2, \cdots$, be a sequence of analytic simplexes covering this set J, and let $E^{(i)*}$, $i = 1, 2, \cdots$, be the corresponding images in $A_n(u)$ under the analytic transformation (7). Then the image J^* of J under the analytic transforma-

[3] See Knopp and Schmidt [3], p. 379.

tion (7) is evidently the sum of all the sets $E^{(i)*}$, $i = 1, 2, \cdots$. Since each $E^{(i)*}$ is a nowhere dense set in $A_n(u)$, it follows that the complement of J^* in $A_n(u)$ is an everywhere dense set. This means that there exists in every neighborhood of the origin of $A_n(u)$ at least a point (e) which is not in J^*. Then for this point $(u) = (e)$ the equations (7) will have no common solution in the set J; or, in other words, we have $J(z, w) \neq 0$ at every common solution in R of the equations (7) for $(u) = (e)$. This implies that every intersection point of E_r and $T_{(e)}E_s$ corresponds to a point in C^{2r+2s} at which $J(z, w) \neq 0$; hence E_r and $T_{(e)}E_s$ are regular with respect to each other. Thus our assertion is also proved for the case $r + s = n$.

It remains to show that the set Ω is open in P; that is, we have to show that if E_r and E_s are regular with respect to each other, then E_r and TE_s are also regular with respect to each other for all T in a sufficiently small neighborhood of the identity I. For $r + s < n$, this is obvious; for the E_r and E_s are disjoint closed subsets in S_n, and TE_s varies continuously with T. To prove the assertion for $r + s = n$, let a projective transformation T be given by the equations

$$x'_i = (a_{i0} + \sum_{j=1}^{n} a_{ij}x_j)/(a_{00} + \sum_{j=1}^{n} a_{0j}x_j), \qquad i = 1, \cdots, n.$$

If the matrix (a_{ij}) is sufficiently near the identity, then the analytic simplex TE_s will still lie in the affine space $A_n(x)$ and is given by the one-to-one analytic mapping

$$x_i = g'_i(w_1, \cdots, w_s) = [a_{i0} + \sum_{j=1}^{n} a_{ij}g_j(w_1, \cdots, w_s)]/[a_{00} + \sum_{j=1}^{n} a_{0j}g_j(w_1, \cdots, w_s)], \qquad i = 1, \cdots, n,$$

of the topological simplex C^{2s} in the region R''. Consider the Jacobian determinant

$$J^T(z, w) = | \partial f_i/\partial z_j, \partial g'_i/\partial w_k |,$$

and the n functions

$$F_i^T(z, w) = f_i(z_1, \cdots, z_r) - g'_i(w_1, \cdots, w_s), \qquad i = 1, \cdots, n,$$

which are all defined in the region R. It is clear that these functions are also continuous functions of the transformation T in a neighborhood of the identity I. Now, it is easily seen that the two analytic simplexes E_r and TE_s are regular with respect to each other if and only if the following equations

$$(8) \qquad \begin{aligned} J^T(z, w) &= 0, \\ F_i^T(z, w) &= 0, \qquad i = 1, \cdots, n, \end{aligned}$$

have no common solution in the topological simplex C^{2r+2s} in the region R. Let Δ^T be the set of all the common solutions of the equations (8) in R; then Δ^T is closed in R and varies continuously with T. Since E_r and E_s are assumed to be regular with respect to each other, the intersection $C^{2r+2s} \cap \Delta^I$ is empty. Now, if $C^{2r+2s} \cap \Delta^T$ is not empty for all T in a sufficiently small neighborhood of I, then there is a sequence of $T^{(i)}$, $i = 1, 2, \cdots$, converging toward I such that $C^{2r+2s} \cap \Delta^{T^{(i)}}$ is not empty. Since C^{2r+2s} is a compact set, the sequence of sets $C^{2r+2s} \cap \Delta^{T^{(i)}}$ has at least one limit point in C^{2r+2s}; and it is easily seen that this limit point is a point of $C^{2r+2s} \cap \Delta^I$, which is a contradiction. Therefore, the intersection $C^{2r+2s} \cap \Delta^T$ is empty for every T in a sufficiently small neighborhood of the identity I; and this means that E_r and TE_s are regular with respect to each other for all such T. This concludes the proof of Theorem I.

From Theorem I we can deduce a similar theorem for analytic varieties. We shall say that a point is a regular intersection of two analytic complexes K_r and K_{n-r} if it is a regular point of both K_r and K_{n-r} and if the tangent spaces of K_r and K_{n-r} at this point are transversal to each other.

THEOREM II. *Let W_r and W_{n-r} be two analytic varieties in S_n, and let Ω be the set of all the transformations T in P such that W_r and TW_{n-r} have only regular intersection points. Then the set Ω is everywhere dense in P.*

Proof. Let $\{E\}$ and $\{F\}$ be two enumerable sets of analytic simplexes which cover the analytic varieties W_r and W_{n-r} respectively. Then the set $\{G\} = \{(E, F)\}$ of all the pairs of analytic simplexes, one from each of the two sets $\{E\}$ and $\{F\}$, is also enumerable and hence can be arranged in a sequence $G^{(i)} = (E^{(i)}, F^{(i)})$, $i = 1, 2, \cdots$. For each $i = 1, 2, \cdots$, let $\Omega^{(i)}$ be the set of all transformations T in P such that $E^{(i)}$ and $TF^{(i)}$ are regular with respect to each other. It is easily seen that the intersection of all the sets $\Omega^{(i)}$, $i = 1, 2, \cdots$, consists of exactly those transformations T such that W_r and TW_{n-r} have only regular intersection points. Since, according to Theorem I, each set $\Omega^{(i)}$ is an everywhere dense open set in P, it follows that Ω is also an everywhere dense set in P.

3. Intersection of analytic varieties. There is a "natural" orientation for an analytic simplex E_r which can be extended to any analytic complex K_r. Let E_r be an analytic simplex in S_n defined by the one-to-one analytic mapping (3) of the topological $2r$-simplex C^{2r} in $A_r(z)$. Let $z_j = z'_j + iz''_j$, $j = 1, \cdots, r$; then the ordered set of $2r$ real coordinates $z'_1, z''_1, \cdots, z'_r,$

z''_r of the space $A_r(z)$ determines a definite orientation of this space and consequently also an orientation of the simplex C^{2r}. This orientation of C^{2r} determines an orientation of E_r which we shall call its natural orientation. It can be easily shown that the natural orientation is independent of the choice of the analytic parameters and is thus an intrinsic property of an analytic simplex. The set of all regular points $K_r - (\dot{K}_r + \tilde{K}_r)$ of an analytic complex K_r can be covered by a set of analytic simplexes of r dimensions. It can be shown that if each analytic simplex of the set is given its natural orientation, then all these orientations are coherent; and if we orient each topological $2r$-simplex of K_r in concordance with the natural orientation of any one analytic simplex E_r contained in it, we obtain a topological $2r$-chain on K_r with boundary in the sub-complex $\dot{K}_r$, or a $2r$-cycle mod $\dot{K}_r$. Thus an analytic complex K_r is an orientable pseudo-manifold, and it has a natural orientation determined by its analytical structure. This applies in particular also to the projective space S_n itself, since it is also an analytic complex. Furthermore, if K_r is a finite complex, its natural orientation will make it a topological $2r$-chain in S_n with boundary in the subset $\dot{K}_r$. Hence a compact analytic variety W_r, if oriented with the natural orientation, is a topological $2r$-cycle in S_n. We shall from now on assume that all analytic complexes, including the space S_n itself, are oriented with the natural orientation, so that a finite analytic complex K_r corresponds to a uniquely determined topological $2r$-chain in S_n. The significance of the natural orientation is given by the following well-known result:

LEMMA 1. *The topological intersection multiplicity of a regular intersection point of two finite analytic complexes K_r and K_{n-r} is $+1$.*

For the proof of this lemma, which is very simple, we refer to the literature.[4]

As to the topological properties of S_n, it is well known that the $2r$-th homology group of S_n is cyclic and the class determined by a linear analytic variety L_r of r dimensions is a generator of this group. From this we can derive at once the following consequences: (1) To each $2r$-cycle in S_n there is associated an integer g called its degree; it is the total topological intersection number of this $2r$-cycle with a linear variety L_{n-r}; (2) two $2r$-cycles are homologous if and only if they have the same degree; (3) the total topological intersection number of a $2r$-cycle of degree g and a $2(n-r)$-cycle of degree h is equal to gh. In view of Lemma 1 we can conclude that the

[4] For the topological properties of analytic varieties, we refer once for all to Lefschetz [5], Ch. VIII, § 3 and van der Waerden [9].

degree of a compact analytic variety W_r is always positive; for it is equal to the number of intersection points of W_r with a linear variety L_{n-r} which has only regular intersections with W_r.

We shall need the following lemma concerning the multiplicity of an intersection point. Let B^s and B^{2n-s} be two topological chains in S_n, and let (p) be an isolated point of $B^s \cap B^{2n-s}$ which does not lie in either $\dot{B}^s$ or $\dot{B}^{2n-s}$. Then this point (p) has a uniquely determined multiplicity as an intersection of the two chains B^s and B^{2n-s}. For the determination of this multiplicity we have the following criterion:

Lemma 2. *Let R be a neighborhood of (p) such that the closure of R does not intersect $\dot{B}^s$ or $\dot{B}^{2n-s}$ or any other component of the intersection $B^s \cap B^{2n-s}$, and let A^s and A^{2n-s} be two chains which are d-homologous to the chains B^s and B^{2n-s} respectively, where d is a sufficiently small positive number depending on B^s, B^{2n-s} and R. Then each component of the intersection $A^s \cap A^{2n-s}$ lies either in R or outside of R, and the intersection multiplicity of (p) is equal to the sum of the intersection multiplicities of those components of $A^s \cap A^{2n-s}$ which lie in R.*

In the above lemma the notion of d-homology is defined as follows: Let Σ be a d-neighborhood of B^s and Σ' be a closed d-neighborhood of $\dot{B}^s$, then a chain A^s is said to be d-homologous to B^s if it is contained in Σ and is homologous to B^s mod Σ' on Σ. This includes in particular the case when A^s is a d-deformation of B^s. The proof of Lemma 2 follows easily from the well known results about intersection of chains,[5] so that we can omit it here.

Let W_r and W_{n-r} be two compact analytic varieties of degrees g and h respectively, and let (p) be an isolated intersection point of them. Let R be a neighborhood of the point (p) in S_n, and let d be the number such that the Lemma 2 holds for all d-deformations of W_r and W_{n-r}. According to Theorem II, there exists a projective transformation such that TW_{n-r} is a d-deformation of W_{n-r} and the two analytic varieties W_r and TW_{n-r} have only regular intersection points. Since according to Lemma 1 each of these intersection points has exactly the multiplicity $+1$, it follows that the intersection $W_r \cap TW_{n-r}$ consists of exactly gh points. If μ of these gh points lie in the neighborhood R, then μ is the multiplicity of (p) as an intersection of W_r and W_{n-r}. Applying this in particular to a regular point of W_r, we obtain the following result: Let E_r be an analytic simplex in S_n and (p) be an interior point of E_r. If E_r is an element of a compact analytic variety W_r, then there exists a positive number N such that if W_{n-r} is a

[5] See Lefschetz [5], Ch. IV, § 3.

compact analytic variety of degree h and if (p) is an isolated intersection point of E_r with W_{n-r}, then the multiplicity of this intersection is not greater than hN. In fact, we can take N to be any number not less than the degree of W_r. In the next section we shall see that this fact is characteristic of an analytic simplex which is an element of an algebraic variety of r dimensions. However, in order to state this result in the definitive form which we shall use in the next section, we introduce the concept of an analytic cycle. An analytic cycle Z_r of r dimensions in S_n is a topological $2r$-cycle in S_n which can be expressed as a sum of multiples of compact analytic varieties. Thus an analytic cycle Z_r is a finite set of compact analytic varieties $W_r^{(1)}, \cdots, W_r^{(s)}$, to each of which is assigned an integer n_i (positive or negative) called its multiplicity. If g_i is the degree of $W_r^{(i)}$ respectively, then the integer $\sum_{i=1}^{s} n_i g_i$ is evidently the degree of the analytic cycle Z_r. An analytic cycle Z_r is called positive if the multiplicities of its component varieties are all positive; it is called algebraic if all of its component varieties are algebraic. It is a simple matter to show that what we have proved above for compact analytic varieties can be generalized to positive analytic cycles. Thus we have the following theorem:

THEOREM III. *Let E_r be an analytic simplex in S_n and (p) be an interior point of E_r. If E_r is an element of a compact analytic variety W_r, then there exists a positive number N such that if Z_{n-r} is a positive analytic cycle of degree h and if (p) is an isolated intersection point of E_r with Z_{n-r}, then the multiplicity of this intersection is not greater than hN.*

We have stated the above theorem only for a regular analytic element, as this is the only case we shall need for our purpose. It is however clear that the theorem holds in general for any analytic element, whether regular or not. Incidentally we should like to remark that it is well known [6] that an isolated intersection of two analytic varieties, in the usual sense as defined in section 1, has always a positive multiplicity. This property is fundamental for many applications of topological methods to algebraic geometry, but we have no direct use for it in the present paper. In fact, we do not know whether this property is true at all for the analytic varieties as we have defined here in the previous section; for the proof of it depends on the fact that the neighborhood of each point of an analytic variety consists of a finite number of analytic elements. What is essential for our present purposes is not to rule out the possibility of a zero multiplicity for an isolated intersection point, but to

[6] See van der Waerden [9], or Lefschetz [5], Ch. VIII, § 4.

rule out the possibility of a negative multiplicity for any connected component (of any dimension) of the intersection. That the latter is true for intersections of positive analytic cycles in S_n can be easily deduced from our results, but it is not generally true for intersections of positive analytic cycles in an arbitrary analytic or algebraic manifold.

4. Proof of the main theorem. We begin with two lemmas.

LEMMA 3. *Let $f(x_1,\cdots,x_r)$ be a power series in r variables $x_1,\cdots,x_r$. Given any positive integer N, we can always find an integer M with the following property: For every integer $m > M$, there exists a polynomial $F(x_1,\cdots,x_r,x_{r+1})$ in $r+1$ variables $x_1,\cdots,x_{r+1}$ of degree m such that the power series $F(x_1,\cdots,x_r,f(x_1,\cdots,x_r))$ contains no terms of degree $\leq mN$.*

Proof. Let $F(x_1,\cdots,x_{r+1})$ be the general polynomial of degree m in $r+1$ variables $x_1,\cdots,x_{r+1}$. If ${}^h\mathcal{C}_j$ denotes the binomial coefficient $h!/j!(h-j)!$, then F contains ${}^{r+1+m}\mathcal{C}_{r+1}$ terms and hence that many indeterminate coefficients (c). The power series $F(x_1,\cdots,x_r,f(x_1,\cdots,x_r))$ contains at most ${}^{r+mN}\mathcal{C}_r$ terms of degrees $\leq mN$, the coefficients of which are linear forms $L_\lambda(c)$ in the (c). The ${}^{r+mN}\mathcal{C}_r$ linear equations $L_\lambda(c)=0$ will have a non-trivial solution in the (c) if ${}^{r+1+m}\mathcal{C}_{r+1} > {}^{r+mN}\mathcal{C}_r$. This will be the case if we take m greater than $M=(N+1)^r(1+r)$; for we have then

$$(m+1) > (1+r)m^r(N+1)^r/m^r$$
$$> (r+1)[(mN+r)\cdots(mN+1)]/[(m+r+1)\cdots(m+2)],$$

hence $[(m+r+1)\cdots(m+1)]/[r+1] > (mN+r)\cdots(mN+1)$,
hence ${}^{r+1+m}\mathcal{C}_{r+1} > {}^{r+mN}\mathcal{C}_r$.

Before proceeding to the next lemma we shall recall here some well-known facts about the intersection of two algebraic cycles Z_r and Z_s in S_n. While in general any two arbitrary cycles C' and C'' in S_n have a "topological intersection" consisting of a homologous family of cycles about their geometric intersection $C'\cap C''$, it is possible in case of the algebraic cycles Z_r and Z_s to define an intersection cycle $Z_r\cdot Z_s$ provided all the components of the geometric intersection $Z_r\cap Z_s$ are proper (i. e. have the dimension $r+s-n$). Let $U^{(1)},\cdots,U^{(\lambda)}$ be the components of $Z_r\cap Z_s$; they are all algebraic varieties of dimension $r+s-n$. Then we can assign in a unique manner to each component $U^{(i)}$ a positive integer n_i such that the algebraic cycle $\sum_{i=1}^{\lambda} n_iU^{(i)}$ is a member of the homologous family of cycles constituting the "topological intersection" of Z_r and Z_s. This cycle $\sum_{i=1}^{\lambda} n_iU^{(i)}$ is called the intersection cycle $Z_r \cdot Z_s$ of Z_r and Z_s, and the numbers $n_1, \cdots, n_\lambda$ are called the intersection multiplicities of $U^{(1)}, \cdots, U^{(\lambda)}$ respectively. The degree of $Z_r \cdot Z_s$ is the product of the degrees of Z_r and Z_s. Corresponding to the Lemmas 1 and 2 of the preceding section, we have here the following result about the intersection cycle $Z_r \cdot Z_s$: (1) If V_r and V_s are component varieties of multiplicity one of Z_r and Z_s respectively, and if U is a regular component of $V_r \cap V_s$ and is not contained in any other component of Z_r *or* Z_s, then the multiplicity of U in the cycle $Z_r \cdot Z_s$ is equal to $+1$. We shall say the U is a regular component of $V_r \cap V_s$ if it contains a point which is regular for both V_r and V_s and if the join of the tangent spaces of V_r and V_s at this point has the dimension n. (2) Given any positive number d there is a positive number e (depending of course also on Z_r and Z_s) such that if the algebraic cycles Z'_r and Z'_s are e-homologous to Z_r and Z_s respectively, then the cycle $Z'_r \cdot Z'_s$ is d-homologous to the cycle $Z_r \cdot Z_s$. All these hold in fact (with slight modifications) also for the intersection of analytic cycles and complexes, but we shall not need them here.

LEMMA 4. *Let E_r be an analytic simplex in S_n, with center at (p) and let L_{n-r+1} be a linear variety containing the point (p) such that the intersection of E_r with L_{n-r+1} is an analytic simplex E_1 with center at (p). If a positive algebraic cycle Z_{n-1} of degree m has an isolated intersection with E_1 at the center (p) with a multiplicity μ, then the intersection cycle of Z_{n-1} and L_{n-r+1} is a positive algebraic cycle Z_{n-r} of degree m which has an isolated intersection of multiplicity μ at the point (p) with E_r.*

Proof. The linear variety L_{n-r+1} does not lie in any component hypersurface of Z_{n-1}; for otherwise the entire analytic simplex E_1 would lie in Z_{n-1}, in contradiction to our assumption that Z_{n-1} has an isolated intersection with E_1 at the point (p). Therefore every component of the intersection $L_{n-r+1} \cap Z_{n-1}$ is an algebraic variety of dimension $n-r$ and hence is proper. Therefore the intersection cycle of L_{n-r+1} and Z_{n-1} is a positive algebraic cycle Z_{n-r} of degree m, and it remains only to show that this cycle Z_{n-r} intersects E_r at the point (p) with the multiplicity μ. It is obviously enough to prove the assertion for the case when Z_{n-1} is an irreducible hypersurface, for otherwise we can apply the same argument to each irreducible component of Z_{n-1}. The assumption that Z_{n-1} intersects E_1 at the center (p) with the multiplicity μ implies that there is a projective transformation arbitrarily near to the identity and such that TZ_{n-1} has exactly μ regular intersections with E_1 which lie in any given sufficiently small neighborhood R of (p). For such a T the components of $TZ_{n-1} \cap L_{n-r+1}$ are all proper and hence there is an intersection cycle $Z^T{}_{n-r} = TZ_{n-1} \cdot L_{n-r+1}$; and since each of the μ regular intersection points of TZ_{n-1} with E_1 is contained in exactly one regular component of $TZ_{n-1} \cap L_{n-r+1}$, it is contained in exactly one component of multiplicity one of the cycle $Z^T{}_{n-r}$. Let the positive number d be chosen with respect to the chains Z_{n-r} and E_r and the neighborhood R of the point (p) as in Lemma 2, and let the positive number e be chosen with respect to the number d and the intersection cycle $Z_{n-r} = Z_{n-1} \cdot L_{n-r+1}$ as in the remark (2) immediately preceding this Lemma. Since for T sufficiently near to the identity the cycle TZ_{n-1} is an e-deformation of Z_{n-1}, it follows that for such a T the intersection cycle $Z^T{}_{n-r}$ is d-homologous to Z_{n-r}. It is clear that the part of the intersection $Z^T{}_{n-r} \cap E_r$ which lies in R consists of the same μ intersection points of TZ_{n-1} and E_1. Since each of these points is contained in exactly one component variety of multiplicity one of $Z^T{}_{n-r}$, and since it is also a regular intersection of this component variety with E_r, therefore it is an intersection of multiplicity one of $Z^T{}_{n-r}$ and E_r. It follows then from Lemma 2 that the multiplicity of the intersection (p) of Z_{n-r} and E_r is equal to μ.

With the help of these two lemmas we can prove the following theorem:

THEOREM IV. *Let E_r be an analytic simplex in S_n. Suppose there exists a positive integer N such that if any positive algebraic cycle Z_{n-r} of degree m in S_n has an isolated intersection with E_r at its center, then the multiplicity of this intersection is at most equal to mN. Then E_r is an element of an algebraic variety V_r in S_n.*

Proof. The theorem is evidently equivalent to the following statement: If E_r is not an element of an algebraic variety V_r in S_n, then given any positive integer N there exists a positive algebraic cycle Z_{n-r} of degree m such that Z_{n-r} has an isolated intersection with E_r of multiplicity greater than mN at the center of E_r.

Let D_r be the analytic element containing E_r. Let the affine space $A_n(x)$ be so chosen that the analytic element D_r is given by the set of equations

$$x_{r+1} = f_{r+1}(x_1, \cdots, x_r), \cdots, x_n = f_n(x_1, \cdots, x_r),$$

where the $f_i(x_1, \cdots, x_r)$, $i = r+1, \cdots, n$, are convergent power series in a region in the affine space $A_r(x)$ of the r complex variables $x_1, \cdots, x_r$. We can assume without any loss of generality that E_r is the image of a simplicial $2r$-simplex C^{2r} in $A_r(x)$ and that the center of E_r is the origin of $A_n(x)$.

Since the analytic simplex E_r, hence also the analytic element D_r is not algebraic, at least one of the $n-r$ power series $x_{r+1}=f_{r+1}(x_1,\cdots,x_r),\cdots,$ $x_n=f_n(x_1,\cdots,x_r)$, say the power series $x_{r+1}=f_{r+1}(x_1,\cdots,x_r)$, is not an algebraic element over the field $k(x_1,\cdots,x_r)$. According to Lemma 3, there exists to any given positive integer N a polynomial $F(x)=F(x_1,\cdots,x_{r+1})$ of degree m $(>M)$ in the $r+1$ variables $x_1,\cdots,x_{r+1}$, such that the power series $P(x_1,\cdots,x_r)=F(x_1,\cdots,x_r,f_{r+1}(x_1,\cdots,x_r))$ contains no terms of degree $\leq mN$. Since the element $f_{r+1}(x_1,\cdots,x_r)$ is not algebraic over $k(x_1,\cdots,x_r)$, the power series $P(x_1,\cdots,x_r)$ is not identically zero. Hence, after a suitably chosen linear transformation in $A_r(x)$ if necessary, we can assume that the power series $P(x_1)=P(x_1,0,\cdots,0)$ is not identically zero. It is clear that this power series $P(x_1)$ contains no terms of degree $\leq mN$. Let L_{n-r+1} be the linear variety in S_n defined by the $r-1$ linear equations $x_2=0,\cdots,x_r=0$, then the intersection of L_{n-r+1} with D_r is a regular analytic element D_1 given by the equations $x_2=0,\cdots,x_r=0$, $x_{r+1}=f_{r+1}(x_1,0,\cdots,0),\cdots,x_n=f_n(x_1,0,\cdots,0)$; and the intersection of L_{n-r+1} with E_r is an analytic simplex E_1 contained in D_1, which is the image of the topological 2-simplex C^2 obtained from C^{2r} by intersection with the subspace $x_2=x_3=\cdots=x_r=0$ in $A_r(x)$. (Thus C^2 is a 2-simplex in the x_1-plane.)

Consider now the homogeneous algebraic equation

$$G(x_0,x_1,\cdots,x_{r+1})=x_0{}^mF(x_1/x_0,\cdots,x_{r+1}/x_0)=0;$$

it defines a positive algebraic cycle Z_{n-1} in S_n consisting of as many hypersurfaces as the distinct irreducible factors on the left side of this equation, each taken with its multiplicity. The fact that $P(x_1)=F(x_1,0,\cdots,0,$ $f_{r+1}(x_1,0,\cdots,0))$ vanishes for $x_1=0$ but not identically shows that the point $(x)=(0)$ is an isolated intersection of Z_{n-1} and E_1. It only remains to prove that the multiplicity of this intersection is greater than mN, for then our theorem will follow from Lemma 4. Here again we can assume that the algebraic cycle Z_{n-1} is an irreducible hypersurface, which means that the form $G(x_0,x_1,\cdots,x_{r+1})$ and hence also the polynomial $F(x_1,\cdots,x_{r+1})$ is irreducible; for otherwise we can apply the same argument to each component hypersurface of Z_{n-1}. According to Lemma 2, it is sufficient to show that given any neighborhood R of the point $(x)=(0)$ in $A_n(x)$ there exists a linear transformation $T_{(e)}$,

$$x'_i=x_i+e_i, \qquad i=1,\cdots,n$$

arbitrarily near to the identity, such that $T_{(e)}Z_{n-1}$ has more than mN regular intersections with E_1 in R.

Consider the power series in x_1 with coefficients in $k(u) = k(u_1, \cdots, u_{r+1})$

$$\Phi(x_1, (u)) = F(x_1 + u_1, u_2, \cdots, u_r, u_{r+1} + f_{r+1}(x_1, 0, \cdots, 0)),$$

which is convergent for all sufficiently small x_1 for any given (u). The intersection of $T_{(e)}Z_{n-1}$ and E_1 is then given by the zeros of the power series $\Phi(x_1, (-e))$ in C^2, and it is well known that a simple zero $x_1 = b_1$ corresponds to a point $(b) = (b_1, 0, \cdots, 0, f_{r+1}(b_1, 0, \cdots, 0), \cdots, f_n(b_1, 0, \cdots, 0))$ on E_1 which is a regular intersection of $T_{(e)}Z_{n-1}$ and E_1. Now we have the fact that $\Phi(x_1, (0)) = P(x_1)$ has a zero of order $\mu > mN$ at the point $x_1 = 0$. Hence, by the Weierstrass Preparation Theorem, there is a neighborhood R' of $x_1 = 0$ and a neighborhood R'' of $(u) = (0)$ such that

$$\Phi(x_1, (u)) = g(x_1, (u))Q(x_1, (u)),$$

where $Q(x_1, (u))$ is a distinguished polynomial of degree μ in x_1 with center at $x_1 = 0$, $(u) = (0)$, and $g(x_1, (u))$ is a power series which does not vanish for x_1 in R' and (u) in R''. The distinguished polynomial $Q(x_1, (u))$ might be reducible as a polynomial in x_1 over the field of all quotients of power series in (u), but it cannot have multiple factors. In fact, if any power series $q(x_1, (u))$ with the property $q(0, (0)) = 0$ is a multiple factor of $Q(x_1, (u))$, then the power series $q(u) = q(0, (u))$ is a multiple factor of $Q(0, (u))$ and hence also a multiple factor of the polynomial $\Phi(0, (u)) = F(u)$. One observes that the power series $q(u)$ is not a constant, for $q(u)$ is not identically zero and we have $q(0) = 0$. Hence the equation $q(u) = 0$ defines a finite number of analytic elements of r dimensions in $A_{r+1}(u)$ with center at $(u) = (0)$. If one of the variables (u), say u_1, is actually involved in the polynomial $F(u)$, then $\partial F(u)/\partial u_1$ is also a polynomial in (u). It follows then that both polynomials $F(u)$ and $\partial F(u)/\partial u_1$ are divisible by the power series $q(u)$. This means that the two positive algebraic cycles of r dimensions in $A_{r+1}(u)$ defined by the two equations $F(u) = 0$ and $\partial F(u)/\partial u_1 = 0$ have at least an analytic element of r dimensions in common and consequently also at least one component variety in common. Since $F(u)$ is an irreducible polynomial, the algebraic cycle $F(u) = 0$ is an irreducible algebraic variety and hence must be a component variety of the algebraic cycle $\partial F(u)/\partial u_1 = 0$. This means that the polynomial $F(u)$ is a factor of the polynomial $\partial F(u)/\partial u_1$, both being now considered as elements of the polynomial ring $k[u]$. But this is impossible, for the polynomial $\partial F(u)/\partial u_1$ has a lower degree than $F(u)$. Thus we have shown that the distinguished polynomial $Q(x_1, (u))$ has no multiple factors.[7]

---

[7] The last part of the argument consists essentially in showing that an irreducible

Since $Q(x_1, (u))$ has no multiple factors, its discriminant $H(u)$ is not identically zero in R''. Hence for any point (e) in R'' such that $H(e) \neq 0$, the polynomial $Q(x_1, (e))$ has exactly μ simple zeros and consequently also the power series $\Phi(x_1, (e))$ has exactly μ simple zeros in R'. For all points (e) sufficiently near to the point $(u) = (0)$ with $H(e) \neq 0$, the μ simple zeros will lie in any given neighborhood of the point $x_1 = 0$ and hence the corresponding μ intersection points on E_1 will lie in any given neighborhood of the point $(x) = (0)$. This completes the proof of the theorem.

Combining Theorem III and Theorem IV, we obtain immediately the main theorem:

THEOREM V. *A compact analytic variety in S_n is an algebraic variety.*

Proof. Let E_r be an analytic simplex contained in the compact algebraic variety W_r. According to Theorem III and Theorem IV, there is an algebraic variety V_r which also contains the analytic simplex E_r. By the principle of analytic continuation, any two (irreducible) compact analytic varieties of r dimensions which have an analytic simplex E_r in common must coincide. Hence we have $W_r = V_r$, and the theorem is proved.

It is perhaps not without interest to point out here the analogy between Theorem IV and the well known criterion for an algebraic number by means of diophantine approximation: A number ξ is algebraic if and only if there exists a positive integer N such that no form $\sum_{i=0}^{n} z_i \xi^i$ has a proper approximation t^{-N} with respect to the variables $z_0, z_1, \cdots, z_n$. In other words, if $Z = \max(|z_0|, |z_1|, \cdots, |z_n|)$, then the number ξ is algebraic if and only if for each n the diophantine inequality

$$| \sum_{i=0}^{n} z_i \xi^i | < Z^{-N}$$

has only a finite number of solutions for which the left hand side is not zero. Hence, for each n there exists a positive number Γ_n such that the diophantine inequality

$$| \sum_{i=0}^{n} z_i \xi^i | < (\Gamma_n Z)^{-N}$$

has no solution for which the left hand side is not zero. Taking the negative of the logarithm of both sides, we can write the inequality as follows:

---

polynomial $F(u_1, \cdots, u_{r+1})$ cannot have multiple factors in the ring of power series $k\{u_1, \cdots, u_{r+1}\}$. This is a special case of a theorem proved by Chevalley, see [2], p. 11, Theorem 1, which asserts that any prime ideal in $k(u_1, \cdots, u_{r+1})$ is the intersection of prime ideals in $k\{u_1, \cdots, u_{r+1}\}$.

$$-\log \left| \sum_{i=0}^{n} z_i \xi^i \right| > N(\log Z + \log \Gamma_n).$$

In the analogy between algebraic number field and algebraic function field of one variable we can regard the number ξ as an "analytic branch" and the equation $\sum_{i=0}^{n} z_i y^i = 0$, for a given (z), as an "algebraic curve." The number $\log Z + \log \Gamma_n$ can then be regarded as representing the "degree" of the "algebraic curve" and the left hand side of the above inequality as the "intersection multiplicity" of the "analytic branch" with the "algebraic curve." Carrying this over to the algebraic function field of one variable, we obtain the statement that an analytic branch E_1 in S_2 is algebraic if and only if there is a number N such that no algebraic curve of degree m can have an isolated intersection with E_1 with a multiplicity greater than mN. Thus our Theorem IV can be considered as the extension of this criterion to analytic elements in space of higher dimensions.

Finally, we should like to remark that though Theorem IV is stated and proved only for a regular analytic element, it is not difficult to see that it holds in fact for any analytic element, whether regular or singular. Thus this theorem expresses a property which is both necessary and sufficient for any analytic element to be algebraic. Though our proof is partly algebraic and partly topological, as is necessary for our present purpose, the theorem itself can be expressed in purely algebraic terms, using only formal power series. In fact, one could probably obtain a purely algebraic proof of this result by means of the intersection theory for algebroid varieties as developed by Chevalley.[8]

5. **Meromorphic transformations.** We begin with an almost obvious generalization of Theorem V to a multiply projective space. For the sake of convenience we shall restrict ourselves to the case of a doubly projective space $S_m \times S_n$, though the results hold obviously for the general case. We can define an analytic variety W_r in $S_m \times S_n$ in exactly the same way as we have done in **2** for S_n. Now the space $S_m \times S_n$ can be mapped by a bi-regular birational transformation onto a non-singular algebraic variety V_{m+n} in a suitably chosen projective space S_T. Under this transformation any analytic or algebraic variety in $S_m \times S_n$ is carried into an analytic or algebraic variety in V_{m+n} respectively, and conversely. Therefore, if W_r is a compact analytic variety in $S_m \times S_n$, then its image W_r^* is also a compact analytic variety in

[8] Chevalley [2].

V_{m+n} and hence also in S_T. By Theorem V, W_r^* is an algebraic variety, hence W_r is also an algebraic variety.

THEOREM VI. *A compact analytic variety in a multiply projective space is an algebraic variety.*

In fact, we can extend this theorem to all the so-called "extended spaces" of Osgood,[9] but we shall not go into details here. Instead we shall derive from Theorem VI an important result concerning meromorphic transformation of an analytic variety. A mapping of an analytic variety W_r in S_n into the space S_m is called a meromorphic transformation if in a sufficiently small neighborhood R of any point (a) of W_r and for a suitable choice of the affine spaces $A_n(x)$ and $A_m(y)$ in S_n and S_m respectively, the mapping is defined by a set of equations

$$g(x)y_i = f_i(x), \qquad i = 1, \cdots, m, \tag{9}$$

where the functions $g(x), f_1(x), \cdots, f_m(x)$ are holomorphic in R and $g(x)$ does not vanish for all points of $W_r \cap R$. These equations are to be understood with the following stipulations: For a point $(x) = (p)$ in $W_r \cap R$ such that $g(p) \neq 0$, the image under the mapping is the point $y_i = f_i(p)/g(p)$, $i = 1, \cdots, m$, in S_m. We shall call such a point (p) a regular point of the mapping. For a point $(x) = (p)$ such that $g(p) = 0$, the image under the mapping is to be the set of all points in S_n which are limits of the images of a sequence of regular points approaching (p). Thus a meromorphic transformation, just like the rational transformation in case of an algebraic variety, is not a mapping in the strict sense of the word; it assigns only to a sufficiently "general" point of W_r a unique image, while to the others (which correspond to the fundamental points of a rational transformation) it assigns certain subsets as images.

THEOREM VII. *A meromorphic transformation of a compact analytic (hence algebraic) variety W_r in S_n into a projective space S_m is a rational transformation of W_r, and the image is an algebraic variety in S_m.*

Proof. The graph of the meromorphic transformation in the product space $S_n \times S_m$ is a point set G, the projection of which in S_n is the variety W_r. Let $((a), (b))$ be any point of G, and let R be a sufficiently small neighborhood of $((a), (b))$ with R' and R'' as its projections in S_n and S_m respectively. The variety W_r being algebraic, it is defined in the neighborhood

[9] See Osgood [7], Ch. III, § 32.

R' of the point (a) by a set of algebraic equations $\phi_i(x)=0$, $i=1,\cdots,s$. Then the set $G \cap R$ will be defined by the set of equations

$$(10) \qquad \begin{aligned} g(x)y_i &= f_i(x), & i&=1,\cdots,m,\\ \phi_j(x) &= 0, & j&=1,\cdots,s, \end{aligned}$$

with the stipulation that any component variety lying entirely in the hypersurface $g(x)=0$, if it is an isolated one, should be deleted from the set. It is well known [10] from the theory of power series ideals that this can be achieved by the adjunction of a finite number of suitably chosen analytic equations to the equations (10), if the neighborhood R is chosen sufficiently small. Thus the set $G \cap R$ consists of all the common solutions of a set of analytic equations in R. Since $((a),(b))$ is any point of G, this means that G is an analytic variety in $S_n \times S_m$.

It remains to show that G is a compact set. Let $((a^{(i)}),(b^{(i)}))$, $i=1,2,\cdots$, be a sequence of points in G converging to a point $((a),(b))$ in $S_n \times S_m$; we have to show that $((a),(b))$ is a point of G. Since W_r is compact and the point (a) is the limit point of the sequence $(a^{(i)})$, $i=1,2,\cdots$, it follows that (a) is a point of W_r. In a sufficiently small neighborhood of (a) the given meromorphic transformation will be represented by a set of equations (9), with the corresponding stipulations. We can assume without loss of generality that for all large i the points $(a^{(i)})$ are regular, i. e. $g(a^{(i)}) \neq 0$. For, since each point $((a^{(i)}),(b^{(i)}))$, with $g(a^{(i)})=0$ is by our stipulation the limit of a sequence of regular points, we can always replace each point $((a^{(i)}),(b^{(i)})$, for sufficiently large i, by a regular point sufficiently near to it so that the resulting sequence will have the same limit point $((a),(b))$. Now, if $g(a)\neq 0$, then we have

$$b_j = \lim_{i\to\infty} b_j{}^{(i)} = \lim_{i\to\infty} f_j(a^{(i)})/g(a^{(i)}) = f_j(a)/g(a), \qquad j=1,\cdots,m;$$

and hence $((a),(b))$ is a point of G. On the other hand, if $g(a)=0$, then the point (b) is by our stipulation a point of G. Therefore G is a compact point set.

Thus we have shown that G is a compact analytic variety in $S_n \times S_m$; hence, by Theorem VI, G is an algebraic variety. It follows then that the projection of G in S_m is also an algebraic variety W^* and the given meromorphic transformation is really an (irreducible) algebraic correspondence between W_r and W^*. Since this algebraic correspondence assigns to a generic point of W_r a unique image point in W^* (and since our ground field has

---

[10] See Rückert [8], or Bochner and Martin [1], Ch. X.

the characteristic 0), therefore it is a rational transformation of W_r onto W^*. Thus the proof is completed.

The concept of a meromorphic transformation evidently includes that of a meromorphic function as the special case $m=1$. Hence we have the corollary:

COROLLARY. *An everywhere meromorphic function on a compact analytic variety is a rational function.*

So far as we know, even this corollary has been proved up to now only for a few special cases such as the projective space and the space of analysis (the product of projective lines).

THE JOHNS HOPKINS UNIVERSITY

REFERENCES.

1. S. Bochner and W. T. Martin, *Several Complex Variables*, Princeton, 1948.
2. C. Chevalley, "Intersections of algebraic and algebroid varieties," *Transactions of the American Mathematical Society*, vol. 57 (1945), pp. 1-85.
3. K. Knopp and R. Schmidt, "Funktionaldeterminanten und Abhaengigkeit von Funktionen," *Mathematische Zeitschrift*, vol. 25 (1926), pp. 373-381.
4. B. O. Koopman and A. B. Brown, "On the covering of analytic loci by complexes," *Transactions of the American Mathematical Society*, vol. 34 (1932), pp. 231-251.
5. S. Lefschetz, *Topology*, New York, 1930.
6. ——— and J. H. C. Whitehead, "On analytical complexes," *Transactions of the American Mathematical Society*, vol. 35 (1933), pp. 510-517.
7. W. F. Osgood, *Lehrbuch der Funktionentheorie*, Bd. II, 1. Leipzig, 1929.
8. W. Rückert, "Zum Eliminationsproblem der Potenzreihenideale," *Mathematische Annalen*, vol. 107 (1932), pp. 259-281.
9. B. L. van der Waerden, "Topologische Begründung des Kalküls der abzählenden Geometrie," *Mathematische Annalen*, vol. 102 (1929), pp. 337-362.

ANNALS OF MATHEMATICS
Vol. 50, No. 1, January, 1949

ON THE GEOMETRY OF ALGEBRAIC HOMOGENEOUS SPACES

BY WEI-LIANG CHOW

(Received December 22, 1947)

CHAPTER I. INTRODUCTION.

1. Four Classes of Homogeneous Spaces

The object of the present paper is a study of those symmetric homogeneous spaces (in the sense of E. Cartan) which can be represented as algebraic varieties. We shall call these spaces, in short, algebraic homogeneous spaces. Apart from certain exceptional types, which we shall not consider here, they consist of the following four classes of spaces:[1]

(1) *The Grassmann space of all r-dimensional linear subspaces* [r] *of a projective space* S_n *of n dimensions.* The basic group of transformations consists of those induced by the collineations (and, in case $n = 2r + 1$, also correlations) in S_n. By a collineation is meant a one-to-one transformation of S_n (as a point space) onto itself such that the lines are carried into lines. It is well known that such a collineation might involve an automorphism of the ground field. In case no automorphism of the ground field is involved, or in other words only the identity automorphism is involved, we have what is usually called a projective transformation.

(2) *The space of the self-conjugate* [r] *of a null system in the projective space* S_n ($n = 2r + 1$). The basic group of transformations consists of those induced by all the collineations in S_n which leave the null system invariant. The number r is called the order of the space. It is well known that a null system can always be transformed by a projective transformation (in any ground field) into the normal form $\sum_{i=0}^{r} (y_{r+1+i}x_i - y_i x_{r+1+i}) = 0$; this means that to each point $Y = (y_0, y_1, \cdots, y_n)$ of S_n the corresponding hyperplane under the null system is given by this equation. Unless otherwise stated, we shall assume that the null system is already taken in this normal form. This space is sometimes called the space of symmetric matrices, and the basic transformations are then called the semi-symplectic transformations.

(3) *The space of the self-conjugate* [r] *of a polar system in the projective space* S_n ($n = 2r + 1$). The basic group of transformations consists of those induced by all the collineations in S_n which leave the polar system invariant. The number r is also called the order of the space. Unless otherwise stated, we shall assume that the polar system is in the normal form $\sum_{i=0}^{r} (y_{r+1+i}\, x_i + y_i\, x_{r+1+i}) = 0$, which, though not always possible in any ground field, is always possible in a suitably chosen finite extension of it. In fact, the assumption that the polar system can be transformed into this normal form by a projective transformation in the ground field is equivalent to the assumption that there exist at least two

[1] See Cartan [4]. Numbers in brackets refer to the Reference at the end of the paper.

disjoint self-conjugate [r] in the ground field (i.e. determined by points with coordinates in the ground field).

We observe that in contrast to the preceding case, the space of a polar system is not irreducible, but consists of two equivalent irreducible components. We shall call any one of the irreducible component the irreducible space of a polar system, and the group of those basic transformations which map this irreducible component onto itself is called the basic group of transformations of the irreducible space. It is this irreducible space of a polar system which is the real object of our study. This space is sometimes called the space of skew-symmetric matrices, and its basic group is then called the group of semi-orthogonal transformations.

(4) *The hyperquadric Q_{n-1} of the projective space S_n.* The basic group of transformations consists of those collineations of S_n which leave a hyperquadric Q_{n-1} invariant. Unless stated otherwise, we shall assume that the hyperquadric Q_{n-1} is in the normal form $\sum_{i=0}^{r} x_i x_{r+1+i} = 0$ in case of odd $n = 2r + 1$, or $\sum_{i=0}^{r} x_i x_{r+1+i} + cx_{2r+2}^2 = 0$ in case of even $n = 2r + 2$. Just as in the preceding case, the normal form means simply the assumption that there exist on the hyperquadric Q_{n-1} at least two disjoint [r] in the ground field.

We have framed the above definitions in such a way that they hold not only for the field of complex numbers (as is usually the case), but also for an arbitrary ground field K. It is for this reason that the particular normal forms are taken in (3) and (4), so that we shall have in any ground field a space of sufficient elements as well as sufficient homogeneity. Since we are concerned in this paper mainly with the algebro-geometrical properties of these spaces, we shall keep our ground field K as general as possible. Only at the end of Chapter III, when we are dealing with analytic transformations, is the ground field K specialized to the field of complex numbers.

2. Two Problems

Our main problem is to characterize the basic groups of transformations of these homogeneous spaces, both from the projective and from the birational points of view. As the prototypes of the theorems we intend to prove, we state here in each case a well known theorem concerning the semi-projective transformations, which are in fact the basic transformations of the simplest of the algebraic homogeneous spaces, the projective space S_n.

(A) *Projective characterization of the basic group of transformations.* Here we have the so-called fundamental theorem of projective geometry, which for our purpose may be formulated thus:[2] Any collineation of the projective space S_n $(n > 1)$ is a semi-projective transformation, that is, a linear homogeneous transformation of the $(n + 1)$-vector space over the ground field K, plus an automorphism of the field K. It is also a well known result in projective

[2] See Brauer [3], where this theorem is proved for an arbitrary ground field. His definition of collineation differs somewhat from ours, but is equivalent, in view of the theorem quoted in footnote 3.

geometry[3] that any one-to-one incidence preserving transformation of the linear subspaces of any two dimensions is enough to determine a collineation in S_n. Thus the semi-projective transformations of S_n are characterized by the fact that they preserve the incidence relations between the linear subspaces, and it is natural to ask whether in the more general algebraic homogeneous spaces the invariance of some kind of incidence is also enough to characterize the basic group of transformations among all the one-to-one transformations. We shall answer this question in the affirmative in Chapter II, where similar fundamental theorems are proved for all the four classes of spaces. The methods used are those of projective geometry, and the results hold for an arbitrary ground field K. Our results are, in fact, so general, that they include not only the above four classes of spaces, but also such spaces as the space of "Hermitian" matrices and the space of self-conjugate $[r]$ of a polar system in an S_n with even $n = 2r + 2$.

(B) *Birational characterization of the basic group of transformations.* Here we have the following well known theorem for the projective space S_n : Any regular birational transformation of the projective space S_n onto itself is a projective transformation. A birational transformation of an algebraic variety (onto another algebraic variety, which may of course be itself) is called *regular*, if it is one-to-one for all points of the variety without exception. By hypothesis, an algebraic homogeneous space can be represented as an algebraic variety, which we shall call the homogeneous algebraic variety associated with the space. We shall choose this homogeneous algebraic variety in such a way that the basic group of transformations of the space is represented as a group of regular birational transformations of the variety. The theorem we want to prove is that any regular birational transformation of the homogeneous algebraic variety associated with one of the four classes of homogeneous spaces onto itself is a transformation of the basic group. On the basis of the results of Chapter II, the proof of this main theorem is carried through in Chapter III by the methods of algebraic geometry. The proof is very simple in its general conception, and it holds uniformly for all four classes of varieties, though with some unavoidable divergence in minor details.

Finally, we shall show at the end of Chapter III that in case the ground field is the field of all complex numbers, the above birational characterization can be generalized into an analytic characterization. In other words, any regular analytic transformation of the variety onto itself is a transformation of the basic group. The usual procedure is to reduce this case to the birational case by means of a general theorem given in Osgood,[4] that any regular analytic transformation of the so-called extended space is a birational transformation. Since the proof of this general theorem is a rather cumbrous affair, it is desirable to have a direct proof of the analytic characterization without recourse to this theorem. Our proof is based on direct topological considerations and is about as simple as could be expected.

[3] See Bertini [1], p. 52.

[4] Osgood [14], p. 295.

3. Remarks

As to literature, we should like to mention the recent papers of Hua,[5] in which similar problems have been considered. Approaching the problems from a somewhat different point of view and using an entirely different method, Hua has obtained solutions for a part of these problems, under the assumption that the ground field is the field of complex numbers. His main emphasis is on the space of a null system (2), or the space of symmetric matrices, for which he proved a weaker form of our Theorem II (under the assumption of continuity) and Theorem XI, both for the ground field of complex numbers.[6] For the irreducible space of a polar system (3), or the space of skew-symmetric matrices, Hua has proved a theorem similar to our Theorem VII, under the following restrictive assumptions:[7] (a) The ground field is the field of complex numbers, (b) the transformation is continuous, (c) the invariance of the distances, (d) the invariance of "harmonic separation", and finally (e) the order r of the space is odd. Our proof of the Theorem VII shows that all these assumptions are really not necessary except a part of (c), namely the invariance of the distance 2. Our geometric method not only enables us to solve these problems completely in all generality with a minimum of hypothesis, but also has the advantage of being substantially simpler and more conceptual.

In conclusion, we should like to remark that the irreducible space of a polar system (3) is relatively the most complicated of all cases; the Grassmann space (1) and the space of a null system (2) are both much simpler in comparison. Though it might appear from the definitions in §1 that the spaces (2) and (3) are similar in nature and somewhat different from the space (1), and though this is true to a certain extent, if we consider the entire (reducible) space (3) (as we will show in the first part of Chapter II), a deeper study shows that the two spaces (1) and (2) are essentially similar and are of the same order of difficulty, while the irreducible space (3) presents generally a quite different problem and calls for much greater effort in solution. The reason for this is not difficult to find, for the varieties $G_{n,r}$ and N_r (and for that matter also Q_{n-1}) are ruled varieties, that is, they contain through each point a system of lines on the variety; while, on the other hand, the variety P_r contains not a single line, and is in fact ruled by a system of conic sections. In this connection, we should like to remark that the varieties $G_{n,r}$, N_r, Q_{n-1} are all "minimum models" in the sense of Severi,[8] that is, they cannot be transformed into varieties of lower order

[5] Hua [8], [9], [10], [11].

[6] Hua proved our Theorem II in [8], pp. 441–459, under the assumptions: (a) The ground field is the field of complex numbers, (b) the transformation is continuous, (c) the invariance of the distances, and (d) the invariance of "harmonic separation." In later papers, [9] and [10] pp. 229–238, he was able to remove the last assumption (d).

[7] Hua [8], pp. 459–477. It is to be noticed that in introducing the "inhomogeneous" skew-symmetric matrices, Hua has implicitly restricted himself to one of the two irreducible components of the space of "homogeneous" skew-symmetric matrices. This explains the apparent contradiction between our Theorem V and his results.

[8] Severi [15], p. 91.

by a regular birational transformation; while the variety P_r is not a minimum model. This latter fact is intimately connected with the existence of the spinor representation of the orthogonal group and the representation of P_r in the spinor spaces.[9] We hope to be able to come back to these questions on some future occasions.

CHAPTER II

PROJECTIVE CHARACTERIZATION OF THE BASIC GROUP. THE FUNDAMENTAL THEOREMS.

1. The Grassmann Space

We begin with some simple definitions. We consider a base projective space S_n of dimension n over an arbitrary ground field K. All geometrical entities, points and spaces, considered in this chapter are understood to be rational over this ground field K. For the sake of brevity, we shall often use the word "element" to denote a linear subspace or a point of the base space S_n. An element of dimension s will be denoted by the symbol $[s]$, and various subscripts and superscripts will be used to differentiate between the different elements of the same dimension. This definition also includes the case $s = -1$, the element $[-1]$ denotes simply the empty set. We shall use the symbols $[s] \cap [t]$ and $[s] \cup [t]$ to denote the intersection and join respectively of the two elements $[s]$ and $[t]$; and the symbol $[s] \supset [t]$ means that $[s]$ contains $[t]$. Two elements $[s]$ and $[t]$ are said to be u-incident, if their intersection $[s] \cap [t]$ is an element of dimension u, or (which is the same thing) if their join is an element of dimension $s + t - u$. Two elements $[s]$ and $[t]$ are simply called incident if one of them is contained in the other.

Two elements $[s]_a$ and $[s]_b$ of the same dimension s are said to be *adjacent* if they are $(s - 1)$-incident; they are said to have the *distance* u, if u is the least integer with the property that there exists a sequence of $u + 1$ elements $[s]_1 = [s]_a, [s]_2, \cdots, [s]_{u+1} = [s]_b$ of dimension s, each two consecutive ones of which are adjacent. It is easily seen that $[s]_a$ and $[s]_b$ have the distance u if and only if they are $(s - u)$-incident. The distance of any two elements of dimension s is at most $s + 1$; this limit is actually reached for two disjoint $[s]$ in case $s \leq (n - 1)/2$. (Two elements are called disjoint if their intersection is $[-1]$). A set of $[s]$, each two elements of which are adjacent, is called maximal, if it is not a proper subset of another set with the same property. By an elementary theorem of projective geometry,[10] a maximal set of mutually adjacent $[s]$ must be either the set of all $[s]$ incident with one common $[s + 1]$ or the set of all $[s]$ incident with one common $[s - 1]$, but it evidently cannot be both.

Consider the Grassmann space of all the $[r]$ in the base space S_n. Our purpose is to study the one-to-one transformation of this Grassmann space onto

[9] See Givens [7].
[10] See Bertini [1], p. 14.

itself. However, it is more convenient (here as well as later in similar connections) to consider a somewhat more general problem. Let S'_n be another base projective space of the same dimension n over the same ground field K, and consider the Grassman space of all the $[r]'$ in S'_n. We shall study the one-to-one transformations of the Grassmann space of all the $[r]$ in S_n onto the Grassmann space of all the $[r]'$ in S'_n. This includes, of course, the first problem as a special case where we have $S_n = S'_n$, but it is technically more convenient to keep the two spaces separate, even when they are the same space. However, as we are interested in this paper only in the transformations of a space onto itself, we shall state all the theorems only for this special case, though the arguments are actually carried through for the more general problem.

Suppose now we have a one-to-one transformation of the Grassmann space of all the $[r]$ in S_n onto the Grassmann space of all the $[r]'$ in S'_n, and suppose that this transformation carries every pair of adjacent $[r]$ into a pair of adjacent $[r]'$, i.e. the adjacence of two $[r]$ is preserved. It follows then that the transformation must also carry every maximal set of mutually adjacent $[r]$ in S_n into a maximal set of mutually adjacent $[r]'$ in S'_n, and vice versa. Since, by the elementary theorem of projective geometry mentioned before, a maximal set of mutually adjacent $[r]$ is either the set of all $[r]$ incident with one common $[r-1]$ or the set of all $[r]$ incident with one common $[r+1]$, and since the same assertion evidently also holds for a maximal set of mutually adjacent $[r]'$, it follows that the given transformation induces a one-to-one transformation of the set of all $[r-1]$ and $[r+1]$ in S_n onto the set of all $[r-1]'$ and $[r+1]'$ in S'_n, such that if an $[r]$ is incident with an element of the former set, then the transformed $[r]'$ is incident with the transformed element in the latter set. We maintain now that if any one $[r-1]$ is transformed into an $[r-1]'$ (or an $[r+1]'$), then the same must be true for all $[r-1]$. It is only necessary to prove this for two adjacent $[r-1]$, since any two $[r-1]$ can be joined by a sequence of consecutively adjacent elements. But for two adjacent $[r-1]$ the assertion follows from the observation that two adjacent $[r-1]$ can have only one $[r]$ incident with both, which is their join; while an $[r-1]'$ and an $[r+1]'$ having one $[r]'$ incident with both, will have at least one further $[r]'$ with the same property, e.g. any one $[r]'$ joining the $[r-1]'$ with a point of the $[r+1]'$ not lying on the given $[r]'$. Similarly, it can be shown by an exactly similar argument that if any one $[r+1]$ is transformed into an $[r-1]'$ (or an $[r+1]'$), then the same must be true for all $[r+1]$. It follows from these results that there is either a one-to-one transformation of the set of all $[r-1]$ onto the set of all $[r-1]'$ (and the set of all $[r+1]$ onto the set of all $[r+1]'$), or a one-to-one transformation of the set of all $[r-1]$ onto the set of all $[r+1]'$ (and the set of all $[r+1]$ onto the set of all $[r-1]'$); and, by virtue of the preservation of the incidence of these elements with the $[r]$ or the $[r]'$ respectively, these transformations all preserve the adjacence, hence also the distance of any two elements of each set.

Now, in case $r \neq (n-1)/2$, it is impossible to have a one-to-one transformation of the set of all $[r-1]$ onto the set of all $[r+1]'$ which preserves the distance. For, in case $r < (n-1)/2$, there exist two $[r+1]'$ having at most one point in common and hence a distance at least equal to $r+1$, while the distance of any two $[r-1]$ is at most equal to r. Similarly, in case $r > (n-1)/2$, there exist two $[r-1]$ having at most a $[2r-n-1]$ in common and hence a distance at least equal to $n-r$, while the distance of any two $[r+1]'$ is at most equal to $n-r-1$. Thus we have shown that, in case $r \neq (n-1)/2$, there is a one-to-one transformation of the set of all $[r-1]$ in S_n onto the set of all $[r-1]'$ in S'_n, and that this transformation, together with the original transformation of the set of all $[r]$ onto the set of all $[r]'$, carries every pair of incident $[r-1]$ and $[r]$ into a pair of incident $[r-1]'$ and $[r]'$. By the well known result in the projective geometry[11] already mentioned in Chapter I, §2, such a transformation must be one induced by a collineation of the base space S_n onto the base space S'_n. In case $r = (n-1)/2$, both of the possibilities mentioned above can actually happen, and we have either a one-to-one transformation of the set of all $[r-1]$ onto the set of all $[r-1]'$ or a one-to-one transformation of the set of all $[r-1]$ onto the set of all $[r+1]'$. It is easily seen that the latter case will lead to a correlation between the base space S_n and the base space S'_n. Hence, in case $r = (n-1)/2$, the given transformation must be either one induced by a collineation between S_n and S'_n or one induced by a correlation between S_n and S'_n. Thus (by setting $S_n = S'_n$) we have proved the following fundamental theorem for the Grassman space:

THEOREM I. *Any one-to-one adjacence preserving transformation of the Grassmann space of all the $[r]$ of S_n $(n-1 > r > 0)$ onto itself is a transformation of the basic group of the space.*

2. The Space of an Involutoric Correlation

Let there be given an involutoric correlation Δ in the base space S_n, which we shall call the basic correlation. To each element $[s]$ in S_n there corresponds an $[n-s-1]$ under the basic correlation Δ; we shall call this $[n-s-1]$ the conjugate of the $[s]$ and shall denote it by $\overline{[s]}$. An element $[s]$ is called invariant (with respect to Δ), if it is incident with its conjugate $\overline{[s]}$. It is evident that the highest possible dimension r of an invariant element which is contained in its conjugate is $r = (n-1)/2$ for odd n, $r = (n-2)/2$ for even n. In the first case, an invariant $[r]$ is obviously the conjugate of itself; while in the second case, the conjugate $\overline{[r]}$ is an $[r+1]$ which (as we shall see later) contains no other invariant points than those of $[r]$. For the sake of convenience, we introduce a number $\bar{r}$ defined as follows: $\bar{r} = r$ for odd n, $\bar{r} = r+1$ for even n. Thus the conjugate $\overline{[r]}$ of an $[r]$ has in any case the dimension $\bar{r}$. Since Δ is a correlation, we have evidently $\overline{[s]\cap[t]} = \overline{[s]} \cup \overline{[t]}$ and $\overline{[s]\cup[t]} = \overline{[s]} \cap \overline{[t]}$, for any two $[s]$, $[t]$ in S_n. It is also evident that any element contained in an invariant $[s]$ is also invariant, provided $s \leqq r$; and similarly, any element containing an invariant

[11] See Bertini [1], p. 52.

$[s]$ is also invariant, provided $s \geqq \bar{r}$. For every element $[s]$, the basic correlation Δ induces a correlation $[s] \to \{\overline{[s]}\}$ between the space $[s]$ (as a point space) and the bundle $\{\overline{[s]}\}$ with the center $\overline{[s]}$. In view of this basic duality, it is usually sufficient to confine our arguments to the elements [s] of dimension $s \leqq r$.

We assume now that the basic correlation Δ has the following properties: (a) *there exist at least two disjoint invariant* $[r]$ *in* S_n (rational over the ground field K), (b) *not all invariant* $[r]$ *lie in a proper subspace of* S_n. Then the set of all the invariant $[r]$ in S_n is called the *space of the involutoric correlation* Δ. The number r is called the *order* of the space, and *we shall always assume that* $r > 0$. The first assumption (a) is fundamental for our arguments, for otherwise we would not have sufficient elements in the space for our constructions; but the second assumption (b) is made purely as a matter of convenience. In fact, for odd n, the assumption (b) follows obviously from (a), for already the two disjoint invariant $[r]$ cannot lie in a subspace of S_n. On the other hand, in case of even n, if all the invariant $[r]$ lie in a subspace $[n-1]$, then they are also invariant with respect to an involutoric correlation Δ_1 in $[n-1]$; so that the given space can also be considered as the space of the involutoric correlation Δ_1 in the base space $S_{n-1} = [n-1]$, and the problem is reduced to the previous case.[12] The involutoric correlation Δ_1 in $[n-1]$ can be defined by the following correspondence: $[s] \to \overline{[s]} \cap [n-1]$ for any element $[s]$ in $[n-1]$; it is not difficult to see that an element in $[n-1]$ is invariant with respect to Δ_1 if and only if it is invariant with respect to Δ.

It is clear that any collineation of S_n onto itself which leaves the basic correlation Δ invariant, will carry an invariant element into an invariant element; hence, in particular it induces an incidence preserving transformation of the space of the correlation Δ onto itself. The group of all these transformations will be called the basic group of the space,[13] and we intend to characterize this group of transformations by the incidence properties. It is evident that both the space of a null system and the space of a polar system (in the normal form stated in Chapter I, §1) are special cases of such a space. We shall generalize our original problem by proving the fundamental theorem for the space of any involutoric correlation Δ (under the assumptions (a) and (b)), though we shall see in course of our proof that the case of a null system is much simpler than the other cases. On the other hand, the space of a polar system has some special properties, which demand some further investigations and lead to a refinement of the fundamental theorem for this space.

It will be shown later (§4, Remark (1)) that, as a consequence of our assump-

---

[12] That the assumption (b) is not superfluous in case of an even $n = 2r + 2$, is shown by the following example. Over a ground field of characteristic 2, all the invariant points of the polar system defined by the equation $\sum_{i=0}^{r} (y_{r+1+i}x_i + y_i x_{r+1+i}) + y_{2r+2}x_{2r+2} = 0$ in S_n are contained in the hyperplane $x_{2r+2} = 0$.

[13] It is easily seen that, in case of an odd $n = 2r + 1$, those correlations which leave the basic correlation Δ invariant induce the same set of transformations in the space as the collineations.

tions (a) and (b), the base space S_n contains a fundamental set of $n + 2$ invariant points. A set of $n + 2$ points in S_n is called *fundamental*, if no $n + 1$ points of the set lie on an $[n - 1]$ in S_n. From this it follows that if two collineations of S_n onto itself induce the same transformation of the invariant elements, then they must be identical. In fact, they can only differ by an automorphism of the ground field; but since they must be concordant on an invariant $[r]$ (note that $r > 0$), they must involve the same automorphism of the ground field. Thus there is an isomorphism between the group of collineations in S_n which leave the correlation Δ invariant and the basic group of the space of the correlation Δ.

3. The Invariant Elements

We begin with a general method for the construction of invariant elements:

If $[s]$ and $[t]$ $(s, t \leqq r)$ are invariant elements and $\overline{[s]} \supset [t]$, then the joint $[s] \cup [t]$ is also an invariant element of dimension $\leqq r$.

In fact, from $\overline{[s]} \supset [t]$ it follows that also $\overline{[t]} \supset [s]$. Since $\overline{[s]} \supset [s]$ and $\overline{[t]} \supset [t]$, it follows that $\overline{[s]} \supset [s] \cup [t]$ and $\overline{[t]} \supset [s] \cup [t]$; and hence $\overline{[s] \cup [t]} = \overline{[s]} \cap \overline{[t]} \supset [s] \cup [t]$. Thus $[s] \cup [t]$ must be an invariant element of dimension $\leqq r$.

An immediate consequence of this is that $\overline{[r]}$ contains no other invariant point than those contained in $[r]$. For, if $[0]$ were such a point, then the element $[0] \cup [r] = \overline{[r]}$ would then be an invariant element of dimension $\leqq r$, which is a contradiction.

An important special case of the above construction is the following: Given any two invariant elements $[s]$ and $[t]$ $(s, t \leqq r)$, the element $[s] \cup (\overline{[s]} \cap [t])$ is always an invariant element of dimension $\leqq r$. In case $t = r$, the invariant element $[s] \cup (\overline{[s]} \cap [r])$ has always the dimension r. For, let $[u] = [s] \cap [r]$, then we have also $[s] \cap \overline{[r]} = [u]$, and consequently the element $\overline{[s]} \cup [r]$ has the dimension $n - u - 1$. This means that the intersection $\overline{[s]} \cap [r]$ has the dimension $(n - s - 1) + r - (n - u - 1) = r + u - s$. Since $[s] \cap (\overline{[s]} \cap [r]) = [s] \cap [r] = [u]$, we conclude that $[s] \cup (\overline{[s]} \cap [r])$ has the dimension $(r + u - s) + s - u = r$. Finally, it is easily seen that, if $[0]$ is any invariant point not in $\overline{[t]}$, then the invariant element $[0] \cup (\overline{[0]} \cap [t])$ has the dimension t.

As a further consequence of the above construction, we shall prove the following lemma:

Lemma 1. *If every two elements of a set of invariant $[s]$ $(s \leqq r)$ are $(s - 1)$-incident, then all the elements of the set must be incident with either one common invariant $[s + 1]$ or one common invariant $[s - 1]$. In case $s = r$, all the elements of the set must be incident with a common $[r - 1]$.*

Proof. By the elementary theorem of projective geometry mentioned before, all the elements of the set must be incident with either one common $[s + 1]$ or one common $[s - 1]$. In the latter case, the common $[s - 1]$ must be invariant, and the lemma is already proved. We have to prove that in the former case, the common $[s + 1]$ is also invariant. Let $[s]_1$, $[s]_2$ be two elements of the set, then we have $[s + 1] = [s]_1 \cup [s]_2$. By hypothesis, there exists at least one $[s]_3$ of the set which is not incident with $[s-1] = [s]_1 \cap [s]_2$; then we have also $[s+1] = [s]_3 \cup [s-1]$. Since $\overline{[s-1]} = \overline{[s]}_1 \cup \overline{[s]}_2 \supset [s]_1 \cup [s]_2 = [s+1] \supset [s]_3$, it follows that $[s+1]$ is an invariant element of dimension $\leqq r$. In case $s = r$, the last condition is clearly impossible; hence all the elements of the set must be incident with one common $[r-1]$.

We shall now investigate the consequences of our assumption (a) that there exist at least two disjoint invariant $[r]_a$, $[r]_b$ in S_n (we shall not need the assumption (b) until the end of the next section). First, it is clear that any invariant $[s]$ ($s \leqq r$) is contained in at least one invariant $[r]$; we can, in fact, take $[r] = [s] \cup (\overline{[s]} \cap [r]_a)$. Furthermore, given any invariant $[r]_1$, there always exists an invariant $[r]_2$ disjoint to $[r]_1$. For, let $[t]_a = [r]_a \cap [r]_1$; we can assume that $t \geqq 0$, as otherwise we can evidently take $[r]_2 = [r]_a$. Then we have $\overline{[t]}_a \cup [r]_b \supset \overline{[r]}_a \cup [r]_b = S_n$, and consequently the invariant element $\overline{[t]}_a \cap [r]_b$ has the dimension $(n - t - 1) + r - n = r - t - 1$. Hence there exists a subspace $[t]_b$ of $[r]_b$, which is disjoint to $\overline{[t]}_a$. We take $[r]_2 = [t]_b \cup (\overline{[t]}_b \cap [r]_a)$. Since the join $[r]_1 \cup [r]_2$ contains the elements $[r]_1$, $[r]_a = [t]_a \cup (\overline{[t]}_b \cap [r]_a)$, $[t]_b$, and since $[t]_b$ and $[r]_a \cup [r]_1$ are disjoint (the latter is contained in $\overline{[t]}_a$) and have the dimensions t and $2r - t$ respectively, we conclude that the join $[r]_1 \cup [r]_2$ must have the dimension $2r + 1$; hence $[r]_1$ and $[r]_2$ are disjoint.

After these remarks we can prove the following lemmas:

Lemma 2. *If two invariant* $[s]_1$, $[s]_2$ ($s \leqq r$) *have an* $[u]$ *as their intersection, then there exist two invariant* $[r]_1$, $[r]_2$ *containing* $[s]_1$, $[s]_2$ *respectively and having also* $[u]$ *as their intersection.*

Proof. Consider the invariant element $[t] = [s]_1 \cup (\overline{[s]}_1 \cap [s]_2)$. Let $[r]_a$ be an invariant element containing $[t]$ and let $[r]_b$ be an invariant element disjoint to $[r]_a$. We take $[r]_1 = [s]_1 \cup (\overline{[s]}_1 \cap [r]_b)$, which obviously contains $[s]_1$. Since $[r]_1 \cap [s]_2 \subset \overline{[s]}_1 \cap [s]_2 \subset [r]_a$, and since $[r]_1 \cap [r]_a = [s]_1$, it follows that $[r]_1 \cap [s]_2 = [r]_a \cap [r]_1 \cap [s]_2 = [s]_1 \cap [s]_2 = [u]$. Now, consider the invariant element $[r]_c = [s]_2 \cup (\overline{[s]}_2 \cap [r]_1)$, and let $[r]_d$ be an invariant element disjoint to $[r]_c$. We take $[r]_2 = [s]_2 \cup (\overline{[s]}_2 \cap [r]_d)$, which obviously contains $[s]_2$. Since $[r]_1 \cap [r]_2 \subset [r]_1 \cap \overline{[s]}_2 \subset [r]_c$ and since $[r]_2 \cap [r]_c = [s]_2$, it follows that $[r]_1 \cap [r]_2 = [r]_1 \cap [r]_2 \cap [r]_c = [r]_1 \cap [s]_2 = [u]$. Thus the lemma is proved.

Remark. In Lemma 2, if the $[u]$ is contained in an $[2r+1]$ which contains two disjoint invariant $[r]$, then the invariant elements $[r]_1$ and $[r]_2$ can also be taken in this $[2r+1]$.

Lemma 3. *If two invariant* $[s]_a$, $[s]_b$ ($s \leqq r$) *have an* $[s-u]$ *as their intersection, then there exists a sequence of* $u + 1$ *invariant elements* $[s]_1 = [s]_a, [s]_2, \cdots, [s]_{u+1} = [s]_b$, *all incident with the* $[s-u]$, *such that each two consecutive elements are* $(s-1)$-*incident.*

Proof. Assuming that the $[s]_1 = [s]_a, [s]_2, \cdots, [s]_i$ have already been constructed, such that $[s]_i$ intersects $[s]_b$ in an $[s-u+i-1]_i$. Let $[0]_i$ be a point of $[s]_b$ not on $[s-u+i-1]_i$, then the invariant element $[s]_{i+1} = [0]_i \cup (\overline{[0]}_i \cap [s]_i)$ is $(s-1)$-incident with $[s]_i$ and intersects $[s]_b$ in an $[s-u+i]_{i+1} = [0]_i \cup [s-u+i-1]_i$. The lemma follows after u steps.

As we are interested in the invariant elements only, we shall remodel our definition of the adjacence and distance in such a way that they will be expressed in terms of the invariant elements alone. This leads to a certain refinement of our concepts, for the two previously equivalent definitions of the adjacence are now no longer equivalent. Two invariant elements $[s]_a$ and $[s]_b$ $(s \leqq r)$ are said to be *adjacent*, if their intersection is an invariant $[s-1]$; and they are said to have the *distance* u, if u is the least integer with the property that there exists a sequence of $u+1$ invariant elements $[s]_1 = [s]_a, [s]_2, \cdots, [s]_{u+1} = [s]_b$, each two consecutive ones of which are adjacent. On the other hand, two invariant elements $[s]_a$ and $[s]_b$ $(s \leqq r)$ are said to be *co-adjacent*, if their join is an invariant $[s+1]$; and they are said to have the *co-distance* u, if u is the least integer with the property that there exists a sequence of $u+1$ invariant elements $[s]_1 = [s]_a, [s]_2, \cdots, [s]_{u+1} = [s]_b$, each two consecutive ones of which are co-adjacent. For the invariant elements of dimension $\geqq \bar{r}$, the definitions are to be reversed: Two invariant elements $[s]_a$ and $[s]_b$ $(s \geqq \bar{r})$ are said to be *adjacent*, if their join is an invariant $[s+1]$; while they are said to be *co-adjacent*, if their intersection is an invariant $[s-1]$. The distance and co-distance are then defined in the same way as above. It is clear from the Lemma 3 (and its dual for invariant elements of dimension $\geqq \bar{r}$) that the new definitions of adjacence and distance are concordant with the ones previously given, when applied to the invariant elements. The co-adjacence and co-distance are essentially new notions, dependent on the basic correlation Δ; however, they play only an auxiliary role in our arguments. It is clear that a pair of co-adjacent invariant elements are also adjacent, so that co-adjacence is a stronger condition than adjacence. From this it follows that the co-distance of any two invariant elements is greater than or equal to their distance.

4. The Space of a Null System

We proceed now with the proof of the fundamental theorem. Let Δ be an involutoric correlation in S_n and Δ' be an involutoric correlation in another base space S'_n, both correlations satisfying the conditions (a), (b) of the previous section. Let Γ be a one-to-one transformation of the space of Δ onto the space of Δ', such that any pair of adjacent invariant $[r]$ in S_n (with respect to Δ) is carried into a pair of adjacent invariant $[r]'$ in S'_n (with respect to Δ') and vice versa; in other words, the transformation Γ preserves the $(r-1)$-incidence. Then all the s-incidence (for $s = -1, 0, 1, \cdots, r-1$) are also preserved. For, according to Lemma 3, the s-incidence of any two invariant $[r]$ is equivalent to the fact that they can be connected by a sequence of $r - s + 1$ consecutively adjacent invariant $[r]$, and this property is evidently invariant under the transformation Γ. Consider now the maximal set of all the invariant $[r]$ incident with one common (invariant) $[r-1]$. The set of transformed $[r]'$ under Γ must be also a maximal set of mutually adjacent elements in S'_n; hence, by Lemma 1, they must all be incident with one common invariant $[r-1]'$. If we let this $[r-1]'$ correspond to the original $[r-1]$, we shall then have a one-to-one transformation of the set of all invariant $[r-1]$ in S_n onto the set of all invariant $[r-1]'$ in S'_n, which together with the given transformation carries every pair of incident invariant $[r]$, $[r-1]$ into a pair of incident invariant $[r]'$, $[r-1]'$ and vice versa. Thus we shall consider the given transformation Γ to have been extended one dimension downward to the invariant elements of dimension $r-1$. We maintain that the (so extended) transformation Γ carries also every pair of s-incident invariant $[r-1]$ into a pair of s-incident invariant $[r-1]'$ (for $s=-1, 0, 1, \cdots, r-2$), and vice versa. In fact, let $[r-1]_1$, $[r-1]_2$ be a pair of s-incident invariant elements and let $[r-1]'_1$, $[r-1]'_2$ be their transformed elements respectively under Γ. According to Lemma 2, there exist two invariant $[r]_1$, $[r]_2$ containing $[r-1]_1$, $[r-1]_2$ respectively, which are also s-incident. Let $[r]'_1$, $[r]'_2$ be the transformed elements of $[r]_1$, $[r]_2$ respectively under Γ, then we have also $[r]'_1 \supset [r-1]'_1$ and $[r]'_2 \supset [r-1]'_2$. Since the transformation Γ preserves the s-incidence between any two invariant $[r]$, it follows that $[r]'_1$ and $[r]'_2$ must also be s-incident, and hence the elements $[r-1]'_1$, $[r-1]'_2$ are at most s-incident. Applying the same argument to the inverse transformation Γ^{-1} shows that the elements $[r-1]'_1$, $[r-1]'_2$ must be at least s-incident; therefore they are s-incident, which proves our assertion.

Consider now the maximal set of all the invariant $[r-1]$ incident with one common (invariant) $[r-2]$. The set of transformed elements $[r-1]'$ under Γ must also be a maximal set of mutually adjacent elements in S'_n; and since they obviously cannot all be incident with one common invariant $[r]'$, they must all be, according to Lemma 1, incident with one common invariant $[r-2]'$. If we let this $[r-2]'$ correspond to the original $[r-2]$, we shall then have a one-to-one transformation of the set of all invariant $[r-2]$ in S_n onto the set of all invariant $[r-2]'$ in S'_n, which together with the transformation Γ carries every pair of incident invariant $[r-1]$, $[r-2]$ into a pair of incident invariant $[r-1]'$, $[r-2]'$, and vice versa. Thus we can extend our transformation Γ one dimension further downward to the invariant elements of dimension $r-2$. In similar manner as before, we can also prove that the so extended transformation Γ also carries a pair of s-incident invariant $[r-2]$ into a pair of s-incident invariant $[r-2]'$ (for $s=-1, 0, \cdots, r-3$), and vice versa.

Thus we can proceed in this manner until we have extended the transformation Γ down to including the invariant points. Finally, the transformation Γ can be extended to include all the invariant elements of dimension $\geqq \bar{r}$ by the stipulation that the transformed element of an invariant $[t]$ $(t \geqq \bar{r})$ is the element $[t]'$, if $\overline{[t]}$ is carried by Γ into $\overline{[t]}'$. It is then clear from our construction that the so extended transformation Γ not only preserves all the incidence relations among the invariant elements, but also carries the basic correlations Δ and Δ' into each other.

We notice that in case both correlations Δ and Δ' are null systems, we have already reached our goal. For, in a null system all points and all hyperplanes are invariant elements; hence the extended transformation Γ contains in particular also a one-to-one incidence preserving transformation of the points and hyperplanes in S_n onto the points and hyperplanes in S'_n respectively. As we have mentioned before, such a transformation defines a collineation of S_n onto S'_n ; and it is easily seen that this collineation carries also every other invariant element into its transformed element under Γ, and hence carries the correlation Δ into the correlation Δ'. In case $S_n = S'_n$ and $\Delta = \Delta'$, this means that Γ is a transformation of the basic group. Thus we have proved the following fundamental theorem for the space of a null system:

THEOREM II. *Any one-to-one adjacence preserving transformation of the space of a null system onto itself is a transformation of the basis group of the space.*

Before we proceed further with the proof of the fundamental theorem in the general case, we shall make a few remarks. We note that this is the only place where the assumption (b) is used.

(1) *There exists a fundamental set of $n + 2$ invariant points in S_n.* Let $[r]_1$ and $[r]_2$ be two disjoint invariant elements, and set $[2r + 1]_0 = [r]_1 \cup [r]_2$. Let $[0]_a$ be any point on $[r]_1$, and $[0]_b$ be any point of $\overline{[0]}_a \cap [r]_2$; then the line $[1] = [0]_a \cup [0]_b$ is invariant. Since a line contains at least three points, there is an invariant point $[0]_0$ on $[1]$ outside of both $[r]_1$ and $[r]_2$. Let $[0]_a, [0]_1, \cdots, [0]_{r+1}$ be a fundamental set of points on $[r]_1$, and let $[0]_b, [0]_{r+2}, \cdots, [0]_{2r+2}$ be a fundamental set of points on $[r]_2$. Then the points $[0]_0, [0]_1, \cdots, [0]_{2r+2}$ form a fundamental set of invariant points on $[2r + 1]_0$. In case of an odd $n = 2r + 1$, we have $S_n = [2r + 1]_0$ and the assertion is already proved. In case of an even $n = 2r + 2$, let $[0]_{2r+3}$ be an invariant point outside of $[2r + 1]_0$. We can obviously choose the points $[0]_i$ $(i = 1, \cdots, 2r + 2)$ in such a way that one of them, say the point $[0]_1$ lies on the hyperplane $\overline{[0]}_{2r+3}$; then the line $[1] = [0]_1 \cup [0]_{2r+3}$ is invariant and there is an invariant point $[0]$ on it different from $[0]_1$ and $[0]_{2r+3}$. It is then easily seen that the set $[0]_0, [0], [0]_2, \cdots, [0]_{2r+3}$ is a fundamental set of invariant points in S_n.

(2) *Every invariant hyperplane $[n - 1]$ is the join of all the invariant points contained in it.* In other words, every invariant $[n - 1]$ is determined by its invariant points. Let $[r]_1$ and $[r]_2$ be two disjoint invariant elements, and set $[2r + 1]_0 = [r]_1 \cup [r]_2$. Since the invariant $[n - 1]$ cannot contain both $[r]_1$ and $[r]_2$ (for then both $\overline{[r]}_1$ and $\overline{[r]}_2$, and hence also both $[r]_1$ and $[r]_2$ must contain the invariant point $[n - 1]$), we can assume that one of them, say $[r]_1$ does not lie in $[n - 1]$. If $[n - 1]$ contains $[r]_2$, then the element $([n - 1] \cap [r]_1) \cup [r]_2$ has the dimension $2r$ and we have therefore

$$[n - 1] \cap [2r + 1]_0 = ([n - 1] \cap [r]_1) \cup [r]_2 .$$

If $[n - 1]$ does not contain $[r]_2$, then the intersection $[n - 1] \cap [r]_2$ has the dimension $r - 1$. Let $[0]_1$ be a point in $[r]_1$ which lies outside of $[n - 1]$ and is different from the intersection point of $[r]_1$ with $\overline{[n - 1] \cap [r]_2}$. Since $\overline{[0]}_1$ does not contain $[n - 1] \cap [r]_2$, it follows that $\overline{[0]}_1 \cap [r]_2$ does not contain $[n - 1] \cap [r]_2$; and since both $\overline{[0]}_1 \cap [r]_2$ and $[n - 1] \cap [r]_2$ have the same dimension $r - 1$, this means that $[n - 1] \cap [r]_2$ also does not contain $\overline{[0]}_1 \cap [r]_2$. Therefore there exists a point $[0]_2$ in $\overline{[0]}_1 \cap [r]_2$ which lies outside of $[n - 1]$. The line $[1] =$

$[0]_1 \cup [0]_2$ is evidently invariant and intersects $[n-1]$ in one point $[0] = [1] \cap [n-1]$. This point $[0]$ does not lie in the $(2r-1)$-dimensional element $([n-1] \cap [r]_1) \cup ([n-1] \cap [r]_2)$; for, the line $[1]$ is the only line through $[0]$ which intersects both $[r]_1$ and $[r]_2$, and this line $[1]$ is not in $[n-1]$, much less in $([n-1] \cap [r]_1) \cup ([n-1] \cap [r]_2)$. It follows that the element $([n-1] \cap [r]_1) \cup ([n-1] \cap [r]_2) \cup ([n-1] \cap [1])$ has the dimension $2r$ and hence we have $[n-1] \cap [2r+1]_0 = ([n-1] \cap [r]_1) \cup ([n-1] \cap [r]_2) \cup ([n-1] \cap [1])$. Thus we have shown that $[n-1] \cap [2r+1]_0$ is in any case the join of all the invariant points contained in it. In case of an odd $n = 2r+1$, we have $[2r+1]_0 = S_n$ and our assertion is already proved. In case of an even $n = 2r+2$, let $[n-1]_1$ be an invariant hyperplane such that the point $\overline{[n-1]}_1$ is not in $[2r+1]_0$. Since, as we have just shown, the element $[n-1]_1 \cap [2r+1]_0$ is the join of all its invariant points, and since $[n-1]_1$ is the join of $[n-1]_1 \cap [2r+1]_0$ and the invariant point $\overline{[n-1]}_1$, it follows that $[n-1]_1$ is also the join of all its invariant points. Now, if $[n-1]$ contains an invariant point $[0]$ of $[n-1]_1$ outside of $[2r+1]_0$, then we have $[n-1] = ([n-1] \cap [2r+1]_0) \cup [0]$, and our assertion is proved. On the other hand, if all invariant points of $[n-1] \cap [n-1]_1$ are contained in $[n-1]_1 \cap [2r+1]_0$, then we must have $[n-1] \cap [n-1]_1 = [n-1]_1 \cap [2r+1]_0$; for, if all invariant points of $[n-1] \cap [n-1]_1$ are contained in an $[s]$ with $s < n-2$, then all invariant points of $[n-1]_1$ must be contained in $\overline{[n-1]}_1 \cup [s]$, which is an element of dimension $s+1 < n-1$; and this is in contradiction to what we have just shown above. We have then evidently also $[n-1] \cap [2r+1]_0 = [n-1] \cap [n-1]_1$. It follows then that the three points $\overline{[2r+1]}_0$, $\overline{[n-1]}_1$ and $\overline{[n-1]}$ are collinear. Since $\overline{[n-1]}_1$ is in $[n-1]_1$ and $\overline{[2r+1]}_0$ is not in $[n-1]_1$ (for $\overline{[n-1]}_1$ is not in $[2r+1]_0$), therefore $\overline{[n-1]}$ is not in $[n-1]_1$, hence also not in $[n-1] \cap [n-1]_1 = [n-1] \cap [2r+1]_0$, hence also not in $[2r+1]_0$. It follows that $[n-1] = ([n-1] \cap [2r+1]_0) \cup \overline{[n-1]}$, which proves our assertion. By duality, we can deduce that *every invariant point is the intersection of all the invariant hyperplane containing it.*

(3) In case the original transformation Γ is one induced by a collineation Θ of S_n onto S'_n, then the same collineation must also induce the entire extended transformation Γ and carry the correlation Δ into the correlation Δ'. It is clear that the collineation Θ must carry every invariant element of dimension $\leqq r$ into the transformed element under Γ, for our method of extending the transformation by means of the adjacence as given in this section is purely projective in nature. For the invariant elements of dimension $\geqq \bar{r}$, we note that since an invariant $[n-1]$ is the join of all the invariant points contained in it, the collineation Θ must also carry every invariant $[n-1]$ into its transformed $[n-1]'$ under Γ. Since all invariant elements of dimension $\geqq \bar{r}$ can be obtained from the invariant $[n-1]$ by purely projective constructions, it follows then that the collineation Θ also carries every such invariant element into the transformed elements under Γ. Since the transformation Γ carries the correlation Δ into the correlation Δ', or in other words, the transformations Γ and $\Delta^{-1}\Gamma\Delta'$ are

the same, we conclude that the collineations Θ and $\Delta^{-1}\Theta\Delta'$ induce the same transformation Γ of the invariant elements. Since these two collineations must carry a fundamental set of invariant points in S_n into the same fundamental set in S'_n, and since they must also be concordant on an invariant $[r]$ (which requires that they must involve the same automorphism of the ground field), it follows that they must be identical. In case $S_n = S'_n$ and $\Delta = \Delta'$, this means that the collineation Θ must leave the correlation Δ invariant.

5. The Fundamental Theorem in the General Case

We continue now with the proof of the fundamental theorem. According to Remark (3) in the last section, we only need to construct a collineation of S_n onto S'_n which carries every invariant $[r]$ into the transformed element $[r]'$ under Γ. From now on we shall assume that $r > 1$; as the fundamental theorem is not generally true for $r = 1$, this assumption is not superfluous.

Consider an invariant $[r]$ and its transformed $[r]'$ under Γ. Since every $[t]$ in $[r]$ is transformed into an $[t]'$ in $[r]'$ and vice versa, and since this transformation preserves incidence, it follows (note that $r > 1$) that Γ induces a collineation of $[r]$ onto $[r]'$, which we shall denote by $[r] \to [r]'$. Since every invariant $[t]$ $(t \leqq r)$ can be imbedded in an invariant $[r]$, it follows that there is also an induced collineation $[t] \to [t]'$ of $[t]$ onto $[t]'$. It is evident that this induced collineation $[t] \to [t]'$ has the property that for any invariant $[s]$ we have $[s] \cap [t] \to [s]' \cap [t]'$. For every two adjacent invariant $[r]_a$, $[r]_b$ the collineations $[r]_a \to [r]'_a$ and $[r]_b \to [r]'_b$ must involve the same automorphism σ of the ground field K, for it is the same automorphism as that of the collineation $[r]_a \cap [r]_b \to [r]'_a \cap [r]'_b$. Since, according to Lemma 3, every two invariant $[r]$ can be connected by a sequence of consecutively adjacent invariant $[r]$, and since every invariant $[t]$ can be imbedded in an invariant $[r]$, it follows that all the induced collineations $[t] \to [t]'$ $(t \leqq r)$ must involve the same automorphism σ of the ground field.[14] In the following we shall use the well known theorem that two collineations of two subspaces (onto two other subspaces respectively), which are concordant on their intersection, can be extended in a unique way to a collineation of their join. This theorem is usually proved for projective transformations[15] (i.e. collineations with identity automorphism), but it obviously holds also when all the collineations concerned have the same automorphism. Since we have shown above that all collineations $[t] \to [t]'$ involve the same automorphism, the application of this theorem in our case is justified.

---

[14] Dually, for every invariant t $(t \geqq \bar{r})$ there is also an induced collineation $\{[t]\} \to \{[t]'\}$ of the bundle $\{[t]\}$ with the center $[t]$ onto the bundle $\{[t]'\}$ with the center $[t]'$; and we have also $\{[s] \cup [t]\} \to \{[s]' \cup [t]'\}$ for any invariant $[s]$. It can be similarly shown that all the collineations $\{[t]\} \to \{[t]'\}$ $(t \geqq \bar{r})$ must also involve the same automorphism $\bar{\sigma}$ of the ground field. From the correspondences $\{[t]\} \xrightarrow{\Delta} \overline{[t]} \xrightarrow{\Gamma} \overline{[t]}' \xrightarrow{\Delta} \{[t]'\}$, it follows that $\bar{\sigma} = \delta^{-1}\sigma\delta$, where δ is the (involutoric) automorphism associated with the basic correlation Δ. However, we shall not need these facts.

[15] See Bertini [1], p. 60, where the theorem is proved in the dual form for the bundles.

Consider now any two disjoint invariant elements $[r]_1$ and $[r]_2$. Let $[0]_1$ be any point on $[r]_1$, and let $[0]_2$ be any point on $[0]_1 \cap [r]_2$, then the line $[1] = [0]_1 \cup [0]_2$ is an invariant line which intersects both $[r]_1$ and $[r]_2$. Let $[r]_1'$, $[r]_2'$, $[1]'$ be the transformed elements of $[r]_1$, $[r]_2$, $[1]$ respectively under the transformation Γ, and set $[2r + 1]_0 = [r]_1 \cup [r]_2$ and $[2r + 1]_0' = [r]_1' \cup [r]_2'$. Then there is a collineation Φ of $[2r + 1]_0$ onto $[2r + 1]_0'$ which is concordant with the collineations $[r]_1 \to [r]_1'$, $[r]_2 \to [r]_2'$, $[1] \to [1]'$ induced by Γ. In fact, since the collineations $[r]_1 \to [r]_1'$ and $[1] \to [1]'$ are concordant with each other on their intersection $[0]_1 \to [0]_1'$., there is a collineation Φ_1 of $[r]_1 \cup [1]$ onto $[r]_1' \cup [1]'$ which is concordant with both. Similarly, there is a collineation Φ_2 of $[r]_2 \cup [1]$ onto $[r]_2' \cup [1]'$ which is concordant with the collineations $[r]_2 \to [r]_2'$ and $[1] \to [1]'$. Since the two collineations Φ_1 and Φ_2 are concordant on their intersection $[1] \to [1]'$, there is a collineation Φ of $[2r + 1]_0$ onto $[2r + 1]_0'$ which is concordant with both.[16]

The collineation Φ constructed above is independent of the choice of the invariant line $[1]$. In other words, if $[1]_0$ is another invariant line which intersects both $[r]_1$ and $[r]_2$, and if $[1]_0'$ is the transformed element of $[1]_0$ under Γ, then the collineation Φ is also concordant with the collineation $[1]_0 \to [1]_0'$ induced by Γ. To prove this, let $[0]_a = [r]_1 \cap [1]_0$ and $[0]_b = [r]_2 \cap [1]_0$., and let $[0]_a'$ and $[0]_b'$ be the transformed points of $[0]_a$ and $[0]_b$ respectively under Γ. Let $[0]_c$ be any point in $[0]_2 \cap [0]_b \cap [r]_1$ (such a point exists since we have $r > 1$), and let $[0]_c'$ be its transformed point under Γ. Consider the invariant plane $[2] = [0]_1 \cup [0]_2 \cup [0]_c$; it is clear that the plane $[2]' = [0]_1' \cup [0]_2' \cup [0]_c'$ is the transformed element of $[2]$ under Γ. Since the collineation Φ is concordant with the collineations $[0]_1 \cup [0]_2 \to [0]_1' \cup [0]_2'$ and $[0]_1 \cup [0]_c \to [0]_1' \cup [0]_c'$, it follows that it is also concordant with the collineation $[2] \to [2]'$ induced by Γ, hence also concordant with the collineation $[0]_2 \cup [0]_c \to [0]_2' \cup [0]_c'$ induced by Γ. Applying the same argument successively to the invariant planes $[0]_2 \cup [0]_c \cup [0]_b$ and $[0]_c \cup [0]_b \cup [0]_a$, we conclude that the collineation Φ is concordant with the collineation $[1]_0 \to [1]_0'$.

Let $[n - 1]$ be any invariant hyperplane in S_n., and let $[n - 1]'$ be the transformed element in S_n' under Γ. We maintain that the collineation carries the element $[n - 1] \cap [2r + 1]_0$ into the element $[n - 1]' \cap [2r + 1]_0'$. To prove this we consider first the case where $[n - 1]$ contains one of the two elements $[r]_1$ and $[r]_2$, say $[r]_1$; then the element $[n - 1]'$ contains also $[r]_1'$. We have in this case $[n - 1] \cap [2r + 1]_0 = [r]_1 \cup ([n - 1] \cap [r]_2)$ and $[n - 1]' \cap [2r + 1]_0' = [r]_1' \cup ([n - 1]' \cap [r]_2')$. Since the collineation Φ carries the elements $[r]_1$ and

[16] The collineation Φ can also be constructed easily as follows. Let $X_0 = [0]_1$, X_1, $\cdots$, X_{r+1} be a fundamental set of points of $[r]_1$ (i.e. a set of $r + 2$ points such that no $r + 1$ of them lie on an $[r - 1]$), and $Y_0 = [0]_2$, Y_1, $\cdots$, Y_{r+1} be a fundamental set of points of $[r]_2$, and $[0]$ be any point on $[1]$ different from $[0]_1$ and $[0]_2$. Let X_i', Y_i' and $[0]'$ be the corresponding points under Γ. Then the $n + 2 = 2r + 3$ pairs of points $[0] \to [0]'$, $X_i \to X_i'$, $Y_i \to Y_i'$ $(i = 1, \cdots, r + 1)$, together with the automorphism σ of the ground field, determine uniquely the collineation Φ.

$[n-1] \cap [r]_2$ into the elements $[r]_1'$ and $[n-1]' \cap [r]_2'$ respectively, it must also carry $[n-1] \cap [2r+1]_0$ into $[n-1]' \cap [2r+1]_0'$. In case $[n-1]$ contains neither $[r]_1$ nor $[r]_2$, we shall choose the invariant line $[1]$ so that its two intersection points $[0]_1$ and $[0]_2$ with $[r]_1$ and $[r]_2$ respectively are outside of the hyperplane $[n-1]$. As we have shown in the proof of Remark (2) in the preceding section, this can always be done and we have in this case $[n-1] \cap [2r+1]_0 = ([n-1] \cap [r]_1) \cup ([n-1] \cap [r]_2) \cup ([n-1] \cap [1])$. Since the element $[1]'$ evidently also satisfies the same condition with respect to $[r]_1'$, $[r]_2'$ and $[n-1]'$, we have also $[n-1]' \cap [2r+1]_0' = ([n-1]' \cap [r]_1') \cup ([n-1]' \cap [r]_2') \cup ([n-1]' \cap [1]')$. Since the collineation Φ carries the element $[n-1] \cap [r]_1$, $[n-1] \cap [r]_2$ and $[n-1] \cap [1]$ into the elements $[n-1]' \cap [r]_1'$, $[n-1]' \cap [r]_2'$ and $[n-1]' \cap [1]'$ respectively, it must also carry $[n-1] \cap [2r+1]_0$ into $[n-1]' \cap [2r+1]_0'$.[16a]

Now, let $[0]$ be an invariant point in $[2r+1]_0$, and let $[0]'$ be the transformed point under Γ. Let $[n-1]_i$ be the set of all invariant hyperplanes containing the point $[0]$ (or a suitably chosen finite number of them), then according to §4, Remark (2), the point $[0]$ is the intersection of all the $[n-1]_i$, hence also the intersection of all the elements $[n-1]_i \cap [2r+1]_0$. Let $[n-1]_i'$ be the transformed elements of the $[n-1]_i$ under Γ; as the transformation Γ preserves the incidence, it follows that the point $[0]'$ is also the intersection of all the $[n-1]_i'$. Since the collineation Φ carries each element $[n-1]_i \cap [2r+1]_0$ into the element $[n-1]_i' \cap [2r+1]_0'$, therefore it must also carry the point $[0]$ into a point in the intersection of all the elements $[n-1]_i' \cap [2r+1]_0'$, which can only be the point $[0]'$. This shows in particular that the point $[0]'$ lies in $[2r+1]_0'$. Since every invariant $[s]$ $(s \leqq r)$ is the join of all the invariant points contained in it, we conclude that for every invariant $[s]$ $(s \leqq r)$ in $[2r+1]_0$, the transformed element $[s]'$ under Γ lies also in $[2r+1]_0'$, and that the collineation Φ is concordant with the collineation $[s] \to [s]'$. More generally, it follows easily from this that for any invariant $[s]$ $(s \leqq r)$ and its transformed $[s]'$ under Γ, the element $[s] \cap [2r+1]_0$ is transformed under Γ into the element $[s]' \cap [2r+1]_0'$, and the collineation Φ is concordant with the collineation $[s] \cap [2r+1]_0 \to [s]' \cap [2r+1]_0'$.

In case of an odd $n = 2r+1$, we have $S_n = [2r+1]_0$ and $S_n' = [2r+1]_0'$, and the assertion we have just proved contains already the fundamental theorem. In case of an even $n = 2r+2$, we can proceed as follows. Consider an invariant element $[r]_0$ not contained in $[2r+1]_0$, then the transformed element $[r]_0'$ under Γ is also not contained in $[2r+1]_0'$. Then, since the collineations Φ and $[r]_0 \to [r]_0'$ are concordant on their intersection $[r]_0 \cap [2r+1]_0 \to [r]_0' \cap [2r+1]_0'$, there is a collineation Θ of S_n onto S_n' which is concordant with both. We maintain that this collineation Θ is independent of the choice of the element $[r]_0$. In fact, let $[r]_\alpha$ be any invariant element adjacent to $[r]_0$, such that their intersection $[r-1]_0$ is not contained in $[2r+1]_0$, and let $[r]_\alpha'$ and $[r-1]_0'$ be the transformed

[16a] [This paragraph and the one above it have been revised in proof (Nov. 24, 1948); they constitute a somewhat simplified version of our original argument.]

elements of $[r]_\alpha$ and $[r-1]_0$ respectively under Γ. Then the collineation of S_n onto S_n' determined by the two concordant collineations Φ and $[r]_\alpha \to [r]_\alpha'$ must be identical with the collineation Θ, for they are both uniquely determined already by the two concordant collineations Φ and $[r-1]_0 \to [r-1]_0'$ (note that on account of our assumption $r > 1$, we have $r - 1 > 0$). On the other hand, any two invariant $[r]_a$ and $[r]_b$ not contained in $[2r+1]_0$ can be connected by a sequence of invariant $[r]$, each two consecutive elements of which are not only adjacent but also have at least one common point outside of $[2r+1]_0$. This assertion follows immediately from Lemma 3, if the $[r]_a$ and $[r]_b$ already have a common point outside of $[2r+1]_0$. Otherwise let $[0]_a$ be a point of $[r]_a$ and $[0]_b$ be a point of $[r]_b$, both outside of $[2r+1]_0$, and consider two disjoint invariant $[r]_c$, $[r]_d$ containing the points $[0]_a$, $[0]_b$ respectively. Since $\overline{[r]_d \cap [2r+1]_0}$ cannot contain $[r]_c$, there is a point $[0]_c$ of $[r]_c$ outside of $[2r+1]_0$, such that $\overline{[0]_c} \cap [r]_d$ is not contained in $[2r+1]_0$; hence, $[r]_e = [0]_c \cup (\overline{[0]_c} \cap [r]_d)$ has a point in common with $[r]_d$ outside of $[2r+1]_0$. The sequence $[r]_a$, $[r]_c$, $[r]_e$, $[r]_d$, $[r]_b$ has the property that every two consecutive elements have a common point outside of $[2r+1]_0$; and our assertion follows then from Lemma 3.

The fact that the collineation Θ is independent of the choice of $[r]_0$ shows that it carries every invariant $[r]_0$ not entirely contained in $[2r+1]_0$ into the transformed element $[r]_0'$ under Γ. Since the collineation Θ also carries every invariant $[r]$ in $[2r+1]_0$ into the transformed $[r]'$ under Γ, it therefore carries every invariant $[r]$ into its transformed $[r]'$ under Γ. Thus everything is proved.

If we set $S_n = S_n'$ and $\Delta = \Delta'$, we obtain the following fundamental theorem for the space of any involutoric correlation:

THEOREM III. *Any one-to-one adjacence preserving transformation of the space of an involutoric correlation onto itself is a transformation of the basic group of the space, provided the order r of the space is greater than* 1.

From this theorem we can deduce almost immediately a fundamental theorem for the invariant points (under the same assumptions (a) and (b)), which will be useful to us later:

THEOREM IV. *Any one-to-one transformation of the set of invariant points of an involutoric correlation onto itself, which carries a pair of co-adjacent points into another such pair, is a collineation of S_n (leaving the correlation invariant), provided* $n > 4$.

In fact, an invariant $[r]$ can be characterized as a maximal set of mutually co-adjacent invariant points; and the adjacence of two (distinct) invariant $[r]$ can be characterized by the fact that every pair of points in their join has a co-distance $\leqq 2$. Then the given one-to-one co-adjacence preserving transformation of the invariant points will induce a one-to-one adjacence preserving transformation of the invariant $[r]$, and Theorem IV follows then from Theorem III. On the other hand, the construction of the extended transformation given in the preceding section shows that, conversely, Theorem III can also be deduced from Theorem IV, so that the two theorems are in a sense equivalent.

6. Applications. The Space of a Polar System and the Hyperquadric

Theorem III includes of course Theorem II as a special case (except for $r = 1$), where the basic correlation is a null system. When the basic correlation is taken to be a polar system, we have not only a fundamental theorem for the space of a polar system as defined in Chapter I, §1 for odd $n = 2r + 1$, but also a similar theorem for the case of even $n = 2r + 2$. Besides these two well-known cases of involutoric correlations, there is a third one which may be called a "Hermitian" system, in case the ground field admits an involutoric automorphism.[17] Such a correlation is given by a "Hermitian" or "skew-Hermitian" bilinear form $\sum_{i,j=0}^{n} h_{ij}x_i\delta(y_j)$, where δ is an involutoric automorphism of the ground field. For example, if we take the form $\sum_{i=0}^{r} (\delta(y_{r+1+i})x_i - \delta(y_i)x_{r+1+i})$ $(n = 2r + 1)$, we obtain the so-called space of "Hermitian" matrices; and Theorem III yields in this case a generalization of a theorem proved by Hua[18] under the following restrictive conditions: (a) The ground field is the field of complex numbers, (b) the transformation is continuous, (c) the invariance of all distances, (d) the invariance of "harmonic separation", and (e) the invariance of "signature". Thus our results show that all these conditions are unnecessary except a part of (c), namely the invariance of the distance 1.

Theorem IV contains a fundamental theorem for the hyperquadric Q_{n-1} of S_n, when the basic correlation is taken to be a polar system. As already mentioned in Chapter I, §1, the assumption of the existence of two disjoint invariant $[r]$ of the polar system is equivalent to the assumption of the normal forms of Q_{n-1}. In case of a "Hermitian" system, we have here a new result, which may be called the fundamental theorem of the "Hermitian" hyperquadric. This theorem can be regarded as an extension into higher dimensions of a theorem proved by E. Cartan[19] for the "hypersphere", which is the case $n = 2$ of the "Hermitian" hyperquadric for the field of complex numbers. Our theorem, however, does not include Cartan's result, for the case $n = 2$ (hence $r = 0$) is excluded in Theorem IV. There remains the gap $n = 3, 4$, which it is probably not very difficult to close. However, as this case of "Hermitian" system lies somewhat aside from our main subject, we shall abstain from going into further details.

We add some further remarks about the case of a polar system. Though both Theorems III and IV are restricted to $n > 4$, it is quite easy to see that they also hold for $n = 4$ in case of a polar system. In fact, it is well known[20] that any hyperquadric Q_3 (in the normal form) can be regarded as the associated algebraic variety of the space of a null system of order 1, and our assertion for Theorem IV

---

[17] See Brauer [3], p. 253.

[18] Hua [8], p. 480.

[19] Cartan [5], p. 161.

[20] From the representation of the Grassmann space by the hyperquadric $p_{12}p_{34} + p_{13}p_{24} + p_{14}p_{23} = 0$ (see footnote 22), we get the space of null system by setting $p_{13} = p_{24}$. Let $p_{12} = z_0$, $p_{14} = z_1$, $p_{34} = z_2$, $p_{23} = z_3$, $p_{13} = p_{24} = z_4$, we get the hyperquadric $z_0z_1 + z_2z_3 + z_4^2 = 0$ in normal form.

follows then from Theorem IX in Chapter III, which is derived from Theorem II and Lemma 5. It might be not without interest to remark here that this Q_3 represents also the Lie geometry of circles, when the ground field is taken to be the field of real numbers, and our result in this case is exactly what is usually known as the fundamental theorem of this geometry.[21] (It is on account of this that the hyperquadric is sometimes designated as the geometry of complex spheres.) Since, as we have shown in §5, the Theorems III and IV are equivalent, it follows that Theorem III also holds for $n = 4$ in case of a polar system. Similarly, from the well known fact[22] that any hyperquadric Q_4 (in normal form) can be regarded as the associated algebraic variety of the Grassmann space of lines in S_3, we can also deduce from Theorem VIII in Chapter III the validity of Theorems III and IV for $n = 5$ in case of a polar system. On the basis of these two cases of lowest possible dimensions, for which the theorems still hold, it is possible then to give a simple inductive proof for Theorems III and IV in case of a polar system. However, since we have already given the proof of the general case, we shall not burden this paper with such details which are only of secondary interest.

For later reference, we shall state here again our results for the special cases of the space of a polar system (for odd $n = 2r + 1$) and the hyperquadric.

THEOREM V. *Any one-to-one adjacence preserving transformation of the space of a polar system onto itself is a transformation of the basic group, provided the order r of the space is greater than* 1.

THEOREM VI. *Any one-to-one transformation of the hyperquadric Q_{n-1} onto itself, which carries a pair of points on a line on Q_{n-1} into another such pair, is a transformation of its basic group, provided $n > 3$.*

7. The Irreducible Space of a Polar System

Theorem V for the space of a polar system is a natural form of the fundamental theorem from the projective point of view. However, from the point of view of algebraic geometry, some further refinement is necessary. The reason lies in the well known fact, already mentioned in Chapter I, §1, that all the invariant $[r]$ of a polar system in S_n (for odd $n = 2r + 1$) do not form one single algebraic system, but are divided into two disjoint systems (which are projectively equivalent). Hence, if we want our homogeneous space to be represented by an irreducible algebraic variety, we have to restrict ourselves to one of the two systems. In other words, we must study the irreducible space of a polar system, and characterize its basic group of transformations. Now the condition of $(r - 1)$-incidence in Theorem V involved necessarily elements of different systems, since only elements of different systems can have $(r - 1)$-

[21] See Blaschke [2], p. 211.

[22] The Grassmann space of lines in S_3 is represented through the Pluecker coordinates $p_{ij} = -p_{ij}$ $(i, j = 1, 2, 3, 4)$ by the hyperquadric Q_4 : $p_{12}p_{34} + p_{13}p_{24} + p_{14}p_{23} = 0$. Setting $p_{12} = z_0$, $p_{13} = z_1$, $p_{14} = z_2$, $p_{34} = z_3$, $p_{24} = z_4$, $p_{23} = z_5$, we have the normal form $z_0z_3 + z_1z_4 + z_2z_5 = 0$.

incidence; hence this theorem cannot be regarded as a satisfactory projective characterization for the irreducible space of a polar system. For, one would naturally require that the transformations of the basic group of the irreducible space be characterized by incidence conditions involving only the elements of the space itself. We shall achieve this aim by showing in the following that the preservation of $(r-2)$-incidence, which involves only the elements of one system of invariant $[r]$ alone, is sufficient to characterize the basic group of the irreducible space of a polar system, provided $r > 3$.

We shall henceforth restrict ourselves to one of the two systems of invariant $[r]$ of a polar system, and shall denote by $[r]^*$ any invariant element of the other system which comes into consideration during our argument. Two invariant $[r]$ (of the one system) are now called "adjacent", if they are $(r-2)$-incident, and the "distance" of two such invariant $[r]$ is then defined in the same way in terms of this concept of "adjacence". The method is to refer the present case back to Theorem V; the arguments are similar to those of the preceding cases, though somewhat more complicated, mainly on account of the fact that the set of all invariant $[r]$ incident with a common $[r-2]$ is now not a maximal set of mutually "adjacent" invariant elements. Instead of the Lemma 1, we have here the following lemma concerning sets of mutually "adjacent" invariant $[r]$:

LEMMA 4. *If every two elements of a set of invariant* $[r]$ $(r > 2)$ *are* $(r-2)$-*incident, then all the elements of the set must be either* $(r-1)$-*incident with a common invariant* $[r]^*$ *or incident with a common invariant* $[r-3]$.

PROOF. We can assume that not all elements of the set are incident with a common $[r-2]$, for otherwise they would obviously all be incident with any $[r-3]$ contained in it and there would be nothing more to prove. Let $[r]_1$, $[r]_2$, $[r]_3$ be three elements of the set, which do not have a common $[r-2]$, and let $[r-2]_1 = [r]_2 \cap [r]_3$, $[r-2]_2 = [r]_3 \cap [r]_1$, $[r-2]_3 = [r]_1 \cap [r]_2$. Then the intersection $[r]_1 \cap [r]_2 \cap [r]_3$ must have the dimension $r-3$. For, if $[r-2]_1 \cap [r-2]_2 = [r]_1 \cap [r]_2 \cap [r]_3$ has the dimension $r-4$ (which is obviously the only other possibility), then $[r-2]_1 \cup [r-2]_2$ must have the dimension r and hence coincide with $[r]_3$. But then we would have $[r]_3 = [r-2]_1 \cup [r-2]_2 \subset [r]_1 \cup [r]_2$, and hence $[r]_3 \supset [r]_1 \cap [r]_2$, which is a contradiction. We therefore set $[r-3] = [r]_1 \cap [r]_2 \cap [r]_3$. Since the intersection of the two $(r-1)$-dimension elements $[r-2]_1 \cup [r-2]_3$ and $[r-2]_2 \cup [r-2]_3$ is exactly $[r-2]_3$, their join, which is the element $[r-2]_1 \cup [r-2]_2 \cup [r-2]_3$, must have the dimension r. This latter element is invariant, since $\overline{[r-2]_1 \cup [r-2]_2} = ([r]_2 \cup [r]_3) \cap ([r]_1 \cup [r]_3) \supset [r]_1 \cap [r]_2 = [r-2]_3$; and as its intersection with $[r]_3$ is the $(r-1)$-dimensional element $[r-2]_1 \cup [r-2]_2$, it belongs to the other system. We therefore set $[r]^* = [r-2]_1 \cup [r-2]_2 \cup [r-2]_3$. We shall now show that, if any one element of the set does not contain the $[r-3]$, then all the elements of the set must be $(r-1)$-incident with the $[r]^*$. For proof, assume that $[r]_4$ is an element of the set which does not contain the $[r-3]$, and set $[r-3]_i = [r-2]_i \cap [r]_4$ $(i = 1, 2, 3)$. Applying the same argument as above to the $[r-3]_i$ instead of the $[r-2]_i$, we deduce that the join $[r-3]_1 \cup [r-3]_2 \cup [r-3]_3$ has the dimension $r-1$.

Since $[r]_4 \cap [r]^* = [r-3]_1 \cup [r-3]_2 \cup [r-3]_3$, this means that $[r]_4$ is $(r-1)$-incident with $[r]^*$. Thus every element of the set must either contain the $[r-3]$ or be $(r-1)$-incident with the $[r]^*$. Now, applying the same argument to any three of the four elements $[r]_i$ $(i = 1, 2, 3, 4)$ and any other element $[r]_5$ of the set, we conclude that $[r]_5$ must either be $(r-1)$-incident with the $[r]^*$ or contain the common intersection of any three of the four elements $[r]_i$ $(i = 1, 2, 3, 4)$. But the latter case is impossible, for $[r]_5$ would then contain $[r-3] \cup [r-3]_1 \cup [r-3]_2 \cup [r-3]_3 = [r-3] \cup ([r]_4 \cap [r]^*) = [r]^*$. Thus the lemma is proved.

It follows immediately from the above lemma that a maximal set of mutually "adjacent" invariant $[r]$ must be either the set of all invariant $[r]$ having an $(r-1)$-incidence with a common $[r]^*$, or the set of all invariant $[r]$ containing a common invariant $[r-3]$. We shall say that the former is a maximal set pertaining to $[r]^*$ and the latter is a maximal set pertaining to $[r-3]$. Furthermore, the above proof has also established the following result, which we shall make use of presently: Let $[r]^* \supset [r-3]$, and consider the elements common to the maximal set pertaining to $[r]^*$ and the maximal set pertaining to $[r-3]$. *Then any maximal set of mutually "adjacent" invariant* $[r]$ *which contains all these elements, must be one of these two maximal sets themselves.* In fact, let $[r-1]_1$, $[r-1]_2$, $[r-1]_3$ be three elements in $[r]^*$ such that $[r-1]_1 \cap [r-1]_2 \cap [r-1]_3 = [r-3]$; and let $[r]_1$, $[r]_2$, $[r]_3$ be the invariant elements which intersect $[r]^*$ in the elements $[r-1]_1$, $[r-1]_2$, $[r-1]_3$ respectively. The given maximal set must then contain $[r]_1$, $[r]_2$, $[r]_3$, and the latter part of the above proof shows that all elements of the set must either contain $[r-3]$ or be $(r-1)$-incident with $[r]^*$.

Suppose now we have a one-to-one transformation of the irreducible space of a polar system (in the base space S_n) onto another irreducible space of a polar system (in the base space S'_n), and suppose that this transformation carries every pair of "adjacent" invariant $[r]$ into a pair of "adjacent" invariant $[r]'$. It follows then that the transformation must also carry every maximal set of mutually "adjacent" invariant $[r]$ into a maximal set of mutually "adjacent" invariant $[r]$, and vice versa. Since, according to Lemma 4 (for $r > 2$), a maximal set of mutually "adjacent" invariant $[r]$ is either a maximal set pertaining to an $[r]^*$ or a maximal set pertaining to an invariant $[r-3]$, and since the same assertion evidently must also hold for a maximal set of mutually "adjacent" invariant $[r]'$, it follows that the given transformation induces a one-to-one transformation of the set of all invariant $[r-3]$ and $[r]^*$ onto the set of all invariant $[r-3]'$ and $[r]'^*$, such that if an invariant $[r]$ is incident or $(r-1)$-incident with an element of the former set, then the transformed $[r]'$ is incident or $(r-1)$-incident with an element of the latter set. We maintain now that if any one $[r]^*$ is transformed into an $[r]'^*$ (or an $[r-3]'$), then the same must be true for all $[r]^*$. Again it is only necessary to prove this for two "adjacent" $[r]^*$, i.e. those having an $[r-2]$ as intersection, since any two $[r]^*$ can be joined by a sequence of consecutively "adjacent" elements (Lemma 3, where we delete all the $[r]_i$ with even subscripts). Now, the maximal set pertaining to two "adjacent" $[r]^*$ have as common elements the set of all invariant $[r]$ containing the intersection $[r-2]$ of these two $[r]^*$, and this latter set is contained in every maximal set pertaining to an $[r]^*$ incident with $[r-2]$. Since there exist[23] at least three different $[r]^*$ incident with any given invariant $[r-2]$, there is at least one other maximal set of mutually "adjacent" invariant $[r]$ which contains the common elements of the two given maximal sets. On the other hand, two maximal sets pertaining to an $[r]'^*$ and an invariant $[r-3]'$ respectively, if they have more than one element in common (and consequently $[r]'^*$ must contain $[r-3]'$), will have the property mentioned in the remark after Lemma 4 that no other maximal set can contain all the common elements of the two sets. Thus it is impossible for two maximal sets pertaining to an $[r]'^*$ and an $[r-3]'$ respectively to be the transformed sets of two maximal sets pertaining to two $[r]^*$. Similarly, it can be shown that if any one invariant $[r-3]$ is transformed into an $[r]'^*$ (or an $[r-3]'$), then the same must be true for all invariant $[r-3]$. The argument is exactly the same, starting with two co-adjacent invariant $[r-3]$. It follows then from these results that there is either a one-to-one transformation of the set of all $[r]^*$ onto the set of all $[r]'^*$ (and the set of all invariant $[r-3]$ onto the set of all invariant $[r-3]'$), or a one-to-one transformation of the set of all $[r]^*$ onto the set of all invariant $[r-3]'$ (and the set of all invariant $[r-3]$ onto the set of all $[r]'^*$); and it is clear from our construction that the former transformation carries a pair of "adjacent" $[r]^*$ into a pair of "adjacent" $[r]'^*$ (and a pair of co-adjacent invariant $[r-3]$ into a pair of co-adjacent invariant $[r-3]'$), and the latter transformation carries a pair of "adjacent" $[r]^*$ into a pair of co-adjacent invariant $[r-3]'$ (and a pair of co-adjacent invariant $[r-3]$ into a pair of "adjacent" $[r]'^*$). In the former case, we would have extended our original transformation to a one-to-one transformation of the entire space of a polar system (in S_n) onto another entire space of a polar system (in S'_n), and it is clear that this extended transformation preserves the adjacence. It follows then from Theorem V that the transformation must be one induced by a collineation of S_n onto S'_n which carries the one polar system into the other. On the other hand, the latter case is only possible for $r \leqq 3$. For the "distance" between any two $[r]^*$ cannot be more than $(r+1)/2$, half of their (ordinary) distance, while there exists (as we shall see presently) a pair of invariant $[r-3]'$ which have a co-distance $\geqq r-1$. The inequality $r-1 \leqq (r+1)/2$ shows that $r \leqq 3$.

To complete our proof, we calculate the co-distance of two invariant $[r-3]_1$, $[r-3]_2$ with the property that $\overline{[r-3]}_1$ and $[r-3]_2$ are disjoint. In a shortest possible sequence of co-adjacent invariant $[r-3]$ connecting the $[r-3]_1$ and $[r-3]_2$, let $[r-3]_a$ and $[r-3]_b$ be the elements co-adjacent to $[r-3]_1$ and

[23] This can be seen as follows. Let $[r]$ be an invariant element (of the one system) which contains the $[r-2]$, and let $[1]$ be a line in $[r]$ disjoint to the $[r-2]$. Let $[0]_1$, $[0]_2$, $[0]_3$ be three points on $[1]$, and let $[r-1]_1 = [0]_1 \cup [r-2]$, $[r-1]_2 = [0]_2 \cup [r-2]$, $[r-1]_3 = [0]_3 \cup [r-2]$. Then the three $[r]_1^*$, $[r]_2^*$, $[r]_3^*$ which intersect the $[r]$ in the elements $[r-1]_1$, $[r-1]_2$, $[r-1]_3$ respectively, all contain the $[r-2]$.

$[r-3]_2$ respectively. Then $[r-3]_a$ and $[r-3]_b$ can intersect at most in a point. For, if the intersection $[r-3]_a \cap [r-3]_b$ is an $[u]$ $(u > 0)$, then this $[u]$ must intersect both $[r-3]_1$ and $[r-3]_2$; and since $[u] \subset [r-3]_a \subset \overline{[r-3]_1}$, therefore $\overline{[r-3]_1}$ must also intersect $[r-3]_2$, which is in contradiction to our assumption. It follows then that the distance of $[r-3]_a$ and $[r-3]_b$ is at least $r-3$, and consequently their co-distance is also at least $r-3$. This means that the co-distance of $[r-3]_1$ and $[r-3]_2$ is not less than $r-3+2=r-1$.

Thus (by setting $S_n = S'_n$) we have proved the following fundamental theorem for the irreducible space of a polar system:

THEOREM VII. *Any one-to-one "adjacence" preserving transformation of the irreducible space of a polar system onto itself is a transformation of the basic group of the space, provided the order r of the space is greater than* 3.

The theorem evidently cannot hold (at least in this form) for $r = 1, 2$, for in these cases any two elements of the space are "adjacent", and consequently the hypothesis of the theorem become meaningless. The case $r = 3$ is a real exception and deserves some attention. In fact, Lemma 4 still holds for this case, and so do all the other arguments of this section except the very last step. Here a real difference sets in, for it is actually possible to construct, in case $r = 3$, a one-to-one transformation between the $[3]^*$ and the invariant points which carries a pair of "adjacent" $[3]^*$ into a pair of co-adjacent invariant points and conversely, so that we have a sort of "correlation" between certain invariant elements of the polar system. It is easily seen that every such "correlation" of one irreducible system of invariant [3] will induce an "adjacence" preserving transformation in the other irreducible system, which is not a transformation of the basic group. With the help of Lemma 4 and the other arguments of this section, it is not difficult to determine all such "correlations". In fact, it is sufficient to construct one such "correlation", for our arguments show that all other such "correlations" can be obtained by composing this one "correlation" with the transformations of the basic group. However, as this is somewhat aside from our main interest, we refrain from going into details.

CHAPTER III

BIRATIONAL CHARACTERIZATION OF THE BASIC GROUP

1. Representation as Algebraic Varieties

Except in the case of hyperquadrics, our homogeneous spaces are up to now only defined as certain sets of linear subspaces $[r]$ in the projective space S_n which satisfy certain conditions. As such they are, strictly speaking, not yet represented as algebraic varieties. In order to be able to define birational transformations without ambiguity, it is necessary to represent these spaces in an algebraical manner as algebraic varieties. This means that we have to introduce for the $[r]$ a set of coordinates, which are univalent algebraic functions of the points X^i $(i = 0, 1, \cdots, r)$ determining the $[r]$, in such a way that the set of all $[r]$ constituting any one of the homogeneous spaces are in one-to-one correspondence with the points of a certain algebraic variety in a suitably chosen projective space. This variety, which we can regard as the homogeneous algebraic variety associated with the homogeneous space, is of course not uniquely determined, but any two such varieties representing the same homogeneous space are equivalent under a one-to-one algebraic transformation. For our purpose, the most convenient coordinates to introduce are the well known Pluecker coordinates of the linear subspace $[r]$, which are the set of all determinants of $(r + 1)^{\text{th}}$ order derived from the $(r + 1, n + 1)$-matrix (x_j^i) $(i = 0, 1, \cdots, r; j = 0, 1, \cdots, n)$ the $r + 1$ rows of which are the coordinates of the $r + 1$ points X^i determining the $[r]$ in S_n. The Grassmann space is then represented by what is usually known as the Grassmann variety $G_{n,r}$ of dimension $d = (r + 1)(n - r)$ in the projective space S_m of dimension $m = \binom{n+1}{r+1} - 1$. The two spaces of correlations can be represented as subvarieties of the Grassmann variety $G_{n,r}$ (where now $n = 2r + 1$) in the following manner. An involutoric correlation in S_n induces a one-to-one correspondence of the $[r]$ in S_n, which can be represented as an involutoric collineation in the space S_m leaving the Grassman variety $G_{n,r}$ invariant. The set of the invariant $[r]$ of the correlation in S_n is then represented by the intersection variety of one of the double spaces of the involutoric collineation in S_m with the Grassman variety $G_{n,r}$. Since both null system and polar system do not involve any automorphism of the ground field, the same is true of the corresponding collineations in S_m, and hence their double spaces are linear subspaces of S_m. The space of a null system, being irreducible, is evidently the intersection of one such double space with $G_{n,r}$, the other double space of the collineation having no intersection with $G_{n,r}$. We shall denote this variety by N_r, and its dimension is easily found to be $(r + 1)(r + 2)/2$. The space of a polar system consists of two projectively equivalent varieties, each one being the intersection of $G_{n,r}$ with one double space of the corresponding collineation in S_m. We shall denote any one of the two irreducible varieties by P_r, and its dimension is easily found to be $r(r + 1)/2$. In the following we have often occasion to treat all the cases together; we shall then denote generally any one of the four classes of homogeneous algebraic varieties $G_{n,r}$, N_r, P_r, Q_{n-1} (Q_{n-1} being the hyperquadric in S_n) by M, its dimension by d, and its ambient space by S_k.

It is easily seen that with the above chosen representation of each of the homogeneous spaces as algebraic variety M in S_k, the basic group of the homogeneous space can be represented as a group of collineations of S_k leaving the variety M invariant. This follows from the well known fact that any collineation of S_n induces a collineation in S_m, and hence also a collineation in any linear subspace which is invariant with respect to it. We shall call this group of collineations the basic group of the variety M. We shall soon see that this group contains in fact every collineation of S_k leaving the variety M invariant.

2. Two Fundamental Lemmas

We begin with two fundamental lemmas, which connect the incidence properties of the $[r]$ in S_n with certain properties of the corresponding points on the Grassmann variety $G_{n,r}$ in S_m.

LEMMA 5. *Two $[r]$ of S_n are $(r - 1)$-incident if and only if the corresponding points of $G_{n,r}$ lie on a line on $G_{n,r}$.*

PROOF. The necessity of the condition is evident, so we need only to show that if two $[r]$ of S_n are not $(r - 1)$-incident, then the corresponding points on $G_{n,r}$ cannot lie on a line on $G_{n,r}$. Now, the set of all $[r]$ of S_n incident with a given $[s - 1]$ is represented in S_m by a subvariety of $G_{n,r}$ which is projectively equivalent to the variety $G_{n-s,r-s}$. Hence it is only necessary to prove that two points of $G_{n,r}$ corresponding to two disjoint $[r]$ of S_n $(r < n/2)$ cannot lie on a line on $G_{n,r}$, except when $r = 0$. Now, all pairs of disjoint $[r]$ in S_n are projectively equivalent; consequently, the corresponding pairs of points on $G_{n,r}$ are all equivalent under the basic group of $G_{n,r}$. Hence, if one such pair of points lie on a line on $G_{n,r}$, every such pair must also lie on a line on $G_{n,r}$; and in particular, a generic pair of points of $G_{n,r}$ (which evidently corresponds to a pair of disjoint $[r]$ in S_n) must then also lie on a line on $G_{n,r}$. Since the condition of lying on a line on $G_{n,r}$ is an algebraic condition, it follows that every pair of points of $G_{n,r}$ must lie on a line on $G_{n,r}$. But this means that $G_{n,r}$ is a linear space, and this is impossible except when $r = 0$.

We note here the following immediate consequence of Lemma 5: The arithmetic distance between two $[r]$ in S_n is equal to the least integer s such that there is a sequence of s consecutively intersecting lines on $G_{n,r}$ which connects the two corresponding points.

LEMMA 6. *Two $[r]$ of S_n are at least $(r - 2)$-incident if and only if the corresponding points of $G_{n,r}$ lie on a conic section on $G_{n,r}$.*

PROOF. By virtue of the remark made in the proof of the preceding lemma, we need only to show that two points of $G_{n,r}$ corresponding to two disjoint $[r]$ of S_n $(r < n/2)$ can lie on a conic section on $G_{n,r}$ if and only if $r < 2$. The condition is sufficient. For, in case $r = 1$ (there is nothing to prove in case $r = 0$), every two disjoint lines in S_n will determine an S_3, and the set of all lines of S_3 is represented on $G_{n,1}$ by a subvariety which is projectively equivalent to $G_{3,1}$, which is a hyperquadric Q_4 of S_5. That two points on a Q_4 of S_5 lie on a conic section on Q_4 (hence on $G_{n,1}$) is quite evident. To prove that the condition is necessary, consider two points Z_1, Z_2 of $G_{n,r}$ corresponding to two disjoint $[r]_1$, $[r]_2$ in S_n. Let C be a conic section on $G_{n,r}$ containing the two points Z_1, Z_2, and let Z_0 be a point on C such that the corresponding $[r]_0$ in S_n is disjoint to $[r]_1$ and $[r]_2$. Such a point evidently exists (though not necessarily a point in the ground field K if K is not algebraically closed), since a generic point of C has this property. Consider a pencil of $[r]$ containing $[r]_0$; the generic element $[r]_u$ of this pencil must of course be disjoint to both $[r]_1$ and $[r]_2$. We assume that $[r]_u \cap ([r]_1 \cup [r]_2) = [r]_0 \cap ([r]_1 \cup [r]_2)$ in case $[r]_0$ is not entirely contained in

$[r]_1 \cup [r]_2$; in the latter case we assume that $[r]_u$ is also entirely contained in $[r]_1 \cup [r]_2$. (It can be easily seen that such a pencil exists.) Then there is a projective transformation in S_n which carries $[r]_1$, $[r]_2$, $[r]_0$ into $[r]_1$, $[r]_2$, $[r]_u$ respectively. (We shall soon prove this assertion.) Now, the pencil of $[r]$ in S_n corresponds to a line L on $G_{n,r}$ through the point Z_0, and the generic element $[r]_u$ of the pencil corresponds to a generic point Z_u of the line L. The above mentioned projective transformation in S_n will then correspond to a projective transformation Θ_u in S_m which leaves $G_{n,r}$ invariant and carries the points Z_1, Z_2, Z_0 into the points Z_1, Z_2, Z_u respectively; hence Θ_u also carries the conic section C into a conic section $\Theta_u(C)$ on $G_{n,r}$. Let Z_v be a generic point of C, then the generic point $\Theta_u(Z_v)$ (of transcendence degree 2 over K) will determine a surface in the S_3 determined by the conic section C and the line L. This surface must be a quadric Q_2, since its intersection with a generic plane of a pencil is a conic section $\Theta_u(C)$. Hence Z_1 and Z_2, as two points on a Q_2 on $G_{n,r}$, can be connected by a sequence of at most two intersecting lines on Q_2, hence on $G_{n,r}$. But, according to the remark made after Lemma 5, this means that the corresponding $[r]_1$, $[r]_2$ in S_n have a distance at most equal to 2; since $[r]_1$, $[r]_2$ are disjoint, this means that $r + 1 \leqq 2$, or $r \leqq 1$.

To complete the proof, we shall show that the above mentioned projective transformation in S_n exists. If $[r]_0$ is not entirely contained in $[r]_1 \cup [r]_2$ and s is the dimension of their intersection, let $[r - s - 1]_0$ be a subspace of $[r]_0$ disjoint to $[r]_1 \cup [r]_2$, and $[r - s - 1]_u$ be a similar subspace of $[r]_u$. Then any projective transformation which leaves every point of $[r]_1 \cup [r]_2$ invariant and carries $[r - s - 1]_0$ into $[r - s - 1]_u$ will serve the purpose. In case $[r]_0$ (and hence also $[r]_u$) is entirely contained in $[r]_1 \cup [r]_2$, $[r]_0$ defines a perspectivity Φ_0 between $[r]_1$ and $[r]_2$, and $[r]_u$ also defines a perspectivity Φ_u between $[r]_1$ and $[r]_2$. Let $X_0, X_1, \cdots, X_{r+1}$ be a fundamental set of points of $[r]_1$, and let $Y_0^0, Y_1^0, \cdots, Y_{r+1}^0$ and $Y_0^u, Y_1^u, \cdots, Y_{r+1}^u$ be the corresponding sets on $[r]_2$ under the perspectivities Φ_0 and Φ_u respectively. Then the $n + 2 = 2r + 3$ pairs of points $X_i \to X_i$, $Y_i^0 \to Y_i^u$ $(i = 0, 1, \cdots, r + 1)$, and $[r]_0 \cap (X_0 \cup Y_0^0) \to [r]_u \cap (X_0 \cup Y_0^u)$ determine a projective transformation which leaves $[r]_1$, $[r]_2$ invariant and carries $[r]_0$ into $[r]_u$. Thus the lemma is completely proved.

The importance of the above two lemmas lies in the fact that they enable us to express the fundamental theorems of the homogeneous spaces in Chapter II as corresponding theorems of the homogeneous algebraic varieties associated with them, thus paving the way to the birational characterization. Combining Lemma 5 with Theorem I and Theorem II, and Lemma 6 with Theorem VII, we obtain the following results, which together with Theorem VI may be regarded as the fundamental theorems for the homogeneous algebraic varieties.

THEOREM VIII. *Any one-to-one transformation of the variety $G_{n,r}$ $(n - 1 > r > 0)$ onto itself, which carries any pair of points on a line on $G_{n,r}$ into another such pair, is a transformation of the basic group of $G_{n,r}$.*

THEOREM IX. *Any one-to-one transformation of the variety N_r $(r > 0)$ onto itself, which carries any pair of points on a line on N_r into another such pair, is a transformation of the basic group of N_r.*

THEOREM X. *Any one-to-one transformation of the variety P_r onto itself, which carries any pair of points on a conic section on P_r into another such pair, is a transformation of the basic group of P_r, provided $r > 3$.*

The above theorems hold for any ground field K in the sense that the one-to-one transformation need be defined only for those points of the variety which are in the ground field K, that is, with coordinates in the ground field K. It then follows as a consequence of our theorems that such a transformation, being a transformation of the basic group, can always be extended to all points of the variety.

Though it is contained in the main theorem of birational characterization proved later in §4, we observe here already as a consequence of the above theorems that any collineation of the ambient space S_k leaving the variety M invariant is a transformation of the basic group of M.[24] It is then easily seen that this establishes an isomorphism between the group of collineations of S_k leaving M invariant and the group of collineations (and also correlations in case of $G_{2r+1,r}$) of S_n which corresponds to the basic group of M. This is a generalization of the well known isomorphisms of the orthogonal groups[25] of orders 5 and 6, which are special cases of our results for the varieties N_1 and $G_{3,1}$ respectively.

3. Birational Characterization. Extension of Ground Field

Before proceeding to the problem of birational characterization, we shall make a few preliminary remarks. By its very nature the problem of birational characterization is independent of any extension of the ground field K. For a regular birational transformation remains evidently regular under any extension of the ground field, while on the other hand a transformation of the basic group, being a collineation in S_k, is certainly a transformation over the ground field K (i.e. with coefficients in K), if it carries every point of M with coordinates in K into another such point. Hence, if the main theorem of birational characterization holds for the algebraic closure of K, it holds also for K, so that we can without any loss of generality assume that the ground field K has already been extended to its algebraic closure. Since we are dealing here only with birational transformations, and since an automorphism of the field is not a birational transformation in the usual sense, it is clear that our characterization really concerns only the subgroup of the basic group consisting of all those transformations which do not involve any automorphism of K. In other words, in the problem of birational characterization, only the projective transformations of the basic group come into consideration.

On the other hand, the introduction of the alegbraic closure of K removes the necessity of taking the correlations and hyperquadrics in the special form

[24] Though the Theorems VI, VIII–X involve some restrictions on the n and r, the considerations in the next section show that this assertion holds in fact for all cases except possibly P_3.

[25] See van der Waerden [18], §7, p. 18.

specified in Chapter I, §1 in the definitions of the spaces (3) and (4), hence also in the definitions of the varieties P_r and Q_{n-1}. All that is necessary is that the variety M contains at least $k + 2$ points in the ground field K, no $k + 1$ of them lie in a $[k - 1]$ of S_k. For, by extending K to its algebraic closure, we can always transform any polar system or hyperquadric into any normal form by a projective transformation of S_n; while the existence of the above mentioned $k + 2$ points in K shows that any transformation of the basic group carrying them into points in K must be a transformation over K. In this connection, it is essential to bear in mind that our definition of a regular birational transformation demands that the transformation is one-to-one for *all* points of M (in the algebraic closure of K), not only for those points in K.

Though our proof of the main theorem in the next section is based on the Theorems VI, VIII–X, we should like to remark here that in fact we do not need these theorems in this strong form. The following weaker form, which is of course a consequence of these theorems, is quite sufficient: *Any one-to-one transformation of the variety M onto itself, which carries the lines (in case of $G_{n,r}$, N_r, Q_{n-1}) or conic sections (in case of P_r) on M into lines or conic sections (respectively), is a transformation of the basic group of M.* It is this weaker form which is really the exact analogy of the fundamental theorem of projective geometry, and includes it as a special case for $G_{n,0}$ $(n > 1)$. The stronger form has in fact no meaning for S_n, since every two points of S_n always lie on a line. In this weaker form, the theorem holds also for P_2, for this variety is "isomorphic" to an S_3. This can be seen as follows. The variety P_2 represents one of the two systems of planes on a Q_4 in S_5. It is well known[26] that the variety $G_{3,1}$ is also a Q_4 and that in this representation of the Grassmann space of lines in S_3, there is a one-to-one correspondence between the points of S_3 and one of the two systems of planes on Q_5 (and also between the planes of S_3 and the other system of planes). The resulting correspondence between P_2 and S_3 is such that a conic section of P_2 is mapped into a line of S_3 and vice versa. The theorem for P_2 then follows from the fundamental theorem for S_3. The case of P_3 is, however, a real exception, as can be easily seen from the considerations in Chapter II, §7.

For the hyperquadric Q_{n-1} we can have a still weaker form of the fundamental theorem, if we use both the lines and the conic sections on it. In this weak form the theorem can be easily proved independently (at least in case of an algebraically closed field). In fact, the fact that the lines on Q_{n-1} are carried into lines implies that the tangent hyperplanes of Q_{n-1} are carried into tangent hyperplanes, and hence the invariance under the given transformation of the polarity with respect to Q_{n-1}. Now, the [2] of S_n are in one-to-one correspondence with the conic sections (including the degenerate conic sections) on Q_{n-1}, and through the invariant polarity the $[n - 3]$ of S_n are also in one-to-one correspondence with the Q_{n-4} on Q_{n-1}. Hence the given transformation induces

[26] See van der Waerden [18], §7, p. 19.

a one to-one incidence preserving transformation of the [2] and $[n-3]$ of S_n, and, as mentioned before, such a transformation is a collineation. The proof fails for $n = 5$, but since the Q_4 is equivalent to $G_{3,1}$, the theorem for this case is already contained in Theorem VIII. Thus the theorem holds for all $n > 2$. Since this weak form is sufficient for the proof in the next section, we see that Theorem VI is really not necessary for the birational characterization of Q_{n-1}.

4. The Main Theorem of Birational Characterization

We begin by establishing a special birational transformation Ω between the variety M and the projective space S_d of d dimensions. In fact, this birational transformation Ω is a projection of the variety M from a suitably chosen S_{k-d-1} onto a suitably chosen S_d in S_k, though the fact that Ω is a projection is of no particular importance for us. The essential properties of this projection are: (1) It is one-to-one between M and S_d for all points outside of a certain hyperplane H in S_k, and (2) this hyperplane H intersects M in an *irreducible* variety W, possibly with a certain multiplicity s (in the sense of Bezout Theorem). For the hyperquadric Q_{n-1} (here we have $S_k = S_n$), the stereographic projection from any point on Q_{n-1} onto an S_{n-1} not incident with this point obviously has these two properties, provided $n > 3$. The hyperplane H is in this case the tangent hyperplane of the point of projection, the variety W is a degenerate Q_{n-2} (quadric cone), and $s = 1$. For the other three varieties $G_{n,r}$, N_r, P_r, we start by choosing the hyperplane H in S_k to be the one given by $z_0 = 0$, where z_0 is the Pluecker coordinate $|\, x_j^i \,|$ $(i, j = 0, 1, \cdots, r)$. It is then easily seen that any point of M not in H corresponds to an $[r]$ in S_n which can be defined by the $r + 1$ points X^i $(i = 0, 1, \cdots, r)$, given by the $r + 1$ rows of the $(r + 1, n + 1)$-matrix $(I, (u_j^i))$, where I is the identity matrix and (u_j^i) is any $(r + 1, n - r)$-matrix in case of $G_{n,r}$, or a symmetric $(r + 1)$-matrix in case of N_r, or a skew-symmetric $(r + 1)$-matrix in case of P_r. Conversely, it is obvious that any point in S_m corresponding to such an $[r]$ is a point of M outside of H. Now the d different (non-zero) u_j^i are themselves the values of the d Pluecker coordinates $z_{p,q} = |\, x_j^i \,|$, $(i = 0, 1, \cdots, r; j = 0, \cdots, p - 1, q, p + 1, \cdots, r)$, where $p = 0, 1, \cdots, r$ and $q = r + 1, \cdots, n$ in case of $G_{n,r}$; or $p = 0, 1, \cdots, r$ and $q = r + 1 + p, \cdots, n$ in case of N_r; or $p = 0, 1, \cdots, r$ and $q = r + 2 + p, \cdots, n$ in case of P_r. The other Pluecker coordinates of the $[r]$ (except z_0 which has the value 1) are simply the minors of higher orders of the matrix (u_j^i), hence polynomials of the u_j^i. We take S_d to be the subspace determined by the $d + 1$ Pluecker coordinates z_0, $z_{p,q}$, and S_{k-d-1} to be the subspace determined by the remaining $k - d$ coordinates of S_k. Then it is easily seen that the projection of M from S_{k-d-1} onto S_d is a birational transformation which is one-to-one for all points outside of H. In fact, through this projection a point of M is represented on S_d by (the d different non-zero elements of) the matrix (u_j^i), when the coordinate z_0 is taken to be 1; while, on the other hand, once the matrix (u_j^i) is given, all the other Pluecker coordinates of the corresponding point on M, being polynomials in the u_j^i, are uniquely determined.

It remains to prove that the intersection W of the hyperplane $z_0 = 0$ with the variety M is (up to a multiplicity) irreducible. Since any permutation (in case of $G_{n,r}$), or any product of permutations of the type (x_j, x_{r+1+j}) $(j = 0, 1, \cdots, r)$ in case of N_r, or any product of even number of permutations of the type (x_j, x_{r+1+j}) in case of P_r, of the coordinates $(x_0, x_1, \cdots, x_n)$ of the base space S_n will induce a transformation of the basic group of M, it follows that the coordinate z_0 can be carried over by such a transformation (which is a projective transformation in S_k) into any other Pluecker coordinate in case of $G_{n,r}$, or any other Pluecker coordinate derived from z_0 by substituting a number of columns $(\overset{0}{x}_j, \overset{1}{x}_j, \cdots, \overset{r}{x}_j)$ by $(\overset{0}{x}_{r+1+j}, \overset{1}{x}_{r+1+j}, \cdots, \overset{r}{x}_{r+1+j})$ in case of N_r, or any other Pluecker coordinate derived from z_0 by substituting an even number of columns $(\overset{0}{x}_j, \overset{1}{x}_j, \cdots, \overset{r}{x}_j)$ by $(\overset{0}{x}_{r+1+j}, \overset{1}{x}_{r+1+j}, \cdots, \overset{r}{x}_{r+1+j})$ in case of P_r. Hence we need only to prove our assertion for any one Pluecker coordinate z_v of the described type. Now, the intersection of any one such coordinate hyperplane $z_v = 0$ with M is, except for any possible component contained in the hyperplane $z_0 = 0$, mapped by the birational transformation Ω into a variety $|\overset{i}{u}_j|_v = 0$ in S_d, where $|\overset{i}{u}_j|_v$ is any minor of the rectangular matrix $(\overset{i}{u}_j)$ in case of $G_{n,r}$, or any principal minor of the symmetric matrix $(\overset{i}{u}_j)$ in case of N_r, or any principal minor of even order of the skew-symmetric matrix $(\overset{i}{u}_j)$ in case of P_r. Since the determinant of a general square matrix or symmetric matrix is an irreducible function of its elements, and that of a skew-symmetric matrix the square of an irreducible function of its elements,[27] our assertion will be proved, if at least one of the coordinate hyperplanes $z_v = 0$ intersects M in a subvariety having no $(d - 1)$-dimensional component in the hyperplane $z_0 = 0$. But it is impossible that every coordinate hyperplane of the type $z_v = 0$ contains a $(d - 1)$-dimensional subvariety of M in common with the hyperplane $z_0 = 0$. For, in view of the equivalence of all such coordinates z_v under the basic group of M, this can only happen when all the hyperplanes $z_v = 0$ (including $z_0 = 0$) contain one common $(d - 1)$-dimensional subvariety of M; and we shall show that any point of M which has zero as value for all the coordinates z_v (including z_0) must have the value zero for all other Pluecker coordinates, hence impossible.

The assertion is obvious for $G_{n,r}$, so that we can restrict ourselves to the varieties N_r and P_r. We note first that if z' is any Pluecker coordinate which is derived from z_0 by applying an odd number of substitutions of the type $(x_j) \to (x_{r+1+j})$, then we have $z' = 0$ for every point of P_r. For, a generic point of P_r is given by the Pluecker coordinates of the $[r]$ defined by the matrix $(I, (\overset{i}{\mu}_j))$, where $(\overset{i}{\mu}_j)$ is a generic skew-symmetric $(r + 1)$-matrix. Since the determinant of a skew-symmetric matrix of odd order is always zero, it follows that $z' = 0$ for this generic point, hence also for every point of P_r. If we now denote by z_v any one of the Pluecker coordinates (including z_0) which are derived from z_0 by applying any number of substitutions of the type $(x_j) \to (x_{r+1+j})$ then we

[27] See MacDuffee [13], p. 14.

can state our assertion as follows: If an $(r+1, 2r+2)$-matrix $T = (T_1, T_2)$ satisfies the condition $T_1T_2' \pm T_2T_1' = 0$ (where T_1' and T_2' are the transposed matrices of T_1 and T_2 respectively and the sign is + or − according as the variety is P_r or N_r) and if all the determinants z_v (including z_0) vanish for T, then T is a singular matrix. To prove this, we recall the well known fact from the theory of matrices that we can subject the rows of the matrix T_1 to a suitable non-singular linear transformation and the columns of T_1 to a suitable permutation so that T_1 will have the form

$$T_1 = \begin{pmatrix} I_t & U \\ 0 & 0 \end{pmatrix},$$

where I_t is the identity matrix of order $t < r + 1$ (for $z_0 = | T_1 | = 0$). If we subject the matrix T_2 also to the same linear transformation of the rows and the same permutation of the columns, then it is easily seen that the resulting matrix T will also satisfy the same conditions as the original one; hence, we can assume that T_1 has already the above form. Setting now

$$T_2 = \begin{pmatrix} A & C \\ B & D \end{pmatrix},$$

and evaluating the equation $T_1T_2' \pm T_2T_1' = 0$, we obtain in particular the equation $B = -DU$. Since the determinant

$$\begin{vmatrix} I_t & C \\ 0 & D \end{vmatrix} = | D |$$

is the value of one of the z_v for the matrix T, we have $| D | = 0$. The two equations $B = -DU$ and $| D | = 0$ show that the last $r + 1 - t$ rows of T are not linearly independent; therefore the matrix T is singular.[27a]

Before proceeding further, we observe that by virtue of the existence of the basic group of projective transformations of the ambient space S_k which is transitive on M, the variety M is evidently free of singularities.[28] For the singularity character of a point is preserved under a projective transformation, and an irreducible variety cannot have all its points singular. Hence, we may apply the theory of linear equivalence to the $(d - 1)$-dimensional subvarieties or divisors of M. The birational transformation Ω between M and S_d has the property that it is one-to-one between all points of M outside of the irreducible subvariety W and all points of S_d outside of a hyperplane. This means that the fundamental points on M of the birational transformation Ω can only be on W. Since every $(d - 1)$-dimensional subvariety in S_d is a hypersurface, that is, represented by one equation, it follows that every $(d - 1)$-dimensional subvariety of M can be represented as a partial intersection of a hypersurface F of

---

27a [This one paragraph was added in proof (Nov. 24, 1948) for the convenience of the reader; in our original manuscript the proof of this assertion was left to the reader himself.]

28 For the results of algebraic geometry used in the following, we refer once for all to van der Waerden [19].

S_k with M, with the remainder part of the intersection consisting of a certain multiple of W, say W^t. This means that every $(d - 1)$-dimensional subvariety V of M is equivalent to a multiple of W, and we shall call this multiplicity the index g of V. In fact, if the order of F is f, and $V + W^t$ is the complete intersection of F with M, the index of V is $g = fs - t$. It is easily seen that the index is independent of the choice of the hypersurface F. Since every two $(d - 1)$-dimensional subvarieties of M with the same index g are equivalent, the set of all such subvarieties form one linear system $| W^g |$, which is evidently complete. Conversely, it is also clear that every complete linear system of $(d - 1)$-dimensional subvarieties *without base points* must be one of the $| W^g |$. The linear system $| W^{fs} |$ evidently contains the system cut out on M by the hypersurfaces of order f. Now, for $g > h$, the linear system $| W^g |$ certainly contains the system $| W^h |$, and the dimension of $| W^g |$ is greater than or equal to that of $| W^h |$; the equality can only occur when $| W^g |$ differs from the $| W^h |$ only by a fixed component, which means that the two systems are essentially the same. Thus the set of all distinct systems $| W^g |$ contains already all the complete linear systems without base points of M, and no two of these systems can have the same dimension.

Suppose there is given a regular birational transformation of the variety M onto itself. Then this transformation must carry every linear system $| W^g |$ into itself. For, a regular birational transformation must carry a complete linear system without base points into a complete linear system without base points, and we have seen above that two such complete systems cannot have the same dimension. Consider a line (or a conic section) on M; it intersects every element of the linear system $| W^s |$ in exactly one point (or two points[29]), if it is not entirely contained in this element. This line (or conic section) is carried by the given regular birational transformation into a curve which must also intersect every element of $| W^s |$ in exactly one point (or two points), if not contained in it. Since the system $| W^s |$ contains in particular the system cut out by all the hyperplanes of S_k, it follows that the transformed curve intersects every hyperplane of S_k in one point (or two points); hence this curve must also be a line (or conic section). Thus any regular birational transformation of M onto itself must carry the lines (or conic sections) on M into lines (or conic sections) on M; hence, according to Theorems VI, VIII–X, it is a transformation of the basic group of M.

THEOREM XI. *Any regular birational transformation of the Grassmann variety $G_{n,r}$ onto itself is a transformation of the basic group of $G_{n,r}$.*

THEOREM XII. *Any regular birational transformation of the variety N_r onto itself is a transformation of the basic group of N_r.*

[29] By this we mean either two points or one point with multiplicity 2. The multiplicity of an intersection is invariant under regular birational transformation, see van der Waerden [17], p. 148; this can also be easily shown directly in our case without recourse to this general result. Moreover, our reasoning will hold just as well, if we replace "two points" by "at most two points". The result will be "line or conic section" instead of "conic section"; but, as the variety P_r does not contain any line at all, the conclusion will still be the same.

THEOREM XIII. *Any regular birational transformation of the variety P_r onto itself is a transformation of the basic group of P_r, except when $r = 3$.*

THEOREM XIV. *Any regular birational transformation of the hyperquadric Q_{n-1} onto itself is a transformation of the basic group of Q_{n-1}.*

Though the Theorem VI, VIII–X as well as our proof in this section involve some restrictions on the n and r, it is not difficult to see that the above theorems hold as stated without any restriction, except for the case of P_3. In fact, the discussions in §3 show that our proof holds also for $G_{n,0}$, $G_{n,n-1}$ $(n > 1)$ and P_2. The case of $G_{1,0} = N_0 = S_1$ is well known; the cases of Q_1 and P_1 then follow from the well known isomorphisms[30] of the orthogonal group of orders 3 and 4, and the observation that both Q_1 and P_1 are equivalent to S_1 under regular birational transformations. The P_0 and Q_0 are trivial, consisting of one point and two points respectively. The Q_2 is a slightly exceptional case, since it is equivalent to the product of two S_1; but it is not difficult to establish the theorem in this case by simple direct considerations.

The real exception is P_3. From the remarks at the end of Chapter II, §7 it is not difficult to see that the transformation induced by a "correlation" is birational and regular, and we have seen there that such a transformation cannot be a transformation of the basic group. The situation can be better understood from the viewpoint of the representation of P_r in the spinor space, mentioned at the end of Chapter I, §3. However, we shall not go into the details here.

5. Field of Complex Numbers. Analytic Characterization

In case the ground field K is the field of all complex numbers, the proof in the preceding section can be so modified as to include all regular analytic transformations of the variety M onto itself. We observe first that M as a d-dimensional algebraic variety without singularity is a $2d$-dimensional orientable analytical manifold.[31] The special birational transformation Ω between the variety M and the projective space S_d given in §4 shows that any $2(d-1)$-cycle of M can be deformed into the variety W, which as an irreducible algebraic subvariety is an irreducible $2(d-1)$-cycle of M. This means that every $2(d-1)$-cycle of M is homotopic, hence also homologous to a multiple of W. Since the algebraic cycle W cannot be homologous to zero even with division, the $2(d-1)^{\text{th}}$ homology group of M is therefore an infinite cyclic group with the class of W as the generating class[32]. Now it is a fundamental fact of an algebraic manifold that for an intersection point of two algebraic cycles the algebraic multiplicity coincides with the topological multiplicity (Kronecker index)[33]. It follows then that all sections of M by hypersurfaces of a fixed order f belong to

[30] See van der Waerden [18], §7.

[31] For the general topological properties of analytic or algebraic manifolds, see Lefschetz [12], Ch. VIII, or van der Waerden [16].

[32] This is contained in the general results of Ehresmann [6], where the topological properties of homogeneous algebraic varieties are systematically investigated.

[33] See van der Waerden [16], §6. The proof there is for the projective space, but can be easily extended to any algebraic manifold (without singularity) by using the Severi's definition of multiplicity (see [17], p. 148).

one homology class $|W^{fs}|$, for they all have the same number of intersections (algebraically, hence topologically) with a fixed algebraic curve on M (for example, a line or a conic section on M), and the "index" fs is already determined by this number.[34] In particular, all the hyperplane sections or hyperquadric sections of M belong to one class $|W^s|$ or $|W^{2s}|$ respectively. Hence, a line on M, which is a 2-cycle, intersects the class $|W^s|$ with the topological multiplicity 1, and a conic section on M intersects the class $|W^s|$ or $|W^{2s}|$ with the topological multiplicity 2 or 4 respectively.

Let there be given a regular analytic transformation of M onto itself, i.e. a transformation which effects a one-to-one analytic mapping of a neighborhood of each point onto a neighborhood of the image point. Since this transformation is also a topological transformation, it induces an isomorphism of every homology group onto itself. For the $2(d - 1)^{\text{th}}$ homology group, since it is an infinite cyclic group, this isomorphism can only be the identity mapping. This means that the given analytic transformation carries every class $|W^g|$, in particular the class $|W^s|$ or $|W^{2s}|$ into itself. Now a line on M is transformed by the given analytic transformation into an analytic curve, which must also intersect the class $|W^s|$ with the topological multiplicity 1. Since the topological multiplicity of any intersection point of two analytic cycles is always positive[35], it follows that this analytic curve intersects every hyperplane section of M, hence every hyperplane in S_k in exactly one point, if it is not entirely contained in the hyperplane. But this means that the curve must be a line, for it is contained in the common line of the bundle of hyperplanes of S_k containing any two fixed points of the curve and hence must coincide with this line. Similarly, a conic section on M is transformed by the given analytic transformation into an analytic curve, which must intersect the class $|W^s|$ or $|W^{2s}|$ with the topological multiplicity 2 or 4 respectively; and it follows then that this analytic curve intersects every hyperplane or hyperquadric of S_k in at most two or four points respectively, if it is not entirely contained in the hyperplane or hyperquadric. This means that the curve is a conic section, for it is contained in the common plane of the bundle of all the hyperplanes of S_k containing three fixed points of the curve, and is also contained in any hyperquadric of S_k containing five points of the curve. Referring to the Theorems VI, VIII, IX, X and the remarks at the end of §4, we have the following:

THEOREM XV. *In case the ground field is the field of complex numbers, the Theorems XI–XIV hold also for any regular analytic transformation of the variety onto itself.*

THE INSTITUTE FOR ADVANCED STUDY

REFERENCES

[1] BERTINI, E.: Einfuehrung in die projektive Geometrie mehr-dimensionaler Raeume. Vienna, 1924.

[34] This follows also from the general theorem that linear equivalence implies homology.

[35] See van der Waerden [16], §5, or Lefschetz [12], p. 382.

[2] Blaschke, W.: Vorlesungen ueber Differentialgeometrie III. Berlin, 1929.

[3] Brauer, R.: *A Characterization of Null Systems in Projective Space.* Bull. Amer. Math. Soc., vol. 42 (1936), pp. 247–254.

[4] Cartan, E.: *Sur certaines formes riemanniennes remarquables des géométries à groupe fondamental simple.* Ann. École Norm., 3e série, t. 44 (1927), pp. 345–467.

[5] Cartan, E.: *Sur le groupe de la géométrie hypersphérique.* Comment. Math. Helv., vol. 4 (1932), pp. 158–171.

[6] Ehresmann, C.: *Sur la topologie de certains espaces homogènes.* Annals of Math., vol. 35 (1934), pp. 396–443.

[7] Givens, J. W.: *Tensor Coordinates of Linear Spaces.* Annals of Math., vol. 38 (1937), pp. 355–385.

[8] Hua, L. K.: *Geometries of Matrices. I. Generalizations of von Staudt's Theorem.* Trans. Amer. Math. Soc., vol. 57 (1945), pp. 441–481.

[9] Hua, L. K.: *Geometries of Matrices.* I_1. *Arithmetical Constructions.* Trans. Amer. Math. Soc., vol. 57 (1945), pp. 482–490.

[10] Hua, L. K.: *Geometries of Matrices. III. Fundamental Theorems in the Geometries of Symmetric Matrices.* Trans. Amer. Math. Soc., vol. 61 (1947), pp. 229–255.

[11] Hua, L. K.: *On the Extended Space of Several Complex Variables (1): The Space of Complex Spheres.* Quarterly Journal of Math., vol. 17 (1946), pp. 214–222.

[12] Lefschetz, S.: Topology. New York, 1930.

[13] MacDuffee, C. C.: The Theory of Matrices. Berlin, 1933.

[14] Osgood, W. F.: Lehrbuch der Funktionentheorie. II_1. Leipzig, 1929.

[15] Severi, F.: *Sulla varietà che rappresenta gli spazî subordinati di data dimensione, immersi in uno spazi lineare.* Annali di Mate. serie III, t. 24 (1915), pp. 89–120.

[16] van der Waerden, B. L.: *Topologische Begruendung des Kalkuels der abzaehlenden Geometrie.* Math. Ann., Bd. 102 (1929), pp. 337–362.

[17] van der Waerden, B. L.: *Zur algebraischen Geometrie. VI. Algebraische Korrespondenzen und rationale Abbildungen.* Math. Ann., Bd. 110 (1934), pp. 134–160.

[18] van der Waerden, B. L.: Gruppen von linearen Transformationen, Berlin, 1935.

[19] van der Waerden, B. L.: Einfuehrung in die algebraische Geometrie. Berlin, 1939.

Separatdruck aus vol. 23, fasc. 1 (1949) der
COMMENTARII MATHEMATICI HELVETICI

Über die Lösbarkeit gewisser algebraischer Gleichungssysteme

Von Wei-Liang Chow, Princeton (USA.)

In einer vor kurzem erschienenen Arbeit[1]) hat Herr W. Habicht den folgenden interessanten Satz bewiesen:

K sei ein Körper; $f_1, \ldots, f_n$ seien Formen aus dem Ringe $K[x_1, \ldots, x_n]$; zwischen ihnen bestehe die Relation $\sum_{i=1}^{n} x_i f_i = 0$; es seien nicht gleichzeitig n gerade und die Grade sämtlicher f_i gleich 1. Dann besitzen die Formen f_i eine gemeinsame (nichttriviale) Nullstelle (in der algebraisch-abgeschlossenen Hülle $\overline{K}$ von K).

Wie Herr Habicht bemerkt hat, können wir im obigen Satz ohne Beschränkung der Allgemeinheit annehmen, daß die sämtlichen Formen f_i von demselben Grade sind. Wir werden in dieser Note zeigen, daß dieser Satz von Habicht einen Spezialfall eines allgemeineren Satzes bildet, der wie folgt lautet:

Satz. *K sei ein Körper; V sei eine $(n-1)$-dimensionale algebraische Mannigfaltigkeit im m-dimensionalen projektiven Raume S_m; $f_1(x), \ldots, f_n(x)$ bzw. $g_1(x), \ldots, g_n(x)$ seien Formen h-ten bzw. k-ten Grades aus dem Ringe $K[x_1, \ldots, x_{m+1}]$; zwischen ihnen bestehe die Relation $\sum_{i=1}^{n} f_i(\xi)\, g_i(\xi) = 0$ für jeden Punkt (ξ) in V; es seien nicht gleichzeitig n gerade und $h = k$. Dann besitzen entweder die Hyperflächen $f_i(x) = 0$ $(i = 1, \ldots, n)$ oder die Hyperflächen $g_i(x) = 0$ $(i = 1, \ldots, n)$ einen gemeinsamen Schnittpunkt in $\overline{K}$ mit der Mannigfaltigkeit V.*

Der Satz von Habicht folgt daraus, indem man $m = n - 1$, $V = S_{n-1}$ und $g_i(x) = x_i$ $(i = 1, \ldots, n)$ setzt. Unsere Beweismethode, die wesentlich anders ist als die von Habicht, ist auch etwas einfacher und weniger rechnerisch; dabei haben wir allerdings etwas stärkere Hilfsmittel aus der algebraischen Geometrie herangezogen.

---

[1]) *W. Habicht,* Über die Lösbarkeit gewisser algebraischer Gleichungssysteme. Comm. Math. Helv., vol. 18, S. 154—175.

Beweis des Satzes: Die Voraussetzung über V bedeutet, daß V die Gesamtheit der gemeinsamen Nullstellen von einem System endlichvieler Formen aus $K[x_1, \ldots, x_{m+1}]$ bildet, und daß $n-1$ unabhängige allgemeine Hyperebenen in S_m mit V endlichviele gemeinsame Schnittpunkte haben. Ohne Beschränkung der Allgemeinheit können wir auch annehmen, daß die Mannigfaltigkeit V irreduzibel (in bezug auf K) ist; denn sonst können wir die V durch irgendeinen ihren irreduziblen $(n-1)$-dimensionalen Bestandteil ersetzen. Es sei $k \leqslant h$; und wir nehmen an, daß die Hyperflächen $g_i(x) = 0 \quad (i = 1, \ldots, n)$ keinen gemeinsamen Schnittpunkt mit der Mannigfaltigkeit V besitzen. Wir haben dann zu beweisen, daß die Hyperflächen $f_i(x) = 0 \quad (i = 1, \ldots, n)$ mindestens einen gemeinsamen Schnittpunkt mit V besitzen.

Es seien $F_1(u^1, x), \ldots, F_n(u^n, x)$ n unabhängige allgemeine Formen vom Grade h, wo die n Koeffizientenreihen $(u^1), \ldots, (u^n)$ unabhängige Unbestimmte sind. Faßt man die sämtlichen Koeffizienten $(u^1, \ldots, u^n)$ zusammen als einen Punkt eines affinen Raumes auf, so wird jedes System von n Formen h-ten Grades (wovon einige identisch verschwinden können) durch einen Punkt in diesem Raume dargestellt. Die Bedingung, daß die Relation $\sum_{i=1}^{n} g_i(\xi) F_i(u^i, \xi) = 0$ für jeden Punkt (ξ) in V bestehe, ist offenbar linear in bezug auf die Koeffizienten $(u^1, \ldots, u^n)$; daraus folgt, daß die Gesamtheit aller Systeme von n Formen h-ten Grades, die diese Bedingung erfüllen, durch eine lineare und folglich irreduzible Mannigfaltigkeit in dem affinen Raume von den $(u^1, \ldots, u^n)$ dargestellt wird. Es sei nun $(a^1, \ldots, a^n)$ ein allgemeiner Punkt dieser irreduziblen Mannigfaltigkeit. Wir können dann $f_i(x) = F_i(\alpha^i, x) \quad (i = 1, \ldots, n)$ setzen, wo $(\alpha^1, \ldots, \alpha^n)$ eine (relationstreue) Spezialisierung[2]) von $(a^1, \ldots, a^n)$ ist. Unsere Behauptung wird bewiesen sein, wenn wir zeigen, daß die n Hyperflächen $F_i(a^i, x) = 0 \quad (i = 1, \ldots, n)$ mit V einen gemeinsamen Schnittpunkt besitzen. Denn, das Vorhandensein eines solchen gemeinsamen Schnittpunktes ist bekanntlich gleichbedeutend mit dem Verschwinden der Resultanten $R_j(u^1, \ldots, u^n)$, die man aus den n Gleichungen $F_i(u^i, x) = 0 \quad (i = 1, \ldots, n)$ und den definierenden Gleichungen der Mannigfaltigkeit V durch Elimination der Variabeln (x) gewinnt; und aus $R_j(a^1, \ldots, a^n) = 0$ folgt offenbar $R_j(\alpha^1, \ldots, \alpha^n) = 0$.

[2]) $(\alpha^1, \ldots, \alpha^n)$ heißt eine Spezialisierung von $(a^1, \ldots, a^n)$, wenn aus $f(a^1, \ldots, a^n) = 0$ immer $f(\alpha^1, \ldots, \alpha^n) = 0$ folgt, wo $f(u^1, \ldots, u^n)$ ein Polynom aus dem Ringe $K[u^1, \ldots, u^n]$ ist. Dabei wird es zugelassen, daß einige von den (α^i) oder eventuell auch alle (α^i) ganz verschwinden. Dasselbe gilt im folgenden auch für die Spezialisierung von $(a^1, \ldots, a^{n-1})$. Dagegen wird unter einem Punkt in S_m oder einer Nullstelle von Formen oder einer Spezialisierung davon, immer ein nichttrivialer Punkt oder Nullstelle verstanden.

Wir betrachten nun das System von $n-1$ allgemeinen Hyperflächen $F_1(u^1, x) = 0, \ldots, F_{n-1}(u^{n-1}, x) = 0$. Bekanntlich hat dieses System von Hyperflächen mit V eine endliche Anzahl von gemeinsamen Schnittpunkten $(y^1), \ldots, (y^s)$, die zusammen ein vollständiges System von konjugierten Punkten über $K(u^1, \ldots, u^{n-1})$ bilden. Bei der Spezialisierung $(u^1, \ldots, u^{n-1}) \to (a^1, \ldots, a^{n-1})$ seien etwa die Punkte $(\eta^1), \ldots, (\eta^s)$ eine Spezialisierung der Punkte $(y^1), \ldots, (y^s)$; die Punkte $(\eta^1), \ldots, (\eta^s)$ sind natürlich auch gemeinsame Schnittpunkte von den Hyperflächen $F_1(a^1, x) = 0, \ldots, F_{n-1}(a^{n-1}, x) = 0$ mit V. Liegt einer von diesen Punkten, etwa (η^1) nicht in der Hyperfläche $g_n(x) = 0$, dann muß er auch der Bedingung

$$F_n(a^n, \eta^1) = \frac{-\sum_{i=1}^{n-1} g_i(\eta^1) F_i(a^i, \eta^1)}{g_n(\eta^1)} = 0$$

genügen. Folglich ist (η^1) auch einen gemeinsamen Schnittpunkt von den n Hyperflächen $F_1(a^1, x) = 0, \ldots, F_n(a^n, x) = 0$ mit V. Wir brauchen also nur zu zeigen, daß nicht die sämtlichen Punkte $(\eta^1), \ldots, (\eta^s)$ in der Hyperfläche $g_n(x) = 0$ liegen können.

Nehmen wir für einen Augenblick an, daß es ein System von $n-1$ Hyperflächen $F_1(\beta^1, x) = 0, \ldots, F_{n-1}(\beta^{n-1}, x) = 0$ gibt, wo $(\beta^1, \ldots, \beta^{n-1})$ eine Spezialisierung von $(a^1, \ldots, a^{n-1})$ ist, mit der Eigenschaft, daß dieses System mit V einen isolierten, nicht in der Hyperfläche $g_n(x) = 0$ liegenden gemeinsamen Schnittpunkt (ζ) besitzt. Da dann der Punkt (ζ) eine isolierte Spezialisierung irgendeines von den Punkten $(y^1), \ldots, (y^s)$ bei der Spezialisierung $(u^1, \ldots, u^{n-1}) (\to (\beta^1, \ldots, \beta^{n-1})$ ist, so muß[3]) er in jeder Spezialisierung der Punkte $y^1), \ldots, (y^s)$ bei der Spezialisierung $(u^1, \ldots, u^{n-1}) \to (\beta^1, \ldots, \beta^{n-1})$ immer vorkommen (zwar genau so oft wie seine Multiplizität, die immer positiv ist). Nun ist jeder Spezialisierung von den Punkten $(\eta^1), \ldots, (\eta^s)$ bei der Spezialisierung $(a^1, \ldots, a^{n-1}) \to (\beta^1, \ldots, \beta^{n-1})$ *a fortiori* auch eine Spezialisierung von den Punkten $(y^1), \ldots, (y^s)$ bei der Spezialisierung $(u^1, \ldots, u^{n-1}) \to (\beta^1, \ldots, \beta^{n-1})$; folglich muß mindestens einer von den Punkten (η^i), etwa der (η^1) bei der Spezialisierung $(a^1, \ldots, a^{n-1}) \to (\beta^1, \ldots, \beta^{n-1})$ in den Punkt (ζ) übergehen. Da der Punkt (ζ) nicht in der Hyperfläche $g_n(x) = 0$ liegt, so kann (η^1) auch nicht in ihr liegen.

Das erwünschte System von Hyperflächen

$$F_1(\beta^1, x) = 0, \ldots, F_{n-1}(\beta^{n-1}, x) = 0$$

[3]) Siehe etwa *A. Weil*, Foundation of Algebraic Geometry, S. 62, Theorem 4.

kann man wie folgt konstruieren. Im Falle von einem ungeraden $n = 2r + 1$, setzt man

$$F_{2j-1}(\beta^{2j-1}, x) = \quad g_{2j}(x)\, p(x) \ ,$$
$$F_{2j}(\beta^{2j}, x) \quad = -\, g_{2j-1}(x)\, p(x) \qquad (j = 1, \ldots, r) \ ,$$

wo $p(x)$ eine Form $(h - k)$-ten Grades ist, deren Nullstellen die endlichvielen gemeinsamen Schnittpunkte von V mit den Hyperflächen $g_1(x) = 0, \ldots, g_{2r}(x) = 0$ nicht enthalten. Man kann dann für den Punkt (ζ) irgendeinen von diesen endlichvielen gemeinsamen Schnittpunkten nehmen. Im Falle von einem geraden $n = 2r + 2$, definiert man die ersten $2r$ Formen genau so wie oben, wo $p(x)$ jetzt eine Form $(h - k)$-ten Grades bedeutet, deren Nullstellen die endlichvielen gemeinsamen Schnittpunkte von V mit den Hyperflächen $g_1(x) = 0, \ldots, g_{2r}(x) = 0$, $g_n(x) = 0$ nicht enthalten. Für die $(2r + 1)$-te Form setzt man dann

$$F_{2r+1}(\beta^{2r+1}, x) = g_n(x)\, q(x) \ ,$$

wo $q(x)$ eine Form $(h - k)$-ten Grades ist, deren Nullstellen weder die endlichvielen gemeinsamen Schnittpunkte von V mit den Hyperflächen $g_1(x) = 0, \ldots, g_{2r}(x) = 0$, $g_n(x) = 0$, noch die endlichvielen gemeinsamen Schnittpunkte von V mit den Hyperflächen $g_1(x) = 0, \ldots, g_{2r}(x) = 0$, $p(x) = 0$ enthalten. Dabei ist es zu beachten, daß, laut unserer in diesem Falle gemachten Voraussetzung $h - k > 0$, die Gleichung $q(x) = 0$ (und ebenso die Gleichung $p(x) = 0$) wirklich eine Hyperfläche darstellt. Man kann dann für den Punkt (ζ) irgendeinen von den endlichvielen gemeinsamen Schnittpunkten von V mit den Hyperflächen $g_1(x) = 0, \ldots, g_{2r}(x) = 0$, $q(x) = 0$ nehmen.

(Eingegangen den 28. Juni 1948.)

DRUCK: ART. INSTITUT ORELL FÜSSLI AG., ZÜRICH

ON THE GENUS OF CURVES OF AN ALGEBRAIC SYSTEM

BY

WEI-LIANG CHOW

Professor C. Chevalley has recently pointed out to me the desirability of having a simple algebraic proof of the following theorem: The genus of any curve of an algebraic system of algebraic curves is not greater than the genus of the generic curve of the system. The theorem seems to be well known, but I have sought in vain for a proof of it in the literature(1). We shall give here a purely algebraic proof of this theorem, based on the idea of the associated form of an algebraic variety, which was introduced by van der Waerden and myself some years ago(2).

Let $\mathfrak{C}$ be an irreducible algebraic curve of order p in the projective space $\mathfrak{S}_m$ defined by the associated form

$$F(C; U, V) = \sum C_{i_1 \cdots i_p, j_1 \cdots j_p} U_{i_1} \cdots U_{i_p} V_{j_1} \cdots V_{j_p}, \tag{1}$$

where the (C) are the coefficients, which can be taken as the coordinates of the curve $\mathfrak{C}$, and the (U), (V) are indeterminates (they are, in fact, the coefficients of the generic hyperplanes $\sum_{i=0}^{m} U_i Y_i = 0$, $\sum_{i=0}^{m} V_i Y_i = 0$ in $\mathfrak{S}_m$). The curve $\mathfrak{C}$ can also be defined by a set of equations

$$f_j(B, Y) = 0 \tag{2}$$

derived from the associated form (1) by rational process(3), so that the coefficients (B) in (2) are forms in (C). It is essential to notice that the equations (2) remains valid for any value of (C) corresponding to an associated form (1), so that in particular the relation between (1) and (2) is preserved under any specialization of (C). The same remark applies to all succeeding steps of our argument. A necessary and sufficient condition for the genus $g(\mathfrak{C})$ of $\mathfrak{C}$ to be less than or equal to g' is that for every (or infinitely many) n there exists a linear series of order not greater than n and dimension not less than $n-g'$. We shall show that this is an algebraic condition in terms of the (C).

Consider the variety $\mathfrak{C}^n$ formed by the totality of all groups of n points

Presented to the Society, June 19, 1948; received by the editors December 10, 1947.

(1) After the above note was written, it came to my attention that B. Segre, in a paper *On limits of algebraic varieties*, Proc. London Math. Soc. vol. 47 (1942) pp. 351–403, has also given a proof of this theorem. His proof, quite different from ours, is however not purely algebraic, as it uses the limit process.

(2) Chow and van der Waerden: *Zur algebraischen Geometrie*. IX. *Ueber zugeordnete Formen und algebraische Systeme von algebraischen Mannigfaltigkeiten*, Math. Ann. vol. 113 (1937) pp. 692–704.

(3) Chow and van der Waerden, loc. cit. p. 699, Satz 3.

on $\mathfrak{C}$. By a group of n points is meant a set of points on $\mathfrak{C}$, to each of which is assigned a certain positive integer, called its multiplicity, such that the sum of all such multiplicities is equal to n. If we take as coordinates of such a group of n points the coefficients (P) of its associated form $P(V) = \sum P_{j_1 \cdots j_n} V_{j_1} \cdots V_{j_n}$, where the summation extends over all possible combinations (repetitions included) of the integers $0, 1, 2, \cdots, m$ in sets of n elements, the variety $\mathfrak{C}^n$ is then represented as an algebraic variety. In fact, the condition that $P(V)$ represents a group of n points on $\mathfrak{C}$ is expressed by the assertion that there exist n solutions(4) (Y^i), $i=1, 2, \cdots, n$, of (2) and a quantity $\lambda \neq 0$, such that the following equation

$$P(V) = \sum P_{j_1 \cdots j_n} V_{j_1} \cdots V_{j_n} = \lambda \sum_{i=1}^{n} \left(\sum_{j=0}^{m} Y_j^i V_j \right)$$

holds identically in (V). Equating the coefficients of the corresponding terms on both sides of this equation, we have

$$P_{j_1 \cdots j_n} = \lambda \sum Y_{j_1}^1 \cdots Y_{j_n}^n,$$

where the summation extends over all the permutations of the set $(j_1, \cdots, j_n)$. Since $\lambda \neq 0$, this is equivalent to a system of equations of the form

$$(3) \qquad P_{j_1 \cdots j_n} \sum Y_{k_1}^1 \cdots Y_{k_n}^n = P_{k_1 \cdots k_n} \sum Y_{j_1}^1 \cdots Y_{j_n}^n.$$

Eliminating the (Y^i) between the equations (3) and the equations (2) for (Y^i), that is, the equations $f_j(B, Y^i) = 0$, we arrive at a set of equations in the (P)

$$(4) \qquad K_j^n(P) = 0,$$

which evidently define the variety $\mathfrak{C}^n$.

Let $\mathfrak{H}$ be a linear system of hypersurfaces of order h in $\mathfrak{S}_m$, with the property that no hypersurface of the system contains the curve $\mathfrak{C}$ and not all hypersurfaces of the system contain any one fixed point of $\mathfrak{C}$. The system $\mathfrak{H}$ will cut out on the curve $\mathfrak{C}$ a linear series of order $q = hp$, and we assume that h is so large that $q > n$. Two point groups $P(V)$, $Q(V)$ on $\mathfrak{C}$, both of order n, are called quasi-equivalent with respect to the system $\mathfrak{H}$, if there exists a group of $q-n$ points on $\mathfrak{C}$, with the associated form $R(V)$, such that the point groups $R(V)P(V)$, $R(V)Q(V)$ are two groups of the linear series cut out on $\mathfrak{C}$ by $\mathfrak{H}$. This condition of quasi-equivalence between the point groups $P(V)$ and $Q(V)$ (with respect to $\mathfrak{H}$) can also be expressed as an algebraic condition between the coordinates (P), (Q) of the two corresponding points on the variety $\mathfrak{C}^n$. To show this, let $H(W, V)$ be the associated form of the

(4) Here as well as in the following, the trivial solution, consisting of all zeroes, is always excluded.

linear series $\mathfrak{H}$, where the (W) are the coefficients determining the individual groups of the series and the (V) are as before the coefficients of a generic hyperplane in $\mathfrak{S}_m$. The condition can then be expressed thus: There exist (W'), (W''), $R(V)$ such that

$$H(W', V) = \lambda R(V)P(V), \qquad H(W'', V) = \mu R(V)Q(V) \qquad (\lambda, \mu \neq 0).$$

One concludes just as before that this is equivalent to a system of equations of the form

$$\begin{aligned} H'_{j_1\cdots j_q}\sum P_{i_1\cdots i_n}R_{i_{n+1}\cdots i_q} &= H'_{i_1\cdots i_q}\sum P_{j_1\cdots j_n}R_{j_{n+1}\cdots j_q}, \\ H''_{j_1\cdots j_q}\sum Q_{i_1\cdots i_n}R_{i_{n+1}\cdots i_q} &= H''_{i_1\cdots i_q}\sum Q_{j_1\cdots j_n}R_{j_{n+1}\cdots j_q}, \end{aligned} \tag{5}$$

where the (H'), (H'') are forms in (W'), (W'') respectively. Eliminating the (W'), (W''), (R) between the equations (5) and the equations $K_j^{q-n}(R)=0$, we arrive at a set of equations

$$J_j(P, Q) = 0, \tag{6}$$

which is the algebraic condition for the quasi-equivalence with respect to $\mathfrak{H}$.

For each fixed point $P(V)$ of $\mathfrak{C}^n$, the set of $Q(V)$ satisfying the equations (6) and the equations $K_j^n(Q)=0$ is evidently an algebraic variety, which is the subvariety of $\mathfrak{C}^n$ consisting of all points of $\mathfrak{C}^n$ which are quasi-equivalent (with respect to $\mathfrak{H}$) with $P(V)$. The condition that this variety has a dimension not less than $n-g'$ is expressed by eliminating the (Q) between the equations (6), the equations $K_j^n(Q)=0$, and the $n-g'$ generic linear equations $\sum_j U_j^{(i)}Q_j=0$ $(i=1, \cdots, n-g')$, where the $U_j^{(i)}$ are again indeterminates. The resultants of this elimination are forms in the indeterminates $U_j^{(i)}$. Setting each coefficient of these forms to zero, we arrive at a set of equations

$$T_j(P) = 0 \tag{7}$$

which is the condition that the set of points $Q(V)$ of $\mathfrak{C}^n$ quasi-equivalent (with respect to $\mathfrak{H}$) with the point $P(V)$ forms a variety of dimension not less than $n-g'$. Lastly, the condition that such a point $P(V)$ exists on $\mathfrak{C}^n$ is found by eliminating (P) from (7) and (4), and we get a set of equations

$$G_j(C) = 0 \tag{8}$$

in the coefficients (C) of the curve $\mathfrak{C}$, which is necessary and sufficient for the existence of a point group of order n on $\mathfrak{C}$, such that the set of all point groups of order n on $\mathfrak{C}$ which are quasi-equivalent (with respect to $\mathfrak{H}$) with it forms a variety of dimension not less than $n-g'$.

We shall now show that the existence of a "quasi-equivalent system" of order n and dimension r implies the existence of a linear series on $\mathfrak{C}$ of order n and dimension r. For two point groups $P(V)$, $Q(V)$ are quasi-equivalent with respect to $\mathfrak{H}$ if and only if there exist two integral divisors $\overline{H}'$, $\overline{H}''$ of order at most equal to q and containing only prime divisors corresponding to the multiple points of $\mathfrak{C}$, such that the divisor $\overline{H}'\overline{P}$ is equivalent to $\overline{H}''\overline{Q}$, where $\overline{P}$, $\overline{Q}$ now stand for any (of the finite number of possible) divisors corresponding to the point groups $P(V)$, $Q(V)$ respectively. This means that every $\overline{Q}$ is equivalent to a divisor $(\overline{H}'/\overline{H}'')\overline{P}$ of order n. Since there are only a finite number of such divisors $\overline{H}'/\overline{H}''$, we conclude that the entire "quasi-equivalent system" determined by $P(V)$ dissolves into a finite number of equivalent systems of divisors or linear series of order n, at least one of which must of course have the same dimension r. Hence our assertion is proved.

Consider now an irreducible algebraic correspondence in which to each point X of a variety $\mathfrak{X}$ there corresponds an irreducible curve $\mathfrak{C}_x$ in a variety $\mathfrak{Y}$ in S_m. Let $F(C_x;\ U,\ V)$ be the associated form of $\mathfrak{C}_x$; we have then a correspondence between the point X and the coordinates (C_x) of the associated form. Let (C_a) be a specialization of (C_x) corresponding to the specialization $X\rightarrow a$; then the form $F(C_a;\ U,\ V)$ is the associated form of a curve $\mathfrak{C}_a$ in $\mathfrak{Y}$. We assume that the curve $\mathfrak{C}_a$ is irreducible. In applying the above considerations to the curve $\mathfrak{C}_x$, we take for $\mathfrak{H}$ the linear system of all hypersurfaces of order h in $\mathfrak{S}_m$ which do not contain the curve $\mathfrak{C}_a$ and hence also do not contain $\mathfrak{C}_x$. For sufficiently high h the linear system $\mathfrak{H}$ contains a subsystem which cuts out on $\mathfrak{C}_x$ a complete series of order n, which must have a dimension not less than $n-g(\mathfrak{C}_x)$. Then the "quasi-equivalent system" determined by any point group (corresponding to any integral divisor) of the complete series must obviously have at least this same dimension. Hence the (C_x) satisfy the equation (8) with $g'=g(\mathfrak{C}_x)$; and consequently the (C_a) satisfy also the equations (8), also with $g'=g(\mathfrak{C}_x)$, and hence $g(\mathfrak{C}_a)\leqq g(\mathfrak{C}_x)$.

To complete the proof we must show that the so chosen linear system $\mathfrak{H}$, for sufficiently high h, contains a subsystem which cuts out on $\mathfrak{C}_x$ a complete series of order n. This follows from the fact that already the subsystem of $\mathfrak{H}$ consisting of all the hypercones passing through a generic $\mathfrak{S}_{m-3}$ has this property. This can be seen most easily by projecting the curves $\mathfrak{C}_x$ and $\mathfrak{C}_a$ from the generic $\mathfrak{S}_{m-3}$ into plane curves $\mathfrak{C}_x'$ and $\mathfrak{C}_a'$ in a generic plane; the system of hypercones passing through $\mathfrak{S}_{m-3}$ will then be projected into the system of all curves of order h not containing the curve $\mathfrak{C}_a'$ (and hence not containing the curve $\mathfrak{C}_x'$) in the plane. Now the dimension of such a system of curves depends only on the order h and the order of the curve $\mathfrak{C}_a'$, and since $\mathfrak{C}_x'$ and $\mathfrak{C}_a'$ certainly have the same order, it follows that this system must also consist of all the curves of order h not containing $\mathfrak{C}_x'$. That this system of curves, for sufficiently high h, contains a subsystem which cuts out on $\mathfrak{C}_x'$ a complete series of order n is quite obvious. Thus everything is proved.

THE INSTITUTE FOR ADVANCED STUDY,
PRINCETON, N. J.

ON THE DEFINING FIELD OF A DIVISOR IN AN ALGEBRAIC VARIETY[1]

WEI-LIANG CHOW

In the system of algebraic geometry as developed by A. Weil in his recent book, *Foundations of algebraic geometry*, a variety U in the n-space is defined as the set of all equivalent couples (k, P), each consisting of a field k and a point P in the n-space such that the field $k(P)$ is a regular extension of k. Two such couples (k', P') and (k'', P'') are called equivalent if every finite specialization of P' over k' is also one of P'' over k'' and conversely. Any field k which enters into such a couple is called a field of definition of the variety U. It has been shown by Weil in his book that among all the fields of definition of a variety U there is a smallest one which is contained in all of them, which we shall call the defining field of the variety U. A d-cycle G in a variety U of dimension r is a finite set of simple subvarieties of dimension d in U, to each of which is assigned an integer called its multiplicity; a cycle is called positive if the multiplicity of each of its component varieties is positive. Let K be a field of definition of U. Then the G is said to be rational over K if it satisfies the following conditions: (1) each component variety of G is algebraic over K; (2) if a variety is a component of G, then all the conjugate varieties over K are also components of G with the same multiplicity; (3) the multiplicity of each component of G is a multiple of its order of inseparability. The question arises whether there is a smallest one among all the fields over which the cycle G is rational. If such a smallest field exists, we shall call it the defining field of the cycle G. One observes that since by definition every field over which the cycle G is rational must be a field of definition of the variety U, it follows that the defining field of G, if it exists, must contain the defining field of U.

The following simple example shows that in general a cycle does not have a defining field. Consider the variety U_0 consisting of the one point $(x^{1/p}, y^{1/p})$ in the 2-space S_2, where x and y are independent variables over a field k of characteristic p; this variety U_0 is a simple subvariety of S_2. The 0-cycle pU_0 is rational over both fields $k(x, y^{1/p})$ and $k(x^{1/p}, y)$; but it is not rational over the field $k(x, y^{1/p})\cap k(x^{1/p}, y) = k(x, y)$.

In a recent discussion, Weil has communicated to me the conjecture

Received by the editors September 20, 1949.

[1] This note is essentially an extract from a letter of the author to Professor André Weil, dated February 3, 1949.

that *in case of a divisor in a variety there always exists a defining field.* I shall give here a proof of this conjecture, based on the method of the associated forms.[2] Let U be a variety of dimension r in the n-space and let k be its defining field. Let $F(Z, Z^{(1)}, \cdots, Z^{(r)})$ be the associated form of U, where the (Z), $(Z^{(1)})$, $\cdots$, $(Z^{(r)})$ are $r+1$ sets of $n+1$ indeterminates each. It can be easily shown that this form $F(Z, Z^{(1)}, \cdots, Z^{(r)})$ is rational over k and absolutely irreducible. We can restrict ourselves to positive cycles in U, since every cycle rational over a field K is the difference of two positive cycles which are rational over K. To each positive d-cycle G in U (since it is also a positive d-cycle in the n-space) there corresponds an associated form $f(Z, Z^{(1)}, \cdots, Z^{(d)})$; let (f) be the set of coefficients of this form. Then the field $k((f))$ is contained in every field over which the cycle G is rational.[3] It is clear that if the cycle G is rational over $k((f))$, then this field is its defining field. Hence, to prove our conjecture, it is necessary only to show that in case $d=r-1$ the divisor G is rational over the field $k((f))=K$.

Without loss of generality we can assume that the form $f(Z, Z^{(1)}, \cdots, Z^{(r-1)})$ is irreducible over K; for otherwise we need only to repeat the same argument for each irreducible factor of this form over K. Over the algebraic closure of K the form $f(Z, Z^{(1)}, \cdots, Z^{(r-1)})$ will then be the p^eth power of a product of distinct irreducible forms, all conjugate with respect to each other over K; this p^e is by definition the multiplicity of each component of the divisor G. Let (x) be a generic point of one of the components of G over K (it does not matter which, as they are all conjugates over K). We have to prove that the field $K(x)$ has the order of inseparability $\leqq p^e$ over K.

Let $z_j^{(i)}$, $i=1, \cdots, r$, $j=1, \cdots, n$, be nr independent variables over $K(x)$, and set $z_0^{(i)} = -\sum_{j=1}^{n} x_j z_j^{(i)}$. Since the point (x) has the dimension $r-1$, the $r(n+1)-1$ elements $z_j^{(i)}$, for all i, j except $i=r$, $j=0$, are independent variables over K. Hence, the form $f(Z, z^{(1)} \cdots, z^{(r-1)})$ is exactly the p^eth power of a product of distinct linear forms. Let $f(Z_0)$ be the polynomial obtained from $f(Z, z^{(1)}, \cdots, z^{(r-1)})$ by

[2] See Chow and van der Waerden, Math. Ann. vol. 113 (1937) pp. 692–704. Strictly speaking, the concept of the associated form is defined only for varieties and positive cycles in a projective space. However, since the affine n-space can be extended to a projective space and every variety in the former can be extended to a variety in the latter in a unique way, it is easily seen that our application of this concept to varieties and positive cycles in the affine n-space is justified.

[3] If $(f)=(f_0, f_1, \cdots, f_t)$ is a set of quantities, not all zero, then the field $K((f))$ is the extension of K obtained by the adjunction of all the ratios of any two non-vanishing elements of the set (f).

setting $Z_j = z_j^{(r)}$, $j=1, \cdots, n$; then $f(Z_0)$ is the p^eth power of a separable polynomial. Since $z_0^{(r)}$ is a root of this polynomial $f(Z_0)$, it follows that $z_0^{(r)}$ has the order of inseparability p^e over the field $K(z^{(1)}, \cdots, z^{(r-1)}; z_1^{(r)}, \cdots, z_n^{(r)})$. Denote by K^* the field obtained from K by the adjunction of the nr element $z_j^{(i)}$, $i=1, \cdots, r$, $j=1, \cdots, n$. Then our result shows that $K(z^{(1)}, \cdots, z^{(r)})$ has the order of inseparability not greater than p^e over K^*. Consider now the form $F(Z, z^{(1)}, \cdots, z^{(r)})$; it dissolves over the algebraic closure of $K(z^{(1)}, \cdots, z^{(r)})$ into a product of distinct linear forms each of which corresponds to an intersection of U with the linear variety defined by the equations $\sum_{j=0}^{n} z_j^{(i)} X_j = 0$, $i=1, \cdots, r$. For, the point (x) is a simple point of U (by the definition of a divisor) and the linear variety is a generic one containing the point (x). Since the point (x) is one of these intersections, therefore the linear form $\sum_{j=0}^{n} x_j Z_j$ is a simple factor of the form $F(Z, z^{(1)}, \cdots, z^{(r)})$; it is, in fact, the only factor independent of the variables $z_j^{(i)}$, $i=1, \cdots, r$; $j=1, \cdots, n$. Therefore, the point (x) is rational over the field $K(z^{(1)}, \cdots, z^{(r)})$, and hence $K^*(x) = K(z^{(1)}, \cdots, z^{(r)})$. This shows that $K^*(x)$ has the order of inseparability not greater than p^e over K^*; and since $K(x)$ and K^* are linearly disjoint over K, it follows that $K(x)$ has also the order of inseparability not greater than p^e over K.

THE JOHNS HOPKINS UNIVERSITY

ALGEBRAIC SYSTEMS OF POSITIVE CYCLES IN AN ALGEBRAIC VARIETY.*

By WEI-LIANG CHOW.

1. Introduction. In a paper[1] some years ago van der Waerden and the author introduced the idea of the associated form of a positive cycle. For a positive cycle of dimension d and degree n in the projective space S_m of m dimensions the associated form is a form of $d+1$ sets of $m+1$ indeterminates each, which has the degree n in each set of the indeterminates. If we arrange all the monomials of this type in an arbitrary but fixed order $\omega_0, \omega_1, \cdots, \omega_t$, then each form of this type can be expressed in a unique way as a linear combination $\sum_{i=0}^{t} c_i\omega_i$ of the monomials. By assigning to each such form the point in the projective space S_t with the (homogeneous) coordinates $(c_0, c_1, \cdots, c_t)$, we obtain a one-to-one correspondence between all the forms of this type and all the points of S_t. Applying this in particular to the associated form, we can then assign to each positive cycle of dimension d and degree n in S_m a point in S_t; we shall call the coordinates of this point, which are the coefficients of the associated form, the natural coordinates of the positive cycle. In this way the totality of all the positive cycles of dimension d and degree n in S_m (or in any variety contained in S_m) is represented by a set of points in S_t. The main theorem about this representation of positive cycles in S_m by points in S_t, proved in the above mentioned paper,[2] asserts that this set of points constitutes a bunch of varieties in S_t. An important consequence of this main theorem is that it enables us to give a precise definition of an (irreducible) algebraic system of positive cycles in S_m: If V is a variety contained in the above mentioned bunch of varieties in S_t, then the set of all the positive cycles in S_m which correspond to the points in V is called an algebraic system of positive cycles. The variety V is then called the associated variety of the algebraic system. The associated variety gives a representation of the algebraic system of positive cycles in the sense that there is a one-to-one correspondence between the

* Received January 20, 1949.

[1] Chow and van der Waerden [5]. Following Weil [8], we shall use the expression "positive cycle" instead of "reducible variety."

[2] Chow and van der Waerden [5], p. 698 and p. 700.

points of V and the positive cycles of the algebraic system, and that this correspondence is of an algebraic nature. This latter statement means that there is an algebraic correspondence T between the variety V and the carrier variety U of the algebraic system [3] such that to each point of V there correspond under T exactly the points of the corresponding positive cycle. This algebraic correspondence T is called the associated correspondence of the algebraic system of positive cycles. The significance of such a representation of an algebraic system of positive cycles as an algebraic variety is easily seen, for it enables us to apply the concepts and results in the theory of algebraic varieties to the study of the algebraic systems. Thus, for example, the set of all linear subspaces (i. e. positive cycles of degree 1) of a given dimension in a projective space is represented by means of their natural coordinates, which in this special case are the well-known Pluecker coordinates, as a certain variety called the Grassmann variety, and various algebraic systems of such linear subspaces are then represented as subvarieties of the Grassmann variety. It is well-known that the study of these algebraic systems of linear subspaces is greatly facilitated by their representation as such subvarieties.

However, except in very special cases such as the Grassmann variety and its subvarieties just mentioned, the usefulness of the associated variety of an algebraic system has been hitherto rather limited. The main reason for this lies in the fact that we know very little about the properties of the associated variety besides its algebraic nature and the fact that it represents the system of positive cycles in a one-to-one manner. In fact, if we consider an associated variety in general without any restriction, it cannot have by itself any special properties; for it is easily seen that any variety whatsoever can be the associated variety of a suitably chosen algebraic system of positive cycles of any given dimension. Therefore we must look for whatever special properties that might exist in the relation between the carrier variety U and the associated variety V under the associated correspondence T; in other words, we must study the special properties of the associated correspondence T and the manner in which it connects the properties of U with those of V. In particular, it is important for the purpose of application to know under what conditions about the variety U and the correspondence T can we be sure that the associated variety V is non-singular or at least simple at a given point. For, most of the deeper concepts and results in the geometry of an algebraic variety can be applied without serious restrictions only to a

[3] The carrier variety of an algebraic system is defined as the smallest variety which contains all the cycles of the system.

non-singular variety or only to a simple point or subvariety of a variety. For example, the entire intersection theory has been up to now developed only for the case when the ambient variety is simple at the intersection point or variety, and there are strong indications that such an intersection theory cannot be extended to a substantially more general class of varieties without a radical change of its character. Thus the usefulness of the Grassmann variety in the study of the geometry of systems of linear subspaces is due by no mean measure to its being non-singular. A particular case in point is the problem of the Jacobian variety of an algebraic curve, a problem which we have studied in some detail in a recent paper.[4] We have shown there that each class of equivalent positive divisors of degree $n > 2g - 2$ on a non-singular curve C (where g is the genus of C) can be represented as a subvariety of dimension $n - g$ on the n-fold-symmetric product C^n of the curve C, and that the set of all such subvarieties forms an algebraic system. The associated variety of this algebraic system would then be a Jacobian variety of the curve C, if we could prove that it is non-singular; and this requirement of being non-singular is essential, for without this property a Jacobian variety would not be of much use in the study of the geometry of the divisor classes. In the quoted paper we have solved the problem by proving that any derived normal model of the associated variety is a Jacobian variety; however, as we have indicated there, the entire results would be much more satisfactory if we could prove that the associated variety itself is already a Jacobian variety. That this is so will follow from the results of the present paper; in fact, the attempt to prove this particular result about the Jacobian variety is the original problem from which we started out and which leads us finally to the general problem considered here.

Before we go over to the exact formulation of the problem and results, we shall give a short summary of the main definitions so that there will be no ambiguity about the terminology we are going to use in this paper. This is the more necessary in view of the fact that there has been so much divergence in terminology in the recent works in the field of algebraic geometry. Although this is a rather unfortunate state of affairs, yet it is in a sense not entirely avoidable; for the different terminologies of different authors are actually more suitable for their particular purposes, and it is difficult to adopt a uniform terminology without its being too cumbersome to be convenient. In fact, we ourselves are forced to adopt in the present paper a somewhat different terminology from that of our recent paper on the Jacobian variety; for, in our present problem the consideration of "relative" varieties

[4] Chow [4].

over a given ground field is natural and in a sense essential, while in the other paper we have only to deal with the absolute varieties.

We consider a field $\mathfrak{K}$ called the "universal domain" which is an algebraically closed field of infinite degree of transcendency over the prime field contained in it. The set of all subfields K of the universal domain $\mathfrak{K}$ such that $\mathfrak{K}$ has an infinite degree of transcendency over K, has evidently the property that the intersection of any number of fields of the set is also a field of the set, and that any subfield of $\mathfrak{K}$ which is a finite extension of any field of the set is also a field of the set. In this paper, whenever we speak of "a field," we shall always mean a field of this set; on the few occasions when we have to consider a field not in this set, we shall expressly say so and refer to such a field as an "abstract field." Any ordered set of $m+1$ elements $(x) = (x_0, x_1, \cdots, x_m)$ in the universal domain $\mathfrak{K}$, not all zero, is called a point of the projective space S_m of dimension m, and two points are regarded as identical if their coordinates are proportional. Given any point (x) in S_m and a field K, the field $K((x))$ is the extension of K generated by the adjunction of the ratios of the $m+1$ coordinates $x_0, x_1, \cdots, x_m$ of the point (x). We can regard this field $K((x))$ as the field generated by the adjunction of the point (x); it serves the same purpose as the ordinary field extension by adjunction of coordinates in the case of a point in the affine space. In similar way we can define the extension $K((x), (y), \cdots)$ generated by the adjunction of several points $(x), (y), \cdots$ in the same or different projective spaces. The degree of transcendency of the field $K((x))$ over K is called the dimension of the point (x) over K; the point (x) is called algebraic over K if its dimension over K is zero, and it is called rational over K if $K((x)) = K$. Beside the elements of the universal domain $\mathfrak{K}$, we shall consider also elements which are not in $\mathfrak{K}$, mainly as symbols by means of which we can express polynomials and functions. Such an element is called an indeterminate, and an ordered set of $m+1$ indeterminates $(X) = (X_0, X_1, \cdots, X_m)$ which are independent variables over $\mathfrak{K}$ is called a set of indeterminates in S_m. We shall always denote the indeterminates by such capital Latin letters as X, Y, Z.

Let K be a field and let (X) be a set of indeterminates in S_m. Any homogeneous prime ideal $\mathfrak{J}$ in the ring $K[X] = K[X_0, X_1, \cdots, X_m]$ is said to define a variety U/K in S_m over the field K. A point (x) is said to be in the variety U/K if all the polynomials of the ideal $\mathfrak{J}$ vanish for $(X) = (x)$. A variety W/K' over any other field K' is said to be contained in the variety U/K if every point of W/K' is in U/K; if we have furthermore $K = K'$, then the variety W/K is said to be a subvariety of U/K. Two varieties are identical if they are subvarieties of each other. Any point (x) in S_m determines over any field K a homogeneous prime ideal consisting of all the polynomials of $K[X]$ which vanish for $(X)=(x)$, and hence it determines a variety over K. A point (x') is called a specialization of the point (x) over K if it is contained in the variety determined by the point (x) over K; and if the point (x) is at the same time also a specialization of (x') over K, then the two points are called generic specializations of each other over K. Any variety U/K contains a point (x) such that the variety determined by (x) over K is identical with U/K; such a point (x) is called a generic point of U/K. Any two generic points (x) and (x') of a variety U/K are equivalent over K in the sense that the fields $K((x))$ and $K((x'))$ are K-isomorphic under the correspondence $(x) \leftrightarrow (x')$, i. e. the ratios of the corresponding coordinates are to correspond to each other. In particular, all generic points of a variety U/K have the same dimension over K; this dimension is called the dimension of the variety U/K. A finite set of varieties is called a bunch of varieties; and any point or variety contained in any one of the varieties of the bunch is said to be contained in the bunch.

Let K be a field containing K. If U/K is a variety such that its defining ideal $\mathfrak{J}$ in $K[X]$ generates also a prime ideal $\check{\mathfrak{J}}$ in $\check{K}[X]$, then the variety defined by $\check{\mathfrak{J}}$ over K is called the extension U/K of U/K over K. The extension of a variety is in a sense essentially the same as the original variety, for they not only are contained in each other but also can be defined by the same set of equations.[5] This is the reason why we denote the extension by the same symbol U as the original variety. If a subvariety W/K of U/K has also an extension W/K over $\check{K}$, then $W/\check{K}$ is also a subvariety of $U/\check{K}$. In order that a variety U/K has an extension $U/\check{K}$ over $\check{K}$, it is necessary and sufficient that the fields $\check{K}$ and $K((x))$ are linearly disjoint[6] over K, where (x) is a generic point of U/K, free with respect to $\check{K}$ over K.

Two varieties in S_m are called equivalent if they are both extensions of a third variety. It is easily seen that equivalent varieties are contained in each other and can be defined by the same set of equations; and that two varieties over the same field can only be equivalent if they are identical. We shall consider equivalent varieties as essentially the same and shall denote all of them by the same symbol over different fields such as $U/K, U/K', \cdots$. A variety over K which has an extension over any field containing K is called

---

[5] Following Weil [8], we shall say that a variety is defined by a set of equations $f_i(X)=0$, if the polynomials $f_i(X)$ generate the corresponding prime ideal in $K[X]$.

[6] See Weil [8], p. 15, Theorem 3. We observe here that the prime ideal determined by the point (x) in $\check{K}[X]$ has a basis in $K[X]$ if and only if the prime ideal determined by (x) in $K[X]$ generates a prime ideal in $\check{K}[X]$.

absolutely irreducible. If a variety is absolutely irreducible, then any variety equivalent to it is also absolutely irreducible. A complete set of mutually equivalent absolutely irreducible varieties is called an absolute variety, and any variety in the set is to be regarded as a representative of absolute variety over the particular field. If an absolute variety U has a representative U/K over a field K, then it is said to be defined over K. It is known [7] that among all the fields over which a given absolute variety U is defined, there is a smallest one which is contained in all of them; this smallest field is called the defining field of U. As an absolute variety is uniquely determined by any one of its representatives, there is little danger of confusion in identifying it sometimes with one of the latter. The absolute varieties are of great importance for us, for it is by means of them that we are going to define the concept of a cycle in S_m. Before doing this, we shall introduce the associated form of a variety.

Let U/K be a variety of dimension d in S_m. As we have shown elsewhere,[8] there is associated with U/K a K-irreducible form $F(Z, Z^{(1)}, \cdots, Z^{(d)})$ of the same degree n in each of the $d+1$ sets of indeterminates $(Z), (Z^{(1)})$, $\cdots, (Z^{(d)})$ in S_m with coefficients in K. This form is called the associated form of the variety U/K, and the number n is called the degree of the variety. Conversely, given any K-irreducible form F of this type satisfying a certain set of algebraic conditions, there is a uniquely determined variety U/K of which F is the associated form. It is easily seen that equivalent varieties have the same associated form; the converse of this statement is also true in case the varieties are separably generated or (which is the same thing) the associated form has no multiple factors. A variety is absolutely irreducible if and only if its associated form is absolutely irreducible. It follows then that an absolute variety is uniquely determined by its associated form; this is the reason why the associated forms are particularly suitable for the study of absolute varieties and cycles.

A cycle of dimension d or d-cycle in S_m is an element of the free Abelian group, the generators of which are the set of all absolute varieties of dimension d in S_m. Thus a d-cycle is a finite set of absolute varieties U_j, called its components, to each of which is assigned an integer n_j which is called the multiplicity of U_j in the d-cycle. If the degrees of the varieties U_j are s_j respectively, then the number $n = \sum_j s_j n_j$ is called the degree of the d-cycle. A cycle is said to be contained in a variety if each of its components is contained in the variety. A cycle is called positive if all the multiplicities n_j

[7] See Weil [8], pp. 70-1, Theorem 1, Corollary 3.

[8] Chow and van der Waerden [5], pp. 693-694.

are positive. If F_j are the associated forms of the varieties U_j respectively, then the form $F = \prod_j F_j^{n_j}$ is called the associated form of the positive cycle. A positive cycle is said to be rational over a field K if its associated form is rational over K, i. e. the ratios of its coefficients are rational over K; and if furthermore the associated form is irreducible over K, then the positive cycle is said to be prime rational [9] over K. Since such a K-irreducible form determines a variety over K and vice versa, it folows that there is a one-to-one correspondence between the varieties over K and the prime rational cycles over K. It is clear that every rational cycle over K can be factorized in a unique way into a sum of prime rational cycles over K corresponding to the factorization of its associated form into a product of irreducible forms over K. Therefore every rational cycle over K can also be regarded as a sum of varieties over K, each with a certain multiplicity; we shall call these varieties the K-component of the cycle.

Let (x) be a generic point of a variety U/K, then the field $K((x))$, which is uniquely determined up to an isomorphism by the variety, is called the field of U/K. Let (ξ) be any point of U/K, then the ring of all the elements of $K((x))$ of the form $f(x)/g(x)$, where $f(x)$ and $g(x)$ are two homogeneous polynomials of (x) of the same degree and $g(\xi) \neq 0$, is called the quotient ring $\mathfrak{U} = Q(U/K, (\xi))$ of U/K at the point (ξ). It is clear that if (ξ') is a generic specialization of (ξ) over K, then $Q(U/K, (\xi)) = Q(U/K, (\xi'))$, so that the ring $\mathfrak{U}$ depends only on the subvariety W/K in U/K determined by the point (ξ) or (ξ') over K, not on the particular choice of the generic point (ξ) or (ξ') of W/K. Therefore we shall also call the ring $\mathfrak{U}$ the quotient ring of U/K at the subvariety W/K, and shall sometimes write $\mathfrak{U} = Q(U/K, W)$ to indicate its dependence on W/K. The quotient ring $\mathfrak{U}$ is a local ring in the sense of Krull,[10] i. e. a Noetherian ring in which the set of all non-units is an ideal, and its maximal prime ideal $\mathfrak{u}$

---

[9] This is not the "correct" way to define a rational cycle, but is more convenient for our present purpose. The "correct" definition is as follows. A cycle is said to be rational over a field K if it satisfies the following conditions: (1) Each component of the cycle is algebraic over K; (2) if an (absolute) variety occurs in the cycle, then all the conjugate varieties over K also occur in it with the same multiplicity; (3) the multiplicity of each component is a multiple of its order of inseparability. See Weil [8], Chapter VII, § 6. Our definition differs from this "correct" definition in that the multiplicity of a component will in general be a multiple of a smaller power of the characteristic of K than the order of inseparability of this component over K.

[10] Krull [7]. Our Appendix, On the Extensions of Local Domains, contains a detailed treatment of many of the concepts and results concerning local rings which are used in the text.

consists of all elements $f(x)/g(x)$ such that $f(\xi)=0$. It is also known [11] that the completion $\mathfrak{U}^*$ of the local ring $\mathfrak{U}$ has no nilpotent elements; in other words, the zero ideal in $\mathfrak{U}^*$ is an intersection of a finite number of prime ideals. Let V/K be another variety, and let $\mathfrak{B}$ be the quotient ring $Q(V/K, (\eta))$ of V/K at one of its points (η); if W/K is the subvariety in V/K determined by the point (η) over K, then $\mathfrak{B}$ is also the quotient ring $Q(V/K, W')$ of V/K at the subvariety W'/K. We shall say that the varieties U/K and V/K are analytically equivalent [12] at the points (ξ) and (η) respectively, or at the subvarieties W/K and W'/K respectively, if the completions $\mathfrak{U}^*$ and $\mathfrak{B}^*$ of the two local rings $\mathfrak{U}$ and $\mathfrak{B}$ are K-isomorphic. Local properties of a variety at a point or a subvariety which are invariant under analytical equivalences are called locally analytic properties. It is known [13] that the property of a variety being simple at a point or subvariety in the absolute sense is a locally analytic property.

For an absolute variety U the concept of a quotient ring can only be defined with respect to a given field of definition. However, in order that this concept has a significance independent of the choice of the particular field of definition, it is advisable to restrict ourselves to the case where the subvariety W is also an absolute variety and the field K is a field of definition for both U and W. Then the ring $Q(U/K, W)$ is called the quotient ring of the absolute variety U at the absolute subvariety W over the field K. This includes in particular the case when W has the dimension 0 and hence consists of a point which is rational over K. The advantage for imposing these restrictions regarding W and K lies in the fact that if K' is any field containing K, then the quotient ring $Q(U/K', W)$ is the extension [14] of the quotient ring $Q(U/K, W)$ over K', so that the ring $Q(U/K', W)$ is uniquely determined by the ring $Q(U/K, W)$ and the field K'. Furthermore, the completion of $Q(U/K', W)$ is the completion of the extension of the completion of $Q(U/K, W)$ over K', so that the completion of $Q(U/K', W)$ is also uniquely determined by the completion of $Q(U/K, W)$ and the field K'. Let V be another absolute variety and W' be an absolute subvariety in V; let K_0 be the smallest field over which all the varieties U, V, W, W' are defined. Then the absolute varieties U and V are said to be analytically

---

[11] See Chevalley [3], p. 11, Theorem 1.

[12] In this definition of analytical equivalence we are following Zariski [10], p. 49, Definition 4.

[13] See Zariski [10], pp. 49-51, Theorem 15, where this theorem is proved for the most general case.

[14] See Appendix.

equivalent at the subvarieties W and W' respectively, if the completions of the two local rings $Q(U/K_0, W)$ and $Q(V/K_0, W')$ are K_0-isomorphic. It follows then from what we have just said above that if K is any field containing K_0, then the completions of the two local rings $Q(U/K, W)$ and $Q(V/K, W')$ are also K-isomorphic; or, in other words, the (relative) varieties U/K and V/K are analytically equivalent at the subvarieties W/K and W'/K respectively.

Let U/K and V/K be two varieties in the space S_m and S_t respectively, and let (x) be a generic point of U/K. If V/K has a generic point (y) which is rational over $K((x))$, then the correspondence $(x) \to (y)$ defines a rational transformation T of U/K onto V/K; and the pair (x), (y) is said to be a pair of corresponding generic points of U/K and V/K respectively under the rational transformation T, and we shall write $(y) = T(x)$. For every point (ξ) of U/K, its image under the rational transformation T consists of all the points (η) such that (ξ, η) is a specialization of (x, y) over the specialization $(x) \to (\xi)$ over K. If for a suitably chosen linear form $\sum_{j=0}^{t} a_j y_j$, the elements $y_i / \sum_{j=0}^{t} a_j y_j$, $i = 0, 1, \cdots, t$, are all in the quotient ring $Q(U/K, (\xi))$, then the transformation T is said to be regular at the point (ξ). In this case the image of (ξ) under T is a uniquely determined point of V/K. The rational transformation T is called birational, if conversely the point (x) is also rational over $K((y))$, i. e. if we have $K((x)) = K((y))$; the resulting rational transformation of V/K onto U/K is called the inverse T^{-1} of the transformation T. If a birational transformation T is regular at a point (ξ) and the inverse transformation T^{-1} is also regular at the image point (η) of (ξ) under T, then the transformation is said to be biregular at this point (ξ) or at the pair of points (ξ), (η). Two varieties U/K and V/K which are in biregular correspondence at the two points (ξ) and (η) respectively, have evidently the same quotient rings $Q(U/K, (\xi))$ and $Q(V/K, (\eta))$; hence they are also analytically equivalent at these two points. These concepts can be carried over to the case of two absolute varieties, just as we have done above for the definitions of quotient ring and analytical equivalence; we have then to replace the points (ξ) and (η) by two absolute subvarieties W and W' in U and V respectively, and the field K must be a field of definition for all the varieties U, V, W, W'.

2. Statement of the problem and result. Let U/K be a variety of r dimensions in the projective space S_m. As we have indicated at the beginning of the previous section, any positive d-cycle of degree n in U/K can be represented through its associated form by a point in the projective space S_t, and the set of all such points in S_t constitutes exactly the points of a bunch of varieties over K. Let V/K be any variety contained in this bunch, then the set of all the positive cycles in U/K corresponding to the points in V/K is called an algebraic system of positive cycles in U/K, and the variety V/K is called the associated variety of the algebraic system. Let (y) be a generic point of V/K, then the cycle $G(y)$ determined by (y) is called a generic cycle of the algebraic system and we shall denote the algebraic system itself by $|G(y)|$. We shall assume that the generic cycle $G(y)$ is prime rational over the field $K((y))$, and that a generic point (x) of $G(y)$ over $K((y))$ is also a generic point of U/K. This last assumption means simply that U/K is the carrier variety of the algebraic system $|G(y)|$. The pair of points (x), (y) then determines an algebraic correspondence T between the varieties U/K and V/K, which we shall call the associated correspondence of the algebraic system $|G(y)|$.

The problem with which we are concerned in this paper can be described as follows. Let $\bar{T}$ be any (K-irreducible) algebraic correspondence between the variety U/K and another variety $\bar{V}/K$ (which may of course lie in a different space). If $(\bar{y})$ is a generic point of $\bar{V}/K$, then the set of all points in U/K which correspond to $(\bar{y})$ under the correspondence $\bar{T}$ constitutes a variety or prime rational cycle $G(\bar{y})$ in U/K over the field $K((\bar{y}))$. If this cycle $G(\bar{y})$ has the dimension d and the degree n, then there corresponds to $G(\bar{y})$ by means of its natural coordinates a point (y) in the projective space S_t; let V/K be the variety in S_t determined by the generic point (y) over the field K. This variety V/K then determines an algebraic system of positive cycles in U/K, of which the cycle $G(y) = G(\bar{y})$ is a generic member, and there is an algebraic correspondence T between U/K and V/K associated with this algebraic system. Thus every algebraic correspondence $\bar{T}$ between U/K and a variety $\bar{V}/K$ induces in U/K an algebraic system of positive cycles, but in general the variety $\bar{V}/K$ is of course different from the associated variety V/K of the algebraic system and the correspondence $\bar{T}$ is different from the associated correspondence T. In fact, under the correspondence $\bar{T}$ between U/K and $\bar{V}/K$, the points of $\bar{V}/K$ will not in general be in one-to-one correspondence with the positive cycles of the algebraic system without exception; that is, to some points of $\bar{V}/K$ there might correspond more than one or even an infinite number of cycles of the algebraic system.[15] The question arises as to whether there are any other properties which distinguish

[15] For a more detailed description of the relations between the correspondences T and $\bar{T}$, see Chow and van der Waerden [5], § 2.

the associated variety V/K from the rather arbitrarily chosen variety $\bar{V}/K$, apart from the fact that V/K represents in a one-to-one manner the cycles of the algebraic system $|G(y)|$. One would naturally ask for those properties which are invariant under a biregular birational transformation of the associated variety V/K. In terms of local geometry we can formulate the question as follows: Let $(\bar{\eta})$ be a point of $\bar{V}/K$ to which there corresponds a uniquely determined cycle $G(\bar{\eta})$ in U/K and let (η) be the corresponding point in V/K so that we have $G(\eta) = G(\bar{\eta})$. Then both varieties V/K and $\bar{V}/K$ can be said to represent through the correspondences T and $\bar{T}$ respectively the algebraic system of positive cycles $|G(y)| = |G(\bar{y})|$ in a one-to-one manner in the neighborhoods of the points (η) and $(\bar{\eta})$ respectively. The question is whether there are any locally birational or analytic properties which distinguish the associated correspondence T in the neighborhood of the pair (η), $G(\eta)$ from the correspondence $\bar{T}$ in the neighborhood of the pair $(\bar{\eta})$, $G(\bar{\eta})$. In this paper we shall give an answer to this question in the special case where the correspondence $\bar{T}$ is a rational transformation of U/K onto $\bar{V}/K$ and consequently the associated correspondence T is also a rational transformation of U/K onto V/K. We shall show that in this case the local analytic properties of the variety V/K in the neighborhood of the point (η) has a very simple connection with the local properties of the variety U/K in the neighborhood of the cycle $G(\eta)$, which is not true in general for the variety $\bar{V}/K$ in the neighborhood of $(\bar{\eta})$.

We shall from now on restrict ourselves to the case where the associated correspondence T of the algebraic system $|G(y)|$ is a rational transformation of U/K onto V/K. It is clear in this case that the variety V/K has the dimension $r-d$. We shall call such an algebraic system of positive cycles an involutional system, on account of the fact that in case $d=0$ such a system is essentially what is usually called an involution in the classical algebraic geometry. For the sake of convenience, we shall assume that the ground field K contains infinitely many elements. Our main result is the following theorem:

THEOREM. *Let U/K be a variety; let $|G(y)|$ be an involutional system of positive cycles in U/K with the associated variety V/K and the associated transformation T, where (y) is a generic point of V/K. Let (η) be a rational point of V/K and (ξ) be a generic point of a K-component G'/K of $G(\eta)$; we assume that the field $K((\xi))$ is analytically disjoint*[16] *with respect to the quotient ring $Q(V/K, (\eta))$ over K; this implies in particular that*

[16] For the definition of analytical disjointness, see Appendix.

$K((y))$ and $K((\xi))$ are linearly disjoint over K, so that the product variety W/K of V/K and G'/K is defined. If the transformation T is regular at the point (ξ) and the variety G'/K is a simple K-component of $G(\eta)$ and is separably generated, then the varieties U/K and W/K are analytically equivalent at the points (ξ) and (η, ξ) respectively.

For the sake of greater generality, we have stated our theorem above for the "relative" varieties; this is the reason why we have to make the assumptions that the point (η) is rational and the field $K((\xi))$ is analytically disjoint with respect to the quotient ring $Q(V/K, (\eta))$ over K. In the applications, however, we are mostly concerned with the case where both varieties U/K and V/K are absolutely irreducible; in other words, we are concerned with an involutional system of positive cycles in an absolute variety U with an absolute variety V as the associated variety. It is then natural to assume that G' is an absolute variety, i. e. an (absolute) component of the positive cycle $G(\eta)$, and that the field K is the smallest field over which all the varieties U, V, G', (η) are defined. Since the field $K((\xi))$ is in this case a regular extension of K, it follows (Appendix, Theorem 4) that $K((\xi))$ is analytically disjoint with respect to any local domain with K as a basic field, hence in particular with respect to the local domain $Q(V/K, (\eta))$ over K. Recalling our definition of the analytical equivalence between two absolute varieties, we can state our theorem for absolute varieties as follows:

THEOREM. *Let U be an absolute variety; let $|G(y)|$ be an involutional system of positive cycles in U with the associated (absolute) variety V and the associated transformation T. Let (η) be any point of V and let G' be a component of $G(\eta)$. If G' is a simple component of $G(\eta)$ and the transformation T is regular at the subvariety G', then the variety U and the product variety $W = V \times G'$ are analytically equivalent at the subvarieties G' and $(\eta) \times G'$ respectively.*

From this theorem we can derive as an immediate corollary a criterion for simple points or subvarieties. Since the product variety W is simple at the subvariety $(\eta) \times G'$ if and only if both V and G' are simple at the subvarieties (η) and G' respectively, and since the variety G' is certainly simple at the variety G' itself, i. e. at its generic point, it follows that under the conditions of the above theorem the variety U is simple at the subvariety G' if and only if the variety V is simple at the point (η). It is evident that the application of this criterion to the problem of Jacobian variety gives an affirmative answer to the question we have posed at the beginning of the previous section.

Finally, we should like to add that the significance of our theorem goes beyond a mere criterion for simple points; it gives us a certain information about the analytic structure of U in the neighborhood of a cycle of an algebraic system. In order to express this more clearly, we introduce yet another definition. An involutional system $|G(y)|$ in an absolute variety U is called a fiber system if the transformation T is regular at every point of U and the cycle $G(\eta)$ is an absolute variety (i. e. consisting of one component with multiplicity 1) for every point (η) of V. It is evident that in this case each point of U belongs to exactly one cycle of the system, so that the entire variety U is fibered by the algebraic system of cycles. Our result can then be expressed roughly as follows: If $|G(y)|$ is a fiber system in U with the associated variety V and the associated transformation T, then U has properties similar to those of a fiber space in the topology, with V as the base space and T as the projection. In fact, there seems to be more than a formal analogy between this algebraic "fiber space" and the fiber space in topology; we hope to be able to come back to these questions on some future occasions.

Before we proceed with the proof of our theorem, we shall make a change in our terminology. It is clear that our main theorem is of a purely local nature, and as such it can be stated just as well for varieties in affine spaces instead of projective spaces. In fact, this is the form in which we shall prove the theorem in the next two sections; for, as is usually the case with a problem of purely local nature, it is very convenient to be able to work with the affine coordinates. It is easily seen that most of the definitions and concepts as developed and indicated in this and the previous sections can be carried over with suitable modifications to the varieties and points in affine spaces, and we can replace such field extensions as $K((x))$ by the ordinary field extension $K(x)$. Thus we shall from now on consider only varieties and points in a suitably chosen affine space in the projective space S_m; we shall denote this affine space also by S_m and represent a point in it by the affine coordinates $(x) = (x_1, \cdots, x_m)$. We shall assume that the affine S_m has been so chosen that the d-cycle $G(\eta)$ is finite; this implies of course that the variety U/K as well as the generic cycle $G(y)$ of the system are also finite. A variety is said to be finite with respect to an affine space if every generic point of it is finite (i. e. in the affine space), and a cycle is said to be finite if all its components are finite. Similarly, we can assume that the affine S_t has been so chosen that the point (η) is finite, which implies that the variety V/K is also finite; but for the sake of convenience later in our proof, we shall make a more definite choice of this affine space S_t with respect to a given choice of the affine coordinate system in the affine space S_m.

We set $\omega_0 = (Z_0 Z^{(1)}{}_1 \cdots Z^{(d)}{}_d)^n$ in the ordered set of monomials $\omega_0, \omega_1, \cdots, \omega_t$ in the $d+1$ set of indeterminates $(Z), (Z^{(1)}), \cdots, (Z^{(d)})$ as defined in the beginning of the previous section. Then the affine S_t is defined as the set of all points $(c_0, c_1, \cdots, c_t)$ with $c_0 \neq 0$, and such a point will be denoted by its affine coordinates $(c_1/c_0 \cdots, c_t/c_0)$. In order to insure that the point (η) is finite with respect to such a choice of the affine S_t, we shall assume that the affine coordinate system in the affine space S_m has been so chosen that the linear variety defined by the d equations $X_i = v_i$, $i = 1, \cdots, d$, where the $v_1, \cdots, v_d$ are independent variables over K, intersects the d-cycle $G(\eta)$ properly and finitely, i. e. the intersection consists of a finite number of finite points. Then the coefficient of the term ω_0 in the associated form $F_{(\eta)}(Z, Z^{(1)}, \cdots, Z^{(d)})$ of $G(\eta)$ cannot be zero. For, let $v^{(i)}{}_0 = -v_i$, $v^{(i)}{}_i = 1$, $v^{(i)}{}_j = 0$, for $i = 1, \cdots, d$, and $j \neq i$, then the form $F_{(\eta)}(Z, v^{(1)}, \cdots, v^{(d)})$ is the associated form of the intersection of $G(\eta)$ with the linear variety. It is easily seen that the coefficient of the term $Z_0{}^n$ in the form $F_{(\eta)}(Z, v^{(1)}, \cdots, v^{(d)})$ is exactly the coefficient of the term ω_0 in the form $F_{(\eta)}(Z, Z^{(1)}, \cdots, Z^{(d)})$. If this coefficient were zero, then the form $F_{(\eta)}(Z, v^{(1)}, \cdots, v^{(d)})$ must contain as factor at least one linear form without the term Z_0, and this linear form would then correspond to an infinite intersection point of $G(\eta)$ with the linear variety, in contradiction to our assumption. Thus the coefficient of the term ω_0 in the form $F_{(\eta)}(Z, Z^{(1)}, \cdots, Z^{(d)})$ is not zero, and this means that the point (η) is in the affine space S_t. From now on we shall assume that the associated form has been so normalized that the coefficient of the term ω_0 is equal to 1; the coefficients of the rest of the terms will then be the affine coordinates of the corresponding cycle.

3. Two lemmas. We begin by recalling some well known facts concerning the semi-local rings. A ring $\mathfrak{o}$ is called a semi-local ring,[17] if it is a Noetherian ring and contains only a finite number of maximal prime ideals. We shall consider $\mathfrak{o}$ also as a topological space in the following way. Since $\mathfrak{o}$ is in particular a group with respect to the addition, we can define a topology in $\mathfrak{o}$ by specifying a fundamental system of neighborhoods of the element 0. Let $\mathfrak{a}$ be the product of all the maximal prime ideals of $\mathfrak{o}$, then the sets $\mathfrak{a}^s$, for $s = 1, 2, \cdots$, will be taken as a fundamental system of neighborhoods of 0. It can be shown that the usual conditions are satisfied; they are all trivial except the one asserting that $\bigcap_{s=1}^{\infty} \mathfrak{a}^s = 0$, which is the well known

[17] Chevalley [1].

theorem due to Krull. Thus $\mathfrak{o}$ is a topological group and consequently a regular space; moreover, not only the addition but also the multiplication is a continuous function in $\mathfrak{o}$.

Let $\mathfrak{o}'$ be another semi-local ring which contains $\mathfrak{o}$ as a sub-ring, and let $\mathfrak{o}'$ be topologized in a similar way by means of the product $\mathfrak{a}'$ of its maximal prime ideals. The question arises as to the relation between the topologies of the two rings $\mathfrak{o}$ and $\mathfrak{o}'$. In general, they are not concordant; in other words, the sets $\mathfrak{a}'^s \cap \mathfrak{o}$ do not in general constitute a fundamental system of neighborhoods of 0 in $\mathfrak{o}$. However, in the important special case when $\mathfrak{o}'$ is a finite $\mathfrak{o}$-module (and this implies itself that $\mathfrak{o}'$ is a semi-local ring) and no non-zero element of $\mathfrak{o}$ is a zero-divisor in $\mathfrak{o}'$, it can be shown [18] that the sets $\mathfrak{a}'^s \cap \mathfrak{o}$ also constitute a fundamental system of neighborhoods of 0 in $\mathfrak{o}$, so that $\mathfrak{o}'$ contains $\mathfrak{o}$ not only as a sub-ring, but also as a sub-space. Explicitly this means that, in this case, there exist two sequences of positive integers $l(s)$ and $m(s)$, with $l(s) \to \infty$ and $m(s) \to \infty$ as $s \to \infty$, such that the relations $\mathfrak{a}^{l(s)} \subset \mathfrak{a}'^s \cap \mathfrak{o} \subset \mathfrak{a}^{m(s)}$ hold for all s.

Let U/K and V/K be two varieties of the same dimension r in the affine spaces S_m and S_t respectively. Let T be a rational transformation of U/K onto V/K, and let (x) and $(y) = T(x)$ be a pair of corresponding generic points of U/K and V/K respectively. Then $K(x)$ is a finite algebraic extension of $K(y)$; let n be the degree of the extension. To the generic point (y) of V/K there correspond in the inverse transformation T^{-1} n (not necessarily distinct) points $(x^{(1)}) = (x), (x^{(2)}), \cdots, (x^{(n)})$ of U/K, which are all generic points of U/K and which together constitute a complete set of conjugates over $K(y)$. We shall say that the inverse transformation T^{-1} is defined at a point (η) of V/K, if there is a finite specialization $(x^{(1)}, \cdots, x^{(n)}) \to (\xi^{(1)}, \cdots, \xi^{(n)})$ over the specialization $(y) \to (\eta)$, such that every specialization of $(x^{(1)}, \cdots, x^{(n)})$ over the specialization $(y) \to (\eta)$ coincides with this one except possibly for a reordering of the points $(\xi^{(1)}), \cdots, (\xi^{(n)})$. It is clear that if the rational transformation T is one which is associated with an involutional system of positive 0-cycles on U/K, so that V/K is the associated variety of the system, then the inverse transformation T^{-1} is defined at every point of the variety V/K, for a suitable choice of the affine spaces S_m and S_t.

LEMMA I. *Let T be a rational transformation of a variety U/K onto a variety V/K, both of the same dimension r. Let (ξ) be a point of U/K*

---

[18] See Chevalley [1], p. 699, Proposition 7. For the properties of semi-local rings used in the following, we refer once for all to this paper of Chevalley.

such that T is regular at (ξ) and the inverse transformation T^{-1} is defined at the image point $(\eta) = T(\xi)$. Then the quotient ring $\mathfrak{U} = Q(U/K, (\xi))$ contains the quotient ring $\mathfrak{V} = Q(V/K, (\eta))$ as a subspace.

Proof. Let $\mathfrak{r} = K[y]$ be the coordinate ring of V/K, and let $\mathfrak{p}$ be the prime ideal in $\mathfrak{r}$ determined by the point (η), and let $\mathfrak{s} = \mathfrak{r} - \mathfrak{p}$ be the multiplicatively closed system of all the elements of $\mathfrak{r}$ not in $\mathfrak{p}$. Then the quotient ring of $\mathfrak{r}$ with respect to $\mathfrak{s}$ is a local ring $\mathfrak{o}$, and the ideal $\mathfrak{m} = \mathfrak{o}\mathfrak{p}$ is the maximal prime ideal of $\mathfrak{o}$. It is easily seen that $\mathfrak{o}$ is the quotient ring $\mathfrak{V} = Q(V/K, (\eta))$ of V/K at (η). Let $\mathfrak{r}' = K[x]$ be the coordinate ring of U/K, and let $\mathfrak{p}'_1, \cdots, \mathfrak{p}'_n$ be the n prime ideals (not necessarily distinct) in $\mathfrak{r}'$ determined by the n points $(\xi^{(1)}), \cdots, (\xi^{(n)})$, and let $\mathfrak{s}' = \bigcap_{i=1}^{n} (\mathfrak{r}' - \mathfrak{p}'_i)$ be the multiplicatively closed system of all elements of $\mathfrak{r}'$ not in any one of the ideals $\mathfrak{p}'_1, \cdots, \mathfrak{p}'_n$. Let $\mathfrak{o}'$ be the quotient ring of $\mathfrak{r}'$ with respect to $\mathfrak{s}'$. Then $\mathfrak{o}'$ is a semi-local ring, and its maximal prime ideals are the ideals $\mathfrak{m}'_i = \mathfrak{o}'\mathfrak{p}'_i$ $(i = 1, \cdots, n)$, which are not necessarily distinct. Let $\mathfrak{o}'_i$ be the quotient ring of $\mathfrak{o}'$ with respect to the multiplicatively closed system $\mathfrak{o}' - \mathfrak{m}'_i$; each ring $\mathfrak{o}'_i$ is a local ring with the maximal prime ideal $\mathfrak{o}'_i\mathfrak{m}'_i$. It is clear that $\mathfrak{o}'_1$ is the quotient ring $\mathfrak{U} = Q(U/K, (\xi))$ of U/K at (ξ). We have to show that if $\mathfrak{o}'_1 \supset \mathfrak{o}$, then $\mathfrak{o}$ is a sub-space of $\mathfrak{o}'_1$. We shall achieve this by constructing a bigger ring which contains both $\mathfrak{o}'_1$ and $\mathfrak{o}$ as sub-spaces.

Consider the ring $\mathfrak{R} = K[x^{(1)}, \cdots, x^{(n)}]$, and for each i $(i = 1, \cdots, n)$, let $\mathfrak{P}_{ij}$, $j = 1, \cdots, a$, be the distinct prime ideals determined by the specializations of $(x^{(1)}, \cdots, x^{(n)})$ over the specialization $(x^{(i)}) \rightarrow (\xi^{(1)})$ over K. Since there is an automorphism of the field $k(x^{(1)}, \cdots, x^{(n)})$ over $K(y)$ which carries any one of the points $(x^{(i)})$ into any other, it follows that we can arrange the ideals $\mathfrak{P}_{ij}$ in such a way that the ideals $\mathfrak{P}_{1j}, \cdots, \mathfrak{P}_{nj}$ are conjugate to each other for every $j = 1, 2, \cdots, a$. Let $\mathfrak{S} = \bigcap_{i,j} (\mathfrak{R} - \mathfrak{P}_{ij})$ be the multiplicatively closed system of all elements of $\mathfrak{R}$ not contained in any one of the prime ideals $\mathfrak{P}_{ij}$, and let $\mathfrak{O}$ be the quotient ring of $\mathfrak{R}$ with respect to $\mathfrak{S}$. The ring $\mathfrak{O}$ is then a semi-local ring, and its maximal prime ideals are the ideals $\mathfrak{M}_{ij} = \mathfrak{O}\mathfrak{P}_{ij}$. Let $\bar{\mathfrak{O}}$ be the integral closure of $\mathfrak{O}$; since $\bar{\mathfrak{O}}$ is a finite $\mathfrak{O}$-module, it is also a semi-local ring. For every i and j, let $\bar{\mathfrak{M}}_{ijk}$ $(k = 1, \cdots, b)$ be the distinct maximal prime ideals of $\bar{\mathfrak{O}}$ such that $\bar{\mathfrak{M}}_{ijk} \cap \mathfrak{O} = \mathfrak{M}_{ij}$; and we shall arrange these ideals in such a way that the ideals $\bar{\mathfrak{M}}_{1jk}, \cdots, \bar{\mathfrak{M}}_{njk}$ are conjugate to each other, for every j and k. Now, since every element of $\mathfrak{o}$ is finite over any finite specialization of the elements of $\mathfrak{O}$, it is integrally dependent on $\mathfrak{O}$; hence, we have $\bar{\mathfrak{O}} \supset \mathfrak{o}$. On the other hand, since every element of $\bar{\mathfrak{O}}$ is also finite over any finite specialization of the elements of $\mathfrak{o}$ and is consequently integrally dependent on $\mathfrak{o}$, we conclude that $\bar{\mathfrak{O}}$ is the integral closure of $\mathfrak{o}$ in the field $K(x^{(1)},\cdots,x^{(n)})$. From this it follows that $\bar{\mathfrak{O}}$ is a finite $\mathfrak{o}$-module, and hence $\bar{\mathfrak{O}}$ contains $\mathfrak{o}$ as a sub-space, and we have $\bar{\mathfrak{M}}_{ijk}\cap\mathfrak{o}=\mathfrak{m}$.

It is clear from our construction that $\mathfrak{R}\supset\mathfrak{r}'$; and, as all the prime ideals $\mathfrak{P}_{ij}$ of $\mathfrak{R}$ contract in $\mathfrak{r}'$ to one of ideals $\mathfrak{p}'_1,\cdots,\mathfrak{p}'_n$, we have $\mathfrak{r}'\cap\mathfrak{S}\subset\mathfrak{s}'$, from which it follows that $\mathfrak{O}\supset\mathfrak{o}'$. The ideals $\mathfrak{P}_{1j}$ $(j=1,\cdots,a)$ are evidently the only maximal prime ideals of $\mathfrak{R}$ which contract to $\mathfrak{p}'_1$ in $\mathfrak{r}'$. It follows that the ideals $\mathfrak{M}_{1j}$ $(j=1,\cdots,a)$ consist of all the maximal prime ideals of $\mathfrak{O}$ which contract to $\mathfrak{m}'_1$ in $\mathfrak{o}'$. Hence the ideals $\bar{\mathfrak{M}}_{1jk}$ are all the maximal prime ideals of $\bar{\mathfrak{O}}$ which contract to $\mathfrak{m}'_1$ to $\mathfrak{o}'$. Let $\bar{\mathfrak{O}}_1$ be the quotient ring of $\mathfrak{O}$ with respect to the multiplicatively closed system $\bigcap_{j,k}(\bar{\mathfrak{O}}-\bar{\mathfrak{M}}_{1jk})$. Since $\bar{\mathfrak{M}}_{1jk}\cap\mathfrak{o}'=\mathfrak{m}'_1$, the ring $\mathfrak{o}'_1$ is contained in $\bar{\mathfrak{O}}_1$. On the other hand, it is easily seen that every element of $\bar{\mathfrak{O}}_1$ is finite over any finite specialization of the elements of $\mathfrak{o}'_1$ and hence is integrally dependent on $\mathfrak{o}'_1$. Therefore, $\bar{\mathfrak{O}}_1$ is the integral closure of $\mathfrak{o}'_1$ in the field $K(x^{(1)},\cdots,x^{(n)})$, from which it follows that $\bar{\mathfrak{O}}_1$ is a finite $\mathfrak{o}'_1$-modul and consequently contains $\mathfrak{o}'_1$ as a subspace.

We shall now show that $\bar{\mathfrak{O}}_1$ contains also $\mathfrak{o}$ as a sub-space. Let $\mathfrak{A}_i=\prod_{j,k}\bar{\mathfrak{M}}_{ijk}$ and $\mathfrak{A}=\prod'_{i,j,k}\bar{\mathfrak{M}}_{jik}$, where the product $\prod'$ extends over all the distinct ideals only among the $\bar{\mathfrak{M}}_{ijk}$. Since $\bar{\mathfrak{O}}$ contains $\mathfrak{o}$ as a sub-space, and since the sets $\mathfrak{A}^s$ $(s=1,2,\cdots)$ constitute by definition a fundamental system of neighborhoods of 0 in $\bar{\mathfrak{O}}$, we have the relations $\mathfrak{a}^{l(s)}\subset\mathfrak{A}^s\cap\mathfrak{o}\subset\mathfrak{a}^{m(s)}$ with $l(s)\to\infty$ and $m(s)\to\infty$, as $s\to\infty$. Since any two distinct ones of the ideals $\bar{\mathfrak{M}}_{ijk}{}^s$ are relatively prime to each other, we have

$$\mathfrak{A}^s=\prod_{i,j,k}{}'\bar{\mathfrak{M}}_{ijk}{}^s=\bigcap_{i,j,k}\bar{\mathfrak{M}}_{ijk}{}^s=\bigcap_i\left(\bigcap_{j,k}\bar{\mathfrak{M}}_{ijk}{}^s\right)=\bigcap_i\left(\prod_{j,k}\bar{\mathfrak{M}}_{ijk}{}^s\right)$$
$$=\bigcap_i\mathfrak{A}_i{}^s.$$

Hence we have $\mathfrak{a}^{l(s)}\subset\bigcap_i(\mathfrak{A}_i{}^s\cap\mathfrak{o})\subset\mathfrak{a}^{m(s)}$. Now, since the ideals $\mathfrak{A}_i{}^s$ are all conjugate to each other over $K(y)$, we have evidently $\mathfrak{A}_1{}^s\cap\mathfrak{o}=\cdots=\mathfrak{A}_n{}^s\cap\mathfrak{o}$. Hence we have the relations $\mathfrak{a}^{l(s)}\subset\mathfrak{A}_1{}^s\cap\mathfrak{o}\subset\mathfrak{a}^{m(s)}$, which means that the sets $\mathfrak{A}_1{}^s\cap\mathfrak{o}$ constitute a fundamental system of neighborhoods of 0 in $\mathfrak{o}$. Now, the sets $(\bar{\mathfrak{O}}_1\mathfrak{A}_1)^s$ constitute by definition a fundamental system of neighborhoods of 0 in $\bar{\mathfrak{O}}_1$, and we have from the theory of quotient rings [19]

[19] We observe that since the $\bar{\mathfrak{M}}_{1jk}$ are maximal prime ideals in $\mathfrak{O}$, we have $\mathfrak{A}_1{}^s=\bigcap_{j,k}\bar{\mathfrak{M}}_{1jk}{}^s$, and each $\bar{\mathfrak{M}}_{1jk}{}^s$ is a primary ideal. Therefore we have $\bar{\mathfrak{O}}_1\bar{\mathfrak{M}}_{1jk}{}^s\cap\bar{\mathfrak{O}}=\bar{\mathfrak{M}}_{1jk}{}^s$ and hence also $\bar{\mathfrak{O}}_1\mathfrak{A}_1{}^s\cap\bar{\mathfrak{O}}=\mathfrak{A}_1{}^s$.

that $(\mathfrak{O}_1\mathfrak{A}_1)^s \cap \mathfrak{O} = \mathfrak{A}_1{}^s$. Therefore we have $(\mathfrak{O}_1\mathfrak{A}_1)^s \cap \mathfrak{o} = (\mathfrak{O}_1\mathfrak{A}_1)^s \cap \mathfrak{O} \cap \mathfrak{o} = \mathfrak{A}_1{}^s \cap \mathfrak{o}$, which shows that $\mathfrak{o}$ is a sub-space of $\mathfrak{O}_1$.

Since the ring $\mathfrak{O}_1$ contains both $\mathfrak{o}'_1$ and $\mathfrak{o}$ as sub-spaces, and since (on account of the regularity of the rational transformation T at $(\xi^{(1)})$) $\mathfrak{o}'_1$ contains $\mathfrak{o}$, it follows that $\mathfrak{o}'_1$ contains $\mathfrak{o}$ as a sub-space. Thus the lemma is proved.

[Note added in proof (March 10, 1950): Professor Zariski has kindly called our attention to the fact that our Lemma I is closely related to a result in his recent note in the *Proceedings of the National Academy of Sciences*, vol. 35 (1949), pp. 62-66, Theorem 3. In fact, if the variety V/K is analytically irreducible at the point (η), our Lemma I is an immediate consequence of Zariski's theorem. In case V/K is not analytically irreducible, it is also possible to deduce our Lemma I from Zariski's theorem, if we make use of Zariski's theory of normal varieties. This can be done in the following way, as communicated to us by Professor Zariski.

Let $\bar{U}/K$ and $\bar{V}/K$ be derived normal models of U/K and V/K respectively. Let $(\bar{\xi}^{(1)}), (\bar{\xi}^{(2)}), \cdots, (\bar{\xi}^{(g)})$ be the points of $\bar{U}/K$ which correspond to (ξ) and let similarly $(\bar{\eta}^{(1)}), (\bar{\eta}^{(2)}), \cdots, (\bar{\eta}^{(h)})$ be the points of $\bar{V}/K$ which correspond to (η). Since (η) is the only point of V/K which corresponds to (ξ) under T, the points of $\bar{V}/K$ which correspond to any $(\bar{\xi}^{(i)})$ (in the rational transformation $\bar{T}$ of $\bar{U}/K$ onto $\bar{V}/K$) are among the points $(\bar{\eta}^{(j)})$ and hence are finite in number. Since $\bar{U}/K$ is normal it follows that to each point $(\bar{\xi}^{(i)})$ there corresponds a unique point, say $(\bar{\eta}^{(\mu_i)})$, and that $\bar{T}$ is regular at $(\bar{\xi}^{(i)})$. Recalling the fact that a normal variety is analytically irreducible at any one of its points (see Zariski, *Annals of Mathematics*, vol. 49 (1948), pp. 352-361), it follows from Zariski's theorem that

(1) the quotient ring of $(\bar{\eta}^{(\mu_i)})$ is a sub-space of the quotient ring of $(\bar{\xi}^{(i)})$ $(i = 1, 2, \cdots, g)$.

Now let us assume that in every specialization $(x^{(1)}, x^{(2)}, \cdots, x^{(n)}) \rightarrow (\xi^{(1)}, \xi^{(2)}, \cdots, \xi^{(n)})$ over the specialization $(y) \rightarrow (\eta)$ the point (ξ) occurs at least once among the $(\xi^{(i)})$ (this assumption is weaker than the assumption that the inverse transformation T^{-1} is defined at (η)). We have for each $j = 1, 2, \cdots, h$ the specialization $(y, \bar{y}) \rightarrow (\eta, \bar{\eta}^{(j)})$ (where $(\bar{y})$ is the point $\bar{V}/K$ which corresponds to the general point (y) of V/K), and if we apply the above assumption to the specializations of $(x^{(1)}, x^{(2)}, \cdots, x^{(n)})$ which are over the specialization $(y, \bar{y}) \rightarrow (\eta, \bar{\eta}^{(j)})$ we conclude at once that

(2) each of the h points $(\bar{\eta}^{(j)})$ occurs at least once among the points $(\bar{\eta}^{(\mu_i)})$, $i = 1, 2, \cdots, g$.

From (1) and (2) it follows that the intersection of the quotient rings $\mathfrak{o}(\bar{\eta}^{(j)})$ of the h points $(\bar{\eta}^{(j)})$ is a sub-space of the intersection of the quotient rings $\mathfrak{o}'(\bar{\xi}^{(i)})$ of the g points $(\bar{\xi}^{(i)})$. Since $\bar{V}/K$ is a derived normal model of V/K, the quotient ring $\mathfrak{o}$ of the point (η) is a sub-space of $\bigcap_j \mathfrak{o}(\bar{\eta}^{(j)})$. *Hence* $\mathfrak{o}$ *is a sub-space of* $\bigcap_i \mathfrak{o}'(\bar{\xi}^{(i)})$. Since also $\bar{U}/K$ is a derived normal model of U/K, *the quotient ring* $\mathfrak{o}'_1$ *of the point* $(\xi) = (\xi^{(1)})$ *is a sub-space of* $\bigcap_i \mathfrak{o}'(\bar{\xi}^{(i)})$. From these conclusions and from the fact that the space $\mathfrak{o}$ is a *subset* of the space $\mathfrak{o}'_1$ (since T is regular at (ξ)), our Lemma I follows.]

Before we proceed to the second lemma, we shall make a few remarks about the coefficients of the associated form. Let U/K be a variety in S_m. Consider an involutional system of positive 0-cycles of degree n in U/K with the associated variety V/K and the associated transformation T, and let (x) and $y = T(x)$ be a pair of corresponding generic points of U/K and V/K respectively. Let $F_{(y)}(Z)$ be the associated form of the generic 0-cycle $G(y)$ of the system. We have then

$$F_{(y)}(Z) = \prod_{i=1}^{n} (Z_0 + \sum_{j=1}^{m} Z_j x^{(i)}{}_j),$$

where the n points $(x^{(1)}) = (x)$, $(x^{(2)}), \cdots, (x^{(n)})$ form a complete set of conjugates over $K(y)$. The coordinates of the point (y) are the coefficients of the form $F_{(y)}(Z)$, and they are by hypothesis rational functions of (x). Let $f_k(Z_0, Z_k)$ be the form obtained from $F_{(y)}(Z)$ by setting $Z_j = 0$ for all $j \neq 0, k$; we have then evidently

$$f_k(Z_0, Z_k) = \prod_{i=1}^{n} (Z_0 + Z_k x^{(i)}{}_k), \qquad k = 1, \cdots, m.$$

Thus, for each $k = 1, \cdots, m$, the coefficients of the form $f_k(Z_0, Z_k)$ are the elementary symmetric functions of the n elements $x^{(1)}{}_k, \cdots, x^{(n)}{}_k$. It is clear that the coefficients of the m forms $f_k(Z_0, Z_k)$, $k = 1, \cdots, m$, are all part of the coefficients of the form $F_{(y)}(Z)$; that is, they all occur among the coordinates of the point (y). Furthermore, let

$$g_k(Z_0, Z_k) = \prod_{i=2}^{n} (Z_0 + Z_k x^{(i)}{}_k), \qquad k = 1, \cdots, m$$

so that we have $f_k(Z_0, Z_k) = (Z_0 + Z_k x_k) g_k(Z_0, Z_k)$, $k = 1, \cdots, m$. It is well known that for each $k = 1, \cdots, m$, the coefficients of $g_k(Z_0, Z_k)$, being the elementary symmetric functions of the $n - 1$ elements $x^{(2)}{}_k, \cdots, x^{(n)}{}_k$, can be expressed as polynomials in the coefficients of the form $f_k(Z_0, Z_k)$ and the element x_k. It follows then that the coefficients of the forms $g_k(Z_0, Z_k)$, $k=1,\cdots,m$, are all elements of the ring $K[x,y]$. We shall make use of this remark presently.

LEMMA II. *Let U/K be a variety; let $|G(y)|$ be an involutional system of positive 0-cycles in U/K with the associated variety V/K and the associated transformation T. Let (ξ) be a point of U/K such that the transformation T is regular at (ξ), and that the corresponding point $(\eta)=T(\xi)$ on V/K determines an 0-cycle $G(\eta)$ which contains (ξ) as a simple component. Then the maximal prime ideal $\mathfrak{u}$ of the quotient ring $\mathfrak{U}=Q(U/K,(\xi))$ has a basis consisting of elements in the quotient ring $\mathfrak{B}=Q(V/K,(\eta))$.*

Proof. Let (x) and $(y)=T(x)$ be a pair of corresponding generic points of U/K and V/K respectively. The hypothesis that the point (ξ) is a simple component of the special cycle $G(\eta)$ implies that the generic cycle $G(y)$ can only have simple components; hence the field $K(x)$ must be a separable algebraic extension of $K(y)$ and the n points $(x^{(1)})=(x)$, $(x^{(2)})$, $\cdots$, $(x^{(n)})$ of the generic cycle $G(y)$ are distinct. Let $(x^{(1)},\cdots,x^{(n)})\to(\xi^{(1)},\cdots,\xi^{(n)})$ be any specialization over the specialization $(x)\to(\xi)$ over K; the points $(\xi^{(1)}),\cdots,(\xi^{(n)})$ are then evidently the n points of the cycle $G(\eta)$. Since the point (ξ) occurs only once in the cycle $G(\eta)$, we have $(\xi^{(i)})\neq(\xi)$ for all $i\neq 1$. Therefore, we can assume that the affine coordinate system in S_m has been so chosen that we have $\xi^{(i)}{}_j\neq\xi_j$ for all $i\neq 1$ and all j. It is easily seen that this condition can always be obtained by means of a suitably chosen affine transformation with coefficients in K, the field K being assumed to contain infinitely many elements. Let $\lambda=(i_1,\cdots,i_m)$ be any ordered set of m positive integers not greater than n, we shall set $(\xi^\lambda)=(\xi^{(i_1)}{}_1,\cdots,\xi^{(i_m)}{}_m)$. Then, as λ runs through all the n^m possible ordered sets $(i_1,\cdots,i_m)$ of positive integers not greater than n, we obtain a set of n^m (not necessarily distinct) points $\{(\xi^\lambda)\}$. Since we have $\xi^{(i)}{}_j\neq\xi_j$ for all $i\neq 1$ and all j, it is clear that the point (ξ) occurs only once among the n^m points of the set (ξ^λ). Similarly, we set $(x^\lambda)=(x^{(i_1)}{}_1,\cdots,x^{(i_m)}{}_m)$ and obtain a set of n^m points $\{(x^\lambda)\}$. It is clear that for every λ the specialization $(x^\lambda)\to(\xi^\lambda)$ is the only specialization over the specialization $(x^{(1)},\cdots,x^{(n)})\to(\xi^{(1)},\cdots,\xi^{(n)})$ over K.

Let $\psi_j(X)$ $(j=1,\cdots,h)$ be a basis of the prime ideal determined by the point (ξ) in the polynomial ring $K[X]=K[X_1,\cdots,X_m]$. It is easily seen that the h elements $\psi_j(x)$, $j=1,\cdots,h$, constitute then also a basis of the ideal $\mathfrak{u}$ in the local ring $\mathfrak{U}$. Since the point (ξ) occurs only once among the set $\{(\xi^\lambda)\}$, it follows that for every $\lambda\neq(1,\cdots,1)$ the h elements

$\psi_j(\xi^\lambda)$, $j=1,\cdots,h$, are not all zero. There exist then h linearly independent combinations

$$\phi_j(X)=\sum_{k=1}^{h}a_{jk}\psi_k(X), \qquad j=1,\cdots,h,$$

of the polynomials $\psi_j(X)$ with coefficients a_{jk} in K, such that for every $\lambda\neq(1,\cdots,1)$ none of the h elements $\phi_j(\xi^\lambda)$, $j=1,\cdots,h$, vanishes. The elements $\phi_j(x)$, $j=1,\cdots,h$, constitute of course also a basis of the ideal $\mathfrak{u}$ in $\mathfrak{U}$; and we have for each λ and j the unique specialization $\phi_j(x^\lambda)\to\phi_j(\xi^\lambda)$ over the specialization $(x^{(1)},\cdots,x^{(n)})\to(\xi^{(1)},\cdots,\xi^{(n)})$ over K.

Now we set

$$\Phi_j=\prod_{\lambda}\phi_j(x^\lambda), \qquad j=1,\cdots,h,$$

where the product runs over all the n^m points of the set $\{(x^\lambda)\}$. We prove first that these h elements Φ_j, $j=1,\cdots,h$, are elements of the ring $K[y]$; this evidently implies that they are also elements of the quotient ring $\mathfrak{B}$. To show this, we set $\phi_j(X)=\phi_j^{(0)}(X_1,\cdots,X_m)$, $j=1,\cdots,h$; we can then write

$$\Phi_j=\prod_{i_m=1}^{n}\cdots\prod_{i_1=1}^{n}\phi_j^{(0)}(x^{(i_1)}{}_1,\cdots,x^{(i_m)}{}_m), \qquad j=1,\cdots,h.$$

Consider the polynomials

$$\phi_j^{(1)}(X_2,\cdots,X_m)=\prod_{i_1=1}^{n}\phi_j^{(0)}(x^{(i_1)}{}_1,X_2,\cdots,X_m), \qquad j=1,\cdots,h;$$

we have evidently

$$\Phi_j=\prod_{i_m=1}^{n}\cdots\prod_{i_2=1}^{n}\phi_j^{(1)}(x^{(i_2)}{}_2,\cdots,x^{(i_m)}{}_m), \qquad j=1,\cdots,h.$$

Since the coefficients of the polynomials $\phi_j^{(1)}(X_2,\cdots,X_m)$, $j=1,\cdots,h$, are integral rational symmetric functions of the n elements $x^{(1)}{}_2,\cdots,x^{(n)}{}_2$ they can be expressed as integral rational functions of the elementary symmetric functions of these n elements. This means that the coefficients of the polynomials $\phi_j^{(1)}(X_2,\cdots,X_m)$, $j=1,\cdots,h$, are integral rational functions of the coefficients of the form $f_1(Z_0,Z_1)$; and since the coefficients of $f_1(Z_0,Z_1)$ are themselves some of the coordinates of the point (y), it follows that the coefficients of $\phi_j^{(1)}(X_2,\cdots,X_m)$, $j=1,\cdots,h$, are elements of $K[y]$. Consider next the polynomials

$$\phi_j^{(2)}(X_3,\cdots,X_m)=\prod_{i_2=1}^{n}\phi_j^{(1)}(x^{(i_2)}{}_2,X_3,\cdots,X_m), \qquad j=1,\cdots,h;$$

we have evidently

$$\Phi_j = \prod_{i_m=1}^{n} \cdot \cdot \cdot , \prod_{i_3=1}^{n} \phi_j^{(2)} (x^{(i_3)}{}_3, \cdot \cdot \cdot , x^{(i_m)}{}_m), \qquad j = 1, \cdot \cdot \cdot , h.$$

Since the coefficients of the polynomials $\phi_j^{(2)}(X_3, \cdot \cdot \cdot , X_m)$, $j = 1, \cdot \cdot \cdot , h$, are integral rational symmetric functions of the n elements $x^{(1)}{}_2, \cdot \cdot \cdot , x^{(n)}{}_2$ with coefficients in $K[y]$, it follows that they can be expressed as integral rational functions of the elementary symmetric functions of these n elements also with coefficients in $K[y]$. This means that the coefficients of the polynomials $\phi_j^{(2)}(X_3, \cdot \cdot \cdot , X_m)$, $j = 1, \cdot \cdot \cdot , h$, are integral rational functions of the coefficients of the form $f_2(Z_0, Z_2)$, with coefficients in $K[y]$; and since the coefficients of the form $f_2(Z_0, Z_2)$ are themselves some of the coordinates of the point (y), it follows that the coefficients of $\phi_j^{(2)}(X_3, \cdot \cdot \cdot , X_m)$, $j = 1, \cdot \cdot \cdot , h$, are elements of $K[y]$. Thus, proceeding successively in this manner, we conclude after m steps that the elements Φ_j, $j = 1, \cdot \cdot \cdot , h$, are elements of $K[y]$.

We maintain that the h elements Φ_j, $j = 1, \cdot \cdot \cdot , h$, constitute a basis of the ideal $\mathfrak{u}$ in the ring $\mathfrak{U}$. Since we have $\Phi_j = (\Phi_j/\phi_j(x))\phi_j(x)$, $j = 1, \cdot \cdot \cdot , h$, and since the h elements $\phi_j(x)$, $j = 1, \cdot \cdot \cdot , h$, constitute a basis of $\mathfrak{u}$ in $\mathfrak{U}$, it is sufficient to prove that the elements $\Phi_j/\phi_j(x)$, $j = 1, \cdot \cdot \cdot , h$, are units of the ring $\mathfrak{U}$. We first show that these elements are elements of the ring $\mathfrak{U}$. In fact, indicating by $\prod_{(k)}^{(m)}{}'$ the product over all the $m - k + 1$ indices $i_k, \cdot \cdot \cdot , i_m$ with the exception of the one combination $i_k = \cdot \cdot \cdot = i_m = 1$, we obtain by a simple calculation the following:

$$\begin{aligned} \Phi_j/\phi_j(x) &= \prod_{(1)}^{(m)}{}' \phi_j^{(0)} (x^{(i_1)}{}_1, \cdot \cdot \cdot , x^{(i_m)}{}_m) \\ &= \prod_{i_1=2}^{n} \phi_j^{(0)} (x^{(i_1)}{}_1, x_2, \cdot \cdot \cdot , x_m) \prod_{(2)}^{(m)}{}' \phi_j^{(1)} (x^{(i_2)}{}_2, \cdot \cdot \cdot , x^{(i_m)}{}_m) \\ &= \cdot \cdot \cdot \\ &= \prod_{k=1}^{m} \Big(\prod_{i_k=2}^{n} \phi_j^{(k-1)} (x^{(i_k)}{}_k, x_{k+1}, \cdot \cdot \cdot , x_m) = \prod_{k=1}^{m} \theta_j^{(k)} (x_{k+1}, \cdot \cdot \cdot , x_m), \\ &\qquad (j = 1, \cdot \cdot \cdot , h), \end{aligned}$$

where in the last step we have set

$$\theta_j^{(k)} (X_{k+1}, \cdot \cdot \cdot , X_m) = \prod_{i_k=2}^{n} \phi_j^{(k-1)} (x^{(i_k)}{}_k, X_{k+1}, \cdot \cdot \cdot , X_m); \quad k = 1, \cdot \cdot \cdot , m; j = 1, \cdot \cdot \cdot , h.$$

Thus it is only necessary to prove that for each $k = 1, \cdot \cdot \cdot , m$ and each $j = 1, \cdot \cdot \cdot , h$, the element $\theta_j^{(k)}(x_{k+1}, \cdot \cdot \cdot , x_m)$ is an element of $\mathfrak{U}$. Since the coefficients of the polynomial $\theta_j^{(k)}(X_{k+1}, \cdot \cdot \cdot , X_m)$ are integral rational symmetric functions of the $n - 1$ elements $x^{(2)}{}_k, \cdot \cdot \cdot , x^{(n)}{}_k$ with coefficients in $K[y]$, they can be expressed as integral rational functions of the elementary symmetric functions of these $n-1$ elements, also with coefficients in $K[y]$. This means that the coefficients of $\theta_j^{(k)}(X_{k+1},\cdots,X_m)$ are integral rational functions of the coefficients of the form $g_k(Z_0,Z_k)$, with coefficients in $K[y]$; and since the coefficients of $g_k(Z_0,Z_k)$ are themselves elements of $K[x,y]$, it follows that the coefficients of $\theta_j^{(k)}(X_{k+1},\cdots,X_m)$ are also elements of $K[x,y]$. Therefore, the element $\theta_j^{(k)}(x_{k+1},\cdots,x_m)$ is also an element of $K[x,y]$; and since the transformation T is regular at the point (ξ), the element $\theta_j^{(k)}(x_{k+1},\cdots,x_m)$ is in the quotient ring $\mathfrak{U}$. Thus we have shown that the elements $\Phi_j/\phi_j(x)$, $j=1,\cdots,h$, are all elements of the ring $\mathfrak{U}$.

Finally, we show that the elements $\Phi_j/\phi_j(x)$, $j=1,\cdots,h$, are units of the ring $\mathfrak{U}$. In fact, we have

$$\Phi_j/\phi_j(x)=\prod_\lambda{}'\phi_j(x^\lambda), \qquad j=1,\cdots,h,$$

where the product $\prod_\lambda'$ runs over all the n^m points of the set $\{(x^\lambda)\}$ except the point (x). For every λ and j, we have the unique specialization $\phi_j(x^\lambda)\to\phi_j(\xi^\lambda)$ over the specialization $(x^{(1)},\cdots,x^{(n)})\to(\xi^{(1)},\cdots,\xi^{(n)})$ over K; hence we have also the unique specialization

$$\Phi_j/\phi_j(x)=\prod_\lambda{}'\phi_j(x^\lambda)\to\prod_\lambda{}'\phi_j(\xi^\lambda), \qquad j=1,\cdots,h,$$

over the specialization $(x^{(1)},\cdots,x^{(n)})\to(\xi^{(1)},\cdots,\xi^{(n)})$ over K. Since the elements $\Phi_j/\phi_j(x)$, $j=1,\cdots,h$, are elements of $\mathfrak{U}$, it follows that the specialization

$$\Phi_j/\phi_j(x)\to\prod_\lambda{}'\phi_j(\xi^\lambda), \qquad j=1,\cdots,h,$$

is the uniquely determined specialization of these h elements over the specialization $(x)\to(\xi)$ over K. Now, we have shown above that $\phi_j(\xi^\lambda)\neq 0$ for every $\lambda\neq(1,\cdots,1)$ and every j; hence we have also

$$\prod_\lambda{}'\phi_j(\xi^\lambda)\neq 0, \qquad j=1,\cdots,h.$$

This means that the h elements $\Phi_j/\phi_j(x)$, $j=1,\cdots,h$, are not in $\mathfrak{u}$ and hence units of $\mathfrak{U}$. This concludes the proof of Lemma II.

4. Proof of the theorem. We restate here our theorem with slight modifications for affine varieties:

THEOREM. *Let U/K be a variety; let $|G(y)|$ be an involutional system of positive cycles in U/K with the associated variety V/K and the associated transformation T, where (y) is a generic point of V/K. Let (η) be a* *rational point of V/K such that the cycle $G(\eta)$ is finite, and let (ξ) be a generic point of a K-component G'/K of $G(\eta)$. We assume that the field $K(\xi)$ is analytically disjoint with respect to the ring $Q(V/K, (\eta))$ over K; this implies in particular that the fields $K(\xi)$ and $K(y)$ are linearly disjoint over K, so that the product variety $W/K = V/K \times G'/K$ is defined. If the transformation T is regular at the point (ξ) and if the variety G'/K is a simple K-component of $G(\eta)$ and is separably generated, then the varieties U/K and W/K are analytically equivalent at the points (ξ) and (η, ξ) respectively.*

Proof. As before, let (x) and $(y) = T(x)$ be a pair of corresponding generic points of U/K and V/K respectively. We shall first prove the theorem for the case when the dimension d of the positive cycles of the involutional system is zero and then extend it by a well known device to the general case.

(a) The case $d = 0$. The assumption that G'/K is a simple K-component of $G(\eta)$ and is separably generated implies that the point (ξ) is a simple component of the 0-cycle $G(\eta)$; hence the hypotheses of both Lemma I and Lemma II are satisfied. Let $\mathfrak{U} = Q(U/K, (\xi))$ be the quotient ring of U/K at the point (ξ) and $\mathfrak{u}$ be the maximal prime ideal in $\mathfrak{U}$; let $\mathfrak{B} = Q(V/K, (\eta))$ be the quotient ring of V/K at the point (η); and let $\mathfrak{W} = (W/K, (\eta, \xi))$ be the quotient ring of W/K at the point (η, ξ). Let $\mathfrak{U}^*$ be the completion of the local ring $\mathfrak{U}$ and let $\mathfrak{u}^*$ be its maximal prime ideal; let $\mathfrak{B}^*$ be the completion of the local ring $\mathfrak{B}$ and $\mathfrak{v}^*$ be its maximal prime ideal; let $\mathfrak{W}^*$ be the completion of the local ring $\mathfrak{W}$. Our aim is to prove that the two rings $\mathfrak{U}^*$ and $\mathfrak{W}^*$ are K-isomorphic. We observe first that $\mathfrak{W}$ is also the quotient ring $Q(V/K(\xi), (\eta))$ of the extension variety $V/K(\xi)$ of V/K over $K(\xi)$ at the point (η); which means that $\mathfrak{W}$ is the extension of $\mathfrak{B}$ over the field $K(\xi)$. Furthermore, since $K(\xi)$ is a finite extension of K and since $\mathfrak{B}$ contains K as a coefficient field, it follows that $\mathfrak{W}^* = K(\xi) \times \mathfrak{B}^*$. According to Lemma I, $\mathfrak{U}$ contains $\mathfrak{B}$ as a sub-space; hence $\mathfrak{U}^*$ contains $\mathfrak{B}^*$ as a subring. Since $K(\xi)$ is separable over K, $\mathfrak{U}$ contains K as a basic field; hence the completion $\mathfrak{U}^*$ contains a coefficient field K' over K, which is K-isomorphic to the residue ring $\mathfrak{U}^*/\mathfrak{u}^*$ and hence also K-isomorphic to $K(\xi)$. The field K' is not in general a subfield of the universal domain $\mathfrak{K}$ and is hence an "abstract" field in our terminology; however, since K' is K-isomorphic with the field $K(\xi)$, we can without loss of generality identify the two fields by setting $K' = K(\xi)$ through the given K-isomorphism. If we now set

$R = \mathfrak{U}^*$, $\mathfrak{o} = \mathfrak{B}$, $L = K'$ in the Theorem 7 of the Appendix, it is easily seen that all the conditions in the hypothesis are fulfilled; we can therefore conclude that the subring $K'\mathfrak{B}^*$ in $\mathfrak{U}^*$ is an $(K', \mathfrak{B}^*)$-isomorphic image of $\mathfrak{W}^* = K' \times \mathfrak{B}^*$. This shows in particular that $K'\mathfrak{B}^*$ is a complete local ring and its maximal prime ideal is the ideal $K'\mathfrak{v}^*$, which is contained in the ideal $\mathfrak{u}^*$; it follows then that the subring $K'\mathfrak{B}^*$ is equal to its own adherence in $\mathfrak{U}^*$. Therefore, to complete our proof, we only need to show that $\mathfrak{U}^* = K'\mathfrak{B}^*$; for then the two rings $\mathfrak{U}^*$ and $\mathfrak{W}^*$ will be K'-isomorphic and *a fortiori* also K-isomorphic. To prove this equation $\mathfrak{U}^* = K'\mathfrak{B}^*$, it is sufficient to show that the elements of $K'\mathfrak{B}$ are everywhere dense in the local ring $\mathfrak{U}^*$. This means that given any element α of $\mathfrak{U}^*$ and a positive integer s, there is an element β_s of $K'\mathfrak{B}$ such that $\alpha \equiv \beta_s \pmod{\mathfrak{u}^{*s}}$. Now, according to Lemma II, the ideal $\mathfrak{u}$ of $\mathfrak{U}$ has a basis Φ_j, $j = 1, \cdots, h$, which are elements of $\mathfrak{B}$, hence also of $K'\mathfrak{B}$. It is well known that the elements Φ_j, $j = 1, \cdots, h$, constitute then also a basis of the ideal $\mathfrak{u}^*$ in $\mathfrak{U}^*$. Since the elements of K' constitute a complete set of representatives of the residue field $\mathfrak{U}^*/\mathfrak{u}^*$, there exists an element α_1 in K' such that $\alpha \equiv \alpha_1 \pmod{\mathfrak{u}^*}$. We set $\beta_1 = \alpha_1$. Assuming that the element β_{s-1} has already been constructed, we proceed to construct the element β_s. Since the element $\alpha - \beta_{s-1}$ is an element in the ideal $\mathfrak{u}^{*s-1}$, we can express it as a form of degree $s-1$ in the basis elements Φ_j, $j = 1, \cdots, h$, with coefficients in $\mathfrak{U}^*$. Now, by the case $s = 1$, each of these coefficients is congruent to an element of K' modulo $\mathfrak{u}^*$. If we now substitute every coefficient of the form by the corresponding element in K', we obtain an element α_s in $K'\mathfrak{B}$ such that $\alpha - \beta_{s-1} \equiv \alpha_s \pmod{\mathfrak{u}^{*s}}$. The element $\beta_s = \alpha_s + \beta_{s-1}$ is in the ring $K'\mathfrak{B}$ and satisfies the condition $\alpha \equiv \beta_s \pmod{\mathfrak{u}^{*s}}$. Thus the proof for the case $d = 0$ is complete.

(b) The case $d > 0$. As we have mentioned at the end of Section 2, we shall assume that the affine coordinate system of S_m has been so chosen that the $(m-d)$-dimensional linear variety defined by the equations $X_j = v_j$, $j = 1, \cdots, d$, where the $v_1, \cdots, v_d$ are independent variables over K, intersects the d-cycle $G(\eta)$ properly and finitely. Then, if G_1/K is any K-component of degree n_1 of $G(\eta)$, a generic point (ζ) of G_1/K can be chosen such that the first d coordinates $\zeta_1, \cdots, \zeta_d$ are any given set of d independent variables over K, and the field $K(\zeta_1, \cdots, \zeta_m)$ is an algebraic extension of degree n_1 over the field $K(\zeta_1, \cdots, \zeta_d)$. Since the first d coordinates $x_1, \cdots, x_d$ of the generic point (x) of U/K are in this case independent variables over K, we can therefore set $\zeta_j = x_j$ for $j = 1, \cdots, d$. In other words, each K-component of $G(\eta)$ has a generic point with the first d coordinates $x_1, \cdots, x_d$. This holds in particular for the K-component G'/K

of $G(\eta)$, and we have a generic point (ξ) with $\xi_j = x_j$ for $j = 1, \cdots, d$. Since the variety G'/K is separably generated, we may without loss of generality assume that the affine coordinate system of S_m has been so chosen that $K(\xi)$ is separably algebraic of degree n' over field $K(x_1, \cdots, x_d)$.

Let $\bar{K} = K(x_1, \cdots, x_d)$, and let $U/\bar{K}$ and $V/\bar{K}$ be the extensions of the varieties U/K and V/K respectively over $\bar{K}$. Let $L/\bar{K}$ be the linear variety of dimension $m - d$ defined by the d linear equations $X_j = x_j$, $j = 1, \cdots, d$. Then the intersection of $U/\bar{K}$ with $L/\bar{K}$ is a variety $\bar{U}/\bar{K}$ over $\bar{K}$, and the point (x) is also a generic point of $\bar{U}/\bar{K}$. The point (ξ) is evidently also a point of $\bar{U}/\bar{K}$, and it is well known[20] that the quotient ring $Q(\bar{U}/\bar{K}, (\xi))$ coincides with the quotient ring $\mathfrak{U} = Q(U/K, (\xi))$. Since the point (x) is a generic point of $\bar{U}/\bar{K}$ and the point (y) is also a point of $V/\bar{K}$, the rational transformation $(y) = T(x)$ of U/K onto V/K can also be regarded as a rational transformation $\bar{T}$ of the variety $\bar{U}/\bar{K}$ into the variety $V/\bar{K}$. In fact, this rational transformation T is also a transformation of $\bar{U}/\bar{K}$ *onto* $V/\bar{K}$; in other words, the point (y) is not only a generic point of V/K, but also a generic point of $V/\bar{K}$. To prove this, we only need to show that every generic point (y') of V/K is a specialization of the point (y) over $\bar{K}$. In fact, since $G(y')$ is a generic d-cycle of the involutionary system $|G(y)|$, it has at least a finite intersection point $(x') = (x_1, \cdots, x_d, x'_{d+1}, \cdots, x'_m)$ with the linear variety $L/\bar{K}$. The pair of points (x'), (y') is then a specialization of the pair (x), (y) over K; hence, the pair $(x_1, \cdots, x_d)$, (y') is also a specialization of the pair $(x_1, \cdots, x_d)$, (y) over K, and this means that the point (y') is a specialization of (y) over $\bar{K}$.

It is evident from the definition of the transformation $\bar{T}$ that to the point (y) of $V/\bar{K}$ correspond in the inverse transformation $\bar{T}^{-1}$ exactly the points of the intersection of $L/\bar{K}$ with the d-cycle $G(y)$. Since each K-component G_1/K of $G(\eta)$ has a generic point of the type $(\zeta) = (x_1, \cdots, x_d, \zeta_{d+1}, \cdots, \zeta_m)$ such that the field $K(\zeta)$ is an algebraic extension of degree n_1 over $\bar{K}$, it follows that the intersection of G_1/K (considered as a d-cycle) with $L/\bar{K}$ is a finite 0-cycle $\bar{G}_1$ of degree n_1, prime rational over $\bar{K}$; thus we have a 0-dimensional variety $\bar{G}_1/\bar{K}$, and the point (ζ) is also a generic point of $\bar{G}_1/\bar{K}$. This holds in particular for the intersection $\bar{G}'/\bar{K}$ of $G'/\bar{K}$ with $L/\bar{K}$; and since the field $\bar{K}(\xi) = K(\xi)$ is separable over $\bar{K}$, this 0-dimensional variety $\bar{G}'/\bar{K}$ is also separably generated over $\bar{K}$. Summing over all the K-components of the cycle $G(\eta)$, we conclude that the intersection of $G(\eta)$ with $L/\bar{K}$ is a finite 0-cycle $\bar{G}(\eta)$ of degree n, rational over $\bar{K}$, which

[20] See Zariski [10], p. 8, Lemma 1.

contains the variety $\bar{G}'/\bar{K}$ as a simple $\bar{K}$-component. It is well known that in such a case the intersection of the generic cycle $G(y)$ with $L/\bar{K}$ is also a finite 0-cycle $\tilde{G}(y)$ of degree n, prime rational over $\bar{K}(y)$, and the cycle $\tilde{G}(\eta)$ is the unique specialization of the cycle $\tilde{G}(y)$ over the specialization $(y) \rightarrow (\eta)$ over $\bar{K}$. Thus the rational transformation $\check{T}$ induces in $\bar{U}/\bar{K}$ an involutional system of 0-cycles $|\tilde{G}(y)|$, and the inverse transformation $\check{T}^{-1}$ is defined at the point (η). Since the transformation $\check{T}$ is evidently regular at the point (ξ), it follows then from Lemma I that the quotient ring $\mathfrak{U} = Q(\bar{U}/\bar{K}, (\xi))$ contains the quotient ring $\check{\mathfrak{V}} = Q(V/\bar{K}, (\eta))$ as a sub-space.

The variety $V/\bar{K}$ is, however not the associated variety of the involutional system $|\tilde{G}(y)|$. To find the associated variety of this system, we only need to observe that the associated form of the generic cycle $\tilde{G}(y)$ is the form $\bar{F}(Z) = F_{(y)}(Z, u^{(1)}, \cdots, u^{(d)})$, where the form $F_{(y)}(Z, Z^{(1)}, \cdots, Z^{(d)})$ is the associated form of the d-cycle $G(y)$ and the $(u^{(i)})$, $i = 1, \cdots, d$, are defined as follows: For each $i = 1, \cdots, d$, we set $u^{(i)}{}_j = -x_i$, 1, or 0, according as $j = 0$, $j = i$, or $j \neq 0, i$, respectively. As we have mentioned at the end of Section 2, the form $F_{(y)}(Z, Z^{(1)}, \cdots, Z^{(d)})$ is assumed to be so normalized that the coefficient of the term ω_0 is 1. It follows then the coefficient of the term $Z_0{}^n$ in the form $\bar{F}(Z)$ is also equal to 1, so that $\bar{F}(Z)$ is already in the normalized form. Let $(\bar{y})$ be the set of coefficients (except the coefficient 1 of the term $Z_0{}^n$) of this form $\bar{F}(Z)$, ordered in some arbitrary but fixed way, then the variety $\bar{V}/\bar{K}$ determined by the point $(\bar{y})$ over $\bar{K}$ is the associated variety of the involutional system $|\tilde{G}(y)|$; and we can now write $\tilde{G}(\bar{y})$ instead of $\tilde{G}(y)$ to indicate its dependence on the point $(\bar{y})$. It is clear that the coordinates of $(\bar{y})$ are linear combination of the coordinates of (y); this means that there is a projection P of $V/\bar{K}$ onto $\bar{V}/\bar{K}$ and that this projection P is regular at every finite point of $V/\bar{K}$, hence in particular at the point (η). We shall denote by $(\bar{\eta})$ the corresponding point of (η) under this projection P, and write $\tilde{G}(\bar{\eta})$ instead of $\tilde{G}(\eta)$ to indicate its dependence on the point $(\bar{\eta})$ of $\bar{V}/\bar{K}$. Since the hypothesis of Lemma I is evidently satisfied by P at the point (η), it follows that the quotient ring $\check{\mathfrak{V}} = Q(V/\bar{K}, (\eta))$ contains the quotient ring $\bar{\mathfrak{V}} = Q(\bar{V}/\bar{K}, (\bar{\eta}))$ of $\bar{V}/\bar{K}$ at the point $(\bar{\eta})$ as a sub-space; and this means that $\bar{\mathfrak{V}}$ is also a sub-space of the local ring $\mathfrak{U}$.

Let $\bar{T}$ be the associated rational transformation of the involutional system $|\tilde{G}(\bar{y})|$; it is easily seen that $\bar{T}$ is the product of the rational transformation $\check{T}$ of $\bar{U}/\bar{K}$ onto $V/\bar{K}$ and the projection P of $V/\bar{K}$ onto $\bar{V}/\bar{K}$. Since the transformation $\check{T}$ is regular at the point (ξ) and the image point on $V/\bar{K}$ is the point (η), and since the projection P is regular at the point (η) and the image point on $\bar{V}/\bar{K}$ is the point $(\bar{\eta})$, it follows that the associated transformation $\bar{T}$ is regular at the point (ξ) and the image point on $\bar{V}/\bar{K}$ is the point $(\bar{\eta})$. Moreover, the 0-dimensional variety $\bar{G}'/\bar{K}$ is a simple $\bar{K}$-component of the cycle $\bar{G}(\bar{\eta})$ and is separably generated; hence the hypothesis of Lemma II is fulfilled by the involutional system $|\bar{G}(\bar{y})|$ on the variety $\bar{U}/\bar{K}$ with the associated variety $\bar{V}/\bar{K}$ and the associated transformation $\bar{T}$. It follows then that the maximal prime ideal $\mathfrak{u}$ in the quotient ring $\mathfrak{U}=Q(\bar{U}/\bar{K}, (\xi))$ has a basis consisting of elements in the subring $\bar{\mathfrak{V}}$ and hence also in the subring $\mathfrak{V}$ of $\mathfrak{U}$.

Finally, let $\mathfrak{W}=Q(W/K, (\eta,\xi))$ be the quotient ring of W/K at the point (η,ξ). It is clear that $\mathfrak{W}$ is also the quotient ring $Q(V/K(\xi), (\eta))$ of the extension variety $V/K(\xi)$ of V/K over $K(\xi)$ at the point (η), and the variety $V/K(\xi)$ is also the extension variety of $V/\bar{K}$ over $K(\xi)$; this means that the local ring $\mathfrak{W}$ is the extension of the local ring $\breve{\mathfrak{V}}$ over the field $K(\xi)$. Since $K(\xi)$ is a finite extension of $\bar{K}$ and since $\breve{V}$ contains $\bar{K}$ as a coefficient field, it follows that $\mathfrak{W}^*=K(\xi)\times\breve{\mathfrak{V}}^*$. Furthermore, since the field $K(\xi)$ is analytically disjoint with respect to the quotient ring $\mathfrak{V}=Q(V/K, (\eta))$ over the field K, and since $\breve{\mathfrak{V}}$ is the extension of $\mathfrak{V}$ over the subfield $\bar{K}$ of $K(\xi)$, it follows (Appendix, Theorem 6) that $K(\xi)$ is also analytically disjoint with respect to $\breve{\mathfrak{V}}$ over the field $\bar{K}$.

We are now ready to apply the method of (a). Consider the completion $\mathfrak{U}^*$ of $\mathfrak{U}$; it contains the completion $\breve{\mathfrak{V}}^*$ of $\breve{\mathfrak{V}}$ as a subring. Since $K(\xi)$ is separable over $\bar{K}$, the local ring $\mathfrak{U}$ contains $\bar{K}$ as a basic field; hence the completion $\mathfrak{U}^*$ contains a coefficient field $\bar{K}'$ over $\bar{K}$, which is $\bar{K}$-isomorphic to the field $K(\xi)$. Here again the field $\bar{K}'$ is an "abstract" field, but we can just as before identify it with $K(\xi)$ by setting $\bar{K}'=K(\xi)$ through the given $\bar{K}$-isomorphism. If we set $R=\mathfrak{U}^*$, $\mathfrak{o}=\breve{\mathfrak{V}}$, $L=\bar{K}'$, $K=\bar{K}$ in Theorem 7 of the Appendix, then all the condition in the hypothesis are fulfilled; we can therefore conclude that the subring $\bar{K}'\breve{\mathfrak{V}}^*$ in $\mathfrak{U}^*$ is a $(\bar{K}',\breve{\mathfrak{V}}^*)$-isomorphic image of $\mathfrak{W}^*=\bar{K}'\times\breve{\mathfrak{V}}^*$. Observing that the maximal prime ideal in $\mathfrak{U}^*$ has a basis consisting of elements in $\breve{\mathfrak{V}}$, we can repeat the same argument as in the last part of (a) and conclude that $\mathfrak{U}^*=\bar{K}'\breve{\mathfrak{V}}^*$. Therefore the two rings $\mathfrak{U}^*$ and $\mathfrak{W}^*$ are $\bar{K}'$-isomorphic and *a fortiori* also K-isomorphic with each other. This completes the proof of our theorem.

APPENDIX.*

On the Extensions of Local Domains.

In this appendix we shall develop some notions and results concerning the extension of a local domain over a field, which are used in the text, but are not directly related to the problem considered there, so that we consider it more appropriate to present them separately. Theorems 1-6 are not essentially new; in fact, most of them are contained implicitly in the work of Chevalley, to which we shall refer for some of the proofs. However, for our present purpose, we have to develop the subject in a somewhat different way. The one new result is Theorem 7, which plays an essential part in the proof of the main theorem in the text. As a matter of terminology, we stress here that the expression "field" will be used in this appendix in its usual general sense as meaning an abstract field, not necessarily a subfield of a certain "universal domain."

Let R_1 and R_2 be two (commutative) rings which contain a field K as common subring. We shall consider the Kronecker product $R_1 \times R_2$ of R_1 and R_2 over the field K, which is a ring containing both R_1 and R_2 as subrings. The Kronecker product $R_1 \times R_2$ has the following characteristic property: Let R' be any ring containing K, and let R'_1 and R'_2 be two subrings of R', both containing K; then if R'_1 and R'_2 are K-isomorphic to R_1 and R_2 respectively, then the subring[21] $R'_1R'_2$ in R' is a K-homomorphic image of $R_1 \times R_2$. Furthermore, this homomorphism will be an isomorphism if and only if R'_1 and R'_2 are linearly disjoint over K. We recall here that R'_1 and R'_2 are linearly disjoint over K, if every set of linearly independent elements of R'_1 over K is still such over R'_2; and that when this is so, then every set of linearly independent elements of R'_2 over K is also still such over R'_1. If we identify the rings R'_1 and R'_2 with the rings R_1 and R_2 respectively, as we shall often do, then the homomorphism of $R_1 \times R_2$ onto the subring R_1R_2 in R' is not only a K-homomorphism, but also an (R_1, R_2)-homomorphism in the sense that both R_1 and R_2 are left invariant by the homomorphism.

Let L and M be two fields which contain a common subfield K. We shall say that L and M are *algebraically disjoint* over K, if the prime ideal in the polynomial ring $K[X_1, \cdots, X_n]$ determined by every set of n elements

* Added January 20, 1950.

[21] We shall denote by $R'_1R'_2$ the subring of R' consisting of all the sums of products of elements in R'_1 and R'_2. Thus, in case of two subfields L and M of a larger field, the ring LM is in general *not* the compositum of L and M.

in L generates a prime ideal in the polynomial ring $M[X_1, \cdots, X_n]$. It is easily seen that in case both L and M are subfields of a larger field such that L and M are independent over K, then they are algebraically disjoint over K if and only if they are linearly disjoint over K. On the other hand, given any two fields L and M having exactly a common subfield K, there always exists a field N which contains L and a subfield M', K-isomorphic to M, such that L and M' are independent over K. If L and M are algebraically disjoint over K, then L and M' are also algebraically disjoint over K and hence also linearly disjoint over K; and since in this case the intersection of L and M' is exactly the field K, we can identify M' with M, so that N contains both L and M as subfields. Thus there is no essential difference between algebraical disjointness and linear disjointness; the former is a natural generalization of the latter to two arbitrary fields over K and hence more suitable to use in those investigations in which it is not possible or convenient to impose a fixed "universal domain" in advance.

The above described relation between algebraical disjointness and the linear disjointness enables us to carry over most of the results about the linear disjointness to algebraical disjointness. Thus, in particular, we can conclude that algebraical disjointness is a symmetrical relation between the two fields. Furthermore, we can define a field L as a regular extension of K, if L and $\bar{K}$ (the algebraical closure of K) are algebraically disjoint over K. It follows then from a well known result [22] that if L is a regular extension of K and M is any field containing K, then L and M are algebraically disjoint over K.

The definition of algebraical disjointness can be extended to two integral domains. We shall say that two integral domains R_1 and R_2, both containing the field K as subring, are algebraically disjoint over K, if their corresponding fields of quotients L and M are algebraically disjoint over K.

THEOREM 1. *Let R_1 and R_2 be two integral domains which both contain the field K as subring. Then the Kronecker product $R_1 \times R_2$ of R_1 and R_2 over K is an integral domain if and only if R_1 and R_2 are algebraically disjoint over K.*

Proof. If $R_1 \times R_2$ is an integral domain, then the fields L and M will be subfields in the field of quotients of $R_1 \times R_2$, and since R_1 and R_2 are linearly disjoint over K, it follows that L and M are also linearly disjoint over K and hence algebraically disjoint over K. Conversely, if L and M are algebraically disjoint over K, let N be a field containing both L and M as

[22] See Weil [8], p. 18, Theorem 5.

subfields, such that L and M are linearly disjoint over K. Then the subring LM in N is a homomorphic image of $L \times M$; and since L and M are linearly disjoint over K, this homomorphism is an isomorphism. Since LM is an integral domain, it follows that $L \times M$ is an integral domain, and hence $R_1 \times R_2$ is also an integral domain.

THEOREM 2. *Let R be an integral domain containing the field K as subring, and let L be an algebraic extension of K. If L and R are algebraically disjoint over K, then any R-homomorphism of $L \times R$ is an isomorphism.*

Proof. In fact, if M is the field of quotients of R, then any R-homomorphism of $L \times R$ can be extended to an M-homomorphism of the ring of quotients of $L \times R$ with respect to the multiplicatively closed system R, which is the ring $L \times M$. Since L is algebraic, $L \times M$ is a field, and hence the homomorphism must be an isomorphism.

Following Cohen,[23] we shall define a generalized local ring $\mathfrak{o}$ as a (commutative) ring with an element 1 in which (1) the set $\mathfrak{m}$ of all non-units is an ideal with finite basis, and (2) $\bigcap_{s=1}^{\infty} \mathfrak{m}^s = (0)$. If a generalized local ring is a Noetherian ring, then it is a local ring in the sense of Krull. It is known [24] that a complete generalized local ring is a local ring, so that in any case the completion $\mathfrak{o}^*$ of $\mathfrak{o}$ is a local ring. We shall speak of a local domain or generalized local domain if the ring has no zero-divisors.

If a generalized local ring $\mathfrak{o}$ contains a field K as subring, then no element of K is contained in the maximal prime ideal $\mathfrak{m}$ of $\mathfrak{o}$, and hence the residue ring $\mathfrak{o}/\mathfrak{m}$ can be considered as an extension field of K. Furthermore, the completion $\mathfrak{o}^*$ of $\mathfrak{o}$ also contains the field K as subring, and the residue ring $\mathfrak{o}^*/\mathfrak{o}^*\mathfrak{m}$, which is K-isomorphic to $\mathfrak{o}/\mathfrak{m}$, can also be considered as an extension field of K. Any field K' in the ring $\mathfrak{o}$ which is a complete system of representatives of the residue ring $\mathfrak{o}/\mathfrak{m}$ is called a *coefficient field* in $\mathfrak{o}$; and if the field K' contains the field K as a subfield, then K' is said to be a *coefficient field over K* in $\mathfrak{o}$. It has been shown by Cohen [25] that while the ring $\mathfrak{o}$ in general does not contain a coefficient field, the completion $\mathfrak{o}^*$ of $\mathfrak{o}$ always contains a coefficient field, provided $\mathfrak{o}/\mathfrak{m}$ has the same characteristic as $\mathfrak{o}$; however, even in $\mathfrak{o}^*$ there need not always exist a coefficient field over any given field K. We shall say that a generalized local ring $\mathfrak{o}$ contains a *basic field* [26] K if $\mathfrak{o}^*$ contains a coefficient field over K.

---

[23] Cohen [6].

[24] See Cohen [6], p. 61, Theorem 3.

[25] See Cohen [6], p. 72, Theorem 9.

[26] It is to be noticed that we use the expression "basic field" in a quite different sense from that of Chevalley in [2].

THEOREM 3. *Let U/K be a variety in S_m, and let (ξ) be a point in U/k such that $K(\xi)$ is a separably generated extension of K. Then the quotient ring $Q(U/K, (\xi))$ contains K as a basic field.*

Proof. Let r be the degree of transcendency of $K(\xi)$ over K. It is well known that we can always find a suitable affine transformation in S_m, such that after such a transformation the first r coordinates $\xi_1, \cdots, \xi_r$ of (ξ) are independent variables over K, and the field $K(\xi)$ is separably algebraic over $K(\xi_1, \cdots, \xi_r)$. This implies that if $(x) = (x_1, \cdots, x_m)$ is a generic point of U/K, then the $x_1, \cdots, x_r$ are independent variables over K and the quotient ring $Q(U/K, (\xi))$ contains the field $K(x_1, \cdots, x_r)$. The residue ring of $Q(U/K, (\xi))$ over its maximal prime ideal, being isomorphic to the field $K(\xi)$, is then a separably algebraic extension of $K(x_1, \cdots, x_r)$; and it is well known [27] that in this case the completion of $Q(U/K, (\xi))$ contains a coefficient field over $K(x_1, \cdots, x_r)$, which is then evidently also a coefficient field over K.

Let $\mathfrak{o}$ be a generalized local domain with a basic field K, $\mathfrak{m}$ be the maximal prime ideal in $\mathfrak{o}$, and M be the quotient field of $\mathfrak{o}$. Let L be any field containing K, such that both L and M as well as L and $\mathfrak{o}/\mathfrak{m}$ are algebraically disjoint over K. According to Theorem 1, the Kronecker product $L \times \mathfrak{o}$ of L and $\mathfrak{o}$ over K is an integral domain, and it contains $\mathfrak{o}$ and L as two linearly disjoint subrings over K. It can be easily shown that $(L \times \mathfrak{o})\mathfrak{q} \cap \mathfrak{o} = \mathfrak{q}$ for every ideal $\mathfrak{q}$ in $\mathfrak{o}$, and that $\bigcap_i (L \times \mathfrak{o})\mathfrak{q}_i = (L \times \mathfrak{o})(\bigcap_i \mathfrak{q}_i)$ for every set of ideals $\mathfrak{q}_i$ in $\mathfrak{o}$. In particular, we have the relations $\bigcap_{s=1}^{\infty} (L \times \mathfrak{o})\mathfrak{m}^s = (L \times \mathfrak{o})(\bigcap_{s=1}^{\infty} \mathfrak{m}^s) = (0)$ and $(L \times \mathfrak{o})\mathfrak{m}^s \cap \mathfrak{o} = \mathfrak{m}^s$; which shows that if we consider $L \times \mathfrak{o}$ as a (generalized) $(L \times \mathfrak{o})\mathfrak{m}$-adic ring in the sense of Zariski,[28] then $L \times \mathfrak{o}$ contains $\mathfrak{o}$ as a subspace. Moreover, it is easily seen that the residue ring $L \times \mathfrak{o}/(L \times \mathfrak{o})\mathfrak{m}$ is isomorphic to the ring $L \times (\mathfrak{o}/\mathfrak{m})$ which is according to Theorem 1 an integral domain; this shows that $(L \times \mathfrak{o})\mathfrak{m}$ is a prime ideal in $L \times \mathfrak{o}$. Let $\mathfrak{o}_L$ be the quotient ring of $L \times \mathfrak{o}$ with respect to the multiplicatively closed system $L \times \mathfrak{o} - (L \times \mathfrak{o})\mathfrak{m}$; then $\mathfrak{o}_L\mathfrak{m}$ is the ideal of non-units in $\mathfrak{o}_L$ and has obviously a finite basis. We shall show now that $\mathfrak{o}_L$ is a generalized local domain and contains $\mathfrak{o}$ as a subspace; to do this it is sufficient to show that $\bigcap_{s=1}^{\infty} \mathfrak{o}_L\mathfrak{m}^s = (0)$ and $\mathfrak{o}_L\mathfrak{m}^s \cap \mathfrak{o} = \mathfrak{m}^s$.

[27] See Chevalley [2], p. 701, Proposition 3.

[28] See Zariski [9]; we add the word "generalized" to indicate the fact that the ring is not necessarily Noetherian.

In case K is a coefficient field of $\mathfrak{o}$, the residue ring $L \times \mathfrak{o}/(L \times \mathfrak{o})\mathfrak{m}$ is isomorphic to the field L; this implies that the prime ideal $(L \times \mathfrak{o})\mathfrak{m}$ is maximal in $L \times \mathfrak{o}$ and hence the ideals $(L \times \mathfrak{o})\mathfrak{m}^s$ are all primary. It follows then that $\bigcap_{s=1}^{\infty} \mathfrak{o}_L\mathfrak{m}^s = \mathfrak{o}_L(\bigcap_{s=1}^{\infty} (L \times \mathfrak{o})\mathfrak{m}^s) = \mathfrak{o}_L(\bigcap_{s=1}^{\infty} \mathfrak{m}^s) = (0)$ and $\mathfrak{o}_L\mathfrak{m}^s \cap \mathfrak{o} = \mathfrak{o}_L\mathfrak{m}^s \cap L \times \mathfrak{o} \cap \mathfrak{o} = (L \times \mathfrak{o})\mathfrak{m}^s \cap \mathfrak{o} = \mathfrak{m}^s$. If K is not a coefficient field of $\mathfrak{o}$, then the completion $\mathfrak{o}^*$ of $\mathfrak{o}$ contains a coefficient K' over K. Since K' is K-isomorphic to $\mathfrak{o}/\mathfrak{m}$, the fields L and K' are algebraically disjoint over K. Therefore the Kronecker product $L \times K'$ of L and K' over K is an integral domain; let L' be the field of quotients of $L \times K'$. Since K' is a coefficient field of $\mathfrak{o}^*$ and L' contains K', we can apply to L' and $\mathfrak{o}^*$ over the field K' the case we have just proved and conclude that $\bigcap_{s=1}^{\infty} (\mathfrak{o}^*)_{L'}\mathfrak{m}^s = (0)$ and $(\mathfrak{o}^*)_{L'}\mathfrak{m}^s \cap \mathfrak{o}^* = \mathfrak{o}^*\mathfrak{m}^s$. Since the ring $(\mathfrak{o}^*)_{L'}$ contain $\mathfrak{o}_L$ as a subring, it follows then that $\bigcap_{s=1}^{\infty} \mathfrak{o}_L\mathfrak{m}^s \subset \mathfrak{o}_L \cap (\bigcap_{s=1}^{\infty}(\mathfrak{o}^*)_{L'}\mathfrak{m}^s) = (0)$ and $\mathfrak{o}_L\mathfrak{m}^s \cap \mathfrak{o} \subset (\mathfrak{o}^*)_{L'}\mathfrak{m}^s \cap \mathfrak{o} = \mathfrak{o}^*\mathfrak{m}^s \cap \mathfrak{o} = \mathfrak{m}^s$, and hence $\bigcap_{s=1}^{\infty} \mathfrak{o}_L\mathfrak{m}^s = (0)$ and $\mathfrak{o}_L\mathfrak{m}^s \cap \mathfrak{o} = \mathfrak{m}^s$.

Thus we have shown that in any case the ring $\mathfrak{o}_L$ is a generalized local domain and contains the generalized local domain $\mathfrak{o}$ as a subspace. We shall call $\mathfrak{o}_L$ the *extension* of $\mathfrak{o}$ over the field L. We remark further that in case L is a finitely generated extension of K (which is the only case of interest for our present purpose), it can be shown that $\mathfrak{o}_L$ is a Noetherian ring if $\mathfrak{o}$ is such; in other words, in this case $\mathfrak{o}_L$ is a local domain if $\mathfrak{o}$ is a local domain. However, this additional restriction is not necessary for our considerations here. It is easily seen that if a variety U/K has an extension over a field L containing K and if (ξ) is a point of U/K such that $K(\xi)$ and L are linearly disjoint over K, then the quotient ring $Q(U/L, (\xi))$ is the extension of the quotient ring $Q(U/K, (\xi))$ over the field L.

Consider the completion $(\mathfrak{o}_L)^*$ of the generalized local domain $\mathfrak{o}_L$; $(\mathfrak{o}_L)^*$ is a local ring and contains the local ring $\mathfrak{o}^*$ as a subring. Let $\mathfrak{p}_1, \cdots, \mathfrak{p}_n$ be the minimal (unimbedded) prime divisors of the zero ideal in $\mathfrak{o}^*$. Then for each $i = 1, \cdots, n$, the residue ring $\mathfrak{o}^*/\mathfrak{p}_i$ is a complete local domain, and the residue ring $(\mathfrak{o}_L)^*/(\mathfrak{o}_L)^*\mathfrak{p}_i$ is a complete local ring and contains $\mathfrak{o}^*/\mathfrak{p}_i$ as a subring. We shall say that the field L is *analytically disjoint* with respect to the generalized local domain $\mathfrak{o}$ over the field K, if the ideals $(\mathfrak{o}_L)^*\mathfrak{p}_i$, $i = 1, \cdots, n$, are all prime ideals in $(\mathfrak{o}_L)^*$, so that the rings $(\mathfrak{o}_L)^*/(\mathfrak{o}_L)^*\mathfrak{p}_i$, $i = 1, \cdots, n$, are all complete local domains.

Let $\mathfrak{q}$ be any ideal in $\mathfrak{o}^*$; we are concerned with the relations between the complete local ring $(\mathfrak{o}_L)^*/(\mathfrak{o}_L)^*\mathfrak{q}$ and the complete local ring $\mathfrak{o}^*/\mathfrak{q}$. Let K' be a coefficient field over K in the complete local ring $\mathfrak{o}^*$, and let L' be the field of quotients of the Kronecker product $L \times K'$ over K. Then the Kronecker product $L' \times \mathfrak{o}^*$ of L' and $\mathfrak{o}^*$ over K' is an $\mathfrak{m}$-adic ring, and the residue ring $L' \times \mathfrak{o}^*/(L' \times \mathfrak{o}^*)\mathfrak{m}$ is isomorphic with the subfield L' of $L' \times \mathfrak{o}^*$. It is then easily seen that every element in the set $L' \times \mathfrak{o}^* - (L' \times \mathfrak{o}^*)\mathfrak{m}$ is $\equiv 1 \pmod{(L' \times \mathfrak{o}^*)\mathfrak{m}}$, from which it follows [29] that the $(L' \times \mathfrak{o}^*)\mathfrak{m}$-adic ring $L' \times \mathfrak{o}^*$ and the local ring $(\mathfrak{o}^*)_{L'}$ have the same completion. Since the completion of $(\mathfrak{o}^*)_{L'}$ is evidently equal to $(\mathfrak{o}_L)^*$, we have therefore $(\mathfrak{o}_L)^* = (L' \times \mathfrak{o}^*)^*$. If now $\mathfrak{q}$ is any ideal $(\neq \mathfrak{o}^*)$ in $\mathfrak{o}^*$, then the residue ring $L' \times \mathfrak{o}^*/(L' \times \mathfrak{o}^*)\mathfrak{q}$ contains L' and $\mathfrak{o}^*/\mathfrak{q}$ as subrings and is $(L', \mathfrak{o}^*/\mathfrak{q})$-isomorphic with the Kronecker product $L' \times (\mathfrak{o}^*/\mathfrak{q})$ over K'. Since the completion of $L' \times \mathfrak{o}^*/(L' \times \mathfrak{o}^*)\mathfrak{q}$ is the residue ring $(L' \times \mathfrak{o}^*)^*/(L' \times \mathfrak{o}^*)^*\mathfrak{q} = (\mathfrak{o}_L)^*/(\mathfrak{o}_L)^*\mathfrak{q}$, we conclude therefore that the two complete local rings $(\mathfrak{o}_L)^*/(\mathfrak{o}_L)^*\mathfrak{q}$ and $(L' \times (\mathfrak{o}^*/\mathfrak{q}))^*$ are $(L', \mathfrak{o}^*/\mathfrak{q})$-isomorphic with each other. In other words, the complete local ring $(\mathfrak{o}_L)^*/(\mathfrak{o}_L)^*\mathfrak{q}$ contains the Kronecker product $L' \times (\mathfrak{o}^*/\mathfrak{q})$ of the two subrings L' and $\mathfrak{o}^*/\mathfrak{q}$ over K' as a subspace and coincides with the adherence of the latter.

The following two theorems (in somewhat different form) are due to Chevalley:

THEOREM 4. *If $\mathfrak{o}$ is a generalized local domain with a basic field K and if the residue field of $\mathfrak{o}$ is a finitely generated extension of K, then any regular extension of K is analytically disjoint with respect to $\mathfrak{o}$ over K.*

THEOREM 5. *If $\mathfrak{o}$ is a generalized local domain with a coefficient field K, and if $\bar{K}$ is analytically disjoint with respect to $\mathfrak{o}$ over K, then any field containing K is analytically disjoint with respect to $\mathfrak{o}$ over K.*

For proof we shall refer to Chevalley [2], § 4. It is sufficient to observe that if we set $(L' \times (\mathfrak{o}^*/\mathfrak{p}_i))^*$, $\mathfrak{o}^*/\mathfrak{p}_i$, L', K' equal to $\mathfrak{O}$, $\mathfrak{o}$, Z, K in the notation of Chevalley [2], p. 77, then all the conditions there are fulfilled. If L is a regular extension of K (i. e. K is "strongly algebraically closed" in L, in the terminology of Chevalley), then L' is also a regular extension of K', and Theorem 4 then follows from [2], p. 78, Proposition 9a and the remark immediately after it. Theorem 5 follows from [2], p. 80, Proposition 10b.

We shall also mention the following rather trivial theorem:

---

[29] See Zariski [9], p. 183. That our ring $L' \times \mathfrak{o}^*$ is not necessarily Noetherian is immaterial here.

THEOREM 6. *Let $\mathfrak{o}$ be a generalized local domain containing a field K, and let L and L' be two fields such that $L \supset L' \supset K$. If L is analytically disjoint with respect to $\mathfrak{o}$ over K, then $\mathfrak{o}_{L'}$ is defined and L is analytically disjoint with respect to $\mathfrak{o}_{L'}$ over L'.*

For, we have $(\mathfrak{o}_L)^* \supset (\mathfrak{o}_{L'})^* \supset \mathfrak{o}^*$ and $(\mathfrak{o}_L)^*\mathfrak{p}_i \cap (\mathfrak{o}_{L'})^* = (\mathfrak{o}_{L'})^*\mathfrak{p}_i$ for every i; hence if the ideals $(\mathfrak{o}_L)^*\mathfrak{p}_i$ are prime in $(\mathfrak{o}_L)^*$, the ideals $(\mathfrak{o}_{L'})^*\mathfrak{p}_i$ must also be prime in $(\mathfrak{o}_{L'})^*$.

Let $\mathfrak{o}$ be a generalized local domain with a coefficient field K, and let L be a field containing K such that the extension $\mathfrak{o}_L$ of $\mathfrak{o}$ over L is defined. The residue of $L \times \mathfrak{o}/(L \times \mathfrak{o})\mathfrak{m}$ is in this case isomorphic to L, and hence every element in the multiplicatively closed system $L \times \mathfrak{o} - (L \times \mathfrak{o})\mathfrak{m}$ is $\equiv 1 \pmod{(L \times \mathfrak{o})\mathfrak{m}}$. It follows then that the $(L \times \mathfrak{o})\mathfrak{m}$-adic ring $L \times \mathfrak{o}$ and the local ring $\mathfrak{o}_L$ have the same completion; that is, we have $(L \times \mathfrak{o})^* = (\mathfrak{o}_L)^*$ or also $(L \times \mathfrak{o}^*)^* = (\mathfrak{o}_L)^*$, since it is evident that $(L \times \mathfrak{o})^* = (L \times \mathfrak{o}^*)^*$. In case L is a finite algebraic extension of K, we can even write $(\mathfrak{o}_L)^* = L \times \mathfrak{o}^*$; for, in this case the $\mathfrak{m}$-adic ring $L \times \mathfrak{o}^*$ is already complete and hence coincides with $(L \times \mathfrak{o}^*)^*$.

THEOREM 7. *Let $\mathfrak{o}$ be a generalized local domain with a coefficient field K, such that its completion $\mathfrak{o}^*$ has no nilpotent elements. Let R be a ring containing $\mathfrak{o}^*$ as a subring, and let L be a field in R such that L is a finite algebraic extension of K. If L is analytically disjoint with respect to $\mathfrak{o}$ over K, then the subring $L\mathfrak{o}^*$ in R is $(L, \mathfrak{o}^*)$-isomorphic to $(\mathfrak{o}_L)^*$.*

Proof. It is clear that the subring $L\mathfrak{o}^*$ in R is an $(L, \mathfrak{o}^*)$-homomorphic image of $(\mathfrak{o}_L)^* = L \times \mathfrak{o}^*$. Therefore, in order to prove our theorem, it is only necessary to show that the subrings L and $\mathfrak{o}^*$ in R are linearly disjoint over K. Let $\mathfrak{p}_1, \cdots, \mathfrak{p}_n$ be the prime divisors of the zero ideal in $\mathfrak{o}^*$; the assumption that L is analytically disjoint with respect to $\mathfrak{o}$ implies that each ring $L \times \mathfrak{o}^*/(L \times \mathfrak{o}^*)\mathfrak{p}_i$ is an integral domain. Since the ring $L \times \mathfrak{o}^*/(L \times \mathfrak{o}^*)\mathfrak{p}_i$ is $(L, \mathfrak{o}^*/\mathfrak{p}_i)$-isomorphic to the ring $L \times (\mathfrak{o}^*/\mathfrak{p}_i)$, therefore $L \times (\mathfrak{o}^*/\mathfrak{p}_i)$ is also an integral domain; it follows then from Theorem 2 that any $(\mathfrak{o}^*/\mathfrak{p}_i)$-homomorphism of $L \times (\mathfrak{o}^*/\mathfrak{p}_i)$ is an isomorphism. Since every element in $\mathfrak{p}_i$ is annulled by every element in $\bigcap_{j \neq i} \mathfrak{p}_j$, it follows that every element in $(L\mathfrak{o}^*)\mathfrak{p}_i \cap \mathfrak{o}^*$ is also annulled by every element in $\bigcap_{j \neq i} \mathfrak{p}_j$; and since $(L\mathfrak{o}^*)\mathfrak{p}_i \cap \mathfrak{o}^* \supset \mathfrak{p}_i$ and since the ring $\mathfrak{o}^*$ has no imbedded prime divisors of the zero ideal, it follows that $(L\mathfrak{o}^*)\mathfrak{p}_i \cap \mathfrak{o}^* = \mathfrak{p}_i$. It follows then that the residue ring $L\mathfrak{o}^*/(L\mathfrak{o}^*)\mathfrak{p}_i$ contains L and $\mathfrak{o}^*/\mathfrak{p}_i$ as subrings, and it is an $(L, \mathfrak{o}^*/\mathfrak{p}_i)$-homomorphic image of $L \times (\mathfrak{o}^*/\mathfrak{p}_i)$ and hence also an $(L, \mathfrak{o}^*/\mathfrak{p}_i)$-

isomorphic image of $L \times (\mathfrak{o}^*/\mathfrak{p}_i)$. Thus each $L\mathfrak{o}^*/(L\mathfrak{o}^*)\mathfrak{p}_i$ is a complete local domain and contains L and $\mathfrak{o}^*/\mathfrak{p}_i$ as two linearly disjoint subrings over K.

Let $u_1, \cdots, u_m$ be a set of linearly independent elements over K in $\mathfrak{o}^*$, and denote by $(u_1, \cdots, u_m)$ the linear space determined by this set over K. If l_1 is the dimension of the sub-space of $(u_1, \cdots, u_m)$ which is contained in the ideal $\mathfrak{p}_1$, then we can replace the set $u_1, \cdots, u_m$ by a linearly equivalent set over K, such that the space $(u_1, \cdots, u_{l_1})$ is in $\mathfrak{p}_1$ and the space $(u_{l_1+1}, \cdots, u_m)$ contains no elements in $\mathfrak{p}_i$. If l_2 is the dimension of the subspace of $(u_1, \cdots, u_{l_1})$ which is contained in $\mathfrak{p}_2$, then we can replace the set $u_1, \cdots, u_{l_1}$ by a linearly equivalent set over K, such that the space $(u_1, \cdots, u_{l_2})$ is in $\mathfrak{p}_1 \cap \mathfrak{p}_2$ and the space $(u_{l_2+1}, \cdots, u_{l_1})$ contains no elements in $\mathfrak{p}_2$. Thus we can proceed until in the final stage we have a set $u_1, \cdots, u_{l_{n-1}}$ such that $(u_1, \cdots, u_{l_{n-1}})$ is contained in $\mathfrak{p}_1 \cap \cdots \cap \mathfrak{p}_{n-1}$; here the process must stop and the space $(u_1, \cdots, u_{l_{n-1}})$ contains no elements in $\mathfrak{p}_n$, for we have $\bigcap_{i=1}^{n} \mathfrak{p}_i = (0)$ on account of the absence of nilpotent elements in $\mathfrak{o}^*$. Adding for the sake of convenience an element $u_0 = 0$, our final results will be a set $u_0, \cdots, u_m$, which is linearly equivalent over K to the original set, but has the property that for each $i = 0, \cdots, n$, the space $(u_0, \cdots, u_{l_i})$ is contained in $\mathfrak{p}_1 \cap \cdots \cap \mathfrak{p}_i$ and the space $(u_{l_i+1}, \cdots, u_{l_{i-1}})$ contains no elements in $\mathfrak{p}_i$ (where we set $l_0 = m$ and $l_n = 0$). We shall now show that if there is a linear relation $\sum_{j=1}^{m} c_j u_j = 0$ with coefficients c_j in L, then all the c_j must vanish. Assume that it has already been proved that $c_j = 0$ for all $j > l_{i-1}$, we shall show that $c_j = 0$ for all $j > l_i$. In fact, if not all $c_{l_i+1}, \cdots, c_{l_{i-1}}$ are zero, then the relation $\sum_{j=1}^{l_{i-1}} c_j u_j = 0$ will induce a linear relation $\sum_{j=l_i+1}^{l_{i-1}} c_j u_j = 0$ in $L\mathfrak{o}^*/(L\mathfrak{o}^*)\mathfrak{p}_i$ between the $u_{l_i+1}, \cdots, u_{l_{i-1}}$, considered as elements in $\mathfrak{o}^*/\mathfrak{p}_i$. Since in $L\mathfrak{o}^*/(L\mathfrak{o}^*)\mathfrak{p}_i$ the subrings L and $\mathfrak{o}^*/\mathfrak{p}_i$ are linearly disjoint over K, it follows that the $u_{l_i+1}, \cdots, u_{l_{i-1}}$ must be linearly dependent over K in $\mathfrak{o}^*/\mathfrak{p}_i$; this means that there is a linear combination of the $u_{l_i+1}, \cdots, u_{l_{i-1}}$ with coefficients in K which is an element in $\mathfrak{p}_i$, in contradiction to the fact that the space $(u_{l_i+1}, \cdots, u_{l_{i-1}})$ contains no elements in $\mathfrak{p}_i$. Thus we have shown by induction that all the $c_1, \cdots, c_m$ must vanish, and hence the elements $u_1, \cdots, u_m$ are linearly independent over L. This concludes the proof of our theorem.

THE JOHNS HOPKINS UNIVERSITY.

REFERENCES.

1. C. Chevalley, "On the theory of local rings," *Annals of Mathematics*, vol. 44 (1943), pp. 690-708.

2. ———, "Some properties of ideals in rings of power series," *Transactions of the Americal Mathematical Society*, vol. 55 (1944), pp. 68-84.

3. ———, "Intersections of algebraic and algebroid varieties," *ibid.*, vol. 57 (1945), pp. 1-85.

4. Wei-Liang Chow, "The Jacobian variety of an algebraic curve," to appear in this JOURNAL soon.

5. ——— and B. L. van der Waerden, "Zur algebraischen Geometrie. IX. Ueber zugeordnete Formen und algebraische Systeme von algebraischen Mannigfaltigkeiten," *Mathematische Annalen*, Band 113 (1937), pp. 692-704.

6. I. S. Cohen, "On the structure and ideal theory of complete local rings," *Transactions of the American Mathematical Society*, vol. 59 (1946), pp. 54-106.

7. W. Krull, "Dimensionstheorie in Stellenringen," *Journal für die reine und angewandte Mathematik*, vol. 179 (1938), pp. 204-226.

8. A. Weil, *Foundations of algebraic geometry*, New York, 1946.

9. O. Zariski, "Generalized semi-local rings," *Summa Brasiliensis Mathematicae*, vol. I, fasc. 8 (1946), pp. 169-195.

10. ———, "The concept of a simple point of an abstract algebraic variety," *Transactions of the American Mathematical Society*, vol. 62 (1947), pp. 1-52.

ON THE QUOTIENT VARIETY OF AN ABELIAN VARIETY

BY WEI-LIANG CHOW

THE JOHNS HOPKINS UNIVERSITY, BALTIMORE, MD.

Communicated by O. Zariski, October 30, 1952

Let A and B be two Abelian varieties and let F be a homomorphism of A onto B; then the kernel of the homomorphism F is a subgroup in A which consists of a bunch of subvarieties in A. The question naturally arises whether conversely for any given such subgroup X in A there exist an Abelian variety B and a homomorphism of A onto B with X as its kernel. In other words, the question is whether the quotient group A/X of A over X can also be represented by an Abelian variety $A(X)$, such that the canonical homomorphism of the group A onto A/X is represented by a homomorphism of the Abelian variety A onto $A(X)$. It is the purpose of this note to show that such a quotient variety $A(X)$ always exists.

We shall begin with a few definitions.[1] A subgroup X in an Abelian variety A is said to be an *algebraic subgroup* if X is a bunch of subvarieties in A, i.e., if X is the union of a finite number of subvarieties in A. An algebraic subgroup X is said to be *algebraic* or *normally algebraic* over a field K if as a bunch of subvarieties it is algebraic or normally algebraic over K; X is said to be *separable* over K, if each component variety in the bunch X is defined over a separable algebraic extension of K. It is known[2] that all component varieties of an algebraic subgroup X in A are disjoint to one another, and that the one and only component variety X_0 containing the unit element is an Abelian variety and the other component varieties are obtained from X_0 by translations with respect to some elements in A. The dimension of X_0, which is also the dimension of all other component varieties in X, is called the *dimension* of X. It is clear that if X is normally algebraic over K, then X_0 is defined over a purely inseparable extension of K, and if, furthermore, X is also separable over K, then X_0 is defined over K.

We recall here the fact[3] that the aggregate of all positive cycles of a given dimension r and a given degree d (i.e., the degree in the ambient projective space) in an algebraic variety V can be represented in a one-to-one and algebraic manner by the points of a bunch of varieties, normally algebraic over K, which we shall denote by $M(V; r, d)$. We shall call this representation the *canonical mapping* Γ of the positive r-cycles into $M(V; r, d)$; this mapping has the property that if a positive r-cycle Z of degree d in V is rational over a field K', then the point $\Gamma(Z)$ in $M(V; r, d)$ is also rational over K'.

Let F be a rational transformation of an algebraic variety V and let K be a field of definition for both V and F, then F is said to be *separable* if $K(x)$ is a separably generated extension of $K(F(x))$, where x is a generic point of V over K. Since $K(x)$ is separably generated over K, this definition of separability is independent of the choice of the field K.

Let F be a homomorphism of an Abelian variety A into an Abelian variety B, and let K be a field over which A, B, F are all defined. Then the kernel of F, being the projection in A of the intersection of $A \times (0)$ with graph of F in the product variety $A \times B$, is an algebraic subgroup in A, normally algebraic over K; if furthermore the homomorphism F is separable, then this algebraic subgroup is also separable over K.

THEOREM 1. *Let A be an Abelian variety of dimension n and let X be an algebraic subgroup of dimension r in A. Then the quotient group A/X can be represented by an Abelian variety $A(X)$ of dimension n-r, such that the canonical homomorphism of the group A onto A/X is represented by a separable homomorphism F of the Abelian variety A onto $A(X)$. Furthermore, if H is any homomorphism of A into an Abelian variety B with X as the kernel, then H is the product of F and a one-to-one homomorphism of $A(X)$ into B. If K is a field such that A is defined over K and X is normally algebraic and separable over K, then both $A(X)$ and F are defined over K.*

Proof: Since X is a bunch of varieties of dimension r, normally algebraic and separable over K, it determines a rational positive r-cycle Z over K in A, which contains each variety in X as a simple component. Let x be a generic point of A over K, and let T_x be the bi-regular birational transformation of A onto itself given by the translation attached to x; then $T_x(Z)$ is a rational positive r-cycle over $K(x)$ in A, which of course also contains each variety in $T_x(X)$ as a simple component. If d is the degree of $T_x(Z)$ as a positive r-cycle in the ambient projective space of A, then the canonical mapping Γ of the positive r-cycles of degree d in A onto $M(A; r, d)$ induces a rational transformation $F(x) = \Gamma(T_x(Z))$ of A into $M(A; r, d)$. It is clear that the image $F(A)$ of A is a variety $A(X)$ defined over K; we shall show that $A(X)$ has the properties stated in our theorem.

Let W be the graph of the law of composition in A; it is known[4] that for any point x_0 in A, we have $x_0 \times T_{x_0}(Z) = pr_{13}[W\cdot(x_0 \times Z \times A)]$. This shows at once the $T_x(Z)$ has the unique specialization $T_{x_0}(Z)$ over the specialization $x \to x_0$ for any point x_0 in A. It follows then that $F(x)$ has a unique specialization $F(x_0) = \Gamma(T_{x_0}(Z))$ over every specialization $x \to x_0$, and since A is non-singular, this implies that the rational transformation $F(x)$ is defined everywhere in A. If x_1 is another point in A, then we have evidently $F(x_0) = F(x_1)$ if and only if $T_{x_0}(Z) = T_{x_1}(Z)$. In particular, for the generic point $y = F(x)$ in $A(X)$ over K, the inverse image $F^{-1}(y)$ is a prime rational cycle over $K(y)$ which contains the same component varieties as the cycle $T_x(Z)$; and since $T_x(Z)$ is rational over $K(y)$ and has only simple components, it follows that $F^{-1}(y) = T_x(Z)$ and hence also $F^{-1}(y_0) = T_{x_0}(Z)$ for every x_0 and $y_0 = F(x_0)$. According to a result of ours proved elsewhere,[5] all this implies that the variety

$A(X)$ is a non-singular variety; furthermore, it is clear that the rational transformation F is separable.

Next we shall show that $A(X)$ is an Abelian variety. Let y_1 and y_2 be two independent generic points of $A(X)$ over K, and let x_1 and x_2 be generic points of the prime rational cycles $F^{-1}(y_1)$ and $F^{-1}(y_2)$ over $K(y_1, y_2)$, respectively, independent with respect to each over $K(y_1, y_2)$. The point $F(x_1 + x_2)$ is then rational over $K(y_1, y_2, x_1, x_2)$. Let x_1' and x_2' be generic points of $F^{-1}(y_1)$ and $F^{-1}(y_2)$, respectively, over $K(y_1, y_2)$, independent with respect to each other as well as to the points x_1 and x_2 over $K(y_1, y_2)$; then the point $F(x_1' + x_2')$ is rational over $K(y_1, y_2, x_1', x_2')$. Since x_1' is in $T_{x_1}(Z)$ and x_2' is in $T_{x_2}(Z)$, the point $x_1' + x_2'$ is in $T_{x_1+x_2}(Z)$; hence we have $F(x_1 + x_2) = F(x_1' + x_2')$. This shows that the point $F(x_1 + x_2)$ is rational over $K(y_1, y_2, x_1, x_2) \cap K(y_1, y_2, x_1', x_2') \subset K(y_1, y_2)$ and is independent of the choice of the generic points x_1 and x_2. Since x_1 and x_2 are separably generated over $K(y_1, y_2)$, the point $F(x_1 + x_2)$ is also separably generated over $K(y_1, y_2)$; and since $F(x_1 + x_2)$ is uniquely determined by the points y_1 and y_2, it follows that $F(x_1 + x_2)$ is rational over $K(y_1, y_2)$. We shall now define $y_1 + y_2$ to be the point $F(x_1 + x_2)$ and show that this defines a law of composition in $A(X)$, defined everywhere in $A(X)$. The group properties of this law of composition follow easily from the same properties in A, and can be indicated as follows: $y_1 + (y_2 + y_3) = F(x_1 + (x_2 + x_3)) = F((x_1 + x_2) + x_3) = (y_1 + y_2) + y_3$, $-y = F(-x)$, and the unit element in $A(X)$ is evidently $F(O)$. It only remains to show that the function $y_1 + y_2$ is defined everywhere in the product variety $A(X) \times A(X)$; since $A(X)$ is non-singular, we only need to show that $y_1 + y_2$ has a unique specialization over every specialization $y_1 \to \eta_1$, $y_2 \to \eta_2$. Now, every such specialization $y_1 + y_2 \to \eta_3$ can be extended to a specialization of $y_1 + y_2 \to \eta_3$, $x_1 \to \xi_1$, $x_2 \to \xi_2$ over the specialization $y_1 \to \eta_1$, $y_2 \to \eta_2$, and in this specialization the relations $y_1 = F(x_1)$, $y_2 = F(x_2)$, $y_1 + y_2 = F(x_1 + x_2)$ go over into the relations $\eta_1 = F(\xi_1)$, $\eta_2 = F(\xi_2)$, $\eta_3 = F(\xi_1 + \xi_2)$. Since $F(\xi_1 + \xi_2)$ remains fixed for every choice of the points ξ_1 and ξ_2 in $F^{-1}(\eta_1)$ and $F^{-1}(\eta_2)$ respectively, the point η_3 is therefore uniquely determined by η_1 and η_2.

Consider now the homomorphism H of A into B; let K' be a field over which A, B, $A(X)$, F and H are defined, and let x be a generic point of A over K'. Then $H(A)$ is an Abelian subvariety in B, defined over K', and the point $z = H(x)$ is a generic point of $H(A)$ over K'. Since F is separable and is defined over K', $K'(x)$ is a separably generated extension of $K'(y)$, and hence $K'(y, z)$ is also a separably generated extension of $K'(y)$. Let y_0 be a point in $A(X)$, and let $x \to x_0$, $z \to z_0$ be specializations over the specialization $y \to y_0$ over K'; then the relations $F(x) = y$, $H(x) = z$ go over into the relations $F(x_0) = y_0$, $H(x_0) = z_0$. The fact that H has X as its kernel implies that $H(x_0) = z_0$ is a uniquely determined point for all x_0 satisfying the relation $F(x_0) = y_0$, so that z has a unique specialization over every specialization of y over K'. This implies not only that z is algebraic over $K'(y)$ but also that z is rational over a purely inseparable extension of $K'(y)$; and since $K'(y, z)$ is a separably generated extension of $K'(y)$, it follows that z is rational over $K'(y)$. The rational function $z = H_1(y)$ then defines a rational transformation of $A(X)$ into B, defined over K', which can be easily seen to be a homomorphism, and we have evidently the equation $H(x) = H_1F(x)$. This concludes the proof of our theorem.

Theorem 1 gives a satisfactory answer to our question from the point of view of "absolute" algebraic geometry. However, one is often interested in the properties of a variety with respect to a given field of definition, and the consideration of such relative properties can be significant even in cases where one's final objective is "absolute." From the relative point of view it is quite natural to take as K a field of definition of A over which the algebraic subgroup X is normally algebraic, as otherwise a quotient variety with the specified properties cannot possibly be defined over K; but the additional condition that X be separable over K seems to be an undesirable restriction. It is therefore not without some interest to show (Theorem 2) that there exists a relative quotient variety $A(X, K)$ over any field of definition K of A over which X is normally algebraic; this relative quotient variety $A(X, K)$ has similar properties as $A(X)$, though naturally in a somewhat restricted form.

We shall begin with two simple lemmas and a few definitions.

LEMMA 1. *Let K' be a finitely generated extension of K, and let $\mathfrak{L}$ be an aggregate of subfields in K' which all contain K. If $\mathfrak{L}$ has the property that the compositum of every two fields in $\mathfrak{L}$ is again a field in $\mathfrak{L}$, then the subfield in K' generated by all fields in $\mathfrak{L}$ is also a field in $\mathfrak{L}$.*

Proof: Let L_1 be a field in $\mathfrak{L}$ which has the maximum possible dimension over K; then for any field L in $\mathfrak{L}$, the compositum LL_1, which by assumption is in $\mathfrak{L}$, is a finite algebraic extension of L_1. Since $K' \cap \bar{L}_1$ is a finite algebraic extension of L_1 and since LL_1 is contained in $K' \cap \bar{L}_1$, the degree of LL_1 over L_1 is bounded for all L in $\mathfrak{L}$. Let L_2 be a field in $\mathfrak{L}$ such that L_2 contains L_1 and has the maximum possible degree over L_1; then for every field L in $\mathfrak{L}$, the fields LL_2 and L_2 must have the same degree over L_1 and hence must coincide. This shows that L_2 is the subfield in K' generated by all the fields in $\mathfrak{L}$.

If an aggregate $\mathfrak{L}$ of subfields in K' has the property stated the preceding lemma, then the subfield in K' generated by all the subfields in $\mathfrak{L}$ is called the *maximal field* of $\mathfrak{L}$.

A field L containing K is said to be an *Abelian field over* K, if it is the function field of an Abelian variety defined over K. Such a field L is of course a finitely generated extension, and the Abelian variety A associated with it is uniquely determined up to an isomorphism over K. If L_1 is an Abelian subfield over K in L and if A_1 is the Abelian variety associated with L_1, then there is associated with L_1 a homomorphism F of A onto A_1, defined over K, which is determined up to isomorphisms of A and A_1.

LEMMA 2. *Let L be an Abelian field over K, and let L_1 and L_2 be two Abelian subfields over K in L. Then the compositum L_1L_2 is also an Abelian field over K; furthermore, if L_1 is associated with a homomorphism of A with the kernel X_1 and L_2 is associated with a homomorphism of A with the kernel X_2, then L_1L_2 is associated with a homomorphism of A with the kernel $X_1 \cap X_2$.*

Proof: Let A, A_1, A_2 be the Abelian varieties associated with L, L_1, L_2, respectively, and let F_1 and F_2 be the associated homomorphisms of A onto A_1 and A_2, respectively, with X_1 and X_2 as the kernels, respectively. Then the rational transformation $F_1 \times F_2$ is a homomorphism of A into the product variety $A_1 \times A_2$, and the image of A under this homomorphism is then an Abelian variety A_3 over K whose function field is isomorphic to L_1L_2. It is clear that the homomorphism $F_1 \times F_2$ of A onto A_3 has the kernel $X_1 \cap X_2$.

THEOREM 2. *Let A be an Abelian variety of dimension n, defined over K, and let X be an algebraic subgroup of dimension r in A, normally algebraic over K. Then the quotient group A/X can be represented by an Abelian variety $A(X, K)$, defined over K, such that the canonical homomorphism of the group A onto A/X is represented by a homomorphism F of the Abelian variety A onto $A(X, K)$, also defined over K; furthermore, if H is any homomorphism of A into an Abelian variety B with X as the kernel and if both B and H are defined over K, then H is the product of F and a one-to-one homomorphism of $A(X, K)$ into B, also defined over K.*

Proof: Since X is normally algebraic over K, there exists a positive integer λ such that X is separable over K^{-p^λ}; then according to Theorem 1 the Abelian variety $A(X)$ is defined over the field K^{-p^λ}. Let y be a generic point of $A(X)$ over K^{-p^λ}, and consider the field $K(y^{p^\lambda})$, where y^{p^λ} denotes the point obtained from y by raising all coordinates of y to the p^λth power. Since $K^{-p^\lambda}(y)$ and $K(y^{p^\lambda})$ are isomorphic under mapping of raising each element of the former to its p^λth power, and since $K^{-p^\lambda}(y)$ is an Abelian field of dimension n-r over K^{-p^λ}, it follows that $K(y^{p^\lambda})$ is an Abelian field of dimension n-r over K. Also from $K^{-p^\lambda}(x) \supset K^{-p^\lambda}(y)$ follows $K(x^{p^\lambda}) \supset K(y^{p^\lambda})$ and hence $K(x) \supset K(y^{p^\lambda})$.

Consider now the aggregate $\mathfrak{L}$ of all Abelian subfields over K in $K(x)$ with the property that each subfield has an associated homomorphism with X as its kernel; this aggregate $\mathfrak{L}$ is not empty as $K(y^{p^\lambda})$ is one such subfield. Furthermore, it is easily seen that the compositum of two fields in $\mathfrak{L}$ is also a field in $\mathfrak{L}$; in fact, according to Lemma 2 this compositum is also an Abelian subfield over K and it has an associated homomorphism with $X \cap X = X$ as its kernel. It follows then from Lemma 1 that $\mathfrak{L}$ has a maximal field L, and we can define $A(X, K)$ to be the Abelian variety determined by L and define F to be the corresponding homomorphism of A onto $A(X, K)$. It is clear that the so-defined $A(X, K)$ and F have the properties stated in our theorem.

[1] We shall assume that the reader is familiar with the contents of the book by A. Weil, *Variétés abéliennes et courbes algébriques*, Paris, 1948, and we shall use the terminology of Weil as developed in this book, with the exception that all our varieties are understood to be embedded in some projective spaces. Since a point x in a projective space is given by a system of homogeneous coordinates, we shall denote by $K(x)$ the field obtained from K by the adjunction of all the ratios (with non-zero denominators) of the coordinates of x.

[2] Weil, *loc. cit.*, No. 24, Proposition 8, p. 39.

[3] Chow, W. L., and van der Waerden, B. L., "Zur algebraischen Geometrie. IX. Ueber zugeordnete Formen und algebraische Systeme von algebraischen Mannigfaltigkeiten," *Math. Ann.*, Band 113, 692–704 (1937).

[4] Weil, *loc. cit.*, No. 13, Théorème 4, p. 22.

[5] Chow, W. L., "Algebraic Systems of Positive Cycles in an Algebraic Variety," *Am. J. Math.*, **72**, 247–283 (1950).

ON PICARD VARIETIES.*

By Wei-Liang Chow.

1. **Introduction.** In a recent paper [1] J. Igusa has developed a theory of the Picard variety attached to a non-singular algebraic variety V. Let $\mathfrak{G}_c(V)$ be the group of all divisors in V which are continuously equivalent to zero, and let $\mathfrak{G}_l(V)$ be the group of all divisors in V which are linearly equivalent to zero; let A be the Abelian variety determined by the periods of the Abelion integrals of the first kind on V, and let P be the dual Abelian variety of A in the sense of Weil (see section 3). The variety P is called the Picard variety attached to V, and Igusa's main result is that the group $\mathfrak{G}_c(V)$ can be mapped homomorphically onto the Picard variety P such that the kernel of this homomorphism is precisely the group $\mathfrak{G}_l(V)$. A. Weil [2] has taken up the same problem (for the possibly more general case of a Hodge variety) by a quite different method and has obtained among others also this result of Igusa, which he calls the first duality theorem of Igusa. There is however one important point concerning this theorem which is not settled in Igusa's paper; this is the question whether the mapping of $\mathfrak{G}_c(V)$ onto P is an analytic mapping in the sense (to be defined more precisely later in section 2) that it is analytic for every analytic system of divisors in $\mathfrak{G}_c(V)$. The importance of this question has been recognized by Weil, who considers the solution of this question, which he has obtained by a very elegant method, as the Main Theorem of his theory.[3] In this paper we shall give a new, simple proof of Igusa's duality theorem, which will at the same time take care of this additional question of analyticity almost without any further effort. Our method, like that of Igusa, is based on the existence of a "sufficiently general" curve in V, but we shall make systematic use of the concept of

* Received February 4, 1952.

[1] J. Igusa, "On the Picard varieties attached to algebraic varieties," *American Journal of Mathematics*, vol. 74 (1952), pp. 1-22.

[2] A. Weil, "On Picard varieties," *American Journal of Mathematics*, vol. 74 (1952), pp. 865-894.

[3] A. Weil, *loc. cit.*, § III. Weil's proof of this theorem holds only for the case when the parameter variety is an analytic manifold (i. e. a variety without multiple points); however, it can be extended to the general case of an arbitrary analytic parameter variety by using some of the concepts and results developed in our (unpublished) paper cited in footnote 10.

duality between Abelian varieties introduced by Weil.[4] Our theory holds only for an algebraic variety and is hence possibly less general than that of Weil; as a compensation, we are able to formulate and prove the analyticity of the mapping in a somewhat stronger sense than that given by Weil's theory.[5] In the last section we shall deal with the more algebraic aspects of the problem; on the basis of some results on the Jacobian varieties proved by us elsewhere,[6] we investigate the fields of definition of some of the varieties and transformations introduced in the previous section. These results enable us to construct in a very simple way a system of divisors which represents the group $\mathfrak{G}_c(V)/\mathfrak{G}_l(V)$ in a "generally" one-to-one manner. This is however a weaker result than that obtained by Weil,[7] who has constructed a system of divisors which represents $\mathfrak{G}_c(V)/\mathfrak{G}_l(V)$ in an exactly one-to-one manner. Finally, we should like to mention that the method used in this last section has far greater potentialities than the immediate objective there indicates; in fact, this method is part of an idea which can be developed into a theory of Picard varieties for algebraic varieties over an arbitrary ground field, as will be shown in a forthcoming paper of ours on this subject.

We add here a few remarks concerning terminology. In this paper, the expression "analytic" shall always mean complex-analytic; and, unless explicitly stated to the contrary, the dimension is always the complex dimension, which is twice the topological dimension. By a cycle is always meant an algebraic cycle, except when explicitly stated to be a topological cycle, in which case the dimension is of course understood to be topological. When we are dealing with mappings of group varieties into one another, the expressions homomorphism, isomorphism, automorphism, etc. shall always mean analytic or algebraic homomorphism, isomorphism, automorphism, etc., according as we are dealing with analytic or algebraic varieties. The possibility of any confusion on this score is excluded by the theorem [8] that any (closed) analytic subvariety of an algebraic variety or analytic correspondence of an algebraic variety into another is an algebraic subvariety or an algebraic correspondence respectively.

---

[4] A. Weil, *loc. cit.*, § I.

[5] We shall prove that the mapping is holomorphic in the sense defined in section 2, while Weil's method (extended as indicated in footnote 3) yields only the somewhat weaker result that the mapping is analytic.

[6] W. L. Chow, "The Jacobian variety of an algebraic curve," to appear in print soon.

[7] A. Weil, *loc. cit.*, § II.

[8] W. L. Chow, "On compact complex analytic varieties," *American Journal of Mathematics*, vol. 71 (1949), pp. 893-914, Theorem V and Theorem VII.

2. **Analytic mapping of cycles.** By means of the method of associated forms [9] the aggregate of all positive cycles of a given dimension d and a given degree m (i. e. the degree in the ambient projective space) in an algebraic variety V can be represented in a one-to-one and algebraic manner by the points of an (in general reducible) algebraic variety, which we shall denote by $M(V; d, m)$. In fact, the (homogeneous) coordinates of a point in $M(V; d, m)$ can be taken to be the coefficients of the associated form of the positive cycle corresponding to it. We shall call this representation the canonical mappings of the positive cycles into $M(V; d, m)$. The canonical mapping has the property that if an analytic system of positive cycles $|Z|$ of dimension d and degree m in V is parametrized by an analytic variety W, i. e. if there is a closed analytic correspondence [10] between W and V such that to each point in W corresponds (with a suitable definition of multiplicity) a positive cycle of dimension d and degree m, then this parametrization induces an everywhere single-valued analytic transformation of W into $M(V; d, m)$. In case this last transformation is a holomorphic transformation of W into V, then $|Z|$ is said to be regularly parametrized by W. It is easily seen that if W is a manifold (i. e. a variety without multiple points), then the parametrization is always regular. If $|Z_1|$ and $|Z_2|$ are two analytic systems of positive cycles of the same dimension d and of degrees m_1 and m_2 respectively, both parametrized by the same analytic variety W, then the difference system $|Z_1 - Z_2|$, consisting of all

---

[9] W. L. Chow and B. L. van der Waerden, "Zur algebraischen Geometrie. IX. Ueber zugeordnete Formen und algebraische Systeme von algebraischen Mannigfaltigkeiten," *Mathematische Annalen*, Band 113 (1937), pp. 692-704; see also B. L. van der Waerden, *Einfuehrung in die algebraische Geometrie*, §§ 36-7, pp. 153-162. More precisely, it is proved there that there is an algebraic correspondence (which is in general reducible) between $M(V; d, m)$ and V such that to each point of $M(V; d, m)$ there corresponds under this correspondence a positive cycle of dimension d and degree m in V, the multiplicities of the components of the cycle being determined by the associated form corresponding to the point in $M(V; d, m)$, and that in this way all positive cycles of dimension d and degree m in V can be obtained. This is what is meant by the statement that the positive cycles of dimension d and degree m in V are represented by the points of $M(V; d, m)$ in a one-to-one and algebraic manner.

[10] We shall use in this section some of the concepts developed in our paper, "On complex analytic varieties," to appear soon in this Journal. We refer to that paper for the definition of a complex analytic variety in abstracto and the definition of a holomorphic function on such a variety. We mention here that a closed analytic correspondence of an analytic variety W into an analytic variety V is defined by a closed analytic subvariety in the product variety $W \times V$, and that such a correspondence is called an everywhere single-valued analytic transformation, if to each point in W corresponds exactly one point in V. A single-valued analytic transformation of W into V is said to be a holomorphic transformation, if every holomorphic function on V is carried by the inverse transformation into a holomorphic function on W.

cycles $Z_1 - Z_2$ as the common parameter runs through the variety W, is said to be an analytic system of cycles of dimension d and degree $m = m_1 - m_2$, and W is its parameter variety. The parametrization of $| Z_1 - Z_2 |$ by W is said to be regular, if both $| Z_1 |$ and $| Z_2 |$ are regularly parametrized by W.

Suppose we have a mapping $f(Z)$, which assign to each element of an aggregate $\{Z\}$ of cycles of a given dimension d in V a point in an analytic variety U. We shall say that this mapping $f(Z)$ is analytic, if for any analytic system of cycles in $\{Z\}$ parametrized by an analytic variety W, the mapping $f(Z)$ induces an (everywhere single-valued) analytic transformation of W into U. The mapping $f(Z)$ is said to be holomorphic (or regularly analytic) if this induced transformation is holomorphic whenever W is a regular parametrization of the analytic system. Assume that U is an Abelian variety and that the mapping $f(Z)$ is a homomorphism of a subgroup $\{Z\}$ of the group of all cycles of dimension d in V into U; then the mapping $f(Z)$ is essentially determined by the mapping of the positive cycles in $\{Z\}$. It is then easily seen that the mapping $f(Z)$ is holomorphic, if for the positive cycles of degree m in $\{Z\}$ (for any m) the mapping $f(Z)$ is the product of the canonical mapping of Z into $M(V; d, m)$ and a holomorphic transformation of a subvariety of $M(V; d, m)$ into U.

Let Y be an algebraic subvariety of dimension $r - d$ in V (r being the dimension of V) which is the intersection of V with a linear subspace L of the proper dimension in the ambient space, and let $\{Z\}$ be the aggregate of all positive cycles of dimension d and degree m in V which has a proper intersection with Y. Let $f(Z)$ be the mapping of $\{Z\}$ in $M(Y; 0, m)$ which is the product of the mapping $Z \to Z \cdot Y$ and the canonical mapping of $Z \cdot Y$ into $M(Y; 0, m)$; then $f(Z)$ is a holomorphic mapping. In fact, the mapping $Z \to Z \cdot Y$ induces a mapping f' of an open subset[11] in $M(V; d, m)$ into

[11] This subset is not only open in the sense of the ordinary topology, but also in the sense of the topology where the closed subsets are the (reducible or irreducible) algebraic subvarieties. To show this, we recall the fact that if $F(u^{(0)}, u^{(1)}, \cdots, u^{(d)})$ is the associated form of a positive cycle Z of dimension d and degree m and if $v^{(i)} = (v^{(i)}_0, v^{(i)}_1, \cdots, v^{(i)}_n)$, $i = 0, 1, \cdots, d$, are the hyperplane coordinates of any $d + 1$ hyperplanes in the ambient projective space, then we have $F(v^{(0)}, v^{(1)}, \cdots, v^{(d)}) = 0$ if and only if these $d + 1$ hyperplanes have a common intersection point with Z. See, e. g., B. L. van der Waerden, *loc. cit.*, p. 157. Now, if we let $v^{(1)}, \cdots, v^{(d)}$ be the hyperplane coordinates of any d hyperplane whose intersection is the linear subspace L, then $F(u^{(0)}, v^{(1)}, \cdots, v^{(d)})$ will vanish identically in the variables $u^{(0)}$ if and only if the cycle Z does not intersect L and hence also Y properly. The condition that $F(u^{(0)}, v^{(1)}, \cdots, v^{(d)})$ vanishes identically in the variables $u^{(0)}$ can obviously be expressed as a system of homogeneous algebraic equations in the coefficients of the form $F(u^{(0)}, u^{(1)}, \cdots, u^{(d)})$, which are the coordinates of the point in $M(V; d, m)$ representing the

$M(Y; 0, m)$; and since $f(Z)$ is evidently the product of the canonical mapping of Z into $M(V; d, m)$ and f', it is sufficient to show that this mapping f' is a holomorphic transformation. Now, the coordinates of the point in $M(V; d, m)$ representing a cycle Z are the coefficients of the associated form of Z, and the associated form of $Z \cdot Y$ can be obtained by substituting for certain sets of variables in the associated form of Z the coefficients of the linear equations defining the linear subspace L. Furthermore, by a well known property of the associated form mentioned in footnote 11, the so obtained associated form of $Z \cdot Y$ cannot vanish identically so long as Z intersects Y properly. As the coefficients of the associated form of $Z \cdot Y$ are the coordinates of the point in $M(Y; 0, m)$ representing $Z \cdot Y$, it follows that the mapping f' is a holomorphic transformation.

3. Dual pairing of complex tori. Following Weil we shall introduce the concept of duality between complex tori of a given dimension; a dual pair of complex tori is characterized by the properties that the one is the character group of the fundamental group of the other and that the automorphism of the covering group of the one corresponding to the multiplication by $\sqrt{-1}$ induces the same automorphism of the covering group of the other. We can describe this concept in non-invariant terms as follows. Let S be the complex vector space of dimension p, and let a coordinate system be chosen as that a vector in S is represented by a point $(z) = (z_1, \cdots, z_p)$ in the p-dimensional complex number space. Let $\mathfrak{P}$ be a $(2p, p)$-matrix whose rows represent $2p$ real-linearly independent vectors, and let $[\mathfrak{P}]$ be the subgroup of S generated by these $2p$ vectors. Then the quotient group $U = S/[\mathfrak{P}]$ is a complex torus of dimension p. Let S' be another complex vector space of dimension p, and let a coordinate system be chosen so that a vector in S' is also represented by a point $(z') = (z'_1, \cdots, z'_p)$ in the p-dimensional complex number space. We shall consider S' to be invariantly associated with S in such a way that if S undergoes a coordinate transformation $(z) \to (z)\mathfrak{T}$, then S' undergoes the coordinate transformation $(z') \to (z')\,{}^t\bar{\mathfrak{T}}^{-1}$. Let $\mathfrak{P}'$ be the $(2p, p)$-matrix such that $(\mathfrak{P}'\bar{\mathfrak{P}}')^t(\mathfrak{P}\bar{\mathfrak{P}}) = 1$; then the quotient group $U' = S'/[\mathfrak{P}']$ is a complex torus of dimension p which is invariantly associated with U and which can be considered as the character group of $[\mathfrak{P}]$ by assigning to each point in U', represented by a point (z') mod. $[\mathfrak{P}']$, the character

$$\frac{1}{2}(\bar{z}')^t(z) + \frac{1}{2}(z')^t(\bar{z}) \qquad \text{(mod. 1)} \tag{1}$$

---

cycle Z; it follows then that the set of all points in $M(V; d, m)$ which represents cycles having improper intersections with Y is an (in general reducible) algebraic subvariety in $M(V; d, m)$.

for all points (z) in $[\mathfrak{P}]$. It is clear that this relation between U and U' is mutual; they are called the dual of each other, and we shall say that they are dually paired by the form (1).

The duality between U and U' can be extended to the (closed, connected) analytic subgroups and the corresponding factor groups, which are also all complex tori. Let U_0 be an analytic subgroup of dimension $p-q$ in U; then the component of the trace of U_0 in S which contains the zero vector is also an analytic subgroup and hence must be a linear subspace S_0 of dimension $p-q$, so that we have $U_0 = S_0/S_0 \cap [\mathfrak{P}]$ and hence U_0 is a complex torus. Without any loss of generality, we can assume the coordinate system of S so chosen that S_0 is the subspace defined by $z_1 = \cdots = z_q = 0$; let S_1 be the subspace defined by $z_{q+1} = \cdots = z_p = 0$. Similarly, let S'_0 be the subspace of S' defined by $z'_1 = \cdots = z'_q = 0$, and let S'_1 be the subspace defined by $z'_{q+1} = \cdots = z'_p = 0$. The fact that $S_0/S_0 \cap [\mathfrak{P}]$ is a compact manifold implies that the group $S_0 \cap [\mathfrak{P}]$ contains $2p-2q$ real-linearly independent vectors; hence, by a well known property of Abelian groups, it must possess a base which can be extended to a base of $[\mathfrak{P}]$. By a change of base of $[\mathfrak{P}]$ if necessary (which corresponds to the multiplication of $\mathfrak{P}$ by a unimodular integral matrix on the left), we can therefore assume that the last $2p-2q$ rows of $\mathfrak{P}$ constitute a base of $S_0 \cap [\mathfrak{P}]$. The matrix $\mathfrak{P}$ will then have the form $\begin{pmatrix} \mathfrak{P}_1 & * \\ 0 & \mathfrak{P}_0 \end{pmatrix}$, so that we have $S_0 \cap [\mathfrak{P}] = [\mathfrak{P}_0]$, and the corresponding matrix $\mathfrak{P}'$ will have the form $\begin{pmatrix} \mathfrak{P}'_1 & 0 \\ * & \mathfrak{P}'_0 \end{pmatrix}$. The relation $(\bar{\mathfrak{P}}'\mathfrak{P}')\,{}^t\mathfrak{P}\bar{\mathfrak{P}}) = 1$ implies then the relations $\bar{\mathfrak{P}}'_1{}^t\mathfrak{P}_1 + \mathfrak{P}'_1{}^t\bar{\mathfrak{P}}_1 = 1$ and $\bar{\mathfrak{P}}'_0{}^t\mathfrak{P}_0 + \mathfrak{P}'_0{}^t\bar{\mathfrak{P}}_0 = 1$; this shows that the rows of each of the matrices $\mathfrak{P}_1$, $\mathfrak{P}'_0$, $\mathfrak{P}'_1$ are also real-linearly independent and hence the groups $U_1 = S_1/[\mathfrak{P}_1]$, $U'_0 = S'_0/[\mathfrak{P}'_0]$, $U'_1 = S'_1/[\mathfrak{P}'_1]$ are all complex tori (of dimensions q, $q-p$, q respectively). Furthermore, these relations show that both U_0, U'_0 and U_1, U'_1 can be made into dual pairs, if we define the character relations in each pair by the same form (1) as restricted to the pairs of subspaces S_0, S'_0 and S_1, S'_1 respectively. The fact that the rows of $\mathfrak{P}'_0$ are real-linearly independent shows that $S'_1 \cap [\mathfrak{P}'] = [\mathfrak{P}'_1]$, and hence $U'_1 = S'_1/S'_1 \cap [\mathfrak{P}']$ can be considered as an analytic subgroup of U'. It is also clear that U_1 is isomorphic to the factor group U/U_0 and U'_0 is isomorphic to the factor group U'/U'_1, so that we can set $U_1 = U/U_0$ and $U'_0 = U'/U'_1$ and thus obtain the dual pairs U_0, U'/U'_1 and U/U_0, U'_1. Finally, we observe that U'_1, considered as a subgroup of the character group U' of $[\mathfrak{P}]$, is precisely the annihilator of $[\mathfrak{P}_0]$, and hence can be considered as the character group of $[\mathfrak{P}]/[\mathfrak{P}_0]$ under the canonical homomorphism of $[\mathfrak{P}]$ onto $[\mathfrak{P}]/[\mathfrak{P}_0]$; and that if we identify $[\mathfrak{P}]/[\mathfrak{P}_0]$ with $[\mathfrak{P}_1]$ by projecting S onto the subspace S_1, then for each element in U'_1 represented by a point (z') mod. $[\mathfrak{P}'_1]$, this induced character is defined by the form (1) for all (z) in $[\mathfrak{P}_1]$. Thus the dual pairing of U/U_0 and U'_1 is an analytic extension of the well known group-theoretic pairing of $[\mathfrak{P}]/[\mathfrak{P}_0]$ and U'_1, and we shall express this fact by saying that the dual pairing of U/U_0 and U'_1 is subordinate to the given dual pairing of U and U'. Similarly, the dual pairing of U_0 and U'/U'_1 if of course also subordinate to the dual pairing of U and U'.

A complex torus U is said to be an Abelian variety, if U is an algebraic variety, i. e., more precisely, if there is a one-to-one bi-holomorphic transformation of U onto an algebraic variety in some projective space. It follows from this definition and the theorem [8] quoted in section 1 that any analytic subgroup U_0 in an Abelian variety U is also an Abelian variety. Furthermore, it is known from the theory of Riemann matrices that the dual of an Abelian variety is also an Abelian variety; it follows then from this and the above paragraph that also the factor group U/U_0 is an Abelian variety.[12]

4. The Jacobian varieties. We shall give in this section a brief resumé of some properties of the Jacobian varieties which will be useful to us later. Let C be a non-singular algebraic curve of genus p. Let $(\alpha) = (\alpha_1, \cdots, \alpha_{2p})$ be a base of its first Betti group $\mathcal{B}(C)$, and let I_α be the intersection matrix of this base. Let $(\int\Phi) = (\int\Phi_1, \cdots, \int\Phi_p)$ be a system of p independent Abelian integrals of the first kind on C, and let Ω be the $(2p, p)$-matrix whose i-th row is the vector $(\int_{\alpha_i}\Phi_1, \cdots, \int_{\alpha_i}\Phi_p)$. The complex torus $J = S/[\Omega]$ is called the Jacobian variety of C, and the dual complex torus $J' = S'/[\Omega']$ is called the dual Jacobian variety of C; it is known that J is an Abelian variety and that J' is isomorphic to J and hence is also an Abelian variety. In fact, if we set $\mathfrak{R} = {}^t\bar{\Omega} I_\alpha^{-1}\Omega$, then we will have from the known properties

[12] It can be shown by the method of associated forms that the factor group U/U_0 of an Abelian variety U over an Abelian subvariety (or a finite subgroup) U_0 is also an Abelian variety. More precisely, we have the following theorem: Let T be a homomorphism of an Abelian variety U into an Abelian variety U', and let K be a field over which U, U' and T are all defined; then the kernel U_0 of T is a rational cycle over K and the factor group U/U_0 is an Abelian variety which can be represented by an algebraic variety defined over K; furthermore, the homomorphism T induces then an isomorphism of the so represented Abelian variety U/U_0 into U', which is also defined over K. The proof of this theorem will be given in our forthcoming paper mentioned in section 1.

of the Riemann matrix Ω the relation $I_\alpha^{-1}\Omega = \Omega'\mathfrak{N}$; this shows that the linear transformation $(z) = (z')\mathfrak{N}$ of S' onto S induces an isomorphism of J' onto J, which we shall denote by N.

Let x_0 be any fixed point in C; then for any point x in C the mapping

$$f(x) = (\int_{x_0}^{x_1} \Phi_1, \cdots, \int_{x_0}^{x_1} \Phi_p) \qquad (\text{mod. } [\Omega])$$

is a holomorphic transformation of C into J, which can be extended to any 0-cycle $Z = \sum n_i x_i$ in C by the stipulation $f(Z) = \sum n_i f(x_i)$. This mapping $f(Z)$ is holomorphic. To show this we can without any loss of generality restrict ourselves to the case where Z is a positive 0-cycle of a given degree m in C. Then the corresponding variety $M(C; 0, m)$ is simply the m-th symmetric product of the curve C and hence is a non-singular algebraic variety.[13] The mapping $f(Z)$ induces an everywhere single-valued analytic transformation of $M(C; 0, m)$ into J, which on account of the non-singular nature of $M(C; 0, m)$ must be a holomorphic transformation. As $f(Z)$ is evidently the product of the canonical mapping of Z into $M(C; 0, m)$ and this last holomorphic transformation, our assertion is therefore proved.

Let $\int\Psi$ be the Abelian integral of the third kind on C which has the 0-cycle $x - x_0$ as its residue divisor and whose periods are all pure imaginary. We can then define a character of $\mathfrak{B}(C)$ by assigning to each element α in $\mathfrak{B}(C)$ the real number $\frac{1}{2\pi\sqrt{-1}}\int_\alpha \Psi (\text{mod. } 1)$. If we identify $\mathfrak{B}(C)$ with $[\Omega]$ by means of the isomorphic mapping $\sum_{i=1}^{2p} m_i\alpha_i \rightarrow (m)\Omega$, then this character is represented by a point $g(x)$ in J'. In fact, we have the formula

$$g(x) = \frac{1}{2\pi\sqrt{-1}}(\int_{\alpha_1}\Psi, \cdots, \int_{\alpha_{2p}}\Psi)\Omega' \qquad (\text{mod. } [\Omega']),$$

and this mapping $g(x)$ of C into J' can be extended to any 0-cycle $Z = \sum n_i x_i$ in C by the stipulation $g(Z) = \sum n_i g(x_i)$. It is clear that $g(Z) = 0$ (mod. $[\Omega']$) if and only if Z is a principal divisor, so that $g(Z)$ induces an isomorphic mapping of $\mathfrak{G}_c(C)/\mathfrak{G}_l(C)$ into J'. Furthermore, this mapping $g(Z)$ is also holomorphic. In fact, let $\int\phi_1, \cdots, \int\phi_{2p}$ be the harmonic integrals on C corresponding to the topological 1-cycles $\alpha_1, \cdots, \alpha_{2p}$ respectively, and let $\mathfrak{M}$ be a $(p, 2p)$-matrix such that

$$(\int\phi_1, \cdots, \int\phi_{2p}) = (\int\bar{\Phi}_1, \cdots, \int\bar{\Phi}_p, \int\Phi_1, \cdots, \int\Phi_p)\begin{pmatrix}\bar{\mathfrak{M}}\\ \mathfrak{M}\end{pmatrix}.$$

[13] See B. L. van der Waerden, "Divisorklassen in algebraischen Funktionenkoerpern," *Commentarii Mathematici Helvetici*, vol. 20 (1947), pp. 68-109, § 13.

Then we have

$$g(x) = (\int_{x_0}^{x} \phi_1, \cdots, \int_{x_0}^{x} \phi_{2p})\Omega'$$

$$= (\int_{x_0}^{x} \bar{\Phi}_1, \cdots, \int_{x_0}^{x} \bar{\Phi}_p, \int_{x_0}^{x} \Phi_1, \cdots, \int_{x_0}^{x} \Phi_p) \begin{pmatrix} \bar{\mathfrak{M}} \\ \mathfrak{M} \end{pmatrix} \Omega'$$

$$= (\int_{x_0}^{x} \bar{\Phi}_1, \cdots, \int_{x_0}^{x} \bar{\Phi}_p, \int_{x_0}^{x} \Phi_1, \cdots, \int_{x_0}^{x} \Phi_p)\, {}^t(\Omega'\bar{\Omega}') I_\alpha \Omega'$$

$$= (\int_{x_0}^{x} \Phi_1, \cdots, \int_{x_0}^{x} \Phi_p) \mathfrak{N}^{-1} \qquad (\text{mod. } [\Omega']);$$

this shows that $g(x) = N^{-1}f(x)$, which proves our assertion. Thus in the isomorphism N of J' onto J, the mapping $g(x)$ is carried over into the mapping $f(x)$; this shows that these two mappings are essentially the same, being given here two different interpretations as mappings into a pair of dual Abelian varieties.

5. The Picard varieties. Let V be a non-singular algebraic variety of dimension r embedded in a projective space, and let C be a non-singular algebraic curve in V such that any topological 1-cycle in V is homologous to a topological 1-cycle in C. Such a curve C certainly exists; we can take, for example, the intersection of V with a sufficiently general linear subspace of the complementary dimension in the ambient space of V.[14] If $2q$ is the rank of the first Betti group $\mathcal{B}(V)$ of V, then there exist (according to a theorem of Hodge) exactly q independent Abelian integrals of the first kind $\int\Theta_1, \cdots, \int\Theta_q$ in V. Let $(\gamma) = (\gamma_1, \cdots, \gamma_{2q})$ be a base of $B(V)$, and let Ω_1 be the $(2q, q)$-matrix whose i-th row is the vector $(\int_{\gamma_i} \Theta_1, \cdots, \int_{\gamma_i} \Theta_q)$. Since the rows of Ω_1 are real-linearly independent, the quotient group $S_1/[\Omega_1]$ is a complex torus of dimension q, which we shall call the Albanese variety A attached to V. The dual complex torus P of A is then called the Picard variety attached to V. Let x_0 be any fixed point in V, which we shall assume for the sake of convenience to be also a point in C and take as the reference

[14] That such an intersection has the indicated property is usually given a very sketchy proof in the literature. For a more detailed treatment, see W. L. Chow, "On the fundamental group of an algebraic variety," *American Journal of Mathematics*, vol. 74 (1952), pp. 726-736, Theorem 1.

point in the definition of the mappings $f(x)$ and $g(x)$ in the previous section; then, for any point x in V the mapping

$$F(x) = (\int_{x_0}^{x} \Theta_1, \cdots, \int_{x_0}^{x} \Theta_q) \qquad (\text{mod. } [\Omega_1])$$

is a holomorphic transformation of V into A, which can be extended to any 0-cycle $Z = \Sigma\, n_i x_i$ in V by the stipulation $F(Z) = \Sigma\, n_i F(x_i)$. It is easily seen that $F(Z)$ is an analytic mapping of the 0-cycles in V into A; however, in contrast to the case of a curve, we do not know whether this mapping is also holomorphic.

The inclusion mapping of C into V induces a homomorphism L of $\mathfrak{B}(C)$ into $\mathfrak{B}(V)$, which can be expressed by the relation $(\alpha) \approx (\gamma)\mathfrak{L}$ on V, where $\mathfrak{L}$ is an integral $(2q, 2p)$-matrix. The assumption that every topological 1-cycle in V is homologous to a topological 1-cycle in C implies that this homomorphism L is onto, i. e. there exixsts an integral $(2p, 2q)$-matrix $\mathfrak{L}'$ such that $\mathfrak{L}\mathfrak{L}' = 1$. Since the integrals $\int \Theta_1, \cdots, \int \Theta_q$ are also Abelian integrals of the first kind on the curve C, there exists a (p, q)-matrix $\mathfrak{T}$ such that the relation $(\int \Theta) = (\int \Phi)\mathfrak{T}$ holds on the curve C. Integrating both sides over the base (α), we obtain the equation ${}^t\mathfrak{L}\Omega_1 = \Omega\mathfrak{T}$ and hence the equation $\Omega_1 = {}^t\mathfrak{L}'\Omega\mathfrak{T}$; we have then the equation $[\Omega\mathfrak{T}] = [\Omega_1]$ and hence the mapping $(z) \to (z)\mathfrak{T}$ induces a homomorphism T of J onto A. The kernel J_0 of T is an analytic subgroup of dimension $p - q$ in J; furthermore, if we denote by S_0 the linear subspace of dimension $p - q$ in S defined by the equation $(z)\mathfrak{T} = 0$, then we have evidently $J_0 = S_0/S_0 \cap [\Omega]$ and hence J_0 is an Abelian variety. The homomorphism T induces an isomorphism of the factor group J/J_0 onto A, which we shall also denote by T; since J/J_0 is an Abelian variety, it follows that A and consequently also P are Abelian varieties. Let the subgroup J'_1 in J' be the annihilator of $S_0 \cap [\Omega]$; we have shown before that J'_1 is the dual of J/J_0 and that the dual pairing of J/J_0 and J'_1 is subordinate to the dual pairing of J and J'. The isomorphism T of J/J_0 onto A then induces an isomorphism T' of P onto J'_1, such that the dual pairing of A and P is carried by these isomorphisms into the dual pairing of J/J_0 and J'_1.

Let D be a divisor in V which is homologous to zero. According to the Weil's generalization of a theorem of Lefschetz,[15] there exists an Abelian integral $\int \Psi$ of the third kind on V with D as its residue divisor. Adding a linear combination of the Abelian integrals of the first kind if necessary,

[15] A. Weil, "Sur la théorie des formes différentielles attachées à une variété analytique complexe," *Commentarii Mathematici Helvetici*, vol. 20 (1947), pp. 110-116.

we can assume that the periods of $\int\Psi$ are all pure imaginary; then the integral $\int\Psi$ is uniquely determined by the divisor D up to an additive constant. We can then define a character of $\mathfrak{B}(V)$ by assigning to each element γ in $\mathfrak{B}(V)$ the real number $\frac{1}{2\pi\sqrt{-1}}\int_{\gamma}\Psi \pmod{1}$. If we identify $\mathfrak{B}(V)$ with $[\Omega_1]$ by means of the isomorphic mapping $\sum_{i=1}^{2q} m_i\gamma_i \to (m)\Omega_1$, then this character is represented by a point $G(D)$ in P. Thus we have a homomorphism $G(D)$ of the group $\mathfrak{G}_h(V)$ of all divisors in V which are homologous to zero, into the Abelian variety P. By considering the multiplicative function $\exp(\int_{x_0}^{x}\Psi)$ on V, it can be easily seen that the kernel of this homomorphism is precisely the group $\mathfrak{G}_l(V)$, so that $G(D)$ induces an isomorphism of $\mathfrak{G}_h(V)/\mathfrak{G}_l(V)$ into P. We shall show that this mapping $G(D)$ is holomorphic. We observe first that if the intersection 0-cycle $C\cdot D$ is defined, then the integral $\int\Psi$ can also be considered as an integral of the third kind on the curve C with $C\cdot D$ as its residue; and as such it defines a character of $\mathfrak{B}(C)$ which is the same as that induced by the corresponding character of $\mathfrak{B}(V)$ under the homomorphism L of $\mathfrak{B}(C)$ onto $\mathfrak{B}(V)$. We have then the equation $G(D) = T'^{-1}g(C\cdot D)$; we observe here that the mapping $T'^{-1}g(C\cdot D)$ is defined for any divisor D (not necessarily homologous to zero) such that $C\cdot D$ is defined. Now, let $D(w) = D_1(w) - D_2(w)$ be an analytic system of divisors in V, (regularly) parametrized by an analytic variety W, where $D_1(w)$ and $D_2(w)$ are positive divisors such that $D_1(w) \sim D_2(w)$ for all w. If w_0 is any point in W, we can choose the curve C such that the intersection 0-cycle $C\cdot D(w_0)$ is defined; then the intersection 0-cycle $C\cdot D(w)$ is defined for all points w in a sufficiently small neighborhood of w_0. Furthermore, we can assume that C is the intersection of V with a suitably chosen linear subspace in the ambient space. Then we have the equation

$$G(D(w)) = T'^{-1}g(C\cdot D(w)) = T'^{-1}g(C\cdot D_1(w)) - T'^{-1}g(C\cdot D_2(w))$$

for all points w in a neighborhood of w_0. The mapping $T'^{-1}g(C\cdot D_1(w))$ is obviously the product of the canonical mapping of $D_1(w)$ into $M(V; r-1, m)$ (m being the degree of $D_1(w)$), the holomorphic transformation of $M(V; r-1, m)$ into $M(C; 0, m)$ induced by the mapping $D \to C\cdot D$ (see the last paragraph of section 1), the holomorphic transformation of $M(C; 0, m)$ into J' induced by $g(Z)$, and the isomorphism T'^{-1} of J'_1 onto P; it follows then that $T'^{-1}g(C\cdot D_1(w))$ is an analytic (holomorphic) transformation of a neighborhood of w_0 into P, and similarly also $T'^{-1}g(C\cdot D_2(w))$

is an analytic (holomorphic) transformation. Thus we have shown that $G(D(w))$ is an analytic (holomorphic) transformation of a neighborhood of w_0 into P; and since w_0 is any point in W, it follows that $G(D(w))$ is an analytic (holomorphic) transformation of W into P. This proves our assertion.

In order to complete the proof of Igusa's duality theorem, it only remains to show that $\mathfrak{G}_h(V) = \mathfrak{G}_c(V)$ and that the mapping $G(D)$ of $\mathfrak{G}_h(V)$ into P is onto. Since $\mathfrak{G}_c(V)$ is a subgroup of $\mathfrak{G}_h(V)$, both assertions will follow if we show that $G(D)$ maps the group $\mathfrak{G}_c(V)$ onto P; and this can be done by the construction of the so-called Poincaré family of divisors. However, we shall construct in the next section an analytic system of divisors which represents the group $\mathfrak{G}_c(V)/\mathfrak{G}_l(V)$ in a more precise manner.

6. Representative system of divisors. We begin by observing (what we have implicitly assumed up to now) that the field $\mathfrak{K}$ of all complex numbers is the universal domain, in the sense of Weil,[16] of the algebraic geometry we are concerned with here. We shall assume that the Abelian variety A is represented by an arbitrarily chosen, but fixed non-singular algebraic variety in some projective space; it follows then that the mapping $F(x)$ of V into A is a rational transformation, defined (in the sense of Weil) for all points of V. Let K be a field (i. e., in the terminology of Weil, a subfield of $\mathfrak{K}$ such that $\mathfrak{K}$ has infinite degree of transcendency over K) over which the algebraic varieties V and A, as well as the rational transformation $F(x)$ and the group composition in A, are all defined; furthermore, we shall assume that the point x_0, the reference point in the definition of $F(x)$, is rational over K. We can, in fact, take K to be a finitely generated extension of the rational field. Consider an $(r-1)$-dimensional linear system $|C|$ of curves on V, which is cut out on V by a system of linear subspaces in the ambient space; by choosing the system of linear subspaces in a suitable manner we can assume that the system $|C|$ has the following properties: (1) the system $|C|$ is defined over K, i. e. the associated variety W of $|C|$ in $M(V; 1, m)$, m being the degree of $|C|$, is defined over K; for any point w in W, we shall denote by $C(w)$ the curve of the system corresponding to it; (2) the point x_0 is a base point of $|C|$, and a generic point of V over K is contained in exactly one curve of the system $|C|$; (3) for any generic point u of W over K, the curve $C(u)$ is a non-singular curve, whose genus

[16] A. Weil, *Foundations of Algebraic Geometry*, American Mathematical Society Colloguium Publication no. 29 (1946). We shall use in this section the algebraic-geometric concepts and terminology developed in this book.

will be denoted by p. According to the theorem cited in footnote 14, the curve $C(u)$ will have the property (stated at the beginning of section 5) that every topological 1-cycle in V is homologous to a topological 1-cycle in $C(u)$. We can therefore apply the results of sections 4 and 5 to the curve $C(u)$; let $J(u)$ be the Jacobian variety of $C(u)$ and let T_u be the homomorphism of $J(u)$ onto A defined in section 5. Since $C(u)$ is defined over $K(u)$, it follows from a result proved elsewhere [17] that we can consider $J(u)$ as represented by a non-singular algebraic variety, such that both the variety $J(u)$ as well as the group composition in $J(u)$ are defined over $K(u)$. Moreover, since the point x_0 is a rational divisor of degree 1 in $C(u)$ over $K(u)$, the "canonical mapping" f_u of $C(u)$ into $J(u)$, defined with x_0 as the reference point, is also a rational transformation defined over $K(u)$. Then the homomorphism T_u is a rational transformation of $J(u)$ onto A; we maintain that it is also defined over $K(u)$. In fact, the rational transformation $F(x)$ of V into A induces a rational transformation of $C(u)$ into A, which is evidently defined over $K(u)$. Since A is an Abelian variety, this rational transformation has a uniquely determined linear extension,[18] which is a rational transformation of $J(u)$ into A, also defined over $K(u)$; and it can be easily seen, in view of our choice of the reference point x_0 for both $F(x)$ and f_u, that this rational transformation is precisely the homomorphism T_u.

The fact that T_u is defined over $K(u)$ is the basis for the construction of what is usually called the Poincaré family of divisors,[19] which is "generically" parametrized by A. Though this family does contain a representative for each element in $\mathfrak{G}_c(V)/\mathfrak{G}_l(V)$, it has the disadvantage that this representative is in general not uniquely determined, even for a "generic" element in $\mathfrak{G}_c(V)/\mathfrak{G}_l(V)$, there being a finite number of them in the family; this is due to the fact that A can be represented as a factor group $J(u)/J_0(u)$ of $J(u)$, but not necessarily as a subgroup of $J(u)$. In order to get a "generally" one-to-one representative system of divisors for $\mathfrak{G}_c(V)/\mathfrak{G}_l(V)$, we shall show that the isomorphism T'_u of P into $J'(u)$, as defined in section 5, is also defined over $K(u)$, provided we represent both P and $J'(u)$ by suitably chosen algebraic varieties. In fact, since $J(u)$ and $J'(u)$ are

[17] See the paper cited in footnote 6.

[18] See A. Weil, *Variétés abéliennes et courbes algébriques*, p. 77, Théorème 21.

[19] We have in mind here the method of construction due to F. Enriques, assuming the existence of q independent Abelian integrals of the first kind in V. See O. Zariski, *Algebraic Surfaces*, Ergebnisse der Mathematik und ihrer Grenzgebiete, vol. 3, no. 5 (1935), p. 123 and the references cited there.

isomorphic, we can represent both of them by the same algebraic variety and thus identify $J'(u)$ with $J(u)$ by means of the isomorphism N defined in section 4; we can then consider $J'_1(u)$ as an Abelian subvariety of $J(u)$ and we have evidently $g_u = f_u$. The construction of the algebraic variety representing P will be given in course of the proof below; we shall see that the so represented variety P is defined over K.[20]

Let X be a cycle of dimension q in $J(u)$, rational over $K(u)$; then for a generic point v of A over $K(u)$, the intersection $X \cdot T_u^{-1}(v)$ is a rational 0-cycle over $K(u, v)$. If for any 0-cycle Z in an Abelian variety we denote (with Weil[21]) by $\mathfrak{S}(Z)$ the element in the Abelian variety which is the sum of all points in Z, each point being added or subtracted as often as its (positive or negative) multiplicity indicates, then the point $\mathfrak{S}(X \cdot T_u^{-1}(v))$ in $J(u)$ is rational over $K(u, v)$; hence, the mapping $v \rightarrow \mathfrak{S}(X \cdot T_u^{-1}(v))$ is a rational transformation of A into $J(u)$, defined over $K(u)$. The mapping $H_u: v \rightarrow \mathfrak{S}(X \cdot T_u^{-1}(v)) - \mathfrak{S}(X \cdot T_u^{-1}(0))$, is then a homomorphism of A into $J(u)$, defined over $K(u)$, and hence the image $H_u(A)$ is an Abelian subvariety in $J(u)$, also defined over $K(u)$. It is clear that if s is the degree of the 0-cycle $X \cdot T_u^{-1}(v)$, then we have the equation $T_u H_u(v) = \mathfrak{S}(sv)$; this shows that $H_u(A)$ has the dimension q, and that the kernel A_0 of H_u is a finite subgroup of A and hence is a positive 0-cycle, rational over $K(u)$. Since the Abelian variety A is defined over K, the positive 0-cycle A_0, being a finite subgroup in A, must be algebraic over K; and as $K(u)$ is a regular extension of K, it follows that A_0 is rational over $K(u) \cap \bar{K} = K$. According to the theorem cited in footnote 12, the factor group A/A_0 is then an Abelian variety which can be represented by a non-singular algebraic variety, defined over K, and the homomorphism H_u of A into $J(u)$ induces an isomorphism T'_u of A/A_0 into $J(u)$, which is also defined over $K(u)$. We shall now show that $T'_u(A/A_0) = J'_1(u)$, so that A/A_0 is isomorphic to P and hence can be taken as a representative algebraic variety of P. In order to prove this, we only need to show that for a generic point y of A/A_0 over $K(u)$, there exists a divisor $D(y)$ in $G_c(V)$ such that $f_u(D(y) \cdot C(u)) = T'_u(y)$. We recall that $M(C; 0, p)$ is a non-singular (irreducible) algebraic variety of dimension p, defined over K; in fact, it is the symmetric p-th product of the curve $C(u)$.

---

[20] The first part of the proof below, namely the construction of the Abelian variety $H_u(A)$, is essentially the same as Castelnuovo's proof for Poincaré's "complete reducibility theorem." See G. Castelnuovo, "Sugli integrali semplici appartenenti ad una superficie irregolare," *Rend. della R. Accad. dei Lincei*, ser. V, vol. XIV (1905); compare also A. Weil, *loc. cit.*, p. 94, Théorème 26.

[21] A. Weil, *loc. cit.*, p. 28.

The mapping f_u of the cycles in $C(u)$ into $J(u)$ induces a birational transformation h_u of $M(C;0,p)$ onto $J(u)$, defined over $K(u)$; hence, since $T'_u(y)$ is a generic point of $T'_u(A/A_0)$ over $K(u)$, the inverse image $h_u^{-1}(T'_u(y))$ is a rational point of $M(C;0,p)$ over $K(u,y)$ and represents a positive 0-cycle $Z(u,y)$ of degree p in $C(u)$, which is rational over $K(u,y)$. Over the field $K(y)$, the mapping $u \rightarrow Z(u,y)$ defines an algebraic correspondence of W into V, and it is easily seen that the projection of this correspondence in V is a positive divisor $Z(y)$ in V, rational over $K(y)$, whose intersection with $C(u)$ is precisely $Z(u,y)$, apart from a possible fixed component Z_0. Let $Z(0)$ be a specialization of $Z(y)$ over the specialization $y \rightarrow 0$ over K, such that $C(u) \cdot Z(0) = px_0 + Z_0$; if K' is an extension of K over which the divisor $Z(0)$ is rational, then $D(y) = Z(y) - Z(0)$ is a divisor in $\mathfrak{G}_c(V)$ such that

$$\begin{aligned} f_u(D(y)\cdot C(u)) &= f_u(Z(y)\cdot C(u)) - f_u(Z(0)\cdot C(u)) \\ &= f_u(Z(u,y)) - f_u(px_0) \\ &= f_u(h_u^{-1}(T'_u(y))) - f_u(px_0) = T'_u(y). \end{aligned}$$

This proves our assertion.

If we now identify P with A/A_0, then the divisor $D(y)$ will generate an analytic system of divisors in $\mathfrak{G}_c(V)$, "generically" parametrized by P in the sense that P is the parameter variety of this system for all generic points of P over K'; furthermore, if $D(y_0)$ is any specialization of $D(y)$ over any specialization $y \rightarrow y_0$ over K', then we have $G(D(y_0)) = y_0$. However, this system gives a one-to-one representation of the group $\mathfrak{G}_c(V)/\mathfrak{G}_l(V)$ only "in general," since it is possible that for some special point y_0 in P there exist more than one specializations of $D(y)$ over the specialization $y \rightarrow y_0$ over K'.

Finally, going over once more the arguments in this section, it will be readily seen that we have in fact proved the following result: If H is a homomorphism of any Abelian variety B into $J(u)$ and if B is defined over K and H is defined over $K(u)$, then $H(B)$ is contained in $J'_1(u)$. Thus we can characterize the Picard variety P as the "largest Abelian subvariety" in $J(u)$ which is defined over K. As mentioned in section 1, this result is part of a purely algebraic theory of Picard varieties for algebraic varieties over an arbitrary ground field, which will be developed in a forthcoming paper.

THE JOHNS HOPKINS UNIVERSITY.

VOL. 38, 1952 MATHEMATICS: CHOW AND KODAIRA 319

ON ANALYTIC SURFACES WITH TWO INDEPENDENT MEROMORPHIC FUNCTIONS

BY WEI-LIANG CHOW AND KUNIHIKO KODAIRA

JOHNS HOPKINS UNIVERSITY AND INSTITUTE FOR ADVANCED STUDY

Communicated by Oscar Zariski, January 24, 1952

Let V be a compact complex analytic variety of complex dimension r, defined *in abstracto* without any embedding space.[1] According to a recent result of Chow,[2] the field $\mathfrak{F}(V)$ of all meromorphic functions on V is a finite algebraic function field whose degree of transcendency s is at most equal to r. There exists therefore an algebraic variety V^* of dimension s in a projective space, uniquely determined up to a birational transformation, and a meromorphic transformation Φ of V onto V^*, such that every meromorphic function on V is the image of a meromorphic (and hence rational) function on V^* under the inverse transformation Φ^{-1}. Thus the meromorphic transformation Φ induces an isomorphism between the two fields $\mathfrak{F}(V)$ and $\mathfrak{F}(V^*)$, so that we can identify the two and set $\mathfrak{F}(V) = \mathfrak{F}(V^*) = \mathfrak{F}$. The variety V^* will be called the *algebraic* equivalent of V. Now, we shall say that the analytic variety is *algebraic* (in an extended sense), if its algebraic equivalent V^* can be so chosen that the transformation Φ is bi-regular between V and V^*. Since from the abstract point of view two bi-regularly equivalent analytic varieties can be considered as not essentially distinct, it follows that in this case we can identify V with V^*, so that our definition of an algebraic variety is a natural extension of the usual one for a variety in a projective space. It is obvious that an algebraic variety V of dimension r possesses the maximum possible number r of algebraically independent meromorphic functions. The question naturally arises whether the converse is also true: *If a compact analytic variety has as many algebraically independent meromorphic functions as its dimension, is it then necessarily algebraic?* So far as we know, this question has not yet been decided except for some special cases. It is therefore probably not without some interest to present here an answer to this question for the case of a *compact analytic manifold of complex dimension 2 with a Kählerian metric,* which we shall call a Kählerian surface. In the following we shall prove:

THEOREM. *Any Kählerian surface with two algebraically independent meromorphic functions is an algebraic surface without singularities.*

We begin with the observation that the so-called quadratic Cremona transformation in algebraic geometry can be applied also to an analytic surface V. In fact, it can be described in terms of local coordinates as follows: let p be a point on V and (x_1, x_2) be a system of local coordinates with the center p. We consider the direct product $U_p \times S$ of a neighborhood U_p of p and a projective line S and, by means of the homogeneous coordinates (t_0, t_1) on S, construct the subvariety

$$W_p = \{(x_1, x_2;\ t_0, t_1) \,|\, t_0x_2 - t_1x_1 = 0\} \tag{1}$$

of $U_p \times S$. Obviously W_p is an analytic surface without singularities and contains the projective line $p \times S$ which will be denoted by the same symbol S: $S = p \times S \subset W_p$. The mapping

$$P_p': (x_1, x_2;\ t_0, t_1) \rightarrow (x_1, x_2)$$

is a regular mapping of W_p onto U_p which maps S onto p and $W_p - S$ bi-regularly onto $U_p - p$. Hence, replacing U_p by W_p, we get from V a new analytic surface $V_1 = (V - U_p) \cup W_p$ having the following properties: *There exists a regular mapping P_p of V_1 onto V such that $S = P_p^{-1}(p)$ is a projective line and that P_p is bi-regular between $V_1 - S$ and $V - p$ (P_p is* a "natural" extension of P_p'). The inverse mapping $Q_p = P_p^{-1}$ of P_p will be called the *quadratic transformation with the center p*. It is easy see that the *quadratic transform* $V_1 = Q_p(V)$ of V is determined uniquely by p (and V) and is independent of the choice of the local coordinates (x_1, x_2) appearing in the above construction. Every meromorphic function $F = F(x)$ on V induces a meromorphic function $F(P_p)$ on $Q_p(V)$. Conversely every meromorphic function F_1 on $Q_p(V)$ induces a meromorphic function F on V such that $F(P_p) = F_1$. In fact, it is obvious that F_1 induces on $V - p$ a meromorphic function F such that $F(P_p) = F_1$. To prove that F is meromorphic also at p, consider the meromorphic function $x_1^{-m}F_1$ in a neighborhood of S, m being the multiplicity of the component S in the divisor of F_1. Then, on the curve S, $x_1^{-m}F_1$ equals a rational function $R(t)$ in $t = t_1/t_0$, so that $R(x_2/x_1)^{-1}x_1^{-m}F_1$ is constant on S. This shows that $R(x_2/x_1)^{-1}x_1^{-m}F_1$ induces a holomorphic function $H(x_1, x_2)$ in a neighborhood of p, proving that $F = x_1^mR(x_2/x_1)H(x_1, x_2)$ is meromorphic at p. The mapping $F \rightarrow F(P_p)$ gives therefore an isomorphism between $\mathfrak{F}(V)$ and $\mathfrak{F}(Q_p(V))$. Thus *the quadratic transformation Q_p does not affect the field of all meromorphic functions on V: $\mathfrak{F}(Q_p(V)) = \mathfrak{F}(V)$.*

LEMMA 1. *If V is algebraic, then $Q_p(V)$ is also algebraic.*

Proof is well known.[3]

LEMMA 2. *If V is a Kählerian surface, then $Q_p(V)$ is also a Kählerian surface.*

Proof: Using the system of local coordinates (x_1, x_2) with the center p, the Kählerian metric on V can be expressed in a neighborhood U_p of p as

$$ds^2 = \sum_{j,\,k=1}^{2} (\partial^2K/\partial x_j\partial \bar{x}_k)dx_j\, d\bar{x}_k,$$

K being a function of class C^∞ defined in U_p. Assuming that U_p is the unit sphere $\{(x_1, x_2) \,|\, |x_1|^2 + |x_2|^2 < 1\}$ and choosing a function $\psi(\rho)$ of class C^∞ such that

$$\psi(\rho) = \begin{cases} 1, & \text{for } 0 \leqq \rho \leqq 1/3, \\ 0, & \text{for } \rho \geqq 2/3, \end{cases}$$

we construct a new function

$$\tilde{K} = K + \epsilon \cdot \psi(|x_1|^2 + |x_2|^2) \cdot \log(|x_1|^2 + |x_2|^2),$$

ϵ being a sufficiently small positive constant, and put

$$d\tilde{s}^2 = \sum_{j,\, k=1}^{2} (\partial^2 \tilde{K}/\partial x_j \partial \bar{x}_k) dx_j\, d\bar{x}_k, \qquad (\text{for } |x_1|^2 + |x_2|^2 > 0).$$

Then, considering (x_1, x_2) as the coordinates of the point $Q_p(x_1, x_2)$ on $W_p = Q_p(U_p)$, we infer readily that $d\tilde{s}^2$ is a Kählerian metric in $W_p - S$. Putting $t = t_1/t_0 = x_2/x_1$, we get

$$d\tilde{s}^2 = ds^2 + \epsilon |dt|^2/(1 + |t|^2)^2, \qquad \text{for } 0 < |x_1|^2 + |x_2|^2 < 1/3,$$

while $\epsilon |dt|^2/(1 + |t|^2)^2$ is a Kählerian metric on S; thus, for $0 < |x_1|^2 + |x_2|^2 < 1/3$, $d\tilde{s}^2$ coincides with the "natural" metric in the Cartesian product $U_p \times S$. This shows that $d\tilde{s}^2$ can be extended to a Kählerian metric in the whole of W_p. On the other hand, $d\tilde{s}^2$ coincides with ds^2 for $|x_1|^2 + |x_2|^2 \geqq 2/3$. Hence $d\tilde{s}^2$ can be extended to a Kählerian metric in the whole of $Q_p(V)$ simply by setting $d\tilde{s}^2 = ds^2$ in $Q_p(V) - W_p$. Thus $Q_p(V)$ is a Kählerian surface.

LEMMA 3. *If $Q_p(V)$ is an algebraic surface, then V is also an algebraic surface.*

Proof: It is sufficient to show that $S = Q_p(p)$ is an *exceptional curve of the first kind.*[4] In fact, if this is the case, there exists a regular mapping P^* of $Q_p(V)$ onto an algebraic surface V^* without singularities such that $P^*(S)$ is a point p^* on V^* and that P^* is bi-regular between $Q_p(V) - S$ and $V^* - p^*$. Then P^*Q_p is a bi-regular mapping of $V - p$ onto $V^* - p^*$ and therefore it must be a bi-regular mapping of V onto V^*. Hence V is algebraic.

Now, in order to prove that $S = Q_p(p)$ is an exceptional curve of the first kind, it is sufficient to show that the intersection number $I(S, S)$ equals -1.[5] Let (x_1, x_2) be the local coordinates on V with the center p. Then x_2 can be considered as a holomorphic function defined in the neighborhood W_p of $S = Q_p(p)$ whose divisor (x_2) consists of S and a curve C such that $I(S, C) = 1$. Now, choose a meromorphic function F on $Q_p(V)$ having S as a polar curve of multiplicity 1 and put $D = (F) + S$, (F) being the divisor of F. Then, since x_2F induces a meromorphic function on S with the divisor $(x_2F) = DS + CS$, we have

$$I(D, S) + I(C, S) = \text{degree of } (x_2F) = 0$$

and therefore

$$I(S, S) = I(D, S) = -I(C, S) = -1,$$

q. e. d.

Now, let V be an arbitrary Kählerian surface such that its algebraic equivalent V^* has the dimension 2; as was mentioned above, we may assume that V and V^* have the same meromorphic functions field $\mathfrak{F} = \mathfrak{F}(V) = \mathfrak{F}(V^*)$. Let $(X_0, X_1, \ldots, X_n)$ be the homogeneous coordinates of a generic point X of V^*. Then, by means of meromorphic functions $F_k(x) = X_k/X_0$ of a "variable" point x on V, the meromorphic mapping Φ of V onto V^* can be represented as

$$\Phi: x \rightarrow \Phi(x) = (1, F_1(x), \ldots, F_n(x)).$$

Denote by C_λ the inverse image $\Phi^{-1}(C_\lambda^*)$ of the hyperplane section C_λ^* of V^* cut out by any hyperplane $\lambda_0X_0 + \lambda_1X_1 + \ldots + \lambda_nX_n = 0$. Then the set $|C_\lambda|$ of all such curves C_λ constitutes a *linear system* on V having no fixed component and the meromorphic mapping $x \rightarrow \Phi(x)$ is *regular* except for a finite number of *base points* of this linear system $|C_\lambda|$. By the *multiplicity* of a base point p of $|C_\lambda|$ we shall mean the integer

$$m(p) = \min_{\lambda, \mu} I_p(C_\lambda, C_\mu),$$

where $I_p(C_\lambda, C_\mu)$ denotes the intersection multiplicity of C_λ and C_μ *at* p. Obviously $m(p)$ is always not smaller than 1 (or $m(p)$ equals zero if and only if p is *not* a base point). Now we shall show that these base points can be eliminated by means of quadratic transformations (if any base point ever exists). Take a base point p of $|C_\lambda|$ and construct the quadratic transform $V' = Q_p(V)$. The mapping Φ induces a mapping Φ': $(x') \rightarrow \Phi'(x') = \Phi Q_p^{-1}(x')$ of V' onto V^* which corresponds to the linear system $|C_\lambda'|$ of curves $C_\lambda' = Q_p(C_\lambda)$ on V'. This new system $|C_\lambda'|$ might have a base point p' on $S = Q_p(p)$, but the *multiplicity* $m(p')$ *of* p' *is strictly smaller than* $m(p)$, since we have

$$I_{p'}(C_\lambda', C_\mu') = I_p(C_\lambda, C_\mu) - m_\lambda \cdot m_\mu,$$ [6]

m_λ and m_μ being the multiplicities of the curves C_λ and C_μ at p, respectively. This shows clearly that, by applying a finite number of suitable quadratic transformations $Q_p, Q_{p'}, \ldots$ to V successively, we get an analytic surface

$$\tilde{V} = \ldots Q_{p''}Q_{p'}Q_p(V) \tag{2}$$

such that the mapping $\tilde{\Phi} = \Phi Q_p^{-1}Q_{p'}^{-1}Q_{p''}^{-1} \ldots$ of $\tilde{V}$ onto *its algebraic equivalent* V^* is a regular mapping. By virtue of Lemma 2, $\tilde{V}$ *is also a Kählerian surface.*

The inverse mapping $\tilde{\Phi}^{-1}$ has possibly a finite number of fundamental points and branch curves and, except for these "singular" points, $\tilde{\Phi}^{-1}$ is locally bi-regular. Hence, for any point p^* on V^* which is not "singular" with respect to $\tilde{\Phi}^{-1}$, the set $\tilde{\Phi}^{-1}(p^*)$ consists of a finite number d of points on $\tilde{V}$ (the number d is independent of p^*). Now we shall show that $d = 1$. Let E_h^* be the section of V^* cut out by a "general" hypersurface of order h containing no fundamental point of $\tilde{\Phi}^{-1}$ and put $E_h = \tilde{\Phi}^{-1}(E_h^*)$. Then, since V^* is the algebraic equivalent of $\tilde{V}$, the complete linear systems $|E_h|$ on $\tilde{V}$ and $|E_h^*|$ on V^* have the same dimension: $\dim|E_h| = \dim |E_h^*|$. For sufficiently large h the system $|E_h^*|$ is *regular* and therefore it follows from *Riemann-Roch's theorem*[7] that

$$\dim |E_h^*| = {}^1/_2 I(E_h^*, E_h^*) - {}^1/_2 I(K^*, E_h^*) + p_a^*,$$

where K^* is the canonical divisor on V^* and p_a^* is the arithmetic genus of V^*, while, using *Riemann-Roch's theorem on Kählerian surfaces* proved recently by Kodaira,[8] we get

$$\dim |E_h| \geqq {}^1/_2 I(E_h, E_h) - {}^1/_2 I(K, E_h) + p_a - i,$$

where K is the canonical divisor on $\tilde{V}$, p_a is the arithmetic genus of $\tilde{V}$ and $i = \dim |K - E_h| + 1$. Putting $\nu = I(E_1^*, E_1^*)$, $k^* = I(K^*, E_1)$, $k = I(K, E_1)$, we get

$$\dim |E_h^*| = {}^1/_2 \nu \cdot h^2 - {}^1/_2 k^* \cdot h. + p_a^*$$

and, since $I(E_h, E_h) = d \cdot I(E_h^*, E_h^*)$,

$$\dim |E_h| \geqq {}^1/_2 d \cdot \nu \cdot h^2 - {}^1/_2 k \cdot h + p_a - \dim |K| - 1.$$

Combined with $\dim |E_h^*| = \dim |E_h|$, the above two formulae show clearly that d *must be equal to* 1. It follows from $d = 1$ that $\tilde{\Phi}^{-1}$ has no branch curve; thus $\tilde{\Phi}^{-1}$ *is regular and single-valued except for a finite number of fundamental points.* $\tilde{\Phi}$ *is therefore bi-regular on* $\tilde{V}$ *except for a finite number of fundamental curves.*

Now we shall show that these fundamental curves can be eliminated by applying quadratic transformations to V^*. Let p^* be a fundamental point on V^*. Then $L = \tilde{\Phi}^{-1}(p^*)$ is a (possibly reducible) fundamental curve on $\tilde{V}$. Let $L = L_1 + L_2 + \ldots + L_n$ be the decomposition of L into a sum of irreducible components L_j. Now, let F_1, F_2 be two meromorphic functions in $\mathfrak{F} = \mathfrak{F}(\tilde{V}) = \mathfrak{F}(V^*)$ such that (F_1, F_2) constitutes a system of local coordinates on V^* with the center p^*. Then the curves B_λ of zeros of the meromorphic functions $\lambda_1 F_1 + \lambda_2 F_2$ on $\tilde{V}$ constitute a linear pencil $|B_\lambda|$ on $\tilde{V}$. Apparently the part of the fixed component B_0 of $|B_\lambda|$ lying on L is written as $B_0 = \sum k_j L_j$ so that we get

$$B_\lambda = \sum k_j L_j + B_{\lambda 1}.$$

The same argument as in the case of fundamental curves on algebraic surfaces[9] shows that the curve $B_{\lambda 1}$ meets with one irreducible component, say L_1, of L in one "variable" point with intersection multiplicity 1 (i.e., $I(B_{\lambda 1}, L_1) = 1$) but does not meet other components $L_2, L_3, \ldots$ of L. This implies that, at each point p on L, the ratio F_2/F_1 can be expressed as

$$F_2/F_1 = H_{2p}/H_{1p}$$

by means of two holomorphic functions H_{1p}, H_{2p} in a neighborhood of p not vanishing simultaneously and that H_{2p}/H_{1p} is a meromorphic function on L_1 whose value $\lambda = H_{2p}/H_{1p}$ corresponds one-to-one to the intersection point $B_{\lambda 1}L_1$ on L_1 while H_{2p}/H_{1p} is constant on other components L_2, $L_3, \ldots$. Now, consider the mapping $Q_{p*}\tilde{\Phi}$ of $\tilde{V}$ onto the quadratic transform $Q_{p*}(V^*)$ which is also an algebraic surface, as Lemma 1 shows. By means of the local coordinates $(x_1^*, x_2^*; t_0^*, t_1^*)$ on $Q_{p*}(V^*)$ introduced in (1), this mapping $Q_{p*}\tilde{\Phi}$ can be represented in a neighborhood of L as

$$(x_1^*, x_2^*; t_0^*, t_1^*) = (F_1, F_2; H_{p1}, H_{p1});$$

hence $Q_{p*}\tilde{\Phi}$ is a regular mapping. $Q_{p*}\tilde{\Phi}$ maps L_1 onto $S^* = Q_{p*}(p^*)$ one-to-one. Consequently L_1 is not a fundamental curve of $Q_{p*}\tilde{\Phi}$, whereas $L_2, L_3, \ldots$ are still fundamental curves of $Q_{p*}\tilde{\Phi}$; again it is obvious that no new fundamental curve is created by Q_{p*}. Thus the quadratic transformation Q_{p*} eliminates one irreducible component L_1 of the fundamental curve L. This shows that, by applying a finite number of suitable quadratic transformations $Q_{p*}, Q_{p*'}, \ldots$ to V^*, we get an algebraic surface

$$V_1^* = \ldots Q_{p*'}Q_{p*}(V^*)$$

such that the mapping $\tilde{\Phi}_1 = \ldots Q_{p*'}Q_{p*}\tilde{\Phi}$ of $\tilde{V}$ onto V_1^* is bi-regular. This proves that $\tilde{V}$ is an algebraic surface. Now $\tilde{V}$ is obtained from V by a finite number of quadratic transformations, as (2) shows. Hence, using Lemma 3, we conclude that *the given Kählerian surface V is an algebraic surface,* q. e. d.

[1] For the definition of a complex analytic variety *in abstracto*, see Chow, W. L., "On Complex Analytic Varieties," to appear in the *Am. J. Math.* For the purpose of this paper we can restrict ourselves to complex analytic manifolds in the usual sense.

[2] See Chow, W. L., *loc. cit.*, Theorem 4.6. We mention here that a closed analytic correspondence Φ of an analytic variety V into a compact analytic variety U is a meromorphic transformation, if it is single-valued in general and if any meromorphic function (in the large) on U whose polar variety does not contain $\Phi(V)$ is carried by Φ^{-1} into a meromorphic function on V. In case V is an algebraic variety embedded in a projective space, this is equivalent to saying that Φ can be represented by setting the inhomogeneous coordinates of a point in V equal to a system of meromorphic functions on U.

[3] See, for example, van der Waerden, B. L., *Infinitely Near Points*, Koninklijke Nederlandse Akademie van Wetenschappen, Vol. LIII, 401–410, § 3 (1950).

[4] Cf. Zariski, *Algebraic Surfaces*, Berlin, pp. 36–41 (1935).

[5] Using Riemann-Roch's theorem it can be shown that a rational curve S without singularities is an exceptional curve of the first kind if $I(S, S) = -1$. See Zariski,

loc. cit., pp. 71–72. Strictly speaking this criterion holds only for an algebraic surface in the usual sense, i.e., embedded in a projective space. However, it can be easily extended to an algebraic V in the extended sense by observing that if V is mapped by a bi-regular transformation Φ onto an algebraic surface V' in a projective space, then any curve S in V is mapped by Φ bi-regularly onto a compact analytic curve S' in the projective space, which must then necessarily be an algebraic curve. See Chow, W. L., "On Compact Complex Analytic Varieties," *Am. J. Math.*, **71,** 893–914 (1949), Theorem V. It is clear that if S is a rational curve without singularities in V such that $I(S, S) = -1$, then S' is also a rational curve without singularities in V' such that $I(S', S') = -1$. It follows then that S' is an exceptional curve of the first kind in V', i.e., there exists a bi-rational transformation Ψ of V' which has S' as a fundamental curve and is bi-regular everywhere else. Then the "product" transformation $\Psi\Phi$ is a bi-rational transformation of V which has S as a fundamental curve and is bi-regular everywhere else.

[6] See van der Waerden, *loc. cit.*, § 3.

[7] Cf. Zariski, *loc. cit.*, pp. 66–71.

[8] Kodaira, K., "The Theorem of Riemann-Roch on Compact Analytic Surfaces," *Am. J. Math.*, **73,** 813–875 (1951), Theorem 5.3.

[9] See Zariski, *loc. cit.*, pp. 36–41.

ON THE FUNDAMENTAL GROUP OF AN ALGEBRAIC VARIETY.*

By WEI-LIANG CHOW.

It is well known that any 1-cycle in an algebraic surface can be deformed into a 1-cycle lying in a generic plane section of the surface.[1] The usual proof of this theorem, which can be easily generalized from a surface to any non-singular algebraic variety, is topological and consists of a simple construction of the deformation chain. In the transcendental theory there is a generalization of this theorem, at least in its homology aspect, which can be stated as follows: [2] There exist exactly $2p$ independent 1-cycles in an algebraic surface which are not homologous to 1-cycles belonging to a generic curve of an irrational pencil of genus p. In this paper we shall show that this theorem is a special case of a more general theorem about the fundamental group of an algebraic variety under a rational transformation. Our method of proof will be purely topological; the essential idea is that although a rational transformation is not in general a fibre mapping, the covering homotopy theorem is nevertheless true, in a somewhat modified form, for the mapping of a 1-simplex.

In section **1** the notion of a fibre system is introduced, and certain subsystems of an algebraic system are shown to be (or can be considered as) fibre systems. The notion of a fibre system is a generalization of that of a fibre space, and just as in the case of a fibre space we have also here as a fundamental property the validity of the covering homotopy theorem, which must now be formulated in a somewhat modified form. This notion of a fibre system is a very useful tool in the study of the topology of algebraic varieties; in this paper we shall limit ourselves strictly to the particular problem in question, but we hope to show in a later paper that the method is applicable also to other similar problems in algebraic geometry. In section **2** the results of section **1** will be used to prove two theorems; one of them (Theorem 2) is the theorem mentioned above, the other (Theorem 1) is a theorem concerning the deformation of 1-cycles into a member of an algebraic system with at least

* Received August 20, 1951.

[1] See, e. g., O. Zariski, *Algebraic Surfaces*, p. 108.

[2] See O. Zariski, *Algebraic Surfaces*, p. 144.

one base point, which can also be regarded as a (partial) generalization of a result of Severi[3] (proved by transcendental methods).

1. Let U and V be topological spaces, and let $G(y)$ be a function which assigns to each point y in V a subset $G(y)$ in U. We shall say that the system of subsets $G(y)$ defines a fibre system in U, if there exists an open covering $N = \{N\}$ of V such that for each set N there exists a continuous function $\phi_N(x, y)$, defined for all points $x \subset G(N)$, $y \subset N$ in the product space $U \times V$ with values in U, with the following properties:

$$\phi_N(x, y) \subset G(y), \qquad (x \subset G(N), y \subset N),$$

$$\phi_N(x, y) = x \qquad (x \subset G(y), y \subset N).$$

The space V is called the base space of the fibre system, and the open sets N and the corresponding functions $\phi_N(x, y)$ are called the slicing neighborhoods and the slicing functions respectively. Let $f(z)$ and $g(z)$ be continuous mappings of a topological space Z into U and V respectively, such that for each point $z \subset Z$ we have $f(z) \subset G(g(z))$, and let $g(z, t)$, $0 \leq t \leq 1$, be a homotopy of the mapping $g(z)$ in V. Then a homotopy $f(z, t)$, $0 \leq t \leq 1$, of $f(z)$ in U is said to cover the homotopy $g(z, t)$, or a covering homotopy of $g(z, t)$, if we have $f(z, t) \subset G(g(z, t))$ for all z, t. The covering homotopy theorem (in the weak form) asserts that if V is a normal Hausdorff space and if Z is compact, then there exists always a covering homotopy; furthermore, if $g(z, t)$ leaves a point $z_0 \subset Z$ fixed, we can assume that $f(z, t)$ also leaves z_0 fixed. That this covering homotopy theorem is true for any fibre system in U can be seen as follows. We observe first that in case the $G(y)$ is the inverse function $\pi^{-1}(y)$ of a continuous mapping $\pi(x)$ of U onto V, then we have a (generalized) fibre space as defined by S. T. Hu[4]; and, as has been observed by Hu, the covering homotopy theorem is true for such a fibre space. The general case can be reduced to this special case by considering the graph W of the function $G(y)$, i. e. the set of all points $w = x \times y$ in $U \times V$ satisfying the condition $x \subset G(y)$. Let $\tau(w)$ and $\pi(w)$ be the

---

[3] F. Severi, "Intorno al teorema d'Abel sulle superficie algebriche ed alla riduzione a forma normale degl'integrali di Picard," *Rendiconti del Circolo Matematico di Palermo*, vol. 21 (1906), p. 261, Teorema I.

[4] Sze-Tsen Hu, "On generalizing the notion of fibre spaces to include the fibre bundles," *Proceedings of the American Mathematical Society*, vol. 1 (1950), pp. 756-762. It is convenient to use this generalized notion of a fibre space, though the particular fibre systems used in the present paper can all be "derived" from fibre bundles which are also at the same time fibre spaces in the sense of Hurewicz-Steenrod.

mappings of W into U and V induced by the projections of $U \times V$ into U and V respectively. It is clear that $\pi(w)$ is a mapping of W onto V, and that we have $\tau(\pi^{-1}(y)) = G(y)$ for each $y \subset V$. The space W can then be made into a fibre space with respect to the mapping $\pi(w)$ if we define for each N the slicing function by the formula:

$$\phi_N(w, y) = \phi_N(\tau(w), y) \times y \qquad (w \subset \pi^{-1}(N), y \subset N).$$

If we set $h(z) = f(z) \times g(z)$, then $h(z)$ is a continuous mapping of Z into W such that $h(z) \subset \pi^{-1}(g(z))$. Since the covering homotopy theorem holds for the fibre mapping $\pi(w)$, there exists a homotopy $h(z, t)$ of $h(z)$ in W such that $h(z, t) \subset \pi^{-1}(g(z, t))$. Then we have $\tau h(z, t) \subset \tau\pi^{-1}(g(z, t)) = G(g(z, t))$, so that $\tau h(z, t)$ is a homotopy of $f(z)$ in U which covers the homotopy $g(z, t)$ in V.

[*Note added in proof* (June 6, 1952). Professor Beno Eckmann has recently called my attention to the fact that essentially the same concept as that of a fibre system has been introduced by him under the name "retrahierbare Ueberdeckung" in his paper "Zur Homotopietheorie gefaserter Raeume," *Commentarii Mathematici Helvetici*, vol. 14 (1941), pp. 141-192. In the first part of that paper the covering homotopy theorem was proved for a "retrahierbare Ueberdeckung", and an application was made of this theorem to the covering of a sphere by its great spheres.]

Let U be a non-singular algebraic variety of dimension r, and let Φ be an irreducible algebraic correspondence of dimension t between U and an algebraic variety V of dimension s. Then for a generic point η in V the set $\Phi^{-1}(\eta)$ is an irreducible algebraic variety of dimension $d = t - s$ in U, and we can consider $\Phi^{-1}(\eta)$ as a generic element of an algebraic system of d-cycles in U. For any point y in V the variety $\Phi^{-1}(y)$ is the carrier of the set of all d-cycles which are specializations of the d-cycle $\Phi^{-1}(\eta)$ over the specialization $\eta \to y$. A point y in V is said to be semiregular with respect to the correspondence Φ (or rather the inverse correspondence Φ^{-1}), if there is a uniquely determined specialization cycle of $\Phi^{-1}(\eta)$ over the specialization $\eta \to y$, and if this specialization cycle has no multiple components. It is easily seen that a point y in V is semi-regular with respect to Φ if and only if the variety $\Phi^{-1}(y)$ has the dimension d and the same degree (in the ambient projective space) as the variety $\Phi^{-1}(\eta)$, so that we can consider $\Phi^{-1}(y)$ itself as the specialization cycle of $\Phi^{-1}(\eta)$ over the specialization $\eta \to y$.

Since U is a differentiable manifold (of class C^∞), we can introduce in U a Riemannian metric. In fact, let $M = \{M_i\}$ be a locally finite system of coordinate neighborhood covering U, and let the set of differentiable functions $\{e_i(x)\}$ be a partition of unity subordinate to this covering; then, if we denote by ds_i^2 the Euclidean metric of the coordinate neighborhood M_i, the differential form $ds^2 \doteq \sum_i e_i(x) ds_i^2$ defines a Riemannian metric on U. We observe that by means of a suitable choice of the covering neighborhood M_i and the partition functions $e_i(x)$ we can make the Riemannian metric ds^2 in a sufficiently small neighborhood of any given point equal to the Euclidean metric with respect to any given coordinate system around this point. This fact will be convenient for us later. Let y' be a semi-regular point in V, and let R be a compact subset in $\Phi^{-1}(y')$ consisting of only simple points in $\Phi^{-1}(y')$. If $p(x)$ is any differentiable function which assigns to each point x in R a $(2r - 2d)$-dimensional direction element which is transversal to the tangent space of $\Phi^{-1}(y')$ at the point x, then there exists a differentiable system of $(2r - 2d)$-dimensional geodesic surfaces (or geodesic $(2r - 2d)$-surfaces) $P(x)$ $(x \subset R)$, such that for each point x in R the surface $P(x)$ has the tangential direction $p(x)$ at x. If N is a sufficiently small neighborhood of y' in V, then for every point y in N and every point x in R, the intersection $\Phi^{-1}(y) \cap P(x)$ consists of exactly one point which is simple in $\Phi^{-1}(y)$, and the mapping $x \to \Phi^{-1}(y) \cap P(x)$ is a homeomorphism of R onto a compact subset $R(y)$ in $\Phi^{-1}(y)$. Thus each point x in $\sum_{y \subset N} R(y)$ is contained in exactly one geodesic $(2r - 2d)$-surface of the system, which we shall also denote by $P(x)$; the function $\phi_N(x, y) = P(x) \cap R(y)$ $(x \subset \sum_{y \subset N} R(y), y \subset N)$ is then a slicing function for the system $R(y)$ $(y \subset N)$.

In case the variety $\Phi^{-1}(y')$ is non-singular, we can set $R = \Phi^{-1}(y')$ and hence $R(y) = \Phi^{-1}(y)$ for all y in N, so that the system of varieties $\Phi^{-1}(y)$ $(y \subset N)$ defines a fibre system in U. Now, if the generic variety $\Phi^{-1}(\eta)$ *is* non-singular, then there exists a proper subvariety H in V such that every point y in $V - H$ is semi-regular with respect to Φ and the variety $\Phi^{-1}(y)$ is non-singular. It follows then that the system of varieties $\Phi^{-1}(y)$ $(y \subset V - H)$ defines a fibre system in U. We shall now show that in case $d = 1$ this assertion is also true (with a suitable definition of H) even if $\Phi^{-1}(\eta)$ has singular points. We shall say that a point y' in V is regular with respect to Φ, if y' is semi-regular and if the variety $\Phi^{-1}(y')$ has the same singularities as the generic variety $\Phi^{-1}(\eta)$, i. e. each singular branch of the curve $\Phi^{-1}(y')$ is the specialization of a singular branch of the same order of the curve $\Phi^{-1}(\eta)$. It is easily seen that the set of all points in V which are not regular with respect to Φ is a proper subvariety in V, which we shall also denote by H. Let y' be a point in $V - H$, and let x' be a singular point in $\Phi^{-1}(y')$; then there is a coordinate neighborhood M of x' in U and a suitably chosen system of coordinates $u_1, \cdots u_r$ in M with origin at x', such that for every point y in a sufficiently small neighborhood N of y' in V, the curve $\Phi^{-1}(y) \cap M$ is a regular analytic covering space of a fixed degree g over a neighborhhood M_1 of the origin in the complex u_1-plane with a unique branch point over the origin $u_1 = 0$. Let M'_1 be a circular region $|u_1| < \epsilon$ $(\epsilon > 0)$ such that its closure $\bar{M}'_1$ is contained in M_1, and let $P(\alpha)$ $(\alpha \subset M_1)$ be the "hyperplane" in M defined by the equation $u_1 = \alpha$. Then, for each $u_1 \neq 0$ in $\bar{M}'_1$, there exist g disjoint circular domains $P_i(u_1)$, $i = 1, \cdots, g$, in $P(u_1)$, such that for each y in N and for each $i = 1, \cdots, g$, the intersection $\Phi^{-1}(y) \cap P_i(u_1)$ consists of exactly one point, while for $u_1 = 0$, the intersection $\Phi^{-1}(y) \cap P(0)$ itself consists of exactly one point, namely the branch point of $\Phi^{-1}(y)$ over $u_1 = 0$. For each y in N, we set $L(y) = \sum_{u_1 \subset M'_1} \Phi^{-1}(y) \cap P(u_1)$ and $\bar{L}(y) = \sum_{u_1 \subset \bar{M}'_1} \Phi^{-1}(y) \cap P(u_1)$ so that $\bar{L}(y)$ is the closure of the domain $L(y)$ in $\Phi^{-1}(y)$. We can then define a slicing function for the system $\bar{L}(y)$ $(y \subset N)$ by setting $\phi_N(x, y) = \Phi^{-1}(y) \cap P_i(u_1)$, $(x \subset \sum_{y \subset N} \bar{L}(y), y \subset N)$, where $P_i(u_1)$ is the one of the g domains in $P(u_1)$ which contains the point x. Furthermore, if we choose our Riemannian metric in U in such a way that it coincides with the Euclidean metric in the coordinate neighborhood M, then each $P_i(u_1)$ is a geodesic $(2r - 2)$-surface, and for each point $x \neq x'$ in $\bar{L}(y')$ the $P_i(u_1)$ passing through it has a tangential direction element $p(x)$ which is transversal to the tangential space of $\Phi^{-1}(y')$ at x. In particular this function $p(x)$ is defined on the boundary $\bar{L}(y') - L(y')$ curve of the domain $L(y')$, and it is also differentiable. Let now $x^{(i)}$, $i = 1, \cdots, a$, be the singular points of $\Phi^{-1}(y')$, and let $M^{(i)}$, $i = 1, \cdots, a$, be suitably chosen (disjoint) coordinate neighborhoods of the points $x^{(i)}$, $i = 1, \cdots, a$, respectively in U, and let the Riemannian metric in U be so chosen that it coincides with the Euclidean metric in each $M^{(i)}$. If we choose the neighborhood N of y' sufficiently small, then we can define a fibre system $\bar{L}^{(i)}(y)$ $(y \subset N)$ and a slicing function $\phi^{(i)}{}_N(x, y)$ $(x \subset \sum_{y \subset N} \bar{L}^{(i)}(y), y \subset N)$ in each $M^{(i)}$, and we have also the function $p^{(i)}(x)$ defined in $\bar{L}^{(i)}(y') - L^{(i)}(y')$. If we set $Q(y) = \sum_{i=1}^{a} L^{(i)}(y)$ and $\bar{Q}(y) = \sum_{i=1}^{a} \bar{L}^{(i)}(y)$, then we can consider all the functions $\phi^{(i)}{}_N(x, y)$ together as a single slicing function $\phi_N(x, y)$ $(x \subset \sum_{y \subset N} \bar{Q}(y), y \subset N)$ for the fibre system $\bar{Q}(y)$ $(y \subset N)$, and also all the functions $p^{(i)}(x)$ together as a single function $p(x)$ defined on the boundary $\bar{Q}(y') - Q(y')$ of the domain $Q(y')$. For each $y \subset N$, let $R(y) = \Phi^{-1}(y) - Q(y)$; it is clear that $\bar{Q}(y)$

and $R(y)$ are two complementary closed domains in $\Phi^{-1}(y)$ with the curve $\bar{Q}(y) - Q(y)$ as their common boundary. In particular the boundary of $R(y')$ is $\bar{Q}(y') - Q(y')$, and the function $p(x)$ is defined and differentiable on the boundary $\bar{Q}(y') - Q(y')$. If we assign to each point x in $R(y')$ the set of all $(2r-2)$-dimensional direction elements at x which are transversal to the tangent space of $\Phi^{-1}(y')$ at x, then we obtain a differentiable fibre bundle over the space $R(y')$, in which the fibre is topologically a $(4r-4)$-cell. It follows then that the differentiable function $p(x)$, which is defined on the boundary of $R(y')$, can be extended to a differentiable function $p'(x)$ in the entire space $R(y')$. If we denote by $P'(x)$ $(x \subset R(y'))$ the system of geodesic $(2r-2)$-surfaces corresponding to the function $p'(x)$, then for each y in N, provided N is taken sufficiently small, the mapping $x \to \Phi^{-1}(y) \cap P'(x)$ is a homeomorphism of $R(y')$ onto a closed domain in $\Phi^{-1}(y)$ whose boundary is $\bar{Q}(y) - Q(y)$ and which approaches $R(y')$ as y approaches y'; hence this domain must be $R(y)$. If we denote by $P'(x)$, for any point x in $\sum_{y \subset N} R(y)$, the one geodesic $(2r-2)$-surface of this system which contains the point x, then the function

$$\phi'_N(x, y) = R(y) \cap P'(x) = \Phi^{-1}(y) \cap P'(x) \quad (x \subset \sum_{y \subset N} R(y), y \subset N)$$

is a slicing function of the system $R(y)$ $(y \subset N)$. Since, for

$$x \subset (\sum_{y \subset N} R(y)) \cap (\sum_{y \subset N} \bar{Q}(y)), \; y \subset N,$$

we have

$$\phi_N(x, y) = \Phi^{-1}(y) \cap P_i(u_1) = \Phi^{-1}(y) \cap P'(x) = \phi'_N(x, y),$$

the two slicing function $\phi_N(x, y)$ and $\phi'_N(x, y)$ are concordant whenever both are defined and hence can be considered together as one slicing function for the system of curves $\Phi^{-1}(y)$ $(y \subset N)$. Thus we have shown that the system of curves $\Phi^{-1}(y)$ $(y \subset V - H)$ is a fibre system.

2. THEOREM 1. *Let U be a non-singular algebraic variety of dimension r, and let a subvariety G of dimension d be a member of an irreducible algebraic system $G(y)$ $(y \subset V)$, with U as its carrier variety, which has at least one base point, and whose generic member is irreducible. If $f(z)$ is a continuous mapping of the unit interval I into U, with $f(0) = f(1) = x^{(0)}$ in G, then a finite power of this mapping is homotopic* rel. $z = 0, 1$, *to a continuous mapping of I into G. If the system $G(y)$ is involutional, then $f(z)$ itself is homotopic* rel. $z = 0, 1$, *to a continuous mapping of I into G.*

Remark. The m-th power of $f(z)$ is the mapping $\bar{f}(z)$ of I defined by setting $\bar{f}(z) = f(mz - i)$ for $i/m \leq z \leq (i+1)/m$, $i = 0, 1, \cdots, m-1$. The system $G(y)$ is said to be involutional, if it is induced by a rational transformation of U onto V.

Proof. Let the algebraic system be defined by an irreducible correspondence Φ between U and the variety V, and let $y^{(0)}$ be the point in V such that $\Phi^{-1}(y^{(0)}) = G(y^{(0)}) = G$. It is sufficient to prove our theorem for the case where $y^{(0)}$ is any point in an everywhere dense subset in V; this follows from the fact that for any point y in V the variety $\Phi^{-1}(y)$ is a neighborhood deformation retract in U. We begin with the special case where the system $G(y)$ is the linear system cut out on U by the system of linear subspaces of dimension $n - r + d$ $(d \geq 1)$ in the ambient space S_n, which all pass through a sufficiently general $S_{n-r+d-1}$. Since in this case $\Phi^{-1}(\eta)$ is non-singular for a generic point η in V, there exists a subvariety H in V such that each point y in $V - H$ is semi-regular and $\Phi^{-1}(y)$ is non-singular. We can assume without any loss of generality that $y^{(0)}$ is a point in $V - H$; furthermore, we can also assume that $x^{(0)}$ is a point in G outside of $\Phi^{-1}(H)$. Since $\Phi^{-1}(H)$ is a proper subvariety in U and hence topologically a subcomplex of dimension $\leq 2r - 2$ in the $2r$-dimensional topological manifold U, we can assume that, after a suitable homotopy (rel. $z = 0, 1$) if necessary, $f(z)$ is a mapping of I into $U - \Phi^{-1}(H)$. Then the mapping $g(z) = \Phi(f(z))$ of I into $V - H$ is well defined, and we have evidently $g(0) = g(1) = y^{(0)}$ and $f(z) \subset G(g(z))$ for all z. Let $x^{(1)}$ be any point in $G \cap S_{n-r+d-1}$, and let $h(z)$ be any continuous mapping of I into G such that $h(0) = x^{(0)}$ and $h(1) = x^{(1)}$. We set

$$f'(z) = \begin{cases} f(4z) & (0 \leq z \leq \frac{1}{4}), \\ h(4z-1) & (\frac{1}{4} \leq z \leq \frac{1}{2}), \\ x^{(1)} & (\frac{1}{2} \leq z \leq \frac{3}{4}), \\ h(4-4z) & (\frac{3}{4} \leq z \leq 1), \end{cases}$$

and

$$g'(z) = \begin{cases} g(4z) & (0 \leq z \leq \frac{1}{4}), \\ y^{(0)} & (\frac{1}{4} \leq z \leq \frac{1}{2}), \\ g(3-4z) & (\frac{1}{2} \leq z \leq \frac{3}{4}), \\ y^{(0)} & (\frac{3}{4} \leq z \leq 1); \end{cases}$$

it is clear that $f'(z) \simeq f(z)$ rel. $z = 0, 1$, and $g'(z) \simeq 0$ rel. $z = 0, 1$. Since $f'(z) \subset G(g'(z))$ for all z, it follows from the covering homotopy theorem that there is a homotopy of $f'(z)$ rel. $z = 0, 1$, which deforms $f'(z)$ into a mapping of I into G; hence $f(z)$ is also homotopic rel. $z = 0, 1$, to a mapping of I into G.

Turning to the general case, we observe first that we can assume without any loss of generality the following: (1) For a generic point ξ in U the variety $\Phi(\xi)$ consists of a finite number m of points; for otherwise we can replace V by its intersection with a suitably chosen linear subspace (passing through $y^{(0)}$) in its ambient space. It is clear that we have $m=1$ in case $G(y)$ is an involutional system. (2) For a generic point η in V the variety $\Phi^{-1}(\eta)$ is a curve, i. e. $d=1$; for otherwise we can replace U by its intersection with a suitably chosen S_{n-d+1} (passing through a base point of the system $G(y)$) in S_n, and we have just shown above that any continuous mapping $f(z)$ of I into U, with $f(0)=f(1)$ in $U \cap S_{n-d+1}$, is homotopic rel. $z=0,1$, to a continuous mapping of I into $U \cap S_{n-d+1}$. Now, let H be the subvariety in V containing all points which are not regular with respect to Φ, so that the system of curves $G(y)$ $(y \subset V-H)$ is a fibre system; let T be the subvariety in U such that for every point x in $U-T$ the set $\Phi(x)$ consists of m distinct points outside of H. Without any loss of generality we can assume that $y^{(0)}$ is a point in $V-H$ such that $G=\Phi^{-1}(y^{(0)})$ is not entirely in T, and that $x^{(0)}$ is a point in $G-T$; then $\Phi(x^{(0)})$ consists of m points, one of which is the point $y^{(0)}$. Since $T+\Phi^{-1}(H)$ is a proper subvariety in U and hence topologically a subcomplex of dimension $\leq 2r-2$ in the $2r$-dimensional manifold U, we can assume that, after a suitable homotopy (rel. $z=0,1$) if necessary, $f(z)$ is a mapping of I into $U-T-\Phi^{-1}(H)$. Then the image $\Phi(f(z))$ consists of m distinct mappings of I into $V-H$, one (and only one) of which will be a mapping $g(z)$ such that $g(0)=y^{(0)}$. The point $g(1)$ is one of the m points in the set $\Phi(x^{(0)})$, though not necessarily the point $y^{(0)}$. It is easily seen that if $\bar{f}(z)$ is the m-th power of $f(z)$, then the image $\Phi(\bar{f}(z))$ of $\bar{f}(z)$ will consist of m distinct mappings of I into $V-H$, one of which will be a mapping $\bar{g}(z)$ such that $\bar{g}(0)=\bar{g}(1)=y^{(0)}$. Let $x^{(1)}$ be a base point of the system $G(y)$, and let $h(z)$ be a continuous mapping of I into G such that $h(0)=x^{(0)}$ and $h(1)=x^{(1)}$. If we now define $f'(z)$ and $g'(z)$ again as before, replacing $f(z)$ and $g(z)$ by $\bar{f}(z)$ and $\bar{g}(z)$ respectively, we can repeat exactly the same argument and conclude that $\bar{f}(z)$ is homotopic rel. $z=0,1$, to a mapping of I into G. This concludes the proof of Theorem 1.

In the following we shall denote by $F(U)$ the fundamental group of a topological space U, considered as a group of mapping classes with some one fixed reference point. If W and X are two subsets in U, then the identity mapping of $W \cap X$ into W will induce a homomorphism of $F(W \cap X)$ into $F(W)$, the reference point in both groups being one and the same point in $W \cap X$; we shall then denote by $F(W, X)$ the subgroup of $F(W)$ which is the image of $F(W \cap X)$ under this homomorphism.

THEOREM 2. *Let Φ be a rational transformation of a non-singular algebraic variety U of dimension r onto a non-singular algebraic variety V of dimension s, with the properties*: (1) *For a generic point η in V the variety $\Phi^{-1}(\eta)$ is irreducible, and* (2) *the set H of all points which are not semi-regular with respect to Φ is a subvariety of dimension $\leq s-2$ in V. Then there is a homomorphism of $F(U)$ onto $F(V)$, and the kernel of this homomorphism is the subgroup $F(U,\Phi^{-1}(y^{(0)}))$, where $y^{(0)}$ is any sufficiently general point in V.*

Proof. Let T be the fundamental variety of Φ in U; it is well known that T has a dimension $\leq r-2$. It is clear that the variety $\Phi^{-1}(\eta)$ has the dimension $d=r-s$; let c be the degree of $\Phi^{-1}(\eta)$ considered as a variety in the ambient space S_n of U. Let $y^{(0)}$ be a point in V such that $\Phi^{-1}(y^{(0)})$ is not contained in T, and let $x^{(0)}$ be a point in $\Phi^{-1}(y^{(0)})-T$. We shall take $x^{(0)}$ as the reference point of the groups $F(U)$, $F(U-T)$, $F(\Phi^{-1}(y^{(0)}))$, $F(U,\Phi^{-1}(y^{(0)}))$, and $F(U-T,\Phi^{-1}(y^{(0)}))$, and take $y^{(0)}$ as the reference point of the group $F(V)$. The identity mapping of $U-T$ into U induces a homomorphism θ of $F(U-T)$ into $F(U)$; since T is topologically a subcomplex of dimension $\leq 2r-4$ in the $2r$-dimensional manifold U, it is easily seen that θ is an isomorphism of $F(U-T)$ onto $F(U)$, and that the image of $F(U-T,\Phi^{-1}(y^{(0)}))$ under θ is precisely $F(U,\Phi^{-1}(y^{(0)}))$. Since Φ is a continuous mapping of $U-T$ into V, there is an induced homomorphism ϕ of $F(U-T)$ into $F(V)$. We set $\Delta=\phi\theta^{-1}$, so that Δ is a homomorphism of $F(U)$ into $F(V)$. We shall prove our theorem by showing that (provided the point $y^{(0)}$ is sufficiently general) Δ is a homomorphism of $F(U)$ onto $F(V)$ and its kernel is $F(U,\Phi^{-1}(y^{(0)}))$.

Let L be a linear subspace of dimension $n-r+s$ in the ambient space S_n of U, such that $U\cap L$ is a non-singular variety of dimension s. The rational transformation Φ will then induce a rational transformation $\bar{\Phi}$ of $U\cap L$ onto V, and we have $\bar{\Phi}^{-1}(y)\subset\Phi^{-1}(y)\cap L$ for every point y in V. Since, for a general point η in V, the set $\bar{\Phi}^{-1}(\eta)=\Phi^{-1}(\eta)\cap L$ consists of c distinct points in $U-T$, there exists a proper subvariety K in V such that for every point y in $V-K$ the set $\bar{\Phi}^{-1}(y)$ consists of c distinct points in $U-T$ (which are then necessarily simple points in the variety $\Phi^{-1}(y)$). We shall also assume that $y^{(0)}$ is a point in $V-K$ and $x^{(0)}$ is a point in $\bar{\Phi}^{-1}(y^{(0)})$.

Let $g(z)$ be a continuous mapping of the unit interval I into V such that $g(0)=g(1)=y^{(0)}$. Since K is topologically a subcomplex of dimension $\leq 2s-2$ in the topological manifold V of dimension $2s$, we can assume, after a homotopy (rel. $z=0,1$) if necessary, that $g(z)$ is a mapping of I into $V-K$. Since $\bar{\Phi}^{-1}(V-K)$ is a c-fold regular covering space of $V-K$, the inverse image $\bar{\Phi}^{-1}(g(z))$ consists of c distinct mappings of I into $U-T$, one of which will be a mapping $f(z)$ such that $f(0)=x^{(0)}$. Let $x^{(1)}=f(1)$, and let $f_1(z)$ be any continuous mapping of I into $\Phi^{-1}(y^{(0)})-T$ such that $f_1(0)=x^{(1)}$ and $f_1(1)=x^{(0)}$. If we set

$$f'(z)=\begin{cases} f(2z) & (0\leq z\leq \frac{1}{2}),\\ f_1(2z-1) & (\frac{1}{2}\leq z\leq 1),\end{cases}$$

then we have evidently a continuous mapping of I into $U-T$ such that $\Phi(f'(z))\simeq g(z)$ rel. $z=0,1$. This shows that Δ is a homomorphism of $F(U)$ onto $F(V)$.

Now let $f(z)$ be a continuous mapping of I into U such that $f(0)=f(1)=x^{(0)}$. According to Theorem 1, $f(z)$ is homotopic rel. $z=0,1$ to a mapping of I into $U\cap L$, and hence also homotopic rel. $z=0,1$ to a mapping of I into $\bar{\Phi}^{-1}(V-K)$. We can therefore assume that $f(z)$ is already a mapping of I into $\bar{\Phi}^{-1}(V-K)$; then the mapping $g(z)=\Phi(f(z))$ is defined, and we have $g(0)=g(1)=y^{(0)}$. In order to prove that the kernel of Δ is $F(U,\Phi^{-1}(y^{(0)}))$, we have only to show that if the mapping $g(z)$ is homotopic to zero rel. $z=0,1$ in V, then $f(z)$ is homotopic rel. $z=0,1$ to a mapping of I into $\Phi^{-1}(y^{(0)})$. Without loss of generality we can restrict ourselves to the case $d=1$; for otherwise we can replace U by its intersection with a suitably chosen linear subspace of dimension $n-r+s+1$ which contains the space L. It is then easily seen that we can assume the subvariety K so chosen that every point in $V-K$ is regular with respect to Φ; then the system of curves $\Phi^{-1}(y)$ $(y\subset V-K)$ is a fibre system. Furthermore, there exists a subvariety H' of dimension $\leq s-2$ in V, which contains H and is contained in K, such that for every point y in $V-H'$ the set $\bar{\Phi}^{-1}(y)$ consists of at most c points, each of which is either a simple point in $\Phi^{-1}(y)$ or a singular point of a "generic" nature.

Since K is topologically a subcomplex of dimension $\leq 2s-2$, and H' is topologically a subcomplex of dimension $\leq 2s-4$ in the $2s$-dimensional manifold V, there exists in $V-H'$ a homotopy $g(z,t)$ of the mapping $g(z)$ rel. $z=0,1$ with the following properties: (1) $g(z,t)\subset V-K$ for all z and t except a finite number m of points $g((2i-1)/2m,1)$, $i=1,\cdots,m$, which all belongs to $K-H'$; (2) $g(i/m,1)=y^{(0)}$ for $i=1,\cdots,m$, and $g(z,1)=g((2i-1-mz)/m,1)$ for $(i-1)/m\leq z\leq i/m$. Since $\bar{\Phi}^{-1}(V-H')$ is an analytic covering space (with branch points) of $V-H'$, and since moreover $\bar{\Phi}^{-1}(V-K)$ is a regular covering space of $V-K$, there corresponds to the homotopy $g(z, t)$ a uniquely determined homotopy $f(z, t)$ rel. $z = 0, 1$ in $\bar{\Phi}^{-1}(V - H')$, with $f(z, 0) = f(z)$ and $f(z, t) \subset \bar{\Phi}^{-1}(g(z, t))$. Therefore, to prove our assertion, it is sufficient to consider the following problem. Let $g(z)$ be a continuous mapping of I into $V - H'$ with the properties: $g(0) = g(1) = y^{(0)}$; $g(z) = g(1 - z)$ for all z; $g(z) \subset V - K$ for all $z \neq \frac{1}{2}$, and $g(\frac{1}{2})$ belongs to $K - H'$. Let $f(z)$ be a continuous mapping of I into $\bar{\Phi}^{-1}(V - H')$ such that $f(z) \subset \bar{\Phi}^{-1}(g(z))$ for all z; then we have to show that $f(z)$ is homotopic rel. $z = 0, 1$ in U to a mapping of I into $\Phi^{-1}(y^{(0)})$.

Let $x^{(1)} = f(\frac{1}{2})$ and $y^{(1)} = g(\frac{1}{2})$, and let R be a compact neighborhood of $x^{(1)}$ in $\Phi^{-1}(y^{(1)})$ which contains no singular points of $\Phi^{-1}(y^{(1)})$ except possibly the point $x^{(1)}$ itself. Then, as we have shown in the preceding section, there exists a neighborhood N of $y^{(1)}$, and to every point y in N there corresponds a compact subset $R(y)$ in $\Phi^{-1}(y)$ (with $R(y^{(1)}) = R$), such that the system $R(y)$ $(y \subset N)$ is a fibre system; furthermore, there is a neighborhood M of $x^{(1)}$ in U such that $M \cap \Phi^{-1}(y) \subset R(y)$ for all y in N. Let I_e $(e > 0)$ denote the interval $|z - \frac{1}{2}| \leq e$; and let e be taken so small that $f(z) \subset M$ and $g(z) \subset N$ for all z in I_e. Since the mapping $g(z)$ of I_e into N is evidently homotopic rel. $z = \frac{1}{2} - e, \frac{1}{2} + e$ in N to a mapping of I_e into the point $y^{(2)} = g(\frac{1}{2} - e) = g(\frac{1}{2} + e)$, it follows from the covering homotopy theorem that the mapping $f(z)$ of I_e into $M \cap \Phi^{-1}(N)$ is also homotopic rel. $z = \frac{1}{2} - e, \frac{1}{2} + e$ to a mapping $h(z)$ of I_e into $\Phi^{-1}(y^{(2)})$. If we set

$$f'(z) = \begin{cases} f(z) & (|z - \frac{1}{2}| \geq e), \\ h(z) & (|z - \frac{1}{2}| \leq e), \end{cases}$$

$$g'(z) = \begin{cases} g(z) & (|z - \frac{1}{2}| \geq e), \\ y^{(2)} & (|z - \frac{1}{2}| \leq e), \end{cases}$$

then we have evidently $f(z) \simeq f'(z)$ rel. $z = 0, 1$ in U and $g'(z) \simeq 0$ rel. $z = 0, 1$ in $V - K$. Since $f'(z) \subset \Phi^{-1}(g'(z))$ for all z, it follows then from the covering homotopy theorem that $f'(z)$ is homotopic rel. $z = 0, 1$ to a mapping of I into $\Phi^{-1}(y^{(0)})$, and hence $f(z)$ is also homotopic rel. $z = 0, 1$ to a mapping of I into $\Phi^{-1}(y^{(0)})$. This concludes the proof of Theorem 2.

THE JOHNS HOPKINS UNIVERSITY.

THE JACOBIAN VARIETY OF AN ALGEBRAIC CURVE.* †

By Wei-Liang Chow.

1. Preliminary remarks. We begin with a few remarks concerning terminology and notations. We shall follow in general the terminology and notations of A. Weil, as developed in his books [8], [9], with the modification that we shall operate throughout with points and (complete) varieties in projective spaces only. The projective space of dimension m will be denoted by $\mathfrak{S}_m$. If $(x) = (x_0, x_1, \cdots, x_m)$ is a point in $\mathfrak{S}_m$, the x_i being its homogeneous coordinates, and if K is a field, we shall denote by $K((x))$ the field generated over K by the adjunction of the mutual ratios of the x_i; similarly, if $(x), (y), \cdots$, are points in projective spaces, we denote by $K((x), (y), \cdots)$ the field generated over K by the adjunction of the mutual ratios of the homogeneous coordinates of every one of the points. The dimension of a point (x) over K is then the degree of transcendency of $K((x))$ over K, and (x) is said to be algebraic or rational over K if $K((x))$ is algebraic over K or coincides with K respectively; we shall also say that a homogeneous polynomial or form is rational over K, if the mutual ratios of its coefficients are all in K. We shall use the capital letters U, V, W, X, Y, Z (and these only) to denote indeterminates; for convenience, a system consisting of $m+1$ indeterminates $(X) = (X_0, X_1, \cdots, X_m)$ will be called a system of indeterminates in $\mathfrak{S}_m$.

We shall assume that the reader is familiar with the theory of associated forms of positive cycles in a projective space, as developed in Chow-van der Waerden [4]. For a positive cycle $\mathfrak{Z}$ of dimension r and degree d in a projective space $\mathfrak{S}_m$, the associated form is a form $f(V^0, \cdots, V^r)$ of degree d in each system of indeterminates $(V^i) = (V^i{}_0, V^i{}_1, \cdots, V^i{}_m)$, $i = 0, 1, \cdots, r$; this form is characterized by the following property: For any set of r hyperplanes in $\mathfrak{S}_m$ given by the equations $\sum_{j=0}^{m} v^i{}_j X_j = 0$, $i = 1, \cdots, r$, the form

* Received June 29, 1953.

† This paper was written in 1948; a brief summary of the results contained in it appeared under the same title in the Abstract of the addresses given at the Conference on Algebraic Geometry and Algebraic Number Theory, University of Chicago, 1949, pp. 25-27. It is published here with some minor revisions, mainly in the first two sections.

$f(V^0, v^1, \cdots, v^r)$ does not vanish identically if and only if $\mathfrak{Z}$ has a finite number of common intersections with these r hyperplanes, and when such is the case, the form $f(V^0, v^1, \cdots, v^r)$ dissolves into a product of d linear forms, giving all the common intersections with the proper multiplicities. If we regard the set of coefficients (f) of the form $f(V^0, \cdots, V^r)$, arranged in an arbitrary but fixed order, as a point in a projective space $\mathfrak{S}_t$, and if (f') is a specialization of (f) over any field K, then it follows from the main theorem of associated form ([4], Satz 2) that the point (f') corresponds also to the associated form $f'(V^0, \cdots, V^r)$ of a positive cycle $\mathfrak{Z}'$ of dimension r and degree d in $\mathfrak{S}_m$; this positive cycle $\mathfrak{Z}'$ is said to be a specialization of $\mathfrak{Z}$ over K. A property or relation concerning a point (x) in $\mathfrak{S}_m$ and a point (y) in $\mathfrak{S}_n$ is said to be algebraic over a field K, if there is a set of forms $F_j(X, Y)$ in $K[X, Y]$, homogeneous in each of the systems of indeterminates $(X) = (X_0, X_1, \cdots, X_m)$ and $(Y) = (Y_0, Y_1, \cdots, Y_n)$, such that the property holds for any two points (x) and (y) in $\mathfrak{S}_m$ and $\mathfrak{S}_n$ respectively if and only if the equations $F_j(x, y) = 0$ hold; we then say that the property in question can be expressed as an algebraic condition (over K) in terms of the points (x) and (y). In case the set $F_j(X, Y)$ contains forms in (X) or (Y) alone, the point (x) or (y) can be restricted to the points of a certain bunch of varieties in $\mathfrak{S}_m$ or $\mathfrak{S}_n$ respectively; in such a case the property in question is often defined only for the points of certain bunches of varieties in $\mathfrak{S}_m$ and $\mathfrak{S}_n$, so that the equations defining these bunches of varieties are already implicitly contained in the definition of the property. The following fact is important for us later: The property that a given algebraic relation over K holds for a point (y) in $\mathfrak{S}_n$ and *every* point (x) in a positive cycle $\mathfrak{Z}$ in $\mathfrak{S}_m$ can itself be expressed as an algebraic condition over K in terms of (y) and the point (f) in $\mathfrak{S}_t$. To show this, we refer to the proof of Satz 2 in [4] and consider the equations (2), (3), (5) listed there; in our present notations these can be writen as a set of equations

$$G_j(V^1, \cdots, V^r, X^1, \cdots, X^d, Z) = 0,$$

where $(X^1), \cdots, (X^d)$ are d distinct systems of indeterminates in $\mathfrak{S}_m$ and (Z) is a system of indeterminates in $\mathfrak{S}_t$. Let $R_j(V^1, \cdots, V^r, Y, Z) = 0$ be the set of equations which we obtain by eliminating the indeterminates $(X^1), \cdots, (X^d)$ from the set of equations $F_j(X^1, Y) = 0, \cdots, F_j(X^d, Y) = 0$, $G_j(V^1, \cdots, V^r, X^1, \cdots, X^d, Z) = 0$, and let $A_j(Y, Z)$ be the set of all coefficients of the forms $R_j(V^1, \cdots, V^r, Y, Z)$, considered as forms in $(V^1), \cdots, (V^r)$ only; then it is easily seen from the cited proof in [4] that the property in question holds for (y) and $\mathfrak{Z}$ if and only if $A_j(y,f)=0$. It is clear that this result remains true if more than one point (y) or more than one cycle $\mathfrak{Z}$ are involved, and it is also clear that it remains true even if the original algebraic property does not actually involve the point (y) at all, so that it is simply an algebraic property for the points of the positive cycle or cycles. For later reference we shall state our result as a lemma:

LEMMA 1. *Let $\mathfrak{Z}^1,\cdots,\mathfrak{Z}^s$ be positive cycles in projective spaces and let $(f^1),\cdots,(f^s)$ be the systems of coefficients of the associated forms of $\mathfrak{Z}^1,\cdots,\mathfrak{Z}^s$ respectively; let $(y^1),\cdots,(y^h)$ be points in projective spaces. Then the property that a given algebraic relation over a field K holds for the system of points $(y^1),\cdots,(y^h)$ and every system of points $(x^1),\cdots,(x^s)$, where each (x^i) runs through all the points in $\mathfrak{Z}^i$, can be expressed as an algebraic condition over K in terms of $(y^1),\cdots,(y^h)$ and $(f^1),\cdots,(f^s)$.*

It is easily seen from the definition of the associated form that if the positive cycle $\mathfrak{Z}$ is rational over a field K, then its associated form $f(V^0,\cdots,V^r)$ is also rational over K, i. e. $K((f))=K$. We shall now show that the converse of this statement is also true, provided that $\mathfrak{Z}$ contains no component with a multiplicity which is divisible by the characteristic of the universal domain; more precisely, we shall now prove the following lemma:

LEMMA 2. *Let $\mathfrak{Z}$ be a positive cycle in a projective space and let (f) be the set of coefficients of the associated form of $\mathfrak{Z}$; if $\mathfrak{Z}$ contains no component with a multiplicity which is divisible by the characteristic of the field, then $\mathfrak{Z}$ is rational over the fields $k_0((f))$, where k_0 is the prime field.*

Proof. Without any loss of generality we can assume that the form $f(V^0,\cdots,V^r)$ is irreducible over the field $K_0=k_0((f))$, for otherwise we can apply the same argument to each irreducible factor of $f(V^0,\cdots,V^r)$ over K_0; then our assumption about the multiplicities of the components in $\mathfrak{Z}$ implies that the form $f(V^0,\cdots,V^r)$ has no multiple factors over $\bar{K}_0$. It follows that $f(V^0,\cdots,V^r)$ is the product of a number s of absolutely irreducible forms, all distinct and conjugate to each other over K_0, and each one of these forms is rational over a finite extension of degree s over K_0. In order to prove our assertion, it is therefore sufficient to restrict ourselves to the case where the form $f(V^0,\cdots,V^r)$ is absolutely irreducible and the positive cycle $\mathfrak{Z}$ is a variety defined over a field K containing K_0, and to show in this case the variety $\mathfrak{Z}$ is also defined over K_0. Consider the equation $f(V^0,\cdots,V^r)=0$; since the form $f(V^0,\cdots,V^r)$ is absolutely irreducible, this equation defines a variety $\mathfrak{T}$ of dimension $(r+1)m-1$ in the $(r+1)$-fold product space $S_m \times \cdots, \times S_m$, defined over the field K_0. Let (x) be a generic point of $\mathfrak{Z}$ over K, and let $(v^0), \cdots, (v^r)$ be $r+1$ independent generic solutions of the linear equation $\sum_{j=0}^{m} x_j V_j = 0$ over the field $K((x))$; if $r' \ (\geq r)$ is the dimension of (x) over K_0, then the point $(v^0, \cdots, v^r)$ in $\mathfrak{T}$ has the dimension

$$(r+1)(m-1) + r' = (r+1)m - 1 + (r' - r) \geq (r+1)m - 1$$

over K_0, from which it follows that $r = r'$ and that $(v^0, \cdots, v^r)$ is a generic point of $\mathfrak{T}$ over K_0. Since the point $(v^0, \cdots, v^r)$ has the same dimension $(r+1)m-1$ over both K and K_0, the fields $K_0((v^0), \cdots, (v^r))$ and K must be independent with respect to each other over K_0, and since $K_0((v^0), \cdots, (v^r))$ is a regular extension of K_0, this means that $K_0((v^0), \cdots, (v^r))$ and K are linearly disjoint with respect to each other over K_0. Without any loss of generality, we can assume that $x_0 \neq 0$ and that $x_m = 0$ in case any one of the coordinates x_i vanishes; then we can take all $v^i{}_j \ (j \neq 0)$ to be independent variables over $K((x))$ and we have $v^i{}_m \neq 0$ for all $i = 0, 1, \cdots, r$. We have then the relations $\sum_{j=1}^{m} (v^i{}_j / v^i{}_m)(x_j / x_0) = - v^i{}_0 / v^i{}_m$, $i = 0, 1, \cdots, r$; the first r equations show ([8], Chapter V, Theorem 1) that (x) is separably algebraic over $K_0((v^0), \cdots, (v^{r-1}))$, and then the last equation shows that (x) is rational over $K_0((v^0), \cdots, (v^r))$. Since $K_0((x))$ is subfield in $K_0((v^0), \cdots, (v^r))$, it is also linearly disjoint with respect to K over K_0; this shows that the variety $\mathfrak{Z}$ is also the locus of (x) over K_0 and hence is defined over K_0.

2. The problem. Let $\mathfrak{C}$ be a curve in the projective space $\mathfrak{S}_m$ of dimension m, defined over a field k, and let h and g be the degree and genus of $\mathfrak{C}$ respectively. A positive divisor $\mathfrak{p}$ of degree n in $\mathfrak{C}$ can be considered as a positive 0-cycle in $\mathfrak{S}_m$ and as such it can be represented by its associated form $p(V)$. Let $\Omega_0(V), \Omega_1(V), \cdots, \Omega_t(V)$ be the set of all power products of degree n in (V), arranged in an order which is arbitrary but fixed once for all; if we set $p(V) = \sum_{j=0}^{t} p_j \Omega_j(V)$, then the set of coefficients $(p) = (p_0, p_1, \cdots, p_t)$, arranged in this fixed order, determines a point in the projective space $\mathfrak{S}_t$ of dimension t, so that every positive divisor of degree n in $\mathfrak{C}$ is represented in a one-to-one manner by a point in $\mathfrak{S}_t$. We shall assume that the curve $\mathfrak{C}$ is free of singularities, so that every 0-cycle in $\mathfrak{S}_m$

which is contained in $\mathfrak{C}$ can also be considered as a divisor in $\mathfrak{C}$; then the set of all points in $\mathfrak{S}_t$ which represent positive divisors of degree n in $\mathfrak{C}$ is a variety $\mathfrak{C}^n$ of dimension n, defined over k. In fact, let $\mathfrak{y}$ be the positive divisor of degree n consisting of n independent generic points $(x^1),\cdots,(x^n)$ of $\mathfrak{C}$ over k, and let (y) be the point in $\mathfrak{S}_t$ representing $\mathfrak{y}$; then, for any positive divisor $\mathfrak{p}$ of degree n in $\mathfrak{C}$, the point (p) is a specialization of the point (y) over k, and conversely, by the main theorem of associated form, every such specialization (p) determines a positive 0-cycle of degree n in $\mathfrak{S}_m$ which is contained in $\mathfrak{C}$ and hence is a positive divisor of degree n in $\mathfrak{C}$. Since $k((y))$ is contained in the regular extension $k((x^1),\cdots,(x^n))$ of k, it is also a regular extension of k, and since $k((x^1),\cdots,(x^n))$ has the degree of transcendency n over k and is an algebraic extension of $k((y))$, the field $k((y))$ must have also the degree of transcendency n over k. It follows that the point (y) has a locus over k which is a variety of dimension n and which evidently coincides with $\mathfrak{C}^n$. As we shall see later, this variety $\mathfrak{C}^n$ is also free of singularities, but we shall not need this fact for the present. It is clear that if a positive divisor $\mathfrak{p}$ is rational over a field K, then the point (p) representing it is also rational over K; the converse of this statement is also true, as we have shown in [3], so that for any point (p) in $\mathfrak{C}^n$ the positive divisor $\mathfrak{p}$ determined by it is rational over $k((p))$.

Two divisors $\mathfrak{p}$, $\mathfrak{q}$ in $\mathfrak{C}$ are said to be linearly equivalent and we shall write $\mathfrak{p}\sim\mathfrak{q}$, if $\mathfrak{p}-\mathfrak{q}$ is the divisor of a rational function on $\mathfrak{C}$, or in other words, if $\mathfrak{p}-\mathfrak{q}$ is a principal divisor in $\mathfrak{C}$. It is well known that the set of all principal divisors in $\mathfrak{C}$ forms a subgroup $\mathfrak{G}_l(\mathfrak{C})$ in the group $\mathfrak{G}(\mathfrak{C})$ of all divisors in $\mathfrak{C}$, and that the quotient group of $\mathfrak{G}(\mathfrak{C})$ over $\mathfrak{G}_l(\mathfrak{C})$ is the group $\mathfrak{D}$ of divisor classes in $\mathfrak{C}$, each class consisting of all divisors which are linearly equivalent to each other. In this group $\mathfrak{D}$ the classes of degree zero, sometimes called the zero classes, again form a subgroup $\mathfrak{D}_0$, while the classes of any one fixed degree n form a coset $\mathfrak{D}_n$ of $\mathfrak{D}_0$ in $\mathfrak{D}$; if we denote by $\mathfrak{G}_a(\mathfrak{C})$ the group of all divisors of degree zero in $\mathfrak{C}$, then $\mathfrak{D}_0$ is the quotient group of $\mathfrak{G}_a(\mathfrak{C})$ over $\mathfrak{G}_l(\mathfrak{C})$. If we let any one fixed divisor class $\mathfrak{O}$ of $\mathfrak{D}_n$ correspond to the principal class of $\mathfrak{D}_0$, then the relation $\mathfrak{D}_n=\mathfrak{O}+\mathfrak{D}_0$ establishes a one-to-one correspondence between the classes of $\mathfrak{D}_n$ and those of $\mathfrak{D}_0$. Through this correspondence the group addition of $\mathfrak{D}_0$ will induce in $\mathfrak{D}_n$ the following operation: Given any two classes $\mathfrak{P}$, $\mathfrak{Q}$ in $\mathfrak{D}_n$, the class $\mathfrak{P}+\mathfrak{Q}-\mathfrak{O}$, also belonging to $\mathfrak{D}_n$, is called the sum of $\mathfrak{P}$ and $\mathfrak{Q}$. We shall call this operation in $\mathfrak{D}_n$ also addition (with respect to the class $\mathfrak{O}$).

A divisor class $\mathfrak{P}$ is said to be rational over a field K, if it contains a divisor $\mathfrak{p}$ which is rational over K. It is well known that the set of positive divisors in the class $\mathfrak{P}$ constitutes a linear system; that is, there exists a linear system of hypersurfaces in $\mathfrak{S}_m$ which cut out on $\mathfrak{C}$, besides a fixed component, exactly the positive divisors of the class $\mathfrak{P}$. If the class $\mathfrak{P}$ is rational over K, then ([8], Chapter VIII, Theorem 10) this linear system of hypersurfaces has a basis consisting of hypersurfaces which are defined over K. We recall that according to the Theorem of Riemann-Roch, the linear system of positive divisors of a class of degree $n > 2g - 2$ has exactly the dimension $n - g$.

A variety $\mathfrak{J}$, defined over a field K containing k, is called a Jacobian variety of the curve $\mathfrak{C}$, if it has the following properties: (1) $\mathfrak{J}$ is an Abelian variety defined over K; (2) there is a homomorphic mapping Φ of $\mathfrak{G}_a(\mathfrak{C})$ onto $\mathfrak{J}$, called the canonical homomorphism, whose kernel is precisely $\mathfrak{G}_l(\mathfrak{C})$; (3) the homomorphism Φ is rational and is defined over K in the following sense: if $\mathfrak{p}_u$ is a rational divisor over $K(u)$, where (u) is any set of elements (in the universal domain), then the point $\Phi(\mathfrak{p}_u)$ is also rational over $K(u)$, and if $\mathfrak{p}_\mu$ is any specialization of $\mathfrak{p}_u$ over a specialization $(u) \rightarrow (\mu)$ over K, then $\Phi(\mathfrak{p}_\mu)$ is also a specialization of $\Phi(\mathfrak{p}_u)$ over the specialization $(u) \rightarrow (\mu)$ over K; (4) the rational homomorphism Φ has the "universal maping" property: if Ψ is any rational homomorphism of $\mathfrak{G}_a(\mathfrak{C})$ into any Abelian variety $\mathfrak{A}$, then Ψ is the product of Φ and a rational homomorphism of $\mathfrak{J}$ into $\mathfrak{A}$. It can be easily seen from the "universal mapping" property (4) that the Jacobian variety $\mathfrak{J}$, it it exists, is uniquely determined up to a birational isomorphism, so that we can sometimes speak of *the* Jacobian variety. However, the question whether such a Jacobian variety exists at all is not a simple one, and this existence problem of the Jacobian variety is one of principal importance not only for the theory of algebraic curves, but also for the theory of Abelian varieties. In this paper we shall offer a general solution of this problem.

In the classical case, where the ground field is the field of complex numbers, a proof of the existence of the Jacobian variety can be obtained from the known results in the literature in the following way. (The terminology and notations in this one paragraph do not follow strictly the conventions set forth in section 1.) Let $I_1, \cdots, I_g$ be a system of g independent Abelian differentials of the first kind on $\mathfrak{C}$, and let $\gamma_1, \cdots, \gamma_{2g}$ be a base of the first Betti group of $\mathfrak{C}$ and we set $p_{ij} = \int_{\gamma_i} I_j$; let $\mathfrak{E}_g$ be the complex linear space of dimension g, considered as an analytic (= complex-analytic) group variety, and let $[P]$ be the discrete subgroup generated by the $2g$ rows of the matrix $P=(p_{ij})$, each row being considered as a point in E_g, and let $\mathfrak{E}_g/[P]$ be the quotient group variety of $\mathfrak{E}_g$ over $[P]$. Let (x_0) be any fixed point in $\mathfrak{C}$ and let ϕ be the analytic mapping of $\mathfrak{C}$ into $\mathfrak{E}_g/[P]$ defined by the formula $\phi(x)=(\int_{x_0}^{x} I_1,\cdots,\int_{x_0}^{x} I_g) \pmod{[P]}$; then, if $\mathfrak{p}=\sum n_i(x_i)$ is any divisor in $\mathfrak{G}_a(\mathfrak{C})$, the correspondence $\mathfrak{p}\rightarrow\sum n_i\phi(x_i)$ defines a homomorphism Φ of $\mathfrak{G}_a(\mathfrak{C})$ onto $\mathfrak{E}_g/[P]$, whose kernel is, by the Abel's Theorem, precisely the group $\mathfrak{G}_l(\mathfrak{C})$. If Ψ is any homomorphism of $\mathfrak{G}_a(\mathfrak{C})$ into an analytic commutative group variety $\mathfrak{A}$, then $\psi(x)=\Psi((x)-(x_0))$ is a mapping of $\mathfrak{C}$ into $\mathfrak{A}$ and we have $\Psi(\mathfrak{p})=\sum n_i\psi(x_i)$; we shall say that the homomorphism Ψ is analytic if the mapping ψ is analytic, and we observe that, in case $\mathfrak{A}$ is algebraic, the homomorphism Ψ is rational if and only if ψ is rational (over the complex ground field). It is clear that the homomorphism Φ defined above is analytic; we shall show that if Ψ is analytic and if $\mathfrak{A}$ is analytically isomorphic to a complex torus, then Ψ is the product of Φ and an analytic homomorphism of $\mathfrak{E}_g/[P]$ into $\mathfrak{A}$. We set $\mathfrak{A}=\mathfrak{E}_r/[Q]$, where $Q=(q_{ij})$ is a $(2r,r)$-matrix such that the matrix $(Q,\bar{Q})$ is non-singular, and let $z_1,\cdots,z_r$ be the coordinates of a variable point in $\mathfrak{E}_r$, so that $dz_1,\cdots,dz_r$ can be considered as Abelian differentials of the first kind on $\mathfrak{E}_r/[Q]$; it is well known that a base $\delta_1,\cdots,\delta_{2r}$ for the first Betti group of $\mathfrak{E}_r/[Q]$ can be so chosen that we have $q_{ij}=\int_{\delta_i} dz_j$. Let $d\psi_1,\cdots,d\psi_r$ be the Abelian differentials of the first kind on $\mathfrak{C}$ induced by $dz_1,\cdots,dz_r$ respectively through the mapping ψ, and let $\psi(\gamma_1),\cdots,\psi(\gamma_{2g})$ be the image cycles of $\gamma_1,\cdots,\gamma_{2g}$ respectively under ψ; then there exist a (g,r)-matrix $M=(m_{ij})$ and an integral $(2g,2r)$-matrix $N=(n_{ij})$ such that we have the relations $d\psi_i=\sum_{j=1}^{g} m_{ji}I_j$, $\psi(\gamma_i)=\sum_{j=1}^{2r} n_{ij}\delta_j$, and

$$PM=\sum_{k=1}^{g} p_{ik}m_{kj}=\sum_{k=1}^{g} m_{kj}\int_{\gamma_i} I_k=\int_{\gamma_i} d\psi_j=\int_{\psi(\gamma_i)} dz_j=\sum_{k=1}^{2r} n_{ik}\int_{\delta_k} dz_j$$
$$=\sum_{k=1}^{2r} n_{ik}q_{kj}=NQ.$$

This shows that the linear mapping of $\mathfrak{E}_g$ into $\mathfrak{E}_r$ determined by the matrix M induces an analytic homomorphism Θ of $\mathfrak{E}_g/[P]$ into $\mathfrak{E}_r/[Q]$, and we have the relation

$$\psi(x)=(\int_{x_0}^{x} d\psi_1,\cdots,\int_{x_0}^{x} d\psi_r)=(\int_{x_0}^{x} I_1,\cdots,\int_{x_0}^{x} I_g)M \pmod{[Q]},$$

or $\psi(x) = \Theta\phi(x)$, from which it follows that $\Psi = \Theta\Phi$. Now, according to a result of Lefschetz ([5], p. 368), there exists an analytic isomorphism σ of $\mathfrak{C}_g/[P]$ onto an Abelian variety $\mathfrak{J}$; if we assume that there is an analytic isomorphism τ of $\mathfrak{C}_r/[Q]$ onto an Abelian variety $\mathfrak{A}$, and if we now write ϕ, Φ, ψ, Ψ, Θ in place of $\sigma\phi$, $\sigma\Phi$, $\tau\psi$, $\tau\Psi$, $\tau\Theta\sigma^{-1}$, respectively, then we have again the relations $\psi = \Theta\phi$ and $\Psi = \Theta\Phi$. Since ϕ, ψ, and Θ are now analytic mappings of an algebraic variety into another algebraic variety, it follows from a theorem of ours proved elsewhere ([2], Theorem VII) that they are all rational transformations, and hence the homomorphisms Φ and Ψ are also rational. This shows that $\mathfrak{J}$ is the Jacobian variety of $\mathfrak{C}$ with Φ as the canonical homomorphism.

In the general case of an arbitrary ground field, it is clear that the above highly transcendental proof cannot be carried over and that new methods of a purely algebraic nature have to be introduced. A. Weil, who first proposed and studied this problem, obtained a solution by a generalization of the concept of an algebraic variety ([9], § V); in fact, the attempt to solve this problem is one of the main reasons which led Weil to introduce his notation of an "abstract" variety, an algebraic variety in abstracto in contrast to the usual ones which are embedded in a projective space. As Weil has been able to extend most of the fundamental results in algebraic geometry to the "abstract" varieties, his "abstract" Jacobian variety proves to be almost just as useful as the projective Jacobian variety in the classical case and thus constitutes a satisfactory solution of the problem. Nevertheless, it remains an important and interesting question whether a Jacobian variety exists as a projective algebraic variety in the usual sense; furthermore, what is probably even more important from the algebraic viewpoint, there remains still unsolved the question of the field of definition for the Jacobian variety and the canonical homomorphism, as has been pointed out by Weil himself ([9], p. 68). Our solution of the problem will provide satisfactory answers to both these questions; we shall show that the Jacobian variety exists as a variety in a projective space, and that both the Jacobian variety and the canonical homomorphism are defined over the defining field of the curve $\mathfrak{C}$. This latter fact, which seems to be new even in the classical case, has important applications in the theory of Picard varieties over arbitrary ground fields, as will be shown in a forthcoming paper of ours.

We shall describe briefly the main idea underlying our construction of the Jacobian variety. Since the positive divisors in $\mathfrak{C}$ can be represented by the points of an algebraic variety, it is natural to try to represent the divisor classes by associating with them certain positive divisors. According to the Riemann-Roch Theorem, each class of $\mathfrak{D}_g$ contains *in general* only one positive divisor; thus each point of $\mathfrak{C}^g$ will represent a class in $\mathfrak{D}_g$, and the representation is one-to-one except for the points of a proper subvariety in $\mathfrak{C}^g$. It is therefore natural that this variety $\mathfrak{C}^g$ has been usually taken as the starting point in the construction of the Jacobian variety; thus Weil's "abstract" Jacobian variety is essentially a collection of birational transforms of $\mathfrak{C}^g$ patched together in a suitable manner. Also van der Waerden, in a paper [7] which deals in reality with certain aspects of the Jacobian variety, has taken this variety $\mathfrak{C}^g$ as the starting point and has constructed by means of the associated forms a variety in a projective space, the points of which are in one-to-one correspondence with the classes in $\mathfrak{D}_g$; however, his results do not show that the so constructed variety is the Jacobian variety, as, for one thing, there is no proof that the variety is non-singular. We observe that in both cases the main difficulty lies in the existence of special divisor classes of degree g which are represented not by points but by subvarieties in $\mathfrak{C}^g$; one has then to cope with the fact that the divisor classes are represented by different types of geometrical entities. If now, instead of $\mathfrak{C}^g$, we consider the variety $\mathfrak{C}^n$ for sufficiently high n (say $n > 2g-2$), then there will be no special divisor classes and every divisor class of degree n will be represented by a subvariety of dimension $n-g$ in $\mathfrak{C}^n$; the method of associated forms then will enable us to represent the divisor classes by ponts of a certain variety, and from the homogeneous nature of the construction one would expect the so constructed variety to be non-singular. This is the underlying idea of the method by which we shall construct the Jacobian variety in the next section; however, the proof that the so constructed variety is non-singular is not so simple and will be given later in section 5 as an application of a general theorem proved by us elsewhere. In the final section we then introduce the canonical homomorphism and complete the proof of the existence of the Jacobian variety.

3. Construction of the variety $\mathfrak{B}$. We start with the variety $\mathfrak{C}^n$ of n dimensions, defined over the field k, for a fixed $n > 2g-2$. For any point (p) of $\mathfrak{C}^n$, such that the corresponding positive divisor $\mathfrak{p}$ is rational over a field K containing k, the Theorem of Riemann-Roch, applied to the curve $\mathfrak{C}$ over the field K, shows that the complete linear system of positive divisors on $\mathfrak{C}$ determined by the divisor $\mathfrak{p}$ has the dimension $n-g$ and is rational over the field K. In other words, the set of all positive divisors in the class $\mathfrak{P}$ determined by $\mathfrak{p}$ consists exactly of those divisors cut out on the cuve $\mathfrak{C}$, apart from a fixed component, by a linear system of $n-g+1$ hypersurfaces $\sum_{j=0}^{n-g} U_j\psi_j(X) = 0$, where the $\psi_j(X)$ are forms of the same degree in $K[X]$ and the $(U) = (U_0, \cdots, U_{n-g})$ is a system of indeterminates; this implies also that no hypersurface of this system contains the curve $\mathfrak{C}$. Let $(u) = (u_0, \cdots, u_{n-g})$ be a set of independent variables over K, and let $\mathfrak{p}_u$ be the divisor cut out on $\mathfrak{C}$, apart from the fixed component, by the generic hypersurface $\sum_{j=0}^{n-g} u_j\psi_j(X) = 0$ of the system; then the divisor $\mathfrak{p}_u$ is rational over $K(u)$ and hence is represented by a rational point (p_u) over $K(u)$ in $\mathfrak{C}^n$. It is well known that over any specialization $(u) \to (\mu)$ over K, the divisor $\mathfrak{p}_u$ specializes into the divisor $\mathfrak{p}_\mu$ cut out on $\mathfrak{C}$, apart from the fixed component, by the hypersurface $\sum_{j=0}^{n-g} \mu_j\psi_j(X) = 0$. It follows that over any specialization $(u) \to (\mu)$ over K, there is a uniquely determined specialization (p_μ) of (p_u), which represents a divisor in $\mathfrak{P}$, and that in this way all positive divisors in $\mathfrak{P}$ can be obtained. Since $K((p_u))$, being a subfield of $K(u)$, is a regular extension of K, this shows that if we denote by $\mathfrak{G}_{\mathfrak{P}}$ the set of all points in $\mathfrak{C}^n$ which represent positive divisors in $\mathfrak{P}$, then $\mathfrak{G}_{\mathfrak{P}}$ is a variety of dimension $n - g$, defined over K, and the point (p_u) is a generic point of $\mathfrak{G}_{\mathfrak{P}}$ over K. Furthermore, the correspondence $u \to (p_u)$ defines a birational transformation of the projective space $\mathfrak{S}_{n-g}$ onto $\mathfrak{G}_{\mathfrak{P}}$, defined over K; in fact, the divisor $\mathfrak{p}_u$ consists of n distinct points $(\xi^1), \cdots, (\xi^n)$, which form a complete set of separable conjugates over $K((p_u))$, and the point (u) is determined rationally by the linear conditions $\sum_{j=0}^{n-g} u_j\psi_j(\xi^i) = 0$, $i = 1, \cdots, n$. We shall denote by $G_{\mathfrak{P}}(W) = G_{\mathfrak{P}}(W^0, \cdots, W^{n-g})$ the associated form of the variety $\mathfrak{G}_{\mathfrak{P}}$, where for each $i = 0, 1, \cdots, n - g$, the $(W^i) = (W^i{}_0, W^i{}_1, \cdots, W^i{}_t)$ is a system of indeterminates in $\mathfrak{S}_t$; since the variety $\mathfrak{G}_{\mathfrak{P}}$ is defined over K, the form $G_{\mathfrak{P}}(W)$ is rational over K.

Let (y) be a generic point of $\mathfrak{C}^n$ over k, and let $\mathfrak{y}$ be the corresponding positive divisor on $\mathfrak{C}$. Since $\mathfrak{y}$ is rational over $k((y))$ and consequently the class $\mathfrak{Y}$ determined by $\mathfrak{y}$ is also rational over $k((y))$, it follows that the associated form $G_{\mathfrak{Y}}(W)$ of the variety $\mathfrak{G}_{\mathfrak{Y}}$ is rational over $k((y))$. Let d be the degree of the variety $\mathfrak{G}_{\mathfrak{Y}}$ (as a subvariety in $\mathfrak{S}_t$), and let $\Delta_0(W)$, $\Delta_1(W), \cdots, \Delta_l(W)$ be the set of all power products of degree d in (W), arranged in an order which is arbitrary but fixed once for all; if we set $G_{\mathfrak{Y}}(W) = \sum_{j=0}^{l} G_j\Delta_j(W)$, then the set of coefficients $(G_{\mathfrak{Y}}) = (G_0, G_1, \cdots, G_l)$, arranged in this fixed order, determines a point in the projective space $\mathfrak{S}_l$

of dimension l, rational over $K((y))$. Since the variety $\mathfrak{C}^n$ is defined over k, it follows that the point $(G_{\mathfrak{Y}})$ defines a variety in $\mathfrak{S}_l$ over k, which we shall denote by $\mathfrak{B}$; and the correspondence $(y) \rightarrow (G_{\mathfrak{Y}})$ determines a rational transformation ϕ of $\mathfrak{C}^n$ onto $\mathfrak{B}$, defined over k. We maintain that the rational transformation has a unique value $\phi(p)$ at every point (p) in $\mathfrak{C}^n$, and that the positive cycle $\mathfrak{G}'_{\mathfrak{P}}$ of dimension $n-g$ and degree d in $\mathfrak{S}_t$ determined by the point $\phi(p)$ consists of exactly the one variety $\mathfrak{G}_{\mathfrak{P}}$, possibly with a certain multiplicity. To show this, we observe ([7], § 14) that the property that any two points (p) and (q) in $\mathfrak{C}^n$ represent linearly equivalent divisors is an algebraic condition over k; it follows then from this and Lemma 1 (section 1) that the property that a positive cycle $\mathfrak{G}$ of dimension $n-g$ and degree d in $\mathfrak{S}_t$ is contained in the variety $\mathfrak{G}_{\mathfrak{P}}$ can be expressed as an algebraic conidtion over k in terms of the point (p) in $\mathfrak{S}_t$ and the point (G) in $\mathfrak{S}_l$ representing $\mathfrak{G}$. In other words, there exists a set of forms $A_j(Y,Z)$ in $k[Y,Z]$, where the (Y) and (Z) are systems of indeterminates in $\mathfrak{S}_t$ and $\mathfrak{S}_l$ respectively, such that the cycle $\mathfrak{G}$ is contained in $\mathfrak{G}_{\mathfrak{P}}$ if and only if the equations $A_j(p, G) = 0$ hold. It is evident that the equations $A_j(y, G_{\mathfrak{Y}}) = A_j(y, \phi(y)) = 0$ hold; if $\phi(p)$ is any specialization of $\phi(y)$ over the specialization $(y) \rightarrow (p)$ over k, then we have the relations $A_j(p, \phi(p)) = 0$. This shows that the positive cycle $\mathfrak{G}'_{\mathfrak{P}}$ determined by $\phi(p)$ can only have component varieties which are contained in $\mathfrak{G}_{\mathfrak{P}}$; and since every component in $\mathfrak{G}'_{\mathfrak{P}}$ has the same dimension $n-g$ as that of $\mathfrak{G}_{\mathfrak{P}}$, it must coincide with $\mathfrak{G}_{\mathfrak{P}}$, so that the positive cycle $\mathfrak{G}'_{\mathfrak{P}}$ is a multiple of the variety $\mathfrak{G}_{\mathfrak{P}}$ (we shall show in section 5 that $\mathfrak{G}'_{\mathfrak{P}}$ actually coincides with $\mathfrak{G}_{\mathfrak{P}}$). This evidently also shows that the specialization $\phi(p)$ is uniquely determined.

Thus we have shown that the points of the variety $\mathfrak{B}$ represent in a one-to-one manner the divisor classes of $\mathfrak{C}$, and that for any two divisors $\mathfrak{p}$ and $\mathfrak{q}$ in $\mathfrak{C}$, we have $\phi(p) = \phi(q)$ if and only if $\mathfrak{p} \sim \mathfrak{q}$. The next step is to define a composition function in $\mathfrak{B}$ which corresponds to the group addition in $\mathfrak{D}_0$, or rather the addition in $\mathfrak{D}_n$ with respect to some one fixed divisor class $\mathfrak{D}$. For this assume the integer n so chosen that there exists a divisor class $\mathfrak{D}$ in $\mathfrak{D}_n$ which is rational over k; it is easily seen that such an integer n exists. The corresponding point (G^0) on $\mathfrak{B}$ is then also rational over k. Let (G^1) and (G^2) be any two points in $\mathfrak{B}$, and let $\mathfrak{P}^1$ and $\mathfrak{P}^2$ be the divisor classes determined by (G^1) and (G^2) respectively; if (G^3) is the point in $\mathfrak{B}$ which represents the divisor class $\mathfrak{P}^3 = \mathfrak{P}^1 + \mathfrak{P}^2 - \mathfrak{D}$, then we define (G^3) as the sum of (G^1) and (G^2), and write $(G^3) = (G^1) + (G^2)$. We shall show that if (G^1) and (G^2) are two independent generic points of $\mathfrak{B}$ over k, then the point (G^3) is rational over the field $K = k((G^1), (G^2))$. We observe first that since both (G^1) and (G^2) are generic points of $\mathfrak{V}$ over k, the positive cycles $\mathfrak{G}^1$ and $\mathfrak{G}^2$ determined by them are subvarieties in $\mathfrak{C}^n$, which by Lemma 2 (section 1), are defined over the fields $k((G^1))$ and $k((G^2))$ respectively, so that both are defined over K. Let (p^1) and (p^2) be independent generic points of the varieties $\mathfrak{G}^1$ and $\mathfrak{G}^2$ respectively over K; since the divisor class $\mathfrak{P}^1$ is rational over $K((p^1))$ and the divisor class $\mathfrak{P}^2$ is rational over $K((p^2))$, the class $\mathfrak{P}^3 = \mathfrak{P}^1 + \mathfrak{P}^2 - \mathfrak{O}$ is rational over the field $K((p^1), (p^2))$ and consequently we have $k((G^3)) \subset K((p^1), (p^2))$. Now, let (q^1) and (q^2) be another pair of such independent generic points of the varieties $\mathfrak{G}^1$ and $\mathfrak{G}^2$ respectively, both independent with respect to the field $K((p^1), (p^2))$ over K; then we must also have $k((G^3)) \subset K((q^1), (q^2))$. Since both $K((p^1), (p^2))$ and $K((q^1), (q^2))$ are regular extensions of K and since they are independent with respect to each other over K, it follows that $K((p^1), (p^2)) \cap K((q^1), (q^2)) = K$. Therefore we conclude that $k((G^3)) \subset K = k((G^1), (G^2))$. In order to prove that the addition on $\mathfrak{V}$ is a rational function, we have only to show that $(G^a) + (G^b)$ is the only specialization of $(G^1) + (G^2)$ over any specialization $(G^1) \to (G^a)$, $(G^2) \to (G^b)$ over k. Here again our assertion will follow immediately from the fact that the relation $(G^c) = (G^a) + (G^b)$ is an algebraic condition over k between the three points. To show this, let $\mathfrak{P}^a$, $\mathfrak{P}^b$, $\mathfrak{P}^c$ be the divisor classes determined by the points (G^a), (G^b), (G^c) respectively, and let $\mathfrak{p}^a$, $\mathfrak{p}^b$, $\mathfrak{p}^c$ be any positive divisors in the classes $\mathfrak{P}^a$, $\mathfrak{P}^b$, $\mathfrak{P}^c$ respectively; let $\mathfrak{o}$ be a rational positive divisor over k in the divisor class $\mathfrak{O}$. We recall again ([7], § 14) that linear equivalence between positive divisors is an algebraic condition over k; since the positive divisor $\mathfrak{o}$ is rational over k and hence is represented by a rational point over k in $\mathfrak{C}^n$, the relation $\mathfrak{p}^a + \mathfrak{p}^b \sim \mathfrak{p}^c + \mathfrak{o}$ can be expressed as an algebraic condition over k in terms of the points (p^a), (p^b), (p^c). It follows then by Lemma 1 (the case where the points are not involved at all) that the relation $(G^c) = (G^a) + (G^b)$ is an algebraic condition over k.

Thus we have shown that the addition in $\mathfrak{D}_n$ (or the addition in $\mathfrak{D}_0$) induces on the variety $\mathfrak{V}$ an addition of points, which is a rational function, defined over the ground field k, on the product variety $\mathfrak{V} \times \mathfrak{V}$ with values in $\mathfrak{V}$. Moreover, this function is not only defined for a generic point of $\mathfrak{V} \times \mathfrak{V}$, but also has a uniquely determined value for any point on $\mathfrak{V} \times \mathfrak{V}$. From this it follows that the points $\mathfrak{V}$ form an abelian group under this addition operation, with the point (G^0) as the unit element. The inverse of any point (G) corresponding to the class $\mathfrak{P}$ is the point $(G)^{-1}$ corresponding to the class $2\mathfrak{O} - \mathfrak{P}$; and in case (G) is a generic point of V over k, the point $(G)^{-1}$ is rational over $k((G))$. From this it follows, by a similar argument as before, that the inverse of a point is a rational function on $\mathfrak{V}$, defined over k. We shall write $(G^1) - (G^2)$ for $(G^1) + (G^2)^{-1}$. Finally, it is also easily seen that, for any fixed point (G^a) on $\mathfrak{V}$, the mapping $(G) \to (G^a) + (G)$ defines a birational transformation of $\mathfrak{V}$ onto itself, which is one-to-one for every point of $\mathfrak{V}$. In order to prove that $\mathfrak{V}$ is an Abelian variety, it remains to shows that the function $(G^1) + (G^2)$ is defined at every point of the product variety $\mathfrak{V} \times \mathfrak{V}$; since a rational function is defined at every simple point for which it has a uniquely determined value, it is sufficient to show that the variety $\mathfrak{V}$ is non-singular. This will be done in section 5, where this result is obtained as an application of a general theorem proved by us elsewhere; in the final section we shall then introduce the canonical homomorphism and show that it has the desired properties, thus completing the proof that $\mathfrak{V}$ is a Jacobian variety of the curve $\mathfrak{C}$. Before we proceed with this, we shall insert a digression in the next section, where we shall show that a derived (absolutely) normal model $\mathfrak{W}$ of the variety $\mathfrak{V}$ is an Abelian variety and that the birational correspondence between $\mathfrak{V}$ and $\mathfrak{W}$ is one-to-one without exception; this result, together with the results of the final section on the canonical homomorphism, gives us an alternate proof of the existence of a Jacobian variety, though not necessarily defined over the original ground field k, except in case k is perfect. Although this alternate proof is now superseded by the more complete result in section 5, it still has the advantage of simplicity; furthermore, the method of proof used here may have possible applications to other similar problems.*

4. The derived normal model $\mathfrak{W}$. Consider a derived (absolutely) normal variety $\mathfrak{W}$ of $\mathfrak{V}$, which is defined over a purely inseparable extension K of the field k. The birational correspondence $\Gamma(\mathfrak{W}) = \mathfrak{V}$ between $\mathfrak{W}$ and $\mathfrak{V}$, also defined over K, has the property that to each point (H) of $\mathfrak{W}$ corresponds exactly one point $\Gamma(H)$ of $\mathfrak{V}$, while conversely to each point (G) of $\mathfrak{V}$ correspond at most a finite number of points $\Gamma^{-1}(G)$ of $\mathfrak{W}$. Let (H^1) and (H^2) be two independent generic points on $\mathfrak{W}$ over K, and let $(G^1) = \Gamma(H^1)$

* [*Added in proof, March* 10, 1954] T. Matsusaka, in a recent paper "On the algebraic construction of Picard variety (II)," *Japanese Journal of Mathematics*, vol. 22, pp. 51-62, has successfully applied this method to prove the existence of the Picard variety over an arbitrary field; the application of this method was made possible by a crucial result, proved by Matsusaka in another paper, concerning the existence of a "regular maximal algebraic family" of divisors. Matsusaka has also shown that it is sufficient to take a derived normal model with reference to k, which can then be shown to be also absolutely normal; this removes the restriction on the field of definition stated above.

and $(G^2) = \Gamma(H^2)$ be the corresponding points on $\mathfrak{V}$ under the correspondence Γ; then (G^1) and (G^2) are also generic and independent with respect to each other over K. The sum $(G^3) = (G^1) + (G^2)$ of the two independent generic points (G^1) and (G^2) on $\mathfrak{V}$, being a rational function on $\mathfrak{V} \times \mathfrak{V}$ defined over K, is carried by the birational cooorespondence Γ into the rational function $(H^3) = \Gamma^{-1}(G^3) = \Gamma^{-1}(\Gamma(H^1) + \Gamma(H^2))$ on $\mathfrak{W} \times \mathfrak{W}$, also defined over K, which we shall define as the sum $(H^1) + (H^2)$ of the two independent generic points (H^1) and (H^2) on $\mathfrak{W}$. Now, what is the behavior of this function $(H^1) + (H^2)$ for a pair of special points (H^a) and (H^b) on $\mathfrak{W}$? By definition, any value of the function $(H^1) + (H^2)$ is a specialization (H^c) of the point (H^3) over the specialization $(H^1) \to (H^a)$, $(H^2) \to (H^b)$ over K. It is well known that any such specialization $(H^1, H^2, H^3) \to (H^a, H^b, H^c)$ can always be extended to a specialization of (G^1, G^2, G^3), while on the other hand it is clear from the above that any such extension of specialization is already uniquely determined by the specialization $(H^1, H^2) \to (H^a, H^b)$ alone. In fact, since the transformation $\Gamma(H)$ has always uniquely determined specialization at every point of $\mathfrak{W}$, we have $(G^1) \to \Gamma(H^a)$, $(G^2) \to \Gamma(H^b)$; from this it follows that since the sum $(G^1) + (G^2)$ has always a uniquely determined specialization for every pair of points on $\mathfrak{V}$, we must have $(G^3) \to \Gamma(H^a) + \Gamma(H^b)$. Since, in particular, the relation $\Gamma(H^3) = (G^3)$, which defines the correspondence Γ, must be preserved under the extended specialization, we have the relation $\Gamma(H^c) = \Gamma(H^a) + \Gamma(H^b)$. It follows then that any specialization (H^c) of (H^3), over the specialization $(H^1, H^2) \to (H^a, H^b)$ over K, must be one of the finite number of points of $\mathfrak{W}$ which are carried over by Γ into the point $\Gamma(H^a) + \Gamma(H^b)$ on $\mathfrak{V}$. In similar manner, we can consider a specialization (H^b) of the point $(H^2) = \Gamma^{-1}(\Gamma(H^3) - \Gamma(H^1))$ over any given specialization $(H^1, H^3) \to (H^a, H^c)$, where the (H^1) and (H^3) are now a pair of independent generic points of $\mathfrak{W}$. Since the specialization can also be extended in a unique way to the specialization $(G^1, G^3, G^2) \to (\Gamma(H^a), \Gamma(H^c), \Gamma(H^c) - \Gamma(H^a))$, we conclude that the relation $\Gamma(H^b) = \Gamma(H^c) - \Gamma(H^a)$ must hold for any such specialization, and hence (H^b) must be one of the finite number of points of $\mathfrak{W}$ which are carried over by Γ into the point $\Gamma(H^c) - \Gamma(H^a)$ of $\mathfrak{V}$.

Thus we have shown that for any specialization $(H^1, H^2, H^3) \to (H^a, H^b, H^c)$ the relation $\Gamma(H^c) = \Gamma(H^a) + \Gamma(H^b)$ always holds, and that in this specialization any two of the three points can be chosen arbitrarily on $\mathfrak{W}$, while the third one is then one of the finite number of possible points determined by this relation. Let now (H^a) be any given point of $\mathfrak{W}$, and let (H) be a generic point of $\mathfrak{W}$ over $K((H^a))$. Since the point $\Gamma(H^a)+\Gamma(H)$ is a generic point of $\mathfrak{B}$ over $K((H^a))$, the relation $\Gamma(H')=\Gamma(H^a)+\Gamma(H)$ determines uniquely a specialization $(H^3)\to(H')$ over the specialization $(H^1,H^2)\to(H^a,H)$ over K, and this point (H') is also a generic point of $\mathfrak{W}$ over $K((H^a))$. Conversely, given a generic point (H') of $\mathfrak{W}$ over $K((H^a))$, the same relation determines uniquely a specialization $(H^2)\to(H)$ over the specialization $(H^1,H^3)\to(H^a,H')$ over K, and the point (H) is also generic over $K((H^a))$. In other words, for any given point (H^a) of $\mathfrak{W}$, the relation $(H^a)+(H)=(H')$ determines a birational correspondence $\Theta_a(H)=(H')$ of the variety $\mathfrak{W}$ onto itself; in fact we have $(H')=\Gamma^{-1}(\Gamma(H^a)+\Gamma(H))$ for any generic (H) over $K((H^a))$ and $(H)=\Gamma^{-1}(\Gamma(H')-\Gamma(H^a))$ for generic (H') over $K((H^a))$. This birational correspondence Θ_a between the normal variety $\mathfrak{W}$ and itself has the property that to each specialization of (H) correspond at most a finite number of specializations of (H'), and vice versa. For, any specialization $(H,H')\to(H^b,H^c)$ of this correspondence Θ_a over K is contained in a specialization $(H^1,H^2,H^3)\to(H^a,H^b,H^c)$ over K, and we have seen above that there is only a finite number of the latter when two of the three points (H^a,H^b,H^c) are given. This means that the birational correspondence Θ_a has no fundamental points on $\mathfrak{W}$, neither for the direct transformation Θ_a nor for the inverse transformation Θ_a^{-1}. Since for a birational transformation, or more generally for a rational transformation of a normal variety, a point is either fundamental or regular ([10], Th. 9 and Th. 10, note the difference in terminology), we conclude that the birational correspondonce Θ_a must be biregular everywhere on $\mathfrak{W}$.

From the biregularity of the correspondence Θ_a for every point (H^a) we can draw some important conclusions. Let (H^c) be a simple point on $\mathfrak{W}$ with the property that it is the only point on $\mathfrak{W}$ which is carried by Γ into the point $\Gamma(H^c)$ on $\mathfrak{B}$; we can take, for example, a generic point of $\mathfrak{W}$ over K. Given any point (H^b), let (H^a) be any point such that $\Gamma(H^c)=\Gamma(H^a)+\Gamma(H^b)$. The birational correspondence Θ_a, determined by the so chosen point (H^a), will then carry the point (H^b) into a point with the image $\Gamma(H^a)+\Gamma(H^b)$ on $\mathfrak{B}$ under the correspondence Γ, which can only be the point (H^c). Since the correspondence Θ_a is biregular and the point (H^c) is simple, it follows that the point (H^b) is also simple. As (H^b) is any point on $\mathfrak{W}$, this shows that the variety $\mathfrak{W}$ is free of singularities. Furthermore, let (H^β) be another point such that $\Gamma(H^\beta)=\Gamma(H^b)$, then the same birational correspondence Θ_a must also carry (H^β), into (H^c), from which it follows, on account of the biregularity of Θ_a, that $(H^\beta) = (H^b)$. This shows that the correspondence Γ between $\mathfrak{W}$ and $\mathfrak{V}$ is one-to-one for all points without exception.

Thus we have shown the existence of an Abelian variety $\mathfrak{W}$ which is in one-to-one birational correspondence without exception with the variety $\mathfrak{V}$; as we have mentioned in our paper [1], this is our original proof of the existence of a Jacobian variety for the curve $\mathfrak{C}$. The underlying idea of the proof is the simple observation that if a variety is homogeneous in the sense that there is a transitive group of rational transformations which are everywhere determined, and if the variety is normal, then it is non-singular. That a variety which is homogeneous in this sense, but not normal, need not be non-singular, is shown by the simple example of a plane cubic curve with a cusp, which is in one-to-one birational correspondence without exception with the projective line. The question naturally arises as to whether the variety $\mathfrak{V}$ is itself also non-singular and is already a Jacobian variety. This question, besides being of some interest in itself, is significant not only on account of the fact that $\mathfrak{V}$ is defined over k, but also on account of the fact, proved in the next section, that for any point (G) in $\mathfrak{V}$ the corresponding variety $\mathfrak{G}$ in $\mathfrak{C}^n$ is defined over the field $k((G))$.

5. Proof that the variety $\mathfrak{V}$ is non-singular. Consider the rational transformation ϕ of $\mathfrak{C}^n$ onto $\mathfrak{V}$, defined over k; if (z) is a generic point of $\mathfrak{V}$ over k, then the inverse image $\phi^{-1}(z)$ is a subvariety $\mathfrak{G}(z)$ of dimension $n-g$ in $\mathfrak{C}^n$, defined over $k((z))$, and for any specialization $(z) \to (\zeta)$ over k, the variety $\mathfrak{G}(z)$, considered as a positive cycle of dimension $n-g$, has a uniquely determined specialization $\mathfrak{G}(\zeta)$, which is a multiple of the variety $\phi^{-1}(\zeta)$. In the terminology introduced in our paper [1], the rational transformation ϕ induces an involutional system of positive cycles $|\mathfrak{G}(z)|$ on $\mathfrak{C}^n$, and the variety $\mathfrak{V}$ is the associated variety of this involutional system; according to the main theorem in that paper ([1], p. 258), the variety $\mathfrak{V}$ is simple at every point (ζ) for which the positive cycle $\mathfrak{G}(\zeta)$ coincides with the variety $\phi^{-1}(\zeta)$ and contains a simple point in $\mathfrak{C}^n$. In order to prove that the variety $\mathfrak{V}$ is non-singular, it is therefore sufficient to show that (1) the variety $\phi^{-1}(\zeta)$ has the same degree (as a subvariety in $\mathfrak{S}_t$) for every point (ζ) in $\mathfrak{V}$, i. e. the degree of the variety $\mathfrak{G}_{\mathfrak{P}}$ is the same for every divisor class $\mathfrak{P}$ of $\mathfrak{C}$, and (2) the variety $\mathfrak{C}^n$ is non-singular.

We shall show that the degree of the variety $\mathfrak{G}_{\mathfrak{P}}$, for any divisor class $\mathfrak{P}$ of degree n, is equal to h^{n-g}. By definition, the degree of $\mathfrak{G}_{\mathfrak{P}}$ is the number of intersections of $\mathfrak{G}_{\mathfrak{P}}$ with $n-g$ independent generic hyperplanes $\sum_{j=0}^{t} w^i{}_j Y_j = 0$ $(i=1,\cdots,n-g)$ in $\mathfrak{S}_t$ over K, where K is a field over which the divisor class $\mathfrak{P}$ is rational. Consider the $n-g$ independent generic hyperplanes $\sum_{j=0}^{m} v^i{}_j X_j = 0$ $(i=1,\cdots,n-g)$ in $\mathfrak{S}_m$ over K, and for each i $i=1,\cdots,n-g$, let $(\xi^{i,1}),\cdots,(\xi^{i,h})$ be the h intersections of the i-th hyperplane with the curve $\mathfrak{C}$; let $\mathfrak{p}^\lambda$, where $\lambda=(m_1,\cdots,m_{n-g})$ runs through all the h^{n-g} possible ordered sets of $n-g$ positive integers $m_i \leq h$, be the positive divisor of the class $\mathfrak{P}$ which contains the $n-g$ points (ξ^{i,m_i}), $i=1,\cdots,n-g$. It is then easily seen that the h^{n-g} points (p^λ) in $\mathfrak{G}_{\mathfrak{P}}$ corresponding to the h^{n-g} positive divisors $\mathfrak{p}^\lambda$ constitute all the intersections of $\mathfrak{G}_{\mathfrak{P}}$ with the $n-g$ special hyperplanes $\sum_{j=0}^{t} \Omega_j(v^i) Y_j = 0$ $(i=1,\cdots,n-g)$ in $\mathfrak{S}_t$. In fact, let (q) be the point in $\mathfrak{C}^n$ corresponding to a positive divisor $\mathfrak{q}$ of degree n consisting of the n points $(\eta^1),\cdots,(\eta^n)$, then we have $\sum_{j=0}^{t} \Omega_j(v^i) q_j = \prod_{k=1}^{n} (\sum_{j=0}^{m} v^i{}_j \eta^k{}_j)$, $i=1,\cdots,n-g$. Hence, all the h^{n-g} points (p^λ) of $\mathfrak{G}_{\mathfrak{P}}$ lie on the $n-g$ special hyperplanes; and conversely, if a point of $\mathfrak{G}_{\mathfrak{P}}$ lies on these $n-g$ special hyperplanes, then the corresponding positive divisors must contain at least one point from each of the $n-g$ sets $(\xi^{i,1}),\cdots,(\xi^{i,h})$, $i=1,\cdots,n-g$, and hence must be one of the h^{n-g} positive divisors $\mathfrak{p}^\lambda$. Therefore, in order to prove our assertion, we need only show that each point (p^λ) is a simple intersection of $\mathfrak{G}_{\mathfrak{P}}$ with these $n-g$ special hyperplanes. Since the point (p^λ) is a generic point of $\mathfrak{G}_{\mathfrak{P}}$ over K (it has the dimension $n-g$ over K), it is a simple point of $\mathfrak{G}_{\mathfrak{P}}$; hence, we need only show that the linear variety of dimension $t-n+g$ in $\mathfrak{S}_t$ defined by the $n-g$ special hyperplanes is transversal to the tangent linear variety of $\mathfrak{G}_{\mathfrak{P}}$ at the point (p^λ). As we shall from now on only deal with the one divisor $\mathfrak{p}^\lambda$ or point (p^λ), for an arbitrary but fixed λ, we can drop the superscript λ and simply write $\mathfrak{p}$ or (p).

Let $(x^1),\cdots,(x^n)$ be the n points contained in the divisor $\mathfrak{p}$, where we can assume that the first $n-g$ points are the points (ξ^{i,m_i}), $i=1,\cdots,n-g$. Since $\mathfrak{p}$ is a generic positive divisor of the class $\mathfrak{P}$ over K, all the n points $(x^1),\cdots,(x^n)$ are generic points of the curve $\mathfrak{C}$ over K and distinct from each other; moreover, any $n-g$ of these points are independent with respect to each other over K and determine the remaining g points uniquely. Hence, there is an automorphism of the field $K((x^1),\cdots,(x^n))$ over the field $K((p))$ which carries any set of $n-g$ of the n points $(x^1),\cdots,(x^n)$ into any other such set. If $\sum_{j=0}^{m} \nu_j X_j = 0$ is a hyperplane in $\mathfrak{S}_m$ containing one of the n points $(x^1), \cdots, (x^n)$, then the hyperplane $\sum_{j=0}^{t} \Omega_j(\nu) Y_j = 0$ in $\mathfrak{S}_t$ evidently contains the point (p); we shall call such a hyperplane in $\mathfrak{S}_t$ a derived hyperplane of the hyperplane in $\mathfrak{S}_m$. The $n-g$ special hyperplanes $\sum_{j=0}^{t} \Omega_j(v^i) Y_j = 0$ $(i = 1, \cdots, n-g)$ considered above are all derived hyperplanes, since each hyperplane $\sum_{j=0}^{m} v^i{}_j X_j = 0$ in $\mathfrak{S}_m$ contains the point (x^i), $i = 1, \cdots, n-g$. Moreover, for each $i = 1, \cdots, n-g$, the hyperplane $\sum_{j=0}^{m} v^i{}_j X_j = 0$ is a generic hyperplane over $K((x^1), \cdots, (x^n))$ of the system of all hyperplanes in $\mathfrak{S}_m$ which contain the point (x^i); and the $n-g$ hyperplanes $\sum_{j=0}^{m} v^i{}_j X_j = 0$ $(i = 1, \cdots, n-g)$, considered as generic hyperplanes of the $n-g$ corresponding systems over $K((x^1), \cdots, (x^n))$, are independent with respect to each other over $K((x^1), \cdots, (x^n))$.

Consider now the set of all derived hyperplanes in $\mathfrak{S}_t$; it can be easily seen that the point (p) is the only point in $\mathfrak{S}_t$ common to all these hyperplanes. For, each point in $\mathfrak{S}_t$ corresponds to a form of degree n in (V), and the form $\sum_{j=0}^{t} p_j \Omega_j(V) = \prod_{i=1}^{n} (\sum_{j=0}^{m} x^i{}_j V_j)$ is obviously the only form of degree n in (V) which vanishe for all the solutions of any one of the n linear equations $\sum_{j=0}^{m} x^i{}_j V_j = 0$, $i = 1, \cdots, n$. Thus there exist t linearly independent derived hyperplanes in $\mathfrak{S}_t$, and consequently there is a set of $n-g$ hyperplanes $\sum_{j=0}^{m} \nu^i{}_j X_j = 0$ $(i = 1, \cdots, n-g)$ in $\mathfrak{S}_m$, each containing a point (x^{l_i}) of the divisor $\mathfrak{p}$, such that the $n-g$ derived hyperplanes $\sum_{j=0}^{t} \Omega_j(\nu^i) Y_j = 0$ $(i = 1, \cdots, n-g)$ define a linear variety of dimension $t-n+g$ transversal to the tangent linear variety of $\mathfrak{G}_{\mathfrak{P}}$ at (p). The $n-g$ points (x^{l_i}), $i = 1, \cdots, n-g$, are distinct; for, otherwise there would be infinitely many positive divisors in the class $\mathfrak{P}$ which contain all the (less than $n-g$ distinct) points (x^{l_i}), $i = 1, \cdots, n-g$, and the corresponding points in $\mathfrak{G}_{\mathfrak{P}}$ would then be contained in the $n-g$ derived hyperplanes $\sum_{j=0}^{t} \Omega_j(\nu^i) Y_j = 0$ $(i = 1, \cdots, n-g)$, and consequently these derived hyperplanes could not define a linear variety transversal to the tangent linear variety of $\mathfrak{G}_{\mathfrak{P}}$ at (p). Thus, for each $i = 1, \cdots, n-g$, the set

(ν^i) is a solution of the linear equation $\sum_{j=0}^{m} x^{l_i}{}_j V_j = 0$. Let $(\bar{v}^i)$, $i = 1, \cdots, n-g$, be respectively generic solutions of the linear equations $\sum_{j=0}^{m} x^{l_i}{}_j V_j = 0$, $i = 1, \cdots, n-g$, over $K((x^1), \cdots, (x^n))$, independent with respect to each other over $K((x^1), \cdots, (x^n))$; then the $n-g$ derived hyperplanes $\sum_{j=0}^{t} \Omega_j(\bar{v}^i) Y_j = 0$ $(i = 1, \cdots, n-g)$ define a linear variety which is also transversal to the tangent linear variety of $\mathfrak{G}_{\mathfrak{P}}$ at (p). For, the set of $n-g$ hyperplanes $\sum_{j=0}^{t} \Omega_j(\nu^i) Y_j = 0$ $(i = 1, \cdots, n-g)$ is a specialization over $K((x^1), \cdots, (x))$ of the set of $n-g$ hyperplanes $\sum_{j=0}^{t} \Omega_j(\bar{v}^i) Y_j = 0$ $(i = 1, \cdots, n-g)$, and the former set defines a linear variety transversal to the tangent linear variety of $\mathfrak{G}_{\mathfrak{P}}$ at (p). Now, there is an automorphism of the field $K((x^1), \cdots, (x^n))$ over $K((p))$, which carries the $n-g$ points $(x^{l_1}), \cdots, (x^{l_{n-g}})$ into the $n-g$ points $(x^1), \cdots, (x^{n-g})$ respectively, and this automorphism can be extended into an isomorphism between the fields $K((x^1), \cdots, (x^n), (\bar{v}^1), \cdots, (\bar{v}^{n-g}))$ and $K((x^1), \cdots, (x^n), (v^1), \cdots, (v^{n-g}))$ by the correspondence $(\bar{v}^i) \leftrightarrow (v^i)$, $i = 1, \cdots, n-g$. In this isomorphism the derived hyperplanes $\sum_{j=0}^{t} \Omega_j(\bar{v}^i) Y_j = 0$ $(i = 1, \cdots, n-g)$ are carried over into the derived hyperplanes $\sum_{j=0}^{t} \Omega_j(v^i) Y_j = 0$ $(i = 1, \cdots, n-g)$; since the former set defines a linear variety of $\mathfrak{G}_{\mathfrak{P}}$ at (p), the same must be true of the latter. This completes the proof of the assertion that the degree of $\mathfrak{G}_{\mathfrak{P}}$ is h^{n-g}.

Next, we shall show that the variety $\mathfrak{C}^n$ is non-singular. We observe first that in a similar manner as above, by taking (p) to be a generic point of $\mathfrak{C}^n$ over k and considering n instead of $n-g$ derived hyperplanes in $\mathfrak{S}_t$, we can show that the degree of the variety $\mathfrak{C}^n$ is equal to h^n. In fact, the proof in this case is somewhat simpler than the above, for there it is not necessary to consider any automorphisms of the field $k((x^1), \cdots, (x^n))$. Now, let (q) be any point in $\mathfrak{C}^n$ and let the corresponding positive divisor $\mathfrak{q}$ be composed of the n (not necessarily distinct) points $(\eta^1), \cdots, (\eta^n)$ in $\mathfrak{C}$. Consider the n linear equations $\sum_{j=0}^{m} \eta^i{}_j V_j = 0$ $(i = 1, \cdots, n)$; let (ρ^i) $(i = 1, \cdots, n)$ be n generic solutions of these n equations, respectively, over $k((\eta^1), \cdots, (\eta^n))$, independent with respect to each other over $k((\eta^1), \cdots, (\eta^n))$; for each $i = 1, \cdots, n$, let $(\eta^{i,1}) = (\eta^i), (\eta^{i,2}), \cdots, (\eta^{i,h})$

be the h intersections of the hyperplane $\sum_{j=0}^{m} \rho^i{}_j X_j = 0$ with the curve $\mathfrak{C}$. Since $\mathfrak{C}$ is non-singular (and the hyperplanes have been chosen as above), the points $(\eta^{i,j})$ $(i = 1, \cdots, n; j = 2, \cdots, h)$ are all distinct from each other and different from the points $(\eta^{i,1})$ $(i = 1, \cdots, n)$. It follows then that all the h^n positive divisors $\mathfrak{q}^\lambda = ((\eta^{1,m_1}), \cdots, (\eta^{n,m_n}))$, where the $\lambda = (m_1, \cdots, m_n)$ runs through all the h^n possible ordered sets of n positive integers $m_i \leq h$ $(i = 1, \cdots, n)$, are distinct from each other. It is easily seen that the h^n corresponding distinct points (q^λ) are precisely the intersections of $\mathfrak{C}^n$ with the linear variety of dimension $t - n$ in $\mathfrak{S}_t$ defined by the hyperplanes $\sum_{j=0}^{t} \Omega_j(\rho^i) Y_j = 0$ $(i = 1, \cdots, n)$. Since the degree of $\mathfrak{C}^n$ is h^n, it follows that each point (q^λ) and hence in particular the point (q) is a simple intersection of $\mathfrak{C}^n$ with this linear variety. This shows that (q) is a simple point in $\mathfrak{C}^n$, and as (q) is an arbitrary point in $\mathfrak{C}^n$, this means that the variety $\mathfrak{C}^n$ is non-singular.

Finally, we observe that in exactly the same manner as above, we can show that there exists a linear variety of dimension $t - n + g$ in $\mathfrak{S}_t$ which intersects simply any given $\mathfrak{G}_{\mathfrak{P}}$ at any given point (q); it follows then that the variety $\mathfrak{G}_{\mathfrak{P}}$ is also non-singular for every $\mathfrak{P}$.

6. The canonical homomorphism. Consider again the rational transformation ϕ of $\mathfrak{C}^n$ onto $\mathfrak{V}$; we shall show that it generates in a natural way (by linear extension, in the terminology of Weil) a rational homomorphism Φ of $\mathcal{G}_a(\mathfrak{C})$ onto the Abelian variety $\mathfrak{V}$ with the subgroup $\mathcal{G}_l(\mathfrak{C})$ as the kernel, and that the so defined homomorphism Φ is defined over k and has the "universal mapping" property. We observe first that if $\mathfrak{p}^1, \cdots, \mathfrak{p}^s$ and $\mathfrak{q}^1, \cdots, \mathfrak{q}^s$ are two sets of positive divisors of degree n in $\mathfrak{C}$, then we have the relation $\sum_{i=1}^{s} \phi(p^i) = \sum_{i=1}^{s} \phi(q^i)$ if and only if $\sum_{i=1}^{s} \mathfrak{p}^i \sim \sum_{i=1}^{s} \mathfrak{q}^i$; in fact, let $\mathfrak{o}$ be any divisor in the class $\mathfrak{D}$ and let $\mathfrak{p}^0$ and $\mathfrak{q}^0$ be respectively positive divisors such that the relations $\sum_{i=1}^{s} \mathfrak{p}^i \sim (s-1)\mathfrak{o} + \mathfrak{p}^0$ and $\sum_{i=1}^{s} \mathfrak{q}^i \sim (s-1)\mathfrak{o} + \mathfrak{q}^0$ hold respectively, then we have, by definition, the relations $\phi(p^0) = \sum_{i=1}^{s} \phi(p^i)$ and $\phi(q^0) = \sum_{i=1}^{s} \phi(q^i)$, and it is clear that $\phi(p^0) = \phi(q^0)$ if and only if $\sum_{i=1}^{s} \mathfrak{p}^i \sim \sum_{i=1}^{s} \mathfrak{q}^i$. We shall call this relation the compatibility condition for the transformation ϕ. Now, let $(\xi^1),\cdots,(\xi^{n-1})$ be $n-1$ independent generic points over k in $\mathfrak{C}$, and let (x) be a generic point of $\mathfrak{C}$ over $k((\xi^1),\cdots,(\xi^{n-1}))$; let (y) be the point in $\mathfrak{C}^n$ representing the divisor consisting of the points $(\xi^1),\cdots,(\xi^{n-1})$ and (x), and let ϕ' be the rational transformation of $\mathfrak{C}$ into $\mathfrak{B}$ determined by the correspondence $(x)\rightarrow\phi(y)$, which is defined over $k((\xi^1),\cdots,(\xi^{n-1}))$. For any divisor $\mathfrak{p}=\sum_{i=1}^{s} n_i(x^i)$ in $\mathfrak{G}_a(\mathfrak{C})$, we define $\Phi(\mathfrak{p})=\sum_{i=1}^{s} n_i\phi'(x^i)$; it is clear that the so defined mapping Φ is a rational homomorphism of $\mathfrak{G}_a(\mathfrak{C})$ into $\mathfrak{B}$, defined over $k((\xi^1),\cdots,(\xi^{n-1}))$. It can be easily seen from the compatibility condition that the kernel of the homomorphism Φ is precisely the subgroup $\mathfrak{G}_l(\mathfrak{C})$, and that this definition of Φ is actually independent of the choice of the points $(\xi^1),\cdots,(\xi^{n-1})$. If $(\eta^1),\cdots,(\eta^{n-1})$ are $n-1$ independent generic points of $\mathfrak{C}$ over k, independent with respect to the points $(\xi^1),\cdots,(\xi^{n-1})$ over k, then the rational homomorphism Φ is also defined over $k((\eta^1),\cdots,(\eta^{n-1})$; it follows then that Φ must be defined over

$$k((\xi^1),\cdots,(\xi^{n-1}))\cap k((\eta^1),\cdots,(\eta^{n-1}))=k.$$

In fact, if $\mathfrak{p}$ is any divisor in $\mathfrak{C}$, rational over any extension K of k, then the point $\Phi(\mathfrak{p})$ is rational over both $K((\xi^1),\cdots,(\xi^{n-1})$ and $K((\eta^1),\cdots,(\eta^{n-1}))$ and hence must be rational over

$$K((\xi^1),\cdots,(\xi^{n-1}))\cap K((\eta^1),\cdots,(\eta^{n-1}))=K.$$

Consider now a rational homomorphism Ψ of $\mathfrak{G}_a(\mathfrak{C})$ into an Abelian variety $\mathfrak{A}$, defined over an extension K of k; let (z) be a generic point of $\mathfrak{B}$ over K, and let (q) be a generic point of the variety $\phi^{-1}(z)$ over $K((z))$ and $\mathfrak{q}$ be the positive divisor represented by (q). Since the divisor class $\mathfrak{D}$ is rational over K, it contains a divisor $\mathfrak{o}$ which is rational over K; the point $\Psi(\mathfrak{q}-\mathfrak{o})$ is then rational over $K((q),(z))$. The correspondence $(q)\rightarrow\Psi(\mathfrak{q}-\mathfrak{o})$ then determines over $K((z))$ a rational transformation of $\phi^{-1}(z)$ into $\mathfrak{A}$; since $\phi^{-1}(z)$ is birationally equivalent to a projective space, it follows from a result of Weil ([9], Th. 8, Cor.) that this rational transformation must be a constant, so that the point $\Psi(\mathfrak{q}-\mathfrak{o})$ is rational over $K((z))$. The correspondence $(z)\rightarrow\Psi(\mathfrak{q}-\mathfrak{o})$ then determines a rational transformation Ψ_1 of $\mathfrak{B}$ into $\mathfrak{A}$, defined over K, which carries the unit element in $\mathfrak{B}$ into the unit element in $\mathfrak{A}$; according to another result of Weil ([9], Th. 9), this rational transformation Ψ_1 must be a homomorphism of $\mathfrak{B}$ into $\mathfrak{A}$. Let $\mathfrak{p}=\sum_{i=1}^{s} n_i(x^i)$ be again any divisor in $\mathfrak{G}_a(\mathfrak{C})$, and for each

$i = 1, \cdots, n$, let $\mathfrak{y}^i$ be the positive divisor of degree n consisting of the points $(\xi^1), \cdots, (\xi^{n-1})$ and (x^i); since $\sum_{i=1}^{s} n_i = 0$, we have evidently the equation $\mathfrak{p} = \sum_{i=1}^{s} n_i(\mathfrak{y}^i - \mathfrak{o})$. We have then

$$\Psi(\mathfrak{p}) = \sum_{i=1}^{s} n_i \Psi(\mathfrak{y}^i - \mathfrak{o}) = \sum_{i=1}^{s} n_i \Psi_1(\phi'(x^i)) = \Psi_1(\sum_{i=1}^{s} n_i \phi'(x^i)) = \Psi_1 \Phi(\mathfrak{p}).$$

Thus we have shown that Φ is the canonical homomorphism, and this completes the proof that $\mathfrak{V}$ is the Jacobian variety of the curve $\mathfrak{C}$.

Once we have obtained the canonical homomorphism Φ, we can derive the rational transformation ϕ from it by the formula $\phi(q) = \Phi(\mathfrak{q} - \mathfrak{o})$, and this formula holds for every positive integer $n > 2g - 2$, provided there exists a rational divisor $\mathfrak{o}$ over k of degree n. In fact, in order that the rational transformation ϕ be defined over k, it is sufficient that $\mathfrak{o}$ is contained in a class $\mathfrak{O}$ such that the variety $\mathfrak{G}_{\mathfrak{O}}$ is defined over k. In general, we shall say that a divisor class $\mathfrak{P}$ in $\mathfrak{D}_n$ $(n > 2g - 2)$ is rationally defined over a field K if the variety $\mathfrak{G}_{\mathfrak{P}}$ is defined over K; and an arbitrary divisor class is said to be rationally defined over K if it can be expressed as the difference of two divisor classes which are rationally defined over K. It is clear that a rational divisor class over K is also rationally defined over K, but the converse of this statement is not generally true, as can be shown by examples. It can be easily shown that, for any positive integer n, if there exists a rationally defined divisor class $\mathfrak{O}$ over k of degree n, then the formula $\phi(q) = \Phi(\mathfrak{q} - \mathfrak{o})$ determines a rational transformation ϕ of $\mathfrak{C}^n$ into $\mathfrak{V}$, defined over k. We maintain now that, conversely, if there is a rational transformation of $\mathfrak{C}^n$ into $\mathfrak{V}$, defined over k, which generates by linear extension the canonical homomorphism Φ, then there exists a rationally defined divisor class over k of degree n in $\mathfrak{C}$. In fact, for every positive integer s, the rational transformation ϕ induces a rational transformation ϕ_s of $\mathfrak{C}^{ns}$ into $\mathfrak{V}$, determined by the formula $\phi_s(p) = \sum_{i=1}^{s} \phi(p^i)$, where (p) is any point in $\mathfrak{C}^{ns}$ and $(p^1), \cdots, (p^s)$ are s points in $\mathfrak{C}^n$ such that we have $\mathfrak{p} = \sum_{i=1}^{s} \mathfrak{p}^i$ for the corresponding positive divisors, the particular choice of the points $(p^1), \cdots, (p^s)$ being immaterial on account of the compatibility condition. It is clear that the rational transformation ϕ_s is defined over k, and that, again by virtue of the compatibility condition, we have the equation $\phi_s^{-1}(\phi^s(p)) = \mathfrak{G}_{\mathfrak{P}}$, where $\mathfrak{P}$ is the divisor class determined by $\mathfrak{p}$. It is also clear that for sufficiently large s, say $s \geq g$, the rational transformation ϕ_s is onto; if $\mathfrak{P}_s$ is the divisor class of degree ns such that $\phi_s^{-1}(0) = \mathfrak{G}_{\mathfrak{P}_s}$, then $\mathfrak{P}_s$ is evidently rationally defined over k. The divisor class $\mathfrak{P}_{s+1} - \mathfrak{P}_s$, for any sufficiently large s, is then a rationally defined divisor class over k of degree n. In case $n = 1$, we have the "canonical function" of Weil, and our result offers an answer to the question raised by Weil ([9], p. 68) as to when a field of definition is "complete" for a curve.

Finally, we shall add a remark concerning the special case when the curve $\mathfrak{C}$ contains a rational divisor over k of degree 1; such is always the case, for example, when the field k is finite, as has been shown by F. K. Schmidt ([6], p. 27). We shall show that in this case every rationally defined divisor class over any field K (containing k) is also rational over K, so that a divisor class $\mathfrak{P}$ is rational over the field $k((G_{\mathfrak{P}}))$. To prove this, let $\mathfrak{o}$ be a divisor of degree 1, rational over k, and let $\mathfrak{O}$ be the class determined by $\mathfrak{o}$. Let $\mathfrak{P}$ be a rationally defined class of degree n over K, and let (p) be a generic point of $\mathfrak{G}_{\mathfrak{P}}$ over K. Then the corresponding positive divisor $\mathfrak{p}$ is rational over $K((p))$, and consequently the class $\mathfrak{P}$ is rational over $K((p))$. There exists an integer r, such that the class $\mathfrak{Q} = \mathfrak{P} - r\mathfrak{O}$ of degree $n - r$ has exactly the dimension 0 and hence contains exactly one positive divisor $\mathfrak{q}$. Since the class $\mathfrak{Q}$ is rational over $K((p))$, the divisor $\mathfrak{q}$ is rational over $K((p))$. Now, let (p') be another generic point of $\mathfrak{G}_{\mathfrak{P}}$ over K, independent with respect to (p) over K; then, by a similar argument, we can conclude that $\mathfrak{q}$ is rational over $K((p'))$. It follows then that the positive divisor $\mathfrak{q}$ is rational over $K((p)) \cap K((p')) = K$, and consequently the class $\mathfrak{Q}$ is rational over K. This implies that the class $\mathfrak{P} = \mathfrak{Q} + r\mathfrak{O}$ is also rational over K.

REFERENCES.

[1] W. L. Chow, "Algebraic systems of positive cycles in an algebraic variety," *American Journal of Mathematics*, vol. 72 (1950), pp. 247-283.

[2] ———, "On compact complex analytic varieties," *American Journal of Mathematics*, vol. 71 (1949), pp. 893-914.

[3] ———, "On the defining field of a divisor in an algebraic variety," *Proceedings of the American Mathematical Society*, vol. 1 (1950), pp. 797-799.

[4] ——— and B. L. van der Waerden, "Zur algebraischen Geometrie. IX. Ueber

zugeordnete Formen und algebraische Systeme von algebraischen Mannigfaltigkeiten," *Mathematische Annalen*, vol. 113 (1937), pp. 692-704.

[5] S. Lefschetz, "On certain numerical invariants of algebraic varieties with application to Abelian varieties," *Transactions of the American Mathematical Society*, vol. 22 (1921), pp. 327-482.

[6] F. K. Schmidt, "Analytische Zahlentheorie in Koerpern der Charakteristik *p*," *Mathematische Zeitschrift*, vol. 33 (1931), pp. 1-32.

[7] B. L. van der Waerden, "Divisorenklassen in algebraischen Funktionenkoerpern," *Commentarii Mathematici Helvetici*, vol. 20 (1947), pp. 68-109.

[8] A. Weil, *Foundations of Algebraic Geometry*, New York, 1946.

[9] ———, *Varietes abeliennes et courbes algebriques*, Paris, 1948.

[10] O. Zariski, "Foundations of a general theory of birational correspondences," *Transactions of the American Mathematical Society*, vol. 53 (1943), pp. pp. 490-542.

REMARKS ON MY PAPER "THE JACOBIAN VARIETY OF AN ALGEBRAIC CURVE."*

By Wei-Liang Chow.

We should like to make two remarks on this paper, which was published in vol. 76 (1954), pp. 453-76, of this Journal.

(1) In section 5, at the bottom of p. 469, we have stated that if $\mathfrak{P}$ is a divisor class of degree $n > 2g$, rational over an extension K of k, and if $(x^1), \cdots, (x^n)$ are the n points contained in a generic positive divisor $\mathfrak{p}$ of $\mathfrak{P}$ over K, then all the n points $(x^1), \cdots, (x^n)$ are generic points of $\mathfrak{C}$ over K, distinct from each other, and any $n-g$ of these n points are independent with respect to each other over K and determine the remaining g points uniquely. On the basis of this statement it can be easily deduced that there is an automorphism of the field $K((x^1), \cdots, (x^n))$ over the field K which carries any set of $n-g$ of the n points $(x^1), \cdots, (x^n)$ into any other such set; furthermore, it is clear that such an automorphism over K is also an automorphism over $K((p))$, so that the Galois group of $K((x^1)), \cdots, (x^n))$ over $K((p))$ is $(n-g)$-fold transitive. Professor J. Igusa has kindly pointed out to us that it is by no means obvious that any $n-g$ points among the n points $(x^1), \cdots, (x^n)$ are independent with respect to each other over K, and that a formal proof of this statement is desirable, although it is sufficient for the purpose we had there to prove this for the special case where $\mathfrak{p}$ is a generic positive divisor of degree n over k in $\mathfrak{C}$, in which case the proof follows easily from the fact that the Galois group of $k((x^1), \cdots, (x^n))$ over $k((p))$ is the entire symmetric group of permutations of the n points $(x^1), \cdots, (x^n)$. We shall give here a simple proof of our statement as follows.

For any point (η) in $\mathfrak{C}$, we denote by $\mathfrak{C}^n(\eta)$ the subset in $\mathfrak{C}^n$ consisting of all points which represent positive divisors of degree n in $\mathfrak{C}$ having the point (η) as a component. It is clear that $\mathfrak{C}^n(\eta)$ is a subvariety of dimension $n-1$ in $\mathfrak{C}^n$, defined over $k((\eta))$; in fact, if (q) is the point in $\mathfrak{C}^n$ which represents the divisor consisting of the point (η) and $n-1$ independent generic points of $\mathfrak{C}$ over $k((\eta))$, then $\mathfrak{C}^n(\eta)$ is the locus of (q)

* Received December 1, 1957.

over $k((\eta))$. Furthermore, it is easily seen that if (x) is a generic point of $\mathfrak{C}$ over k, then $\mathfrak{C}^n(\eta)$ is the unique specialization of $\mathfrak{C}^n(x)$ over the specialization $(x) \to (\eta)$ over k. In order to prove our statement, it is sufficient to show that if the $n - g$ points $(x^1), \cdots, (x^{n-g})$ in $\mathfrak{p}$ are independent generic points of $\mathfrak{C}$ over K, then the same is true for the $n - g$ points $(x^1), \cdots, (x^{n-g-1}), (x^s)$, for any choice of the point (x^s) among the remaining g points in $\mathfrak{p}$ (always under the assumption $n > 2g$). To show this, we shall assume that it is possible to choose the point (x^s) so that it is algebraic over $K((x^1), \cdots, (x^{n-g-1}))$, and we shall show that this leads to a contradiction. Since $(x^1), \cdots, (x^{n-g-1})$ are independent generic points of $\mathfrak{C}$ over K, the set of all positive divisor in the residual class of $\mathfrak{P}$ with respect to the divisor $\sum_{i=1}^{n-g-1} (x^i)$ is a complete linear system of dimension 1, rational over $K((x^1), \cdots, (x^{n-g-1}))$, and hence is represented in $\mathfrak{C}^{g+1}$ by a subvariety $\mathfrak{E}$ of dimension 1, defined over $K((x^1), \cdots, (x^{n-g-1}))$. The assumption that (x^s) is algebraic over $K((x^1), \cdots, (x^{n-g-1}))$ implies that (x^s) is a fixed point of this complete linear system, so that we have the relation $\mathfrak{E} \subset \mathfrak{C}^{g+1}(x^s)$. Consider now the simultaneous specialization $(x^i) \to (x^{n-i+1})$, for $i = 1, \cdots, n - g - 1$, over K, and let $\mathfrak{E}'$ and (ξ) be specializations of the positive 1-cycle $\mathfrak{E}$ and the point (x^s) over this specialization over K; then every component $\mathfrak{E}_0'$ in $\mathfrak{E}'$ is a subvariety of dimension 1 and we have the relation $\mathfrak{E}_0' \subset \mathfrak{C}^{g+1}(\xi)$. If (q) is any point in $\mathfrak{E}_0'$ and $\mathfrak{q}$ is the positive divisor in $\mathfrak{C}$ represented by (q), then the positive divisor $\mathfrak{q} + \sum_{i=1}^{n-g-1} (x^{n-i+1})$ is a specialization over K of a positive divisor in $\mathfrak{P}$ and hence is contained in $\mathfrak{P}$; it follows that every such positive divisor $\mathfrak{q}$ is an element in the residual class of $\mathfrak{P}$ with respect to the divisor $\sum_{i=1}^{n-g-1} (x^{n-i+1})$. Now, the set of all positive divisors in this last mentioned residual class is the complete linear system determined by the positive divisor $\sum_{i=1}^{g+1} (x^i)$ and since $n - g \geqq g + 1$, the points $(x^1), \cdots, (x^{g+1})$ are independent generic points of $\mathfrak{C}$ over K, and hence the divisor $\sum_{i=1}^{g+1} (x^i)$ is a generic positive divisor of degree $g + 1$ over K in $\mathfrak{C}$. This implies in particular that this complete linear system has the dimension 1 and hence is represented in $\mathfrak{C}^{g+1}$ by a subvariety $\mathfrak{E}''$ of dimension 1; since $\mathfrak{E}_0' \subset \mathfrak{E}''$ and both varieties have the same dimension, we conclude that $\mathfrak{E}_0' = \mathfrak{E}''$ and hence $\mathfrak{E}'' \subset \mathfrak{C}^{g+1}(\xi)$. This means that the complete linear system determined by the generic positive divisor $\sum_{i=1}^{g+1} (x^i)$ of degree $g + 1$ over K in $\mathfrak{C}$ has a fixed point, which is evidently impossible.

(2) In the last paragraph of our paper, on p. 475, we have given a proof of the statement that if the curve $\mathfrak{C}$ contains a rational divisor $\mathfrak{o}$ over k of degree 1, then every rationally defined divisor class over any extension K of k is also rational over K. Dr. P. Roquette has kindly called our attention to the fact that the proof given there holds only for the case where $\mathfrak{o}$ is a positive divisor, or rather a prime divisor, so that there is in this sense a gap in our proof. However, this gap can be easily remedied as follows. In fact, it is sufficient to prove the following statement: If $\mathfrak{C}$ contains a rational positive divisor $\mathfrak{o}$ over k of degree s and if $\mathfrak{P}$ is a rationally defined divisor class over K, then the divisor class $s\mathfrak{P}$ is rational over K. Now, if $(x^1), \cdots, (x^s)$ are the s points (which need not be distinct) contained in $\mathfrak{o}$, then, by what we have proved there, there exists a rational divisor $\mathfrak{p}_i$ over $K((x^i))$ in $\mathfrak{P}$ for each $i = 1, \cdots, s$; the divisor $\sum_{i=1}^{s} \mathfrak{p}_i$ is evidently rational over K and is contained in sP.

THE JOHNS HOPKINS UNIVERSITY.

ON ABELIAN VARIETIES OVER FUNCTION FIELDS*

BY WEI-LIANG CHOW

INSTITUTE FOR ADVANCED STUDY AND JOHNS HOPKINS UNIVERSITY

Communicated by O. Zariski, May 3, 1955

In a recent paper[1] on Abelian varieties over function fields, we have shown that to every algebraic system of Abelian varieties, defined over a field K, can be assigned two invariants, called the *K-image* and the *K-trace*. Let $K(u)$ be a primary extension of K, and let A^* be an Abelian variety defined over $K(u)$; then the K-image A of A^* over $K(u)$ is an Abelian variety over K characterized by the existence of a rational homomorphism F (called the *canonical homomorphism*) of A^* onto

A, defined over $K(u)$, such that, if H is a rational homomorphism of A^* into an Abelian variety B, and if B is defined over a separably generated extension K_1 of K, independent with respect to $K(u)$ over K, and H is defined over $K_1(u)$, then H is the product of F and a rational homomorphism of A into B, defined over K_1. The K-trace A' of A^* over $K(u)$ is defined in case $K(u)$ is a regular extension of K, and is an Abelian variety over K characterized by the existence of a rational isomorphism F (called the *canonical isomorphism*) of A' into A^*, defined over $K(u)$, such that if H is a rational homomorphism of an Abelian variety B into A^*, and if B is defined over an extension K_1 of K, independent with respect to $K(u)$ over K, and H is defined over $K_1(u)$, then H is the product of F and a rational homomorphism of B into A', defined over K_1. We observe that, while the K-trace is defined only for the case of a regular extension $K(u)$ of K, it possesses as a compensation the stronger property that the field K_1 involved is any extension of K which is independent with respect to $K(u)$ over K, not necessarily a separably generated one. Furthermore, as we have shown in Paper AF (Theorem 8, Corollary), the K-trace has the following property:

Let $u_1, \ldots, u_m$ be independent generic specializations of u over K, and let $A_1^*, \ldots, A_m^*$ and $F_1, \ldots, F_m$ be the corresponding specializations of A^* and F, respectively; if y is a generic point of A' over $K(u_1, \ldots, u_m)$, then the correspondence $y \rightarrow (F_1(y), \ldots, F_m(y))$ defines a regular isomorphism of A' into the product variety $\prod_{i=1}^{m} A_i^*$, provided that m is taken sufficiently large.

As we have indicated at the end of AF, there naturally arises the question whether a similar result also holds for the K-image A of A^* over $K(u)$ in case $K(u)$ is a regular extension of K; that is, if $(x_1, \ldots, x_m)$ is a generic point of the product variety $\prod_{i=1}^{m} A_i^*$ over $K(u_1, \ldots, u_m)$, whether the correspondence $(x_1, \ldots, x_m) \rightarrow \sum_{i=1}^{m} F_i(x_i)$ defines a regular homomorphism of $\prod_{i=1}^{m} A_i^*$ onto A, provided that m is sufficiently large. The purpose of the present note is to show that the answer to this question is in the affirmative. This result is significant, for in many applications it is advantageous to be able to deal with a regular homomorphism, and we do not know whether the canonical homomorphism F is regular; furthermore, as we shall see later, it follows easily from this result that the field K_1 mentioned above can be any extension of K which is independent with respect to $K(u)$ over K, in case $K(u)$ is a regular extension of K, so that in this case also the K-image is independent of any extension of the ground field K.

We recall that an extension K_1 of K is said to be *primary* if every element in K_1 which is separably algebraic over K is contained in K. We note that, if $K_1, \ldots, K_s$ are independent primary extensions of K, then the compositum $K_1 \ldots K_s$ is also a primary extension of K. In fact, there exist subfields $K_1', \ldots, K_s'$ in $K_1, \ldots, K_s$, respectively, such that each K_i' is a regular extension of K and K_i is a purely inseparable extension of K_i'; since $K_1', \ldots, K_s'$ are independent extensions of K, the compositum $K_1' \ldots K_s'$ is also a regular extension of K, and it is clear that $K_1 \ldots K_s$ is a purely inseparable extension of $K_1' \ldots K_s'$.

THEOREM. *Let $K(u)$ be a regular extension of K, and let A^* be an Abelian variety defined over $K(u)$; let A be the K-image of A^* over $K(u)$, and let F be the canonical* *homomorphism of A^* onto A. Let $u_1, \ldots, u_m$ be independent generic specializations of u over K, and let $A_1^*, \ldots, A_m^*$ and $F_1, \ldots, F_m$ be the corresponding specializations of A^* and F, respectively; if $(x_1, \ldots, x_m)$ is a generic point of the product variety $\prod_{i=1}^{m} A_i^*$ over $K(u_1, \ldots, u_m)$, then the correspondence $(x_1, \ldots, x_m) \to \sum_{i=1}^{m} F_i(x_i)$ defines a regular homomorphism of $\prod_{i=1}^{m} A_i^*$ onto A, provided that m is taken sufficiently large.*

Proof: For convenience, we set $K_m = K(u_1, \ldots, u_m)$ and $A_{(m)}^* = \prod_{i=1}^{m} A_i^*$, and we denote by $F_{(m)}$ the rational homomorphism of $A_{(m)}^*$ onto A defined by the correspondence $(x_1, \ldots, x_m) \to \sum_{i=1}^{m} F_i(x_i)$. We maintain that the kernel $X_{(m)}$ of $F_{(m)}$ is an Abelian subvariety in $A_{(m)}^*$. In fact, according to AF, Theorem 6, Corollary 2, for each $i = 1, \ldots, m$, $K_m(x_i)$ is a primary extension of $K_m(F_i(x_i))$; it follows, then, from a remark we made above that $K_m(x_1, \ldots, x_m)$ is a primary extension of $K_m(F_1(x_1), \ldots, F_m(x_m))$. Since $F_1(x_1), \ldots, F_m(x_m)$ are independent generic points of A over K_m, the field $K_m(F_1(x_1), \ldots, F_m(x_m))$ is a primary extension of $K_m\left(\sum_{i=1}^{m} F_i(x_i)\right)$; it follows, then, that $K_m(x_1, \ldots, x_m)$ is a primary extension of $K_m\left(\sum_{i=1}^{m} F_i(x_i)\right)$, which shows that $X_{(m)}$ is an Abelian subvariety in $A_{(m)}^*$. Since $X_{(m)}$ is evidently normally algebraic over K_m, it follows from AF, Theorem 2, that $X_{(m)}$ is defined over K_m. According to a result of ours on the quotient varieties,[2] the canonical homomorphism $R_{(m)}$ of $A_{(m)}^*$ onto its quotient variety $A_{(m)}^*(X_{(m)})$ relative to $X_{(m)}$ is separable, and there exists a rational isomorphism $T_{(m)}$ of $A_{(m)}^*(X_{(m)})$ onto A, defined over K_m, such that $F_{(m)} = T_{(m)}R_{(m)}$. We set $y_{(m)} = R_{(m)}(x_1, \ldots, x_m)$ and $z_{(m)} = F_{(m)}(x_1, \ldots, x_m)$, so that we have the relation $z_{(m)} = T_{(m)}(y_{(m)})$.

Next we observe that the degree $[K_m(y_{(m)}):K_m(z_{(m)})]$ is a monotone nonincreasing function of the integer m. To show this, denote by X_m the kernel of F_m; according to AF, Theorem 6, X_m is an Abelian subvariety in A_m^*, and, according to Theorem 2 of the same paper, X_m is defined over $K(u_m)$; according to the result on quotient varieties cited above,[2] the canonical homomorphism R_m of A_m^* onto its quotient variety $A_m^*(X_m)$ relative to X_m is separable, and there exists a rational isomorphism T_m of $A_m^*(X_m)$ onto A, defined over $K(u_m)$, such that $F_m = T_mR_m$. We set $y_m = R_m(x_m)$ and $z_m = F_m(x_m)$, so that we have the relations $z_m = T_m(y_m)$ and $z_{(m)} = z_{(m-1)} + z_m$. Since the Abelian subvariety $X_{(m-1)} \times X_m$ in $A_{(m)}^*$ is defined over K_m and since it is evidently contained in $X_{(m)}$, it follows[2] that there is a rational homomorphism of $A_{(m-1)}^*(X_{(m-1)}) \times A_m^*(X_m)$ onto $A_{(m)}^*(X_{(m)})$, defined over K_m, and $(y_{(m-1)}, y_m)$ and $y_{(m)}$ are a pair of corresponding generic points under this homomorphism. We have, therefore, the relation $K_m(y_{(m-1)}, y_m) \supset K_m(y_{(m)})$ and hence the relation $K_m(y_{(m-1)}, y_m) \supset K_m(y_{(m)}, y_m) \supset K_m(z_{(m)}, y_m) = K_m(z_{(m-1)}, y_m)$; since $K_m(y_{(m)}, y_m)$ is (for $m > 1$) a regular extension of $K_m(y_{(m)})$, it follows that $[K_m(y_{(m)}): K_m(z_{(m)})] = [K_m(y_{(m)}, y_m):K_m(z_{(m)}, y_m)] \leq [K_m(y_{(m-1)}, y_m):K_m(z_{(m-1)}, y_m)] = [K_m(y_{(m-1)}):K_m(z_{(m-1)})]$. This proves our assertion. We observe also that the relation $K_m(y_{(m-1)}, y_m) \supset K_m(y_{(m-1)}, y_{(m)})$ implies the existence of a rational homomorphism G of $A^*_m(X_m)$ onto $A^*_{(m)}(X_{(m)})$, defined over $K_m(y_{(m-1)})$; according to AF, Theorem 2, Corollary 1, G is also defined over K_m. On the other hand, the relation $K_m(y_{(m-1)}, z_m) = K_m(y_{(m-1)}, z_{(m)})$ implies the existence of a regular isomorphism G_1 of A onto itself, defined over $K_m(y_{(m-1)})$, which carries z_m into $z_{(m)}$; and according to AF, Theorem 2, Corollary 1, G_1 is defined over K. Finally, the relations $T_{(m)}G(y_m) = z_{(m)} = G_1(z_m) = G_1T_m(y_m)$ show that $T_{(m)}G = G_1T_m$.

From now on we shall assume that the integer m has been taken so large that the order $[K_m(y_{(m)}):K_m(z_{(m)})]$ attains its minimum value, and we have to show that this minimum value is equal to 1, i.e., $T_{(m)}$ is a regular isomorphism.

Now let $u_1', \ldots, u_m'$ be independent generic specializations of u over K_m, and denote by $A^{*\prime}_{(m)}$, $X'_{(m)}$, $F'_{(m)}$, $R'_{(m)}$, $T'_{(m)}$, $y'_{(m)}$, $z'_{(m)}$ the same elements as those without the prime superscript introduced above, with the field K_m now replaced by the field $K_m' = K(u_1', \ldots, u_m')$; similarly, we denote by $A^*_{(2m)}$, $X_{(2m)}$, $F_{(2m)}$, $R_{(2m)}$, $T_{(2m)}$, $y_{(2m)}$, $z_{(2m)}$ the corresponding elements defined with respect to the field $K_{2m} = K(u_1, \ldots, u_m, u_1', \ldots, u_m')$. It is clear that $z_{(2m)} = z_{(m)} + z'_{(m)}$. Since the Abelian subvariety $X_{(m)} \times X'_{(m)}$ in $A^*_{(2m)}$ is defined over K_{2m} and since it is evidently contained in $X_{(2m)}$, it follows[2] that there is a rational homomorphism of $A^*_{(m)}(X_{(m)}) \times A^{*\prime}_{(m)}(X'_{(m)})$ onto $A^*_{(2m)}(X_{(2m)})$, defined over K_{2m}, and $(y_{(m)}, y'_{(m)})$ and $y_{(2m)}$ are a pair of corresponding generic points under this homomorphism. We have, therefore, the relation $K_{2m}(y_{(m)}, y'_{(m)}) \supset K_{2m}(y_{(2m)})$ and hence the relation $K_{2m}(y_{(m)}, y'_{(m)}) \supset K_{2m}(y_{(2m)}, y'_{(m)}) \supset K_{2m}(z_{(2m)}, y'_{(m)}) = K_{2m}(z_{(m)}, y'_{(m)})$; furthermore, by our choice of the integer m, we have the relations $[K_{2m}(y_{(m)}, y'_{(m)}): K_{2m}(z_{(m)}, y'_{(m)})] = [K_{2m}(y_{(m)}): K_{2m}(z_{(m)})] = [K_{2m}(y_{(2m)}):K_{2m}(z_{(2m)})] = [K_{2m}(y_{(2m)}), y'_{(m)}):K_{2m}(z_{(2m)}, y'_{(m)})]$. This shows that $K_{2m}(y_{(m)}, y'_{(m)}) = K_{2m}(y_{(2m)}, y'_{(m)})$, and hence there exists a regular isomorphism H of $A^*_{(m)}(X_{(m)})$ onto $A^*_{(2m)}(X_{(2m)})$ defined over $K_{2m}(y'_{(m)})$; according to AF, Theorem 2, Corollary 1, H is also defined over K_{2m}. Similarly, we conclude that there exists a regular isomorphism H' of $A^{*\prime}_{(m)}(X'_{(m)})$ onto $A^*_{(2m)}(X_{(2m)})$, defined over K_{2m}, so that $H'^{-1}H$ is a regular isomorphism of $A^*_{(m)}(X_{(m)})$ onto $A^{*\prime}_{(m)}(X'_{(m)})$, defined over K_{2m}. According to AF, Theorem 3, Corollary 1$'$, there exists an Abelian variety A_0, defined over K, and a regular isomorphism H_0 of $A^*_{(m)}(X_{(m)})$ onto A_0, defined over K_m; it follows that H_0GR_m is a rational homomorphism of A^*_m onto A_0, defined over K_m, and, according to AF, Theorem 2, Corollary 1, H_0GR_m is also defined over $K(u_m)$. Since A is the K-image of A^*_m over $K(u_m)$, there exists a rational homomorphism H_1 of A into A_0, defined over K, such that $H_0GR_m = H_1F_m$ or $GR_m = H_0^{-1}H_1F_m$; we have, then, the relations $T_{(m)}H_0^{-1}H_1F_m = T_{(m)}GR_m = G_1T_mR_m = G_1F_m$ and hence $T_{(m)}H_0^{-1}H_1 = G_1$, which shows that $T_{(m)}$ is a regular isomorphism. This concludes the proof of our theorem.

COROLLARY. *Let $K(u)$ be a regular extension of K, and let A^* be an Abelian variety defined over $K(u)$; let A be the K-image of A^* over $K(u)$, and let F be the canonical homomorphism of A^* onto A. If K_1 is any extension of K, independent with respect to $K(u)$ over K, then A is also the K_1-image of A^* over $K_1(u)$, and F is also the corresponding canonical homomorphism.*

Proof: If $\bar{A}$ is the K_1-image of A^* over $K_1(u)$ and if we denote by $\bar{F}$ the canonical homomorphism of A^* onto A_1, then, according to AF, Theorem 4, Corollary, there exists a rational isomorphism I of $\bar{A}$ onto A such that $F = I\bar{F}$; if we denote by $\bar{F}_{(m)}$ the rational homomorphism of $A^*_{(m)}$ onto $\bar{A}$ defined by the correspondence

$(x_1, \ldots, x_m) \rightarrow \sum_{i=1}^{m} \bar{F}_i(x_i)$, where $(x_i, \ldots, x_m)$ is now a generic point of $A^*_{(m)}$ over $K_1(u_1, \ldots, u_m)$, then we have the relation $F_{(m)} = I\bar{F}_{(m)}$. Since the homomorphism $F_{(m)}$ is regular for sufficiently large m, we conclude that the isomorphism I must be regular.

* This work was partially supported by the Office of Ordnance Research, United States Army.

1 "Abelian Varieties over Function Fields," *Trans. Am. Math. Soc.*, **78**, 253, 1955. We shall assume that the reader is familiar with the contents of this paper, which we shall cite as "AF."

2 "On the Quotient Variety of an Abelian Variety," these PROCEEDINGS, **38**, 1040, 1952, Theorem 1. We observe that Theorem 1 (as well as Theorem 2) in this paper and its proof hold without any modifications also for the more general case where (in the notation used there) the kernel of H contains X.

ABELIAN VARIETIES OVER FUNCTION FIELDS

BY
WEI-LIANG CHOW

1. **Introduction.** Let K be a field and U be a variety defined over K; let u be a generic point of U over K, so that $K(u)$ is a regular extension of K. Let W be another variety defined over K, and let A_u be a simple subvariety in W which is an Abelian variety defined over $K(u)$; then there exists (Weil [5, Chap. VII, Theorem 12 (iii)]) in the product variety $U\times W$ a subvariety T, defined over K, such that $T\cdot(u\times W)=u\times A_u$. As we shall see later (§3, Lemma 2 and the last part of the proof of Theorem 3), there exists a bunch B of subvarieties in U, normally algebraic over K, such that the intersection $T\cdot(u_0\times W)$ is defined for every point u_0 in $U-B$ and is an Abelian variety defined over $K(u_0)$. Thus the Abelian variety A_u generates over the field K an algebraic system $\{A_u\}$ of Abelian varieties in W, parametrized by the open variety $U-B$, and A_u is a generic element over K of this algebraic system. In this paper we shall deal with the problem of finding geometrical entities which are associated with an algebraic system of Abelian varieties in a birationally invariant way and hence can be regarded as birational invariants of such a system; more specifically, since we are interested mainly in Abelian varieties here, we shall investigate whether there are Abelian varieties which are invariantly associated with such a system. It is clear that any birational invariant of the parameter variety U, or the graph T, or the carrier variety of the system (which is the projection of T into W), is also a birational invariant of the algebraic system $\{A_u\}$; however, since these are invariants of algebraic systems in general, not specifically invariants of algebraic systems of Abelian varieties, we shall not consider them here, though there are some interesting relations between these invariants and those which will be considered. Instead, we shall consider here some invariants which arise specifically from the fact that we are dealing with a system of Abelian varieties. The main purpose of this paper is to show that, to every algebraic system of Abelian varieties, defined over a field K, it is possible to attach in a birationally invariant way two Abelian varieties, which are defined over K and are isogeneous to each other over K. These two Abelian varieties will be called the K-image and the K-trace of the generic Abelian variety A_u of the algebraic system; they will play an important role in our abstract theory of Picard varieties, which will be presented in a forthcoming paper.

We begin with a few remarks on terminology. We shall follow in general the terminology of Weil, as developed in his books [5] and [6], with the following modifications. First, we shall deal only with points and varieties in

Received by the editors October 22, 1953.

projective spaces; thus, in particular, all Abelian varieties can be considered as embedded in some one projective space, so that, for example, we can without any loss of generality replace the variety W in the preceding paragraph by a projective space. In fact, this restriction to varieties in a projective space is used in an essential way only in the proof of Theorem 3; moreover, in view of the fact (which will be proved in our forthcoming paper on Picard varieties) that any Abelian variety is regularly isomorphic to an Abelian variety in a projective space, defined over the same field as the original one, this restriction entails in reality no loss of generality for our results. Next, we shall use the expression "extension" of a field in its usual general sense, not necessarily meaning a finitely generated extension, as is the case with Weil; however, whenever we speak of an extension $K(u)$ or $K(v)$ of a field K, this is to indicate that $K(u)$ or $K(v)$ is the (finitely generated) extension obtained from K by the adjunction of a point u or v in some projective space. We shall say that an extension K^* of K is separably generated if it preserves p-independence, p being the characteristic of K (MacLane [4, §4]); and we shall say that an extension K^* of K is primary, if $K^* \cap \overline{K}$ is a purely inseparable extension of K. An extension which is both separably generated and primary is called regular, and one sees readily that this definition agrees with that of Weil in case of a finitely generated extension. Another modification is that we shall use the expressions "homomorphism" and "isomorphism" in their strictly group theoretic sense. We shall say that a homomorphism H of an Abelian variety A into another Abelian variety B is rational if H is also a rational transformation of A into B (so that it is a homomorphism in the terminology of Weil), and we shall say that a rational homomorphism H is regular if, K being a field of definition for A, B, H, and x being a generic point of A over K, the field $K(x)$ is a regular extension of $K(H(x))$. It is clear that a rational isomorphism is regular if and only if it is birational. Finally, for the sake of convenience, we shall use sometimes the expression "almost every" in the following sense: Let U be a variety defined over a field K, or more generally a prime rational cycle over K, and let u be a generic point of U over K; we shall say that a property holds for almost every specialization of u over K if there is a bunch B of proper subvarieties in U such that the property holds for every point in $U-B$.

We shall assume that the reader is familiar with the theory of associated forms of positive cycles in a projective space, as developed in Chow-van der Waerden [3], and we shall denote by $M(S_n; r, d)$ the bunch of varieties in a projective space which consists of the associated points of all positive cycles of dimension r and degree d in the projective space S_n of n dimensions. We shall also use the terminology and results of our note [2], except that we shall use the expression "Abelian function field" instead of "Abelian field"; this note was originally a part of the present paper, but was published separately on account of its independent interest.

In §2, we shall prove the fact, well known in the classical case, that an algebraic system of Abelian varieties (of positive dimension) cannot exist in an Abelian variety; in §3, we shall show that the generic element A_u of an algebraic system of Abelian varieties, defined over a field K, is regularly isomorphic to an Abelian variety defined over K if and only if every two generic elements in the algebraic system are regularly isomorphic to each other. On the basis of these results, or rather somewhat more general versions of these results proved in §§2 and 3, we shall prove in the next two sections the existence of the K-image and the K-trace. In §4, we shall consider rational homomorphisms of the Abelian variety A_u into Abelian varieties defined over K, and we shall show that among all such rational homomorphisms there exists one which has the "universal mapping" property in the sense that any rational homomorphism of A_u into any Abelian variety defined over K must "go through" this one. The image Abelian variety of A_u under this rational homomorphism is then defined as the K-image of A_u over $K(u)$; it is uniquely determined up to a regular isomorphism over K. In §5, we shall consider dually rational homomorphisms (or isomorphisms) of Abelian varieties defined over K into the Abelian variety A_u, and we shall show that among all such rational homomorphisms there exists one (which is an isomorphism) which has the "universal mapping" property in the sense that any rational homomorphism of any Abelian variety defined over K into A_u must "go through" this one. The Abelian variety defined over K which is mapped by this isomorphism into A_u is called the K-trace of A_u over $K(u)$; it is uniquely determined up to a regular isomorphism over K. The K-trace is independent of any extension of the ground field K, while the K-image remains unchanged up to an isomorphism under any such extension, though the isomorphism involved might not be regular (see however the remarks at the end of §5). These two invariants of the algebraic system $\{A_u\}$, the K-image and the K-trace, are connected by a relation of isogeneity, which is derived from the application of the Poincaré Complete Reducibility Theorem to the Abelian variety A_u and its K-image.

The concepts of the K-image and the K-trace arise in a natural way from the theory of Picard varieties. Let V be a variety of dimension r in a projective space S_n, defined over a field K, which has no singular subvarieties of dimension $r-1$, and consider the algebraic system of curves on V generated over K by the intersection curve C_u of a generic linear subspace (over K) of dimension $n-r+1$ in S_n, where $K(u)$ is a purely transcendental extension of K. Since the curve C_u is defined over $K(u)$, the Jacobian variety J_u of C_u is also defined over $K(u)$, according to a result of ours proved elsewhere. In our paper [1], we have shown that, in the classical case, the K-image of J_u over $K(u)$ is the Albanese variety of V and the K-trace of J_u over $K(u)$ is the Picard variety of V; as we have mentioned there, an independent proof of the existence of the K-image and the K-trace will enable us to dispense with the classical construction of the Albanese and Picard varieties by means of the periods of Abelian integrals and thus serve as a starting point of a purely algebraic theory of the Picard varieties over an arbitrary ground field. This algebraic theory of Picard varieties will be presented in a forthcoming paper.

2. **Algebraic subgroups in an Abelian variety.** Let V be a variety and A be an Abelian variety, both defined over a field K, and let F be a rational transformation of V into A, defined over a purely transcendental extension $K(u)$ of K, where u is a generic point over K of a projective space S. If x is a generic point of V over $K(u)$, then the correspondence $u \rightarrow F(x)$ defines a rational transformation of S into A, defined over $K(x)$; according to [6, Theorem 8, Corollary], the point $F(x)$ is independent of u and hence is rational over $K(x)$. Since $K(x)$ is linearly disjoint with respect to $K(u)$, it follows that $K(F(x))$ is also linearly disjoint with respect to $K(u)$; this shows that the rational transformation F is defined over K and hence the subvariety $F(V)$ in A is also defined over K.

If $K(u)$ is not a purely transcendental extension of K, then the above statement will not be true in general; however, we shall show that if V is also an Abelian variety and if F is a rational homomorphism of V into A, then it is still true that F is defined over K and hence the image Abelian subvariety $F(V)$ in A is also defined over K, provided that $K(u)$ is a primary extension of K. Furthermore, even if the Abelian variety V is only defined over $K(u)$, it is still true that the Abelian subvariety $F(V)$ in A is defined over K, though of course the homomorphism F will be in general defined only over $K(u)$. Both these results are consequences of a general theorem which asserts that any Abelian subvariety in A which is defined over $K(u)$ is also defined over K, a theorem which expresses essentially the fact that an Abelian variety cannot contain a continuous system of Abelian subvarieties or, more generally, algebraic subgroups. In case of complex ground field, this theorem is a simple consequence of the fact that an analytic subgroup of a complex torus is uniquely determined by a subgroup of its fundamental group, which is a discrete group. In case of an arbitrary ground field, this argument is of course not available; however, a substitute for this can be found in the fact that to each Abelian subvariety B in an Abelian variety A there exists an endomorphism of A which maps A onto B (Weil [6, Proposition 25]), and the fact that the module of endomorphisms of A is a finite module over the rational integers (Weil [6, Theorem 37]). We shall give here a simple direct proof of this theorem, which is entirely elementary in character.

THEOREM 1. *Let A be an Abelian variety; let K be a field of definition for A and let K^* be an extension of K. If an algebraic subgroup X in A is normally algebraic over K^*, then it is normally algebraic over $K^* \cap \bar{K}$.*

Proof. Without any loss of generality, we can assume that $K^* = K(u)$ is a finitely generated extension, for otherwise we can replace K^* by $K^* \cap K_1$, where K_1 is a finitely generated extension of K over which X is defined. It is sufficient to show that X is defined over $\overline{K}$, for then X must be normally algebraic over $K^*\cap\overline{K}$; we can therefore assume that K is algebraically closed, so that $K(u)$ is a regular extension of K. Furthermore, we can also assume that X is an Abelian variety, for an algebraic subgroup in A is algebraic over K if and only if the maximal Abelian variety contained in it is algebraic over K. Let U be the variety which is the locus of u over K, and let x be a generic point of X over $K(u)$; then the point $u\times x$ in $U\times A$ has as its locus over K a subvariety T in $U\times A$. If $u'\times x'$ is a generic point of T over K, independent with respect to $u\times x$ over K, then the point x' has a locus X' over $K(u')$, which is an Abelian subvariety in A and has the same dimension d as that of X. Let X'' be the locus of the point $x+x'$ over $K(u, u')$; it is clear that X'' is an Abelian subvariety in A and is the subgroup in A generated by the X and X'. Our theorem would be proved if we show that X'' has a dimension not greater than d, for then we must have $X=X''=X$ and hence X must be defined over $K(u)\cap K(u')=K$. In the product space $U\times U\times A\times A\times A$ we consider the following subvarieties, where y, y', y'' are three independent generic points of A over $K(u, u', x, x')$: The locus M of the point $u\times u'\times x \times x'\times x+x'$ over K, the locus Δ of the point $u\times u'\times x\times y'\times y''$ over K, the locus Δ' of the point $u\times u'\times y\times x'\times y''$ over K, and the locus Γ of the point $u\times u'\times y\times y'\times y+y'$ over K. We have evidently the relations $M=\Delta\cap\Delta'\cap\Gamma$ and $M\cap(u\times u'\times A\times A\times A)=u\times u'\times X\times X'\times X''$. Let N be the projection of M into the partial product space $U\times U\times A$ of the first, second, and fifth factors, and consider the intersection $N\cap(u\times u\times A)$; since $N\cap(u\times u\times A)$ is evidently contained in the projection of $M\cap(u\times u\times A\times A\times A)$ into $U\times U\times A$, every point $u\times u\times x_0''$ in the former is the projection of a point $u\times u\times x_0\times x_0'\times x_0''$ in the latter. The relation $u\times u\times x_0\times x_0'\times x_0''\subset\Delta\cap\Delta'\cap\Gamma$ implies that $x_0\subset X$, $x_0'\subset X$, and $x_0''=x_0+x_0'$, which shows that the point x_0'' is contained in X; it follows then that $N\cap(u\times u\times A)$ is contained in $u\times u\times X$ and hence has a dimension not greater than d. If s is the dimension of U, then the dimension of N cannot be greater than $2s+d$, from which it follows that the dimension of $N\cap(u\times u'\times A)=u\times u'\times X''$ cannot be greater than d; this implies that the dimension of X'' cannot be greater than the dimension d of X.

It might be not without some interest to indicate here an alternate proof of our theorem, which is simple but uses a deeper result of Weil. If l is a positive integer which is prime to the characteristic of the field K, then the subgroup g_l of all elements in X whose orders divide l is a finite group of order l^{2d}, normally algebraic over $K(u)$ (Weil [6, Theorem 33, Corollary 1]). Since each point in g_l, being of finite order, is rational over K (K being algebraically closed), the group g_l is defined over K and hence is normally algebraic over $K(u)\cap K=K$. It follows that for every point x_0 in g_l, the point $u'\times x_0$ is a specialization of the point $u\times x_0$ over K, and since $u\times x_0$ is contained in

T, $u'\times x_0$ is also contained in T; this means that g_l is contained in X', and hence is also contained in $X\cap X'$. Since this is true for every positive integer l prime to the characteristic of K, it follows from the same result quoted above that the dimension of the Abelian variety $X\cap X'$ cannot be less than d and hence must be equal to d. This means that $X=X'$, and hence X must be defined over $K(u)\cap K(u')=K$.

THEOREM 2. *Let A be an Abelian variety; let K be a field of definition for A and let K^* be a primary extension of K. If an Abelian subvariety X in A is normally algebraic over K^*, then it is defined over K.*

Proof. According to Theorem 1, the Abelian variety X is normally algebraic over $K^*\cap\overline{K}$ and hence is defined over a purely inseparable extension K_1 of $K^*\cap\overline{K}$. Since $K^*\cap\overline{K}$ is a primary extension of K, K_1 is also a purely inseparable extension of K; let p^e be the degree of K_1 over K. Let x be a generic point of X over K_1, and denote by x^{p^e} the point obtained from x by raising all the coordinates of x to the p^eth power; since $K_1(x)$ is a regular extension of K_1, $K_1^{p^e}(x^{p^e})$ is a regular extension of $K_1^{p^e}$, and since $K_1^{p^e}$ is contained in K, it follows that $K(x^{p^e})$ is also a regular extension of K. Consider now the point p^ex in X; since the associated point of 0-cycle $p^e(x)$ in A is rational over $K(x^{p^e})$, the point p^ex is rational over $K(x^{p^e})$, i.e., we have $K(p^ex)\subset K(x^{p^e})$ (Weil [6, Theorem 1], the proof holds for the more general situation considered here). This shows that $K(p^ex)$ is a regular extension of K. On the other hand, since $K_1(x)$ is a finite algebraic extension of $K_1(p^ex)$ (Weil [6, Proposition 24]), the point p^ex is also a generic point of X over K_1; and since $K(p^ex)$ is a regular extension of K and K_1 is an algebraic extension of K, the point p^ex has the same specializations over K as it has over K_1. This proves the theorem.

REMARK. For any integer e (positive or negative), the Abelian variety over K^{p^e} defined by the point x^{p^e} is called the *p^eth power* of the Abelian variety A.

COROLLARY 1. *Let A be an Abelian variety, defined over a field K, and let K^* be a primary extension of K. If B is an Abelian variety over K^*, and if H is a rational homomorphism of B into A, also defined over K^*, then the image $H(B)$ is an Abelian subvariety in A which is defined over K. Furthermore, if the Abelian variety B is also defined over K, then H is defined over K.*

Proof. The first part follows immediately from Theorem 2 and the fact that $H(B)$ is an Abelian variety defined over K^*. As to the second part, we observe that the graph of H is an Abelian subvariety in the product variety $B\times A$, and since H is defined over K^* and $B\times A$ is defined over K, it follows from Theorem 2 that H is also defined over K.

COROLLARY 2. *Let A be an Abelian variety, defined over a field K, and let $K^*=K(u)$ be a purely transcendental extension of K. If B is an Abelian variety* *over K^*, and if H is a rational transformation of B into A, defined over K^*, then $H(B)$ is a subvariety in A defined over K; furthermore, if B is also defined over K, then H is defined over K.*

Proof. Since H is defined over $K(u)$, the point $H(0)$ is rational over $K(u)$ and hence the correspondence $u \rightarrow H(0)$ defines a rational transformation of a projective space into A; it is well known that the image point $H(0)$ in this case is independent of u and hence is rational over K. If x is a generic point of B over $K(u)$, then the correspondence $x \rightarrow H(x) - H(0)$ defines a rational homomorphism of B into A, defined over $K(u)$, and the corollary then follows from Corollary 1.

3. **An existence theorem.** Let $K(u)$ be a regular extension of a field K, and let A_u be an Abelian variety defined over $K(u)$; then, as we have explained in §1, the Abelian variety A_u generates over K an algebraic system of Abelian varieties. Let u_1 and u_2 be independent generic specializations of u over K, and let A_1 and A_2 be respectively the specializations of A_u over the specializations $u \rightarrow u_1$ and $u \rightarrow u_2$ over K; in general, the Abelian varieties A_1 and A_2 will not be isomorphic to each other. It is clear that if the "variable" Abelian variety A_u is regularly isomorphic over $K(u)$ to a "fixed" Abelian variety defined over K, then A_1 and A_2 will be regularly isomorphic to each other over $K(u_1, u_2)$. We shall show in this section that the converse of this statement is also true; in other words, we shall show that if A_1 and A_2 are regularly isomorphic over $K(u_1, u_2)$, then there exists an Abelian variety A defined over K, such that A_u is regularly isomorphic to A over $K(u)$ (Theorem 3, Corollary 1'). However, for the purpose of application to the problem of the effect of ground field extension in the next section, we shall prove a somewhat more general result (Theorem 3), where the regular extension $K(u)$ is replaced by a separably generated extension. We shall need the following two lemmas.

LEMMA 1. *Let $K(u)$ be an extension of a field K, and let V be a variety defined over $K(u)$; then, for almost every specialization $u \rightarrow u_0$ over K, the variety V has a uniquely determined specialization V_0, which is a variety defined over $K(u_0)$.*

Proof. Since the associated point y of the variety V is rational over $K(u)$, for almost every specialization $u \rightarrow u_0$ over K the point y has a uniquely determined specialization y_0 which is rational over $K(u_0)$, and hence the variety V has a uniquely determined specialization V_0 which is a rational positive cycle over $K(u_0)$. Furthermore, it is well known that for almost every specialization $u \rightarrow u_0$ over K, the associated form of V, which is absolutely irreducible, will have a uniquely determined specialization which is also absolutely irreducible; one need only to observe that the condition of a form being absolutely reducible can be expressed as a system of algebraic equations in terms of the coefficients. The lemma then follows immediately.

Lemma 2. *Let $K(u)$ be an extension of a field K; let V and W be two varieties defined over $K(u)$, and let F be a rational transformation of V onto W, defined over $K(u)$, which is defined everywhere in V. Then, for almost every specialization $u \to u_0$ over K, there exist uniquely determined specializations V_0, W_0, F_0 of V, W, F respectively, such that V_0 and W_0 are varieties defined over $K(u_0)$ and F_0 is a rational transformation of V_0 onto W_0, defined over $K(u_0)$.*

Proof. That V_0 and W_0 are varieties defined over $K(u_0)$ follows from Lemma 1; furthermore, since F is a subvariety in $V \times W$, it follows from Lemma 1 that F_0 is also a subvariety in $V_0 \times W_0$, defined over $K(u_0)$. It remains to prove that the projection of F_0 onto V_0 is regular at every point of V_0. Let S and S' be the ambient projective spaces of V and W respectively, and let x be a generic point of V over $K(u)$; without any loss of generality, we can assume that K is algebraically closed, so that u has a locus U over K. Let T be the locus of the point $u \times x \times F(x)$ over K in $U \times S \times S'$, and let X be the locus of the point $u \times x$ over K in $U \times S$; then we have the relations $T \cdot (u \times S \times S') = u \times F$ and $X \cdot (u \times S) = u \times V$ (Weil [5, Chap. VII, Theorem 12]). The projection of T into $U \times S$ is X, and the projection of T onto X is regular; it can be shown that for almost every point $u_0 \times x_0$ in X, the projection of T onto X is regular at $u_0 \times x_0$. Since the projection of T onto X is regular at the point $u \times x'$ for every point x' in V, it follows that for almost every point u_0 in U, the projection of T onto X is regular at the point $u_0 \times x_0'$ for every point x_0' in V_0. Since the projection of T onto X is regular at a point $u_0 \times x_0$ if and only if the projection of F_0 onto V_0 is regular at x_0, we conclude that for almost every point u_0 in U, F_0 defines a rational transformation of V_0 onto W_0, defined everywhere in V_0.

Theorem 3. *Let K^* be a primary extension of a field K, and let $K(v)$ be a separably generated extension of K, independent with respect to K^* over K; let L^* be an Abelian function field over K^*, independent with respect to $K^*(v)$ over K^*, and let L be a subfield in $L^*(v)$ which is an Abelian function field over $K(v)$ and is independent with respect to $K^*(v)$ over $K(v)$. Let v' be a generic specialization of v over K, independent with respect to $L^*(v)$ over K, and let L' be the subfield in $L^*(v')$ which is the image of L under the isomorphism between $L^*(v)$ and $L^*(v')$. If for every choice of v' we have $LL' = L(v') = L'(v)$, then $L^* \cap L$ contains an Abelian function field $K(z)$ over K, independent with respect to $K(v)$ over K, such that $L = K(v, z)$.*

Remark. Since K^* and $K(v)$ are linearly disjoint over K, v' is also a generic specialization of v over L^*; hence $L^*(v)$ and $L^*(v')$ are isomorphic over L^*.

Proof. Let A^* and A be Abelian varieties over K^* and $K(v)$ respectively defined by the fields L^* and L respectively, and let F be the rational homomorphism of A^* onto A, defined over $K^*(v)$, associated with the subfield K^*L in $L^*(v)$; if x is a generic point of A^* over $K^*(v)$, then $y = F(x)$ is a generic point of A over $K^*(v)$, and we can set $L^* = K^*(x)$ and $L = K(v, y)$. Without any loss of generality we can assume that both A^* and A are embedded in the same projective space S_n; furthermore, we can also assume that $K(v)$ is the defining field of A, for otherwise we can replace A by the product of A and the point v, and replace S_n by another suitably chosen projective space. If A' and F' are respectively the specializations of A and F over the specialization $v \to v'$ over K^*, then A' is an Abelian variety in S_n, with $K(v')$ as the defining field, and F' is a rational homomorphism of A^* onto A' defined over $K^*(v')$; furthermore, $y' = F'(x)$ is a generic point of A' over $K^*(v')$, and we have $L' = K(v', y')$. Let $v^{(1)}, \cdots, v^{(m)}$ be a set of independent generic specializations of v' over K, independent with respect to $L^*(v)$ over K, such that no two of them are specializations of each other over $K(v)$ and that every specialization of v' over K is a specialization of one of these m points over $K(v)$. For each $i = 1, \cdots, m$, let $A^{(i)}$ and $F^{(i)}$ be respectively the specializations of A' and F' over the specialization $v' \to v^{(i)}$ over K^*; then $A^{(i)}$ is an Abelian variety in S_n, with $K(v^{(i)})$ as the defining field, and $F^{(i)}$ is a rational homomorphism of A^* onto $A^{(i)}$, defined over $K^*(v^{(i)})$. Furthermore, the point $y^{(i)} = F^{(i)}(x)$ is a generic point of $A^{(i)}$ over $K^*(v^{(i)})$ and the field $L^{(i)} = K(v^{(i)}, y^{(i)})$ is the subfield in $L^*(v^{(i)})$ which is the image of L under the isomorphism between $L^*(v)$ and $L^*(v^{(i)})$. We observe that the varieties $A^{(1)}, \cdots, A^{(m)}$ are all distinct. In fact, if we have $A^{(i)} = A^{(j)}$, then the field $K(v^{(i)}) \cap K(v^{(j)})$ is a field of definition for both $A^{(i)}$ and $A^{(j)}$; we would then have $K(v^{(i)}) = K(v^{(i)}) \cap K(v^{(j)}) = K(v^{(j)})$, which can only be possible for $i = j$. The condition $K(v, v^{(i)}, y, y^{(i)}) = LL^{(i)} = L(v^{(i)}) = K(v, v^{(i)}, y)$ is equivalent to the fact that there is a regular isomorphism $H^{(i)}$ of the Abelian variety A onto the Abelian variety $A^{(i)}$, defined over $K(v, v^{(i)})$, such that $y^{(i)} = H^{(i)}(y)$. We shall show presently (in order not to interrupt the argument now) that there exists a specialization $v' \to v_1'$ over K with the following properties:

(1) The point v_1' is separably algebraic over K, and for each conjugate point v_i' $(i = 1, \cdots, d)$ of v_1' over K there exists exactly one point $v^{(m_i)}$ among the points $v^{(1)}, \cdots, v^{(m)}$, such that v_i' is a specialization of $v^{(m_i)}$ over $K(v)$; we observe that the set $v_1', \cdots, v_d'$ contains the complete set of conjugates of each v_i' over $K(v)$. (2) For each $i = 1, \cdots, d$, there exist uniquely determined specializations $A' \to A_i'$ and $F' \to F_i'$ over the specialization $v' \to v_i'$ over K^*, such that A_i' is an Abelian variety over $K(v_i')$ and F_i' is a rational homomorphism of A^* onto A_i', defined over $K^*(v_i')$; furthermore, the varieties $A_1', \cdots, A_d'$ are all distinct subvarieties in S_n. We observe that for each $i = 1, \cdots, d$, A_i' and F_i' are specializations of $A^{(m_i)}$ and $F^{(m_i)}$ respectively over the specialization $v^{(m_i)} \to v_i'$ over $K^*(v)$. (3) For each $i = 1, \cdots, d$, there exists a uniquely determined specialization H_i of $H^{(m_i)}$ over the specialization $v^{(m_i)} \to v_i'$ over $K(v)$, such that H_i is a regular isomorphism of A onto A_i', defined over $K(v, v_i')$, and we have the relation $H_i(y) = F_i'(x)$. We observe that the set $H_1, \cdots, H_d$ contains the complete set of conjugates of each H_i over $K(v)$.

Consider now the set of d distinct points $H_1(y), \cdots, H_d(y)$ in S_n; it determines a positive 0 cycle $Z(y)$ of degree d in S_n, rational over $K(v, y)$. If Γ is the canonical mapping of all positive 0 cycles of degree d in S_n into $M(S_n; 0, d)$, then the mapping $y \to \Gamma(Z(y))$ defines a rational transformation R of A into $M(S_n; 0, d)$, defined over $K(v)$. Let V be the image variety $R(A)$ in $M(S_n; 0, d)$, which is defined over $K(v)$; we maintain that V is also defined over K. In fact, the relations $H_i(y) = F_i'(x)$ show that the 0-cycle $Z(y)$ is also rational over $K^*(x)$; it follows that the mapping $x \to \Gamma(Z(y))$ defines a rational transformation R^* of A^* into $M(S_n; 0, d)$, defined over K^*, and since every specialization of $R(y)$ over $K(v)$ is a specialization of $R^*(x)$ over K^* and vice-versa, we have $R^*(A^*) = R(A) = V$, so that V is defined over K^* and hence also over $K(v) \cap K^* = K$.

Since the 0-cycle $Z(y)$ consists of d distinct points, each point $H_i(y)$ is separably algebraic over $K(z)$, where we set $z = R(y)$; and since furthermore $H_i(y)$ is the only point of $Z(y)$ contained in the variety A_i' and y is the only point in A with the image $H_i(y)$ in A_i' under the isomorphism H_i, it follows that R is a birational transformation of A onto V. Let z_0 be any point in V, and let $Z_0 = (\zeta_1, \cdots, \zeta_d)$ be the positive 0-cycles of degree d determined by z_0; then each point ζ_i is in A_i' and $R^{-1}(z_0)$ must be one of the d points $H_i^{-1}(\zeta_i)$ in A. This shows that R has no fundamental points on V. Let V_1 be a derived normal model of the variety V over K, and let R_1 be the birational transformation of A onto V_1 induced by R; then R_1^{-1} does not have any fundamental points on V_1. It follows then that R_1^{-1} is defined everywhere on V_1; and since R_1 evidently has no fundamental points on A, it follows that R_1 is an everywhere biregular birational transformation of A onto V_1. If z_1 and z_2 are two generic points of V_1 over K, we can define a composition function in V_1 by setting $z_1 + z_2 = R_1(R_1^{-1}(z_1) + R_1^{-1}(z_2))$; it is easily seen that this composition function satisfies all the conditions of an Abelian variety. The function $z_1 + z_2$ is evidently defined over $K(v)$; we maintain that this function is also defined over K. Let R_1^* be the rational transformation of A^* onto V_1 induced by R^*; it is clear that R_1^* is defined over K^*, and the relation $R_1^* = R_1 F$ shows that R_1^* is a rational homomorphism of A^* onto V, whereby V is considered as an Abelian variety defined over $K(v)$. It follows that $R_1^{*-1}(z_1)$ and $R_1^{*-1}(z_2)$ are prime rational cycles in A^* over $K^*(z_1, z_2)$; let w_1 and w_2 be generic points of $R_1^{*-1}(z_1)$ and $R_1^{*-1}(z_2)$ over $K^*(z_1, z_2)$ respectively. Then we have $z_1 + z_2 = R_1^*(w_1 + w_2)$, which shows that $z_1 + z_2$ is rational over $K^*(z_1, z_2, w_1, w_2)$; and as $z_1 + z_2$ is evidently independent of the choice of the generic points w_1 and w_2, it follows that $z_1 + z_2$ is rational over a purely inseparable extension of $K^*(z_1, z_2)$. As $z_1 + z_2$ is also rational over the separably generated extension $K(v, z_1, z_2)$ of $K(z_1, z_2)$, we conclude that $z_1 + z_2$ is rational over $K(z_1, z_2)$. Thus we have shown that V_1 is an Abelian variety over K and hence $K(z)$ is an Abelian function field over K; as we have evidently $L = K(v, z)$, this concludes the proof of our theorem, except for one point mentioned above, to which we shall now turn our attention.

In order to complete the proof of Theorem 3, we shall show the existence of a specialization v_1' of v' with the properties mentioned above. We observe first that the following properties hold for almost every specialization $v' \to v_0'$ over K:

(1) There exists a unique specialization $A' \to A_0'$ such that A_0' is a non-singular variety, defined over $K(v_0')$ (Lemma 1);

(2) the composition function in A', defined everywhere on $A' \times A'$ to A', specializes uniquely into a function on $A_0' \times A_0'$ to A_0', defined over $K(v_0')$, which is defined everywhere; by the permanence of functional relations under specialization, this specialized composition function is also associative (Lemma 2);

(3) the unit element in A', being rational over $K(v')$, specializes uniquely into a point in A_0', rational over $K(v_0')$, which is also a unit for the specialized composition function, by the permanence of functional relations;

(4) the inverse function in A', defined everywhere in A', specializes uniquely into a function in A_0', defined over $K(v_0')$, which is also defined everywhere in A_0' (Lemma 2), and the permanence of functional relations shows that this function is also the inverse function in A_0' with respect to the specialized composition function and the specialized unit element;

(1)–(4) together shows that A_0' is an Abelian variety defined over $K(v_0')$;

(5) over the specialization $v' \to v_0'$ over K^*, the rational homomorphism F' of A^* onto A' specializes uniquely into a rational transformation F_0' of A^* onto A_0', defined over $K^*(v_0')$, which is defined everywhere in A^* (Lemma 2); by the permanence of functional relations, this rational transformation F_0' is a homomorphism of A^* onto A_0';

(6) if v_0' is a specialization of $v^{(m_0)}$ over $K(v)$, then there exists a uniquely determined specialization $H^{(m_0)} \to H_0$ such that H_0 is an everywhere biregular birational transformation of A onto A_0', defined over $K(v, v_0')$, which by the permanence of functional relations must be an isomorphism of A onto A_0' (Lemma 2, applied to both $H^{(m_0)}$ and its inverse); since we have the relation $H^{(m_0)}(y) = F^{(m_0)}(x)$, and since F_0' is also a specialization of $F^{(m_0)}$ over the specialization $v^{(m_0)} \to v_0'$ over $K^*(v)$ and H_0 is also a specialization of $H^{(m_0)}$ over the specialization $v^{(m_0)} \to v_0'$ over $K^*(v)$, it follows that $H_0(y) = F_0'(x)$.

Now, let $w_1, \cdots, w_r$ be a separating transcendental base of $K(v')$; then the following properties hold for almost every specialization $w \to w_0$ over K:

(7) there are only a finite number of specializations $v_1', \cdots, v_d'$ of v' over the specialization $w \to w_0$ over K, forming a complete set of conjugates over $K(w_0)$;

(8) each point v_i' is separably algebraic over $K(w_0)$;

(9) each point v_i' is the specialization of exactly one point $v^{(m_i)}$ over $K(v)$;

(10) each point v_i' satisfies the conditions (1)–(6) above.

The proof of (7), (8), (10) follows immediately from the fact that each is expressed by the condition that a system of algebraic equations over K is not satisfied by the point w_0 or the point v_i'.

Thus we have shown that almost every specialization v_1' of v' over K has all the desired properties except the condition that v_1' be separably algebraic over K. For this last condition, it is clearly sufficient to show that there exists at least one point w_0 in the r-space which is separably algebraic over K and is not contained in a given bunch of varieties, and this follows immediately from the fact that the set of all separably algebraic points over K in the r-space is everywhere dense with respect to the Zariski topology (in which the closed subsets are the bunches of subvarieties). In fact, this is obvious in case of the 1-space, there being more than a finite number of separably algebraic elements over K in the universal domain, and the general case can be easily deduced from this by induction.

COROLLARY 1. *Let $K(v)$ be a regular extension of K, and let L be an Abelian function field over $K(v)$; let v' be a generic specialization of v over K, independent with respect to v over K, and let L' be an Abelian function field over $K(v')$ such that the isomorphism between $K(v)$ and $K(v')$ can be extended to an isomorphism between L and L'. If the field L' can be so chosen that $LL'=L(v')=L'(v)$, then L contains an Abelian function field $K(z)$ over K, independent with respect to $K(v)$ over K, such that $L=K(v, z)$.*

Proof. The corollary follows from Theorem 3 if we take u to be a generic specialization of v over K, independent with respect to L over K, and set $K^*=K(u)$ and take L^* to be an Abelian function field over $K(u)$ such that the isomorphism between $K(v)$ and $K(u)$ can be extended to an isomorphism of L and L^* and that the relation $LL^*=L^*(v)=L(u)$ holds; for we would then have the relation $L^*(v, v')=L^*L(v')=L(u, v')=L'(u, v)$, from which it follows by Theorem 2, Corollary 1, that $L^*(v')=L'(u)$. One observes that the condition "for any choice of v'" is not necessary here, for all such v' are generic specializations of each other over $K(v)$.

In order to exhibit the geometrical meaning of this corollary and also for the sake of convenience in later applications, we shall restate it in terms of Abelian varieties instead of function fields:

COROLLARY 1'. *Let $K(v)$ be a regular extension of K, and let A be an Abelian variety defined over $K(v)$; let v' be a generic specialization of v over K, independent with respect to v over K, and let A' be the specialization of A over the specialization $v\to v'$ over K. If A is regularly isomorphic to A' over $K(v, v')$, then A is regularly isomorphic to an Abelian variety defined over K.*

COROLLARY 2. *Let V be a variety defined over a field K, and let $K(u)$ be a* *regular extension of K. If V is birationally equivalent over $K(u)$ to an Abelian variety defined over $K(u)$, then V is also birationally equivalent over $K(u)$ to an Abelian variety A defined over K; furthermore, this Abelian variety A is uniquely determined up to a regular isomorphism over K, independent of the choice of the regular extension $K(u)$ of K.*

Proof. Let u' be a generic specialization of u over K, independent of u over K, and let x be a generic point of V over $K(u, u')$; then $K(u, x)$ is an Abelian function field over $K(u)$ and $K(u', x)$ is an Abelian function field over $K(u')$, and since $K(u, x, u')=K(u', x, u)$, it follows from Corollary 1 that $K(u, x)=K(u, z)$, where $K(z)$ is an Abelian function field over K, independent with respect to $K(u)$ over K. If A is the Abelian variety over K, determined by $K(z)$, then V is evidently birationally equivalent to A (over $K(u)$); if A_1 is an Abelian variety over K obtained in this manner with respect to another regular extension K_1 of K, then A_1 must also be birationally equivalent to V over K_1 and hence must be regularly isomorphic to A over $K_1(u)$; according to Theorem 2, Corollary 1, then A_1 must be regularly isomorphic to A over K.

4. The K-image.

THEOREM 4. *Let K^* be a primary extension of K, and let A^* be an Abelian variety over K^*. Then there exists an Abelian variety A over K, uniquely determined up to a regular isomorphism over K, and a rational homomorphism F of A^* onto A, defined over K^*, such that if H is a rational homomorphism of A^* into an Abelian variety B and if B is defined over a separably generated extension $K(v)$ of K, independent with respect to K^* over K, and H is defined over $K^*(v)$, then H is the product of F and a rational homomorphism of A into B, defined over $K(v)$.*

Proof. Let x be a generic point of A^* over K^*, and set $L^*=K^*(x)$; if F_0 is any rational homomorphism of A^* onto an Abelian variety A_0 defined over K and if F_0 is defined over K^*, then $L_0=K(F_0(x))$ is an Abelian function field over K which is a subfield in L^* and is linearly disjoint with respect to K^* over K. This field L_0, being the function field of A_0 over K, determines A_0 uniquely up to a regular isomorphism over K, and F_0 is the rational homomorphism associated with the subfield K^*L_0 in L^* and hence is uniquely determined up to regular isomorphisms of A^* and A_0 over K^* and K respectively. Consider now the aggregate $\mathfrak{L}$ of all such subfields $K(F_0(x))$ in L^* as F_0 runs through all such homomorphisms; let L_1 and L_2 be two fields in $\mathfrak{L}$, and let F_1 and F_2 be the corresponding rational homomorphisms of A^* onto the Abelian varieties A_1 and A_2 respectively, so that we have $L_1=K(F_1(x))$ and $L_2=K(F_2(x))$. The rational transformation $F_1\times F_2$ is a homomorphism of A^* into the product Abelian variety $A_1\times A_2$, and the image of A^* in $A_1\times A_2$ under this homomorphism is then an Abelian subvariety A_3, defined over K^*; since $A_1\times A_2$ is defined over K, it follows from Theorem 2, Corollary

1, that the Abelian variety A_3 is defined not only over K^*, but also over K. This means that the compositum $L_1L_2 = K(F_1(x) \times F_2(x))$ is also a field in $\mathfrak{L}$; it follows then from [2, Lemma 1] that there is a maximal field L in $\mathfrak{L}$, and we can define A to be the Abelian variety over K defined by L and define F to be a homomorphism of A^* onto A associated with the subfield L of L^*. Consider now the homomorphism H and the extension $K^*(v)$ of K over which H is defined; then the last assertion of our theorem is equivalent to the statement that if we replace the fields K and K^* by $K(v)$ and $K^*(v)$ respectively, and if we denote by $\mathfrak{L}_v$ the aggregate of all subfields in $L^*(v)$ of the form $K(v, F_0(x))$, where F_0 is any rational homomorphism, defined over $K^*(v)$, of A^* onto an Abelian variety defined over $K(v)$, then the maximal field L_v in $\mathfrak{L}_v$ is the field $L(v)$. To show this, we consider a field K_1 which is the separably-algebraic closure of an infinite transcendental extension of K; it is well known (MacLane [4, Theorem 9(a) and Theorem 15]) that any finitely generated extension of K in K_1 is a separably generated extension of K, and it is easily seen that for any finitely and separably generated extension of K there exists an isomorphic extension of K in K_1 which is independent with respect to it over K. If we now apply the preceding argument with K_1 replacing K as the ground field and if A is the so obtained Abelian variety over K_1 and F is a homomorphism of A^* onto A associated with it, then there is a finitely and separably generated extension of K in K_1 over which both A and F are defined. It is clearly sufficient to prove our assertion for the case where $K(v)$ is such a field, so that A is the Abelian variety over $K(v)$ defined by L_v. Let v' be a generic specialization of v over K such that $K(v')$ is contained in K_1 and is independent with respect to $K(v)$ over K, and let $L_{v'}$ be the maximal field in the aggregate $\mathfrak{L}_{v'}$, where $\mathfrak{L}_{v'}$ is defined similarly as $\mathfrak{L}_v$ with $K(v')$ replacing $K(v)$; similarly, let $L_{(v,v')}$ be the maximal field in the aggregate $\mathfrak{L}_{(v,v')}$. Then it follows from the definition of A that we have the relation $L_{(v,v')} = L_v(v')$, and since $L_{(v,v')} \supset L_vL_{v'} \supset L_v(v')$, we have the relation $L_vL_{v'} = L_v(v')$. It follows then from Theorem 3 that $L^* \cap L_v$ contains an Abelian function field $K(z)$ over K, independent with respect to $K(v)$ over K, such that $L_v = K(z, v)$; since $K(z)$ is evidently a field in $\mathfrak{L}$, we have the relation $L \supset K(z)$ and hence $L(v) \supset K(z, v) = L_v$, which shows that $L(v) = L_v$. This concludes the proof of our theorem.

The Abelian variety A in Theorem 4 is called the *K-image of A^* over K^**, and the rational homomorphism F is called the *canonical homomorphism* of A^* onto A. The K-image A of A^* over K^* is uniquely determined up to a regular isomorphism over K, and the canonical homomorphism F of A^* onto A is uniquely determined up to regular isomorphisms of A^* and A over K^* and K respectively. If K_1 is any separably generated extension of K, independent with respect to K^* over K, then the K-image A of A^* over K^* is also the K_1-image of A^* over K^*K_1, with the same canonical homomorphism F; furthermore, it is easily seen from Theorem 2, Corollary 1, and the proof of

Theorem 4 that if K_1^* is any primary extension of K^*, then the K-image A of A^* over K^* is also the K-image of A^* over K_1^*, with the same canonical homomorphism F.

COROLLARY. *Let K_1 be any extension of K, independent with respect to K^* over K, and let A_1 be the K_1-image of A^* over K_1K^* and F_1 be the canonical homomorphism of A^* onto A_1; then there is a rational isomorphism I of A_1 onto A, defined over K_1, such that $F=IF_1$.*

Proof. Without any loss of generality, we can assume that K_1 is a purely inseparable extension of K, for otherwise we can replace K by its separably-algebraic closure in K_1, and A would still remain the K-image of A^*. Let F_1 be the canonical homomorphism of A^* onto A_1; then according to Theorem 4, F is the product of F_1 and a rational homomorphism I of A_1 onto A, defined over K_1. It follows then that A is also the K-image of A_1 over K_1, and I is the corresponding canonical homomorphism. Since K_1 is purely inseparable over K, there exists a positive integer e such that the p^eth power A_2 of A_1 is an Abelian variety over K and there is a rational isomorphism I_1 of A_1 onto A_2, defined over K_1. According to Theorem 4, I_1 is the product of I and a rational homomorphism of A onto A_2; this shows that I must be an isomorphism.

Theorem 4 enables us to obtain a stronger form of Theorem 2 in our note [2], where the hypothesis "if both B and H are defined over K" is replaced by "if both B and H are defined over a separably generated extension of K." We shall state this stronger result here, as a theorem, mainly on account of its own interest, although we shall also find it convenient to use it in the proof of the next theorem.

THEOREM 5. *Let A be an Abelian variety over a field K, and let X be an algebraic subgroup in A, normally algebraic over K; let $A(X, K)$ be the quotient Abelian variety of A over X relative to K, and let F be the canonical homomorphism of A onto $A(X, K)$. If H is any rational homomorphism of A into an Abelian variety B with X as the kernel and if both B and H are defined over a separably generated extension $K(v)$ of K, then H is the product of F and a rational isomorphism of $A(X, K)$ into B, defined over $K(v)$.*

Proof. Since X is normally algebraic over K, there is a purely inseparable extension K_1 of K such that X is normally algebraic and separable over K_1; according to [2, Theorem 1], the quotient variety $A(X)$ and the canonical homomorphism F_1 of A onto $A(X)$ are both defined over K_1. Let $A(X, K)$ be the K-image of $A(X)$ over K_1 and F_2 be the canonical homomorphism of $A(X)$ onto $A(X, K)$, and set $F=F_2F_1$; since both A and $A(X, K)$ are defined over K, it follows from Theorem 2, Corollary 1, that F is also defined over K. It is then easily seen from Theorem 4 and [2, Theorem 1], that the so defined $A(X, K)$ and F have the property stated in the theorem.

THEOREM 6. *Let K^* be a primary extension of K, and let A^* be an Abelian* *variety over K^*; let K_1 be any extension of K, independent with respect to K^* over K. If there is a rational homomorphism H of an Abelian variety B onto A^*, and if B is defined over K_1 and H is defined over K^*K_1, then the canonical homomorphism F of A^* onto its own K-image A over K^* is an isomorphism.*

Proof. Without any loss of generality, we can assume that K is algebraically closed and $K_1=K$; for otherwise we can replace K by $\overline{K}_1$ and K^* by $K^*\overline{K}_1$, and if the canonical homomorphism F_1 of A^* onto its $\overline{K}_1$-image A_1 over $K^*\overline{K}_1$ is an isomorphism, then according to Theorem 4, Corollary, the canonical homomorphism F is the product of F_1 and a rational isomorphism of A_1 onto A, and hence is also an isomorphism.

Let X be the kernel of H in B; since X is normally algebraic over K^* and since B is defined over K, it follows from Theorem 1 that X is also normally algebraic over K. According to Theorem 5, H is the product of the canonical homomorphism of B onto $B(X, K)$ and a rational isomorphism R of $B(X, K)$ onto A^*, defined over K^*. Let x be a generic point of $B(X, K)$ over K; then $y=R(x)$ is a generic point of A^* over K^* and the field $K^*(x)$ is a purely inseparable extension of $K^*(y)$. Let e be the exponent of $K^*(x)$ over $K^*(y)$, and consider the p^eth power A_2 of $B(X, K)$, whose function field over K^* is isomorphic to $K^*(x^{p^e})$, where x^{p^e} denotes the point obtained from x by raising all coordinates of x to the p^eth power. Since we have $K^*(x)\supset K^*(y)\supset K^*(x^{p^e})$, $K^*(y)$ is evidently a purely inseparable extension of $K^*(x^{p^e})$; it follows that there is a rational isomorphism of A^* onto A_2, which according to Theorem 4 must be a product of F and a rational homomorphism of A onto A_2. This shows that F must be an isomorphism.

REMARK. It is easily seen that the condition in Theorem 6 is not only sufficient but also necessary.

COROLLARY 1. *Let K^* be a primary extension of K, and let A^* be an Abelian variety over K^*; if the canonical homomorphism F of A^* onto its K-image A over K^* has a finite kernel, then F is an isomorphism.*

Proof. According to Weil [6, Theorem 27], there is a rational homomorphism of A onto A^*, defined over K^*, and the corollary then follows from Theorem 6.

COROLLARY 2. *Let K^* be a primary extension of K, and let A^* be an Abelian variety over K^*; then the kernel of the canonical homomorphism F of A^* onto its K-image A over K^* is an Abelian subvariety in A^*.*

Proof. The kernel X of F is an algebraic subgroup in A^*, normally algebraic over K^*; let X_0 be the maximal Abelian variety contained in X, which is also normally algebraic over K^*, and consider the quotient variety $A^*(X_0, K^*)$ and the canonical homomorphism F^* of A^* onto $A^*(X_0, K^*)$, both defined over K^*. According to Theorem 5, F is the product of F^* and a rational homomorphism H of $A^*(X_0, K^*)$ onto A, defined over K^*, and it is easily seen that A is also the K-image of $A^*(X_0, K^*)$ over K^* and H is the canonical homomorphism. Since the kernel of H is isomorphic to the quotient group X/X_0 and hence is finite, it follows from Corollary 1 that H is an isomorphism; this shows that $X=X_0$.

5. **The K-trace.**

THEOREM 7. *Let K^* be a primary extension of K, and let A^* be an Abelian variety over K^*. Then there exists a uniquely determined Abelian subvariety A_0^* in A^*, defined over K^*, with the properties*: (1) *If H is a rational homomorphism of an Abelian variety B into A^*, and if B is defined over any extension $K(v)$ of K, independent with respect to K^* over K, and H is defined over $K^*(v)$, then $H(B)$ is contained in A_0^**, (2) *A_0^* is the smallest Abelian subvariety in A^* which has this property. Furthermore, A_0^* is the only Abelian subvariety in A^* which is isogenous to the K-image of A^* over K^*, and the canonical homomorphism F_0 of A_0^* onto its own K-image A_0 over K^* is an isomorphism.*

Proof. Let K_1 and K_2 be two extensions of K which are both independent with respect to K^* over K, and let B_1 and B_2 be Abelian varieties defined over K_1 and K_2 respectively; let H_1 and H_2 be rational homomorphisms of B_1 and B_2 respectively into A^*, and let A_1^* and A_2^* be the respective image varieties in A^*. Let $K^{*\prime}$ be an extension of K^* which is isomorphic to K^*K_2 over K and is independent with respect to K^*K_1 over K^*, and let K_2' be the subfield in $K^{*\prime}$ which is the image of K_2 under the isomorphism between K^*K_2 and $K^{*\prime}$; then K_2' is independent with respect to K^*K_1 over K, and hence $K_2'K_1$ is independent with respect to K^* over K, and we have the relation $K^*K_2' =K^{*\prime}$. Let B_2' be the Abelian variety defined over K_2' which is the isomorphic image of B_2 under the isomorphism between K_2 and K_2' over K; since K^*K_2 is isomorphic to K^*K_2' over K^*, there is a rational homomorphism H_2' of B_2' into A^* such that $H_2'(B_2')=H_2(B_2)=A_2^*$. The product Abelian variety $B_1\times B_2'$ is then defined over K_1K_2', which is independent with respect to K^* over K, and $H_1pr_1+H_2pr_2$ is a rational homomorphism of $B_1\times B_2'$ into A^* which has as its image the Abelian subvariety in A^* generated by A_1^* and A_2^*. Consider now the aggregate of all Abelian subvarieties in A^* which are rationally homomorphic images of Abelian varieties defined over fields which are independent with respect to K^* over K; it follows from what we have just said that there is an Abelian subvariety A_0^* in this aggregate which contains every variety in the aggregate. It is clear that A_0^* has the properties (1) and (2); furthermore, it follows from Theorem 6 that F_0 is an isomorphism of A_0^* onto A_0. Consider now the K-image A of A^* over K^*; according to the Poincaré Complete Reducibility Theorem (Weil [6, Theorem 26 and Corollary 2]), there exists a homomorphism H of A onto an Abelian subvariety of the same dimension in A^*, and it can be easily seen from the cited proof of this theorem that H is defined over K^*. It follows that $H(A)$ is contained in A_0^*, so that the dimension of A_0^* is not less than that of A. On the other hand, by the same theorem, there exists a homomorphism H_1 of A^* onto A_0^*, defined over K^*; then F_0H_1 is a homomorphism of A^* onto A_0, defined over K^*, and since F_0H_1 is according to Theorem 4 the product of the canonical homomorphism F of A^* onto A and a homomorphism of A onto A_0, it follows that the dimension of A_0 and hence also the dimension of A_0^* cannot be greater than that of A. Thus $H(A)$ and A_0^* have the same dimension and hence we have $H(A)=A_0^*$, so that A and A_0^* are isogenous over K^* (and hence A and A_0 are isogenous over K); and it is clear from our argument that the Abelian subvariety A_0^* is the only Abelian subvariety in A^* which is isogenous to the Abelian variety A.

The Abelian variety A_0^* in Theorem 7 is called the *K-maximal Abelian subvariety over K^* in A^**; it is uniquely determined by the Abelian variety A^* and the fields K^* and K. If K_1 is any extension of K, independent with respect to K^* over K, then the K-maximal Abelian subvariety A_0^* over K^* in A^* is also the K_1-maximal Abelian subvariety over K^*K_1 in A^*; and if K_1^* is any primary extension of K^*, then the K-maximal Abelian subvariety A_0^* over K^* in A^* is also the K-maximal Abelian subvariety over K_1^* in A^*.

THEOREM 8. *Let K^* be a regular extension of K and let A^* be an Abelian variety over K^*. Then there exists an Abelian variety A' over K, uniquely determined up to a regular isomorphism over K, and a rational isomorphism F of A' into A^*, defined over K^*, such that if B is an Abelian variety defined over any extension K_1 of K, independent with respect to K^* over K, and if H is a rational homomorphism of B into A^*, defined over K^*K_1, then H is the product of F and a rational homomorphism of B into A', defined over K_1.*

Proof. Without any loss of generality, we can assume that $K^*=K(u)$ is a finitely generated extension of K; let A be the K-image of A^* over K^* and A_u be the maximal K-Abelian subvariety over K^* in A^*, and let T_u be a rational homomorphism of A onto A_u, defined over $K(u)$. Let $u_1(=u), u_2, \cdots, u_m$ be m independent generic specializations of u over K; if S_n is the ambient projective space of A_u, then there exist Abelian varieties $A_{u_1}, \cdots, A_{u_m}$ in S_n which are respectively the specializations of A_u over the specializations $u_1, \cdots, u_m$ of u over K. If y is a generic point of A over $K(u_1, \cdots, u_m)$, then for each $i=1, \cdots, m$, the point $x_i=T_{u_i}(y)$ is a generic point of A_{u_i} over $K(u_1, \cdots, u_m)$. The correspondence $y\rightarrow(x_1, \cdots, x_m)$ then defines a rational homomorphism T_m of A into the product variety $A_{u_1}\times \cdots \times A_{u_m}$, defined over $K(u_1, \cdots, u_m)$, and the image $T_m(A)$ is an Abelian variety, defined over $K(u_1, \cdots, u_m)$, whose projection into the first factor $A_{u_1}=A_u$ is a rational isomorphism onto A_u, i.e., $pr_1(T_m(A))=sA_u$, where s is the degree of $K(u_1, \cdots, u_m, x_1, \cdots, x_m)$ over $K(u_1, \cdots, u_m, x_1)$.

Since $K(u_1, \cdots, u_m, y)$ is a finite extension of $K(u_1, \cdots, u_m, x_1)$ whose degree is independent of the integer m, and since we have $K(u_1, \cdots, u_m, y)$

$\supset K(u_1, \cdots, u_m, x_1, \cdots, x_m) \supset K(u_1, \cdots, u_m, x_1)$, it follows that the degree s of $K(u_1, \cdots, u_m, x_1, \cdots, x_m)$ over $K(u_1, \cdots, u_m, x_1)$ is bounded for all m. We shall assume that the integer m has been chosen so large that this degree reaches the maximum; let $u_1', \cdots, u_m'$ be m independent generic specializations of u over K, independent with respect to $u_1, \cdots, u_m$, and y over K, and set $x_i' = T_{u_i'}(y)$, $i=1, \cdots, m$. Since $K(u_1, \cdots, u_m, u_1', \cdots, u_m', x_1, \cdots, x_m, x_1', \cdots, x_m')$ contains $K(u_1, \cdots, u_m, u_1', \cdots, u_m', x_1, \cdots, x_m)$ and has the same degree s over $K(u_1, \cdots, u_m, u_1', \cdots, u_m', x_1)$ as the latter, we must have the relation $K(u_1, \cdots, u_m, u_1', \cdots, u_m', x_1, \cdots, x_m, x_1', \cdots, x_m') = K(u_1, \cdots, u_m, u_1', \cdots, u_m', x_1, \cdots, x_m)$ and hence also the relation $K(u_1, \cdots, u_m, u_1', \cdots, u_m', x_1, \cdots, x_m) = K(u_1, \cdots, u_m, u_1', \cdots, u_m', x_1', \cdots, x_m')$. This shows that the Abelian variety $T_m(A)$ satisfies the hypothesis of Theorem 3, Corollary 1′, and hence there exists an Abelian variety A' over K, uniquely determined up to a regular isomorphism over K, such that for every sufficiently large integer m, there is a regular isomorphism R_m of A' onto $T_m(A)$. We set $F = pr_1 R_m$.

Consider now the rational homomorphism H of B into A^*, and let w be a generic point of B over $K_1(u_1, \cdots, u_m)$, where we now assume that $K(u_1, \cdots, u_m)$ is independent with respect to K_1 over K; according to Theorem 7, the point $H(w)$ is contained in A_u, hence there exists a point y_0 in A such that $T_u(y_0) = H(w)$. The correspondence $w \to S[T_u^{-1}(H(w))] - S[T_u^{-1}(0)]$ defines a rational homomorphism of B into A (where the symbol S has the meaning given in Weil [6, pp. 28–29]), defined over $K_1(u)$; according to Theorem 2, Corollary 1, this homomorphism is also defined over K_1 and hence $S[T_u^{-1}(H(w))] - S[T_u^{-1}(0)]$ is a rational point over $K_1(w)$ in A. Furthermore, the points in $T_u^{-1}(0)$ constitute a finite subgroup in A, normally algebraic over $K(u)$; according to Theorem 1, this subgroup is also algebraic over K. It follows then that the point $y_1 = S[T_u^{-1}(H(w))]$ is algebraic over $K_1(w)$. If d is the degree of T_u, then we have $y_1 = S[T_u^{-1}(H(w))] = S[T_u^{-1}(T_u(y_0))] = dy_0$; it follows then that the point y_0 is algebraic over $K_1(y_1)$ and hence also algebraic over $K_1(w)$.

Now, for each $i=1, \cdots, m$, let H_{u_i} be the specialization of H over the specialization $u \to u_i$ over K_1; then H_{u_i} is a rational homomorphism of B into A_{u_i}, and since $K_1(u_i)$ and $K_1(w, y_0)$ are linearly disjoint over K_1, we have the relation $H_{u_i}(w) = T_{u_i}(y_0)$. The correspondence $w \to (H_{u_1}(w), \cdots, H_{u_m}(w)) = (T_{u_1}(y_0), \cdots, T_{u_m}(y_0))$ then defines a rational homomorphism H_m of B into $T_m(A)$, defined over $K(u_1, \cdots, u_m)$; and for sufficiently large m, we have then $H(w) = pr_1 H_m(w) = pr_1 R_m R_m^{-1} H_m(w) = F R_m^{-1} H_m(w)$. Since $R_m^{-1} H_m$ is a rational homomorphism of B into A', defined over $K_1(u_1, \cdots, u_m)$, it is also defined over K_1, according to Theorem 2, Corollary 1. This concludes the proof of our theorem.

The Abelian variety A' in Theorem 8 is called the *K-trace of A^* over K^**, and the rational isomorphism F is called the *canonical isomorphism* of A' into

A^*. The K-trace A' of A^* over K^* is uniquely determined up to a regular isomorphism over K, and the canonical isomorphism F of A' into A^* is uniquely determined up to regular isomorphisms of A' and A_0^* over K and K^* respectively, where A_0^* is the K-maximal Abelian subvariety over K^* in A^*. If K_1 is any extension of K, independent with respect to K^* over K, then the K-trace A' of A^* over K^* is also the K_1-trace of A^* over K^*K_1, with the same canonical isomorphism F; furthermore, it is easily seen from Theorem 2, Corollary 1, and the proof of Theorem 8 that if K_1^* is any regular extension of K^*, then the K-trace A' of A^* over K^* is also the K-trace of A^* over K_1^*, with the same canonical isomorphism F. In fact, in order that the latter statement holds, it is sufficient to assume that K_1^* is a primary extension of K^* and is also a regular extension of K.

COROLLARY. *Let $K^*K(u)$ be a regular extension of K, and let A^* be an Abelian variety over K^*; let A' be the K-trace of A^* over K^*, and let F be the canonical isomorphism of A' into A^*. Let $u_1, \cdots, u_m$ be independent generic specializations of u over K, and let $A_1^*, \cdots, A_m^*$ and $F_1, \cdots, F_m$ be the corresponding specializations of A^* and F respectively; if y is a generic point of A' over $K(u_1, \cdots, u_m)$, then the correspondence $y \to (F_1(y), \cdots, F_m(y))$ defines a regular isomorphism of A' into the product variety $\prod_{i=1}^m A_i^*$, provided the integer m is taken sufficiently large.*

This follows almost immediately from the proof of Theorem 8.

There arises naturally the question whether a result similar to this corollary also holds for the K-image A of A^* over K^*, that is, in the notation of Theorem 4, whether the correspondence $(x_1, \cdots, x_m) \to \sum_{i=1}^m F_i(x_i)$ defines a regular homomorphism of $\prod_{i=1}^m A_i^*$ onto A, if the integer m is taken sufficiently large. If this is true, then it would follow that, in case K^* is a regular extension of K, the K-image of A^* over K^* is also the K_1-image of A^* over K^*K_1 for any extension K_1 of K which is independent with respect to K^* over K. That the answer to this question is in the affirmative will be proved in our forthcoming paper on Picard varieties, as an application of the principle of duality developed there; however, it would be interesting and desirable to have a direct proof of this fact by the methods used in this paper.

Notes. (Added in proof, January 8, 1955.) (1) Professor Weil has kindly communicated to us a new, simpler proof of our Theorem 1 in §1, based on the Theorem 7 and Proposition 25 in his book [6]. Using the notations in the proof of Theorem 1, as well as the simplifying assumptions introduced there, Weil's proof can be summarized as follows. Let W be a subvariety of dimension $r-d$ (r and d being the dimensions of A and X respectively) in A, defined over K, having at 0 a simple point and transversal to X at 0; then, according to [6, Proposition 25], the correspondence $y \to S(X \cdot W_y)$ defines up to a translation a rational homomorphism H of A onto X, defined over $K(u)$. On the other hand, the correspondence $y \times u \to H(y)$ defines a rational transformation F of $A \times U$ into A, defined over K; according to [6, Theorem 7], we have $F(y, u) = F_0(y) + F_1(u)$, where F_0 and F_1 are rational transformations of A and U respectively into A, both defined over K. It follows then that we must have $H(y) = F_0(y) - F_0(0)$, so that H is also defined over K, and this shows that X as the image variety of H is defined over K.

(2) We should like to make a few remarks on the proof of Theorem 3 in §3; for the sake of simplicity, we shall limit ourselves to the case where $K(v)$ is a regular extension of K, i.e. the case dealt with in the Corollary 1, so that the existence of A^* is not needed. We recall that, in the proof of Theorem 3, the following construction is used to obtain an Abelian variety defined over K which is regularly isomorphic over $K(v)$ to a given Abelian variety A defined over $K(v)$, after we have shown that there exists an Abelian variety A_1' defined over an algebraic extension $K(v_1')$ of K which is regularly isomorphic to A over $K(v, v_1')$: Let $v_1', \cdots, v_d'$ be the complete set of conjugates of v_1' over K, and let $A_1', \cdots, A_d'$ be the corresponding specializations of A_1' over K; let y be a generic point of A over $K(v)$, and let $y_1', \cdots, y_d'$ be the corresponding generic points of $A_1', \cdots, A_d'$ over $K(v, v_1'), \cdots, K(v, v_d')$ respectively; finally, let z be the associated point of the 0-cycle $\sum_{i=1}^{d} (y_i')$ as a cycle in the ambient projective space. Since z is rational over $K(v, y)$, it has a locus V over $K(v)$; then, under the additional assumptions of Theorem 3, it is shown that V is an Abelian variety defined over K and is regularly isomorphic to A over $K(v)$. Professor Weil has kindly called our attention to the fact that a similar construction has been used by T. Matsusaka in his paper *Some theorems on Abelian varieties* (Natural Science Report, Ochanomizu University, vol. 4, 1953, pp. 22–35), in the proof of his main theorem (Theorem 3), for a similar purpose; in his case, the variety A is already defined over K, but is assumed to have only a normal law of composition, and the object is to prove that it is birationally equivalent to an Abelian variety defined over K. This method of construction seems to be useful also for other problems where one wishes to narrow down the field of definition of a certain type of varieties; in fact, in a paper to be published soon in the American Journal of Mathematics, Weil has applied this method in his theory of "transformation spaces" and succeeds in generalizing to abstract varieties some of the results of Matsusaka and ourselves.

In this connection, we should like to add the following comments, which will throw some further lights on the relation between the main result of Matsusaka's paper and our Theorem 3. If we remove from both the hypothesis and the conclusion of Theorem 3, or rather Corollary 1 of Theorem 3, the condition that the function fields involved are Abelian function fields, then the theorem will still be true, for we need only to omit in the proof all references to group properties. In fact, in this case, a much simpler proof of the theorem can be given as follows, using the notations in the proof of Theorem 3, whereby we shall again restrict ourselves to the case of Corollary

1, for the sake of simplicity. Let X and W be the loci of (y', v') and v' over K respectively; the condition $K(v, v', y)=K(v, v', y')$ shows that there is a birational correspondence T of the product variety $A\times W$ onto X, defined over $K(v)$. Since $y\times W$ is a subvariety in $A\times W$, defined over $K(v, y)$, and since it contains a point (y, v') which is a generic point of $A\times W$ over $K(v)$, the image variety $T(y\times W)$ is defined and is a prime rational cycle over $K(v, y)$. If z is the associated point of $T(y\times W)$, then the correspondence $y\rightarrow z$ defines a rational transformation R of A, defined over $K(v)$, which can be easily seen to be birational. Since the generic point v of W can be replaced by any other generic point, and since $K(v)$ is assumed to be regular over K, it follows that the image variety $V=R(A)$ must be defined over K; furthermore, if we now add the assumption that A is Abelian, then the group composition in A induces on V a normal law of composition, which can be easily seen to be also defined over K. Thus we have given a new proof of our Theorem 3, with however the weaker conclusion that the variety V has a normal law of composition, instead of being an Abelian variety; this was, in fact, our first (incomplete) proof of this theorem, and it is precisely in order to obtain the stronger conclusion, that V is an Abelian variety, that we introduced an algebraic specialization of A' and applied the method of construction outlined above. Now, the main theorem in Matsusaka's paper (which unfortunately was not known to us at the time when this paper was written) asserts that any variety with a normal law of composition defined over K is birationally equivalent over K to an Abelian variety defined over K, and this is exactly what we need in order to complete the above argument into a proof of our Theorem 3, at least in the case of Corollary 1. This new proof, though in reality longer and more complicated than the one given in the text, considering the length of Matsusaka's proof of his main theorem, has nevertheless the advantage of conceptual simplicity.

(3) As we have observed after Theorem 4, the K-image A of A^* over K^* remains to be the K-image of A^* over K_1^* for any primary extension K_1^* of K^*. Professor Matsusaka has kindly pointed out to us that this is in a sense the exact extent to which the K-image is independent of the choice of the reference field K^*, and that a counter example can be constructed where K_1^* is even a regular extension of K, though of course no longer a primary extension of K^*. A simple special case of the example communicated by Matsusaka is the following. Let K be an algebraically closed field of characteristic zero, let u be a variable over K, and consider the elliptic curve $C(u)$ in the plane defined by the equation $y^2=uf(x)$ over the field $K(u)$, where $f(x)$ is a polynomial of degree 4 (or 3) with coefficients in K and having no multiple roots; since $C(u)$ has evidently a rational point over $K(u)$, it is an Abelian variety over $K(u)$. If u' is another variable over K, independent with respect to u over K, then the curve $C(u')$ cannot be birationally equivalent over $K(u, u')$ to the curve $C(u)$; for, if it were so, there would be a birational transformation

of $C(u)$ onto $C(u')$, defined over $K(u, u')$, which leaves the variable x invariant and hence must transform the variable y by the correspondence $y \to (u'/u)^{1/2} y$, which is in contradiction to the fact that birational transformation is defined over $K(u, u')$. This shows that the K-image of $C(u)$ over $K(u)$ is zero, i.e. reduces to a point, while on the other hand it is obvious that the K-image of $C(u)$ over $K(u^{1/2})$ is the elliptic curve C defined by the equation $y^2 = f(x)$. This situation is further confirmed by the observation, based on some of the results in our forthcoming paper on Picard varieties, that the K-images of $C(u)$ over $K(u)$ and $K(u^{1/2})$ are respectively the Albanese varieties of the surfaces in 3-space defined by the equations $y^2 - zf(x) = 0$ and $y^2 - z^2 f(x) = 0$ over K; the former is a rational surface and hence has a zero Albanese variety, while the latter is a ruled surface, birationally equivalent to the product of the curve C with a projective line.

References

1. W. L. Chow, *On Picard varieties*, Amer. J. Math. vol. 74 (1952) pp. 895–909.

2. ———, *On the quotient variety of an Abelian variety*, Proc. Nat. Acad. Sci. U.S.A. vol. 38 (1952) pp. 1039–1044.

3. W. L. Chow and B. L. van der Waerden, *Zur algebraischen Geometrie*. IX. *Ueber zugeordnete Formen und algebraische Systeme von algebraischen Mannigfaltigkeiten*, Math. Ann. vol. 113 (1937) pp. 692–704.

4. S. MacLane, *Modular fields*, I. *Separating transcendental bases*, Duke Math. J. vol. 5 (1939) pp. 372–393.

5. A. Weil, *Foundations of algebraic geometry*, New York, 1946.

6. ———, *Variétés abéliennes et courbes algébriques*, Paris, 1948.

The Johns Hopkins University,
Baltimore, Md.

Annals of Mathematics
Vol. 64, No. 3, November, 1956
Printed in U.S.A.

ON EQUIVALENCE CLASSES OF CYCLES IN AN ALGEBRAIC VARIETY*

By Wei-Liang Chow

(Received December 30, 1955)

1. It is well-known that there is a close analogy between the intersection theories in the abstract algebraic geometry and in topology; in fact, this similarity is more than an analogy, for, in the algebraic geometry over the complex field, the algebraic cycles are also topological cycles and their algebraic intersections coincide with their topological intersections. However, as has been observed by Weil in the preface to his *Foundations of Algebraic Geometry* [6][1], there is one important difference; that is, in topology the theory of intersection can be extended to an algebra of homology classes of cycles, while in the abstract algebraic geometry, because of the lack of a suitably developed concept of equivalence, one has to restrict oneself to operate with the cycles themselves, with the resulting inconvenience that the intersection-product is not always defined. The purpose of this paper is to show that this restriction can be removed, and that also in abstract algebraic geometry we can operate with classes of cycles (under a suitably chosen concept of equivalence) instead of the cycles themselves, and develop an algebra of equivalence classes closely analogous to the algebra of homology classes in topology.

The idea of replacing cycles by equivalence classes of cycles in intersection theory is not new in algebraic geometry. It was implicitly contained in the Schubert calculus, where numerical equivalence was the main object of study. A more explicit attempt to introduce the concept of equivalence, mainly the algebraic equivalence, was made by Severi with his theory of "virtual varieties" and the "functional equivalent" of an intersection; the ideas of Severi were later developed and amplified by van der Waerden in his paper on intersection theory [5], and in the recent book of Hodge and Pedoe [2] the results of van der Waerden were developed further and presented in detail. All this, however, falls short of a full theory of intersection for equivalence classes of cycles, and as our recent work on the theory of algebraic vector bundles shows the urgent need of such a theory, not only for algebraic equivalence but also for the more restrictive rational equivalence, we propose to develop such a theory here. Just as the treatment in the book of Hodge and Pedoe, our theory is also based on the ideas of Severi and van der Waerden, and in this sense we do not claim any great originality; nevertheless, apart from the fact that it is desirable to have a systematic treatment of a subject which lies at the very foundation of algebraic geometry, a discerning reader will easily perceive that such a systematic

* This work was supported by a research grant of the National Science Foundation, NSF-G1714.

[1] Hereafter, we shall refer to this book as *Foundations*.

treatment does involve substantially more than what can be readily obtained from the papers of Severi and van der Waerden.

We shall follow in general the terminology of Weil's *Foundations*, except for a few modifications which we shall introduce here and in the next section. We shall deal exclusively with varieties embedded in a projective space, so that we shall not always mention this fact specifically; we say that a variety is embedded in a projective space (or any other variety) if it is a subset in the latter, but not necessarily a subvariety, as the variety is not assumed to be complete. More precisely, we shall call a subset V in a projective space S a variety if there exists a complete variety $\bar{V}$ in S (i.e. a variety in the ordinary sense) such that $\bar{V}$ contains V and that $\bar{V} - V$ is a bunch of subvarieties in $\bar{V}$; the projective space is then called the *ambient space* of V. As one will see from the methods used in our proofs, this assumption of projective embedding is an essential limitation of our theory, and the question is still open whether the theory can be extended to all abstract varieties in the sense of Weil. On the other hand, it is a new feature of our theory, as against the previous treatments, that it is not limited to complete varieties; in fact, for an incomplete variety there are two (or three) possible interpretations of our theory, which we shall explain in the next section. All varieties are assumed to be topologized with the Zariski topology, in which the closed subsets are the bunches of subvarieties; this will enable us to use the terminology of topology without any further explanations. If S is the ambient projective space of a variety V, we shall denote by $\bar{V}$ the variety which is the closure of V in S; more generally, if $X = \sum n_i V_i$ is a cycle in V, where V_i are simple subvarieties in V, we shall denote by $\bar{X}$ the cycle $\sum n_i \bar{V}_i$ in S and we shall call this cycle $\bar{X}$ the *closure* of X in S. Also following the terminology of topology, we shall define the *support* of a cycle X in a variety V to be the bunch of subvarieties in V consisting of all the components of X, and we shall denote the support of X by $| X |$.

Let V be a variety of dimension n, and let W_1 and W_2 be bunches of subvarieties of dimensions s and t respectively in V; then a component of $W_1 \cap W_2$ is said to be *proper* in V if it has the dimension $s + t - n$ (we note that, in contrast to Weil's terminology, we do not assume that the component is simple in V), and a component of $W_1 \cap W_2$ is said to be *negligible* in V if it has a dimension less than $s + t - n$. Let W be a subvariety in V and let K be a field of definition for V; we shall say that W is *general* over K for V if W contains a generic point of V over K, and a cycle X in V is said to be *general* over K for V if every component of X is general over K for V. If V' is a variety contained in V (not necessarily a subvariety in V) and if W is a subvariety in V, then the *restriction* $R(W, V')$ of W in V' is defined to be $W \cap V'$ in case $W \cap V'$ is a variety such that $(\overline{W \cap V'}) \cap V = W$, and is defined to be the empty set otherwise; if $X = \sum n_i V_i$ is a cycle in V, then the *restriction* of X in V' is defined by the formula $R(X, V') = \sum n_i R(V_i, V')$, whereby the empty set represents the zero cycle and the summation extends over all i such that $R(V_i, V')$ is simple in V'.

2. Since the theory of equivalence classes of cycles involves in an essential way the concepts of specialization and intersection of cycles, the contents of our theory will vary to a certain extent with the choice of the definitions of these concepts. Therefore, before proceeding to our main subject, we shall make a few observations concerning the various possible choices of the definitions of these concepts. We begin by recalling the definition of a specialization of a cycle in a projective space S. For a positive cycle X in S, the concept of a specialization of X is to be defined as in the theory of associated form [1], which we assume the reader is familiar with; thus, a positive cycle X_1 in S is said to be a specialization of the cycle X over a field K if the associated form of X_1 is a specialization of the associated form of X over K. For an arbitrary cycle X in S, we shall define a specialization X_1 of X over a field K in the following way. We set $X = X' - X''$, where X' and X'' are two positive cycles without any common components, and similarly we set $X_1 = X_1' - X_1''$; then the cycle X_1 is said to be a specialization of the cycle X over K, if there exists a positive cycle Z in S such that the pair of positive cycles $(X_1' + Z, X_1'' + Z)$ is a specialization of the pair (X', X'') over K. Consider now a variety V embedded in S, and let X be a cycle in V; if K is a field of definition for V, then a specialization X_0 of the cycle $\bar{X}$ in S, the sense just defined above, is contained in $\bar{V}$, i.e. $| X_0 |$ is contained in $\bar{V}$. The cycle X_0 in S may however fail to be the closure in S of a cycle in V, either because a component of X_0 is a singular subvariety in $\bar{V}$ or because a component of X_0 lies entirely outside of V, even though it may be a simple subvariety in $\bar{V}$; and in order to obtain from X_0 a cycle in V which one may take as the specialization in V of the cycle X, one may in each of the two cases choose either to ignore such components as happen to appear in X_0 or to restrict oneself to those cases where such components do not appear in X_0. By combining various choices of these alternatives, it is possible to get six different definitions for the concept of a specialization in V of a cycle X, but not all of them prove to be suitable for our present purpose, though they might very well be useful in other connections. It turns out that, for our present purpose, it is not very advisable to introduce any restrictions regarding the possible presence of a component in X_0 which is a singular subvariety in $\bar{V}$; for, any such restriction would entail a corresponding restriction on the definition of the intersection-product in order to preserve its compatibility with the specialization (Lemma 1 below), and this restriction would in turn impair the validity of the crucial Lemma 2 below, with the result that the intersection-product would not be always defined for any two cycle classes. Thus we are left with essentially two possible choices in the definition of a specialization in V, according as we do or do not impose the restriction that X_0 has no component which is simple for $\bar{V}$ and lies outside of V; this gives rise to two different theories, called respectively the relative and the absolute theories, which are somewhat analogous to the homology theory of an open manifold on the one hand and the homology theory of a complex, modulo a subcomplex, on the other hand in topology, though the terms relative and absolute are in our case more appropriately used the other way around.

(α) *The absolute theory.* If X is a cycle in V and if K is a field of definition for V, then a cycle X_1 in V is said to be a specialization in V of X over K, if there exists a cycle X_2 in S, whose components are either singular subvarieties in $\bar{V}$ or subvarieties in $\bar{V} - V$, such that the cycle $\bar{X}_1 + X_2$ is a specialization in S of the cycle $\bar{X}$ over K. With this definition of specialization in V, we can define the intersection-product $X.Y$ in V of two cycles X and Y in V exactly as in *Foundations*; we recall that the intersection-product $X.Y$ in V is defined in this case if and only if every component of $| X | \cap | Y |$ which is simple in V is proper in V. We observe that this definition of a specialization in V is essentially the same as the definition given by G. Shimura in a recent paper [4] for the general case of an abstract variety in the sense of Weil; this shows that the definition is really independent of the particular embedding of V in S (and thereby justifies the caption absolute), a fact which can also be easily shown directly.

(β) *The relative theory.* If X is a cycle in V and if K is a field of definition for V, then a cycle X_1 in V is said to be a specialization in V of X over K, if there exists a cycle X_2 in S, whose components are all singular subvarieties in $\bar{V}$, such that the cycle $\bar{X}_1 + X_2$ is a specialization in S of the cycle $\bar{X}$ over K; in other words, a cycle X_1 in V is a specialization in V of the cycle X over K in the relative sense if and only if $\bar{X}_1$ is a specialization in $\bar{V}$ of the cycle $\bar{X}$ over K in the absolute sense. With this definition of a specialization in V, it is advantageous, for the theory of equivalence classes of cycles, to consider the intersection-product $X.Y$ in V of two cycles X and Y in V as defined only under somewhat stricter conditions than those given in *Foundations*. We shall consider the intersection-product $X.Y$ in V as defined (and then it is defined as in *Foundations*) if and only if every component of $| \bar{X} | \cap | \bar{Y} |$ which is simple in $\bar{V}$ is proper in $\bar{V}$ and is the closure in $\bar{V}$ of a component of $| X | \cap | Y |$; in other words, the intersection-product $X.Y$ in V is defined in the relative sense if and only if the intersection-product $\bar{X}.\bar{Y}$ in $\bar{V}$ is defined in the absolute sense and is the closure in $\bar{V}$ of the intersection-product $X.Y$ in V in the absolute sense. In contrast to the absolute theory, the relative theory is in general not invariant under an everywhere biregular birational transformation of V and hence is dependent on the particular embedding of V in S; however, in view of the relations between the relative theory in V and the absolute theory in $\bar{V}$, it is clear that the relative theory in V depends in reality only on the embedding of V in $\bar{V}$, but is independent of the embedding of the latter in S. Thus the relative theory in V is essentially the theory of the complete variety $\bar{V}$ with special reference to the subset $\bar{V} - V$ in $\bar{V}$, in the sense that no component of a specialization or an intersection-product should be contained in $\bar{V} - V$.

We should like to add that if we are willing to dispense with the condition that the intersection-product of every pair of cycle classes be defined, then it is possible to restore the birational invariance of the relative theory by restricting ourselves to certain special classes of cycles and by a slight modification in the definitions. This leads us to what we should like to call the *restricted theory*, which we shall describe in a later section.

The theory which we shall develop in the following sections holds both in the absolute and in the relative senses, and we shall develop both theories at the same time by using the concepts of specialization and intersection-product without any further qualifications, which can then be interpreted in either sense, except when specifically stated to the contrary. We observe finally that the two theories coincide in case the variety V is complete.

3. For the sake of simplicity, we shall restrict ourselves to the theory of rational equivalence, which we shall define presently, partly because this equivalence has hitherto not been given much attention in the abstract algebraic geometry and also because it plays an important role in the theory of algebraic vector bundles which we shall develop in a forthcoming paper. It is easily seen that our theory holds with hardly any modifications also for algebraic or numerical equivalence.

Let V be a variety, X be a cycle in V, and K be a field of definition for V such that X is rational over K; we shall say that X is *rationally equivalent to zero in V over the field K*, if there exists a cycle X_u in V, rational over a purely transcendental extension $K(u)$ of K, such that both X and the zero cycle are specializations in V of X_u over K; and we shall say that a cycle is *rationally equivalent to zero in V* if it is so over some such field K. Since specialization is compatible with the addition (and subtraction) of cycles and since the compositum of two independent, purely transcendental extensions of a field is another such extension, it is easily seen that the set of all cycles of a given dimension s in V which are rationally equivalent to zero in V is a subgroup $G_r(V, s)$ in the group $G(V, s)$ of all s-cycles in V; in fact, $G_r(V, s)$ is evidently a subgroup in the group $G_a(V, s)$ of all s-cycles in V which are algebraically equivalent to zero. Two cycles X_1 and X_2 in V are said to be *rationally equivalent in V* to each other if the cycle $X_1 - X_2$ is rationally equivalent to zero in V, and the set of all cycles in V rationally equivalent in V to a given cycle X in V is called a *rational cycle class*, or in short a *cycle class*, determined by the cycle X. It is clear that the set of all cycle classes of a given dimension s in V constitutes a group, which is isomorphic to the quotient group $G(V, s)/G_r(V, s)$ in a canonical way.

Among the three standard operations on cycles, namely the product (symbol $\times$), the projection (symbol pr), and the intersection-product (symbol .), there are no difficulties in extending the first operation to the cycle classes, and, in case the varieties involved are complete, also the second operation. In fact, since the specialization of cycles is compatible with the operation product and, in case of complete varieties, also with the operation projection [4] Theorems 24 and 25, and since (in case of the product) the compositum of two independent, purely transcendental extensions of a field is again such an extension, it is easily seen that these two operations preserve the rational equivalence between cycles, so that they can also be considered as operations on the cycle classes. The situation is, however, somewhat different and more complicated with the operation intersection. On the one hand, the intersection-product is not defined for every two cycles in a variety, but only for those which intersect properly in it; in order to be able to define an intersection-product for every two cycle classes, as is the case with the homology classes in topology, we have to show that there always exist cycles in the two classes which intersect properly in the variety. On the other hand, the compatibility of specialization with the operation intersection holds in general only for positive cycles [4] Theorem 23,[2] so that, even if all the cycles involved intersect properly in the variety, it is not at all evident that two pairs of equivalent cycles will have equivalent intersections. Further complications also arise from the fact that the varieties are not assumed to be complete. Nevertheless, we shall show that, in spite of these complications, we can define for every two cycle classes in a variety an intersection-product which is a cycle class containing every intersection-product of any two cycles in the two cycle classes respectively, whenever it is defined.

Before proceeding with the main theory, we shall introduce here the concept of a *latent cycle of a specialization*. Let K be a field of definition for the variety V, and let X_1 be a specialization in V of a cycle X over K; as in Section 2, we set $X = X' - X''$, where X' and X'' are two positive cycles in V without any common components, and similarly we set $X_1 = X_1' - X_1''$. Then, according as we are dealing with the absolute or relative theory, there exists a positive cycle Z in V or $\bar{V}$ respectively such that the pair of cycles $(X_1' + Z, X_1'' + Z)$ or $(\bar{X}_1' + Z, \bar{X}_1'' + Z)$ is a specialization in V or $\bar{V}$ of the pair (X', X'') or $(\bar{X}', \bar{X}'')$ respectively over K. We shall call this positive cycle Z in V or $\bar{V}$ a *latent cycle* of the specialization $X \to X_1$ over K; and if, in case of the relative theory, the relation $Z = \overline{R(Z, V)}$ holds, we shall call the cycle $R(Z, V)$ a *latent cycle in* V of the specialization $X \to X_1$ over K. We observe that the latent cycle Z in this definition is in general not uniquely determined by the given specialization $X \to X_1$, though it can often be made so by adding further specifications to the given specialization. The possible existence of a latent cycle in a specialization is what destroys the general compatibility of specialization with the operation intersection, for, even if both X and X_1 intersect a cycle Y in V properly in V, it does not necessarily follow that the latent cycle Z also intersects Y or $\bar{Y}$ properly in V or $\bar{V}$, so that we cannot conclude that $X_1.Y$ is a specialization in V of $X.Y$ over K. Therefore, instead of the general principle of compatibility, we have for the present at our disposal only the restricted version in the following lemma.

LEMMA 1. *Let V be a variety, let X and Y be two cycles in V such that the intersection-product $X.Y$ in V is defined, and let K be a field of definition for V such that Y is rational over K; if X_1 is a specialization in V of X over K such that the intersection-product $X_1.Y$ in V is defined, and if there exists a latent cycle Z in V of the specialization $X \to X_1$ over K such that the intersection-product $Z.Y$ in V is*

[2] That specialization is not in general compatible with intersection of cycles can be easily shown by simple examples; this fact was first brought to my attention by C. Chevalley.

also defined, then $X_1.Y$ *is a specialization in* V *of* $X.Y$ *over the specialization* $X \to X_1$ *over* K.

PROOF. In view of our assumption concerning the latent cycle Z in V, it is sufficient to prove this lemma for the case where both X and Y (and hence also X_1) are positive cycles. Let S be the ambient projective space of V, and let Q be the union of $\bar{V} - V$ together with all the singular subvarieties in $\bar{V}$ in the absolute theory, or the union of all singular sunvarieties in $\bar{V}$ in the relative theory; then there exists a positive cycle X_2 in S, with support in Q, such that $\bar{X}_1 + X_2$ is a specialization in S of $\bar{X}$ over K. Let Z_1 be a specialization in S of $\overline{X.Y}$ over the specialization $\bar{X} \to \bar{X}_1 + X_2$ over K; then we have the relation $| Z_1 | \subset | \bar{X}_1 + X_2 | \cap | \bar{Y} |$, and since every component of $| R(\bar{X}_1 + X_2, V) | \cap | R(\bar{Y}, V) | = | X_1 | \cap | Y |$ is proper in V and is a component in $X_1.Y$, it follows from the intersection theory in V and the specialization theory in S that $Z_1 = \overline{X_1.Y} + D$, where D is a positive cycle in S with support in $| \bar{X}_1 + X_2 | \cap | \bar{Y} | \cap Q$. This shows that $X_1.Y$ is a specialization in V of $X.Y$ over the specialization $X \to X_1$ over K.

REMARK. The above proof shows also that there exists a cycle D in S, with support in $(| \bar{X}_1 | \cup | X_2 | \cup | \bar{Z} |) \cap | \bar{Y} | \cap Q$, such that $\overline{X_1.Y} + D$ is a specialization in S of $\overline{X.Y}$ over the specialization $\bar{X} \to \bar{X}_1 + X_2$ over K; furthermore, every cycle D in S with this latter property must have its support in $(| \bar{X}_1 | \cup | X_2 | \cup | \bar{Z} |) \cap | \bar{Y} | \cap Q$.

We note that Lemma 1 is not true in general if no restriction is placed on the latent cycle Z; we shall give a generalization of this lemma in a later section (Lemma 1*, Section 5).

4. The following lemma is in a sense fundamental for our theory:

LEMMA 2. *Let* V *be a variety, let* X *be a cycle in* V, *and let* K *be a field of definition for* V *such that* X *is rational over* K; *then there exists a purely transcendental extension* $K(u)$ *of* K, *which can be taken to be independent with respect to any given extension of* K, *and a rational cycle* X_u *over* $K(u)$ *in* V, *with the following properties:* (1) X *is a specialization in* V *of* X_u *over* K; (2) *if* W *is a subvariety in* V *which is defined over any extension* K' *of* K, *independent with respect to* $K(u)$ *over* K, *then every component of* $| X_u | \cap W$ *which is either simple in* V *or not contained in* $| X |$ *is proper in* V; (3) *there exists a latent cycle* Z *in* V *of the specialization* $X_u \to X$ *over* K *such that,* W *being the same as in* (2), *every component of* $| Z | \cap W$ *which is not contained in* $| X |$ *is proper in* V *and every component of* $| Z | \cap W$ *which is simple in* V *and is contained in* $| X |$ *is properly contained in a component of* $| X | \cap W$.

Before proceeding to the proof of this lemma, we shall first deduce from it a few corollaries.

COROLLARY 1. *The situation being as in Lemma* 2, (2), *every component of* $| X_u | \cap W$ *which is either simple in* V *or not contained in* $| X |$ *is general over* K' *for* W.

PROOF: Let W_1 be such a component of $| X_u | \cap W$, and let U be a proper subvariety in W which is defined over K'; then every component of $| X_u | \cap U$ which is either simple in V or not contained in $| X |$ is proper in V, and hence must have a dimension less than that of W_1. On the other hand, a component of $| X_u | \cap U$ which is contained in $| X |$ and is also singular in V certainly cannot contain W_1; therefore, W_1 is not contained in U.

COROLLARY 2. *The situation being as in Lemma* 2, (3), *every component of* $| Z | \cap W$ *which is not contained in* $| X |$ *is general over* K' *for* W.

PROOF: Let W_1 be a component of $| Z | \cap W$ which is not contained in $| X |$, and let U be a proper subvariety in W which is defined over K'; then every component of $| Z | \cap U$ which is not contained in $| X |$ is proper in V, and hence must have a dimension less than that of W_1. On the other hand, a component of $| Z | \cap U$ which is contained in $| X |$ and is simple in V must be a proper subvariety in a component of $| X | \cap W$, and hence must be a proper subvariety in W_1, while a component of $| Z | \cap U$ which is contained in $| X |$ and is singular in V certainly cannot contain W_1; therefore, W_1 is not contained in U.

COROLLARY 3. *If* X'_u *and* Z' *are the cycles obtained by applying Lemma* 2 *to the cycle* $\bar{X}$ *in* $\bar{V}$, *then the restrictions* X_u *and* Z *of* X'_u *and* Z' *respectively in* V *satisfy Lemma* 2 *with respect to the cycle* X *in* V, *and we have the relations* $X'_u = \bar{X}_u$ *and* $Z' = \bar{Z}$.

PROOF: The corollary is obvious by the nature of Lemma 2, except possibly the last relation $Z' = \bar{Z}$, which is equivalent to the assertion that no component of Z' is contained in $\bar{V} - V$. Since every component of $\bar{V} - V$ is defined over $\bar{K}$, this assertion follows immediately from Lemma 2, (3).

COROLLARY 4. *The situation being as in Lemma* 2, (2), *if* W *is simple in* V, *then the intersection-product* $X_u.W$ *in* V *is defined*; *hence, if* Y *is any cycle in* V *which is rational over* K', *then the intersection-product* $X_u.Y$ *in* V *is defined.*

PROOF: In case V is complete, the corollary follows immediately from Lemma 2, (2). If V is not complete, we choose the cycle X_u so that $\bar{X}_u$ satisfies the Lemma 2 with respect to $\bar{X}$ in $\bar{V}$, which is possible by Corollary 3; if U is a component of $(\bar{V} - V) \cap \bar{W}$, then it is a proper subvariety in $\bar{W}$ and is defined over $K'\bar{K}$, and it follows from Corollary 1 that no component of $\bar{X}_u.\bar{W}$ is contained in U. This shows that the intersection-product $X_u.W$ in V is defined.

COROLLARY 5. *The situation being as in Lemma* 2, (3), *if* W *is simple in* V *and if the intersection-product* $X.W$ *in* V *is defined, then the intersection-product* $Z.W$ *in* V *is defined*; *hence, if* Y *is any cycle in* V, *rational over* K', *such that the intersection-product* $X.Y$ *in* V *is defined, then the intersection-product* $Z.Y$ *in* V *is defined.*

PROOF: If a component of $| Z | \cap W$ which is simple in V is contained in $| X |$, then it must be a proper subvariety in a component of $| X | \cap W$ which is simple in V and hence proper in V; since this is evidently impossible, every component of $| Z | \cap W$ which is simple in V cannot be contained in $| X |$ and hence must be proper in V. In case V is complete, this shows that the intersection-product $Z.W$ in V is defined; if V is not complete, we can proceed as in the proof of Corollary 4, using Corollary 2 instead of Corollary 1.

After these corollaries, we now turn to the proof of Lemma 2. We observe first that the assertions (2) and (3) in the lemma can be reformulated in the following form, where W is any subvariety in V which is defined over K': (A) Every component of $|X_u| \cap W$ or $|Z| \cap W$ which is not contained in $|X|$ is proper in V; (B) every component of $|X_u| \cap W$ which is simple in V is proper in V, while every component of $|Z| \cap W$ which is simple in V and is contained in $|X|$ is properly contained in a component of $|X| \cap W$. However, instead of (B), it is sufficient to prove the following weaker statement: (B′) Every component of $|X_u| \cap W$ or $|Z| \cap W$ which is simple in V and is contained in $|X|$ is properly contained in a component of $|X| \cap W$. For the cycle Z (B′) is the same as (B), while for the cycle X_u (B) can be obtained from (B′) by successive applications, whereby the maximum dimension of all components of $|X_u| \cap W$ which are simple in V is reduced at each step, until finally all such components become proper in V.

Next, we shall introduce a geometric construction in the ambient projective space S^m of V; and we observe here that the use of this construction is the main reason for the restriction of our theory to varieties embedded in a projective space. Consider an s-cycle $X = \sum_{i=1}^d n_i U_i$ in a projective space S^m, where $U_1, \cdots, U_d$ are complete varieties of dimension s and $n_1, \cdots, n_d$ are integers, and let K be a field over which X is rational. Let L_v be a generic linear subspace of dimension r ($r < m - s$) over K in S^m, with a purely transcendental extension $K(v)$ of K as its defining field, and let U_i^* be the complete variety which is the union of all the linear subspaces of dimension $r + 1$ obtained by joining L_v with any point in U_i, so that U_i^* is a variety of dimension $s + r + 1$; the $(s + r + 1)$-cycle $X_v^* = \sum_{i=1}^d n_i U_i^*$, which is rational over $K(v)$, is then said to be *obtained from X by projection from L_v*.

Finally, we shall introduce for any two complete varieties U and W in S^m a subvariety $C(U, W)$ and, in case W is a simple subvariety in U, also a subvariety $T(U, W)$ in the Grassmann variety $G_{m,1}$ of all lines in S^m. If K is a field of definition for both U and W, and if x and y are independent generic points of U and W respectively over K, then the line joining x and y is represented in $G_{m,1}$ by a point p which is rational over $K(x, y)$, and the variety $C(U, W)$ is defined to be the locus of p over K; if W is a simple subvariety in U, and if q is the point in $G_{m,1}$ which represents a generic tangent line of U at y over the field $K(y)$, then $K(y, q)$ is a purely transcendental extension of $K(y)$, and the variety $T(U, W)$ is defined to be the locus of q over K. If l and t are the dimensions of U and W respectively, then it is easily seen that the dimension of $C(U, W)$ is at most $l + t$ and the dimension of $T(U, W)$ is at most $l + t - 1$. If x' and y' are any two distinct points in U and W respectively, then the point p' in $G_{m,1}$ representing the line joining x' and y' is a specialization of p over K and hence is contained in $C(U, W)$; similarly, the point in $G_{m,1}$ representing a tangent line of U at a point in W which is simple in U must be contained in $T(U, W)$. If we denote by N_v the subvariety of dimension $r + m - 1$ in $G_{m,1}$ whose points represent all the lines intersecting L_v, and N_v is a generic Schubert variety of the type $[r, m - 1]$ over K and is defined over $K(v)$; it is well known that, since L_v is generic over K, N_v intersects both $C(U, W)$ and $T(U, W)$ properly in $G_{m,1}$, so that every component of $N_v \cap C(U, W)$ has a dimension at most equal to $l + t - m + r + 1$, and every component of $N_v \cap T(U, W)$ has a dimension at most equal to $l + t - m + r$.

After these preliminaries, we come finally to the proof of Lemma 2.

PROOF OF LEMMA 2. Without any loss of generality, we can assume that X is a (simple) subvariety in V, the general case following evidently from this by linearity; furthermore, in view of Corollary 3 above, we can evidently restrict ourselves to the case where V is a complete variety. Let n be the dimension of V and s be the dimension of X; we take the dimension r of L_v to be $m - n - 1$, and consider the variety X_v^* obtained from X by projection from L_v. Let W be a subvariety of dimension t in V, defined over an extension K' of K which is independent with respect to $K(v)$ over K, and let W_1 be a component of $X_v^* \cap W$ which is not contained in X (but *a priori* may contain X); if y_1 is a generic point of W_1 over $K'(v)$, then there exists a point y_0 in X, of course distinct from y_1, such that the line joining y_1 and y_0 meets L_v and hence is represented in $G_{m,1}$ by a point p_1 in $N_v \cap C(X, W)$, rational over $K'(v, y_0, y_1)$. Since W is contained in V and hence does not intersect L_v, the line represented by the point p_1 can only intersect W_1 at a finite number of points; it follows that the dimension of y_1 over $K'(v)$ is at most equal to that of p_1 over $K'(v)$ and hence is at most equal to $s + t - n$. This shows that the dimension of W_1 is at most equal to $s + t - n$ and hence must be equal to $s + t - n$. If we take for W the variety V itself, then every component of $X_v^* \cap V$ has the dimension s, so that the intersection-product $X_v^*.V$ in S^m is defined; furthermore, if we take for W any singular subvariety W_0 in V, then every component of $X_v^* \cap W_0$ has a dimension at most equal to $s - 1$, from which it follows that every component of $X_v^*.V$ is a simple subvariety in V. Finally, if L^s is the tangent space of X at a generic point x over $K(v)$, then the join of L^s and L_v is evidently the tangent space of X_v^* at x, and since L_v is also generic over $K(x)$, the tangent space of V at x must be transversal to the join of L^s and L_v. This shows that X is a simple component in the positive cycle $X_v^*.V$, so that we can set $X_v^*.V = X + X_1 + \cdots + X_e$, where $X_1, \cdots, X_e$ are simple subvarieties in V, all distinct from X. (Actually it can be shown that all the varieties $X_1, \cdots, X_e$ are distinct from each other, but we shall not need this fact.) Since every component of $X_i \cap W$ which is not contained in X or any other X_j $(j \neq i)$ is also a component of $X_v^* \cap W$, it has the dimension $s + t - n$ and hence is proper in V; thus the property (A) holds for each X_i, except possibly for those components of $X_i \cap W$ which are contained in some X_j $(j \neq i)$.

Consider now a component W_1 of $X_i \cap W$ which is simple in V and is contained in X, and let y_1 be a generic point of W_1 over $\overline{K'(v)}$; since y_1 is a point in X_v^*, the join of y_1 and L_v meets X in a point y_0, which may coincide with y_1. If y_0 is distinct from y_1, then the line joining y_1 and y_0 meets L_v and hence is represented in $G_{m,1}$ by a point p_1 in $N_v \cap C(X, W)$; since this line can only intersect W_1 at a finite number of points, it follows that the dimension of y_1 over $\overline{K'(v)}$ is at most equal to that of p_1 over $\overline{K'(v)}$ and hence is at most equal to $s + t - n$. Since W_1 is simple in V, this shows that the dimension of W_1 is equal to $s + t - n$. Assume now that y_0 coincides with y_1, and let z_1 be a generic point of X_i over $\overline{K'(v)}$; then the join of z_1 and L_v will meet X in a point z_0, which is of course different from z_1. As the point z_1 specializes over $\overline{K'(v)}$ to the point y_1, the join of z_1 and L_v specializes to the join of y_1 and L_v, and hence the point z_0 must specialize to the point $y_0 = y_1$. This shows that the point y_1 has a multiplicity at least 2 as an intersection of V with the join of y_1 and L_v; since y_1 is a simple point in V, it follows that there is a tangent line to V at y_1 which meets L_v. If W_0 is a component of $X \cap W$ which contains W_1 and if t_0 is the dimension of W_0, then the variety $T(V, W_0)$ has a dimension at most equal to $n + t_0 - 1$; since L_v is generic over K', the variety N_v intersects $T(V, W_0)$ properly in $G_{m,1}$, so that every component of $N_v \cap T(V, W_0)$ has a dimension at most equal to $n + t_0 - 1 + 2m - n - 2 - (2m - 2) = t_0 - 1$. If q is the point in $G_{m,1}$ representing a tangent line to V at y_1 which meets L_v, then q is a point in $N_v \cap T(V, W_0)$ and hence has a dimension at most equal to $t_0 - 1$ over $\overline{K'(v)}$; since this tangent line can only intersect W_1 at a finite number of points, the dimension of y_1 over $\overline{K'(v)}$ is at most equal to that of q over $\overline{K'(v)}$ and hence is at most equal to $t_0 - 1$. This shows that the dimension of W_1 is at most equal to $t_0 - 1$, and hence W_1 is properly contained in W_0; thus the property (B′) holds for each X_i.

We set $Z = X_1 + \cdots + X_e$; it is clear that the properties (A) and (B′) hold for Z. Let T_u be a generic projective transformation in S^m over $K(v)$, defined over a purely transcendental extension $K(u)$ of K, independent with respect to $K(v)$ over K, and consider the transformed cycle $T_u(X_v^*)$ of X_v^*, which is rational over $K(u, v)$; in view of the generic nature of the transformation T_u, it is easily seen that the intersection-product $T_u(X_v^*).V$ in S^m is defined and is a rational s-cycle over $K(u, v)$ in V, and that the properties (A) and (B′) holds for this cycle. If we set $X_{u,v} = T_u(X_v^*).V - Z$, then X is evidently a specialization of $X_{u,v}$ over K and Z is a latent cycle of this specialization, and the properties (A) and (B′) hold for $X_{u,v}$; if we now change our notation by replacing $K(u, v)$ by $K(u)$, we conclude the proof of Lemma 2.

Remarks on Lemma 2. It follows from the above proof that the cycles X_u and Z in Lemma 2 can be so chosen that the following properties hold:

(1) $\bar{X}$ is a specialization in S of $\bar{X}_u$ over K.

(2) There exists a cycle A in S such that $R(\bar{X}_u, S - |\bar{X}|)$ is the intersection-product $R(A, S - |\bar{X}|).R(\bar{V}, S - |\bar{X}|)$ in $S - |\bar{X}|$, in the absolute sense.

(3) There exists a cycle B in S such that $R(\bar{Z}, S - |\bar{X}|)$ is the intersection-product $R(B, S - |\bar{X}|).R(\bar{V}, S - |\bar{X}|)$ in $S - |\bar{X}|$, in the absolute sense.

We remark that the idea of using the projection of a cycle from L_v in connection with Lemma 2 goes back to Severi [3], and the method of proof used above is essentially due to van der Waerden [5] Section 5; however, our Lemma 2 contains somewhat stronger results than those given by these authors, and these are important for our present purpose. For later convenience, we shall deduce some further results from Lemma 2, which we shall gather together as a lemma.

LEMMA 3. *Let V be a variety, let X and Y be two cycles in V, and let K be a field of definition for V such that both X and Y are rational over K; then there exist cycles X' and Y' in V such that the following properties are satisfied*: (1) *X' and Y' are rationally equivalent in V to X and Y respectively, and are both general over K for V*; (2) *the intersection-product $X'.Y'$ in V is defined, and is general over K for V*; (3) *every component of $|X'| \cap |Y'|$ which is singular in V is contained in $|X| \cap |Y|$*; (4) *if W is any subvariety in V which is defined over $\bar{K}$, then every component of $|X'| \cap W$, $|Y'| \cap W$, or $|X'| \cap |Y'| \cap W$ which is either simple in V or not contained in $|X|$, $|Y|$, or $|X| \cup |Y|$ respectively, is proper in V.*

PROOF: According to Lemma 2 and Corollary 4, there exists a cycle X' in V, rational over a purely transcendental extension $K(u)$ of K, which satisfies the conditions (1) and (4), and for which the intersection-product $X'.Y$ in V is defined. Furthermore, every component of $|X'| \cap |Y|$ which is singular in V is contained in $|X| \cap |Y|$; in fact, if W_1 is such a component of $|X'| \cap |Y|$, then it is contained in a singular subvariety W_2 in V which is defined over $\bar{K}$ and is a proper subvariety in a component Y_1 of Y, and if W_1 is not contained in $|X|$, then by Lemma 2, (2), W_1 must be a proper component of both $|X| \cap W_2$ and $|X| \cap Y_1$, which is impossible. Now, apply Lemma 2 to the cycle Y and for this purpose take a purely transcendental extension $K(v)$ of K which is independent with respect to $K(u)$ over K; then there exists a cycle Y' in V, rational over $K(v)$, which satisfies the conditions (1) and (4), and for which the intersection-product $X'.Y'$ in V is defined (by Corollary 4). This proves (2). Furthermore, by a similar argument as before, every component of $|X'| \cap |Y'|$ which is singular in V is contained in $|X'| \cap |Y|$; and since the situation is symmetric with respect to the two cycles X and Y, we conclude that every component of $|X'| \cap |Y'|$ which is singular in V is contained in $|X| \cap |Y|$. This proves (3). Finally, if W_1 is a component of $|X'| \cap |Y'| \cap W$ which is either simple in V or not contained in $|X| \cup |Y|$, then W_1 is contained in a component W_2 of $|Y'| \cap W$ which is either simple in V or not contained in $|X| \cup |Y|$; it follows then from Lemma 2, (2) that W_2 is a proper component of $|Y'| \cap W$, and since W_2 is defined over $\overline{K(v)}$, which is independent with respect to $K(u)$ over K, we conclude also that W_1 is a proper component of $|X'| \cap W_2$. This proves (4) for $X'.Y'$, and, as in the proof of Lemma 2, Corollary 1, we deduce from this that the cycle $X'.Y'$ is general over K for V. This concludes the proof of the lemma.

5. Consider now two cycle classes $\mathfrak{X}$ and $\mathfrak{Y}$ in V, and let X and Y be cycles in the classes $\mathfrak{X}$ and $\mathfrak{Y}$ respectively such that the intersection-product $X.Y$ in V is defined; according to Lemma 2, Corollary 4, such cycles X and Y certainly exist and one of them can even be taken arbitrarily in its class. We shall show that if X' and Y' are any two other cycles in the classes $\mathfrak{X}$ and $\mathfrak{Y}$ respectively such that the intersection-product $X'.Y'$ in V is defined, then $X'.Y'$ is rationally equivalent to $X.Y$ in V; in view of Lemma 2, Corollary 4, it is evidently sufficient to show this for the case where Y' coincides with Y. Let K be a field of definition for V such that the cycles X, X', and Y are rational over K; let X_u be a cycle in V, rational over a purely transcendental extension $K(u)$ of K, such that both X and X' are specializations of X_u over K, and let Z and Z' be latent cycles of the specializations $X_u \to X$ and $X_u \to X'$ respectively. According to Lemma 2, Corollary 4, there exists a purely transcendental extension $K(v)$ of K, independent with respect to $K(u)$ over K, and a rational cycle Y_v over $K(v)$ in V, such that Y is a specialization of Y_v over $K(u)$ and that the intersection-products $X.Y_v$, $X'.Y_v$, $X_u.Y_v$, $Z.Y_v$, $Z'.Y_v$ in V are all defined (in case of relative theory, the last two should be replaced by the intersections-products $Z.\bar{Y}_v$ and $Z'.\bar{Y}_v$ in $\bar{V}$); furthermore, by Lemma 2, Corollary 5, the specialization $Y_v \to Y$ over $K(u)$ has a latent cycle Z_0 such that the intersection-products $X.Z_0$ and $X'.Z_0$ in V are defined. According to Lemma 1, this implies firstly that $X.Y_v$ is rationally equivalent in V to $X'.Y_v$, and secondly that $X.Y_v$ and $X'.Y_v$ are rationally equivalent in V to $X.Y$ and $X'.Y$ respectively; this shows that $X.Y$ is rationally equivalent in V to $X'.Y$.

We shall examine the above argument more closely and obtain a somewhat stronger result which will be important for us later. The assertion that $X.Y$ and $X'.Y$ are rationally equivalent in V implies by definition that there exist cycles D and D' in the ambient projective space S of V, with supports in the union Q of $\bar{V} - V$ with all the singular subvarieties in $\bar{V}$ in the absolute case (or the union Q of all the singular subvarieties in $\bar{V}$ in the relative case), such that the cycles $\overline{X.Y} + D$ and $\overline{X'.Y} + D'$ are rationally equivalent in S. We shall show that the cycles D and D' can be so chosen that their supports are also contained in $| \bar{Y} |$; in case Y is a prime cycle, this means that $X.Y$ and $X'.Y$ are rationally equivalent not only in V but also in the variety Y. Let X_u, X, X', Z, Z', and Y be the same as before, but we shall find it convenient to use a slightly different procedure than the introduction of the cycle Y_v. Without any loss of generality, we can assume that the intersection-product $X_u.Y$ in V is defined, for otherwise we can replace X_u by the cycle obtained from it by the application of Lemma 2. Let X_1 and X_1' be cycles in S, with supports in Q, such that $\bar{X} + X_1$ and $\bar{X}' + X'_1$ are specializations in S of $\bar{X}_u$ over K, and let Z_1 and Z_1' be latent cycles of the specializations $\bar{X}_u \to \bar{X} + X_1$ and $\bar{X}_u \to \bar{X}' + X_1'$ over K respectively, so that the cycles Z and Z' can be taken to be the restrictions in V or $\bar{V}$ of Z_1 and Z_1' respectively. For the sake of convenience, we shall assume the field K so extended that the cycles X_1, X_1', Z_1, Z_1' are all rational over K. Our assertion is then contained in the following lemma, which is a generalization of Lemma 1:

LEMMA 1*. *There exist cycles D and D' in S, with supports in $| \bar{Y} | \cap Q$, such that $\overline{X.Y} + D$ and $\overline{X'.Y} + D'$ are rationally equivalent in S; furthermore if W_0 is a closed subset in S such that every component of $(| \bar{X} + X_1 | \cup | Z_1 |) \cap | \bar{Y} |$ which is not contained in W_0 and is not negligible in $\bar{V}$ is proper in $\bar{V}$, and such that every component of $| \bar{X}_u | \cap | \bar{Y} |$ which is not contained in W_0 and is not neg-*

ligible in $\bar{V}$ *is proper in* $\bar{V}$ *and is the closure of a simple subvariety in* V, *then* $R(\overline{X.Y} + D, S - W_0)$ *is the uniquely determined specialization in* $S - W_0$ *of* $R(\overline{X_u.Y}, S - W_0)$, *in the absolute sense, over the specialization* $\bar{X}_u \to \bar{X} + X_1$ *over* K; *and a similar statement holds for* $\overline{X'.Y} + D'$.

PROOF: It is sufficient to prove this lemma for the absolute theory, for we can easily obtain the relative theory by applying the absolute theory to the variety $\bar{V} - Q$ (Q being the union of all singular subvarieties in $\bar{V}$), replacing the cycles X_u, X, X', and Y by the restrictions of $\bar{X}_u$, $\bar{X}$, $\bar{X}'$, and $\bar{Y}$ in $\bar{V} - Q$ respectively. If A and B are two bunches of subvarieties of dimensions s and t respectively in V, then we shall denote by $d(V; A, B)$ the maximum dimension of all components of $A \cap B$ which are simple in V, in case such a component exists, otherwise we set $d(V; A, B)$ equal to $s + t - n$, where n denotes the dimension of V; it is clear that $d(V; A, B) \geqq s + t - n$, and that this minimum value $s + t - n$ is reached if and only if every component of $A \cap B$ which is simple in V is proper in V. We shall prove our lemma by induction on the integer $d(V; |Z| \cup |Z'|, |Y|)$; and we observe that in case $d(V; |Z| \cup |Z'|, |Y|)$ has the minimum value (i.e. $s + t - n$, if s and t are the dimensions of X and Y respectively), our lemma follows from Lemma 1 and the Remark after its proof, so that we can assume that $d(V; |Z| \cup |Z'|, |Y|) > s + t - n$. Instead of the cycle Y_v, we shall go back to the projective construction used in the proof of Lemma 2. Let m be the dimension of S, and let L be a generic linear subspace of dimension $m - n - 1$ over $K(u)$ in S; let K' be an extension of K, independent with respect to $K(u)$ over K, such that L is defined over K'. Let $\bar{Y}^*$ be the cycle in S obtained from $\bar{Y}$ by projection from L, and let T_v be a generic projective transformation of S over $K'(u)$, so that $K(v)$ is a purely transcendental extension of K, independent with respect to $K'(u)$ over K; then, as we have shown in the proof of Lemma 2, the cycle $T_v(\bar{Y}^*)$ intersects properly in S every cycle which is rational over $K'(u)$, and the cycle $\bar{V}.\bar{Y}^* - \bar{Y}$ has no component in $|\bar{Y}|$ and is the closure in S of a cycle Z_0 in V; furthermore, if W is any subvariety in V which is defined over $K'(u)$, then we have the inequality $d(V; W, |Z_0|) < d(V; W, |Y|)$ except when $d(V; W, |Y|)$ already has the minimum possible value. According to Lemma 1, the cycles $(\bar{X} + X_1).T_v(\bar{Y}^*)$ and $(\bar{X}' + X_1').T_v(\bar{Y}^*)$ are both specializations in S of $\bar{X}_u.T_v(\bar{Y}^*)$ over $K(v)$ and hence are rationally equivalent in S. Let E be a specialization of $(\bar{X} + X_1).T_v(\bar{Y}^*)$ over the specialization $T_v(\bar{Y}^*) \to \bar{Y}^*$ over K', and let W_1 be the smallest closed subset in S such that every component of $|\bar{X} + X_1| \cap |\bar{Y}^*|$ which is not contained in W_1 is proper in S; then $R(E, S - W_1)$ is the intersection-product $R(\bar{X} + X_1, S - W_1).R(\bar{Y}^*, S - W_1)$ in $S - W_1$, so that we can set $E = E_0 + E_1$, where E_1 is the closure in S of this intersection-product and E_0 is a cycle in S with support in W_1. Since the intersection-product $X.Y$ in V is defined, it follows from the proof of Lemma 2 that the intersection-product $X.Z_0$ in V is also defined and has no component in $|Y|$, and that $R(E, V) = R(E_1, V) = X.Y + X.Z_0$; we can then set $E_1 = \overline{X.Y} + \overline{R(E, S - |\bar{Y}|)} + E_2$, where E_2 is a cycle in S with support in $|\bar{Y}| \cap Q$, and $\overline{R(E, S - |\bar{Y}|)} = \overline{X.Z_0} +$

E_3, where E_3 is a cycle in S with support in Q. Since $R(\bar{Y}^*, S - W_0)$ intersects $R(\bar{X} + X_1, S - W_0)$, $R(Z_1, S - W_0)$, and $R(\bar{X}_u, S - W_0)$ properly in $S - W_0$ and since $R(\bar{X}_u.\bar{Y}^*, S - W_0) = R(\overline{X_u.Y} + \overline{X_u.Z_0}, S - W_0)$, it follows from the intersection theory in $S - W_0$ that $R(E, S - W_0)$ is the unique specialization in $S - W_0$ of $R(\overline{X_u.Y} + \overline{X_u.Z_0}, S - W_0)$, in the absolute sense, over the specialization $\bar{X}_u \to \bar{X} + X_1$ over K'; and since $| \bar{Y} |$ satisfies the conditions for W_0, this implies that $R(E, S - | \bar{Y} |)$ is the unique specialization in $S - | \bar{Y} |$ of $R(\overline{X_u.Z_0}, S - | \bar{Y} |)$, in the absolute sense, over the specialization $\bar{X}_u \to \bar{X} + X_1$ over K'. Thus we have $E = \overline{X.Y} + (E_0 + E_2) + \overline{X.Z_0} + E_3$, where $E_0 + E_2$ has its support in $| \bar{Y} | \cap Q$ (W_1 being evidently contained in $| \bar{Y} | \cap Q$) and E_3 has its support in Q, and where $\overline{X.Z_0} + E_3$ is the unique specialization in $S - | \bar{Y} |$ of $R(\overline{X_u.Z_0}, S - | \bar{Y} |)$, in the absolute sense, over the specialization $\bar{X}_u \to \bar{X} + X_1$ over K'. Similarly, if E' is a specialization of $(\bar{X}' + X_1').T_v(\bar{Y}^*)$ over the specialization $T_v(\bar{Y}^*) \to \bar{Y}^*$ over K', then we have $E' = \overline{X'.Y} + (E_0' + E_2') + \overline{X'.Z_0} + E_3'$, where $E_0' + E_2'$ has its support in $| \bar{Y} | \cap Q$ and E_3' has its support in Q, and where $\overline{X'.Z_0} + E_3'$ is the unique specialization in $S - | \bar{Y} |$ of $R(\overline{X_u.Z_0}, S - | \bar{Y} |)$, in the absolute sense, over the specialization $\bar{X}_u \to \bar{X}' + X_1'$ over K'. Since E and E' are rationally equivalent in S to $(\bar{X} + X_1).T_v(\bar{Y}^*)$ and $(\bar{X}' + X_1').T_v(\bar{Y}^*)$ respectively and since the latter two cycles are rationally equivalent in S, it follows that E and E' are rationally equivalent in S.

Consider now the situation of our lemma with the cycle Y replaced by the cycle Z_0; since $d(V; | Z | \cup | Z' |, | Z_0 |) < d(V; | Z | \cup | Z' |, | Y |)$, we can apply the induction hypothesis. Therefore there exist cycles D_0 and D_0' in S, with supports in $| \bar{Z}_0 | \cap Q$, such that $\overline{X.Z_0} + D_0$ and $\overline{X'.Z} + D_0'$ are rationally equivalent in S; moreover, since W_0 has the properties stated in the lemma also with respect to Z_0, $R(\overline{X.Z_0} + D_0, S - W_0)$ is the unique specialization in $S - W_0$ of $R(\overline{X_u.Z_0}, S - W_0)$, in the absolute sense, over the specialization $\bar{X}_u \to \bar{X} + X_1$ over K'. If we take $| \bar{Y} |$ for W_0, then we have $R(\overline{X.Z_0} + D_0, S - | \bar{Y} |) = R(\overline{X.Z_0} + E_3, S - | \bar{Y} |)$; it follows from this that the cycle $(\overline{X.Z_0} + E_3) - (\overline{X.Z_0} + D_0) = E_3 - D_0$ in S has its support in $| \bar{Y} |$, and since $| E_3 - D_0 | \subset | E_3 | \cup | D_0 | \subset Q$, it follows that $| E_3 - D_0 | \subset | \bar{Y} | \cap Q$. We have then the relation $E - (\overline{X.Z_0} + D_0) = \overline{X.Y} + (E_0 + E_2) + (E_3 - D_0) = \overline{X.Y} + D$, where $D = (E_0 + E_2) + (E_3 - D_0)$ is a cycle in S with support in $| \bar{Y} | \cap Q$; and similarly we have the relation $E' - (\overline{X'.Z_0} + D_0') = \overline{X'.Z} + D'$, where D' is a cycle in S with support in $| \bar{Y} | \cap Q$. Since $E - (\overline{X.Z_0} + D_0)$ and $E' - (\overline{X'.Z_0} + D_0')$ are rationally equivalent in S, it follows that $\overline{X.Y} + D$ and $\overline{X'.Y} + D'$ are rationally equivalent in S. Finally, since $R(E, S - W_0)$ and $R(\overline{X.Z_0} + D_0, S - W_0)$ are unique specializations in $S - W_0$ of $R(\overline{X_u.Y} + \overline{X_u.Z_0}, S - W_0)$ and $R(\overline{X_u.Z_0}, S - W_0)$ respectively, in the absolute sense, over the specialization $\bar{X}_u \to \bar{X} + X_1$ over K', it follows by subtraction that $R(\overline{X.Y} + D, S - W_0)$ is the unique specialization in $S - W_0$ of $R(\overline{X_u.Y}, S - W_0)$, in the absolute sense, over the specialization $\bar{X} \to \bar{X} + X_1$ over K'. This concludes the proof of Lemma 1*.

We shall state the first part of Lemma 1* as a theorem:

THEOREM 1. *Let V be a variety and let X, X', and Y be cycles in V such that the intersections-products $X.Y$ and $X'.Y$ in V are defined; if X and X' are rationally equivalent in V, then $X.Y$ and $X'.Y$ are also rationally equivalent in V. Furthermore, if Y is a simple subvariety in V and if the cycles $X.Y$ and $X'.Y$ are also cycles in Y, then $X.Y$ and $X'.Y$ are also rationally equivalent in Y.*

On the basis of Theorem 1, we can define the intersection product $X.Y$ in V of the two cycle classes $\mathfrak{X}$ and $\mathfrak{Y}$ to be the cycle class in V determined by the cycle $X.Y$, this cycle class having been shown to be independent of the choice of the cycles X and Y. With this definition of the intersection-product of two cycle classes in V as multiplication, we shall now extend the set of all cycle classes in V to a ring in the following way. We consider the direct sum $G(V) = \sum_{s=0}^{n} G(V, s)$ of the $n + 1$ groups $G(V, s)$, and we extend the concept of a cycle in V to include any element in this group $G(V)$; the elements in each group $G(V, s)$ will then be called a homogeneous cycle in V, of the dimension s. We shall consider the direct sum $G_r(V) = \sum_{s=0}^{n} G_r(V, s)$ as a subgroup in $G(V)$ in an obvious way, and an element in $G(V)$ is said to be rationally equivalent to zero in V if and only if it is in $G_r(V)$; it is easily seen that the quotient group $G(V)/G_r(V)$ of cycle classes in V is isomorphic, in a natural way, to the direct sum of the groups $G(V, s)/G_r(V, s)$ of homogeneous cycle classes of the various dimensions in V. Since the intersection-product in V is defined for any two homogeneous cycle classes in V, we can extend this definition by linearity to any two elements in $G(V)/G_r(V)$, and if we take the so defined intersection-product as multiplication, then it can be easily shown, by means of the known commutative, associative and distributive properties in the calculus of cycles in V, that the group $G(V)/G_r(V)$ is thus made into a ring which we shall denote by $H_r(V)$. This ring $H_r(V)$ of cycle classes in V is commutative and is graded (by codimension), and it has a unit element which is the class $\mathfrak{V}$ determined by the variety V itself, considered as a cycle of dimension n.

THEOREM 2. *The set of all cycle classes in a variety V constitutes a ring $H_r(V)$, with the sum and intersection-product of two cycle classes as the sum and product respectively in $H_r(V)$; this ring $H_r(V)$ is commutative and graded (by codimension), and it has a unit element.*

6. We consider now two varieties U and V, and investigate the relations between the rings $H_r(U)$, $H_r(V)$, and $H_r(U \times V)$ of cycle classes in U, V, and $U \times V$ respectively. Since the operation product is compatible with both the addition and intersection of cycles, one sees readily that this operation is compatible with the operations in the rings of cycle classes, so that there is a natural homomorphism of the Kronecker product $H_r(U) \times H_r(V)$ of the two rings $H_r(U)$ and $H_r(V)$, over the ring of intergers, into the ring $H_r(U \times V)$. In case of the operation projection, from $U \times V$ on U or V, we naturally assume that V or U is complete and non-singular respectively in the absolute theory, and that $\bar{V}$ is non-singular and U is complete or $\bar{U}$ is non-singular and V is complete respectively in the relative theory, so that this operation can be considered as an operation on cycle classes in any case; however, even then the projection is in general not compatible with the intersection of cycle classes, so that it induces a homomorphism of the group $G(U \times V)/G_r(U \times V)$ into $G(U)/G_r(V)$ respectively, but in general not a homomorphism of $H_r(U \times V)$ into $H_r(U)$ or $H_r(V)$. If $\mathfrak{F}$ is a homogeneous cycle class of dimension l in $U \times V$, and if we denote by n the dimension of V, then, always under the assumption that V is complete and non-singular in the absolute theory and that U is complete and $\bar{V}$ is non-singular in the relative theory, we can assign to each homogeneous cycle class $\mathfrak{X}$ of dimension s in V a cycle class $\mathrm{pr}_U((\mathfrak{U} \times \mathfrak{X}).\mathfrak{F})$ of dimension $s + l - n$ in U. It is easily seen that the mapping $\mathfrak{X} \to \mathrm{pr}_U((\mathfrak{U} \times \mathfrak{X}).\mathfrak{F})$ is a homomorphism of the group of all cycle classes of dimension s in V into the group of all cycle classes of dimension $s + l - n$ in U, and that this homomorphism can be extended by linearity to a homomorphism of $G(V)/G_r(V)$ into $G(U)/G_r(U)$ which we shall denote by $\mathfrak{F}^{-1}$; however, owing to the lack of compatibility between intersection and projection in general, as we have noted above, this homomorphism $\mathfrak{F}^{-1}$ is in general not a ring homomorphism of $H_r(V)$ into $H_r(U)$. In fact, if $\mathfrak{Y}$ is a homogeneous cycle class of dimension t in V, then $\mathfrak{F}^{-1}(\mathfrak{X})$, $\mathfrak{F}^{-1}(\mathfrak{Y})$, and $\mathfrak{F}^{-1}(\mathfrak{X}.\mathfrak{Y})$ have the dimensions $s + l - n$, $t + l - n$, and $s + t + l - 2n$ respectively; hence, by a counting of the dimensions alone, it is clear that $\mathfrak{F}^{-1}(\mathfrak{X}.\mathfrak{Y})$ can be the intersection-product of $\mathfrak{F}^{-1}(\mathfrak{X})$ and $\mathfrak{F}^{-1}(\mathfrak{Y})$ in U only if the dimension of U is equal to l. We shall show in the following that this condition is essentially also sufficient, if we assume in addition that $\mathfrak{F}$ contains a prime cycle F such that the projection from F to U is regular at every point in U (and, in the relative case, we make the same assumption for the projection from $\bar{F}$ to $\bar{U}$). We observe that the prime cycle F is in this case the graph of a rational transformation of U into V, and the mapping $\mathfrak{F}^{-1}$ is induced by the inverse of this rational transformation, in similar way as a dual homomorphism of the cohomology rings is induced by the inverse of a continuous mapping in topology.

Before proceeding further, we shall make an observation concerning the assumptions we have made about the completeness and non-singularity of the varieties U and V. These assumptions, although convenient for many applications, are somewhat too strong for our present purpose, and it is important for our theory that they can be replaced by less restrictive assumptions on the cycle F, instead of on the varieties U and V. We recall that any homogeneous cycle F in $U \times V$ defines a correspondence between the cycles in U and V if we set $X \to F(X) = \mathrm{pr}_V((X \times V).F)$ for a cycle X in U and set $X \to F^{-1}(X) = \mathrm{pr}_U((U \times X).F)$ for a cycle X in V, provided the expressions involved are defined. Set-theoretically, $|F|$ defines also a correspondence between the subsets in U and V if we set $W \to F(W) = \mathrm{pr}_V((W \times V) \cap |F|)$ for any subset W in U and set $W \to F^{-1}(W) = \mathrm{pr}_U((U \times W) \cap |F|)$ for any subset W in V, whereby the symbols pr_U and pr_V denote here the projection in set-theoretic sense. We observe here that we shall use the same symbol F to denote the cycle on $U \times V$ as well as the correspondence between the cycles and subsets in U

and V; there will be no danger of any confusion on this score, as the context will make clear what we mean in each case. We shall say that the correspondence F (or F^{-1}) *preserves the singularity* of a point x_0 in U (or V) if, whenever x_0 is singular in U (or V), every point in $F(x_0)$ (or $F^{-1}(x_0)$) is singular in V (or U); and we shall say that F (or F^{-1}) preserves the singularities in a subset W in U (or V) if it preserves the singularity of every point in W. We shall say that the correspondence F (or F^{-1}) is *complete* over a point x_0 or a subvariety W in V (or U) if every component of the cycle F is complete over x_0 or W, whenever x_0 or W is in the projection of $| F |$ in V (or U). In the following we shall make no restrictions on the varieties U and V, but instead we shall impose on the correspondences between them some of the properties just described. For simplicity we shall restrict ourselves to the correspondence F^{-1} from V to U, as it is in terms of F^{-1} that we shall find it more convenient to formulate our results. We shall assume that (α) F^{-1} preserves the singularities in V and is complete over all points in U in the absolute case, or (β) $\bar{F}^{-1}$ preserves the singularities in $\bar{V}$ in the relative case, whereby we remark that $\bar{F}^{-1}$ denotes the correspondence from $\bar{V}$ to $\bar{U}$ defined by the cycle $\bar{F}$ in $\bar{U} \times \bar{V}$.

In view of the restrictions we shall impose on the correspondences between U and V, it is advisable to introduce here for correspondences a stricter concept of rational equivalence than that which is given by the rational equivalence of the cycles in $U \times V$ which define them. Let Q be (α) the union of

$$\bar{U} \times \bar{V} - U \times V$$

with all the singular subvarieties in $\bar{U} \times \bar{V}$ in the absolute case, or (β) the union of all singular subvarieties in $\bar{U} \times \bar{V}$ in the relative case; similarly, let Q' be (α) the union of $\bar{U} - U$ with all the singular subvarieties in $\bar{U}$ in the absolute case, or (β) the union of all singular subvarieties in $\bar{U}$ in the relative case. Then our above assumptions on the correspondence F between U and V can be expressed by the condition that $\mathrm{pr}_{\bar{U}}(Q \cap | \bar{F} |)$ is contained in Q', and we shall call a correspondence between U and V *admissible* if it satisfies this condition. If F is any cycle in $U \times V$ and if K is a field of definition for $U \times V$, then a cycle F_1 in $U \times V$ is said to be a *C-specialization* in $U \times V$ of F over K, if there exists a cycle F_2 in the ambient space S of $U \times V$, satisfying the conditions $| F_2 | \subset Q$ and $\mathrm{pr}_{\bar{U}}(Q \cap | F_2 |) \subset Q'$, such that the cycle $\bar{F}_1 + F_2$ is a specialization in S of the cycle $\bar{F}$ over K, and if the specialization $\bar{F} \rightarrow \bar{F}_1 + F_2$ over K has a latent cycle Z such that $\mathrm{pr}_{\bar{U}}(Q \cap | Z |) \subset Q'$. Two cycles in $U \times V$ are said to be rationally *C-equivalent* in $U \times V$ if they are rationally equivalent as defined in section 3 with the concept of specialization used there replaced by that of C-specialization. Two admissible correspondences between U and V are said to be *rationally equivalent* if the cycles in $U \times V$ defining them are rationally C-equivalent. It is easily seen that the so defined concept of rational equivalence is reflexive, symmetric, and transitive among the set of all admissible correspondences between U and V, so that we can speak of an *admissible correspondence class* between U and V. If $\mathfrak{F}$ is an admissible correspondence class between U and V, and if $\mathfrak{X}$ is any cycle class in V, then we shall say that the expression $\mathfrak{F}^{-1}(\mathfrak{X}) = \mathrm{pr}_{\bar{U}}((\mathfrak{U} \times \mathfrak{X}).\mathfrak{F})$ is defined if there exist elements X and F in $\mathfrak{X}$ and $\mathfrak{F}$ respectively such that the expression $F^{-1}(X)$ is defined and if all the so obtained cycles $F^{-1}(X)$ (for all possible choices of X and F) are contained in one cycle class in U; and when such is the case, we shall denote this cycle class in U by $\mathfrak{F}^{-1}(\mathfrak{X})$. We observe here again that our definitions of an admissible correspondence and of an admissible correspondence class are so formulated that they give the desired properties for the correspondence F^{-1} from V to U; if one is interested in the correspondence F from U to V, the roles of U and V have to be interchanged, and we observe that the resulting definitions will in general not be the same.

THEOREM 3. *If $\mathfrak{F}$ is an admissible correspondence class between a variety U and a variety V, then the expression $\mathfrak{F}^{-1}(\mathfrak{X})$ is defined for every cycle class $\mathfrak{X}$ in V and the mapping $\mathfrak{X} \to \mathfrak{F}^{-1}(\mathfrak{X})$ is a homomorphism of $G(V)/G_r(V)$ into $G(U)/G_r(U)$.*

The proof of this theorem is based on the following two lemmas.

LEMMA 4. *Let U and V be varieties, let X be a homogeneous cycle in V, and let F be a homogeneous cycle in $U \times V$; let K be a field of definition for U and V such that both X and F are rational over K, and let X' be a homogeneous cycle in V having the property that if W is any subvariety in V, defined over $\bar{K}$, then every component of $|X'| \cap W$ which is not contained in $|X| \cap W$ (or which is simple in V) is proper in V. Then every component of $|(U \times X')| \cap |F|$ which is not contained in $|(U \times X)| \cap |F|$ and is not negligible in $U \times V$ (or which is simple in $U \times V$), is proper in $U \times V$ and is general over $\bar{K}$ for a component of F.*

PROOF. Without any loss of generality we can assume that F is a prime cycle, general case following evidently from this by linearity; we observe also that, just as in the proofs of Lemma 2, Corollaries 1 and 2, our assumptions imply that every component of $|X'| \cap W$ satisfying our assumptions is general over $\bar{K}$ for W. Let V' be the projection of F in V, and let n, n', l, and s be the dimensions of V, V', F, and X respectively; then V' is a simple subvariety in V, and for every point x_0 in V' every component of $(U \times x_0) \cap F$ has a dimension at least equal to $l - n'$. For each $i = 1, \cdots, n'$, let W_i be the subset in V' such that for every point x_0 in W_i the bunch $(U \times x_0) \cap F$ has a component of dimension at least equal to $l - n' + i$; then each W_i is a closed subset in V' of maximum dimension at most equal to $n' - i - 1$, normally algebraic over K, and we have the relation $W_i \supset W_{i+1}$. Since V' is defined over K, every component of $|X'| \cap V'$ which is not contained in $|X|$ (or which is simple in V) is proper in V and is general over $\bar{K}$ for V', and a similar assertion holds if we replace V' by any component of any W_i.

Consider now a component Y of $|U \times X'| \cap F$ which is not contained in $|U \times X|$ and which has a dimension not less than $s + l - n$ (or which is simple in $U \times V$ and hence has a dimension not less than $s + l - n$), and let Z be its projection in V; then Z is not contained in $|X|$ (or is simple in V). Let K' be an extension of K such that X' is rational over K', and let y and z be a pair of corresponding generic points of Y and Z respectively over $\bar{K}'$; since y is a generic point over $\overline{\ '(z)}$ of a component of $(U \times z) \cap F$, its dimension over $K'(z)$ is at least equal to $l - n'$. If the dimension of y over $K'(z)$ is $l - n' + i$ $(i > 0)$, then z is contained in $|X'| \cap W_i$ and hence has a dimension at most equal to $n' + s - n - i - 1$ over K', so that y has a dimension at most equal to $s + l - n - 1$ over K'; since this is impossible, it follows that y has the dimension $l - n'$ over $K'(z)$, and hence z must have a dimension at least equal to $s + n' - n$ over K'. Since Z is contained in a component of $|X'| \cap V'$ which is not contained in $|X|$ (or which is simple in V) and which therefore must be proper in V, it has a dimension at most equal to $s + n' - n$; this shows that Z has exactly the dimension $s + n' - n$ and hence is a proper component of $|X'| \cap V'$ and is general over $\bar{K}$ for V'. Since y has the dimension $l - n'$ over $K'(z)$, it has the dimension $l + s - n$ over K; this shows that Y is a proper component of $|U \times X'| \cap F$ in $U \times V$ and is general over $\bar{K}$ for F. This proves Lemma 4.

Lemma 5. *Let F be a correspondence between a variety U and a variety V such that* (α) *F^{-1} preserves the singularities in V and is complete over all points in U in the absolute case, or* (β) *$\bar{F}^{-1}$ preserves the singularities in $\bar{V}$ in the relative case; if X is a cycle in V, rationally equivalent to zero in V, such that the expression $F^{-1}(X)$ is defined, then the cycle $F^{-1}(X)$ is rationally equivalent to zero in U.*

Proof. Let Q be (α) the union of $\bar{U} \times \bar{V} - U \times V$ with all the singular subvarieties in $\bar{U} \times \bar{V}$ in the absolute case, or (β) the union of all singular subvarieties in $\bar{U} \times \bar{V}$ in the relative case; similarly, let Q' be (α) the union of $\bar{U} - U$ with all the singular subvarieties in $\bar{U}$ in the absolute case, or (β) the union of all singular subvarieties in $\bar{U}$ in the relative case. Then our assumptions imply that the projection of $Q \cap |\bar{F}|$ in $\bar{U}$ is contained in Q'. Since X is rationally equivalent to zero in V, the cycle $U \times X$ is rationally equivalent to zero in $U \times V$, and hence, by Theorem 1, $(U \times X).F$ is rationally equivalent to zero in $U \times V$; furthermore, if we denoted by S the ambient projective space of V, there exists, according to Lemma 1*, a cycle X_1 in $\bar{U} \times S$, whose support is contained in $Q \cap |\bar{F}|$, such that the cycle $\overline{(U \times X).F} + X_1$ is rationally equivalent to zero in $\bar{U} \times S$. Since S is complete and non-singular (or since S is non-singular and $\bar{U}$ is complete), the cycle $\mathrm{pr}_{\bar{U}}(\overline{(U \times X).F} + X_1)$ is rationally equivalent to zero in $\bar{U}$, and since the cycle $\mathrm{pr}_{\bar{U}}(X_1)$ has its support in Q', the cycle $\mathrm{pr}_{\bar{U}}(\overline{(U \times X).F})$ is also rationally equivalent to zero in $\bar{U}$. Since $F^{-1}(X) = \mathrm{pr}_U((U \times X).F)$ is evidently the restriction of $\mathrm{pr}_{\bar{U}}(\overline{(U \times X).F})$ in U (and since every component of $\mathrm{pr}_{\bar{U}}(\overline{(U \times X).F})$ is the closure in $\bar{U}$ of a component of $\mathrm{pr}_U((U \times X).F)$, on account of the fact that the intersection-product $(U \times X).F$ in $U \times V$ is defined in the relative sense), it follows that the cycle $F^{-1}(X)$ is also rationally equivalent to zero in U. This proves Lemma 5.

Proof of Theorem 3. In view of Lemmas 4 and 5, we only need to show that if F and F' are rationally equivalent admissible correspondences between U and V, and if X and X' are rationally equivalent cycles in V such that the expressions $F^{-1}(X)$ and $F'^{-1}(X')$ are defined, then $F^{-1}(X)$ and $F'^{-1}(X')$ are rationally equivalent in U. Let Q, Q', and S be the same as defined in the proof of Lemma 5 and let S' be the ambient projective space of U; we apply the definition of rational equivalence between F and F', and replace the involved cycles in the ambient space of $U \times V$ by their restrictions to the variety $S' \times S$. Then there exist a cycle F^* in $U \times V$ and cycles F_1, F_1', F_2, F_2' in $S' \times S$, with supports in $\bar{U} \times \bar{V}$, such that $\bar{F} + F_1$ and $\bar{F}' + F_1'$ are both specializations of $\bar{F}^*$ over a suitable field of definition for $U \times V$, and such that F_2 and F_2' are latent cycles of the specializations $\bar{F}^* \to \bar{F} + F_1$ and $\bar{F}^* \to \bar{F}' + F_1'$ respectively, and such that $| F_1 | \subset Q$, $| F_1' | \subset Q$, $\mathrm{pr}_{\bar{U}}(Q \cap | F_1 |) \subset Q'$,

$$\mathrm{pr}_{\bar{U}}(Q \cap | F_1' |) \subset Q', \qquad \mathrm{pr}_{\bar{U}}(Q \cap | F_2 |) \subset Q', \qquad \mathrm{pr}_{\bar{U}}(Q \cap | F_2' |) \subset Q'.$$

Let K be a field of definition for $U \times V$ over which all the above introduced cycles and correspondences are rational, and let X^* be the cycle obtained by applying Lemma 2 to the cycle X with respect to the field K such that $\bar{X}^*$ satisfies the condition of Corollary 3 of Lemma 2. Then, by Lemma 4, the intersection-products $(U \times X^*).F$, $(U \times X^*).F'$, $(U \times X^*).F^*$ in $U \times V$ and the intersection-products $(U \times \bar{X}^*).R(F_2, \bar{U} \times \bar{V})$, $(U \times \bar{X}^*).R(F_2', \bar{U} \times \bar{V})$ in $\bar{U} \times \bar{V}$ are all defined. According to the Remark after Lemma 1, there exist cycles D and D' in $S' \times S$, with supports in $Q \cap (| \bar{F} | \cup | F_1 | \cup | F_2 |)$ and $Q \cap (| \bar{F}' | \cup | F_1' | \cup | F_2' |)$ respectively, such that $\overline{(U \times X^*).F} + D$ and $\overline{(U \times X^*).F'} + D'$ are specializations of $\overline{(U \times X^*).F^*}$ in $S' \times S$. Since $S' \times S$ is complete and non-singular, the cycles $\mathrm{pr}_{S'}(\overline{(U \times X^*).F} + D)$ and $\mathrm{pr}_{S'}(\overline{(U \times X^*).F'} + D')$ are specializations of $\mathrm{pr}_{S'}(\overline{(U \times X^*).F^*})$ in S'; and since $\mathrm{pr}_{S'}(\overline{(U \times X^*).F^*}) = \overline{\mathrm{pr}_U((U \times X^*).F^*)}$, $\mathrm{pr}_{S'}(\overline{(U \times X^*).F}) = \overline{\mathrm{pr}_U((U \times X^*).F)}$, $\mathrm{pr}_{S'}(\overline{(U \times X^*).F'}) = \overline{\mathrm{pr}_U((U \times X^*).F')}$, and since $| \mathrm{pr}_{S'}D | \subset Q'$, $| \mathrm{pr}_{S'}D' | \subset Q'$, it follows that $\mathrm{pr}_U((U \times X^*).F)$ and

$$\mathrm{pr}_U((U \times X^*).F')$$

are specializations of $\mathrm{pr}_U((U \times X^*).F^*)$ in U. This shows that $F^{-1}(X^*)$ and $F'^{-1}(X^*)$ are rationally equivalent in U. Now, since the cycles $U \times X$ and $U \times X^*$ are rationally equivalent in $U \times S$, it follows from Lemma 1* that there exists a cycle X_1 in $\bar{U} \times S$, with support in $Q \cap | \bar{F} |$, such that the cycles

$$\overline{(U \times X).F} + X_1$$

and $\overline{(U \times X^*).F}$ are rationally equivalent in $\bar{U} \times S$; from this it follows then by a similar argument as in the proof of Lemma 5 that $F^{-1}(X)$ and $F^{-1}(X^*)$ are rationally equivalent in U. Similarly, we can show that $F'^{-1}(X')$ and $F'^{-1}(X^*)$ are rationally equivalent in U. This shows that $F^{-1}(X)$ and $F'^{-1}(X')$ are rationally equivalent in U.

We turn now to the case where the admissible correspondence class $\mathfrak{F}$ contains an element F such that (α) F is a rational transformation of U into V, defined at every point of U, in the absolute case, or (β) $\bar{F}$ is a rational transformation of $\bar{U}$ into $\bar{V}$, defined at every point in $\bar{U}$, in the relative case. When this is the case, we shall call F an *admissible transformation* of U into V and $\mathfrak{F}$ an *admissible transformation class* of U into V. We have then the following theorem:

THEOREM 4. *If $\mathfrak{F}$ is an admissible transformation class of a variety U into a variety V, then the expression $\mathfrak{F}^{-1}(\mathfrak{X})$ is defined for every cycle class $\mathfrak{X}$ in V, and the mapping $\mathfrak{X} \to \mathfrak{F}^{-1}(\mathfrak{X})$ is a homomorphism of the ring $H_r(V)$ into the ring $H_r(U)$.*

This theorem will follow immediately from Theorem 3 and the following lemma, which will also be useful to us later.

LEMMA 6. *Let U and V be varieties, let X and Y be cycles in V, and let F be a rational transformation of U into V such that (α) F is defined everywhere in*

$$F^{-1}(|X|)$$

and F^{-1} preserves the singularities in $|X| \cup |Y|$ in the absolute theory, or (β) $\bar{F}$ is defined everywhere in $\bar{F}^{-1}(|\bar{X}|)$, F is complete over all points in $|X|$ (except in case $\bar{F}$ is defined everywhere in $\bar{U}$), and $\bar{F}^{-1}$ preserves the singularities in

$$|\bar{X}| \cup |\bar{Y}|$$

in the relative theory; then there exist cycles X′ and Y′, rationally equivalent in V to X and Y respectively, such that the intersection-product X′.Y′ in V as well as the expressions $F^{-1}(X')$, $F^{-1}(Y')$, $F^{-1}(X'.Y')$ are all defined, and the class determined by $F^{-1}(X'.Y')$ in U is the intersection-product of the classes determined by $F^{-1}(X')$ and $F^{-1}(Y')$ in U; furthermore, if F or $\bar{F}$ is defined everywhere in U or in $\bar{U}$ respectively in the absolute or relative case, then we have the relation

$$F^{-1}(X'.Y') = F^{-1}(X').F^{-1}(Y')$$

in U.

PROOF. Without any loss of generality we can restrict ourselves to the case where both X and Y are homogeneous, the general case following evidently from this by linearity; for simplicity, we shall restrict ourselves to the absolute theory first, and we shall show later how the proof can be modified to take care of the relative theory. Let K be a field of definition for U, V, and F, such that X and Y are rational over K, and let n, l, s and t be the dimensions of V, U (and hence F), X and Y respectively. We apply Lemma 3 and denote by X′ and Y′ the cycles so obtained, so that the intersection-product X′.Y′ in V is defined and the cycles X′, Y′, and X′.Y′ all satisfy the hypothesis of Lemma 4; this implies in particular that the intersection-products $X^* = (U \times X').F$, $Y^* = (U \times Y').F$, and $Z^* = (U \times (X'.Y')).F$ in $U \times V$ are all defined and are general over K for F. Furthermore, by applying Lemma 4 to each component of $|X'| \cap |Y'|$, we conclude that if Z_1^* is a component of $(U \times (|X'| \cap |Y'|)) \cap F$ which has a dimension at least equal to $s + t + l - 2n$ and which is either not contained in $U \times (|X| \cup |Y|)$ or simple in $U \times V$, then Z_1^* is a proper component of $(U \times (|X'| \cap |Y'|)) \cap F$ in $U \times V$ and is general over K for F, and hence is also a proper component of $|U \times (X'.Y')| \cap F$ in $U \times V$ as well as a proper component of $|X^*| \cap |Y^*|$ in F.

We set $X'' = \mathrm{pr}_U X^*$, $Y'' = \mathrm{pr}_U Y^*$, and $Z'' = \mathrm{pr}_U Z^*$. We shall first assume that F is defined everywhere in U, and we shall prove the relation $X''.Y'' = Z''$. Let Z_1'' be a component of $|X''| \cap |Y''|$ which is simple in U; since the projection from F to U is regular along Z''_1, there exists a uniquely determined subvariety Z^*_1 in F which has Z''_1 for its projection in U, and it follows from this uniqueness that Z^*_1 is a component of $| X^* | \cap | Y^* |$; let Z'_1 be the projection of Z^*_1 in V. If Z^*_1 is singular in $U \times V$ and is contained in $U \times (| X | \cup | Y |)$, then Z'_1 would also be singular in V and contained in $| X | \cup | Y |$; since the rational transformation F preserves the singularities in $| X | \cup | Y |$ and since Z''_1 is evidently contained in the projection in U of $(U \times Z'_1) \cap F$, this would imply that Z''_1 is singular in U, which is in contradiction to our assumption. Thus Z^*_1 is either simple in $U \times V$ or not contained in $U \times (| X | \cup | Y |)$; furthermore, since Z''_1 has a dimension at least equal to $s + t + l - 2n$, Z^*_1 has also a dimension at least equal to $s + t + l - 2n$. It follows then from our above observation that Z^*_1 is a proper component of $| X^* | \cap | Y^* |$ in F and is general over K for F, and hence Z''_1 is also a proper component of $| X'' | \cap | Y'' |$ in U and is general over K for U; this imples in particular that the intersection-product $X''.Y''$ in U is defined. It follows from *Foundations*, Ch. VII, Theorem 7, and Ch. VI, Theorem 10, that Z^*_1 is simple in F, and that we have the relation $i(X''.Y'', Z''_1; U) = i(X^*.Y^*, Z^*_1; F)$. Now, according to *Foundations*, Ch. VII, Theorem 11, the intersection-product $(U \times X').(U \times Y')$ in $U \times V$ is defined and is equal to $U \times (X'.Y')$; and since the intersection-products

$$(U \times X').F, \qquad (U \times Y').F,$$

and $(U \times (X'.Y')).F$ in $U \times V$ are all defined, we have by associativity the relations $Z^* = (U \times X').(U \times Y').F = X^*.(U \times Y') = Y^*.(U \times X')$. It follows then from *Foundations*, Ch. VII, Theorem 18 (i), that $i(X^*.Y^*, Z^*_1; F) = i(X^*.(U \times Y'), Z^*_1; U \times V) =$ coefficient of Z^*_1 in Z^*; and since Z^*_1 is the only component of Z^* which has the projection Z''_1 in U, the coefficient of Z^*_1 in Z^* is equal to the coefficient of Z''_1 in Z''. Thus we have shown that for every component Z''_1 of $X''.Y''$ the coefficient of Z''_1 in $X''.Y''$ is equal to the coefficient of Z''_1 in Z'', and since every component of Z'' is evidently also a component of $X''.Y''$, it follows that $X''.Y'' = Z''$.

We turn now to the general case, where F is only assumed to be defined everywhere in $F^{-1}(| X |)$. Since X' is obtained by applying Lemma 2 to X and the field K, we can choose X' so that the conditions in Corollary 3 and Remark (1) after Lemma 2 are satisfied; this implies that $\bar{X}'$ satisfies Lemma 2 with respect to $\bar{X}$ and the field K, and that the specialization $X' \to X$ over K has a latent cycle E in V such that $\bar{E}$ is a latent cycle of the specialization $\bar{X}' \to \bar{X}$ over K (in the ambient space of V), with the properties stated in Lemma 2. Then, by Lemma 4, every component of $| \bar{U} \times \bar{E} | \cap | \bar{F} |$ which is not contained in $| \bar{U} \times \bar{X} |$ and is not negligible in $\bar{U} \times \bar{V}$ is proper in $\bar{U} \times \bar{V}$, so that the intersection-product $R(\bar{U} \times \bar{E}, \bar{U} \times \bar{V} - | \bar{U} \times \bar{X} |).R(\bar{F}, \bar{U} \times \bar{V} - | \bar{U} \times \bar{X} |)$ in $\bar{U} \times \bar{V} - | \bar{U} \times \bar{X} |$ is defined in the absolute sense. Since we have evidently $R(\bar{U} \times \bar{X}, \bar{U} \times \bar{V} - | \bar{U} \times \bar{X} |) = 0$, it follows from Lemma 1 and the Remark after it that if D is any specialization of $\bar{X}^*$ in the ambient space of $U \times V$ over the specialization $\bar{X}' \to \bar{X}$ over K, then we have the relation $| D | \subset | \bar{F} | \cap (| \bar{U} \times \bar{X} | \cup | \bar{U} \times \bar{E} |) \cap (| \bar{U} \times \bar{X} | \cup Q)$, where Q denotes the union of all singular subvarieties in $\bar{U} \times \bar{V}$. Since, again by Lemma 4, every component of $| \bar{U} \times \bar{E} | \cap | \bar{F} |$ which is not contained in $| \bar{U} \times \bar{X} |$ and is not negligible in $\bar{U} \times \bar{V}$ is general for $| \bar{F} |$ over K and hence cannot be contained in Q, we obtain from this the relation $| D | \subset | \bar{F} | \cap | \bar{U} \times \bar{X} |$. If we set $X_0^* = R(D, U \times V)$ and $X_0'' = \mathrm{pr}_U X_0^*$, then we have evidently the relations $| X_0^* | \subset | F | \cap | U \times X |$ and $| X_0'' | \subset F^{-1}(| X |)$; furthermore, since X_0^* is rationally equivalent to X^* in $U \times V$ and since F^{-1} preserves the singularities in $| X |$, it follows by a similar argument as in the proof of Theorem 3 that X_0'' is rationally equivalent to X'' in U. Now, let W be the smallest subset in U such that the transformation F is defined everywhere in $U - W$; it is well known that W is a closed subset in U, normally algebraic over K. Our assumption that $F^{-1}(| X |)$ does not intersect W implies that no component of X_0'' intersects W. Let K_1 be an extension of K such that the cycles X', E, and D are rational over K_1, and let Y' be obtained by applying Lemma 2 to Y and the field K_1 (it is evidently permissible to assume this); then, by Lemma 4, the intersection-product $X_0^*.(U \times Y')$ in $U \times V$ is defined, and every component of $| X_0^* | \cap | U \times Y' |$ which is not contained in $| U \times Y |$ and is not negligible in $U \times V$, or which is simple in $U \times V$, is proper in $U \times V$ and is general for a component of X_0'' over $\bar{K}_1$. Let Z_0'' be a component of $| X_0'' | \cap | Y'' |$ which is simple in U. Since Z_0'' is not contained in W, the projection from F to U is regular along Z_0''; there exists then a uniquely determined subvariety Z_0^* in F which has Z_0'' for its projection in U, and it follows from this uniqueness that Z_0^* is also a component of $| X_0^* | \cap | Y^* |$. By a similar argument as in the first part of the preceding paragraph, replacing $| X | \cup | Y |$ and F by $| Y |$ and $| X_0^* |$ respectively and observing that a component of $| X_0^* | \cap | Y^* |$ is also a component of $| X_0^* | \cap | U \times Y' |$ and vice-versa, we conclude that the intersection-product $X_0''.Y''$ in U is defined (and that every component of $| X_0'' | \cap | Y'' |$ is general over $\bar{K}_1$ for a component of X_0''). Since F is defined everywhere in $U - W$, it follows from our result in the preceding paragraph that

$$R(X'', U - W).R(Y'', U - W) = R(Z'', U - W)$$

in the absolute sense; on the other hand, by Theorem 1, $R(X'', U - W).R(Y'', U - W)$ is rationally equivalent to $R(X_0'', U - W).R(Y'', U - W) = R(X_0''.Y'', U - W)$ in $U - W$ in the absolute sense. It follows then that $R(Z'', U - W)$ and $R(X_0''.Y'', U - W)$ are rationally equivalent in $U - W$ in the absolute sense; however, since neither Z'' nor $X_0''.Y''$ has any component in W, it follows that Z'' and $X_0''.Y''$ are rationally equivalent in U. This concludes the proof of Lemma 6 for the absolute theory. (Remark. Although we are dealing only with the absolute theory in this paragraph, we have in several places added specifically the phrase "in the absolute sense"; the reason for this is that, as we shall see below, the entire argument holds also for the relative theory, but the places where "in the absolute sense" is specifically mentioned should still be understood in the absolute sense even in dealing with the relative theory.)

For the relative theory, we first introduce as above the cycles X' and Y' by applying Lemma 3; and we observe that, by virtue of Lemma 2, Corollary 3, the cycles $\bar{X}'$ and $\bar{Y}'$ satisfy Lemma 3 with respect to $\bar{X}$ and $\bar{Y}$ in $\bar{V}$, and that, by the definition of intersection-product in the relative theory, we have the relation $\overline{X'.Y'} = \bar{X}'.\bar{Y}'$ in $\bar{V}$. Again we attend first to the case where $\bar{F}$ is defined everywhere. Since $\bar{F}^{-1}$ preserves the singularities in $| \bar{X} | \cup | \bar{Y} |$, we can apply what we have proved above for the absolute theory to the cycles $\bar{X}'$, $\bar{Y}'$, and $\overline{X'.Y'}$; we conclude therefore that the expressions $\bar{F}^{-1}(\bar{X}')$, $\bar{F}^{-1}(\bar{Y}')$, $\bar{F}^{-1}(\overline{X'.Y'})$ are all defined and that the relation $\bar{F}^{-1}(\overline{X'.Y'}) = \bar{F}^{-1}(\bar{X}').\bar{F}^{-1}(\bar{Y}')$ hold in $\bar{U}$. Furthermore, the cycles $\bar{F}^{-1}(\bar{X}')$, $\bar{F}^{-1}(\bar{Y}')$, and $\bar{F}^{-1}(\overline{X'.Y'})$ are all general over K for $\bar{U}$; it follows then that the expressions $F^{-1}(X')$, $F^{-1}(Y')$, and $F^{-1}(X'.Y')$ are all defined (in the relative sense), and by restriction from $\bar{U}$ to U, we obtain the relation $F^{-1}(X'.Y') = F^{-1}(X').F^{-1}(Y')$ in U.

For the general case, we observe that the proof given above for the absolute theory holds also here, if we take account of the additional assumption that F is complete over all points in $| X |$. In fact, it follows from this assumption that $\overline{R(D, U \times V)} = R(D, \bar{U} \times \bar{V})$; this shows that $\bar{X}_0^*$ is also a specialization of $\bar{X}^*$ in $\bar{U} \times \bar{V}$, so that X_0^* is rationally equivalent to X^* in $U \times V$ in the relative sense. From this point on, by considering at the same time also the transformation $\bar{F}$, one can see readily that the entire proof holds also for the relative theory. This concludes the proof of Lemma 6.

7. We turn now to the question of birational invariance and, in connection with this problem, we shall introduce the restricted theory and finally obtain some results of a purely local nature.

We observe first that an everywhere biregular birational transformation F of a variety U onto a variety V certainly satisfies the hypothesis of Theorem 3 and hence induces a homomorphism $\mathfrak{F}^{-1}$ of $H_r(V)$ into $H_r(U)$. In case of the absolute theory, the expression $F^{-1}(X)$ is defined for every cycle X in V and the mapping $X \to F^{-1}(X)$ is a one-to-one correspondence between the cycles in V and in U; it follows that in this case $\mathfrak{F}^{-1}$ is an isomorphism between $H_r(V)$ and $H_r(U)$, and this gives a proof of the invariance of the absolute theory. The situation is, however, quite different in the relative theory, for, even when F is an everywhere biregular birational transformation of U onto V, the expression $F^{-1}(X)$ need not be defined in the relative sense for every cycle X in V; in fact, as we shall see by examples later, the homomorphism $\mathfrak{F}^{-1}$ is in this case not necessarily an isomorphism, which shows that the relative theory is in general not invariant under an everywhere biregular birational transformation. We shall now show that this invariance can be restored by restricting ourselves to a special class of cycles and modifying the concept of equivalence accordingly, and this leads us to what we should like to call the *restricted theory*, which we shall proceed to describe.

(γ) *The restricted theory*. If X is a cycle in V and if K is a field of definition for V, then a cycle X_1 in V is said to be a specialization in V of X over K, if there exists a cycle X_2 in the ambient projective space S of V, whose components are all closures of singular subvarieties in V, such that the cycle $X_1 + X_2$ is a specialization in S of the cycle X over K. The intersection-product $X.Y$

in V of two cycles X and Y in V is said to be defined if it is defined in the absolute sense and if every component of $|\bar{X}| \cap |\bar{Y}|$ in $\bar{V}$ is the closure in $\bar{V}$ of a component of $|X| \cap |Y|$. One sees that these definitions are very similar to those in the relative theory, but somewhat stricter, and on account of these restrictions, it is in general not possible to define an intersection-product for every two cycle classes in V, so that the group $G(V)/G_r(V)$ cannot in general be made in this way into a ring; furthermore, just as in case of the relative theory, the restricted theory is in general not invariant under an everywhere biregular birational transformation. In spite of all these disadvantages, this theory turns out to be useful if we restrict ourselves to those cycle classes in V which we shall call complete. A cycle X in V is said to be *complete* if its components are all complete varieties and a cycle class $\mathfrak{X}$ in V (in the absolute, relative, or restricted sense) is said to be *complete* if it contains a complete cycle. It can be easily seen that if X and Y are two cycles in V such that the intersection-product $X.Y$ in V is defined in the absolute sense, and if one of the two cycles X and Y is complete, then the intersection-product $X.Y$ in V is also defined in the relative and restricted senses; it follows from this that the intersection-product $\mathfrak{X}.\mathfrak{Y}$ in V is always defined in the restricted sense if one of the two cycle classes $\mathfrak{X}$ and $\mathfrak{Y}$ is complete, so that in particular the set of all complete cycle classes in V can be made into a ring. In fact, Lemma 1 holds for the restricted theory without any modification, while Lemmas 2 and 3 hold also for the restricted theory if the cycles X and Y are complete; it follows then that Theorem 1 holds for the restricted theory if the cycles X and X' are complete, and that Theorem 2 holds for the set of all complete cycle classes. We shall denote the ring of all complete cycle classes in V by $H'_r(V)$; in case of the absolute and relative theories, $H'_r(V)$ is a subring and an ideal in $H_r(V)$, but it is in the restricted theory, where $H'_r(V)$ is the only ring we have defined, that it is specially significant. Finally, if W is any subset in V, we shall denote by $H'_r(V, W)$ the subset of all complete cycle classes which contains complete cycles with supports in W; it is easily seen that $H'_r(V, W)$ is a subring in $H'_r(V)$.

In order to extend Theorems 3 and 4 to the restricted theory, we shall prove the following lemma, which gives a sufficient condition for the preservation of complete cycle classes under a correspondence.

LEMMA 7 (for restricted theory). *Let X be a complete cycle in V, and let F be a correspondence between a variety U and a variety V such that F is complete over all points in $|X|$; then there exists a cycle X' in V, rationally equivalent in V to X, such that the expression $F^{-1}(X')$ is defined and is rationally equivalent to a complete cycle in U, with support in $F^{-1}(|X|)$.*

PROOF. Let K be a field of definition for U, V, and F such that X is rational over K; let X' be a cycle in V, rational over a purely transcendental extension $K(u)$ of K, which is obtained by applying Lemma 2 to X, and let Z be a latent cycle in V of the specialization $X' \to X$ over K satisfying the conditions of that lemma. According to Lemma 2, Corollary 3, we can choose the cycles X' and Z so that $\bar{X}'$ and $\bar{Z}$ satisfy the Lemma 2 with respect to the cycle $\bar{X}$ in $\bar{V}$, and according to Remark (1) after the proof of Lemma 2, we can choose X' so that $\bar{X}$ is actually a specialization of $\bar{X}'$ in the ambient projective space S of V; this shows that X is a specialization in V of X' over K in the restricted sense. Furthermore, according to Remarks (2) and (3) after the proof of Lemma 2, there exist cycles A and B in S such that $R(\bar{X}', S - |X|)$ is the intersection-product $R(A, S - |X|).R(\bar{V}, S - |X|)$ in $S - |X|$, and that $R(\bar{Z}, S - |X|)$ is the intersection-product $R(B, S - |X|).R(\bar{V}.S - |X|)$ in $S - |X|$, both intersection-products being understood in the absolute sense. If S' denotes the ambient projective space of U, then we have the relations $(\bar{U} \times (|A| \cap \bar{V})) \cap |\bar{F}| = (S' \times (|A| \cap \bar{V})) \cap |\bar{F}| = (S' \times |A|) \cap |\bar{F}|$, and hence every component of $|\bar{U} \times \bar{X}'| \cap |\bar{F}|$ which is not contained in $|X|$ is a component of $|S' \times A| \cap |\bar{F}|$. If s, n, m, and l are the dimensions of X, V, S, and F respectively, then A has the dimension $s + m - n$, and since $S' \times S$ is non-singular, every component of $|S' \times A| \cap |\bar{F}|$ has a dimension at least equal to $s + l - n$; it follows then that every component of $|\bar{U} \times \bar{X}'| \cap |\bar{F}|$ which is not contained in $|X|$ must have a dimension at least equal to $s + l - n$ and hence is non-negligible in $\bar{U} \times \bar{V}$. Similarly, we can show that every component of $|\bar{U} \times \bar{Z}| \cap |\bar{F}|$ which is not contained in $|X|$ is non-negligible in $\bar{U} \times \bar{V}$. Since, by Lemma 2, $\bar{X}'$ satisfies the hypothesis of Lemma 4 and $\bar{Z}$ satisfies the first part of the hypothesis of Lemma 4, it follows that every component of $|\bar{U} \times \bar{X}'| \cap |\bar{F}|$ which is simple in $\bar{U} \times \bar{V}$ or not contained in $|\bar{U} \times X| \cap |\bar{F}|$ is proper and simple in $\bar{U} \times \bar{V}$, and that every component of $|\bar{U} \times Z| \cap |\bar{F}|$ which is not contained in $|\bar{U} \times X| \cap |\bar{F}|$ is proper and simple in $\bar{U} \times \bar{V}$. Since F is complete over all points in $|X|$, the set $|\bar{U} \times X| \cap |\bar{F}|$ is contained in $U \times V$; this shows that the intersection-product $(U \times X').F$ in $U \times V$ is defined in the restricted sense, and hence also the expression $F^{-1}(X')$ is defined in the restricted sense. Consider now a specialization D of the cycle $\overline{(U \times X').F}$ in $S' \times S$ over the specialization $\bar{X}' \to X$ over K; since $\overline{U \times X} = \bar{U} \times X$ is the unique specialization in $S' \times S$ of $\overline{U \times X'}$ over the specialization $\bar{X}' \to X$ over K, it follows from the Remark after the proof of Lemma 1 that $|D|$ is contained in $|\bar{U} \times X| \cap |\bar{F}|$ and hence $|\mathrm{pr}_{S'}D|$ is contained in $F^{-1}(|X|)$. Since $\mathrm{pr}_{S'}D$ is a specialization in S' of $\mathrm{pr}_{S'}(\overline{(U \times X').F})$ over K, $R(\mathrm{pr}_{S'}D, U)$ is a specialization in U of $R(\mathrm{pr}_{S'}(\overline{(U \times X').F}), U) = \mathrm{pr}_U((U \times X').F) = F^{-1}(X')$, at least in the absolute sense; and since $|\mathrm{pr}_{S'}D|$ is contained in $F^{-1}(|X|)$ and hence also in U, it follows that $R(\mathrm{pr}_{S'}D, U)$ is a complete cycle in U and is a specialization in U of $F^{-1}(X')$ over K in the restricted sense. Since $F^{-1}(X')$ is rational over $K(u)$, it is rationally equivalent in U to $R(\mathrm{pr}_{S'}D, U)$; this proves our lemma.

In view of Lemma 7, it is not difficult to see that Lemmas 5 and 6 hold also for the restricted theory, in case the cycles X and Y are complete, whereby we make the same assumptions as in the relative theory except that F^{-1} need only be assumed to preserve the singularities in V or in $|X| \cup |Y|$ respectively. In fact, the proofs of both lemmas for the relative theory hold here with hardly any modifications, and the weaker assumption in case of Lemma 5 is due to the stricter definition of specialization in the restricted theory. For the restricted theory, a correspondence F between a variety U and a variety V is said to be *admissible with respect to a subset* W in V, if (γ) F^{-1} preserves the singularities in W and F is complete over all points in W. If Q and Q' are as defined in the preceding section for the relative theory, then this condition can be expressed by the relations $\mathrm{pr}_U(Q \cap W \cap |F|) \subset Q'$ and $\mathrm{pr}_{\bar{V}}((\bar{U} \times \bar{V} - U \times V) \cap F^{-1}(W) \cap |F|) \subset \bar{V} - V$. A corresponding modification is to be made in the definition of the *rational equivalence* between admissible correspondences and hence also in the definition of an *admissible correspondence class*. Finally, an admissible correspondence class $\mathfrak{F}$ between U and V, admissible with respect to a subset W on V, is said to be an *admissible transformation class* of U into V, if $\mathfrak{F}$ contains an element F which is a rational transformation of U into V, defined everywhere in $F^{-1}(W)$; and when this is the case, any such transformation F is then said to be an *admissible transformation* of U into V, admissible with respect to W. With these modifications, Theorems 3 and 4 then hold also for the restricted theory, and we shall state them as the next two theorems.

THEOREM 5 (for restricted theory). *If $\mathfrak{F}$ is an admissible correspondence class between a variety U and a variety V, admissible with respect to a subset W in V, then the expression $\mathfrak{F}^{-1}(\mathfrak{X})$ is defined for every cycle class $\mathfrak{X}$ in $H'_r(V, W)$ and the mapping $\mathfrak{X} \to \mathfrak{F}^{-1}(\mathfrak{X})$ is a homomorphism of the additive group $H'_r(V, W)$ into the additive group $H'_r(U, F^{-1}(W))$.*

THEOREM 6 (for restricted theory). *If $\mathfrak{F}$ is an admissible transformation class of a variety U into a variety V, admissible with respect to a subset W in V, then the expression $\mathfrak{F}^{-1}(\mathfrak{X})$ is defined for every cycle class $\mathfrak{X}$ in $H'_r(V, W)$ and the mapping $\mathfrak{X} \to \mathfrak{F}^{-1}(\mathfrak{X})$ is a homomorphism of the ring $H'_r(V, W)$ into the ring $H'_r(U, F^{-1}(W))$.*

Finally, we shall deduce from Theorem 6 the following result of a local nature. Let X and Y be two complete cycles in variety V, and let F be a birational transformation of a variety U onto V such that F^{-1} is defined and biregular at every point in $|X| \cup |Y|$; since F evidently satisfies the hypothesis of Theorem 6, it induces a homomorphism $\mathfrak{F}^{-1}$ of $H'_r(V, |X| \cup |Y|)$ into $H'_r(U, F^{-1}(|X| \cup |Y|))$, in the restricted sense. Since the expression $F^{-1}(Z)$ is evidently defined for every complete cycle Z in $|X| \cup |Y|$, in the restricted sense, and the mapping $Z \to F^{-1}(Z)$ is one-to-one between the complete cycles in $|X| \cup |Y|$ and the complete cycles in $F^{-1}(|X| \cup |Y|)$, it follows that $\mathfrak{F}^{-1}$ is an isomorphism. In particular, since the class $\mathfrak{X}.\mathfrak{Y}$ is defined in the restricted sense and is evidently contained in $H'_r(V, |X| \cup |Y|)$, we have the relation $\mathfrak{F}^{-1}(\mathfrak{X}.\mathfrak{Y}) = \mathfrak{F}^{-1}(\mathfrak{X}).\mathfrak{F}^{-1}(\mathfrak{Y})$; this shows that the intersection class $\mathfrak{X}.\mathfrak{Y}$ is invariant under any birational transformation of V which is biregular at every point of $|X| \cup |Y|$. This implies in particular that if V' is any neighborhood of $|X| \cup |Y|$ in V, then the restriction of $\mathfrak{X}.\mathfrak{Y}$ in V' is the intersection class in V' of the restrictions of $\mathfrak{X}$ and $\mathfrak{Y}$ in V'. In other words, the intersection class in the restricted sense of any two complete cycles X and Y in a variety V is determined by an arbitrarily small neighborhood of the support $|X| \cup |Y|$ of the two cycles and hence depends only on the local properties of $|X| \cup |Y|$.

8. We conclude this paper by a few simple examples, illustrating the use of the concepts developed here. We observe first that if a rational variety V is incomplete or is not non-singular, then every point in V is rationally equivalent to zero in V in the absolute sense, and hence every 0-dimensional cycle class is the zero class in the absolute theory in V; in particular, if V is such a surface, then the intersection-product of every two 1-dimensional cycle classes is zero in the absolute theory, so that this theory is rather devoid of contents in this case. The situation is quite different with the relative theory, as long as the closure $\bar{V}$ of V in its ambient projective space is non-singular. For example, if we take for V an affine plane, then the self-intersection of the class determined by a line in V is the class determined by a point in V, which is not the zero class in the relative theory; or, if we take for V the variety obtained from the affine plane by a locally quadratic transformation at the origin, which replaces the origin by a subvariety W of dimension 1 (this transformation is known as the σ-process of Hopf in the theory of complex-analytic manifold), then the self-intersection of the cycle class determined by W is the negative of the cycle class determined by a point in V. We observe that in the second example, since the cycle classes involved are complete (and since $\bar{V}$ is non-singular), the relative theory coincides with the restricted theory, so that the self-intersection class has an invariant meaning independent of the embedding of the variety V. These examples show that under certain circumstances the relative and restricted theories can be more useful than the absolute theory, which justifies the inclusion of these theories in our treatment; they also show that the relative theory is analogous to the homology theory of the variety V as an open manifold, while the absolute theory is more similar to the relative homology theory of $\bar{V}$ modulo the subcomplex $\bar{V} - V$.

Next, we shall cite a simple example to show the dependence of the relative theory in V on the embedding of V in $\bar{V}$. We take again for V an affine plane, and let (ξ, η) be a coordinate system in it; we can embed V in a projective plane, or we can embed V in a product of two projective lines by embedding the ξ-axis and the η-axis in two projective lines separately, and we shall denote these two completions of V by $\bar{V}$ and $\bar{V}'$ respectively. Then, according as we consider the relative theory of V with respect to the embedding $\bar{V}$ or $\bar{V}'$ respectively, the two lines in V defined by the equations $\xi = 0$ and $\xi = \eta$ determine the same or different cycle classes in V; also, in the former case the intersection-product of the two lines defined by the equations $\xi = 0$ and $\xi = 1$ is not defined, while in the latter it is defined and is equal to zero. Or, we can take for V a projective plane with a point x deleted and take for $\bar{V}$ the projective plane itself, while we take for $\bar{V}'$ the variety obtained from the projective plane by applying the σ-process to the point x; then, in the relative theory of V with respect to $\bar{V}$ every two lines determine the same cycle class, while in the relative theory of V with respect to $\bar{V}'$ a line passing through x is not in the same cycle class as a line not passing through x. One may give readily more complicated examples, but these suffice to show that the relative theory definitely depends upon the embedding of V in $\bar{V}$. We observe that, in either of our examples, both $\bar{V}$ and $\bar{V}'$ are non-singular; if we denote V by V or V' according to its embedding in $\bar{V}$ or $\bar{V}'$ respectively, then the birational correspondence between V and V' induces, according to Theorem 4, a homomorphism $\mathfrak{F}$ of $G(V)/G_r(V)$ into $G(V')/G_r(V')$ and a homomorphism $\mathfrak{F}^{-1}$ of $G(V')/G_r(V')$ into $G(V)/G_r(V)$, both in the relative sense. In either of our examples, the homomorphism $\mathfrak{F}$ is injective and the homomorphism $\mathfrak{F}^{-1}$ is surjective, and while the product $\mathfrak{F}^{-1} \circ \mathfrak{F}$ is the identity automorphism of $G(V)/G_r(V)$, the product $\mathfrak{F} \circ \mathfrak{F}^{-1}$ maps $G(V')/G_r(V')$ into a proper subgroup in $G(V)/G_r(V)$.

Finally, we shall cite a simple example which shows the effect of the presence of singularities in $\bar{V}$ on the relative and restricted theories. We consider an affine 3-space A^3 with a coordinate system (ξ, η, ζ); we can embed in a projective space S^3, but we can also embed it in a variety $S^2 \times S^1$, where S^2 is the projective plane completing the (ξ, η)-plane and S^1 is the projective line completing the ζ-axis. We take for V a cylinder in A^3 whose generators are parallel to the z-axis and whose base curve in the (ξ, η)-plane is non-singular also in S^2; such a surface is, for example, defined by the equation $\xi - \eta^2 = 0$. Corresponding to the two different embeddings of A^3 in S^3 and $S^2 \times S^1$, we obtain two different embeddings of V in two complete surfaces which we shall denote by $\bar{V}$ and $\bar{V}'$ respectively; the variety $\bar{V}'$ is non-singular, while the variety $\bar{V}$ has a singularity at the point (0, 0, 1), which is the vertex of the cone in S^3. One sees readily that the cycle class determined by one point in V will be zero or not according as we are dealing with the relative theory of V with respect to the embedding $\bar{V}$ or the embedding $\bar{V}'$, while in the restricted theory this class will not be zero in either case, being in fact independent of the embedding. It can be easily shown that if X is any cycle in V which is equivalent to a generator, the support of its closure $\bar{X}$ in $\bar{V}$ must contain the singular point at the vertex; it follows then that for two cycle classes determined by any two generators in V, the intersection-product in the restricted sense is not defined, while the situation would be evidently different if we replace $\bar{V}$ by $\bar{V}'$. This shows that the restricted theory is not suitable for the intersection theory of cycle classes in general.

THE JOHNS HOPKINS UNIVERSITY

BIBLIOGRAPHY

1. W. L. CHOW and B. L. VAN DER WAERDEN, *Zur algebraischen Geometrie*, IX, *Ueber zugeordnete Formen und algebraische Systeme von algebraischen Mannigfaltigkeiten*, Math. Ann., 113 (1937), pp. 692–704.
2. W. V. D. HODGE and D. PEDOE, Methods of Algebraic Geometry, vol. II, Cambridge, 1952.
3. F. SEVERI, *Ueber die Grundlagen der algebraischen Geometrie*, Abh. Math. Sem. Hamburg Univ., 9 (1933), pp. 335–364.
4. G. SHIMURA, *Reduction of algebraic varieties with respect to a discrete valuation of the basic field*, Amer. J. Math., 77 (1955), pp. 134–176.
5. B. L. VAN DER WAERDEN, *Zur algebraischen Geometrie*, XIV, *Schnittpunktszahlen von algebraischen Mannigfaltigkeiten*, Math. Ann., 115 (1938), pp. 619–642.
6. A. WEIL, Foundations of Algebraic Geometry, New York, 1946.

*ALGEBRAIC VARIETIES WITH RATIONAL DISSECTIONS**

BY WEI-LIANG CHOW

INSTITUTE FOR ADVANCED STUDY AND JOHNS HOPKINS UNIVERSITY

Communicated by O. Zariski, December 21, 1955

Let V be a variety,[1] and let $\{W_i\}$ be a finite set of varieties contained in V (which need not be closed in V and hence are not, in general, subvarieties in V) such that every point in V is contained in one and only one of these varieties; we shall call such a set of varieties $\{W_i\}$ a *dissection* of V, and the dissection is said to be *rational* if each variety W_i is equivalent to an affine space by an everywhere biregular birational transformation. A dissection $\{W_i\}$ of a variety V is said to be *defined* over a field K if K is a field of definition for all the varieties W_i (and hence necessarily also for V); and in case $\{W_i\}$ is a rational dissection, we add the further condition that each W_i is equivalent to an affine space by an everywhere biregular birational transformation which is defined over K. It is evident that if W is an element in a (rational) dissection $\{W_i\}$ of V, then the subset consisting of all W_i contained in the closure $\bar{W}$ or the boundary $\bar{W} - W$ of W in V is also a (rational) dissection of $\bar{W}$ or $\bar{W} - W$, respectively. We shall be concerned in this note only with rational dissections, and we shall call an element W in a rational dissection a *cell*, or an *r-cell* if r is the dimension of W. It is clear that a variety having a rational dissection must be a rational variety; in fact, any dissection of a variety V of dimension d can contain only one element W of dimension d, and it is easily seen that V must coincide with the closure $\bar{W}$ of W in V. The converse of this statement is, however, not true in general; for this, one need only observe that if a variety V has a rational dissection, then (in case the universal domain is the complex field) all the odd-dimensional Betti numbers of V must vanish, and that there exist rational varieties with nonvanishing odd-dimensional Betti numbers.

It has been shown by Ehresmann[2] that certain well-known homogeneous varieties, such as the Grassmann varieties, have rational dissections which can be constructed in a simple geometrical way; using topological methods, Ehresmann was able to derive from these dissections many of the fundamental properties of these varieties, in case the universal domain is the complex field. In particular, it follows from Ehresmann's proofs that if a variety has a rational dissection, then the quotient group $G(V, s)/G_n(V, s)$ (where $G(V, s)$ is the group of all s-cycles in V and $G_n(V, s)$ is the subgroup in $G(V, s)$ consisting of all elements which are numerically equivalent to zero) has a finite base for every dimension s; in other words, the basis theorem for numerical equivalence follows from the existence of a rational dissection. It is natural to ask whether this is also true in abstract algebraic geometry and whether we can make the same assertion also for algebraical or rational equivalence; the purpose of this note is to answer these questions in the affirmative. In the case of the Grassmann varieties, this has been done by Hodge and Pedoe[3] by using an idea previously introduced by van der Waerden in his proof of the Bezout theorem.[4] Our method, though not unrelated to this idea of van der Waerden, is much simpler than that of Hodge and Pedoe and is closely analogous to the topological treatment of Ehresmann; and, since we make use only of the existence of a rational dissection, not of its specific form, our results are valid for any variety with such a dissection.

Before proceeding any further, we add a remark concerning the equivalence between cycles in an algebraic variety. The usual equivalence concept used in the basis theorem is the algebraic equivalence, and this is indeed the proper one for the general form of this theorem; however, for our special case it is useful and important to introduce a more strict concept of equivalence, namely, the rational equivalence.[5] Let V be a variety, Z be a cycle in V, and K be a field of definition for V such that Z is rational over K; we shall say that Z is *rationally equivalent to zero over the field* K if there exists a cycle Z_u in V, rational over a purely transcendental extension $K(u)$ of K, such that both Z and the zero cycle are specializations of Z_u over K; and we shall say that a cycle is *rationally equivalent to zero* if it is so over some such field K. Since specialization is compatible with the addition (and subtraction) of cycles, it is easily seen that the set of all cycles of a given dimension s in V which are rationally equivalent to zero is a subgroup $G_r(V, s)$ in the group

$G(V, s)$ of all s-cycles in V; in fact, $G_r(V, s)$ is evidently a subgroup in the group $G_a(V, s)$ of all s-cycles in V which are algebraically equivalent to zero. Two cycles Z_1 and Z_2 in V are said to be *rationally equivalent* to each other if the cycle $Z_1 - Z_2$ is rationally equivalent to zero; and a finite set of s-cycles in V is said to form a *rational base* for the s-cycles in V if every s-cycle in V is rationally equivalent to a linear combination of the cycles in this set. It is clear that in general a variety does not have a rational base for cycles of any dimension; the purpose of this note is to show that in the special case of a variety with rational dissection such a rational base does exist for every dimension.

As we shall see presently, our problem reduces essentially to the proof of the following lemma, which is the algebraical counterpart of the topological lemma used by Ehresmann:

LEMMA. *Let W be an r-cell in a rational dissection of a variety V, and let $B = \bar{W} - W$ be the boundary of W; then any s-cycle Z ($s < r$) in V with support in $\bar{W}$ is rationally equivalent on $\bar{W}$ to an s-cycle Z_0 whose support is contained in B.*

Remark: The *support* of a cycle Z is the bunch of all varieties which appear as components, with nonvanishing coefficients, in Z.

Proof: It is clearly sufficient to prove the theorem for the case where Z is a prime s-cycle, that is, a variety of dimension s, the general case following obviously by linearity; furthermore, we can evidently assume that the support of Z is not already contained in B, so that the restriction of Z in W is a (nonzero) prime s-cycle X in W, and we have the relation $Z = \bar{X}$. Let F be an everywhere biregular birational transformation of W onto an affine space A^r; let p be a point in W which is not contained in the support of X, and let $x_1, \ldots, x_r$ be a system of affine co-ordinates in A^r, with the origin at the point $F(p)$. Let v be a variable over a field of definition K for $\{W_i\}$ and Z, and let T_v be the linear transformation of A^r defined by the equations $x_i' = x_i/v$; then $F^{-1}T_vF$ is a biregular birational transformation of W onto itself, defined over $K(v)$, so that $F^{-1}T_vF(X) = X_v$ is a prime s-cycle in W, rational over $K(v)$, and hence $Z_v = \bar{X}_v$ is also a prime s-cycle in $\bar{W}$, rational over $K(v)$. Since v is a generic point over K of the affine line A^1, there exists a uniquely determined specialization Z_0 of Z_v over the specialization $v \to 0$ over K; and since Z is evidently the specialization of $\bar{X}_v$ over the specialization $v \to 1$ over K, it follows that Z_0 is rationally equivalent to Z. It remains to show that the support of Z_0 is contained in B. To do this, we imbed A^r in a projective space P^r; then F can be extended to a birational transformation $\bar{F}$ of $\bar{W}$ into P^r, such that the total image of every point in $\bar{W} - W$ is contained in $P^r - A^r$, and conversely. Let $y_0, y_1, \ldots, y_r$ be a system of homogeneous co-ordinates in P^r such that $x_i = y_i/y_0$, $i = 1, \ldots, r$, and let $f(u^{(0)}, u^{(1)}, \ldots, u^{(s)})$ be the associated form of the prime s-cycle $Y = \bar{F}(\bar{X})$ in P^r (the closure of $F(X)$ in P^r), expressed in terms of the system of homogeneous co-ordinates $y_0, y_1, \ldots, y_r$ in P^r; then the form $f_v(u^{(0)}, u^{(1)}, \ldots, u^{(s)})$, obtained from the form f by substituting $vu_0^{(i)}$, for $u_0^{(i)}$, $i = 0, 1, \ldots, s$, is the associated form of the prime s-cycle $Y_v = T_v\bar{F}(\bar{X})$, rational over $K(v)$, whereby T_v is now considered as a projective transformation of P^r onto itself. Over the specialization $v \to 0$ over K, the form f_v specializes to form f_0, which is obtained from f by setting $u_0^{(i)} = 0$, $i = 0, 1, \ldots, s$, and, since the support of the s-cycle Y does not contain the origin in A^r, one sees readily that the form f_0 does not vanish identically;

it follows that f_0 is the associated form of a positive s-cycle in P^r, rational over K, which must be the uniquely determined specialization Y_0 of Y_v over the specialization $v \to 0$ over K. Since the form f_0 does not involve the variables $u_0^{(i)}$, $i = 0, 1, \ldots, s$, any set of s hyperplanes together with the hyperplane $P^r - A^r$, defined by the equation $y_0 = 0$, will have a common intersection with the support of Y_0; this shows that the support of Y_0 is contained in $P^r - A^r$. Now let q be a generic point of $Z_v = \bar{X}_v$ over $K(v)$, which is, of course, also a generic point of X_v over $K(v)$, and consider the prime s-cycle G_v in the product variety $\bar{W} \times P^r$ which is determined by the locus of the point $(q, F(q))$ over $K(v)$; since q and $F(q)$ are generic points of Z_v and Y_v over $K(v)$, respectively, it follows that Z_v and Y_v are the projections of G_v into $\bar{W}$ and P^r, respectively. Furthermore, since $(q, F(q))$ is evidently a point in the graph of F, the support of G_v is contained in the graph of $\bar{F}$, as the latter is the closure of the graph of F in $\bar{W} \times P^r$. If G_0 is the specialization of G_v over the specialization $v \to 0$ over K, then the projections of G_0 into $\bar{W}$ and P^r must be the cycles Z_0 and Y_0, respectively; since the support of G_0 is contained in the graph of $\bar{F}$ and since the support of Y_0 is contained in $P^r - A^r$, it follows that the support of Z_0 is contained in B. This concludes the proof of the lemma.

Consider, now, a rational dissection of a variety V, and let $W_1, \ldots, W_t$ be the set of all s-cells in this dissection; then it is easily seen, by the repeated application of the above lemma, that every s-cycle in V is rationally equivalent to an s-cycle whose support is contained in the bunch $\bigcup_{i=1}^{t} \bar{W}_i$. This means that every s-cycle in V is rationally equivalent to a linear combination of the varieties $\bar{W}_1, \ldots, \bar{W}_t$, considered as prime s-cycles in V, so that these t s-cycles form a rational base for the s-cycles in V. Furthermore, it also follows from the proof of the above lemma that if K is a field of definition for V and the rational dissection such that cycle Z is rational over K, then Z is rationally equivalent over K to a linear combination of the base $\bar{W}_1, \ldots, \bar{W}_t$. In this sense we cay say that the rational base $\bar{W}_1, \ldots, \bar{W}_t$ is defined over any field of definition for the rational dissection of V. Thus we can summarize our results as the following theorem:

THEOREM. *The set of all s-cells in a rational dissection of a variety V constitutes a rational base for the s-cycles in V, which is defined over any field of definition for the rational dissection.*

* This work was partially supported by the Office of Ordnance Research, United States Army.

1 We shall use in general the terminology of Weil's *Foundations of Algebraic Geometry*. By a "variety" we shall mean an "Abstract Variety" in the sense of Weil, unless the contrary is explicitly stated. All varieties are assumed to be topologized with the Zariski topology, in which the closed subsets are the bunches of subvarieties; this will enable us to use the terminology of topology without any further explanation. The concept of a specialization of a cycle (of arbitrary dimension) in a variety can be defined as in the paper of G. Shimura, *Am. J. Math.*, **77,** 134, 1955.

2 C. Ehresmann, *Ann. Math.*, **35,** 396, 1934.

3 W. V. D. Hodge and D. Pedoe, *Methods of Algebraic Geometry*, **2,** 337, 1952.

4 B. L. van der Waerden, *Math. Ann.*, **115,** 619, 1938; van der Waerden attributes this idea to G. Schaake.

5 The concept of an equivalence class of cycles under rational equivalence is important in the algebraic theory of vector bundles, as we shall show in a forthcoming paper on this subject.

On the Projective Embedding of Homogeneous Varieties

Wei-Liang Chow

IN an article [5] elsewhere in this volume, Weil has shown that an idea of Lefschetz on the projective embedding of an Abelian variety over the complex field, which seemingly depends upon the use of theta functions ([1], pp. 368–9), can actually be extended to the case of an abstract Abelian variety over a field of arbitrary characteristic. In this note we shall show that this idea can be further extended to get a projective embedding not only of an arbitrary group variety, but also of any homogeneous variety. We shall say that a variety V can be embedded in a projective space or has a projective embedding, if there is an everywhere biregular birational transformation of V onto a (not necessarily complete) variety contained in a projective space. A group variety G is said to act on a variety V, if there is a subvariety T in $G \times V \times V$, such that for any point a in G, the cycle

$$T_a = pr_{23}(T \cdot (a \times V \times V))$$

is defined and is an everywhere biregular birational transformation of V onto itself, and that for any two points a and b in G, we have the relation $T_{ab} = T_a T_b$; G is said to act transitively on V, if for any two points p and q in V, there exists a point a in G such that $p = T_a(q)$. A variety is said to be *homogeneous* if there is a group variety which acts transitively on it; it is clear that a group variety is homogeneous, since it acts transitively on itself by left translation. A field is said to be a field of definition for a homogeneous variety V, if it is such for the varieties V and G as well as for the subvariety T in $G \times V \times V$.

For the sake of convenience, we shall avail ourselves of the use of topological terminology by introducing the Zariski topology on an algebraic variety V, in which the closed subsets are bunches of subvarieties in V. If k is a field of definition for V, then we shall say that a subset W in V is k-closed if the bunch W is normally algebraic over k, and that a subset is k-open if it is the complement of a k-closed subset. A subset in a projective space is evidently a variety if and only if it is an open subset in a complete variety, and it is defined over a field k if and only if it is a k-open subset. We do not know whether any variety, not necessarily one embedded in a projective space, is also an open subset in a complete variety, but the following Lemma 1 offers in a sense a substitute for this property. This simple lemma is significant for our purpose for the following reason. One of the main difficulties in extending Weil's proof to our more general case is the fact that, for an incomplete variety V, the linear system of all functions f on V which are multiples of a given divisor X in V has in general an infinite dimension and hence does not possess a finite base; since the embedding mapping in the Weil proof is obtained from a finite base of a suitably chosen linear system, one sees readily that the lack of such a base is a serious obstacle. In case V is an open subset in a complete variety $\bar{V}$, any divisor X is the restriction to V of a divisor $\bar{X}$ in $\bar{V}$ and any function f is the restriction of a function $\bar{f}$ on $\bar{V}$; it is then natural to consider only those functions f such that $\bar{f}$ is a multiple of $\bar{X}$ on $\bar{V}$. As we shall see later, the complete variety V' in Lemma 1 below can be used for a similar purpose in the general case.

LEMMA 1. *Let V be a variety and k be a field of definition for V; then there exist a complete normal subvariety V' in a projective space and a birational correspondence F between V' and V, both defined over a purely inseparable extension k' of k, such that F is defined at every point in a k'-open subset W in V' and only at such points, and the image of W under F is V; furthermore, the inverse image of every point in V under F is a closed subset in V'.*

PROOF. Let $V_1, \ldots, V_l$ be the representatives of the variety V and let $B_1, \ldots, B_l$ be the frontiers of $V_1, \ldots, V_l$ respectively; without any loss of generality, we can consider each V_i to be a complete subvariety in a projective space, so that the product variety $V_1 \times \ldots \times V_l$ can also be considered as a complete subvariety in a projective space. Let $x_1, \ldots, x_l$ be a system of corresponding generic points of $V_1, \ldots, V_l$ respectively over k, and let V' be the locus of the point $x_1 \times \ldots \times x_l$ over k; then V' is a complete subvariety in $V_1 \times \ldots \times V_l$, defined over k, and the projection from V' to V_i defines a birational correspondence F_i between V' and $V_i - B_i$, defined over k, which is defined at every point in the k-open subset $W_i = V' \cap (V_1 \times \ldots \times V_{i-1} \times V_i - B_i \times V_{i+1} \times \ldots \times V_l)$; furthermore, it is clear that for any point y in $V_i - B_i$, its inverse image under F_i is the closed subset $V' \cap (V_1 \times \ldots \times V_{i-1} \times y \times V_{i+1} \times \ldots \times V_l)$ in V'. The system $(F_1, \ldots, F_l)$ then defines a birational correspondence F between V' and V, defined over k, which is defined at every point in the k-open subset $W = \bigcup_{i=1}^{l} W_i$ and only at these points; since $V_1, \ldots, V_l$ constitute a and hence the relation $(f) \succ X$, so that a strong multiple of X is also a multiple of X in the usual sense. Also, it is clear that the definition of a strong multiple is independent of the choice of the generic point u of G over K. We shall denote by $L(X)$ the set of all functions on V which are strong multiples of the divisor $-X$; it is easily seen that $L(X)$ is a vector space of finite dimension over the field of constants. In fact, the dimension of $L(X)$ cannot exceed the dimension of the linear system $|F^{-1}(X_u)|$ on V'. For, if $f_1, \ldots, f_d$ is a set of linearly independent functions in $L(X)$, and if we denote by $f'_1, \ldots, f'_d$ the functions induced on V' by the functions $f_1 \cdot T_u^{-1}, \ldots, f_d \cdot T_u^{-1}$ respectively, then $f'_1, \ldots, f'_d$ are evidently linearly independent functions which are multiples of the divisor $F^{-1}(X_u)$, and the number of such functions cannot exceed the dimension of $|F^{-1}(X_u)|$. We shall denote by $|X|$ the set of all positive divisors in V which has the form $(f)+X$, where f is any function in $L(X)$, and we shall call this set the complete linear system determined by X. We shall say that a divisor X in V is *strongly equivalent to zero* (relative to F), and we shall write $X \approx 0$, if there exist a function f on V such that we have both the relations $(f) \succ\succ -X$ and $(1/f) \succ\succ X$, and we shall say that a divisor Y is *strongly equivalent to* X (relative to F) if $X - Y \approx 0$. It is easily seen that we have the relation $X \approx 0$ if and only if there exist a function f on V such that $(f'_u) = F^{-1}(X_u)$, where f'_u is the function induced on V' by the function $f \cdot T_u^{-1}$ and u is a generic point of G over a field of definition for V, V', F and f over which X is rational. The complete linear system $|X|$ contains all positive divisors which are strongly equivalent to X, but it may contain also other 'partial' divisors in case V is not complete. It can be easily seen that the relations $X \approx Y$ and $|X| = |Y|$ imply each other, and that both imply that $L(X)$ and $L(Y)$ have the same dimension.

LEMMA 2. *Let X be a divisor in V, and let K be a field of definition for V, V' and F, such that X is rational over K; then the relation $X \approx 0$ holds if and only if the relation $F^{-1}(X_u) \sim 0$ holds for every generic point u of G over K.*

PROOF. That the condition is necessary follows immediately from what we have just said above, so that we need only to prove the converse statement. Let f' be a function on V', rational over $K(u)$, such that $(f') = F^{-1}(X_u)$, and let f be the function induced by f' on V; if we set $g = f \cdot T_u$, then g is defined over $K(u)$ and we have the relation $(g) = X$. According to a remark we made before, there is a constant element c such that the function $h = cg$ is defined over $\bar{K}$; if h'_u is the function induced on V' by the function $h \cdot T_u^{-1} = cg \cdot T^{-1} = cf$, then we have evidently the relation $h'_u = cf'$ and hence $(h'_u) = (f') = F^{-1}(X_u)$. Since u is a generic point of G over $\bar{K}$, this proves our lemma.

We shall say that a linear system X on a homogeneous variety V is *ample*, if it has no fixed component and satisfies the two conditions (A) and (B) stated in Weil [5]. Similarly as in the case of a complete variety, we shall show that the existence of an ample linear system $|X|$ on V implies the existence of a projective embedding of V, whereby it is already sufficient to assume the condition (B) for a generic point of V. In fact, the condition (A) shows that a base of $L(X)$ will define a rational transformation T of V into a variety U (of the same dimension as V) in a projective space, such that T is defined at every point in V and is one-to-one between the points of V and its image U_0 under T; and the validity of condition (B) at a generic point shows that T is a birational transformation. Let U' be a derived (absolutely) normal model of U, and let R be the birational transformation of U onto U', so that $T' = RT$ is the induced birational transformation of V into U' and $T'' = T'F$ is the induced birational transformation of V' onto U'; let U'_0 be the image of U_0 under R. Since the image of any point in V under T' consists of at most a finite number of points and since V is non-singular, it follows from ([3], Chap. VI, Theorem 13), that T' is defined at every point in V; on the other hand, since the image of any point in U'_0 under T'^{-1} consists of a single point in V and since U' is normal, it follows, from [6], Main Theorem, that T'^{-1} is defined at every point in U'_0. Thus T' is an everywhere biregular birational transformation of V onto U'_0, and it remains to show that U'_0 is an open subset in U'. Since the image of any point in U'_0 under T'^{-1} consists of a single point in V and since the image of a point in V under F^{-1} is a closed subset in V' which is contained in W, it follows that the image of any point in U'_0 under T''^{-1} consists of a non-empty closed subset in V' which does not intersect $V' - W$ and a closed subset in $V' - W$; since U' is normal, it follows from Zariski's Connectedness Theorem ([7], p. 6), that the image of any point in U'_0 under T''^{-1} does not intersect $V' - W$, and hence $U' - U'_0$ is the image of $V' - W$ under T''. Since $V' - W$ is a closed subset in V', $U' - U'_0$ must be a closed subset in U' and hence U'_0 must be an open subset in U'.

In order to prove the possibility of a projective embedding of a homogeneous variety V, it is therefore sufficient (and necessary) to show the existence of an ample complete system of divisors in V. In order to do this, we observe that Lefschetz's idea, as formulated by Weil, can be carried over to our more general case with no essential change, provided we have the relation $3X \approx X_a + X_b + X_{a^{-1}b^{-1}}$ for any divisor X in V and any two points a and b in G. For this, we prove the following lemma, which is known to be true for Abelian varieties (Weil [4], Theorem 30, Corollary 2):

LEMMA 3. *If X is any divisor in a homogeneous variety V, and if a and b are any two points in the group variety G which acts on V, then we have the relation $X_{ab} - X_a - X_b + X \approx 0$.*

PROOF. We shall use the following result from the theory of Picard varieties: For any complete normal variety M in a projective space, there exists an Abelian variety P, called the Picard variety of M, and a rational homomorphism of $G_a(M)$ onto P, called the canonical homomorphism, which has the group $G_l(M)$ as its kernel; here $G_a(M)$ is the group of all divisors in M which are algebraically equivalent to zero and $G_l(M)$ is the group of all divisors in M which are linearly equivalent to zero. For the proof of this fact we refer to Matsusaka [2] as well as a forthcoming paper of ours on Picard varieties over arbitrary ground fields. Let P be the Picard variety of V', and let Φ be the canonical homomorphism of $G_a(V')$ onto P; let K be a field of definition for V, V', F, P and such that the divisor X and the points a and b are all rational over K, and let u and v be independent generic points of G over K. The divisor $F^{-1}(X_{uv}) - F^{-1}(X_v)$ is rational over $K(u, v)$, and is contained in $G_a(V')$; hence the point $\Phi(F^{-1}(X_{uv}) - F^{-1}(X_u))$ is rational over $K(u, v)$; therefore the correspondence $v \to \Phi(F^{-1}(X_{uv}) - F^{-1}(X_u))$ defines a rational transformation ϕ of G into P, defined over $K(u)$, which carries the unit element in G into the unit element in P. According to ([4], § 19), the rational transformation ϕ is a homomorphism of G into P; furthermore, for any point c in G such that uc is a generic point of G over K, we have the relation $\phi(c) = \Phi(F^{-1}(X_{uc}) - F^{-1}(X_u))$. Since the points uab, ua, ub evidently are generic points of G over K, we have then the relation

$$\begin{aligned}\Phi(F^{-1}(X_{uab}) - F^{-1}(X_u)) &= \phi(ab) = \phi(a) + \phi(b)\\ &= \Phi(F^{-1}(X_{ua}) - F^{-1}(X_u)) + \Phi(F^{-1}(X_{ub}) - F^{-1}(X_u))\\ &= \Phi(F^{-1}(X_{ua}) + F^{-1}(X_{ub}) - 2F^{-1}(X_u)).\end{aligned}$$

This means that we have the relation

$$F^{-1}(X_{uab}) - F^{-1}(X_u) \sim F^{-1}(X_{ua}) + F^{-1}(X_{ub}) - 2F^{-1}(X_u)$$

and hence $F^{-1}(X_{uab}) - F^{-1}(X_{ua}) - F^{-1}(X_{ub}) + F^{-1}(X_u) \sim 0$; according to Lemma 2, this implies the relation $X_{ab} - X_a - X_b + X \approx 0$.

If we apply the above lemma to the pairs of points (a, b), (ab, b^{-1}) and $(a, a^{-1}b^{-1})$, we obtain the following relations:

$$X \approx X_a + X_b - X_{ab},$$
$$X \approx X_{ab} + X_{b^{-1}} - X_a,$$
$$X \approx X_a + X_{a^{-1}b^{-1}} - X_{b^{-1}},$$

which, when added together, gives the relation $3X \approx X_a + X_b + X_{a^{-1}b^{-1}}$. From here on we can proceed exactly as in Weil [5], except for a few obvious changes here and there; there is no need to repeat the proof here.

THEOREM. *Any homogeneous variety has a projective embedding.*

JOHNS HOPKINS UNIVERSITY

REFERENCES

[1] S. LEFSCHETZ, *On certain numerical invariants of algebraic varieties, with application to abelian varieties*, Trans. Amer. Math. Soc., 22 (1921), pp. 327–482.

[2] T. MATSUSAKA, *On the algebraic construction of the Picard variety*, Jap. J. Math., 21 (1951), pp. 217–235.

[3] A. WEIL, Foundations of algebraic geometry, New York, 1946.

[4] ——, *Variétés abeliennes et courbes algébriques*, Act. Sci. et Ind., no. 1064, Paris, 1948.

[5] ——, *On the projective embedding of abelian varieties*, this volume, pp. 177–181.

[6] O. ZARISKI, *Foundations of a general theory of birational correspondence*, Trans. Amer. Math. Soc., 53 (1943), pp. 490–542.

[7] ——, *Theory of applications of holomorphic functions on algebraic varieties over arbitrary ground fields*, Memoirs Amer. Math. Soc., no. 5 (1951).

ANNALS OF MATHEMATICS
Vol. 66, No. 1, July, 1957
Printed in U.S.A.

ON THE PRINCIPLE OF DEGENERATION IN ALGEBRAIC GEOMETRY*

BY WEI-LIANG CHOW

(Received December 4, 1956)

(Revised March 29, 1957)

The Principle of Degeneration in algebraic geometry asserts that every specialization (over any field) of a connected positive cycle in a projective space is connected, whereby a positive cycle is said to be connected if its support, i.e., the set of its component varieties, is connected in the sense of the (absolute) Zariski topology. This principle, although more or less "known" for a long time in algebraic geometry and which can be easily proved in the classical case by transcendental methods, has been only recently given a purely algebraic proof by Zariski [7], as a special case of a more general principle called the Principle of Connectedness. In view of the fact that Zariski's proof of the Principle of Connectedness involves the very difficult theory of holomorphic functions on an algebraic variety, it is probably not without some interest to give here a new and very much simpler proof of the Principle of Degeneration. Furthermore, since the concept of a specialization is subsumed under the more general concept of a homomorphism of an integral domain onto its residue ring modulo a prime idea, it is natural to ask for a generalization of the Principle of Degeneration which is valid in this general case. In the present paper we shall give a proof of such a generalization; this will include in particular the case of reduction modulo a prime number of a cycle defined over the field of rational numbers, a case which has of late become increasingly important in number-theoretic applications.

1. The Generalized Principle of Degeneration can be formulated as follows. Let R be a Noetherian domain with the quotient field K, and let $\mathfrak{p}$ be a prime ideal in R. We shall denote by $\bar{R}$ the residue ring $R/\mathfrak{p}$ of R over $\mathfrak{p}$; generally, for any element a in R or a polynomial $F(X)$ in the indeterminates X with coefficients in R, we shall denote by $\bar{a}$ or $\bar{F}(X)$ respectively the corresponding element in $\bar{R}$ or polynomial in $\bar{R}[X]$. Let S^n be the projective space of dimension n in the algebraic geometry over a universal domain[1] which contains K, and let Z be an

* This work was partially supported by a research grant of the National Science Foundation, NSF-G3022.

[1] We shall use some of the terminology in A. Weil [6], such as the terms "universal domain", "variables", "indeterminates", "proper intersections", "cycles", etc.; however, the knowledge of this book is not necessary for the understanding of the present paper, as our methods are entirely elementary. We shall also use the topological concepts based on the Zariski topology, absolute or relative to a given ground field, in which the closed subsets are bunches of varieties (in the terminology of Weil) or (in the relative case) those bunches which are normally algebraic over the given ground field. On the other hand, we shall assume that the reader is familiar with the theory of associated forms, for which see W. L. Chow and B. L. van der Waerden [3]. A summary of some of the properties of the associated forms which are used here will be given at the beginning of Section 4.

r-cycle of degree d in S^n; let $F(U) = F({}_0U, \cdots, {}_rU)$ be the associated form of Z, where ${}_iU = ({}_iU_0, \cdots, {}_iU_n)$, $i = 0, 1, \cdots, r$, are $r + 1$ sets of $n + 1$ indeterminates, and assume that the coefficients in $F(U)$ are all in R but not all in $\mathfrak{p}$. Since the property of being the the associated form of an r-cycle of a given degree d is preserved under a homomorphism, the form $\bar{F}(U)$ is also the associated form of an r-cycle $\bar{Z}$, of the same degree d, in the projective space $\bar{S}^n$ in the algebraic geometry over a universal domain which contains $\bar{R}$; we shall call $\bar{Z}$ the cycle obtained from Z by *reduction modulo* $\mathfrak{p}$. The Principle of Degeneration then asserts:

If the cycle Z is connected, then the cycle $\bar{Z}$ obtained from Z by reduction modulo $\mathfrak{p}$ is also connected.

Before proceeding to the proof, we observe first that it is sufficient to restrict ourselves to the special case where R is the valuating ring of a real discrete valuation of K and $\mathfrak{p}$ is the maximal prime ideal in R. This follows from the fact that *there exists a real discrete valuation ring R' in K such that R' contains R and the maximal prime ideal in R' contracts to $\mathfrak{p}$ in R.* Although this fact is more or less well known, we shall give here a short proof of it for the convenience of the reader.[2]

We can assume that R is a local ring with $\mathfrak{p}$ as its maximal prime ideal. In fact, we can replace R by its quotient ring $R_\mathfrak{p}$ with respect to the ideal $\mathfrak{p}$; since $R_\mathfrak{p}$ is a local ring with the maximal prime ideal $R_\mathfrak{p}\mathfrak{p}$ and since $R_\mathfrak{p}\mathfrak{p}$ contracts to $\mathfrak{p}$ in R, it is evidently sufficient to prove our statement for $R_\mathfrak{p}$. Let r be the dimension of the local ring R, and let $\mathfrak{q}$ be a primary ideal for $\mathfrak{p}$ which is generated by a set of r elements $x_1, \cdots, x_r$, so that $x_1, \cdots, x_r$ form a system of parameters in R; we set $u_i = x_i/x_1$ for $i = 2, \cdots, r$, and consider the ring $R[u] = R[u_2, \cdots, u_r]$. It is clear that the ideal $R[u]\mathfrak{q}$ is generated by the one element x_1. We maintain that the ideal $R[u]\mathfrak{q}$ is not the unit ideal. In fact, if the element 1 is contained in $R[u]\mathfrak{q}$, then there exist monomials $m_i(u) (i = 1, \cdots, s)$ of positive degrees in u_j and elements $a_i (i = 0, 1, \cdots, s)$ in R, such that $(a_0 + \sum_{i=1}^{s} a_i m_i(u))x_1 = 1$. If $t + 1$ is the maximum of the degrees of $m_i(u)$, then we have upon multiplying both sides by x_1^t the relation $a_0 x_1^{t+1} \equiv x_1^t$ or $(1 - a_0 x_1)x_1^t \equiv 0 \bmod (x_2, \cdots, x_r)$ in R. Since R is a local ring, the element $1 - a_0 x_1$ has an inverse in R; hence we have the relation $x_1^t \equiv 0 \bmod (x_2, \cdots, x_r)$ in R, which is in contradiction to the fact that $x_1, \cdots, x_r$ form a system of parameters in R. Now, let $\mathfrak{p}_1$ be an isolated prime divisor of $R[u]\mathfrak{q}$ in $R[u]$, and let R_1 be the quotient ring of $R[u]$ with respect to $\mathfrak{p}_1$; then R_1 is a local ring with the ideal $R_1\mathfrak{p}_1$ as its maximal prime ideal. We observe that since $R_1\mathfrak{q}$ is evidently generated by the one element x_1 and is primary for $R_1\mathfrak{p}_1$, the dimension of R_1 is equal to 1 and hence is a primary domain in the sense of Krull. If we denote by R_2 the integral closure of R_1 in its quotient field, which is the quotient field K of R, then R_2 is the intersection of a finite number of real discrete valuation rings, according to a theorem of Krull [4], and the maximal prime ideal in each of these rings contracts to $R_2\mathfrak{p}_1$ in R_2. Let R' be one of these discrete valuations and let $\mathfrak{p}'$ be its maximal prime ideal;

[2] The proof given below is not essentially different from that given in S. Abhyankar and O. Zariski [1], p. 85, Lemma 1.

since $R \cap \mathfrak{p}'$ is a prime ideal in R and evidently contains $\mathfrak{q}$, it must coincide with $\mathfrak{p}$. This proves our statement.

We shall therefore assume from now on that R is a real discrete valuation ring, and we observe that in this case the coefficients of $F(U)$ can always be so chosen (by multiplying with a suitable element in K) that they are not all in $\mathfrak{p}$, so that the assumption we made above in this respect really imposes no conditions at all on the cycle Z. Futhermore, by going over to the completion of R if necessary, we can assume that the ring R itself is already complete with respect to the given valuation. Now, with this restriction to a complete real discrete valuation ring R, we can formulate a somewhat stronger version of the Principle of Degeneration where the absolute connectedness is replaced by connectedness relative to a given ground field, i.e., connectedness in the Zariski topology over a given ground field. If we denote by $\bar{K}$ the residue ring $\bar{R}$ which is now a field, then this stronger version of the Principle of Degeneration can be stated as follows:

MAIN THEOREM. *Let K be a complete real discrete valuation field and $\mathfrak{p}$ be its maximal prime ideal; let Z be a positive cycle whose associated form is rational over K, and let $\bar{Z}$ be the cycle obtained from Z by reduction modulo $\mathfrak{p}$; if the cycle Z is K-connected, then the cycle $\bar{Z}$ is $\bar{K}$-connected.*

This is the theorem which we shall prove in this paper; to deduce from it the Principle of Degeneration in the absolute case, one need only to replace K by a suitable algebraic extension such that every component in $\bar{Z}$ is defined over $\bar{K}$.

The basic idea of our proof of the Main Theorem arises from the observation that it is essentially a generalization to cycles of higher dimensions of the well known Hensel's Lemma. In fact, if we note that a 0-cycle is connected over a field if and only if it is a multiple of an irreducible cycle over the field, then the Hensel's Lemma is precisely the simplest case of our Main Theorem where Z is a 0-cycle in a projective line. This observation naturally suggests the idea of constructing a proof of our theorem along the lines similar to that of the Hensel's Lemma; in other words, from the given factorization of the form $\bar{F}(U)$ into two "disconnected" factors one should construct a corresponding factorization of the form $F(U)$ by determining successively the terms of two "power series" beginning with the given two factors of $\bar{F}(U)$. That this can be done will be shown in Section 4, which contains the proof of our Main Theorem; in the two preceding sections we shall prove two lemmas which are needed in this proof. The first of the two lemmas is itself a generalization of the Hensel's Lemma to the case of a 0-cycle in an arbitrary projective space, which is probably not new (although we are unable to find it in the literature), while the second one concerns itself with the finer properties of the coefficients which enter in the "power series" development given by the first lemma. This second lemma involves an aspect of the Hensel's Lemma which we believe is new, and it can be considered as the crucial point of our proof.

2. As before, let K be a complete field with respect to a real discrete valuation φ, let R be the valuation ring in K, and let $\mathfrak{p}$ be the maximal prime ideal in R.

For convenience of calculation, we shall represent each element a in K by a "power series" $a = \sum_{i>-\infty}^{\infty} a^{(i)}p^i$, where p is a fixed generator of $\mathfrak{p}$; the $a^{(i)}$ are elements of a fixed system Q of representatives of $\bar{R} = R/\mathfrak{p}$ in R, and the summation extends over only a finite number of negative integers; the choice of the system Q is completely arbitrary, but we shall assume for convenience that Q contains the elements 0 and 1. Thus each element in K is uniquely determined by the sequence of elements $a^{(i)}$ in Q, and two elements a and b in K are equal if and only if $a^{(i)} = b^{(i)}$ for all i, or equivalently $\bar{a}^{(i)} = \bar{b}^{(i)}$ for all i.

If u is a variable over K, we shall extend the valuation φ to $K(u)$ by setting $\varphi(a + a_1u + \cdots + a_du^d) = \text{Min}\,(\varphi(a_0), \cdots, \varphi(a_d))$ for a polynomial in u, and $\varphi(f/g) = \varphi(f) - \varphi(g)$ for the quotient of two polynomials f and g in u; that these stipulations actually define a valuation of $K(u)$, concordant with the given valuation φ of K, follows easily from the well known fact that fg is divisible by $\mathfrak{p}$ if and only if either f or g is divisible by $\mathfrak{p}$. We denote by R_u the valuation ring in $K(u)$, and by R_u^* the completion of R_u, which is then the valuation ring of the completion $K(u)^*$ of $K(u)$; then $\mathfrak{p}_u = R_u\mathfrak{p}$ and $\mathfrak{p}_u^* = R_u^*\mathfrak{p}$ are the maximal prime ideals in R_u and R_u^* respectively, and we have $\bar{R}_u^* = R_u^*/\mathfrak{p}_u^* = R_u/\mathfrak{p}_u = \bar{R}_u$. For simplicity, we shall use the same symbol u to denote the residue class determined by u in $\bar{R}_u^*$, i.e., we identify u with $\bar{u}$; there will be no danger of confusion, as the context will make clear what is meant in every case. We denote by $Q(u)$ the set of all elements in R_u of the form f/g, where f and g are polynomials in u with coefficients in Q, the coefficient of the highest term in g is 1, and f and g have no common factors mod $\mathfrak{p}$; it is easily seen that $Q(u)$ is a system of representatives of $R_u/\mathfrak{p}_u = \bar{R}(u)$ in R_u, and that $Q(u)$ includes the set $Q[u]$ of all polynomials in u with coefficients in Q. With this system $Q(u)$ and with the same element p as the generator of $\mathfrak{p}_u$, every element A in $K(u)^*$ can then be represented in a unique manner by a power series $A = \sum_{i>-\infty}^{\infty} A^{(i)}p^i$, such that two elements A and B are equal if and only if $\bar{A}^{(i)} = \bar{B}^{(i)}$ for all i. We observe that, by successive adjunction, the above results can be extended to a system u of independent variables over K.

Now, let K' be a finite algebraic extension of degree t over K; it is well known that the valuation φ of K can be extended in a unique way to a valuation of K', and that K' is complete. If $e_1 = 1, e_2, \cdots, e_t$ is a base of K' over K, then every element a' in K' can be represented in a unique way in the form $a' = \sum_{j=1}^{t} a_je_j$ with the a_j in K. Consider the extension $K'(u)$ and its completion $K'(u)^*$, where u is a set of independent variables over K'; it can be easily shown that K' and $K(u)^*$ are linearly disjoint subfields in $K'(u)^*$ over K (for an irreducible polynomial over K will remain irreducible over $K(u)^*$),[3] so that $K'(u)^*$ is a finite algebraic extension of degree t over $K(u)^*$, with the elements $e_1, \cdots, e_t$ as a

[3] This can be proved directly by showing that any element in $K(u)^*$ which is algebraic over K is contained in K; it also follows from a general result of C. Chevalley [2], Proposition 7, p. 699, by observing that K' has an integral base over K and that this integral base is also an integral base of $K'(u)$ over $K(u)$.

base. It follows then from the preceding paragraph that every element A' in $K'(u)^*$ can be represented in a unique manner in the form $A' = \sum_{j=1}^{t} A'_j e_j$, where each A'_j is a power series of the form $A'_j = \sum_{i>-\infty}^{\infty} A'^{(i)}_j p^i$, where $A'^{(i)}_j$ are elements in $Q(u)$. From this it follows easily that $R^*_u \cap K'[u] = R[u]$. In fact, an element A' in $K'[u]$ has the form $A' = \sum_{j=1}^{t} A'_j e_j$ with A'_j in $K[u]$, and it is in $K(u)^*$ if and only if $A' = A'_1$; and an element in $K[u]$ is in R^*_u if and only if it is in $R[u]$.

We denote by $K[X] = K[X_1, \cdots, X_n]$ the ring of all polynomials in the n indeterminates $X_1, \cdots, X_n$ with coefficients in K. A polynomial in $K[X]$ is said to be *splitting* if it is the product of linear polynomials with coefficients in some extension of K. It is well known that every polynomial in one indeterminate is splitting, but this is no longer true in case of polynomials of more than one indeterminate.

We now prove a generalization of the well known Hensel's Lemma:

LEMMA 1. *Let K be a complete real discrete valuation field with the valuation ring R and the maximal prime ideal $\mathfrak{p}$, and let $F(X)$ be a splitting polynomial of degree d in $R[X]$; we assume that there exist polynomials $G^{(0)}(X)$ and $H^{(0)}(X)$ with coefficients in Q, of degrees s and $\leqq d - s$ respectively, such that $\bar{G}^{(0)}(X)$ and $\bar{H}^{(0)}(X)$ have no common factor, that $G^{(0)}(X)$ contains the term $a^{(0)}X_1^s$ with a non-vanishing coefficient $a^{(0)}$, and that $\bar{F}(X) = \bar{G}^{(0)}(X)\bar{H}^{(0)}(X)$. Then there exist polynomials $G(X)$ and $H(X)$ in $R[X]$, of degrees s and $d - s$ respectively, such that $\bar{G}(X) = \bar{G}^{(0)}(X)$ and $\bar{H}(X) = \bar{H}^{(0)}(X)$, that $G(X)$ contains the term aX_1^s with any given coefficient a in R satisfying the condition $\bar{a} = \bar{a}^{(0)}$, and that $F(X) = G(X)H(X)$.*

PROOF. Let $v = (v_2, \cdots, v_n)$ be a set of independent variables over K, and consider the polynomials $F(X_1, v) = F(X_1, v_2, \cdots, v_n)$, $G^{(0)}(X_1, v) = G^{(0)}(X_1, v_2, \cdots, v_n)$ and $H^{(0)}(X_1, v) = H^{(0)}(X_1, v_2, \cdots, v_n)$ in $R[v][X_1]$; it is easily seen that $\bar{G}^{(0)}(X_1, v)$ and $\bar{H}^{(0)}(X_1, v)$ are relatively prime to each other. In fact, since $\bar{R}[v]$ is a unique factorization ring and the coefficients of $\bar{G}^{(0)}(X_1, v)$ and $\bar{H}^{(0)}(X_1, v)$ are in $\bar{R}[v]$, any common factor of them in $\bar{R}(v)[X_1]$ can be made into a common factor in $\bar{R}[v][X_1]$, by multiplying with a suitable element in $\bar{R}(v)$; and, by replacing the variables $v_2, \cdots, v_n$ by the indeterminates $X_2, \cdots, X_n$ respectively, we would then obtain a common factor of $\bar{G}^{(0)}(X)$ and $\bar{H}^{(0)}(X)$ in $\bar{R}[X]$. We can therefore apply the Hensel's Lemma and conclude that there exist polynomials $F_1(X_1)$ and $F_2(X_1)$ in $R^*_v[X]$, of degrees s and $d - s$ respectively such that $\bar{F}_1(X_1) = \bar{G}^{(0)}(X_1, v)$ and $\bar{F}_2(X_1) = \bar{H}^{(0)}(X_1, v)$, that $F_1(X_1)$ contains the term aX_1^s, and that $F(X_1, v) = F_1(X_1)F_2(X_1)$. On the other hand, there exists a finite algebraic extension K' of K such that $F(X)$ is the product of d linear polynomials $f_i(X)$ in $K'[X]$. Since both $K'(v)$ and $K(v)^*$ are subfields in $K'(v)^*$ and since the coefficient of X_1^s in $F_1(X_1)$ is the element a in R, which is a unit in $K'[v]$, we can set $F_1(X_1) = c\prod_{i=1}^{s} f_i(X_1, v)$ for a suitable enumeration of the set $f_i(X)$ and a suitable choice of the proportionality factors c in K' and hence also set $F_2(X_1) = 1/c \prod_{i=s+1}^{d} f_i(X_1, v)$. This shows that the coefficients of both $F_1(X_1)$ and $F_2(X_1)$ are all contained in $K'[v]$, and hence in $R^*_v \cap K'[v]$

$= \bar{R}[v]$. There exist therefore polynomials $G(X)$ and $H(X)$ in $R[X]$, of degrees s and $d - s$ respectively, such that $G(X_1, v) = F_1(X_1)$ and $H(X_1, v) = F_2(X_1)$, and it is easily seen that $F(X) = G(X)H(X)$, and that $\bar{G}(X) = \bar{G}^{(0)}(X)$ and $\bar{H}(X) = \bar{H}^{(0)}(X)$. This proves the lemma.

3. We set $F(X) = \sum_{i=0}^{\infty} F^{(i)}(X)p^i$, $G(X) = \sum_{i=0}^{\infty} G^{(i)}(X)p^i$, and $H(X) = \sum_{i=0}^{\infty} H^{(i)}(X)p^i$, where the $F^{(i)}(X)$, $G^{(i)}(X)$, $H^{(i)}(X)$ are polynomials of degrees at most d, s, $d - s$ respectively, with coefficients in Q, and are uniquely determined by $F(X)$, $G(X)$, $H(X)$ respectively, with the given choice of p and Q. It is easily seen from our proof of Lemma 1 and from the construction procedure in the usual proof of the Hensel's Lemma that the polynomials $G(X)$ and $H(X)$ are uniquely determined by the polynomials $F(X)$, $G^{(0)}(X)$, and $H^{(0)}(X)$; in fact, for each i, the polynomials $G^{(i)}(X)$ and $H^{(i)}(X)$ are constructed step by step from $F^{(i)}(X)$ and the polynomials $F^{(j)}(X)$, $G^{(j)}(X)$, $H^{(j)}(X)$, $j = 0, 1, \cdots, i - 1$. The following lemma gives some additional information on this matter, which will be of importance for our present purpose.

LEMMA 2. *We consider the same situation as in Lemma* 1, *except that the field K is to be replaced by the field $K(u)^*$ as constructed at the beginning of the preceding section, that the coefficients in $G^{(0)}(X)$, $H^{(0)}(X)$, and every $F^{(i)}(X)$ are assumed to be in $Q[u]$, and that the element a is replaced by an element $a(u) = \sum_{i=0}^{\infty} a^{(i)}(u)p^i$ in $R[u]$ and that the coefficients of X_1^d in $F(X)$ has the form $a(u)\, b(u)$ with $b(u)$ in $R[u]$; let $E(X) = E(X_2, \cdots, X_n)$ be the resultant of $\bar{G}^{(0)}(X)$ and $\bar{H}^{(0)}(X)$, considered as polynomials in X_1 only with coefficients in $\bar{R}[u; X_2, \cdots, X_n]$, and let $e(u)$ be the common factor in $\bar{R}[u]$ of all the coefficients in $E(X)$, considered as a polynomial in $X_2, \cdots, X_n$. Then the coefficients in $\bar{G}^{(i)}(X)$ and $\bar{H}^{(i)}(X)$ are elements in $\bar{R}(u)$, and if every such coefficient is expressed as the quotient of two elements in $\bar{R}[u]$, without common factors, then every factor of the denominator is also a factor of $e(u)$.*

PROOF. The first assertion is obvious, since $G^{(i)}(X)$ and $H^{(i)}(X)$ have coefficients in $Q(u)$; it remains to prove the assertion about the denominators of these coefficients. Let $P^{(i)}(X)$ be the polynomial with coefficients in $Q(u)$ such that

$$P^{(i)}(X) \equiv \frac{1}{p^i}\left\{\sum_{j=0}^{i} F^{(j)}(X)p^j - \sum_{j=0}^{i-1}\left(\sum_{k=0}^{j} G^{(k)}(X)H^{(j-k)}(X)\right)p^j\right\} - \sum_{j=1}^{i-1} G^{(j)}(X)H^{(i-j)}(X) \pmod{\mathfrak{p}}; \tag{1}$$

it is clear that $P^{(i)}(X)$ has a degree $\leqq d$, and that, by the induction hypothesis, the denominator of every coefficient in $\bar{P}^{(i)}(X)$ has only factors which are also factors of $e(u)$. The polynomials $G^{(i)}(X)$ and $H^{(i)}(X)$ are then uniquely determined by the condition

$$\bar{H}^{(i)}(X)\bar{G}^{(0)}(X) + \bar{G}^{(i)}(X)\bar{H}^{(0)}(X) = \bar{P}^{(i)}(X), \tag{2}$$

and the stipulations that $G^{(i)}(X)$ has the degree s in the indeterminate X_1, and that the coefficient of the term X_1^s in $G^{(i)}(X)$ is equal to $a^{(i)}(u)$. We recall now from the resultant theory (see, e.g., [5], pp. 88–91) that there exist polynomials $A(X)$ and $B(X)$, with coefficients in $\bar{R}[u]$, such that $B(X)\bar{G}^{(0)}(X) + A(X)\bar{H}^{(0)}(X)$

$= E(X)$. We divide $\bar{P}^{(i)}(X)A(X)$ by $(1/(\bar{a}^{(0)}(u)))\bar{G}^{(0)}(X)$, both as polynomials in X_1, and obtain the relation

$$\bar{P}^{(i)}(X)A(X) = \frac{1}{\bar{a}^{(0)}(u)} C(X)\bar{G}^{(0)}(X) + D(X) \tag{3}$$

where $C(X)$ and $D(X)$ are polynomials with coefficients in $\bar{R}(u)$, and $D(X)$ has degree $< s$ in X_1. Furthermore, it is easily seen that the denominator of every coefficient in $C(X)$ and $D(X)$ can only have factors which are also factors of $\bar{a}^{(0)}(u)e(u)$.

A simple calculation shows that

$$\begin{aligned} &\left\{\bar{P}^{(i)}(X)B(X) + \frac{1}{\bar{a}^{(0)}(u)} \bar{H}^{(0)}(X)[C(X) - \bar{a}^{(i)}(u)E(X)]\right\} \bar{G}^{(0)}(X) \\ &\qquad + \left\{D(X) + \frac{\bar{a}^{(i)}(u)}{\bar{a}^{(0)}(u)} E(X)\bar{G}^{(0)}(X)\right\} H^{(0)}(X) = E(X)\bar{P}^{(i)}(X); \end{aligned} \tag{4}$$

and, on multiplying (2) by $E(X)$, we obtain the relation

$$E(X)\bar{H}^{(i)}(X)\bar{G}^{(0)}(X) + E(X)\bar{G}^{(i)}(X)\bar{H}^{(0)}(X) = E(X)\bar{P}^{(i)}(X). \tag{5}$$

Since both $E(X)\bar{G}^{(i)}(X)$ and $D(X) + (\bar{a}^{(i)}(u)/\bar{a}^{(0)}(u))E(X)\bar{G}^{(0)}(X)$ have the degree s in X_1 and the same coefficient for the term X_1^s, it follows from (4) and (5) and the uniqueness property mentioned above that

$$\begin{aligned} E(X)\bar{G}^{(i)}(X) &= D(X) + \frac{\bar{a}^{(i)}(u)}{\bar{a}^{(0)}(u)} E(X)\bar{G}^{(0)}(X), \\ E(X)\bar{H}^{(i)}(X) &= \bar{P}^{(i)}(X)B(X) + \frac{1}{\bar{a}^{(0)}(u)} \bar{H}^{(0)}(X)(C(X) - \bar{a}^{(i)}(u)E(X)). \end{aligned} \tag{6}$$

This shows that every coefficient in $\bar{G}^{(i)}(X)$ and $\bar{H}^{(i)}(X)$ can have in its denominator only factors which are also factors in $\bar{a}^{(0)}(u)e(u)$. If $H^{(0)}(X)$ does not contain the term X_1^{d-s}, then $\bar{a}^{(0)}(u)$ is a factor of $e(u)$ and we have nothing more to prove. On the other hand, if $H^{(0)}(X)$ contains the term $b^{(0)}(u)X_1^{d-s}$, then we can apply the same argument as above with $A(X)$ and $B(X)$ interchanged; we would then conclude that every coefficient in $\bar{G}^{(i)}(X)$ and $\bar{H}^{(i)}(X)$ can have in its denominator only factors which are also factors in $\bar{b}^{(0)}(u)e(u)$. Since every common factor of $\bar{a}^{(0)}(u)$ and $\bar{b}^{(0)}(u)$ is also a factor $e(u)$, this concludes the proof of our lemma.

COROLLARY. *If the coefficients in $F(X)$ are homogeneous of degree d in u, the coefficients in $G^{(0)}(X)$ and $H^{(0)}(X)$ are homogeneous of degrees s and $d - s$ respectively in u, and the element $a(u)$ is homogeneous of degree s in u, then all the coefficients in $\bar{G}^{(i)}(X)$ and $\bar{H}^{(i)}(X)$ are homogeneous of degree s and $d - s$ respectively in u.*

This follows from the formulas (1) to (6) by a simple calculation.

4. Before giving the proof of our Main Theorem, we shall briefly summarize here for the convenience of the reader a few properties of the associated form which we shall need. (1) The associated form $F(U)$ of a positive r-cycle Z of degree d in S^n has the degree d in each set ${}_iU$ of indeterminates, for $i = 0, \cdots, r$; every factor of an associated form is again an associated form, and the factorization of the associated form $F(U)$ into factors corresponds exactly to the representation of the cycle Z as a sum of the positive cycles corresponding to these factors. (2) If we denote by w the set of all coefficients of a general form in U of degree d in each set ${}_iU$, then there exists a finite set of homogeneous algebraic equations in X with coefficients which are themselves forms in w with integral coefficients, such that if we substitute for w the coefficients of the form $F(U)$ and take the integers to be the elements in the prime field of the given universal domain, then the resulting set of equations defines precisely the support of the cycle Z. We shall call such a set of equations "universal" in the sense that it remains valid (as equations of the support) for any positive cycle of the given dimension r and the given degree d in S^n over any universal domain. (3) The associated form $F(U)$ is rational over a field K (i.e., the coefficients of the form, for a suitable choice of the proportionality factor, are contained in K) if and only if the support of the cycle Z can be defined by a set of equations with coefficients which are rational over K; and when such is the case, the cycle Z is not K-connected if and only if the form $F(U)$ is the product of two forms which are rational over K and whose corresponding cycles have disjoint supports. (4) If $\mu = ({}_1\mu, \cdots, {}_r\mu)$ is a set of elements such that the form $F({}_0U, \mu) = F({}_0U, {}_1\mu, \cdots, {}_r\mu)$ is not identically zero, then $F({}_0U, \mu)$ splits completely into linear factors, and if $\sum_{j=0}^{n} {}_0U_j\alpha_j$ is any one of these factors, then the point $(\alpha_0, \alpha_1, \cdots, \alpha_n)$ is contained in the support of Z.

Proof of the Main Theorem. Without any loss of generality, we can restrict ourselves to the case where the associated form $F(U)$ of the cycle Z is irreducible over K. For, on account of the property (2) mentioned above and the "universal" property of the resultant criterion for the existence of a common solution of a system of homogeneous algebraic equations, the condition that the supports of two positive r-cycles have a common point can be expressed by a system of "universal" equations in the coefficients of their associated forms, and it is evident that such a condition is preserved under a homomorphism. We assume now that the cycle $\bar{Z}$ is not $\bar{R}$-connected, and we shall show that this will lead us to a contradiction. By the property (3) mentioned above, we can set $\bar{Z} = \bar{Z}_1 + \bar{Z}_2$, where $\bar{Z}_1$ and $\bar{Z}_2$ are positive r-cycles of degrees s and $d - s$ respectively in $\bar{S}^n$, such that their associated forms are rational over $\bar{R}$ and their supports are disjoint. Let $G^{(0)}(U) = G^{(0)}({}^0U, \cdots, {}_rU)$ and $H^{(0)}(U) = H^{(0)}({}_0U, \cdots, {}_rU)$ be forms with coefficients in Q such that $\bar{G}^{(0)}(U)$ and $\bar{H}^{(0)}(U)$ are the associated forms of $\bar{Z}_1$ and $\bar{Z}_2$ respectively and satisfy the equation $\bar{F}(U) = \bar{G}^{(0)}(U)\bar{H}^{(0)}(U)$. We shall assume for convenience that the coordinate system in (S^n and) $\bar{S}^n$ has been so chosen that every component of $\bar{Z}$ intersects properly the hyperplane L_0 in $\bar{S}^n$ with the Pluecker coordinates $\delta_0 = (1, 0, \cdots, 0)$; this is evidently possible in case the field $\bar{R}$ contains more than a finite number of elements, and we can always obtain this condition by the adjunction of a variable to the field K, which will not affect the problem in any essential way. Then

$\bar{G}^{(0)}(\delta_0, {}_1U, \cdots, {}_rU)$ and $\bar{H}^{(0)}(\delta_0, {}_1U, \cdots, {}_rU)$ do not vanish identically and are the associated forms of the cycles $L_0 \cdot \bar{Z}_1$ and $L_0 \cdot \bar{Z}_2$ respectively.

Let ${}_iu = ({}_iu_0, \cdots, {}_iu_n)$, $i = 1, \cdots, r$, be r independent sets of $n + 1$ independent variables over K (and hence also over $\bar{R}$); we set $F({}_0U, u) = F({}_0U, {}_1u, \cdots, {}_ru)$ and similarly for the other forms, and denote generally by u the set of all $r(n + 1)$ variables ${}_iu_j$. Let $E({}_0U, u) = E({}_0U_1, \cdots, {}_0U_n, u)$ be the resultant of $\bar{G}^{(0)}({}_0U, u)$ and $\bar{H}^{(0)}({}_0U, u)$, considered as polynomials in the indeterminate ${}_0U_0$ alone, and let $e(u)$ be the greatest common divisor in $\bar{R}[u]$ of all the coefficients of $E({}_0U, u)$; we shall show that $e(u) = 1$, apart from a multiplicative constant in $\bar{R}$. Let μ be any specialization of u over R such that $e(\mu) = 0$, but $\bar{F}({}_0U, \mu)$ is not identically zero, and assume first that either $G^{(0)}(\delta_0, \mu)$ or $H^{(0)}(\delta_0, \mu)$ does not vanish, say the former. Since $E({}_0U, \mu) = 0$, and since the coefficient $\bar{G}^{(0)}(\delta_0, \mu)$ of the term ${}_0U_0^s$ in $\bar{G}^{(0)}({}_0U, \mu)$ does not vanish, the forms $\bar{G}^{(0)}({}_0U, \mu)$ and $\bar{H}^{(0)}({}_0U, \mu)$ must have a factor in common, at first as polynomials in the indeterminate ${}_0U_0$ alone with coefficients in $\bar{R}(\mu, {}_0U_1, \cdots, {}_0U_n)$, and consequently also as forms in the indeterminates ${}_0U$. By the property (4) mentioned above, this implies that the forms $\bar{G}^{(0)}({}_0U, \mu)$ and $\bar{H}^{(0)}({}_0U, \mu)$ have a linear factor in common, and since this linear factor yields a point which is in the supports of both $\bar{Z}_1$ and $\bar{Z}_2$, this contradicts our hypothesis that the supports of $\bar{S}_1$ and Z_2 are disjoint. Thus we have shown that the condition $e(\mu) = 0$ implies either $\bar{F}({}_0U, \mu) = 0$ or $\bar{G}^{(0)}(\delta_0, \mu) = \bar{H}^{(0)}(\delta_0, \mu) = 0$, so that a factor in $e(u)$ is either a common factor of all the coefficients in $\bar{F}({}_0U, u)$ as a form in ${}_0U$, or a common factor of the two forms $\bar{G}^{(0)}(\delta_0, u)$ and $\bar{H}^{(0)}(\delta_0, u)$. However, by the property (1) mentioned above, no such factors can exist except for constants in $\bar{R}$; for $\bar{F}(U)$ cannot have a factor which has the degree 0 in ${}_0U$ (any factor must have the same degree in each set ${}_iU$), and $\bar{G}^{(0)}(\delta_0, {}_1U, \cdots, {}_rU)$ and $\bar{H}^{(0)}(\delta_0, {}_1U, \cdots, {}_rU)$ can have a common factor if and only if the cycles $L_0 \cdot \bar{Z}_1$ and $L_0 \cdot \bar{Z}_2$ have a common component, which would be in contradiction to our hypothesis again.

We use induction on the dimension r, the case $r=0$ being Lemma 1 itself. Since $F(\delta_0, u)$ is the associated form of $L_0 \cdot Z$ and since $L_0 \cdot \bar{Z}_1$ and $L_0 \cdot \bar{Z}_2$ are disjoint, it follows from induction hypothesis that there exist forms $g(U)$ and $L(U)$ in $R[{}_1U, \cdots, {}_rU]$ such that $F(\delta_0, U) = g(U)\, L(U)$, $\bar{g}(U) = \bar{G}^{(0)}(\delta_0, U)$, and $\bar{L}(U) = \bar{H}^{(0)}(\delta_0, u)$. Now, since $F({}_0U, u)$ is a splitting polynomial, since

$$\bar{F}({}_0U, u) = \bar{G}^{(0)}({}_0U, u)\bar{H}^{(0)}({}_0U, u),$$

and since $\bar{G}^{(0)}({}_0U, u)$ and $\bar{H}^{(0)}({}_0U, u)$ are relatively prime and have non-vanishing coefficients for the terms ${}_0U_0^s$ and ${}_0U_0^{d-s}$ respectively, we can apply Lemma 1. There exists therefore polynomials $F_1({}_0U)$ and $F_2({}_0U)$ with coefficients in R_u^*, of degrees s and $d - s$ respectively, such that $\bar{F}_1({}_0U) = \bar{G}^{(0)}({}_0U, u)$ and $\bar{F}_2({}_0U) = \bar{H}^{(0)}({}_0U, u)$, that the coefficient of the term ${}_0U_0^s$ in $F_1({}_0U)$ is $g(u)$, and that $F({}_0U, u) = F_1({}_0U)F_2({}_0U)$. We set $F_1({}_0U) = \sum_{i=0}^{\infty} F_1^{(i)}({}_0U)p^i$ and $F_2({}_0U) = \sum_{i=0}^{\infty} F_2^{(i)}({}_0U)p^i$. Since $e(u) = 1$, it follows from Lemma 2 that the coefficients in $F_1^{(i)}({}_0U)$ and $F_2^{(i)}({}_0U)$ are all polynomials in u with coefficients in Q; and since the coefficients in $F({}_0U, u)$, $G^{(0)}({}_0U, u)$, and $H^{(0)}({}_0U, u)$ are all homogeneous in each set ${}_iu$ of degrees d, s, and $d - s$ respectively, it follows from the Corollary of Lemma 2 that the coefficients in $F_1^{(i)}({}_0U)$ and $F_2^{(i)}({}_0U)$ are also all homogeneous in each set ${}_iu$ of degrees s and $d - s$ respectively. This implies that the coefficients in $F_1({}_0U)$ and $F_2({}_0U)$ are forms in $R[u]$ of degrees s and $d - s$ respectively in each set ${}_iu$. There exist therefore forms $G(U)$ and $H(U)$ in $R[U]$, of degrees s and $d - s$ respectively in each set ${}_iU$, such that $G({}_0U, u) = F_1({}_0U)$ and $H({}_0U, u) = F_2({}_0U)$, and from the equation $F({}_0U, u) = G({}_0U, u)H({}_0U, u)$ follows evidently the equation $F(U) = G(U)H(U)$. As this is in contradiction to our hypothesis that the form $F(U)$ is irreducible over K, the proof of our theorem is therefore completed.

The Johns Hopkins University

References

1. S. Abhyankar and O. Zariski, *Splitting of valuations in extensions of local domains*, Proc. Nat. Acad. Sci. U. S. A., vol. 41 (1955), pp. 84–90.
2. C. Chevalley, *On the theory of local rings*, Ann. of Math., vol. 44 (1943), pp. 690–708.
3. W. L. Chow and B. L. van der Waerden, *Ueber zugeordnete Formen und algebraische Systeme von algebraischen Mannigfaltigkeiten*, Math. Ann., vol. 113 (1937), pp. 692–704.
4. W. Krull, *Ein Satz über primäre Integritätsbereiche*, Math. Ann., vol. 103 (1930), pp. 450–465.
5. B. L. van der Waerden, Moderne Algebra, vol. 1, second edition, J. Springer, Berlin, 1937.
6. A. Weil, Foundations of Algebraic Geometry, Amer. Math. Soc. Colloquium Publications, vol. 29, 1946.
7. O. Zariski, *Theory and applications of holomorphic functions on algebraic varieties over arbitrary ground fields*, Mem. Amer. Math. Soc., no. 5, 1951.

ON THE BIRATIONAL EQUIVALENCE OF CURVES UNDER SPECIALIZATION.*

By WEI-LIANG CHOW [1] and SERGE LANG.

Let k be a field with a discrete valuation, let p be the maximal prime ideal in the valuation ring on k, and let $\bar{k}$ be the residue field of k. Let S^n be the projective space of dimension n in the algebraic geometry over a universal domain containing k, and denote by $\bar{S}^n$ the projective space of dimension n over a universal domain containing $\bar{k}$; if Z is any cycle in S^n, rational over k, we shall denote by $\bar{Z}$ the cycle obtained from Z by reduction modulo p, in the sense of Shimura [4], and we shall call $\bar{Z}$ the specialization of Z (with respect to the given valuation of k). This definition applies also in case Z is a variety, which is then to be considered as a prime cycle; the the specialization $\bar{Z}$ is then a positive cycle, but in general not necessarily a variety. According to the generalization of the Principle of Degeneration recently proved by Chow, the specialization of a variety is always connected, but we shall not need this fact here.

Consider now two non-singular curves C_1 and C_2 in S^n, both defined over k, and assume that they are birationally equivalent over k. A theorem of Deuring ([2], Satz 3) can be expressed geometrically by saying that if C_1 and C_2 have genus 1 and if the cycles $\bar{C}_1$ and $\bar{C}_2$ are also non-singular curves of genus 1, then $\bar{C}_1$ and $\bar{C}_2$ are birationally equivalent $\bar{k}$. In this note we shall generalize this result to non-singular curves of arbitrary genus and also to abelian varieties; in fact, we shall first prove the result for abelian varieties, and then deduce the result for curves by embedding them in their Jacobians. Our method is quite different from Deuring's, and is based on the "compatibility" of the Chow construction of the Jacobian and the canonical mapping.

Let A be an abelian variety in S^n, defined over k. We shall say that the specialization $\bar{A}$ of A is *non-degenerate* if $\bar{A}$ is an abelian variety defined over $\bar{k}$ and if the law of composition in $\bar{A}$ is the specialization of the law of

* Received January 24, 1957.

[1] Work of W. L. Chow was supported in part by a National Science Foundation Grant, NSF-3022.

composition in A. Specifically, this means the following. Let G be the graph of the law of composition in A, so that G is a subvariety in $A \times A \times A$; then our condition can be expressed by saying that both $\bar{A}$ and $\bar{G}$ are varieties (in $\bar{S}^n$ and $\bar{S}^n \times \bar{S}^n \times \bar{S}^n$ respectively) and that $\bar{G}$ defines an abelian law of composition in $\bar{A}$. We observe that if e is the unit element in A, then $\bar{e}$ is the unit element in $\bar{A}$; in fact, e is characterized by the condition that $\mathrm{pr}_{32}(e \times A \times A))$ is contained in the diagonal in $A \times A$, and this condition is evidently preserved under the specialization.

THEOREM 1. *Let A and B be abelian varieties in S^n, and let T be a birational isomorphism between A and B; we assume that A, B, and T are all defined over k. If the specializations $\bar{A}$ and $\bar{B}$ are non-degenerate, then the specialization $\bar{T}$ is a variety and defines a birational isomorphism between $\bar{A}$ and $\bar{B}$.*

Proof. It is clear that $\bar{A} \times \bar{B}$ is the specialization of $A \times B$ and is non-degenerate. If we denote the graph of T by the same symbol, then the support $|\bar{T}|$ of $\bar{T}$ is an algebraic subgroup in $\bar{A} \times \bar{B}$; this follows from the fact that T is an abelian subvariety in $A \times B$, and the fact that the law of composition in $\bar{A} \times \bar{B}$ is the specialization of the law of composition in $A \times B$. Let $\bar{T}_0$ be a component in $\bar{T}$ which contains the unit element; then $\bar{T}_0$ is an abelian subvariety in $\bar{A} \times \bar{B}$, and every component in $|\bar{T}|$ and hence also every component in $\bar{T}$ is obtained from $\bar{T}_0$ by a translation. By assumption, we have $\mathrm{pr}_1 T = A$ and $\mathrm{pr}_2 T = B$; hence, by specialization, we have $\mathrm{pr}_1 \bar{T} = \bar{A}$ and $\mathrm{pr}_2 \bar{T} = \bar{B}$. Since all cycles obtained from $\bar{T}_0$ by translations must have the same projections on $\bar{A}$ and $\bar{B}$, we conclude readily that $\mathrm{pr}_1 \bar{T}_0 = \bar{A}$ and $\mathrm{pr}_2 \bar{T} = \bar{B}$, and that $\bar{T}_0$ is the only component in $\bar{T}$ and has the coefficient 1. This shows that $\bar{T} = \bar{T}_0$ defines a birational transformation between $\bar{A}$ and $\bar{B}$, which, since it preserves the unit elements, must be an isomorphism.

We turn now to the case of the curves C_1 and C_2 considered at the beginning of this note. Let C be a non-singular curve of genus $g(C)$ in S^n, defined over k; we shall say that the specialization $\bar{C}$ is *non-degenerate* if $\bar{C}$ is a non-singular curve and hence has the same genus $g(C)$ as C. We assume that there exists a birational correspondence F between C_1 and C_2, defined over k, and that the specialization $\bar{C}_1$ and $\bar{C}_2$ are both non-degenerate. Since the conditions $\mathrm{pr}_1 F = C_1$ and $\mathrm{pr}_2 F = C_2$ imply by specialization the conditions $\mathrm{pr}_1 \bar{F} = \bar{C}_1$ and $\mathrm{pr}_2 \bar{F} = \bar{C}_2$, we conclude that $\bar{F}$ is either a birational correspondence between $\bar{C}_1$ and $\bar{C}_2$, or has the form $x \times \bar{C}_2 + \bar{C}_1 \times y$, for some points x and y in $\bar{C}_1$ and $\bar{C}_2$ respectively. In case $g(C_1) = g(C_2) = 0$, it can be easily seen by examples that both possibilities can occur; however, in this case the assertion of Deuring is trivially true. In the next theorem we shall show that, except in this trivial case, the second possibility cannot occur; this implies that the specialization $\bar{F}$ is itself already such a birational correspondence.

THEOREM 2. *Let C_1 and C_2 be non-singular curves of genus > 0 in S^n, and let F be a birational correspondence between C_1 and C_2; we assume that C_1, C_2, and F are all defined over k. If the specializations $\bar{C}_1$ and $\bar{C}_2$ are non-degenerate, then the specialization $\bar{F}$ is a birational correspondance between $\bar{C}_1$ and $\bar{C}_2$.*

Proof. The theorem being "geometric," we may assume that the curves C_1 and C_2 have rational points over k, and hence the curves $\bar{C}_1$ and $\bar{C}_2$ have rational points over $\bar{k}$. In fact, we can enlarge k by a finite algebraic extension if necessary, and extend the discrete valuation in k to a discrete valuation of this extension; it is easily seen that the validity of our theorem over this extension will imply its validity in the original form.

For $i = 1, 2$, we consider the Jacobian J_i of C_i and the canonical mapping f_i of C_i into J_i with reference to a rational point p_i in C_i (i. e., $f_i(p_i) = e_i$, e_i being the unit element in J_i), both J_i and f_i being constructed as in Chow [1]. It is essential for our purpose that J_i and f_i are constructed in this particular way, which we shall call the Chow construction, not just some arbitrarily chosen projective models of J_i and the corresponding mapping f_i. The reason lies in the fact that the Chow construction is "compatible" with specializations, at least when they are non-degenerate. Specifically, this means that if the specialization $\bar{C}_i$ is non-degenerate, then the specialization $\bar{J}_i$ (which is a cycle in $\bar{S}^{t_i}$ if S^{t_i} is the ambient projective space of J_i) of the abelian variety J_i is also non-degenerate, and $\bar{J}_i$ and $\bar{f}_i$ are respectively the Jacobian and the canonical mapping of $\bar{C}_i$. This fact is a special case of a more general "compatibility" theorem of Igusa [3] (Igusa treats only the algebro-geometric specializations, but it is easily seen that his proof is also valid for the more general specializations considered here at least in the non-degenerate case). For the sake of convenience, we shall assume that $p_2 = F(p_1)$; then there exists a birational isomorphism T between J_1 and J_2, defined over k, such that $T \circ f_1 = f_2 \circ F$. If we denote by f the mapping $f_1 \times f_2$ of $C_1 \times C_2$ into $J_1 \times J_2$, and if we denote the graphs of F and T by the same symbols, then the above relation implies that $f(F)$ is contained in T. Since the specialization $\bar{f} = \bar{f}_1 \times \bar{f}_2$ is a rational mapping of $\bar{C}_1 \times \bar{C}_2$ into $\bar{J}_1 \times \bar{J}_2$, it follows that the cycle $\bar{f}(\bar{F})$ is defined and is contained in $|\bar{T}|$.

By Theorem 1, $\bar{T}$ is an abelian subvariety in $\bar{J}_1 \times \bar{J}_2$ (it is in fact sufficient to know that $|\bar{T}|$ is an algebraic group in $\bar{J}_1 \times \bar{J}_2$) and is a proper subset in $\bar{J}_1 \times \bar{J}_2$. If $\bar{F}$ is not a birational correspondence between $\bar{C}_1$ and $\bar{C}_2$, then by a previous remark we must have $\bar{F} = \bar{p}_1 \times \bar{C}_2 + \bar{C}_1 \times \bar{p}_2$ (we observe that $|\bar{F}|$ must contain the point $\bar{p}_1 \times \bar{p}_2$) and hence

$$\bar{f}(\bar{F}) = \bar{e}_1 \times \bar{f}_2(\bar{C}_2) + \bar{f}_1(\bar{C}_1) \times \bar{e}_2.$$

This shows that $|\bar{f}(\bar{F})|$ generates the entire variety $\bar{J}_1 \times \bar{J}_2$, in contradiction to the fact that $\bar{f}(\bar{F})$ is contained in $\bar{T}$. This proves our theorem.

THE JOHNS HOPKINS UNIVERSITY,
COLUMBIA UNIVERSITY.

REFERENCES.

[1] W. L. Chow, "The Jacobian variety of an algebraic curve," *American Journal of Mathematics*, vol. 76 (1954), pp. 453-476.

[2] M. Deuring, "Die Zetafunktion einer algebraischen Kurve vom Geschlechte Eins (II)," *Nachrichten der Akademie der Wissenschaften in Göttingen*, 1955, pp. 13-42.

[3] J. Igusa, "Fibre systems of Jacobian varieties," *American Journal of Mathematics*, vol. 78 (1956), pp. 171-199.

[4] G. Shimura, "Reduction of algebraic varieties with respect to a discrete valuation of the basic field," *ibid.*, vol. 77 (1955), pp. 131-176.

THE CRITERION FOR UNIT MULTIPLICITY AND A GENERALIZATION OF HENSEL'S LEMMA.*[1]

By WEI-LIANG CHOW.

To Artin on his 60th birthday

The well-known criterion for unit multiplicity in algebraic geometry states that if two cycles $\mathfrak{X}^r$ and $\mathfrak{Y}^s$ in a variety V^n $(n = r + s)$ are transversal to each other at a simple point α in V, then $\mathfrak{X}$ and $\mathfrak{Y}$ intersect properly at α with the multiplicity 1. This is the version of the criterion as formulated in Weil's *Foundations of Algebraic Geometry*,[2] whereby we have restricted ourselves for simplicity to the case $r + s = n$. It is possible to give this criterion another formulation which is somewhat more elementary in the sense that it involves the multiplicity of a specialization rather than that of an intersection. It can be stated as follows: If two cycles $\mathfrak{X}^r$ and $\mathfrak{Y}^s$ are specializations of two subvarieties $\mathfrak{X}'$ and $\mathfrak{Y}'$ respectively in a variety V^n $(n = r + s)$ over a field of definition k for V, if $\mathfrak{X}' \cap \mathfrak{Y}'$ consists of a finite set of (distinct) points $a^{(i)}$, and if $\mathfrak{X}$ and $\mathfrak{Y}$ are transversal to each other at a simple point α in V, then the point α occurs exactly once in every specialization of the set $a^{(i)}$ over the specialization $(\mathfrak{X}', \mathfrak{Y}') \rightarrow (\mathfrak{X}, \mathfrak{Y})$ over k. This version of the criterion for unit multiplicity appeared in the earlier intersection theory of van der Waerden,[3] and played an essential role there in the development of the concept of the intersection-multiplicity itself. However, if one examines very closely the treatment of van der Waerden, one observes that only a part of this criterion, namely the assertion that the specialization-multiplicity is *at most* one, was given a simple proof before the introduction of the concept of the intersection-multiplicity, while the other part, which asserts that this specialization-multiplicity is *at least* one, was proved only later after the concept of the intersection-multiplicity had

* Received January 17, 1958.

[1] This work was partially supported by a research grant of the National Science Foundation.

[2] See Chapter VI, Theorem 6.

[3] B. L. van der Waerden, "Zur algebraischen Geometrie V. Ein Kriterium fuer die Einfachheit von Schnittpunkten," *Math. Annalen*, 110, 128-133, 1934.

already been introduced.[4] In fact, this second part concerning the positivity of this particular specialization-multiplicity was proved by van der Waerden in a rather complicated and indirect way, at first only for cycles in a projective space and later carried over to the general case. As we have recently developed a new theory of intersection which preserves the basic ideas of the original theory of van der Waerden, although in an extensively generalized form, we become naturally interested in finding a proof of the criterion for unit multiplicity which is completely free from any use of the notion of the intersection-multiplicity. It turns out that we are able not only to find a simple proof of this nature, but also at the same time, by virtue of the abstract local form in which we have recast this criterion itself, to obtain a more precise result which contains as a special case a certain generalization of the Hensel's Lemma conjectured by Weil some years ago. Although the subject forms an integral part of our intersection theory, which will be published elsewhere at a later date, we feel that this connection with the Hensel's Lemma justifies its separate publication here as a short note. In keeping with the foundational nature of the subject, we shall make our treatment as elementary as posible, using only a few elementary properties in the theory of local rings. Finally, we shall show in the last section that this connection with Hensel's Lemma can be extended to a more general result in the intersection-theory, namely the invariance of intersection-multiplicity under specialization; it turns out that this invariance can be easily deduced from the Associativity Formula, as generalized in a recent paper[5] of Nagata. Here we shall of course need the more sophisticated results in the theory of local rings.

1. Let K be a field, and let v be a real discrete valuation of K; let K_v be the valuation ring of v in K, let π be a generator of its maximal prime ideal, and let $\bar{K}$ be the residue field of K_v modulo the maximal prime ideal. We denote by S^m and $\bar{S}^m$ the projective spaces over the fields K and $\bar{K}$ respectively, and we shall be interested in the relation between the varieties and cycles contained in them. Let α be a rational point over $\bar{K}$ in $\bar{S}^m$. We choose an affine coordinate system in S^m so that in the corresponding affine coordinate system in $\bar{S}^m$ the point α is at the origin; we introduce

[4] The first part was proved in the paper cited above; the second part is contained implicitly in B. L. van der Waerden, "Zur algebraischen Geometrie, XIV. Schnittpunktszahlen von algebraischen Mannigfaltigkeiten," *Math. Annalen*, 115, 619-642, 1938, see § 4.

[5] M. Nagata, "The Theory of Multiplicity in General Local Rings," *Proceedings of the International Symposium on Algebraic Number Theory*, Tokyo-Nikko 1955, pp. 191-226.

the indeterminates $X_1,\cdots,X_m$ corresponding to the affine coordinates so that we have $\alpha=(0,\cdots,0)$ and consider the ring $K[X]=K[X_1,\cdots,X_m]$ as well as the rings $K_v[X]=K_v[X_1,\cdots,X_m]$ and $\bar{K}[X]=\bar{K}[X_1,\cdots,X_m]$. The canonical homomorphism E of K_v onto $\bar{K}$ can be extended to a homomorphism of the polynomial ring $K_v[X]$ onto the polynomial ring $\bar{K}[X]$, and for any $f(X)$ in $K_v[X]$ we shall denote its image under E in $\bar{K}[X]$ by $\bar{f}(X)$. Let $\mathfrak{Q}(S/\alpha)$ be the set of all quotients $f(X)/g(X)$ of two elements $f(X)$ and $g(X)$ in $K_v[X]$ such that $\bar{g}(\alpha)\neq 0$; it is easily seen that $\mathfrak{Q}(S/\alpha)$ is a local ring and that its maximal prime ideal $\mathfrak{p}(S/\alpha)$ is generated by the $m+1$ elements $\pi, X_1,\cdots,X_m$. Similarly, let $\mathfrak{Q}(\bar{S}/\alpha)$ be the set of all quotients $\phi(X)/\theta(X)$ of two elements $\phi(X)$ and $\theta(X)$ in $\bar{K}[X]$ such that $\theta(\alpha)\neq 0$; it is well known that $\mathfrak{Q}(\bar{S}/\alpha)$ is a regular local ring of dimension m and that it maximal prime ideal $\mathfrak{p}(\bar{S}/\alpha)$ is generated by the m elements $X_1,\cdots,X_m$. It is easily seen that the homomorphism E of $K_v[X]$ onto $\bar{K}[X]$ can be extended to a homomorphism of $\mathfrak{Q}(S/\alpha)$ onto $\mathfrak{Q}(\bar{S}/\alpha)$ and that the kernel of the so extended homomorphism E is the ideal in $\mathfrak{Q}(S/\alpha)$ generated by the element π. This shows, in particular, that $\mathfrak{Q}(S/\alpha)$ has a dimension at least $m+1$, and since $\mathfrak{p}(S/\alpha)$ is generated by the $m+1$ elements $\pi, X_1,\cdots,X_m$, it follows that $\mathfrak{Q}(S/\alpha)$ is a regular local ring of dimension $m+1$.

Let $\mathfrak{X}^r$ be a positive cycle in S^m, rational over K, and let $F(U)=F({}_0U,{}_1U,\cdots,{}_rU)$ be the associated form of $\mathfrak{X}$, where each ${}_iU$ is a system of $m+1$ indeterminates; by a proper choice of the proportionality factor, we can obtain that the coefficients in $F(U)$ are all in K_v, but not all in the maximal prime ideal, and we observe that in this way the coefficients in $F(U)$ are uniquely determined up to a common factor which is a unit in K_v. If we denote by $\bar{F}(U)$ the form obtained from $F(U)$ by replacing every coefficient by its image in $\bar{K}$, then $\bar{F}(U)$ is the associated form of a positive r-cycle $\bar{\mathfrak{X}}$ in $\bar{S}^m$; since the form $\bar{F}(U)$ is uniquely determined up to a factor in $\bar{K}$ by the form $F(U)$, the cycle $\bar{\mathfrak{X}}$ is uniquely determined by the cycle $\mathfrak{X}$. We shall call the cycle $\bar{\mathfrak{X}}$ the *specialization of the cycle* $\mathfrak{X}$ *over the valuation* v. We shall now express this situation in terms of local ideals at a given point in $\bar{\mathfrak{X}}$. Let α be a point in $\bar{\mathfrak{X}}$ (i.e., a point in the support of $\bar{\mathfrak{X}}$), rational over $\bar{K}$, and assume that an affine coordinate system in S^m has been chosen as indicated above. Let $\mathfrak{P}$ be the ideal in $K[X]$ associated with the support of the positive cycle $\mathfrak{X}$, i.e., $\mathfrak{P}$ consists of all polynomials in $K[X]$ which vanish at every point in the support of $\mathfrak{X}$; we set $\mathfrak{P}_v=K_v[X]\cap\mathfrak{P}$ and $\mathfrak{P}_\alpha=\mathfrak{Q}(S/\alpha)\mathfrak{P}_v$. The fact that $\mathfrak{X}$ has the dimension r can be expressed by the property that every minimal prime divisor of $\mathfrak{P}$ has the rank $m-r$;

we recall that the rank of a prime ideal is defined as the maximum length of a descending chain of prime ideals starting with, but not including, the given ideal itself, and the rank of any ideal is defined as the minimum of the ranks of its minimal prime divisors. Since every element in $\mathfrak{P}$ can be made into an element in $\mathfrak{P}_v$ by multiplying with a suitable power of π, it follows that every minimal prime divisor of $\mathfrak{P}_v$ must have the same rank $m-r$; since $\mathfrak{X}$ contains the point α, the ideal $p(S/\alpha)$ contains $\mathfrak{P}_\alpha$, and it follows then from well-known properties of quotient rings that every minimal prime divisor of $\mathfrak{P}_\alpha$ must have the same rank $m-r$ as $\mathfrak{P}_v$. Let $\bar{\mathfrak{P}}_v$ be the image ideal of $\mathfrak{P}_v$ under the homomorphism E; then $\bar{\mathfrak{P}}_v$ is contained in the ideal associated with the support of $\bar{\mathfrak{X}}$, and hence every minimal prime divisor of $\bar{\mathfrak{P}}_v$ must have a rank at most $m-r$. If we denote by $\bar{\mathfrak{P}}_\alpha$ the image ideal of $\mathfrak{P}_\alpha$ under the homomorphism E, then it is easily seen that $\bar{\mathfrak{P}}_\alpha = \mathfrak{Q}(\bar{S}/\alpha)\bar{\mathfrak{P}}_v$; hence $\bar{\mathfrak{P}}_\alpha$ must have a rank at most $m-r$.

We shall say that the positive cycle $\bar{\mathfrak{X}}$ is *simple* at the point α if exactly one component variety $\bar{\mathfrak{X}}_\alpha$ in $\bar{\mathfrak{X}}$, with the coefficient 1, contains the point α, and if this component is simple at α.

LEMMA 1. *If the cycle $\bar{\mathfrak{X}}$ is simple at the point α, then $\bar{\mathfrak{P}}_\alpha$ has a basis consisting of $m-r$ elements which can be extended to a minimal basis for $p(\bar{S}/\alpha)$; and when such is the case, then $\bar{\mathfrak{P}}_\alpha$ is the prime ideal in $\mathfrak{Q}(\bar{S}/\alpha)$ associated with the variety $\bar{\mathfrak{X}}_\alpha$.*

Proof. We assume that $\bar{\mathfrak{X}} = \bar{\mathfrak{X}}_\alpha + \bar{\mathfrak{X}}_\beta$, where $\bar{\mathfrak{X}}_\alpha$ is the variety having a simple point at α and $\bar{\mathfrak{X}}_\beta$ is a positive cycle not containing α at all, and let $\bar{F}(U) = \Phi_1(U)\Phi_2(U)$ be the corresponding factorization of the associated form; since $\bar{\mathfrak{X}}_\alpha$ and hence also $\bar{\mathfrak{X}}_\beta$ are rational over $\bar{K}$, the forms $\Phi_1(U)$ and $\Phi_2(U)$ can also be taken to be rational over $\bar{K}$. Since $\bar{\mathfrak{X}}_\alpha$ is simple at the origin α, we can choose the affine coordinate system so that the equations $X_{r+1} = \cdots = X_m = 0$ define the tangential space of $\bar{\mathfrak{X}}_\alpha$ at α. We set ${}_iu_0 = X_i$, ${}_iu_i = 1$, ${}_iu_j = 0$ $(j \neq 0, i)$, for $i = 1, \cdots, m$, and consider the $m-r$ polynomials $\Psi_k(X) = \Phi_1({}_ku, {}_1u, \cdots, {}_ru)$, $k = r+1, \cdots, m$; if we assume for the moment that $\bar{K}$ is infinite, then for a suitable choice of the affine coordinate system in $\bar{S}^m$ we can obtain that the determinant $|\partial\Psi_k(X)/\partial X_j|$, $j, k = r+1, \cdots, m$, does not vanish at α. We set $\phi_k(X) = \bar{F}({}_ku, {}_1u, \cdots, {}_ru) = \Phi_1({}_ku, {}_1u, \cdots, {}_ru)\Phi_2({}_ku, {}_1u, \cdots, {}_ru)$; since the second factor on the right does not vanish at all at α, it follows that the determinant $|\partial\phi_k(X)/\partial X_j|$, $j, k = r+1, \cdots, m$, also does not vanish at α. This shows that the elements $X_1, \cdots, X_r, \phi_{r+1}(X), \cdots, \phi_m(X)$ form a minimal basis for $p(\bar{S}/\alpha)$. If we now set $f_k(X) = F({}_ku, {}_1u, \cdots, {}_ru)$,

$k = r+1, \cdots, m$, then it is easily seen that the polynomials $f_k(X)$ all vanish at every point of $\mathfrak{X}$ and hence are contained in $\mathfrak{P}_v$ (and hence also in $\mathfrak{P}_\alpha$); since we have evidently $\phi_k(X) = \bar{f}_k(X)$ for every k, this shows that the polynomials $\phi_k(X)$ are all in $\bar{\mathfrak{P}}_\alpha$. Since the polynomials $\phi_k(X)$ generate in $\mathfrak{Q}(\bar{S}/\alpha)$ a prime ideal which has the rank $m-r$ and is contained in $\bar{\mathfrak{P}}_\alpha$, and since $\bar{\mathfrak{P}}_\alpha$ has a rank at most $m-r$, it follows that $\bar{\mathfrak{P}}_\alpha$ must coincide with this prime ideal and hence has the polynomials $\phi_k(X)$ as a basis. Finally, since the prime ideal in $\mathfrak{Q}(\bar{S}/\alpha)$ associated with $\bar{\mathfrak{X}}_\alpha$ has the rank $m-r$ and contains $\bar{\mathfrak{P}}_\alpha$, it must coincide with $\bar{\mathfrak{P}}_\alpha$.

In case $\bar{K}$ is finite, we take the ring $K_v[w]$ obtained from K_v by the adjunction of an auxiliary variable w over K, and consider the quotient ring $K_{v,w}$ of $K_v[w]$ with respect to the set $K_v[w] - K_v[w]\pi$; if we denote by $\bar{w}$ the $K_{v,w}\pi$-residue of w, then $\bar{w}$ is a variable over $\bar{K}$, and $\bar{K}(\bar{w})$ is the residue field of $K_{v,w}$. If we replace K_v by $K_{v,w}$ and apply our lemma, and if we denote by $\mathfrak{Q}(\bar{S}/\alpha, \bar{K}(\bar{w}))$ the local ring defined in a similar way as $\mathfrak{Q}(\bar{S}/\alpha)$ with $\bar{K}(\bar{w})$ replacing $\bar{K}$, then we conclude that the ideal $\mathfrak{Q}(\bar{S}/\alpha, \bar{K}(\bar{w}))\bar{\mathfrak{P}}_\alpha$ is the prime ideal in $\mathfrak{Q}(\bar{S}/\alpha, \bar{K}(\bar{w}))$ associated with $\bar{\mathfrak{X}}_\alpha$; since $\bar{\mathfrak{P}}_\alpha$ is the contraction of $\mathfrak{Q}(\bar{S}/\alpha, \bar{K}(\bar{w}))\bar{\mathfrak{P}}_\alpha$ in $\mathfrak{Q}(\bar{S}/\alpha)$, it follows that the ideal $\bar{\mathfrak{P}}_\alpha$ is the prime ideal in $\mathfrak{Q}(\bar{S}/\alpha)$ associated with $\bar{\mathfrak{X}}_\alpha$.

In the next two section we shall be concerned only with those points in S^m (and in $\bar{S}^m$) which are finite with respect to the given affine coordinate system, so that we shall be really dealing with affine varieties and cycles. For the sake of convenience we shall keep the same notations as used before for the entities in the projective space; there will be no danger of confusion, as the context will make it clear what is meant in each case.

2. Consider now a variety V^n in S^m, defined over K, and assume that the specialization $\bar{V}$ of V over v is simple at the point α. Let $\mathfrak{O}$ be the prime ideal in $K[X]$ associated with V, and set $\mathfrak{O}_v = K_v[X] \cap \mathfrak{O}$, $\mathfrak{O}_\alpha = \mathfrak{Q}(S/\alpha)\mathfrak{O}_v$, $R = \mathfrak{Q}(S/\alpha)/\mathfrak{O}_\alpha$, and $\mathfrak{m} = \mathfrak{p}(S/\alpha)/\mathfrak{O}_\alpha$; then R is a local ring with $\mathfrak{m}$ as the maximal prime ideal, and since $\mathfrak{O}_\alpha$ does not contain any element in K_v, we can embed K_v canonically as a subring in R. Let $\bar{\mathfrak{O}}_\alpha$ be the image ideal of $\mathfrak{O}_\alpha$ under the homomorphism E, and set $\bar{R} = \mathfrak{Q}(\bar{S}/\alpha)/\bar{\mathfrak{O}}_\alpha$; then $\bar{R}$ is a local ring, and since $\bar{\mathfrak{O}}_\alpha$ does not contain any element in $\bar{K}$, we can embed $\bar{K}$ canonically as a subring in $\bar{R}$. It is clear that E induces a homomorphism of R onto $\bar{R}$, which we shall denote by E_R; and if we denote by G and $\bar{G}$ the canonical homomorphisms of $\mathfrak{Q}(S/\alpha)$ onto R and of $\mathfrak{Q}(\bar{S}/\alpha)$ onto $\bar{R}$ respectively, then we have the relation $E_R G = \bar{G} E$. We set $x_i = G(X_i)$ and $\bar{x}_i = \bar{G}(X_i)$ for $i = 1, \cdots, m$; then the point $x = (x_1, \cdots, x_m)$ in S^m is a generic point of V over K and the point $\bar{x} = (\bar{x}_1, \cdots, \bar{x}_m)$ is a generic point over $\bar{K}$ of the component $\bar{V}_\alpha$ of $\bar{V}$ containing α, and it is clear that $\bar{x}$ is the image of x under the homomorphism E_R. According to Lemma 1 above, applied to V as a prime cycle in S^m, the ideal $\bar{\mathfrak{O}}_\alpha$ is the prime ideal in $\mathfrak{Q}(\bar{S}/\alpha)$ associated with $\bar{V}_\alpha$; since $\bar{V}_\alpha$ is simple at α, this means that $\bar{R}$ is a regular local ring of dimension n. If now $y_1, \cdots, y_n$ are elements in R such that their images $\bar{y}_1, \cdots, \bar{y}_n$ under E_R form a minimal basis for the maximal prime ideal $\bar{\mathfrak{m}}$ in $\bar{R}$, then it is easily seen that the $n+1$ elements $\pi, y_1, \cdots, y_n$ will form a basis for the maximal prime ideal $\mathfrak{m}$ in R; since R has the dimension at least $n+1$, this shows that this basis is minimal and that R is a regular local ring of dimension $n+1$.

We consider the (affine) coordinate ring $K[x]$ of V, and the subring $K_v[x]$ contained in it; it is clear that $K[x]$ is the quotient ring of $K_v[x]$ with respect to the multiplicatively closed set of all non-zero elements in K_v. Let $\mathfrak{i}$ be an ideal in $K_v[x]$, and let $\mathfrak{p}_i$, $i = 1, \cdots, c$, be the minimal prime divisors of $\mathfrak{i}$; then, for each $\mathfrak{p}_i$ which does not contain any element in K_v (i. e. $K[x]\mathfrak{p}_i \neq K[x]$), the ideal $K[x]\mathfrak{p}_i$ is a minimal prime divisor of the ideal $K[x]\mathfrak{i}$, and in this way we obtain all the minimal prime divisors of $K[x]\mathfrak{i}$. It is well-known that if the ideal $\mathfrak{i}$ is the contraction of an ideal in $K[x]$, then none of the ideals $\mathfrak{p}_i$ can contain any element in K_v and $K[x]\mathfrak{i}$ is the only ideal in $K[x]$ which contracts to $\mathfrak{i}$ in $K_v[x]$. If we denote by W and W_i the subsets in V (or rather in the part of V which is finite with respect to the given affine coordinate system) defined by the ideals $\mathfrak{i}$ and $\mathfrak{p}_i$ respectively (or by $K[x]\mathfrak{i}$ and $K[x]\mathfrak{p}_i$ respectively), then each W_i is an affine K-variety and we have the equation $W = \bigcup_i W_i$; it is clear that if $\mathfrak{i}$ is the contraction of an ideal in $K[x]$, then none of the sets W_i is empty.

Consider now the coordinate ring $\bar{K}[\bar{x}]$ of the variety $\bar{V}_\alpha$, and denote by E' the homomorphism of $K_v[x]$ onto $\bar{K}[\bar{x}]$ which is the restriction to $K_v[x]$ of E_R (E' is also an extension of the canonical homomorphism E of K_v onto $\bar{K}$); we observe that the kernel of E' is the ideal $K_v[x]\pi$. We denote by $\bar{\mathfrak{i}}$ and $\bar{\mathfrak{p}}_i$ the image ideals of $\mathfrak{i}$ and $\mathfrak{p}$ respectively under E'; we have evidently the relation $\bar{\mathfrak{i}} \subset \bar{\mathfrak{p}}_i$. On the other hand, if f_i are elements in $\mathfrak{p}_i$ such that their images $\bar{f}_i$ under E' all coincide with one element ϕ in $\bigcap_i \bar{\mathfrak{p}}_i$, then the element $\prod_i f_i$ is in the radical of $\mathfrak{i}$; since the image of the radical of $\mathfrak{i}$ under E' is contained in the radical of $\bar{\mathfrak{i}}$, this means that a power of the element ϕ is in the radical of $\bar{\mathfrak{i}}$, and hence ϕ is itself in the radical of $\bar{\mathfrak{i}}$; this shows that $\bigcap_i \bar{\mathfrak{p}}_i$ is contained in the radical of $\bar{\mathfrak{i}}$. If we denote by $\bar{W}_i$ the subsets in $\bar{V}_\alpha$ defined by the ideals $\bar{\mathfrak{p}}_i$, then it follows from what we have just said that the union $\bigcup_i \bar{W}_i$ is precisely the subset in $\bar{V}_\alpha$ defined by the ideal $\bar{\mathfrak{i}}$. Now, for each $\mathfrak{p}_i$ which does not contain any element in K_v (and hence W_i is not empty), the set $\bar{W}_i$ is by definition the specialization of the set W_i over E (or over the valuation v); hence the set $\bar{W} = \bigcup_i{}' \bar{W}_i$, where the symbol $\bigcup'$ indicates that the union is taken over all i such that $\mathfrak{p}_i$ does not contain any non-zero element in K_v, is the specialization of W over E.

Now, the point α in $\bar{V}_a$ is the only zero of the ideal $\bar{K}[\bar{x}] \cap \bar{\mathfrak{m}}$, and $\bar{R}$ is the quotient ring of $\bar{K}[\bar{x}]$ with respect to $\bar{K}[\bar{x}] - (\bar{K}[\bar{x}] \cap \bar{\mathfrak{m}})$; it follows that α is contained in $\bar{W}$ if and only if for at least one i the ideal $\bar{R}\bar{\mathfrak{p}}_i$ is not the unit ideal and the ideal $\mathfrak{p}_i$ does not contain any non-zero element in K_v. Since $\bar{R}\bar{\mathfrak{p}}_i$ is clearly the image of $R\mathfrak{p}_i$ under E_R, it follows that α is contained in $\bar{W}$ if and only if for at least one i the ideal $R\mathfrak{p}_i$ does not contain any non-zero element in K_v, and it is easily seen that this is so if and only if the ideal $R\mathfrak{i}$ does not contain any non-zero element in K_v.

We shall say that the point α is an *isolated* zero of the ideal $\bar{\mathfrak{i}}$ if the ideal $\bar{R}\bar{\mathfrak{i}}$ is primary for the ideal $\bar{\mathfrak{m}}$, and the isolated zero α of $\bar{\mathfrak{i}}$ is said to be a *simple* zero of $\bar{\mathfrak{i}}$ if $\bar{R}\bar{\mathfrak{i}} = \bar{\mathfrak{m}}$.

THEOREM 1. *If the point α is an isolated zero of $\bar{\mathfrak{i}}$ and if $R\mathfrak{i}$ has a rank at most n, then α is contained in $\bar{W}$; furthermore, if K_v is complete and if α is a simple zero of $\bar{\mathfrak{i}}$, then there is exactly one point a in W which specializes over E to α, and this point a is rational over K_v and is a simple zero of $\mathfrak{i}$.*

Proof. To prove that α is contained in $\bar{W}$, we have to show that the ideal $R\mathfrak{i}$ does not contain any non-zero element in K_v; we assume that $\bar{R}\bar{\mathfrak{i}}$ is primary for $\bar{\mathfrak{m}}$ and the $R\mathfrak{i}$ contains a non-zero element in K_v, and we shall show that $R\mathfrak{i}$ must then have the rank $n+1$. In fact, for every i, the prime ideal $R\mathfrak{p}_i \cap K_v$ in K_v is in this case not the zero ideal and hence must coincide with the ideal $K_v\pi$; it follows that every ideal $R\mathfrak{p}_i$ contains the kernel $R\pi$ of E_R and hence must coincide with the inverse image under E_R of the ideal $\bar{R}\bar{\mathfrak{p}}_i$. Since $\bar{R}\bar{\mathfrak{i}}$ is primary for $\bar{\mathfrak{m}}$, every ideal $\bar{R}\bar{\mathfrak{p}}_i$ is primary for $\bar{\mathfrak{m}}$; it follows that every ideal $R\mathfrak{p}_i$ is primary for $\mathfrak{m}$ and hence has the rank $n+1$, and this implies that $R\mathfrak{i}$ has also the rank $n+1$.

If $\bar{R}\bar{\mathfrak{i}} = \bar{\mathfrak{m}}$, then we have the relation $R(\mathfrak{i}, \pi) = \mathfrak{m}$. Consider now the residue ring $T = R/R\mathfrak{i}$, which is a local ring with the maximal prime ideal $\mathfrak{m}/R\mathfrak{i}$; since $R\mathfrak{i}$ does not contain any non-zero element in K_v, we can embed K_v canonically in T as a subring. It is clear that T and K_v have the same residue field K, and the relation $\mathfrak{m}/R\mathfrak{i} = R(\mathfrak{i}, \pi)/R\mathfrak{i}$ shows that the maximal prime ideal of the so embedded ring K_v generates in T the maximal prime ideal of T. If K_v is complete, then it follows from a well-known result that

T coincides with its subring K_v. Since $T = K_v$ is an integral domain, $R\mathfrak{i}$ is a prime ideal, and hence we have $c = 1$ and $R\mathfrak{i} = R\mathfrak{p}_1$; since the quotient ring of $K_v[x]$ with respect to the set $K_v[x] - \mathfrak{p}_1$ coincides with the quotient ring of R with respect to the set $R - R\mathfrak{p}_1 = R - R\mathfrak{i}$, we can identify T with the residue ring $K_v[x]/\mathfrak{p}_1$, so that $K_v[x]/\mathfrak{p}_1$ also coincides with K_v. There exists therefore a homomorphism H of $K_v[x]$ onto K_v which leaves every element in K_v invariant and whose kernel is the ideal $\mathfrak{p}_1$; if we set $a_i = H(x_i)$, $i = 1, \cdots, m$, then the point $a = (a_1, \cdots, a_m)$ in V is clearly the only zero of $\mathfrak{p}_1$, and since the point a has a unique image $\bar{a}$ under E and has the point α as a specialization over E, we must have the equality $\bar{a} = \alpha$. Thus the rational point a is the only point in W which can specialize over E to the point α, and it can specialize over E only to α. Finally, it is clear that $K[x]\mathfrak{p}_1$ is the defining prime ideal of the point a in $K[x]$, and that the quotient ring of $K[x]$ with respect to the set $K[x] - K[x]\mathfrak{p}_1$ coincides with the quotient ring of $K_v[x]$ with respect to the set $K_v[x] - \mathfrak{p}_1$, which in turn coincides with the quotient ring of R with respect to the set $R - R\mathfrak{p}_1$; it follows then from the relation $R\mathfrak{i} = R\mathfrak{p}_1$ that the point a is a simple zero of $\mathfrak{i}$. This concludes the proof of our theorem.

As a corollary of the above theorem we obtain a result which can be considered as a generalization of the well-known Hensel's Lemma:[6]

COROLLARY. *Let K be a field which is complete with respect to a real discerte valuation v, let V be a variety in S^m defined over K, and let $\bar{V}$ be its specialization over v; if α is a point in $\bar{V}$ such that α is rational over $\bar{K}$ and $\bar{V}$ is simple at α, then there exists a point a in V such that a is rational over K_v, V is simple at a, and $\bar{a} = \alpha$.*

In fact, if $y_1, \cdots, y_n$ are elements in R such that their images $\bar{y}_1, \cdots, \bar{y}_n$ under E_R form a basis for $\bar{\mathfrak{m}}$, then we can take for $\mathfrak{i}$ the ideal in $K_v[x]$ generated by the elements $y_1, \cdots, y_n$, which has a rank at most n; our corollary then follows immediately from Theorem 1.

3. Let $\boldsymbol{x} = (x^{(1)}, \cdots, x^{(d)})$ be a system of d points in W, which for the sake of convenience we shall assume to be distinct. A system of d (not necessarily distinct) points $\boldsymbol{\beta} = (\beta^{(1)}, \cdots, \beta^{(d)})$ in $\bar{V}_a$ is said to be a specialization of the system $\boldsymbol{x}$ over E if the mapping $x^{(i)} \to \beta^{(i)}$ defines an extension of E to a homomorphism E_0 of the ring $K_v[\boldsymbol{x}]$ onto the ring $\bar{K}[\boldsymbol{\beta}]$. The number of times the point α appears in the system $\boldsymbol{\beta}$ is then called the

[6] **Such a generalization of Hensel's Lemma was conjectured by A. Weil some years ago; a special case of it, namely, for the Jacobian variety of a curve, has been recently proved by A. P. Mattuck in an unpublished manuscript.**

multiplicity of α in the specialization $\boldsymbol{\beta}$ of $\boldsymbol{x}$ over E. It is clear that in order that this multiplicity can be positive, it is necessary that α is not only a zero of the ideal $\bar{\mathfrak{i}}$, but also contained in the set $\bar{W}$; as we have seen above, this condition is equivalent to the condition that the ideal $R\mathfrak{i}$ does not contain any element in K_v.

THEOREM 2. *If the point α is a simple isolated zero of $\bar{\mathfrak{i}}$, then the multiplicity of α in any specialization over E of any system of points in W is at most equal to* 1.

Proof. If K_v is complete, then this theorem follows immediately from Theorem 1; for then the point a is the only point in W which specializes over E to α, and hence α appears in $\boldsymbol{\beta}$ if and only if a appears in $\boldsymbol{x}$ (observe that, since the points in $\boldsymbol{x}$ are assumed to be distinct, a can only appear once in $\boldsymbol{x}$). In order to prove our theorem it is sufficient to show that the homomorphism E_0 can be extended to a homomorphism of $K_v{}^*[\boldsymbol{x}]$ onto $\bar{K}[\boldsymbol{\beta}]$, where $K_v{}^*$ denotes the completion of K_v. Since $\bar{K}[\boldsymbol{\beta}]$ is an integral domain, the kernel $\mathfrak{r}_0$ of E_0 is a prime ideal, and E_0 can be extended to a homomorphism E_1 of the quotient ring N_1 of $K_v[\boldsymbol{x}]$ with respect to the set $K_v[\boldsymbol{x}]-\mathfrak{r}_0$ onto the field $\bar{K}(\boldsymbol{\beta})$; since the contraction of $\mathfrak{r}_0$ in K_v is evidently the kernel $K_{v\pi}$ of E, it follows that $K_v=N_1\cap K$. Let N_1 be embedded in a complete local domain N_2 which dominates N_1 (i.e. N_2 contains N_1 and its maximal prime ideal contract in N_1 to the maximal prime ideal of N_1) and whose residue field coincides with the residue field $\bar{K}(\boldsymbol{\beta})$ of N_1; for example, we can take N_2 to be the residue ring of the completion of N_1 modulo a minimal prime divisor of its zero ideal. The canonical homomorphism E_2 of N_2 onto its residue field $\bar{K}(\boldsymbol{\beta})$ is then an extension of E_1. Since N_2 contains K_v and is complete, we can embed the completion $K_v{}^*$ of K_v in N_2, and E_2 evidently induces a homomorphism of $K_v{}^*[\boldsymbol{x}]$ onto $\bar{K}[\boldsymbol{\beta}]$ which is an extension of E_0.

We consider now the positive cycle $\mathfrak{X}$ introduced in Section 1, and assume that the variety V contains $\mathfrak{X}$ so that $\mathfrak{P}$ contains $\mathfrak{O}$ and hence also $\mathfrak{P}_\alpha$ contains $\mathfrak{O}_\alpha$. Let $\mathfrak{p}$ be the image ideal of $\mathfrak{P}_\alpha$ under the homomorphism G, and let $\bar{\mathfrak{p}}$ be the image of $\bar{\mathfrak{P}}_\alpha$ under the homomorphism $\bar{G}$; then both ideals $\mathfrak{p}$ and $\bar{\mathfrak{p}}$ will have ranks at most $n-r$, and $\bar{\mathfrak{p}}$ is the image of $\mathfrak{p}$ under the homomorphism E_R. Since $\bar{V}_\alpha$ contains $\bar{\mathfrak{X}}_\alpha$, the tangential space of $\bar{V}_\alpha$ at α contains the tangential space of $\bar{\mathfrak{X}}_\alpha$ at α; by a suitable choice of the affine coordinate system, we can assume that the equations $X_{n+1}=\cdots=X_m=0$ defines the tangential space of $\bar{V}_\alpha$ at α. Then, as in the proof of Lemma 1, the ideal $\bar{\mathfrak{O}}_\alpha$ has a basis $\omega_{n+1}(X),\cdots,\omega_m(X)$, in $\bar{K}[X]$, such that $X_1,\cdots,X_n,$ $\omega_{n+1}(X),\cdots,\omega_m(X)$ form a minimal basis for

$p(\bar{S}/\alpha)$; using the determinant criterion for the minimal basis, one sees readily that the basis $\phi_{r+1}(X), \cdots, \phi_m(X)$ of $\mathfrak{P}_\alpha$ can be so chosen that $\phi_i(X) = \omega_i(X)$ for $i = n+1, \cdots, m$. Going over to the images in $\bar{R}$, under the homomorphism $\bar{G}$ and observing that $\bar{K}$ remains invariant under $\bar{G}$, this means that the ideal $\bar{\mathfrak{p}}$ has a basis $\phi_{r+1}(\bar{x}), \cdots, \phi_n(\bar{x})$ which can be extended to a minimal basis $\bar{x}_1, \cdots, \bar{x}_r, \phi_{r+1}(\bar{x}), \cdots, \phi_n(\bar{x})$ for $\bar{\mathfrak{m}}$. Let $f_k(X)$, $k = r+1, \cdots, m$, be elements in $\mathfrak{P}_\alpha$ such that $\bar{f}_k(X) = \phi_k(X)$ for every k and that $f_{n+1}(X), \cdots, f_m(X)$ are in $\mathfrak{O}_\alpha$; then $f_{n+1}(x) = \cdots = f_m(x) = 0$ and the elements $f_{r+1}(x), \cdots, f_n(x)$ are in $\mathfrak{p}$, with $\phi_{r+1}(\bar{x}), \cdots, \phi_n(\bar{x})$ as their respective images under E_R. Since the elements π, $x_1, \cdots, x_r$, $f_{r+1}(x), \cdots, f_n(x)$ clearly generate the ideal $\mathfrak{m}$, they must form a minimal basis for $\mathfrak{m}$; this shows that the elements $f_{r+1}(x), \cdots, f_n(x)$ generate a prime ideal of rank $n-r$ in R, and since $\mathfrak{p}$ has a rank at most $n-r$, this prime ideal must coincide with $\mathfrak{p}$. We observe also that since the elements $f_{r+1}(X), \cdots, f_m(X)$ can be extended to a minimal basis for $p(S/\alpha)$, by adding the elements π, $X_1, \cdots, X_r$, they generate a prime ideal of rank $m-r$ and hence must form a basis for $\mathfrak{P}_\alpha$.

Now, let $\mathfrak{Y}^s$ $(s = n-r)$ be another positive cycle contained in the variety V, and assume that $\mathfrak{Y}$ is rational over K and that the specialization $\bar{\mathfrak{Y}}$ of $\mathfrak{Y}$ over v is also simple at α; we denote by $\mathfrak{O}$, $\mathfrak{O}_v$, $\mathfrak{O}_\alpha$, and $\mathfrak{q}$ the ideals having the same meanings for the cycle $\mathfrak{Y}$ as the ideals $\mathfrak{P}$, $\mathfrak{P}_v$, $\mathfrak{P}_\alpha$, and $\mathfrak{p}$ respectively for the cycle $\mathfrak{X}$. Then there exists a basis $\theta_{s+1}(X), \cdots, \theta_m(X)$ in $\bar{K}[X]$ for $\bar{\mathfrak{O}}_\alpha$ such that $\theta_i(X) = \omega_i(X)$ for $i = n+1, \cdots, m$, and such that $\theta_{s+1}(\bar{x}), \cdots, \theta_n(\bar{x})$ form a basis for $\bar{\mathfrak{q}}$ which can be extended to a minimal basis for $\bar{\mathfrak{m}}$. The condition that $\bar{\mathfrak{X}}$ and $\bar{\mathfrak{Y}}$ are transversal to each other at α in $\bar{V}$ is equivalent to the condition that the elements $\phi_{r+1}(\bar{x}), \cdots, \phi_n(\bar{x})$, $\theta_{s+1}(\bar{x}), \cdots, \theta_n(\bar{x})$ form a minimal basis for $\bar{\mathfrak{m}}$. If this is the case, and if we take elements $g_{s+1}(X), \cdots, g_n(X)$ in $\mathfrak{O}_\alpha$ such that $\bar{g}_j(X) = \theta_j(X)$ for every j, then $g_{s+1}(x), \cdots, g_n(x)$ form a basis for $\mathfrak{q}$, and the elements π, $f_{r+1}(x), \cdots, f_n(x)$, $g_{s+1}(x), \cdots, g_n(x)$ form a minimal basis for $\mathfrak{m}$; this show that the ideal $(\mathfrak{p}, \mathfrak{q})$ has the rank n (and is in fact a prime ideal). If we now apply Theorems 1 and 2 by taking for $\mathfrak{i}$ the restriction of the ideal $(\mathfrak{p}, \mathfrak{q})$ in $K_v[x]$, we obtain then the following theorem, which is the well-known criterion for unit multiplicity in a somewhat more precise version.

THEOREM 3. *Let V^n be a variety in S^m, defined over K, and let $\mathfrak{X}^r$ and $\mathfrak{Y}^s$ $(r+s=n)$ be two positive cycles in V, rational over K; let v be a real discrete valuation of K, and let $\bar{V}$, $\bar{\mathfrak{X}}$, and $\bar{\mathfrak{Y}}$ be the specializations of V, $\mathfrak{X}$, and $\mathfrak{Y}$ respectively over v. If α is a rational point over $\bar{K}$ in $\bar{V}$ such that $\bar{V}$ is simple at α and that $\bar{\mathfrak{X}}$ and $\bar{\mathfrak{Y}}$ are transversal to each other at α,* *then α has the multiplicity 1 in the specialization over v of every system of points in $|\mathfrak{X}| \cap |\mathfrak{Y}|$ which includes all the isolated points in $|\mathfrak{X}| \cap |\mathfrak{Y}|$. If furthermore K is complete with respect to v, then there is exactly one point a in $|\mathfrak{X}| \cap |\mathfrak{Y}|$ which specializes to α over v; and this point a is rational over K_v (hence $\bar{a} = \alpha$). V is simple at a, and $\mathfrak{X}$ and $\mathfrak{Y}$ are transversal to each other at a.*

4. In the above treatment of the criterion for unit multiplicity, we have kept our arguments on as elementary a level as possible, in consonant with the foundational nature of our purpose. In fact, the greater part of our argument consists of a precise local formulation of the various geometric concepts involved in this criterion; the actual proof of the criterion itself is very short and is contained essentially in the proof of Theorem 1. If we now take a more sophisticated point of view and assume a knowledge of the intersection-theory, then the criterion for unit multiplicity, in the present version as an assertion on the specialization-multiplicity, is really only a special case of a more general theorem on the invariance of intersection-multiplicity under specialization, and there naturally arises the question whether, corresponding to Theorem 3, we can also prove a more precise version of this invariance theorem. In the following we shall show that this can be done, by using the Chevalley-Samuel-Nagata theory of multiplicity in a local ring. We shall need some of the results in the paper of Nagata cited before.[5]

As in Section 1, we consider the positive r-cycle $\mathfrak{X}$ in S^m, except that we shall now no longer assume that $\mathfrak{X}$ is simple at the point α; for the sake of simplicity, we shall assume that $\mathfrak{X}$ is a prime rational cycle over K_v (so that $\mathfrak{P}_v$ and hence $\mathfrak{P}_\alpha$ are prime ideals). Let Ω_0 be a prime rational r-cycle over $\bar{K}$ in $\bar{S}^m$ which contains the point α and whose support is contained in the support of $\bar{\mathfrak{X}}$, and let $\Phi_0(U)$ be the associated form of Ω_0; we set $\bar{\mathfrak{X}} = \Omega_1 + \Omega_2$, where Ω_1 is a positive r-cycle with the same support as Ω_0 and Ω_2 is a positive cycle relatively prime to Ω_0, and let $\bar{F}(U) = \Phi_1(U)\Phi_2(U)$ be the corresponding factorization of $\bar{F}(U)$. Let d_0 and d_1 be the degrees of $\Phi_0(U)$ and $\Phi_1(U)$ respectively; the number d_1/d, which will presently be shown to be an integer, is the coefficient of Ω_0 as a component in $\bar{\mathfrak{X}}$ and will be denoted by $\mu(\Omega_0, \bar{\mathfrak{X}})$. Let $\bar{\mathfrak{A}}$ be the prime ideal in $\mathfrak{Q}(\bar{S}/\alpha)$ associated with Ω_0, and let $\mathfrak{A} = E^{-1}(\bar{\mathfrak{A}})$ be the corresponding prime ideal in $\mathfrak{Q}(S/\alpha)$; then $\mathfrak{A}$ is a minimal prime divisor of the ideal $(\pi, \mathfrak{P}_\alpha)$ in $\mathfrak{Q}(S/\alpha)$, and in this way all the minimal prime divisors of $(\pi, \mathfrak{P}_\alpha)$ can be obtained.

LEMMA 2. $\mu(\Omega_0, \bar{\mathfrak{X}}) = e((\pi, \mathfrak{P}_\alpha)\mathfrak{Q}(S/\alpha)_{\mathfrak{A}}/\mathfrak{P}_\alpha\mathfrak{Q}(S/\alpha)_{\mathfrak{A}})$.

Proof. Let ${}_iw_j$, $i=1,\cdots,r$, and $j=0,1,\cdots,m$, be $r(m+1)$ independent variables over K_v; we set ${}_iw=({}_iw_0,{}_iw_1,\cdots,{}_iw_m)$ and denote by w the set of all $r(m+1)$ variables ${}_iw_j$. We set $F({}_0U,w)=F({}_0U,{}_1w,\cdots,{}_rw)$, and similarly for other associated forms; we denote by $K_{v,w}$ the quotient ring of $K_v[w]$ with respect to the prime ideal generated in it by π, so that $K_{v,w}$ is a real discrete valuation ring; and we observe that if we denote by $\bar{w}$ the image of w in the residue field of $K_{v,w}$, then $\bar{w}$ is a set of independent variable over $\bar{K}$ and $\bar{K}(\bar{w})$ is this residue field. We set $L_i={}_iw_0+\sum_{j=1}^{m}{}_iw_jX_j$ and denote by $\mathfrak{Q}(S/\alpha,K_{v,w})$ the local ring defined in the same way as $\mathfrak{Q}(S/\alpha)$ but with $K_{v,w}$ replacing K_v; we set

$$M=\mathfrak{Q}(S/\alpha)\mathfrak{P}_\alpha,\quad \mathfrak{g}=\mathfrak{A}/\mathfrak{P}_\alpha\quad N=\mathfrak{Q}(S/\alpha,K_{v,w})/(\mathfrak{P}_\alpha,L_1,\cdots,L_r),$$

and

$$\mathfrak{h}=\mathfrak{A}\mathfrak{Q}(S/\alpha,K_{v,w})/(\mathfrak{P}_\alpha,L_1,\cdots,L_r).$$

Then it is easily seen that $N_\mathfrak{h}$ can be obtained from $M_\mathfrak{g}$ by first adjoining the rm variables ${}_iw_j$, $j\neq 0$, and then forming the quotient ring of the so obtained ring with respect to the ideal generated in it by $\mathfrak{g}$. It follows that $e(\pi N_\mathfrak{h})=e(\pi M_\mathfrak{g})=e((\pi,\mathfrak{P}_\alpha)\mathfrak{Q}(S/\alpha)_\mathfrak{A}/\mathfrak{P}_\alpha\mathfrak{Q}(S/\alpha)_\mathfrak{A})$. Since $N_\mathfrak{h}{}^*$ is a finite module over $K_{v,w}{}^*$, it follows by a well-known argument[7] using the decomposition of the ideal $\mathfrak{h}N_\mathfrak{h}{}^*$ that

$$e(\pi N_\mathfrak{h})=e(\pi N_\mathfrak{h}{}^*)=l(N_\mathfrak{h}{}^*/\pi N_\mathfrak{h}{}^*)=[N_\mathfrak{h}{}^*:K_{v,w}{}^*]/[N_\mathfrak{h}/\mathfrak{h}:\bar{K}(\bar{w})].$$

It is easily verified that $[N_\mathfrak{h}/\mathfrak{h}:\bar{K}(\bar{w})]$ is equal to the degree of the form $\Phi_0({}_0U,\bar{w})$ and hence is equal to d_0. In order to complete the proof of our lemma, it remains to show that $[N_\mathfrak{h}{}^*:K_{v,w}{}^*]=d_1$. This can be done by generalizing the usual argument in the ramification theory, first observing that[8] there exists a factorization $F({}_0U,w)=F_1({}_0U)F_2({}_0U)$ in $K_{v,w}{}^*[{}_0U]$ corresponding to the factorization $\bar{F}({}_0U,\bar{w})=\Phi_1({}_0U,\bar{w})\Phi_2({}_0U,\bar{w})$, and then showing that the number $[N_\mathfrak{h}{}^*:K_{v,w}{}^*]$ is equal to the degree of the form $F_1({}_0U)$, which is the same as the degree d_1 of $\Phi_1({}_0U,\bar{w})$.

Let A_i be the components in the positive cycle $\bar{\mathfrak{X}}$ which contain the point α, let $\bar{\mathfrak{A}}_i$ be the prime ideals in $\mathfrak{Q}(\bar{S}/\alpha)$ associated with A_i, and let $\mathfrak{A}_i=E^{-1}(\bar{\mathfrak{A}}_i)$ be the corresponding minimal prime divisor of the ideal $(\pi,\mathfrak{P}_\alpha)$ in $\mathfrak{Q}(S/\alpha)$. If we denote by $\mathfrak{a}_i$ the image ideals of $\mathfrak{A}_i$ in R under the homomorphism G, then it is easily seen that the ideals $\mathfrak{a}_i$ are the

---

[7] See e. g. C. Chevalley, "On the Theory of Local Rings," *Annals of Mathematics*, vol. 44 (1943), pp. 690-708, § IV, Lemma 2.

[8] See Lemma 1 in our recent paper "On the Principle of Degeneration in Algebraic Geometry," *Annals of Mathematics*, vol. 66 (1957), pp. 70-79.

minimal prime divisors of the ideal $(\pi, \mathfrak{p})R$, and we have the equality $e((\pi, \mathfrak{p})R_{\mathfrak{a}_i}/\mathfrak{p}R_{\mathfrak{a}_i}) = e((\pi, \mathfrak{P}_{\mathfrak{a}})\mathfrak{Q}(S/\alpha)_{\mathfrak{A}_i}/\mathfrak{P}_{\mathfrak{a}}\mathfrak{Q}(S/\alpha)_{\mathfrak{A}_i})$. It follows then from Lemma 2 that $\mu(A_i, \bar{\mathfrak{X}}) = e((\pi, \mathfrak{p})R_{\mathfrak{a}_i}/\mathfrak{p}R_{\mathfrak{a}_i})$. For each i, let $\mathfrak{a}_{ij}$ be the minimal prime divisors of $\mathfrak{a}_i R^*$; since $\mathfrak{a}_i$ is analytically unramified (for $R/\mathfrak{a}_i$ is a "geometrical" local ring), we have the equality

$$e((\pi, \mathfrak{p})R^*_{\mathfrak{a}_{ij}}/\mathfrak{p}R^*_{\mathfrak{a}_{ij}}) = e((\pi, \mathfrak{p})R_{\mathfrak{a}_i}/\mathfrak{p}R_{\mathfrak{a}_i}).$$

We consider now, as in Section 3, two positive cycles $\mathfrak{X}^r$ and $\mathfrak{Y}^s$ in a variety V^n $(n = r + s)$, and we shall use the same notation as developed there; while $\bar{V}$ is still assumed to be a variety which is simple at the point α, we shall assume no longer that $\bar{\mathfrak{X}}$ or $\bar{\mathfrak{Y}}$ is simple at α, nor that they are transversal to each other. Instead, we shall only assume that $\bar{\mathfrak{X}}$ and $\bar{\mathfrak{Y}}$ intersect properly at α, which is equivalent to assuming that the ideal $(\bar{\mathfrak{p}}, \bar{\mathfrak{q}})$ is primary for $\mathfrak{m}$; this latter assertion follows from the fact that the radical of the ideal $\bar{\mathfrak{p}}$ (or $\bar{\mathfrak{q}}$) is the ideal associated with the support of $\bar{\mathfrak{X}}$ (or $\bar{\mathfrak{Y}}$), and this fact can be easily proved by a method similar to that used in the proof of Lemma 1. Since $(\bar{\mathfrak{p}}, \bar{\mathfrak{q}})$ has the rank n, the ideal $(\mathfrak{p}, \mathfrak{q})$ must have at least the same rank n, and since R is an unramified regular local ring, the rank of $(\mathfrak{p}, \mathfrak{q})$ must be equal to n. Let $\mathfrak{e}_i$ be the minimal prime divisors of the ideal $(\mathfrak{p}, \mathfrak{q})R^*$. Consider the residue ring $T_i = R^*/\mathfrak{e}_i$; it is clear that $K_v{}^*$ can be embedded canonically in T_i as a subring. In fact, T_i is a finite module over $K_v{}^*$, and we have the equality $[T_i : K_v{}^*] = e((\pi, \mathfrak{e}_i)R^*/\mathfrak{e}_i)$. If we denote by H_i the canonical homomorphism of R^* onto $R^*/\mathfrak{e}_i$ and set $a^{(i)}{}_j = H_i(x_j)$, then it can be shown exactly as in Section 3 that the point $a^{(i)} = (a^{(i)}{}_1, \cdots, a^{(i)}{}_m)$ is an intersection point of $\mathfrak{X}$ and $\mathfrak{Y}$; and if we denote by $\mathfrak{Z}_i$ the 0-cycle consisting of the complete set of conjugates of the point $a^{(i)}$ over $K_v{}^*$, then $\mathfrak{Z}_i$ is evidently rational over $K_v{}^*$ and its specialization $\bar{\mathfrak{Z}}_i$ is the cycle consisting of the point α with the coefficient $e((\pi, \mathfrak{e}_i)R^*/\mathfrak{e}_i)$. We set $\mathfrak{Z} = \sum_i i(\mathfrak{Z}_i; \mathfrak{X} \cdot \mathfrak{Y}) \cdot \mathfrak{Z}_i$; then a point a in $|\mathfrak{X}| \cap |\mathfrak{Y}|$ will specialize to α over the specialization $(\mathfrak{X}, \mathfrak{Y}) \to (\bar{\mathfrak{X}}, \bar{\mathfrak{Y}})$ if and only if the coefficient $\mu(a, \mathfrak{Z})$ of a in $\mathfrak{Z}$ is positive. We shall now show that $\deg \mathfrak{Z} = i(\alpha; \bar{\mathfrak{X}} \cdot \bar{\mathfrak{Y}})$.

Let R' be another copy of the local ring R, but with the same basic ring K_v, and let a prime be used to denote generally the corresponding entities in R'; we consider the complete tensor product J of R and R' (or of R^* and R'^*) over $K_v{}^*$ and denote by $\mathfrak{d}$ the ideal in J generated by the n elements $x_1 - x_1{}', \cdots, x_n - x_n{}'$. It is well-known that the prime ideals $(\mathfrak{d}, \mathfrak{e}_i)J$ are precisely the minimal prime divisors of the ideal $(\mathfrak{d}, \mathfrak{p}, \mathfrak{q}')J$, and hence the ideals $(\mathfrak{d}, \mathfrak{e}_i)J/(\mathfrak{p}, \mathfrak{q}')J$ are the minimal prime divisors of the ideal $(\mathfrak{d}, \mathfrak{p}, \mathfrak{q}')J/(\mathfrak{p}, \mathfrak{q}')J$ in the local ring $J/(\mathfrak{p}, \mathfrak{q}')J$. Applying the Associativity Formula, we obtain the equation

$$
\begin{aligned}
&e((\mathfrak{d},\pi,\mathfrak{p},\mathfrak{q}')J/(\mathfrak{p},\mathfrak{q}')J)\\
&\quad=\sum_i e((\mathfrak{d},\mathfrak{p},\mathfrak{q}')J_{(\mathfrak{d},\mathfrak{e}_i)}/(\mathfrak{p},\mathfrak{q}')J_{(\mathfrak{d},\mathfrak{e}_i)})\cdot e((\mathfrak{d},\pi,\mathfrak{e}_i)J/(\mathfrak{d},\mathfrak{e}_i)J)\\
(1)\quad&\quad=\sum_i e((\mathfrak{d},\mathfrak{p},\mathfrak{q}')J_{(\mathfrak{d},\mathfrak{e}_i)}/(\mathfrak{p},\mathfrak{q}')J_{(\mathfrak{d},\mathfrak{e}_i)})\cdot e((\pi,\mathfrak{e}_i)R^*/\mathfrak{e}_iR^*)\\
&\quad=\sum_i i(\mathfrak{Z}_i;\mathfrak{X}\cdot\mathfrak{Y})\cdot\deg\mathfrak{Z}_i=\deg\mathfrak{Z}.
\end{aligned}
$$

Let B_k be the components in $\bar{Y}$ which contain the point α, and let $\mathfrak{b}_k$ and $\mathfrak{b}_{kl}$ have the same meanings with respect to B_k as $\mathfrak{a}_i$ and $\mathfrak{a}_{ij}$ have with respect to A_i; then it can be easily seen that the ideals $\mathfrak{c}_{ij,kl}=(\mathfrak{a}_{ij},\mathfrak{b}_{kl}')J$ are the minimal prime divisors of the ideal $(\pi,\mathfrak{p},\mathfrak{q}')J$, and we have the relation

$$
\begin{aligned}
e((\pi,\mathfrak{p},\mathfrak{q}')J_{\mathfrak{c}_{ij,kl}}/(\mathfrak{p},\mathfrak{q}')J_{\mathfrak{c}_{ij,kl}})&=e((\pi,\mathfrak{p})R^*_{\mathfrak{a}_{ij}}/\mathfrak{p}R^*_{\mathfrak{a}_{ij}})\cdot e((\pi,\mathfrak{q})R^*_{\mathfrak{b}_{kl}}/\mathfrak{q}R^*_{\mathfrak{b}_{kl}})\\
&=e((\pi,\mathfrak{p})R_{\mathfrak{a}_i}/\mathfrak{p}R_{\mathfrak{a}_i})\cdot e((\pi,\mathfrak{q})R_{\mathfrak{b}_k}/\mathfrak{q}R_{\mathfrak{b}_k}).
\end{aligned}
$$

Applying again the Associativity Formula, with the ideal $\mathfrak{d}$ and πJ interchanged, we obtain the equation

$$
\begin{aligned}
&e((\mathfrak{d},\pi,\mathfrak{p},\mathfrak{q}')J/(\mathfrak{p},\mathfrak{q}')J)\\
&\quad=\sum_{i,j,k,l} e((\pi,\mathfrak{p},\mathfrak{q}')J_{\mathfrak{c}_{ij,kl}}/(\mathfrak{p},\mathfrak{q}')J_{\mathfrak{c}_{ij,kl}})\cdot e((\mathfrak{d},\mathfrak{c}_{ij,kl})J/\mathfrak{c}_{ij,kl})\\
(2)\quad&\quad=\sum_{i,k} e((\pi,\mathfrak{p})R_{\mathfrak{a}_i}/\mathfrak{p}R_{\mathfrak{a}_i})\cdot e((\pi,\mathfrak{q})R_{\mathfrak{b}_k}/\mathfrak{q}R_{\mathfrak{b}_k})\cdot\Big(\sum_{j,l} e((\mathfrak{d},\mathfrak{c}_{ij,kl})J/\mathfrak{c}_{ij,kl})\Big)\\
&\quad=\sum_{i,k}\mu(\mathrm{A}_i,\bar{\mathfrak{X}})\cdot\mu(\mathrm{B}_k,\bar{\mathfrak{Y}})\cdot i(\alpha;\mathrm{A}_i\cdot\mathrm{B}_k)=i(\alpha;\bar{\mathfrak{X}}\cdot\bar{\mathfrak{Y}}).
\end{aligned}
$$

Combining the equations (1) and (2), we obtain the equation $\deg\mathfrak{Z}=i(\alpha;\bar{\mathfrak{X}}\cdot\bar{\mathfrak{Y}})$, which proves our assertion.

We can summarize our results in the following theorem:

THEOREM 4. *Let V^n be a variety in S^m, defined over a field K which is complete with respect to a real discrete valuation v, and let $\mathfrak{X}^r$ and $\mathfrak{Y}^s$ $(r+s=n)$ be two positive cycles in V, rational over K; let $\bar{V}$, $\bar{\mathfrak{X}}$, and $\bar{\mathfrak{Y}}$ be the specializations of V, $\mathfrak{X}$, and $\mathfrak{Y}$ respectively over v. If α is a rational point over $\bar{K}$ in $\bar{V}$ such that $\bar{V}$ is simple at α and that $\bar{\mathfrak{X}}$ and $\bar{\mathfrak{Y}}$ intersect properly at α, then there exists exactly one positive 0-cycle $\mathfrak{Z}$ in V, with support in $|\mathfrak{X}|\cap|\mathfrak{Y}|$, such that $\mathfrak{Z}$ is rational over K and its specialization $\bar{\mathfrak{Z}}$ over v consists of the point α with the coefficient $i(\alpha;\bar{\mathfrak{X}}\cdot\bar{\mathfrak{Y}})$; furthermore, a point a in $|\mathfrak{X}|\cap|\mathfrak{Y}|$ specializes to α over the specialization $(\mathfrak{X},\mathfrak{Y})\rightarrow(\bar{\mathfrak{X}},\bar{\mathfrak{Y}})$ if and only if $\mu(\alpha,\mathfrak{Z})>0$, and when such is the case, we have $i(a;\mathfrak{X}\cdot\mathfrak{Y})=\mu(a,\mathfrak{Z})$.*

THE JOHNS HOPKINS UNIVERSITY.

ON THE THEOREM OF BERTINI FOR LOCAL DOMAINS*

BY WEI-LIANG CHOW

DEPARTMENT OF MATHEMATICS, THE JOHNS HOPKINS UNIVERSITY

Communicated by N. E. Steenrod, August 26, 1957

1. The well-known Theorem of Bertini on reducible linear systems of divisors in an algebraic variety asserts that any such system, assumed to be free from fixed components, is composite with a pencil. A somewhat special form of this theorem, which is actually a special case of it, states that if a linear system of divisors is obtained from a rational transformation of the variety onto a projective

space of dimension >1, then a generic element of the system, apart from a possible fixed component, is absolutely irreducible. In a recent discussion on local algebraic geometry, Professor S. Abhyankar raised the question whether this Theorem of Bertini, at least in the special form, also holds for local varieties, with suitable modifications. In this note we shall answer this question in the affirmative by formulating and proving such a theorem for an arbitrary complete local domain. The theorem we shall prove here can be stated as follows:

THEOREM OF BERTINI (for local domains). *Let* $(\mathfrak{R}, \mathfrak{p})$ *be a complete local domain of dimension* $t > 2$, *let* $(\mathfrak{S}, \mathfrak{q})$ *be a complete regular local domain of dimension t contained in* $\mathfrak{R}$ *such that* $\mathfrak{R}$ *is a finite module over* $\mathfrak{S}$ *and is separable over* $\mathfrak{S}$, *and let* x_1, x_2, x_3 *be three elements in a minimal basis for* $\mathfrak{q}$; *we set* $u(c) = (x_1 + cx_2)/x_3$, *where c is an element in* $\mathfrak{S}$. *Then, for sufficiently general c, the ideal* $\mathfrak{R}[u(c)]\mathfrak{p}$ *is prime in* $\mathfrak{R}[u(c)]$ *and the quotient ring* $\mathfrak{R}_{u(c)}$ *of* $\mathfrak{R}[u(c)]$ *with respect to* $\mathfrak{R}[u(c)]\mathfrak{p}$ *is analytically irreducible.*

We add in explanation that the expression "sufficiently general c" in the above theorem means that the $\mathfrak{q}$-residue of c is different from some one given finite set of elements in the residue field $\bar{\mathfrak{S}}$ of $\mathfrak{S}$.

The above local formulation is somewhat different in spirit from the original version of the Theorem of Bertini, in that it belongs more properly to the relative rather than to the absolute algebraic geometry. This is in a sense natural when one deals with local domains, and we have accordingly replaced the extension to algebraic closure by the extension to the completion of a local domain; furthermore, since the restriction to a local variety diminishes the degree of freedom by 1, the dimensionality condition in the local theorem has naturally to be increased by 1. Nevertheless, our proof of the local theorem will be at least formally similar to the proof given by Zariski[1] of the Theorem of Bertini in the global geometric case. Our proof is based on the following lemma, which may have some interest in itself; the discerning reader will observe that our proof of this lemma, given in the next section, is at least formally similar to the proof of Lemma 5 in the paper of Zariski quoted above, which plays a key role in his proof of the Theorem of Bertini.

LEMMA. *Let* $(\mathfrak{S}, \mathfrak{q})$ *be a complete regular local domain of dimension* $t > 2$, *and let* $x_1, \ldots, x_t$ *be the elements in a minimal basis for* $\mathfrak{q}$; *we set* $u(c) = (x_1 + cx_2)/x_3$, *where c is an element in* $\mathfrak{S}$, *and denote by* $\mathfrak{S}_{u(c)}$ *the quotient ring of* $\mathfrak{S}[u(c)]$ *with respect to the prime ideal* $\mathfrak{S}[u(c)]\mathfrak{q}$. *If a monic polynomial* $F(X)$ *in* $\mathfrak{S}[X]$ *(where X is an indeterminate) is irreducible over* $\mathfrak{S}$, *then it is also irreducible over the completion* $\mathfrak{S}_{u(c)}^*$ *of* $\mathfrak{S}_{u(c)}$ *for sufficiently general c.*

2. Before proceeding to the proof of our lemma, we shall make a few remarks on the "power-series" development of an element in the complete regular local domain $(\mathfrak{S}, \mathfrak{q})$. As before, we denote by $\bar{\mathfrak{S}}$ the residue field of $\mathfrak{S}$, and denote by Q a system of representatives of $\bar{\mathfrak{S}}$ in $\mathfrak{S}$; apart from the condition that every element in $\bar{\mathfrak{S}}$ is the $\mathfrak{q}$-residue of exactly one element in Q, the elements in Q can be chosen arbitrarily, but, for convenience, we shall assume that Q contains the elements 0 and 1. Using the properties of a regular local domain, it can easily be shown that every element in $\mathfrak{S}$ can be represented uniquely by a power series in $x_1, \ldots, x_t$ with coefficients in Q; we shall express this by setting $\mathfrak{S} = Q[[x_1, \ldots, x_t]]$. For a fixed $1 \leq j < t$, we set $v_1 = x_1/x_{j+1}, \ldots, v_j = x_j/x_{j+1}$, and denote the set $(v_1,$

$\ldots, v_j)$ by v; then $\mathfrak{S}[v]\mathfrak{q}$ is a prime ideal in $\mathfrak{S}[v]$, and the quotient ring $\mathfrak{S}_v$ of $\mathfrak{S}[v]$ with respect to $\mathfrak{S}[v]\mathfrak{q}$ is a regular local domain of dimension $t - j$, with the elements $x_{j+1}, \ldots, x_t$ as a minimal basis for its maximal prime ideal. Furthermore, the $\mathfrak{S}_v\mathfrak{q}$-residues $\bar{v}_1, \ldots, \bar{v}_j$ of $v_1, \ldots, v_j$, respectively, are independent variables over $\bar{\mathfrak{S}}$, so that the residue field of $\mathfrak{S}_v$ is a purely transcendental extension $\bar{\mathfrak{S}}(\bar{v})$ of $\bar{\mathfrak{S}}$. We denote by $Q[v]$ the set of all polynomials in v with coefficients in Q, and denote by $Q(v)$ the set of all quotients $f(v)/g(v)$ of two elements in $Q[v]$ such that the $\mathfrak{S}_v\mathfrak{q}$-residues of $f(v)$ and $g(v)$ are relatively prime (as polynomials in $\bar{v}$ with coefficients in $\bar{\mathfrak{S}}$) and that the coefficient of the highest term in $g(v)$ lexicographically (i.e., the term has the highest degree in v_1, then among these has the highest degree in v_2, etc.) is 1; then $Q(v)$ is a system of representatives in $\mathfrak{S}_v$ of its residue field $\bar{\mathfrak{S}}(\bar{v})$, and every element in the completion $\mathfrak{S}_v^*$ can be represented uniquely by a power series in $x_{j+1}, \ldots, x_t$ with coefficients in $Q(v)$, so that we can set $\mathfrak{S}_v^* = Q(v)[[x_{j+1}, \ldots, x_t]]$. Finally, given any power series in $x_{j+1}, \ldots, x_t$ with coefficients in $\mathfrak{S}[v]$, we shall define the *v-excess* (*v_i-excess, $1 \leq i < j$*) of the coefficient of any term in it to be the difference (positive or negative) between the total degree of v (the degree of v_i) in this coefficient and the degree of x_{j+1} in the term. It is easily seen that an element in $\mathfrak{S}_v^*$ is in $\mathfrak{S}$ if and only if in its power-series development the coefficient of every term is an element in $Q[v]$ of non-positive v-excess.

Proof of the Lemma: We assume that there exists an infinity of elements c in $\mathfrak{S}$, with distinct $\mathfrak{q}$-residues, such that $F(X)$ is reducible in $\mathfrak{S}_{u(c)}^*$, and we shall show that this will lead us to a contradiction. Since there is only a finite number of ways in which the polynomial $F(X)$ can split into two non-trivial monic factors, there must exist an infinite sequence of elements c_i, $i = 1, 2, \ldots$, in $\mathfrak{S}$, with distinct $\mathfrak{q}$-residues, such that we have the same factorization $F(X) = F_1(X)F_2(X)$ in each ring $\mathfrak{S}_{u(c_i)}$; we set for convenience $u_i = u(c_i)$, $i = 1, 2, \ldots$. Since the elements c_1 and c_2 have distinct $\mathfrak{p}$-residues, the elements u_1x_3, u_2x_3, $x_3, \ldots, x_t$ also form a minimal basis for $\mathfrak{p}$; in the notations introduced above, we then have the relations $\mathfrak{S}_{u_1}^* = Q(u_1)[[u_2x_3, x_3, \ldots, x_t]]$, $\mathfrak{S}_{u_2}^* = Q(u_2)[[u_1x_3, x_3, \ldots, x_t]]$, and $\mathfrak{S}_{u_1, u_2}^* = Q(u_1, u_2)[[x_3, \ldots, x_t]]$, and we observe that both $\mathfrak{S}_{u_1}^*$ and $\mathfrak{S}_{u_2}^*$ are here already canonically imbedded in $\mathfrak{S}_{u_1, u_2}^*$ as subrings. Let a be any coefficient in $F_1(X)$ or $F_2(X)$, and let $a = \sum a_{n_3 \ldots n_t} x_3^{n_3} \ldots x_t^{n_t}$ be the power-series development of a with coefficients in $Q(u_1, u_2)$. Since a is contained in $\mathfrak{S}_{u_1}^*$, it can be represented by a power series in $u_2x_3, x_3, \ldots, x_t$ with coefficients in $Q(u_1)$; it then follows from the uniqueness of the power-series representation of a that each $a_{n_3 \ldots n_t}$ is a polynomial in u_2 of degree at most n_3. Similarly, interchanging u_1 and u_2, we can show that each $a_{n_3 \ldots n_t}$ is a polynomial in u_1 of degree at most n_3. Thus we have shown that each coefficient $a_{n_3 \ldots n_t}$ is an element in $Q[u_1, u_2]$ of non-positive u_1-excess and non-positive u_2-excess. In order to prove our lemma, it is sufficient to show that the (u_1, u_2)-excess of each $a_{n_3 \ldots n_t}$ is non-positive. For, by a remark made above, the element a would then be contained in $\mathfrak{S}$, and since a is any coefficient in $F_1(X)$ or $F_2(X)$, both $F_1(X)$ and $F_2(X)$ would then be elements in $\mathfrak{S}[X]$; this would be in contradiction to our assumption that $F(X)$ is irreducible in $S[X]$.

We have the relation $u_2 = du_1 + eu_3$, where $d = (c_3 - c_2)/(c_3 - c_1)$ and $e = (c_2 - c_1)/(c_3 - c_1)$ are units in $\mathfrak{S}$; if we substitute for u_2 the expression $du_1 + eu_3$ in $a_{n_3 \ldots n_t}$, we obtain a polynomial $a'_{n_3 \ldots n_t}$ in u_1 and u_3 with coefficients in S, whose degree in u_1 is equal to the total degree of $a_{n_3 \ldots n_t}$ in u_1 and u_2. The expression $a = \sum a_{n_3} \ldots {}_{n_t}' x_3^{n_3} \ldots x_t^{n_t}$ is then a development of a as a power series in $x_3, \ldots, x_t$ with coefficients in $\mathfrak{S}[u_1, u_3]$, and if $x_3^{\mu_3} \ldots x_t^{\mu_t}$ is a term of the lowest degree such that the coefficient $a_{\mu_3} \ldots {}_{\mu_t}$ has a positive (u_1, u_2)-excess, then it is also a term of the lowest degree such that the coefficient $a_{\mu_3} \ldots {}_{\mu_t}'$ has a positive u_1-excess. We observe now that if $f(u_1, u_3)$ is any element in $\mathfrak{S}[u_1, u_3]$ and if we represent each coefficient in $f(u_1, u_3)$ by a power series in $u_1 x_3, u_3 x_3, x_3, \ldots, x_t$ with coefficients in Q, then we obtain a development of $f(u_1, u_3)$ as a power series in $x_3, \ldots, x_t$ with coefficients in $\mathfrak{S}[u_1, u_3]$, but the coefficient of the term of degree zero is an element in $Q[u_1, u_3]$; and it is easily seen that the u_1-excess of the coefficient of each term in this development is at most equal to the degree of $f(u_1, u_3)$ in u_1. If we now, beginning with the terms of the lowest degree, successively substitute in the power series $a = \sum a_{n_3} \ldots {}_{n_t}' x_3^{n_3} \ldots x_t^{n_t}$ the coefficients of the terms of a given degree by power-series developments of the type just described, we shall obtain a development of a as a power series $a = \sum b_{n_3} \ldots {}_{n_t} x_3^{n_3} \ldots x_t^{n_t}$ in $x_3, \ldots, x_t$ with coefficients in $Q[u_1, u_3]$, and it can be easily seen from what we have said above that the u_1-excess of $b_{\mu_3} \ldots {}_{\mu_t}$ must be positive. However, this is impossible; for the expression $a = \sum b_{n_3} \ldots {}_{n_t} x_3^{n_3} \ldots x_t^{n_t}$ is the unique development of a as a power series in $x_3, \ldots, x_t$ with coefficients in $Q[u_1, u_3]$, and by a similar argument as before with u_3 replacing u_2, we can show that the u_1-excess of each coefficient $b_{n_3} \ldots {}_{n_t}$ must be non-positive. This completes the proof of the lemma.

3. *Proof of the Theorem of Bertini:* Since $u(c)$ is the quotient of two elements in a system of parameters in $\mathfrak{R}$, it can be easily shown, by a well-known argument, that the $\mathfrak{R}[u(c)]\mathfrak{p}$-residue of $u(c)$ is a variable over $\bar{\mathfrak{R}}$, so that $\mathfrak{R}[u(c)]/\mathfrak{R}[u(c)]\mathfrak{p}$ is a polynomial ring in one variable over $\bar{\mathfrak{R}}$; this shows that $\mathfrak{R}[u(c)]\mathfrak{p}$ is a prime ideal in $\mathfrak{R}[u(c)]$ and that $\mathfrak{R}[u(c)]\mathfrak{p}$ contracts to $\mathfrak{S}[u(c)]\mathfrak{q}$ in $\mathfrak{S}[u(c)]$. In order to prove our theorem, it is sufficient to show that $\mathfrak{R}\mathfrak{S}_{u(c)}$ is an analytically irreducible local domain; for, since $\mathfrak{R}\mathfrak{S}_{u(c)}$ contains $\mathfrak{R}[u(c)]$ and is contained in $\mathfrak{R}_{u(c)}$ and since the maximal prime ideal in $\mathfrak{R}\mathfrak{S}_{u(c)}$ must then contract to $\mathfrak{R}[u(c)]\mathfrak{p}$ in $\mathfrak{R}[u(c)]$, one sees readily that $\mathfrak{R}_{u(c)} = \mathfrak{R}\mathfrak{S}_{u(c)}$. Since $\mathfrak{R}\mathfrak{S}_{u(c)}$ is a finite module over the local domain $\mathfrak{S}_{u(c)}$, it is a semilocal domain in the sense of Chevalley;[2] it is therefore sufficient to show that the completion $(\mathfrak{R}\mathfrak{S}_{u(c)})^*$ of $\mathfrak{R}\mathfrak{S}_{u(c)}$ is a local domain. Let K and L be the quotient fields of $\mathfrak{R}$ and $\mathfrak{S}$ respectively, and let L^* be the quotient field of the completion $\mathfrak{S}_{u(c)}^*$ of $\mathfrak{S}_{u(c)}$, which we recall is a regular local ring. According to a result of Chevalley,[3] $(\mathfrak{R}\mathfrak{S}_{u(c)})^*$ is a finite module over $\mathfrak{S}_{u(c)}^*$ and no non-zero element in $\mathfrak{S}_{u(c)}^*$ is a zero-divisor in $(\mathfrak{R}\mathfrak{S}_{u(c)})^*$; if we denote by $\mathfrak{M}$ the quotient ring of $(\mathfrak{R}\mathfrak{S}_{u(c)})^*$ with respect to the set of all non-zero elements in $\mathfrak{S}_{u(c)}^*$, then both K and L^* are subrings in $\mathfrak{M}$ and we have the equation $\mathfrak{M} = KL^*$ (KL^* denotes here the ring generated by K and L^*). Moreover, by the same result, every set of linearly independent elements in K over L remains such over L^*; it follows from this that $\mathfrak{M}$ is isomorphic to the tensor product $K \times L^*$ over L. Now, since K is separable over L, there exist a finite number of irreducible monic polynomials $F_i(X)$ in $\mathfrak{S}[X]$ such that any extension field L_1 of L is linearly disjoint with respect to K over L if and only if all $F_i(X)$ remain irreducible in $L_1[X]$. According to our lemma, for sufficiently general c, all $F_i(X)$ will remain irreducible in $\mathfrak{S}_{u(c)}^*[X]$; since $\mathfrak{S}_{u(c)}^*$ is integrally closed, this implies that all $F_i(X)$ will remain irreducible in $L^*[X]$. It follows that L^* is linearly disjoint with respect to K

over L, and hence the tensor product $K \times L^*$ is an integral domain; this implies of course that $(\mathfrak{R}\mathfrak{S}_{u(c)})^*$ is an integral domain. Finally, since $(\mathfrak{R}\mathfrak{S}_{u(c)})^*$ is a finite module over the complete local domain $\mathfrak{S}_{u(c)}^*$ and has no zero-divisors, it must be a complete local domain.[4] This completes the proof of our theorem.

We shall conclude this note by a few remarks on the hypothesis of our theorem. We observe that in our formulation of the local Theorem of Bertini, we have introduced the subring $\mathfrak{S}$ and a system of regular parameters $x_1, \ldots, x_t$ in $\mathfrak{S}$, while in the final assertion of the theorem only the three elements x_1, x_2, x_3 enter into the picture. The question naturally arises whether it is necessary at all to introduce the regular local domain $\mathfrak{S}$ or even the system of parameters $x_1, \ldots, x_t$ in $\mathfrak{R}$. As to the latter, one finds a measure of justification in the fact that, even in the equal-characteristic case where $\mathfrak{R}$ contains a coefficient field k, an arbitrary set of three analytically independent elements x_1, x_2, x_3 may generate in $\mathfrak{R}$ an ideal of rank less than 3 and hence cannot be extended into a system of parameters in $\mathfrak{R}$. This is a well-known situation where a local ring may contain as a subring a local ring of higher dimension, and one can easily obtain examples to show that the local Theorem of Bertini cannot hold for an arbitrary set of three analytically independent elements without some qualifications. As to the introduction of the regular local domain $\mathfrak{S}$, it is evidently not necessary in the equal-characteristic case, since any system of parameters $x_1, \ldots, x_t$ in $\mathfrak{R}$ can be considered as the minimal basis of the maximal prime ideal in the regular local domain $k[[x_1, \ldots, x_t]]$, where k is a coefficient field in $\mathfrak{R}$. However, in the unequal-characteristic case, we do not know whether every system of parameters in $\mathfrak{R}$ can be so imbedded in a regular local domain, and such an imbedding is essential for our present method of proof. The question therefore remains open whether the local Theorem of Bertini holds in general for any three elements in a system of parameters in $\mathfrak{R}$.

* This work was partially supported by a research grant of the National Science Foundation.

[1] O. Zariski, "Pencils on an Algebraic Variety and a New Proof of a Theorem of Bertini," *Trans. Am. Math. Soc.*, **50**, 48–70, 1941.

[2] C. Chevalley, "On the Theory of Local Rings," *Ann. Math.*, **44**, 690–708, 1943.

[3] *Ibid.*, § II, Proposition 7.

[4] *Ibid.*, § III, Proposition 8. The proof of this proposition also holds for a local ring in the sense of Krull.

*COHOMOLOGY THEORY OF VARIETIES OVER RINGS**

BY WEI-LIANG CHOW AND JUN-ICHI IGUSA

DEPARTMENT OF MATHEMATICS, JOHNS HOPKINS UNIVERSITY

Communicated by Oscar Zariski, October 13, 1958

1. *Introduction.*—Some recent developments in algebraic geometry, e. g., the theory of deformations of complex analytic structures by Kodaira and Spencer, investigations on the principles of degeneration and of connectedness by Chow and on elliptic modular functions by Igusa, suggest the importance of the notion of varieties considered not merely over fields but more generally over rings. This

type of varieties gradually took shape in some works of Zariski, and a tentative formulation was given by Nagata rather recently.[1] In this short note, we shall outline a cohomology theory of such varieties and derive, among others, the *principle of upper semicontinuity* as a corollary of a Künneth relation in our cohomology theory.

2. *Models and Cohomology.*—We fix a field Σ and consider a subring A which is Noetherian and which has Σ as its field of fractions. The totality of rings of fractions constructed from A with respect to prime ideals is called a *chunk*. A local ring of Σ is called a *spot*. Two chunks U_1 and U_2 defined, respectively, by A_1 and A_2 may be called coherent if the compositum A_1A_2 is a finite ring extension of A_1 and A_2, hence Noetherian, and if the chunk defined by A_1A_2 is precisely the intersection of U_1 and U_2. A non-empty set X of spots is called a *model* of Σ, or simply a model, if it can be expressed as a union of a finite number of coherent chunks. If f is an element of Σ, the set of spots of X containing f is denoted by D_f. The totality of D_f generates a topology in X, which is called the Zariski topology of X. An open chunk contained in X may be called a *neighborhood.* Observe that the finite number of chunks by which X was defined are neighborhoods. Also, every open set is a union of neighborhoods, and every increasing sequence of open sets is necessarily finite.

Since a model X is a topological space, we can consider sheaves on X. In fact, the algebraic sheaf theory as developed by Serre[2] can be taken over verbatim to the present case. We get the sheaf of local rings, algebraic sheaves, coherent sheaves, locally free sheaves, etc. A crucial lemma, known as Hilbert's "Nullstellensatz," is in our case a triviality, because we are not restricting "points" to those which correspond to maximal ideals. If F is a sheaf on X, the *cohomology group* of X with coefficients in F, in notation $H(X,F)$, is obtained by Čech method: If U is a finite open covering of X, the module of q-cochains of U with coefficients in F is denoted by $C^q(U,F)$. The graded module $C(U,F) = \Sigma\, C^q(U,F)$ is a complex with a coboundary operator δ. The corresponding cohomology group is denoted by $H(U,F)$. One then passes to the direct limit of the groups $H(U,F)$ and one obtains the cohomology group $H(X,F)$. The following theorems are fundamental in our cohomology theory·

THEOREM 1. *Let F be a coherent sheaf on a model X which can be covered by one neighborhood. Then we have $H^q(X,F) = 0$ for $q > 0$.*

THEOREM 2. *Let F be a coherent sheaf on a model X and let U be a finite covering of X by coherent neighborhoods. Then the homomorphism $H(U,F) \to H(X,F)$ is an isomorphism.*

As it follows from Theorem 1 and Theorem 2, if $0 \to F' \to F \to F'' \to 0$ is an exact sequence of sheaves on X such that F' is coherent, the corresponding cohomology sequence

$$\to H^q(X,F') \to H^q(X,F) \to H^q(X,F'') \xrightarrow{\delta^*} H^{q+1}(X,F') \to$$

is exact. The third theorem is more delicate, and we must make some restrictions. Suppose that R is a fixed Noetherian domain such that Σ is a finitely generated extension of its field of fractions, say K. Then a finite ring extension of R with Σ as its field of fractions is Noetherian; hence it defines a chunk. If a model X is covered by this type of neighborhoods, we may call it a *geometric model* over R.

The cohomology groups of such models with coefficients in algebraic sheaves are R-modules. There is a classical way of getting geometric models: Pick a finite number of elements $x_0 = 1, x_1, \ldots, x_r$ of Σ which are different from zero and which generate Σ over K. With respect to such choice of generators, we get finite ring extensions $A_i = R[x_0/x_i, \ldots, x_r/x_i]$ of R for $i = 0, 1, \ldots, r$. The corresponding $r + 1$ chunks are coherent; hence their union X is a model. In fact, X is a geometric model over R, and it is called a *projective model* over R. We can now state our third theorem:

THEOREM 3. *If R is a regular local ring and if X is a projective model over R, the cohomology groups of X with coefficients in coherent sheaves are finite R-modules.*

Again, the proof is almost the same as in Serre. It is necessary, however, to have the Hilbert theorem on "chains of syzygies" in the following generalized form:[3]

LEMMA. *Let A be a ring of polynomials in $r + 1$ letters with coefficients in a regular local ring. Then every graded finite A-module admits a finite free resolution.*

We can apply Theorem 3 to the investigation of cohomology of "fiber systems" over non-singular models. In the next section, we shall discuss the simplest non-trivial case, i. e., the case where the base space is defined by a regular local ring of one dimension.

3. *A Künneth Relation.*—Keeping the notation of the previous section, assume that R is a discrete valuation ring with πR as its ideal of non-units and with k as the corresponding residue field. Let X be a projective model over R. The spots of X containing K or, more suggestively, not on the hyperplane $\pi = 0$, form an open subset Y of X. Observe that Y is a projective model over K. On the other hand, we shall *assume* that the tensor product $Y' = k \otimes_R X$ defined in an obvious way is a model over k. Let F be a coherent sheaf on X. Then, the restriction F_Y of F to Y is a coherent sheaf on Y and the tensor product $F' = k \otimes_R F$ is a coherent sheaf on X, which can be considered also as a coherent sheaf on Y'.

Now, consider the free resolution

$$0 \rightarrow R \xrightarrow{\pi} R \rightarrow k \rightarrow 0$$

of k and tensorize this exact sequence by a coherent sheaf F on X. We shall *assume* that the tensorized sequence is exact. Let U be a finite covering of X by coherent neighborhoods and consider the cochain complex $C(U,F)$. Then the Künneth relation[4] for the above complex over k and $C(U,F)$ takes the following form:

$$0 \rightarrow k \otimes_R H(X,F) \rightarrow H(Y',F') \rightarrow Tor_1^R(k,H(X,F)) \rightarrow 0.$$

Now, in general, if for an R-module M we denote by ${}_\pi M$ its submodule of elements m satisfying $\pi m = 0$, the "Tor" can be identified with ${}_\pi H(X,F)$. For the convenience of the reader, we shall give a proof of the above relation in a few lines: By assumption, the sequence $0 \rightarrow F \xrightarrow{\pi} F \rightarrow F' \rightarrow 0$ is exact; hence we get the corresponding cohomology sequence,

$$\rightarrow H^q(X,F) \xrightarrow{\pi^*} H^q(X,F) \rightarrow H^q(Y',F') \xrightarrow{\delta^*} H^{q+1}(X,F) \rightarrow.$$

Also, if we tensorize the exact sequence $0 \rightarrow R \xrightarrow{\pi} R \rightarrow k \rightarrow 0$ by $H(X,F)$, we get the following exact sequence:

$$0 \to {}_{\pi}H(X,F) \to H(X,F) \xrightarrow{\pi^*} H(X,F) \to k \otimes_R H(X,F) \to 0.$$

The combination of these two exact sequences will give the Künneth relation.

On the other hand, the restriction of F to Y gives rise to an R-homomorphism $H(X,F) \to H(Y,F_Y)$. We shall show that the image generates $H(Y,F_Y)$ over K and the kernel is the torsion submodule of $H(X,F)$: As before, take a finite covering U of X by coherent neighborhoods. Then the restriction U_Y of U to Y is a finite covering of Y by coherent neighborhoods. Moreover, if f is an element of $C(U_Y,F_Y)$, there exists a positive integer m such that $\pi^m f$ is the image of an element, say g, of $C(U,F)$. If f is a cocycle, there exists a positive integer n such that $\pi^n g$ become a cocycle. The first part is thereby proved. The proof of the second part is similar. Therefore, we can identify the tensor product $K \otimes_R H(X,F)$ with $H(Y,F_Y)$. In particular, we have

$$\dim k \otimes_R H^q(X,F) = \dim H^q(Y,F_Y) + \dim {}_{\pi}H^q(X,F).$$

If we combine this with the previous relation, we finally get the following *fundamental relation:*

$$\dim H^q(Y',F') - \dim H^q(Y,F_Y) = \dim {}_{\pi}H^q(X,F) + \dim {}_{\pi}H^{q+1}(X,F).$$

Observe that $H^0(X,F)$ is R-free. At any rate, if χ denotes the Euler-Poincaré characteristic, we have $\chi(Y',F') = \chi(Y,F_Y)$. We shall show in the next section that our fundamental relation implies the principle of upper semicontinuity.

4. *Principle of Upper Semicontinuity.*—We shall assume that $\Sigma = K(x)$ is a regular extension of K. Then $(x) = (x_0, x_1, \ldots, x_r)$ can be considered as a generic point of a projective variety V over K. Let V' be its specialization at the center of R. We shall assume that V' is a variety. The generators $x_0, x_1, \ldots, x_r$ of Σ over K define a projective model over R. We take its derived normal model in Σ and call it X. We note that X is also a projective model over R.[5] Moreover, if we assume that V and V' are locally normal,[6] we can identify the open subset Y of X of spots not on the hyperplane $\pi = 0$ with the totality of local rings of V over K and $Y' = k \otimes_R X$ with the totality of local rings of V' over k. In particular Y' is a projective model over k. Let E be a vector bundle on V defined over K such that its specialization E' at the center of R is a vector bundle on V'. Then we get sheaves of germs of sections of E and E'. If we denote these sheaves by $S(E)$ and $S(E')$, they are locally free sheaves on V and V'. Also E defines a locally free sheaf F on X in an obvious way. As before, we shall denote the restriction of F to Y by F_Y and $k \otimes_R F$ by F'. Then the sequence $0 \to F \xrightarrow{\pi} F \to F' \to 0$ is certainly exact. Moreover, if Ω and Ω' are the universal domains over K and k, by a result of Weil[7] we can identify $H(V,S(E))$ with $\Omega \otimes_K H(Y,F_Y)$ and $H(V',S(E'))$ with $\Omega' \otimes_k H(Y',F')$. In this way, we get the following theorem:

THEOREM 4. *Let V be a projective variety and let V' be its equal or unequal-characteristic specialization. We shall assume that V' is a normal variety. Let E be a vecor bundle on V which is specialized to a vector bundle E' on V' over the above specialization. Let $S(E)$ and $S(E')$ be the sheaves of germs of sections of E and E'. Then $\dim H^q(V',S(E'))$ is at least equal to $\dim H^q(V,S(E))$ and we have $\chi(V',S(E')) = \chi(V,S(E))$. In particular V and V' have the same arithmetic genus.*

We note that the last statement was already proved by a different method some years ago.[8] Also, we note that the "pathology" of unequal-characteristic specializations[9] is explained quantitatively by the fundamental relation in the previous section.

* This work was partially supported by the National Science Foundation.

[1] M. Nagata, "A General Theory of Algebraic Geometry over Dedekind Domains. I," *Am. J. Math.*, **78**, 78–116, 1956.

[2] J.-P. Serre, "Faisceaux algébriques cohérents," *Ann. Math.*, **61**, 197–278, 1955.

[3] This lemma is implicitly proved in Cartan-Eilenberg, *Homological Algebra* (Princeton, N. J.: Princeton University Press, 1956), pp. 154–156.

[4] *Ibid.*, pp. 112–116.

[5] Cf. Nagata, *op. cit.*, p. 104.

[6] Actually, the local normality of V' implies the local normality of V. Cf. J. Igusa, "Arithmetic Genera of Normal Varieties in an Algebraic Family," these PROCEEDINGS, **41**, 34–37, 1955.

[7] A. Weil, *Foundations of Algebraic Geometry* ("Am. Math. Soc. Colloquium Publs.," No. 29 [1946]), p. 239.

[8] Cf. Igusa, *op. cit.*

[9] J. Igusa, "On Some Problems in Abstract Algebraic Geometry," these PROCEEDINGS, **41**, 964–967, 1955. The footnote 12 is incorrect; the surface V_2 is indeed a specialization of V_0. This fact was remarked orally to the second-named author by Mr. Hironaka.

ON THE CONNECTEDNESS THEOREM IN ALGEBRAIC GEOMETRY.*[1]

By Wei-Liang Chow.

To Zariski on his 60th birthday.

1. Introduction. The Connectedness Theorem in abstract algebraic geometry was first formulated and proved by Zariski [14]; it stems from his attempt to prove the Principle of Degeneration over an abstract field and actually contains that principle as a special case. The proof of Zariski makes extensive use of his theory of holomorphic functions on an algebraic variety; in fact, this latter theory was developed by him specially for this purpose, and, as has been pointed out by Zariski himself, it has in a sense the Connectedness Theorem as its principal application. In view of the fundamental importance of the Connectedness Theorem on the one hand and the very complicated nature of Zariski's theory of holomorphic functions on the other, it is clearly very desirable to have for this theorem a proof of a simpler and more elemetnary nature. Furthermore, in consonance with the recent tendencies in algebraic geometry toward increasing arithmetic applications, one would naturally wish to generalize this theorem to the case of a correspondence between the rational transforms of arbitrary local domains (the exact meaning of this statement will be explained later in § 2). In a recent paper [4] we have given a very simple proof of such a generalization for a special case of the theorem, namely the Principle of Degeneration. In the present paper we shall solve the problem completely by giving a simple proof of the Connectedness Theorem in just about as general a version as we know how to formulate it at this stage of the subject.

Let R be a Noetherian domain with the quotient field K, and let S^n be the projective space of dimension n in the algebraic geometry over a universal domain $\mathfrak{K}$ which contains the field K. Let $\mathfrak{p}$ be a prime ideal in R, and denote by $\bar{R}$ the residue field $R/\mathfrak{p}$ of R over $\mathfrak{p}$; generally for any element x in R or a polynomial $f(Y)$ in $R[Y]=R[Y_0, Y_1, \cdots, Y_m]$, where the Y_i are indeterminates, we shall denote by $\bar{x}$ or $\bar{f}(X)$ respectively the corresponding element

* Received April 28, 1958; revised August 28, 1959.

[1] This work was partially supported by a research grant of the National Science Foundation.

in $\bar{R}$ or polynomial in $\bar{R}[Y]$ under the canonical homomorphism of R onto $\bar{R}$. We shall denote by $\bar{S}^n$ the projective space of dimension n in the algebraic geometry over a universal domain Λ which contains the field $\bar{R}$. Let Z be a positive r-cycle in S^n such that its associated point $y=(y_0, y_1, \cdots, y_m)$ in S^m is rational over K. A point η in $\bar{S}^m$ is said to be a *specialization* of the point y at the prime ideal $\mathfrak{p}$ if whenever a form $f(Y)$ satisfies the condition $f(y)=0$, it satisfies also the condition $\bar{f}(\eta)=0$. Since the condition for a point to be an associated point is "universal," every specialization η of y at $\mathfrak{p}$ is the associated point of a positive r-cycle $Z(\eta)$ in $\bar{S}^n$, of the same degree as Z; we shall say that the cycle $Z(\eta)$ is a *specialization* of the cycle Z at $\mathfrak{p}$ or at $R_\mathfrak{p}$, where $R_\mathfrak{p}$ is the quotient ring of R with respect to $\mathfrak{p}$. It is easily seen that the set of all specializations of y at $\mathfrak{p}$ is an $\bar{R}$-closed subset in $\bar{S}^m$, and it follows from this that the union of the supports of all specializations of Z at $\mathfrak{p}$ is also an $\bar{R}$-closed subset in $\bar{S}^n$. Recalling the definition that R is *analytically irreducible* at $\mathfrak{p}$ if the completion of the local ring $R_\mathfrak{p}$ is an integral domain, we can state the Connectedness Theorem as follows:

Let R be a Noetherian domain with the quotient field K, and let Z be a rational positive cycle over K in S^n; if Z is connected and if $\mathfrak{p}$ is a prime ideal in R such that R is analytically irreducible at $\mathfrak{p}$, then union of the supports of all the specializations of Z at $\mathfrak{p}$ is also connected.

Apart from the extension to an arbitrary Noetherian domain (instead of the coordinate ring of an algebraic variety), the above version of the Connectedness Theorem coincides in substance with the original formulation of Zariski. Now, since the Connectedness Theorem concerns itself essentially with the properties of a cycle Z with respect to a prime ideal $\mathfrak{p}$ in R, there will be no loss of generality if we replace the ring R by the quotient ring $R_\mathfrak{p}$, so that we can assume from the beginning that R is already a local domain. Moreover, since the concept of analytic irreducibility involves the completion of R, we may just as well assume further that R is itself a complete local domain. It turns out that, once put in this strictly local form, the Connectedness Theorem can be generalized to a theorem in the "relative" geometry, in a similar way as we have generalized the Principle of Degeneration in [4]. This generalization, which we shall call the General Connectedness Theorem, can be stated as follows:

General Connectedness Theorem. *Let R be a complete local domain with the quotient field K and the residue field $\bar{R}$, and let Z be a rational positive cycle over K in S^n; if Z is K-connected, then the union of the supports of all the specializations of Z at R is $\bar{R}$-connected.*

We remark that here the topological concepts such as "closed," "connected," etc., refer to the well-known Zariski topology, in which the closed subsets are unions of finite number of algebraic varieties, and the relative topological concepts such as "K-closed," "K-connected," etc., refer to the relative Zariski topology, in which the closed subsets are unions of finite number of complete sets of conjugate varieties over K. The use of these topological terminology, though not really necessary, proves to be very convenient. Although it is really only a trivial observation, we should like to point out in this connection that our formulation of the General Connectedness Theorem does not at all imply that the preservation of connectedness under specialization, which holds in the absolute Zariski topology, holds also in the relative Zariski topology over a *given* ground field; on the contrary, a little reflection shows that it is rather the disconnectedness over a ground field that is preserved under a specialization over the same ground field.

The General Connectedness Theorem contains two special cases: The Special Connectedness Theorem, where the cycle Z consists of a single point, so that we are dealing essentially with what corresponds to a birational transformation in the geometric case, and the Principle of Degeneration, where the local domain R is a real discrete valuation ring, so that there is a uniquely determined specialization of Z at R. Conversely, once we have proved these two special cases, we can easily obtain from them the Connectedness Theorem, as we shall show later in section 4. However, in order to prove the General Connectedness Theorem in its full generality as formulated above, we need something more than a simple combination of these two special cases. The reason for this is not difficult to see. For the hypothesis of the General Connectedness Theorem provides only the relative connectedness of the cycle Z over the complete local domain R, while the application of the Principle of Degeneration (in the form formulated in [4]) presupposes that we have some information about the relative connectedness of Z over some complete real discrete valuation ring dominating R, and it is clear that the relative connectedness of a cycle is in general not preserved under such an extension. In order to fill this gap, we need a result which says essentially that if a cycle Z splits into two disconnected parts over every complete real discrete valuation ring dominating R and this is in "continuous" manner, then Z cannot be connected over R. It turns out that these considerations lead us in a natural way to a refinement of the General Connectedness Theorem, which we shall call (for reasons which will be apparent) the Extended Principle of Degeneration. We shall prove this Extended Principle of Degeneration only for the case of a complete regular local ring; whether this Extend Principle holds for the general case of an arbitrary complete local domain, is a question which will not be settled here, although it seems very likely that the answer to it is in the affirmative. However, so far as the General Connectedness Theorem is concerned, this restriction to regular local rings is not a serious point, for a very simple argument (in section 4) shows that the validity of this theorem for a regular local ring implies its validity in the general case. On the other hand, this restriction to regular local rings enables us to obtain, in case of the Special Connectedness Theorem, yet another refinement, which is not generally true without this restriction. We shall say that a point set is *linearly connected* if every two points in the set can be connected by a sequence of rational curves in the set. Then we can prove that in the Special Connectedness Theorem for a regular local ring the set of all specializations is not only connected, but also linearly connected. This stronger result, which we shall call the Linear Connectedness Theorem, is of fundamental importance for the theory of rational equivalence, as we shall show in our forthcoming work on the intersection theory of cycles and equivalence classes of cycles.

Our paper therefore consists of two main results, the Linear Connectedness Theorem and the Extended Principle of Degeneration; the former will be proved in section 3 and the latter will be proved in section 5. Our proof of the Linear Connectedness Theorem is based on an analysis of the properties of the quadratic transforms of a regular local ring, and we shall use in course of the proof some of the well-known results on this subject. Looked upon from this point of view, our proof is related to the recent proofs of Zariski [15] and Murre [10] of the Special Connectedness Theorem for a simple point in the geometric case. On the other hand, our proof of the Extended Principle of Degeneration is of a quite different nature; although the quadratic transforms play also an essential role in this proof, the main idea lies in a different direction and is essentially a generalization of the method used in our proof of the Principle of Degeneration in [4], which centers on the idea of Hensel's Lemma. Only, as we have to deal now with an entire system of specialization cycles at the center, instead of a single cycle, the elementary "resultant" method used in [4] proves to be too cumbersome and rather inadequate; instead, we shall draw more heavily on the theory of local rings, with the result that we are able not only to prove the more general result, but also to gain a better insight into the nature of the problem. Incidentally, we should like to point out that the result in [4] is not used in our present proof, which in this sense can be said to start from the beginning, while on the other hand our recent result on the Local Bertini Theorem [5] plays an essential role in the important process of reduction to the case of dimension 2.

As to terminology, we shall use here some of the terms in Weil's well-known *Foundations of Algebraic Geometry,* such as the terms "universal domain," "variables," "indeterminates," "cycles," etc., which have now become more or less standard; however, we shall not make use of any results in this book, so that a knowledge of it is not at all necessary for the understanding of the present paper. On the other hand, we shall assume that the reader is familiar with the theory of local rings, which is fundamental for our subject; we shall therefore use freely the basic notions and results in this theory, although we endeavor to give explicit references for all results beyond the very elementary ones. We shall also use for convenience the topological concepts based on the Zariski topology, either absolute or relative to a given ground field; the Zariski topology for algebraic varieties is of course involved right at the beginning already in the formulation of our problem, but later we shall introduce it also for sets of algebraic cycles as well as for sets of valuation rings dominating given local domains. Finally, in the last two sections we shall assume some knowledge of the theory of associated forms; it is sufficient for this purpose to know the properties of the associated forms listed at the beginning of § 4 in our paper [4].

2. Rational transforms. We begin by making a few remarks concerning the projective space S^n (or S^m) introduced in the preceding section. Although it is quite possible for us to work within a given universal domain provided we choose it suitably to include all the local rings we shall need in the course of our theory, it is nevertheless more convenient and probably in a sense also more appropriate to consider S^n as the "universal projective space" which can be defined as the logical union of all the projective spaces of dimension n over all possible universal domains. We remark however that we shall never invoke more than a finite number of universal domains at one time, so that we are not really involved in any of the logical difficulties usually connected with the use of the expression "all" in such a context. Without any reference to any universal domains, we can define the *universal projective space* S^n by the stipulations that any ordered system of $n+1$ elements $(z_0, z_1, \cdots, z_n)$ in any field (or integral domain), not all zero, defines a point z in S^n, and that two such systems define the same point in S^n if and only if their elements are contained in some one field and are proportional to each other in this field; the elements z_j are called the (homogeneous) coordinates of the point z. We shall say that the point z is rational over a field K if the ratios of its coordinates z_j are contained in K; when this is the case, we shall always assume (unless stated to the contrary) that the proportionality factor has already been so chosen that the coordinates of z are all in K. In applying the usual algebro-geometrical operations to the entities in this universal projective space S^n we must of course see to it that the entities involved all belong to some one field, while among the entities belonging to different fields only very restricted types of relations can occur.

On the other hand, the projective space $\bar{S}^n$ remains, as it was before, to be the one over a given universal domain Λ, so that we can now consider $\bar{S}^n$ to be contained in the universal projective space S^n. However, this space $\bar{S}^n$ is in a sense the center of our attention, and we shall relate to it various entities in S^n in the following way. We shall say that a local ring $(R, \mathfrak{p})$ is *attached* to Λ if its residue field $\bar{R} = R/\mathfrak{p}$ is identified with a field contained in Λ, whereby the expression "a field in Λ" is to be understood in the sense of Weil's *Foundations* (i. e. Λ has infinite degree of transcendency over $\bar{R}$); the so defined concept of an attached local ring is somewhat more specific than that of a local ring by itself, for two local rings which are isomorphic but attached to distinct fields in Λ are to be considered here as different. *In this paper all the local rings are assumed to be attached to* Λ *in some ways,* so that we shall not always explicitly mention this fact. Furthermore, if a local ring $(R_1, \mathfrak{p}_1)$ contains R such that $R \cap \mathfrak{p}_1 = \mathfrak{p}$, and if R_1 is attached to Λ in such a way that the canonical embedding of $R/\mathfrak{p}$ into $R_1/\mathfrak{p}_1$ as abstract fields is accompanied by a corresponding embedding of the attached residue field of R in Λ into the attached residue field of R_1 in Λ, then we shall say that R_1 *dominates* R. In particular, we shall always assume that the completion R^* of R is attached to Λ in such a way that R^* dominates R, so that they have the same residue field in Λ. Once a local ring R is attached to Λ, the canonical homomorphism of R onto $\bar{R}$ will induce (under certain circumstances) a relation between those geometrical entities in S^n which are defined by means of R and similar entities in $\bar{S}^n$, such as the concept of a specialization of a cycle at $\mathfrak{p}$ described in the preceding section. For the sake of completeness and clarity, we shall begin here anew and repeat all the definitions.

Let $(R, \mathfrak{p})$ be a local domain of dimension t (attached to Λ), with the quotient field K and the residue field $\bar{R}$; for any element a in R or a polynomial $f(Y)$ in $R[Y] = R[Y_0, Y_1, \cdots, Y_m]$, we shall denote by $\bar{a}$ or $\bar{f}(X)$ respectively the corresponding element in $\bar{R}$ or polynomial in $\bar{R}[Y]$ under the canonical homomorphism of R onto $\bar{R}$. Let $y = (y_0, y_1, \cdots, y_m)$ be a point in S^m with coordinates in a field containing K; a point $\eta = (\eta_0, \eta_1, \cdots, \eta_m)$ in $\bar{S}^m$ is said to be a *specialization of* y *at* R, if for every form $f(Y)$ in $R[Y]$ the relation $f(y) = 0$ implies the relation $\bar{f}(\eta) = 0$. We shall denote by $\bar{y}(R)$ the set of all specializations of y at R, and it is easily seen that $\bar{y}(R)$

is an $\bar{R}$-closed subset in $\bar{S}^m$. If $(R_1, \mathfrak{p}_1)$ is a local domain which dominates $(R, \mathfrak{p})$ and if y is a point which is rational over a field containing R_1, then every specialization of y at R_1 is also a specialization of y at R, so that $\bar{y}(R_1)$ is contained in $\bar{y}(R)$. In case y is the associated point of a positive r-cycle Z in S^n, then every specialization of y at R is also the associated point of a positive r-cycle in $\bar{S}^n$, which we shall call a *specialization of* Z *at* R; we shall denote by $\bar{Z}(R)$ the set of all specializations of Z at R and denote by $|\bar{Z}(R)|$ the union of the supports of all the cycles in $\bar{Z}(R)$, and we observe that $|\bar{Z}(R)|$ is an $\bar{R}$-closed subset in $\bar{S}^n$.

We shall be particularly interested in the case where the point y is rational over K, and we shall say in this case that the point y defines a *rational transform* $y(R)$ of the local domain R; as we shall see, this notion of a rational transform of a local domain plays a fundamental role in our theory. The set $\bar{y}(R)$ is then called the *center* of the rational transform $y(R)$. If η is any point in $\bar{y}(R)$, consider the subring $\mathfrak{Q}(y(R)/\eta)$ in K consisting of all the elements of the form $f(y)/g(y)$, where $f(Y)$ and $g(Y)$ are forms of the same degree in $R[Y]$ and $\bar{g}(\eta) \neq 0$; it is easily seen that $\mathfrak{Q}(y(R)/\eta)$ is a local domain with the quotient field K, and that its maximal prime ideal $\mathfrak{p}(y(R)/\eta)$ consists of all the elements $f(y)/g(y)$ with $\bar{f}(\eta) = 0$. Since the residue field of $\mathfrak{Q}(y(R)/\eta)$ is isomorphic to the field $\bar{R}(\eta)$ in Λ, we shall attach $\mathfrak{Q}(y(R)/\eta)$ to Λ by identifying the residue field of $\mathfrak{Q}(y(R)/\eta)$ with $\bar{R}(\eta)$, and we shall call the so attached local domain $\mathfrak{Q}(y(R)/\eta)$ the *specialization ring* of the rational transform $y(R)$ at η. We observe that $\mathfrak{Q}(y(R)/\eta)$ evidently dominates the local domain R, and that if $\eta_j \neq 0$ for any one j, then $\mathfrak{Q}(y(R)/\eta)$ contains also the ring $R[y_0/y_j, \cdots, y_m/y_j]$; furthermore, it is easily seen that $\bar{y}(Q(y(R)/\eta))$ is the $\bar{R}$-variety in $\bar{S}^{m-1}$ consisting of all the specializations of η over $\bar{R}$. The point η is said to be *normal* or *simple* in the rational transform $y(R)$ if $\mathfrak{Q}(y(R)/\eta)$ is integrally closed or regular respectively, and when such is the case, we shall also say that the rational transform $y(R)$ is normal or simple at η respectively. We shall say that the rational transform $y(R)$ is normal or non-singular if it is normal or simple respectively at every point in its center $\bar{y}(R)$.

The concept of a rational transform $y(R)$ of a local domain R is a natural generalization of the concept of a rational transformation T of an algebraic variety V around a given point p of the variety. It belongs to the local geometry in the strict sense that one is concerned with what corresponds to the behavior of the rational transformation T in the infinitesimal neighborhood of p in V, not a finite neighborhood, however small. For this reason the only points on our rational transform $y(R)$ are the points in the center $\bar{y}(R)$, which corresponds in the geometric case to the total image $T(p)$ of the point p under T. However, we should like to stress here that we are really considering the points in $\bar{y}(R)$ not as points in what corresponds to the image variety $T(V)$ of V, but rather as points in what corresponds to the graph of T in the product space $V \times T(V)$, so that $\bar{y}(R)$ should be considered as corresponding to $p \times T(p)$, not to $T(p)$; for our definition of the specialization ring $\mathfrak{Q}(y(R)/\eta)$ at a point η in $\bar{y}(R)$ corresponds to the specialization ring of the graph of T at a point in $p \times T(p)$, not that of $T(V)$ at a point of $T(p)$. The reason for introducing here what corresponds to the graph rather than the image is quite easily understood, for the graph of a rational transformation contains essentially the complete information about the transformation, while the image is merely its projection on one of the factor spaces and hence its introduction is often not necessary, as is the case in our present studies.

The study of rational transforms is of fundamental importance not only for our present purpose, but also for many other problems in abstract local geometry. In this section, we shall limit ourselves to stating and proving a few basic properties of these transforms, which we shall need later. The discerning reader will notice that most of these are merely generalizations to arbitrary local domains of some of the elementary properties given in [11] for the geometric case, and it is therefore not surprising that we have also taken over some of the proofs from that paper.

LEMMA 2.1. *If $y(R)$ is a rational transform of R, then for any two points η and η' in $\bar{y}(R)$, $\mathfrak{Q}(y(R)/\eta)$ contains $\mathfrak{Q}(y(R)/\eta')$ if and only if η' is a specialization of η over $\bar{R}$.*

Proof. The condition is evidently sufficient. To prove the necessity, let $g(Y)$ be any form in $R[Y]$ such that $\bar{g}(\eta') \neq 0$, and let $f(Y)$ be a form in $R[Y]$ of the same degree as $g(Y)$ such that $\bar{f}(\eta) \neq 0$; if $\mathfrak{Q}(y(R)/\eta)$ contains $\mathfrak{Q}(y(R)/\eta')$, then $f(y)/g(y)$ being an element in $\mathfrak{Q}(y(R)/\eta')$ must be contained in $\mathfrak{Q}(y(R)/\eta)$. Since $\bar{f}(\eta) \neq 0$, this implies that $\bar{g}(\eta) \neq 0$. Since this is true for any form $g(Y)$ in $R[Y]$ and since any form in $\bar{R}[Y]$ can be obtained as a form $\bar{g}(Y)$, this shows that η' is a specialization of η over $\bar{R}$.

In the special case when the local domain R is a valuation ring, the center $\bar{y}(R)$ of any rational transform $y(R)$ of R consists of a single point; in fact, there exists in this case some one y_j such that $y_0/y_j, \cdots, y_m/y_j$ are all contained in R, and the point $(\overline{y_0/y_j}, \cdots, \overline{y_m/y_j})$ is evidently the only specialization of y at R. Furthermore, it is easily seen from the well-known maximality properties of a valuation ring that in this case we have $\mathfrak{Q}(y(R)/\bar{y}(R)) = R$.

In the general case of an arbitrary local domain R, we consider a valuation ring M in K dominating R; then, for any rational transform $y(R)$ of R, the point $\bar{y}(M)$ is evidently contained in $\bar{y}(R)$. We shall call this point $\bar{y}(M)$ the *center* of M in the rational transform $y(R)$. We maintain that M dominates $\mathfrak{Q}(y(R)/\bar{y}(M))$. To see this, let y_j be so chosen that the elements $y_0/y_j, \cdots, y_m/y_j$ are all in M and hence $\bar{y}(M) = (\overline{y_0/y_j}, \cdots, \overline{y_m/y_j})$; if $f(Y)$ is any form of degree d in $R[Y]$, then $f(y)/y_j{}^d$ is an element in M, and since M dominates R, it is a non-unit in M if and only if $\bar{f}(\bar{y}(M)) = 0$. It follows easily from this that any element $f(y)/g(y) = (f(y)/y_j{}^d)/(g(y)/y_j{}^d)$ in $\mathfrak{Q}(y(R)/\bar{y}(M))$ is an element in M, and that it is a non-unit in M if and only if it is such in $\mathfrak{Q}(y(R)/\bar{y}(M))$. This result, together with Lemma 2.1, gives us almost immediately the following lemma.

LEMMA 2.2. *Let $y(R)$ be a rational transform of a local domain R, let M be a valuation ring in K which dominates R, and let η be a point in $\bar{y}(R)$; then $\bar{y}(M) = \eta$ if and only if M dominates $\mathfrak{Q}(y(R)/\eta)$.*

It is well-known that any local ring in a field K is dominated by some valuation ring in K. Applying this to the local ring $\mathfrak{Q}(y(R)/\eta)$, we conclude from Lemma 2.2 that given any point η in $\bar{y}(R)$ there exists a valuation ring M in K dominating R such that $\eta = \bar{y}(M)$. If we denote by $\mathfrak{V}(R)$ the *set of all valuation rings in K dominating R*, then the correspondence $M \rightarrow \bar{y}(M)$ defines a mapping $\mathfrak{V}(R)$ *onto* the center $\bar{y}(R)$ of $y(R)$. We now introduce a topology in $\mathfrak{V}(R)$, called its *Zariski topology*, by the stipulation that a subset in $\mathfrak{V}(R)$ is said to be closed if it is the inverse image under this mapping of a closed subset in $\bar{y}(R)$ for some one rational transform $y(R)$ of R, whereby $\bar{y}(R)$ is understood to be endowed with its Zariski topology (induced by the Zariski topology in $\bar{S}^m$). If we consider only the Zariski $\bar{R}$-topology of $\bar{y}(R)$ for every $y(R)$, then we obtain the $\bar{R}$-topology of $\mathfrak{V}(R)$, and we have therefore such notions as $\bar{R}$-closed, $\bar{R}$-connected, etc. That this definition of a closed subset in $\mathfrak{V}(R)$ actually defines a topology in $\mathfrak{V}(R)$ can be easily verified. We observe also that this topology is the least fine among all topologies in $\mathfrak{V}(R)$ for which the mapping $M \rightarrow \bar{y}(M)$ is continuous for every rational transform $y(R)$ of R, both in the absolute case and in the relative case.

In case y is the associated point of a positive cycle Z in S^n, then for every valuation ring M in K dominating R we have a specialization cycle $\bar{Z}(M)$ of Z at R, and in this way we can obtain all specializations of Z at R, so that we can set $\bar{Z}(R) = \{\bar{Z}(M) \mid M \in \mathfrak{V}(R)\}$. If $F(U)$ is the associated form of the cycle Z, then we shall denote the associated form of the specialization cycle $\bar{Z}(M)$ by $\bar{F}_M(U)$. We note that both forms $F(U)$ and $\bar{F}_M(U)$ are defined only up to proportionality factors in K and $\bar{M}$ respectively, and that even when $F(U)$ is taken to be a form in $M[U]$, $\bar{F}_M(U)$ is not necessarily the residue image of $F(U)$ in $\bar{M}[U]$; in order to get $\bar{F}_M(U)$ as the residue form of $F(U)$, one has to normalize $F(U)$ by dividing it by one of its coefficients which have the lowest value under the valuation of M, so that all coefficients in $F(U)$ be in M but not all non-units.

Let $z = (z_0, z_1, \cdots, z_l)$ be a point in S^l which is rational over K, and consider the rational transform $z(R)$; then we can define a birational correspondence T between the two rational transforms $y(R)$ and $z(R)$ as follows: Two points η and ζ in $\bar{y}(R)$ and $\bar{z}(R)$ respectively are said to correspond to each other under T if there exists a valuation ring M in K dominating R such that $\eta = \bar{y}(M)$ and $\zeta = \bar{z}(M)$. In order to avoid possible ambiguities, we shall consider the correspondence T as from $y(R)$ to $z(R)$, so that the inverse correspondence from $z(R)$ to $y(R)$ will be denoted by T^{-1}; if η or ζ is any point in $\bar{y}(R)$ or $\bar{z}(R)$ respectively, then we shall denote by $T(\eta)$ or $T^{-1}(\zeta)$ respectively the set of all points in $\bar{z}(R)$ or $\bar{y}(R)$ which correspond to η or ζ under T or T^{-1} respectively. As the reader will readily perceive, this definition of a birational correspondence is essentially the same as that given in [11]. We shall say that the birational correspondence T is *regular* at a point η in $\bar{y}(R)$, if there exists a point ζ in $\bar{z}(R)$ corresponding to η under T such that $\mathfrak{Q}(y(R)/\eta)$ contains $\mathfrak{Q}(z(R)/\zeta)$. Finally, we introduce the *join* $y \circ z$ of the two points y and z, which is defined as the point in $S^{(m+1)(l+1)-1}$ whose coordinates are the elements $y_i z_j$, for $i = 0, 1, \cdots, m$, and $j = 0, 1, \cdots, l$, arranged in some arbitrary (but fixed) order; the rational transform $y \circ z(R)$ is then said to be the *join* of the rational transforms $y(R)$ and $z(R)$. It is clear that if M is any valuation ring in K dominating R, then we have the relation $\overline{y \circ z}(M) = \bar{y}(M) \circ \bar{z}(M)$; it follows easily from this that the birational correspondence between $y \circ z(R)$ and $y(R)$ is regular at every point in $\overline{y \circ z}(R)$.

LEMMA 2.3. *If the birational correspondence T between $y(R)$ and $z(R)$ is regular at a point η in $y(R)$, then $T(\eta)$ consists of exactly one point.*

Proof. By definition there exist a point ζ in $\bar{z}(R)$ and a valuation ring M in K dominating R such that $\eta = \bar{y}(M)$, $\zeta = \bar{z}(M)$, and $\mathfrak{Q}(y(R)/\eta) \supset \mathfrak{Q}(z(R)/\zeta)$. According to Lemma 2.2, M dominates both $\mathfrak{Q}(y(R)/\eta)$ and $\mathfrak{Q}(z(R)/\zeta)$; it follows that $\mathfrak{Q}(y(R)/\eta)$ dominates $\mathfrak{Q}(z(R)/\zeta)$. If M_1 is any valuation ring in K dominating $\mathfrak{Q}(y(R)/\eta)$, then it must also dominate $\mathfrak{Q}(z(R)/\zeta)$; it follows then by Lemma 2.2 again that $\bar{z}(M_1) = \zeta$, which proves our lemma.

LEMMA 2.4. *If $y(R)$ is normal at a point η and if $T(\eta)$ consists of a finite number of points, then T is regular at η.*

Proof. It is easily seen that, without affecting the validity of our lemma, we can replace $z(R)$ by another rational transform of R which is in biregular correspondence with it; in particular, we can replace $z(R)$ by the rational transform defined by a point whose coordinates are the monomials of the elements $z_0, z_1, \cdots, z_l$ of a given degree, arranged in some arbitrary order. By making such a replacement of $z(R)$ if necessary, we can therefore obtain that if $\zeta^{(1)}, \cdots, \zeta^{(d)}$ are the points contained in $T(\eta)$, then none of the coordinates $\zeta^{(1)}{}_0, \cdots, \zeta^{(d)}{}_0$ vanishes. Then $\mathfrak{Q}(z(R)/\zeta^{(i)})$ contains the ring $R[z_1/z_0, \cdots, z_l/z_0]$ for every $i = 1, \cdots, d$, and is in fact the quotient ring of $R[z_1/z_0, \cdots, z_l/z_0]$ with respect to the ideal

$$\mathfrak{q}_i = \mathfrak{p}(z(R)/\zeta^{(i)}) \cap R[z_1/z_0, \cdots, z_l/z_0].$$

Since each $\mathfrak{Q}(z(R)/\zeta^{(i)})$ is contained in some one valuation ring in $\mathfrak{V}(\mathfrak{Q}(y(R)/\eta))$, it follows that $R[z_1/z_0, \cdots, z_l/z_0]$ is contained in the intersection of all valuation rings in $\mathfrak{V}(\mathfrak{Q}(y(R)/\eta))$. Since $y(R)$ is normal at η, $\mathfrak{Q}(y(R)/\eta)$ coincides with the intersection of all valuation rings in $\mathfrak{V}(\mathfrak{Q}(y(R)/\eta))$ and hence contains $R[z_1/z_0, \cdots, z_l/z_0]$. Since $\mathfrak{q}_i$ is the restriction to $R[z_1/z_0, \cdots, z_l/z_0]$ of the maximal ideal in some one valuation ring in $\mathfrak{V}(\mathfrak{Q}(y(R)/\eta))$, it must contain the ideal

$$\mathfrak{q} = \mathfrak{p}(y(R)/\eta) \cap R[z_1/z_0, \cdots, z_l/z_0],$$

and since $\mathfrak{Q}(y(R)/\eta)$ contain the quotient ring of $R[z_1/z_0, \cdots, z_l/z_0]$ with respect to $\mathfrak{q}$, it must also contain $\mathfrak{Q}(z(R)/\zeta^{(i)})$. This shows that T is regular at the point η.

We make a digresion here to observe that up to now we have not used at all the crucial property of a local domain, that it is Noetherian; in fact, all our definitions (except for the definition of a simple point) and results up to now hold for any (commutative) integrity domain with the property that the set of all non-units is an ideal, i. e., what is sometimes called a quasi-local domain. While any deeper study of the rational transforms must of course involve the Noetherian or other similar properties, it is nevertheless convenient to avail ourselves of the fact that some of the elementary notions and results hold for the more general case. This is the justification for stating Lemma 2.2 above for any arbitrary valuation ring M, without the Noetherian restriction. In this paper we shall only consider rational transforms of (Noetherian) local domains; however; in considering the valuation rings dominating a local domain, we shall not restrict ourselves to the

Noetherian ones. Although it is possible in principle to get along without using any other valuation rings except those which are discrete and of rank 1, as shown by Lemma 2.7 below, it is very desirable to be able to use others whenever it turns out to be convenient to do so. We shall meet later occasions when it is convenient to use discrete valuation rings of rank 2.

From now on the Noetherian property of the local domain R will play an essential role. We observe first that by virtue of this Noetherian property the local domain R has a finite dimension t, which can be defined either as the rank of the maximal prime ideal $\mathfrak{p}$ or the number of elements in a system of parameters.

LEMMA 2.5. *If $y(R)$ is a rational transform of a local domain of dimension t, then any point in $\bar{y}(R)$ has a dimension at most $t-1$ over $\bar{R}$, so that the maximum dimension of the algebraic set $\bar{y}(R)$ is at most $t-1$.*

Proof. Let η be any point in $\bar{y}(R)$, and let M be a valuation ring in K dominating $\mathfrak{Q}(y(R)/\eta)$. According to [2], Theorem 1, the residue field of M has a dimension at most $t-1$ over $\bar{R}$; hence the residue field $\bar{R}(\eta)$ of $\mathfrak{Q}(y(R)/\eta)$ can have a dimension at most $t-1$ over $\bar{R}$, which shows that η has a dimension at most $t-1$ over $\bar{R}$.

LEMMA 2.6. *Let $y(R)$ be a rational transform of a local domain of dimension t; if a point η in $\bar{y}(R)$ has the dimension $t-1$ over $\bar{R}$, then there are only a finite number of valuation rings in K which dominate $\mathfrak{Q}(y(R)/\eta)$, and these valuation rings are real discrete (i.e. regular local domains of dimension 1).*

Proof. Since the residue field $\bar{R}(\eta)$ of $\mathfrak{Q}(y(R)/\eta)$ has the dimension $t-1$ over $\bar{R}$, the dimension of $\mathfrak{Q}(y(R)/\eta)$ must be 1; for, if the dimension of $\mathfrak{Q}(y(R)/\eta)$ were greater than 1, then there would exist a valuation ring in K dominating $\mathfrak{Q}(y(R)/\eta)$ whose residue field has a positive dimension over $\bar{R}(\eta)$ and hence a dimension greater than $t-1$ over $\bar{R}$, in contradiction to [2], Theorem 1. According to [7], the integral closure R_1 of $\mathfrak{Q}(y(R)/\eta)$ is Noetherian and hence the ideal $R_1\mathfrak{p}(y(R)/\eta)$ has only a finite number of (minimal) prime divisors; furthermore, the quotient ring of R_1 with respect to any one of these prime divisors is a real discrete valuation ring whose residue field is a finite algebraic extension of $\bar{R}(\eta)$. Since any valuation ring in K dominating $\mathfrak{Q}(y(R)/\eta)$ must coincide with one of these quotient rings, our lemma follows if we observe in addition that since the residue field of such a valuation ring is a finite algebraic extension of $\bar{R}(\eta)$, there are only a finite number of possible ways to embed it as a field extension of $\bar{R}(\eta)$ in Λ.

LEMMA 2.7. *Let $y(R)$ be a rational transform of an unmixed (= equidimensional) local domain of dimension t, and let η be any point in $y(R)$; then there exists a real discrete valuation ring in K which dominates $\mathfrak{Q}(y(R)/\eta)$ and whose residue field has the dimension $t-1$ over $\bar{R}$.*

Proof. Without the last condition on the dimension over R this lemma holds for any arbitrary local domain R (not necessarily unmixed); one has only to apply to $\mathfrak{Q}(y(R)/\eta)$ the general result that any local domain in K is dominated by a real discrete valuation ring in K, as proved in [1] or [4]. Furthermore, a closer look at the proof of this result in [4] shows that the residue field of the so obtained real discrete valuation ring has a dimension over the residue field $\bar{R}(\eta)$ of $\mathfrak{Q}(y(R)/\eta)$ which is exactly equal to the dimension of $\mathfrak{Q}(y(R)/\eta)$. The last condition in our lemma will then be fulfilled if we have the "dimension formula": $\dim(\mathfrak{Q}(y(R)/\eta)) + \dim(\eta/\bar{R}) = t$. This "dimension formula," in fact in a somewhat more general form, has been proved recently by Nagata [10] under the assumption that R satisfies the "second chain condition," and a previous result of Nagata shows that a unmixed local domain always satisfies this "second chain condition."

As a corollary to Lemma 2.7, we have the following useful lemma.

LEMMA 2.8. *If R is an analytically irreducible local domain, then $\bar{y}(R) = \bar{y}(R^*)$ for every rational transform $y(R)$ of R.*

Proof. Since R^* dominates R, it is clear that $\bar{y}(R^*) \subset \bar{y}(R)$; we have therefore only to show that every point in $\bar{y}(R)$ is also a point in $\bar{y}(R^*)$. Let η be a point in $\bar{y}(R)$ and let M be a real discrete valuation ring in K which dominates $Q(y(R)/\eta)$ and whose residue field has the dimension $t-1$ over $\bar{R}$ (t being the dimension of R). According to [13], Theorem 2, R is a subspace in M, so that the completion R^* of R can be embedded in the completion M^* of M; it is clear that M^* dominates both R^* and M. It follows that $\bar{y}(M^*)$ is a point in $\bar{y}(R^*)$ and also that $\bar{y}(M^*) = \bar{y}(M) = \eta$; this shows that η is a point in $y(R^*)$.

Before proceeding further, we shall make an observation here which will be useful to us also later. Let $v = (v_1, \cdots, v_d)$ be a set of independent variables over R, and consider the quotient ring $R(v)$ of $R[v]$ with respect to the prime ideal $R[v]\mathfrak{p}$; it is easily seen that $R(v)$ is a local domain of dimension t, with the quotient field $K(v)$, and that $R(v)$ is normal or regular if and only if R is such. Furthermore, if $\mathfrak{a}$ and $\mathfrak{b}$ are ideals in R, then we have $R(v)(\mathfrak{a} \cap \mathfrak{b}) = R(v)\mathfrak{a} \cap R(v)\mathfrak{b}$ and $R(v)\mathfrak{a} \cap R = \mathfrak{a}$, and if $\mathfrak{q}$ is a primary ideal in R with the associated prime ideal $\mathfrak{q}'$, then $R(v)$ is primary with

$R(v)\mathfrak{q}'$ as its associated prime. Finally, the $R(v)\mathfrak{p}$-residues $\bar{v}=(\bar{v}_1,\cdots,\bar{v}_d)$ of v are algebraically independent over $\bar{R}$, so that we can attach $R(v)$ to Λ by identifying the $\bar{v}$ with any set of d independent variables over $\bar{R}$; therefore, $R(v)$ dominates R and its residue field $\bar{R}(\bar{v})$ is a purely transcendental extension of dimension d over $\bar{R}$. For proof, we refer to [6], Lemma 14 for a part of it, with the remark that the rest of our assertions follows easily from it. In many problems these properties will enable us to replace the original local domain R by $R(v)$, and such a substitution often turns out to be convenient, mainly on account of the fact that the residue field $\bar{R}$ is thereby extended by the adjunction of an arbitrary number of variables and in particular made infinite. We shall call $R(v)$ the local domain obtained from R by the *adjunction of the variables* v.

Assume now that R is a complete regular local domain of dimension $t>2$, and let $x_1,\cdots,x_t$ be the elements in a minimal base for $\mathfrak{p}$; then $R(v)$ is a regular local domain of dimension t and the elements $x_1,\cdots,x_t$ form also a minimal base for its maximal prime ideal $R(v)\mathfrak{p}$. Let $d\leqq t-2$, and set $u_i=(x_2+v_ix_{i+2})/x_1$ for $i=1,\cdots,d$; denote by R_v the quotient ring of $R(v)[u_1,\cdots,u_d]$ with respect to the prime ideal $R(v)[u_1,\cdots u_d]\mathfrak{p}$. It is easily seen that R_v is a regular local domain of dimension $t-d$, and that the $R_v\mathfrak{p}$-residues $\bar{u}=(\bar{u}_1,\cdots,\bar{u}_d)$ of $u=(u_1,\cdots,u_d)$ are algebraically independent over $\bar{R}(\bar{v})$, so that we can attach R_v to by identifying $\bar{u}$ with any set of d independent variables over $\bar{R}(\bar{v})$. The following lemma, which can be considered as the local equivalent of the well-known "Theorem of Bertini" in algebraic geometry, plays an essential role later in the proof of Theorem 2 in that it enables us to reduce that theorem to the case $t=2$.

LEMMA 2.9. *The quotient field K of R is algebraically closed in the quotient field of the completion R_v^* of R_v.*

This lemma is equivalent to the assertion that if a monic polynomial of one variable with coefficients in R is irreducible over R, then it is also irreducible over the completion R_v^* of R_v. We can evidently without any loss of generality restrict ourselves to the case $d=1$, the general case following then by induction, and for this case the assertion can be proved in exactly the same way as the lemma in [5], except that one replaces the element c there by the element $v=v_1$ and observes that v satisfies the condition of being "sufficiently general" as required there. We therefore refer the proof of this lemma to [5].

3. Quadratic transforms and linear connectedness theorem. Let

$(R, \mathfrak{p})$ be a local domain of dimension t, with the quotient field K and the residue field $\bar{R}$, and let $x_1, \cdots, x_s$ be elements which form a minimal basis for $\mathfrak{p}$; if we denote by x the point $(x_1, \cdots, x_s)$ in S^{s-1}, then the rational transform $x(R)$ of R is called the *quadratic transform* of R. Since any two minimal bases for $\mathfrak{p}$ have the same number of elements and can be obtained each from the other by a linear transformation with coefficients in R and with determinant not in $\mathfrak{p}$, the quadratic transform $x(R)$ of R is unique up to an everywhere biregular transformation. We shall be mainly interested in the case where the local domain R is regular, and we recall that in this case the number s is equal to the dimension t. In this case, it is well-known ([1], Lemma 3, and Corollary 1) that $\bar{x}(R) = \bar{S}^{t-1}$ and every point in $\bar{x}(R)$ is simple in $x(R)$; furthermore ([2], Lemma 10, or as a special case of the dimension formula in [10]), if ξ is any point in $\bar{x}(R)$ and if t' is the dimension of $\mathfrak{Q}(x(R)/\xi)$, then $t' + \dim(\xi/\bar{R}) = t$. In case ξ is a generic point of $\bar{x}(R) = \bar{S}^{t-1}$ over $\bar{R}$, then $\mathfrak{Q}(x(R)/\xi)$ is a real discrete valuation ring, which (considered without its attachment to Λ) is the valuation ring of what is sometimes called the *$\mathfrak{p}$-adic valuation of* R; we shall call this valuation ring the *quadratic valuation ring* of R.

Let $y(R)$ be a rational transform of R in S^m, and let η be a point in $\bar{y}(R)$; let $\mathfrak{P}(\eta)$ be the homogeneous ideal in $R[Y] = R[Y_0, Y_1, \cdots, Y_m]$ generated by all forms $f(Y)$ such that $\bar{f}(\eta) = 0$, and let h be a positive integer such that there exists a basis of this ideal $\mathfrak{P}(\eta)$ consisting of forms of degrees less than $h+1$. Let $b_1(Y), \cdots, b_l(Y)$ be a basis of the R-module of all forms of degree h in $\mathfrak{P}(\eta)$, and denote the point $(b_1(y), \cdots, b_l(y))$ by b; if we set $z = y \circ b$, then the rational transform $z(R)$ is called the *quadratic transform of $y(R)$ centered at η*. This definition is analogous to the one given in [11], II, § 11 (for a monoidal transformation), and by a similar argument as given there, we can show that the quadratic transform $z(R)$ is up to an everywhere biregular birational correspondence independent of the choice of the integer h and the basis $b_1(Y), \cdots, b_l(Y)$, being dependent only on the point η. Furthermore, it is clear that $z(R)$ remains unchanged if we replace the point η by any one of its generic specializations over $\bar{R}$; this shows that the concept of a quadratic transform belongs really to the relative geometry.

Let T be the birational correspondence between $y(R)$ and its quadratic transform $z(R)$ centered at η. It is clear from the definition $z = y \circ b$ that the inverse T^{-1} is everywhere regular in $\bar{z}(R)$, and that T itself is regular at every point in $\bar{y}(R)$ which is not a specialization of η over $\bar{R}$. As to the point η itself, we can take any valuation ring M in K dominating $\mathfrak{Q}(y(R)/\eta)$

and get a point $\bar{z}(M) = \eta \circ \bar{b}(M)$ in $T(\eta)$, and in this way we can obtain every point in $T(\eta)$; it follows that $T(\eta) = \bar{z}(\mathfrak{Q}(y(R)/\eta)) = \eta \circ \bar{b}(\mathfrak{Q}(y(R)/\eta))$. Let $w_1, \cdots, w_\tau$ be the elements of a minimal basis for the maximal prime ideal in $\mathfrak{Q}(y(R)/\eta)$, and denote by w the point $(w_1, \cdots, w_\tau)$ in $S^{\tau-1}$. It is easily seen that the coordinates of b, after division by the form y_i^h if $\eta_i \neq 0$, are linear combinations of the coordinates of w with coefficients in $\mathfrak{Q}(y(R)/\eta)$, and vice-versa; therefore, the coordinates of z, after division by the form y_i^h, are also linear combinations of the coordinates of $y \circ w$ with coefficients in $\mathfrak{Q}(y(R)/\eta)$, and vice-versa. Since both

$$\mathfrak{Q}(z(R)/\bar{z}(M) \text{ and } \mathfrak{Q}(y \circ w(R)/\eta \circ \bar{w}(M))$$

contain $\mathfrak{Q}(y(R)/\eta)$ as a subring, it follows easily from this that

$$\mathfrak{Q}(z(R)/\bar{z}(M)) = \mathfrak{Q}(y \circ w(R)/\eta \circ \bar{w}(M));$$

furthermore, going over to the residue field of M, we conclude that the coordinates of $\bar{z}(M) = \eta \circ \bar{b}(M)$ and the coordinates of $\overline{y \circ w}(M) = \eta \circ \bar{w}(M)$ are linear combinations of each other with coefficients in $\bar{R}(\eta)$. This shows that the birational correspondence between $z(R)$ and $y \circ w(R)$ is biregular between $\bar{z}(\mathfrak{Q}(y(R)/\eta))$ and $\overline{y \circ w}(\mathfrak{Q}(y(R)/\eta))$, and that each $\bar{R}(\eta)$-variety in $T(\eta)$ is biregularly, in fact projectively equivalent over $\bar{R}(\eta)$ to an $\bar{R}(\eta)$-variety in $\eta \circ \bar{w}(\mathfrak{Q}(y(R)/\eta))$.

LEMMA 3.1. *For any valuation ring M in K dominating $\mathfrak{Q}(y(R)/\eta)$, we have the relation $\mathfrak{Q}(z(R)/\bar{z}(M)) = \mathfrak{Q}(w(\mathfrak{Q}(y(R)/\eta))/\bar{w}(M))$.*

Proof. Every element in $\mathfrak{Q}(w(\mathfrak{Q}(y(R)/\eta))/\bar{w}(M))$ is the quotient $f(w)/g(w)$ of two forms of the same degree in w with coefficients in $\mathfrak{Q}(y(R)/\eta)$, with the condition $\bar{g}(\bar{w}(M)) \neq 0$, while every element in $\mathfrak{Q}(y(R)/\eta)$ is the quotient $f'(y)/g'(y)$ of two forms of the same degree in y with coefficients in R, with the condition $\bar{g}'(\eta) \neq 0$; it follows that, by multiplying both $f(w)$ and $g(w)$ with a suitable power of a coordinate y_i or w_j such that the i-th coordinate of η or the j-th coordinate of $\bar{w}(M)$ does not vanish respectively, we can express any element in $\mathfrak{Q}(w(\mathfrak{Q}(y(R)/\eta))\bar{w}(M))$ as a quotient $f''(y \circ w)/g''(y \circ w)$ of two forms of the same degree in $y \circ w$ with coefficients in R, with the condition $g''(\eta \circ \bar{w}(M)) \neq 0$. This shows that

$$\mathfrak{Q}(w(\mathfrak{Q}(y(R)/\eta))/\bar{w}(M))$$

is contained in $\mathfrak{Q}(y \circ w(R)/\eta \circ \bar{w}(M)) = \mathfrak{Q}(z(R)/\bar{z}(M))$. In a similar way we can show that conversely $\mathfrak{Q}(z(R)/\bar{z}(M))$ is also contained in

$$\mathfrak{Q}(w(\mathfrak{Q}(y(R)/\eta))/\bar{w}(M)).$$

If the point η is simple in $y(R)$, then the local ring $\mathfrak{Q}(y(R)/\eta)$ is regular; it follows then from what we have said before about the quadratic transform of a regular local ring that (a) $\bar{w}(\mathfrak{Q}(y(R)/\eta)) = \bar{S}^{\tau-1}$, (b) $w(\mathfrak{Q}(y(R)/\eta))$ is simple at every point $\bar{w}(M)$ in $\bar{w}(\mathfrak{Q}(y(R)/\eta))$, and (c) if τ' is the dimension of the local ring $\mathfrak{Q}(w(\mathfrak{Q}(y(R)/\eta))/\bar{w}(M))$, then $\tau' + \dim(\bar{w}(M)/\bar{R}(\eta)) = \tau$. In view of Lemma 3.1 and the remarks immediately preceding it, we can conclude in this case that (a) $T(\eta)$ consists of a single $(\tau-1)$-dimensional variety, defined over $\bar{R}(\eta)$, and there is a biregular birational transformation Ψ, defined over $\bar{R}(\eta)$, which maps $\bar{w}(\mathfrak{Q}(y(R)/\eta)) = \bar{S}^{\tau-1}$ onto $T(\eta)$; (b) $z(R)$ is simple at every point $\bar{z}(M)$ in $T(\eta)$; and (c) τ' is also the dimension of the local ring $\mathfrak{Q}(z(R)/\bar{z}(M))$ and we have the relation $\tau' + \dim(\bar{z}(M)/\bar{R}(\eta)) = \tau$.

We shall now assume further that the point η (already assumed to be simple in $y(R)$) is contained in a variety Ω in $\bar{y}(R)$ with the following properties: (1) Ω is defined over an extension field Δ of $\bar{R}$ such that if $\sigma - 1$ is the dimension of Ω, then we have the equation $\sigma + \dim(\Delta/\bar{R}) = t$; (2) if ξ is any point in Ω, then $\bar{R}(\xi)$ contains Δ; (3) every point in Ω is simple in $y(R)$; (4) there is a biregular transformation Φ, defined over Δ, which maps the projective space $\bar{S}^{\sigma-1}$ onto Ω. Let ζ be a generic point of Ω over Δ, and let ζ' be a generic point of $T(\eta)$ over $\bar{R}(\eta)$; we shall prove the following lemma, which plays an essential part in the proof of Theorem 1 later.

LEMMA 3.2. *There exists a linearly connected set Γ in $\bar{z}(R)$ with the properties:* (1) *Γ contains both $T(\zeta)$ and ζ', and* (2) *every point in Γ is simple in $z(R)$ and has a dimension at least $t-2$ over $\bar{R}$, and* (3) *Γ consists of a finite number of rational curves.*

Proof. Since the point η is a specialization of the point ζ over $\bar{R}$, in fact over Δ, the local ring $\mathfrak{Q}(y(R)/\zeta)$ is the quotient ring of the local ring $\mathfrak{Q}(y(R)/\eta)$ with respect to a prime ideal $\mathfrak{p}_0$, which must then necessarily be of rank 1 (for $\mathfrak{Q}(y(R)/\zeta)$ has the dimension 1). The residue ring $\mathfrak{Q}(y(R)/\eta)/\mathfrak{p}_0$ is then canonically isomorphic to a subring in the residue field $\bar{R}(\zeta) = \Delta(\zeta)$ of $\mathfrak{Q}(y(R)/\zeta)$; in fact, it is easily seen that this subring is precisely the specialization ring $\mathfrak{Q}(\zeta/\eta, \Delta)$ in $\Delta(\zeta)$ of the variety Ω at the point η. For the sake of convenience, we shall identify the residue ring $\mathfrak{Q}(y(R)/\eta)/\mathfrak{p}_0$ with $\mathfrak{Q}(\zeta/\eta, \Delta)$; this implies in particular that we shall identify the residue field of the local ring $\mathfrak{Q}(y(R)/\eta)/\mathfrak{p}_0$ with $\bar{R}(\eta) = \Delta(\eta)$, and in this way attach $\mathfrak{Q}(y(R)/\eta)/\mathfrak{p}_0$ to Λ. Since Ω is biregularly equivalent over Δ to the projective space $\bar{S}^{\sigma-1}$, and since it is well-known (see [11], Lemma 9, p. 541) that the specialization ring over Δ of $\bar{S}^{\sigma-1}$ at any point is a regular local ring whose dimension is equal to $\sigma - 1$ minus the dimension of the point over Δ, it follows that $\mathfrak{Q}(y(R)/\eta)/\mathfrak{p}_0$ is a regular local ring of dimension $\sigma - 1 - \dim(\eta/\Delta)$. Since $\mathfrak{Q}(y(R)/\eta)$ is itself a regular local ring, it follows from a well-known result ([3], § III, Proposition 9, p. 705) that the basis $w_1, \cdots, w_\tau$ of $\mathfrak{Q}(y(R)/\eta)$ can be so chosen that $\mathfrak{p}_0$ is the principal ideal (w_τ) generated by the one element w_τ, and that the canonical images $\omega_1, \cdots, \omega_{\tau-1}$ of $w_1, \cdots, w_{\tau-1}$ in $\mathfrak{Q}(y(R)/\eta)/\mathfrak{p}_0$ coincide with any given minimal basis for its maximal prime ideals; furthermore, we have evidently the relation $\dim(\eta/\Delta) = \sigma - \tau$, and since $\bar{R}(\eta)$ contains Δ, it follows that $\dim(\eta/\bar{R}) = \dim(\eta/\Delta) + \dim(\Delta/\bar{R}) = t - \tau$. Now, let M be the quadratic valuation ring of the regular local ring $\mathfrak{Q}(y(R)/\eta)/\mathfrak{p}_0$; we shall attach M to Λ in such a way that its residue field M contains $\bar{R}(\eta)$ and is free with respect to $\bar{R}(\zeta')$ over $\bar{R}(\eta)$. If we denote by ω the point in $S^{\tau-1}$ (observe that by our convention $\bar{S}^{\tau-1}$ is contained in $S^{\tau-1}$) with the projective coordinates $(\omega_1, \cdots, \omega_{\tau-1}, 0)$, then the point $\bar{\omega}(\mathrm{M})$ in $\bar{S}^{\tau-1}$ has the form $(\lambda_1, \cdots, \lambda_{\tau-1}, 0)$, where $\lambda_1 = 1$, and $\lambda_2, \cdots, \lambda_{\tau-1}$ are independent variable (in Λ) over $\Delta(\eta)$. Let M be the valuation ring of the composite valuation obtained from the valuation of $\mathfrak{Q}(y(R)/\zeta)$ and the valuation of its residue field determined by M, i.e. the ring M consists of all elements in $\mathfrak{Q}(y(R)/\zeta)$ whose images in the residue field are contained in the ring M; we shall attach M to Λ by identifying its residue field with the residue field of M. It is clear that M dominates $\mathfrak{Q}(y(R)/\eta)$, so that $\bar{z}(M)$ is a point in $T(\eta)$. Furthermore, it follows from the definition of a composite valuation that $\bar{w}(M) = \bar{\omega}(\mathrm{M}) = (\lambda_1, \cdots, \lambda_{\tau-1}, 0)$; since $\bar{R}(\eta, \bar{z}(M)) = \bar{R}(\eta, \bar{w}(M))$, the point $\bar{z}(M)$ has the dimension $\tau - 2$ over $\bar{R}(\eta)$, and since $\bar{R}(\bar{z}(M))$ contains $\bar{R}(\eta)$, by the relation $\eta = T^{-1}(\bar{z}(M))$, the point $\bar{z}(M)$ has the dimension $\tau - 2 + \dim(\eta/\bar{R}) = t - 2$ over $\bar{R}$. If we denote by Γ' the line in $\bar{S}^{\tau-1}$ joining the points $\bar{w}(M)$ and $\Psi^{-1}(\zeta')$, then it is easily seen that every point in Γ' has a dimension at least $\tau - 2$ over $\bar{R}(\eta)$; going over to the variety $T(\eta)$, we obtain a rational curve $\Psi(\Gamma')$ in $T(\eta)$ which contains both $\bar{z}(M)$ and ζ', such that every point on it has a dimension at least $\tau - 2$ over $\bar{R}(\eta)$. Since every point in $\Psi(\Gamma')$ has the point η as its image under the holomorphic rational mapping T^{-1} (of $z(R)$) and since $\dim(\eta/\bar{R}) = t - \tau$, it follows that every point in $\Psi(\Gamma')$ has a dimension at least $t - 2$ over $\bar{R}$; furthermore, every point in $\Psi(\Gamma')$, being a point in $T(\eta)$, is simple in $z(R)$. This shows that the curve $\Gamma_1 = \Psi(\Gamma')$ has the property (2) in our lemma.

We set $\theta = (\theta_1, \cdots, \theta_\sigma) = \Phi^{-1}(\zeta)$ and $\alpha = (\alpha_1, \cdots, \alpha_\sigma) = \Phi^{-1}(\eta)$; we observe that $\mathfrak{Q}(\zeta/\eta, \Delta)$ is also the specialization ring $\mathfrak{Q}(\theta/\alpha, \Delta)$ in $\Delta(\theta) = \Delta(\zeta)$ of $\bar{S}^{\sigma-1}$ at the point α, so that the elements $\omega_1, \cdots, \omega_{\tau-1}$ form also a basis for the maximal prime ideal in $\mathfrak{Q}(\theta/\alpha, \Delta)$. We shall show below that there exists a rational curve Γ'' in $\bar{S}^{\sigma-1}$ containing the points θ and α, with the property that every point in Γ'' except possibly α has a dimension at least $\sigma-2$ over Δ and that the elements ω_i/ω_1, $i=2,\cdots,\tau-1$, considered as function on Γ'', have values at α which are independent variables over $\Delta(\alpha)$. Going over to the variety Ω, we have a rational curve $\Phi(\Gamma'')$ containing the points ζ and η such that every point in $\Phi(\Gamma'')$ except possibly η has a dimension at least $\sigma-2$ over Δ and hence a dimension at least $t-2$ over $\bar{R}$. Since the values of ω_i/ω_1, $i=2,\cdots,\tau-1$, on $\Phi(\Gamma'')$ at the point η are independent variables over $\Delta(\alpha)=\bar{R}(\eta)$, we can take them to be the elements $\lambda_2,\cdots,\lambda_{\tau-1}$ introduced before; then in the "proper transform" of $\Phi(\Gamma'')$ in $\overline{y\circ w}(R)$, i.e. the locus over the defining field of $\Phi(\Gamma'')$ of the image in $\overline{y\circ w}(R)$ of a generic point of $\Phi(\Gamma'')$, the (unque) point corresponding to η is $\eta\circ\bar{w}(M)$. Let Γ_2 be the "proper transform" of $\Phi(\Gamma'')$ in $\bar{z}(R)$; it is clear that Γ_2 contains the point $T(\zeta)$. Since the correspondence T is biregular at every point in $\Phi(\Gamma'')$ except η, it follows that every point in Γ_2 outside of $T(\eta)$ has a dimension at least $t-2$ over $\bar{R}$; on the other hand, since the correspondence between $z(R)$ and $y\circ w(R)$ is biregular at every point in $T(\eta)$, it follows that $\bar{z}(M)$ is the (unique) point on Γ_2 corresponding to η in $\Phi(\Gamma'')$. This shows that Γ_2 contains the points $\bar{z}(M)$ and $T(\zeta)$, and has the property (2) in our lemma. If we set $\Gamma=\Gamma_1+\Gamma_2$, then Γ is clearly linearly connected and has the properties (1), (2) and (3) in our lemma.

It remains therefore to prove the existence of the curve Γ''. For convenience we shall assume that ζ is a generic point of Ω over $\bar{R}(\eta)$, not only over Δ; this does not essentially affect our situation, for if Γ_2 does not contain ζ, but contains another generic point χ of Ω over Δ, then we can take a line Γ''' in $\bar{S}^{\sigma-1}$ joining θ and $\Phi^{-1}(\chi)$ and replace Γ by $\Gamma+T(\Phi(\Gamma'''))$. Also, without any loss of generality, we can assume that $\theta_\sigma=\alpha_\sigma=1$ and that $\alpha_\tau,\cdots,\alpha_{\sigma-1}$ are independent variables over Δ; we observe that since θ is a generic point of $\bar{S}^{\sigma-1}$ over $\Delta(\alpha)$, the elements $\theta_1,\cdots,\theta_{\sigma-1}$ are independent variables over $\Delta(\alpha)$. Finally, we recall that we can choose the basis $\omega_1,\cdots,\omega_{\tau-1}$ in any way to suit our convenience; besides, the choice of this basis does not matter any way in our present context, for any two such bases can be obtained one from the other by a linear transformation with coefficients in $\mathfrak{Q}(\theta/\alpha,\Delta)$ and with determinant which is a unit.

For $i=1,\cdots,\tau-1$, let ω_i be an element in $\Delta[\theta_1,\cdots,\theta_i,\theta_\tau,\cdots,\theta_{\sigma-1}]$ which on substituting $\theta_1=\alpha_1,\cdots,\theta_{i-1}=\alpha_{i-1}$, $\theta_\tau=\alpha_\tau,\cdots,\theta_{\sigma-1}=\alpha_{\sigma-1}$ becomes an irreducible polynomial in $\Delta(\alpha_1,\cdots,\alpha_{i-1},\alpha_\tau,\cdots,\alpha_{\sigma-1})[\theta_i]$ having α_i as a root. It can be easily shown by induction on the dimension $\tau-1$ that the elements $\omega_1, \cdots, \omega_{\tau-1}$ form a minimal basis for the maximal prime ideal in $\mathfrak{Q}(\theta/\alpha, \Delta)$, this being in fact the proof in [11], Lemma 9. If we denote by s_i the inseparability index of α_i over $\Delta(\alpha_1, \cdots, \alpha_{i-1}, \alpha_\tau, \cdots, \alpha_{\sigma-1})$, then we can evidently choose ω_i so that it involves only the s_i-th power of θ_i. Now, by a reordering of the first $\tau - 1$ coordinates in $\bar{S}^{\sigma-1}$ if necessary, we can evidently obtain that the sequence of integers $s_1, \cdots, s_{\tau-1}$ has the property that if $s'_1, \cdots, s'_{\tau-1}$ is another such sequence corresponding to any other ordering of the first $\tau - 1$ coordinates in $\bar{S}^{\sigma-1}$, then the last non-vanishing number in the sequence $s'_1 - s_1, \cdots, s'_{\tau-1} - s_{\tau-1}$ is positive. Then, for $j < i \leqq \tau - 1$, the inseparability index of α_j over $\Delta(\alpha_1, \cdots, \alpha_{j-1}, \alpha_{j+1}, \cdots, \alpha_i, \alpha_\tau, \cdots, \alpha_{\sigma-1})$ cannot be less than s_i; for, if this is not true for one such pair of indices i and j, and if we interchange the i-th and the j-th coordinates, then we would obtain a new sequence $s'_1, \cdots, s'_{\tau-1}$ such that $s_{i+1} = s'_{i+1}, \cdots, s_{\tau-1} = s'_{\tau-1}$, and $s'_i < s_i$, in contradiction to our choice of the sequence $s_1, \cdots, s_{\tau-1}$. Consider now ω_i as a polynomial in one variable θ_j $(j < i)$ with coefficients in $\Delta[\theta_1, \cdots, \theta_{j-1}, \theta_{j+1}, \cdots, \theta_i, \theta_\tau, \cdots, \theta_{\sigma-1}]$; it is clear that in our choice of ω_i we can delete any term whose coefficient vanishes for $\theta = \alpha$. If we delete all such terms from ω_i, then the resulting ω_i will involve only the s_i-th powers of θ_j; for, otherwise, if we set $\theta = \alpha$ in every coefficient in ω_i (considered as a polynomial in θ_j), we would obtain a polynomial in θ_j with coefficients in $\Delta(\alpha_1, \cdots, \alpha_{j-1}, \alpha_{j+1}, \cdots, \alpha_i, \alpha_\tau, \cdots, \alpha_{\sigma-1})$ which has α_j as a root and which is not a polynomial in the s_i-th power of θ_j, in contradiction to the fact that the inseparability index of α_j over $\Delta(\alpha_1, \cdots, \alpha_{j-1}, \alpha_{j+1}, \cdots, \alpha_\tau, \cdots, \alpha_{\sigma-1})$ cannot be less than s_i. If we do this for every $j < i$, we would then obtain that ω_i involves only the s_i-th powers of the elements $\theta_1, \cdots, \theta_i$. Thus we have shown that the elements $\omega_1, \cdots, \omega_{\tau-1}$ can be so chosen that each ω_i on substituting $\theta_\tau = \alpha_\tau, \cdots, \theta_{\sigma-1} = \alpha_{\sigma-1}$ becomes a polynomial in the s_i-th powers of $\theta_1, \cdots, \theta_i$; this implies that over an algebraic (purely inseparable) extension Δ' of $\Delta(\alpha)$, each ω_i will have the form $\omega_i = \phi_i(\theta_1 - \alpha_1, \cdots, \theta_i - \alpha_i)^{s_i} + \rho_i$, where ϕ_i is a ploynomial with coefficients in Δ' and ρ_i is an element in $\Delta'[\theta_1, \cdots, \theta_{\sigma-1}]$ which vanishes for $\theta_\tau = \alpha_\tau, \cdots, \theta_{\sigma-1} = \alpha_{\sigma-1}$. Furthermore, it is clear that $\phi_i(0, \cdots, 0) = 0$ and that the coefficient β_i of the term $(\theta_i - \alpha_i)^{s_i}$ in ϕ_i does not vanish; it follows that there exists a power series $\psi_i(\theta_1 - \alpha_1, \cdots, \theta_{i-1} - \alpha_{i-1})$ in $\theta_1 = \alpha_1, \cdots, \theta_{i-1} - \alpha_{i-1}$ with coefficients in Δ' such that $\phi_i(\theta_1 - \alpha_1, \cdots, \theta_{i-1} - \alpha_{i-1}, \psi_i) = 0$.

Let $\mu_1, \cdots, \mu_{\sigma-1}$ be independent variables over $\Delta'(\theta)$; for convenience in notation, we set $s_i = 1$ and $\psi_i = 0$ for $i = \tau, \cdots, \sigma - 1$. We define a curve Γ'' in $\bar{S}^{\sigma-1}$ by the following recursive parametric equations, where $Y_1, \cdots, Y_{\sigma-1}$

denote the (affine) coordinates in $\bar{S}^{\sigma-1}$ (with $Y_\sigma = 1$) and T denotes the parameter:

$$Y_i = \alpha_i + \mu_i T^{(s/s_i)} - \psi_i(Y_1 - \alpha_1, \cdots, Y_{i-1} - \alpha_{i-1})_s + \nu_i T^{s+1},$$

for $i = 1, \cdots, \sigma - 1$, where the $Y_1, \cdots, Y_{i-1}$ on the right hand side are to be substituted by their expressions in the previous equations, and the subscript s after the ψ_i indicates that all terms of degrees higher than s are to be deleted; furthermore, each ν_i is to be determined by the condition $\nu_i = \theta_i - \alpha_i - \mu_i + [\psi_i(Y_1 - \alpha_1, \cdots, Y_{i-1} - \alpha_{i-1})_s]_{T=1}$. It is easily seen that Γ'' is a rational curve and contains the points α (for $T = 0$) and θ (for $T = 1$). Furthermore, if $\Gamma''(\gamma)$ is the point on Γ'' corresponding to any value $\gamma \neq 0$ of T, then $\Delta'(\mu, \nu)$ is contained in $\Delta'(\gamma, \Gamma''(\gamma), \nu)$, and since $\Delta'(\mu, \nu)$ has the dimension $\sigma - 1$ over $\Delta'(\nu)$, it follows that $\Gamma''(\gamma)$ must have a dimension at least $\sigma - 2$ over $\Delta'(\nu)$. For $T = \infty$, the corresponding point on Γ'' has the homogeneous coordinates $(\nu_1, \cdots, \nu_{\sigma-1}, 0)$ and hence has the dimension $\sigma - 2$ over $\Delta'(\mu)$. Thus every point on Γ'', except possibly the point α, has a dimension at least $\sigma - 2$ over Δ. Finally, a simple calculation shows that each ω_i, considered as a function on Γ'', has at the point α a power series development in T beginning with the term $(\beta_i \mu_i + \delta_i) T^s$, where δ_i is an element in $\Delta'(\theta_1, \cdots, \theta_{i-1})$; this shows that the ratios ω_i/ω_1, $i = 2, \cdots, \tau - 1$, considered as functions on Γ'', have values at α which are independent variables over $\Delta(\alpha)$. This concludes the proof of Lemma 3.2.

Remark. If $\sigma = 2$, then Ω is a rational curve already and the rather complicated construction of the curve Γ'' is not necessary. In this case, the curve Ω can contain only a finite number of fundamental points $\eta^{(1)}, \cdots, \eta^{(d)}$ for T, and each set $T(\eta^{(i)})$ is also a rational curve with the property (2). It is clear that the set $T(\Omega)$ consists of the "proper transform" of Ω and all the curves $T(\eta^{(1)}), \cdots, T(\eta^{(d)})$, and that it is a connected set with the properties (2) and (3).

Let M be a valuation ring in K dominating R. We start with a rational transform $y(R)$ of R, and obtain from it a sequence of rational transforms $z^{(i)}(R)$ of R, $i = 1, 2, \cdots$, such that each $z^{(i)}(R)$ is the quadratic transform of $z^{(i-1)}(R)$ centered at $\bar{z}^{(i-1)}(M)$, whereby we set $z^{(0)}(R) = y(R)$. We shall say that $z^{(i)}(R)$ is the *i-th quadratic transform of* $y(R)$ *along* M. As we have observed above, if $y(M)$ is simple in $y(R)$, then $\bar{z}^{(i)}(M)$ will also be simple in $z^{(i)}(R)$ for every i; and, in this case, if we denote by t_i the dimension of $\mathfrak{Q}(z^{(i)}(R)/\bar{z}^{(i)}(M))$, then we have the relation $t_i + \dim(\bar{z}^{(i)}(M)/\bar{R}(\bar{z}^{(i-1)}(M))) = t_{i-1}$, for every $i = 1, 2, \cdots$, while for $i = 0$ we have the relation

$t_0 + \dim(\bar{z}^{(0)}(M)/\bar{R}) = t$. Since the birational correspondence between $z^{(i-1)}(R)$ and $z^{(i)}(R)$ is everywhere regular in $z^{(i)}(R)$, it follows that $\bar{R}(z^{(i)}(M))$ contains $\bar{R}(\bar{z}^{(i-1)}(M))$; we obtain therefore by addition the relation $t_i + \dim(\bar{z}^{(i)}(M)/\bar{R}) = t$ for every $i = 0, 1, \cdots$. We shall say that the sequence $z^{(i)}(R)$ of quadratic transforms of $y(R)$ along M is *finite* if there exists an index i such that $t_i = 1$ or equivalently $\dim(\bar{z}^{(i)}(M)/\bar{R}) = t - 1$. It is clear that this is possible only when the residue field of M has a dimension at least $t - 1$ over $\bar{R}$; according to [2], Theorem 1, this implies that M is a real discrete valuation ring and that its residue field has exactly the dimension $t - 1$ over $\bar{R}$. It is important for our present purpose that the converse of this statement is also true.

LEMMA 3.3. *If M is a real discrete valuation ring in $\mathfrak{V}(R)$ such that its residue field has the dimension $t - 1$ over $\bar{R}$, and if $y(R)$ is a rational transform of R such that $y(M)$ is simple in $y(R)$, then the sequence of quadratic transforms of $y(R)$ along M is finite.*

A proof of this lemma is given in [2], Proposition 3.

We are now ready to prove the Linear Connectedness Theorem.

THEOREM 1 (Linear Connectedness Theorem). *The center $\bar{y}(R)$ of a rational transform $y(R)$ of a regular local ring R is linearly connected.*

Proof. As before, we denote by t, K, and $\bar{R}$ the dimension, the quotient field, and the residue field respectively of R. Consider the quadratic transform $z^{(1)}(R)$ of R and let T_1 be the birational correspondence between $z^{(1)}(R)$ and $y(R)$; if $\zeta^{(1)}$ is a generic point of $\bar{z}^{(1)}(R)$ over $\bar{R}$ (observe that $\bar{z}^{(1)}(R) = \bar{S}^{t-1}$), then T_1 is regular at $\zeta^{(1)}$ and hence $T_1(\zeta^{(1)})$ is a uniquely determined point in $\bar{y}(R)$, which we shall denote by ζ. To prove our theorem, it is sufficient to show that any point η in $\bar{y}(R)$ is linearly connected in $\bar{y}(R)$ to this point ζ. By Lemma 2.7, there exists a real discrete valuation ring M in $\mathfrak{V}(R)$, with a residue field which has the dimension $t - 1$ over $\bar{R}$, such that $\bar{y}(M) = \eta$. For $1 < i \leqq a$ (the integer a will be determined presently), let $z^{(i)}(R)$ be the $(i-1)$-th quadratic transform of $z^{(1)}(R)$ along M, and let T_i be the birational correspondence between $z^{(i)}(R)$ and $y(R)$. We set $T_{i,j} = T_j^{-1}T_i$ for $i, j = 1, \cdots, a$, so that $T_{i,j}$ is the birational correspondence between $z^{(i)}(R)$ and $z^{(j)}(R)$; we set $\eta^{(i)} = \bar{z}^{(i)}(M)$ for $i = 1, \cdots, a$, and take for each $i = 2, \cdots, a$, a generic point $\zeta^{(i)}$ of the variety $T_{i-1,i}(\eta^{(i-1)})$ over $\bar{R}(\eta^{(i-1)})$. The integer a is to be taken as the smallest integer such that the point $\eta^{(a)} = \bar{z}^{(a)}(M)$ has the dimension $t - 1$ over $\bar{R}$; by Lemma 3.3, such an integer a exists, and since $\eta^{(a)}$ is simple in $z^{(a)}(R)$, it must be a regular point of T_a and we have evidently $\eta = T_a(\eta^{(a)})$. Furthermore, since $\eta^{(a)}$ has the dimension $t-1$ over $\bar{R}$, it must have the maximum possible dimension over $\bar{R}(\eta^{(a-1)})$ and hence is a generic point of $T_{a-1,a}(\eta^{(a-1)})$ over $\bar{R}(\eta^{(a-1)})$; we can therefore take $\eta^{(a)}$ itself as the point $\zeta^{(a)}$, which will be convenient for us later. We maintain now that there exists a connected set $\Gamma^{(i)}$ in $\bar{z}^{(i)}(R)$ with the following properties: (1) $\Gamma^{(i)}$ contains both $\zeta^{(i)}$ and $T_{1,i}(\zeta^{(1)})$, (2) every point in $\Gamma^{(i)}$ is simple in $z^{(i)}(R)$ and has a dimension at least $t-2$ over $\bar{R}$, and (3) $\Gamma^{(i)}$ consists of either a single point or a finite number of rational curves. We shall show this by induction on the index i, beginning by setting $\Gamma^{(1)} = \zeta^{(1)}$, and, assuming the induction hypothesis that the set $\Gamma^{(i-1)}$ has already been constructed, we shall proceed to contruct the set $\Gamma^{(i)}$. If $T_{i-1,i}$ has no fundamental points in $\Gamma^{(i-1)}$ and hence is biregular everywhere in $\Gamma^{(i-1)}$, then the set $T_{i-1,i}(\Gamma^{(i-1)})$ is evidently connected, has the properties (2) and (3), and contains the points $T_{1,i}(\zeta^{(1)}) = T_{i-1,i}(T_{1,i-1}(\zeta^{(1)}))$ and $T_{i-1,i}(\zeta^{(i-1)})$. Assuming now that $T_{i-1,i}$ has some fundamental points in $\Gamma^{(i-1)}$; we observe that this can occur only if $\dim(\eta^{(i-1)}/\bar{R}) = t-2$, in which case $T_{i-1,i}$ can have only a finite number of fundamental points $\eta^{(i-1,1)}, \cdots, \eta^{(i-1,b)}$ in $\Gamma^{(i-1)}$, all of dimension $t-2$ over $\bar{R}^{-(i-1)}$, and each set $T_{i-1,i}(\eta^{(i-1,j)})$ is a rational curve in $\bar{z}^{(i)}(R)$ which has the property (2). If we denote by $\Gamma_1^{(i-1)}, \cdots, \Gamma_c^{(i-1)}$ the distinct component curves in $\Gamma^{(i-1)}$, then by the Remark after Lemma 3.2 each set $T_{i-1,i}(\Gamma_k^{(i-1)})$ is connected and has the properties (2) and (3), and it will contain the curve $T_{i-1,i}(\eta^{(i-1,j)})$ whenever $\Gamma_k^{(i-1)}$ contains the point $\eta^{(i-1,j)}$. It follows then from an elementary combinatorial consideration that the set $T_{i-1,i}(\Gamma^{(i-1)}) = \sum_{k=1}^{c} T_{i-1,i}(\Gamma_k^{(i-1)})$ is connected, and it evidently has the properties (2) and (3), and contains the point $T_{1,i}(\zeta^{(1)})$ and $T_{i-1,i}(\zeta^{(i-1)})$. Now, by Lemma 3.2, there exists a connected set Γ_i in $\bar{z}^{(i)}(R)$ which has the properties (2) and (3) and contains the points $\zeta^{(i)}$ and $T_{i-1,i}(\zeta^{(i-1)})$. The set $\Gamma^{(i)} = T_{i-1,i}(\Gamma^{(i-1)}) + \Gamma_i$ is then evidently connected and has the properties (1), (2), and (3).

For $i = a$, we obtain a linearly connected set $\Gamma^{(a)}$ in $\bar{z}^{(a)}(R)$ which has the properties (2) and (3) and which contains the points $T_{1,a}(\zeta^{(1)})$ and $\bar{z}^{(a)}(M) = \eta^{(a)} = \zeta^{(a)}$. If T_a is regular at every point in $\Gamma^{(a)}$, then T_a induces a continuous mapping of $\Gamma^{(a)}$ into $\bar{y}(R)$, so that $T_a(\Gamma^{(a)})$ is a linearly connected set; since $T_a(\Gamma^{(a)})$ contains the points $\zeta = T_1(\zeta^{(1)}) = T_a(T_{1,a}(\zeta^{(1)}))$ and $\eta = T_a(\eta^{(a)}) = T_a(\zeta^{(a)})$, our theorem would therefore be proved. Otherwise, we proceed as follows. Consider the join $y \circ z^{(a)}(R)$, and let T_a' be the birational correspondence between $z^{(a)}$ and $y \circ z^{(a)}(R)$. By Lemma 2.5, there exist in $\overline{y \circ z^{(a)}}(R)$ only a finite number of points of dimension $t-1$ over $\bar{R}$

such that no two of them are equivalent over $\bar{R}$ (i.e. specializations of each other over $\bar{R}$), and by Lemma 2.6, there exist only a finite number of valuation rings in $\mathfrak{V}(R)$ whose centers in $y \circ z^{(a)}(R)$ coincide with any one of these points, and these valuation rings are all real discrete. There exist therefore a finite number of real discrete valuation rings $M_1, \cdots, M_d$ in $\mathfrak{V}(R)$ such that any valuation ring in $\mathfrak{V}(R)$ whose center in $y \circ z^{(a)}(R)$ has the dimension $t-1$ over $\bar{R}$ coincides with one of them up to attachments to Λ. If T_a is not regular at a point ξ in $\Gamma^{(a)}$, then T_a' also cannot be regular at ξ; since ξ is simple and hence normal in $z^{(a)}(R)$, it follows from Lemma 2.4 that $T_a'(\xi)$ has at least one component of positive dimension. There exists therefore a valuation ring N in $\mathfrak{V}(R)$ such that $\bar{z}^{(a)}(N) = \xi$ and such that the point $\overline{y \circ z^{(a)}}(N)$ has a positive dimension over $\bar{R}(\xi)$. Since $T_a'^{-1}$ is regular everywhere, $\bar{R}(\overline{y \circ z^{(a)}}(N))$ must contain $\bar{R}(\xi)$ as subfield, and since ξ has the dimension $t-2$ over $\bar{R}$, it follows that $\overline{y \circ z^{(a)}}(N)$ has the dimension $t-1$ over $\bar{R}$. From what we have just said, N must coincide up to attachment to Λ with one of the valuation rings M_i, say M_1, so that $z^{(a)}(M_1)$ is equivalent to ξ over $\bar{R}$ and hence is also simple in $z^{(a)}(R)$. Now, let $z^{(i)}(R)$, for $a < i \leqq a_1$ (the integer a_1 will be determined presently), be the $(i-a)$-th quadratic transform of $z^{(a)}(R)$ along M_1, and let T_i be as before the birational correspondence between $z^{(i)}(R)$ and $y(R)$; also as before, we set $T_{i,j} = T_j^{-1}T_i$, so that $T_{i,j}$ is the birational correspondence between $z^{(i)}(R)$ and $z^{(j)}(R)$. If we set $\Gamma^{(i)} = T_{i-1,i}(\Gamma^{(i-1)}) = T_{a,i}(\Gamma^{(a)})$, then by exactly the same induction argument as before we can show that $\Gamma^{(i)}$ is a linearly connected set with the properties (2) and (3), and it is evident that $\Gamma^{(i)}$ contains the points $T_{1,i}(\zeta^{(1)}) = T_{a,i}(T_{1,a}(\zeta^{(1)}))$ and $\bar{z}^{(i)}(M) = T_{a,i}(\eta^{(a)})$. By Lemma 3.3, there exists a smallest integer $a_1 > a$ such that the point $\bar{z}^{(a_1)}(M_1)$ has the dimension $t-1$ over $\bar{R}$, so that T_{a_1} is regular at $\bar{z}^{(a_1)}(M_1)$; this determines a_1. If T_{a_1} is regular at every point in $\Gamma^{(a_1)}$, then T_{a_1} induces a continuous mapping of $\Gamma^{(a_1)}$ into $\bar{y}(R)$ and hence $T_{a_1}(\Gamma^{(a_1)})$ is a linearly connected set; since $T_{a_1}(\Gamma^{(a_1)})$ contains the points $\zeta = T_1(\zeta^{(1)}) = T_{a_1}(T_{1,a_1}(\zeta^{(1)}))$ and $\eta = T_{a_1}(\bar{z}^{(a_1)}(M))$, this would prove our theorem. If T_{a_1} is not regular at some one point $\xi^{(1)}$ in $\Gamma^{(a_1)}$, then by the same argument as before we can show that there exists a valuation ring N_1 in $\mathfrak{V}(R)$ such that $\bar{z}^{(a_1)}(N_1) = \xi^{(1)}$ and such that the point $\overline{y \circ z^{(a_1)}}(N_1)$ has the dimension $t-1$ over $\bar{R}$. Since $\bar{z}^{(a)}(N_1)$ is a point in $\Gamma^{(a)}$, it must have the same dimension $t-2$ over $\bar{R}$ as the point $\bar{z}^{(a_1)}(N_1)$; it follows that the point $\bar{z}^{(a_1)}(N_1)$ is algebraic over $\bar{R}(\bar{z}^{(a)}(N_1))$, and hence the point $\overline{y \circ z^{(a_1)}}(N_1)$ must be algebraic over $\bar{R}(\overline{y \circ z^{(a)}}(N_1))$. This shows that the point $\overline{y \circ z^{(a)}}(N_1)$ has the dimension $t-1$ over $\bar{R}$, and it follows from what we have said before that N_1 must coincide up to attachments to Λ with one of the valuation rings M_i, say M_2. We can then proceed in a similar way with $z^{(a)}(R)$, $\Gamma^{(a_1)}$, M_2 as we did before with $z^{(a)}(R)$, $\Gamma^{(a)}$ M_1. Thus we proceed in this way step by step until, after a finite number π of steps, we have obtained a rational transform $z^{(\pi)}(R)$ of R with the following properties, whereby T_π and $T_{1,\pi}$ have similar meanings as before: There exists a linearly connected set $\Gamma^{(\pi)}$ in $\bar{z}^{(\pi)}(R)$ which has the properties (2) and (3) and which contains the points $T_{1,\pi}(\zeta^{(1)})$ and $\bar{z}^{(\pi)}(M)$, and for every M_i the point $\bar{z}^{(\pi)}(M_i)$ either has the dimension $t-1$ over $\bar{R}$ or is not equivalent over $\bar{R}$ to a point in $\Gamma^{(\pi)}$. From what we have said above, this implies that T_π is regular at every point in $\Gamma^{(\pi)}$. Therefore T_π induces a continuous mapping of $\Gamma^{(\pi)}$ into $\bar{y}(R)$, so that $T_\pi(\Gamma^{(\pi)})$ is a linearly connected set; since $T_\pi(\Gamma^{(\pi)})$ contains the points $\zeta = T_1(\zeta^{(1)}) = T_\pi(T_{1,\pi}(\zeta^{(1)}))$ and $\eta = T_\pi(\bar{z}^{(\pi)}(M))$, this concludes the proof of Theorem 1.

COROLLARY 1. *In case $t=2$, there exists a finite quadratic transform $z(R)$ of R such that the birational correspondence between $z(R)$ and $y(R)$ is regular at every point in $\bar{z}(R)$.*

Proof. Since in this case every point in $\bar{z}^{(\pi)}(R)$ must have a dimension over $\bar{R}$ which is either $t-1$ or $t-2=0$, it follows from the proof above that T_π is regular at every point in $\bar{z}^{(\pi)}(R)$. We can therefore take $z(R)$ to be $z^{(\pi)}(R)$.

COROLLARY 2. *The topological space $\mathfrak{V}(R)$ is connected.*

This follows immediately from Theorem 1 and the definition of topology in $\mathfrak{V}(R)$ given in section 2.

4. General connectedness theorem and extended principle of Degeneration. We begin by recalling the well-known Hensel's Lemma: If a splitting form $F(X)$ with coefficients in a complete real discrete valuation ring R is irreducible over R (or over the quotient field K of R, which in this case amounts to the same thing), then the residue form $\bar{F}(X)$, if it does not vanish identically, is not the product of two relatively prime forms in $\bar{R}[X]$. Geometrically, this means that if a positive 0-cycle Z is rational and irreducible over K, then its specialization $\bar{Z}(R)$ at R is $\bar{R}$-connected. Our purpose here is to generalize this Hensel's Lemma to the case where R is an arbitrary complete local domain. In a sense something of this nature has already been done in the literature; in fact, it is well-known that the Hensel's Lemma, at least in the case of a form of two variables, holds for any complete local ring R (see e.g. [6], Theorem 4), for the usual proof of the Hensel's Lemma for a valuation ring can be carried over without any difficulties to the general case. However, this generalization is essentially a trivial one and in reality deals only with a highly specialized situation; for the condition that the residue form $\bar{F}(X)$ does not vanish identically, while easily obtainable in case of a valuation ring by a suitable choice of the factor of proportionality, imposes a very severe restriction on the form $F(X)$ in the general case. Expressed in geometrical terms, this condition means that the cycle Z must have a uniquely determined specialization at R, so that the set $\bar{Z}(R)$ consists of only a single cycle; while this is always true in case of a valuation ring, it evidently represents a severe restriction in the general case. In fact, the fact that $\bar{Z}(R)$ is in general an algebraic set of cycles instead of a single cycle, can be considered as the principal new feature in local algebraic geometry which one encounters in going over from valuation rings to arbitrary local domains. A true generalization of the Hensel's Lemma should therefore take account of this important point, and in lieu of the $\bar{R}$-connectedness of a single residue cycle, it should supply us with some information about the $\bar{R}$-connectedness properties of the entire set of cycles $\bar{Z}(R)$. The General Connectedness Theorem, as formulated in § 1, is a theorem of just such a nature and can therefore very appropriately be considered as a generalization of the Hensel's Lemma, or rather, since it deals with cycles of arbitrary dimensions, as a generalization of the Principle of Degeneration as formulated and proved in [4], which is itself a generalization of the Hensel's Lemma to cycles of higher dimensions while retaining the assumption of a valuation ring. However, it is possible to formulate a theorem which is an even truer generalization of the Principle of Degeneration in the sense that it deals with the $\bar{R}$-connectedness properties not of the point set $|\bar{Z}(R)|$, but rather of the individual cycles in $\bar{Z}(R)$. This leads us to what we shall call the Extended Principle of Degeneration, from which one can easily obtain the General Connectedness Theorem as a simple corollary, as we have already mentioned in § 1.

We shall speak of a *continuous system* of positive r-cycles in $\bar{S}^n$ *parametrized* by a local domain R, if there is a continuous mapping $N \to \Xi(N)$ $(N \in \mathfrak{V}(R))$ of the set $\mathfrak{V}(R)$ of all valuation rings in K dominating R into the set $\mathfrak{Z}(\bar{S}^n)$ of all positive r-cycles in $\bar{S}^n$ such that for every N the cycle $\Xi(N)$ is rational over thet residue field $\bar{N}$ of N. We shall denote such a continuous system of cycles by the symbols $\Xi(R)$. The continuity condition here in the definition refers to the Zariski $\bar{R}$-topologies of both $\mathfrak{V}(R)$ and $\mathfrak{Z}(\bar{S}^n)$; we recall that the topology of $\mathfrak{V}(R)$ is defined by means of the system of Zariski $\bar{R}$-topologies of the centers of all rational transforms of R, as described after Lemma 2.2 in section 2, while the topology of $\mathfrak{Z}(\bar{S}^n)$ is that of the (discrete) union of an infinite sequence of spaces, each one of which is the set of all positive r-cycles of a given degree in $\bar{S}^n$ and is being endowed with its Zariski $\bar{R}$-topology by means of the associated points. In particular, this implies that if N and N' are two elements in $\mathfrak{V}(R)$ such that N' is the quotient ring of N, then $\Xi(N)$ is a specialization of $\Xi(N')$ over $\bar{R}$. In terms of the associated forms, the continuous system of cycles $\Xi(R)$ can be given by a *continuous system* of forms $\Phi_R(U) = \{\Phi_N(U) \mid N \in \mathfrak{V}(R)\}$ such that each $\Phi_N(U)$ is an associated form and is rational over $\bar{N}$, whereby we observe that these forms are determined only up to proportionality factors and that rationality of a form refers to the ratios of the coefficients.

Let R_1 be a local domain dominating R, not necessarily contained in the quotient field K of R; since every valuation ring in the quotient field of R_1 dominating R_1 contracts in K to a valuation ring dominating R, there is defined in this way a canonical mapping of $\mathfrak{V}(R_1)$ into $\mathfrak{V}(R)$, which is easily seen to be continuous. If we set $\Xi(N) = \Xi(K \cap N)$ for every N in $\mathfrak{V}(R_1)$, then the mapping $N \to \Xi(N)$ $(N \in \mathfrak{V}(R_1))$ defines evidently a continuous system parametrized by R_1, which we shall denote by $\Xi(R_1)$ and call the *subsystem* of $\Xi(R)$ determined by R_1.

We remark that in general the cycles in a continuous system $\Xi(R)$ can very well have different degrees; in particular, for some N the cycle $\Xi(N)$ may have the degree 0, i.e. $\Xi(N)$ is the zero r-cycle. If every cycle in a continuous system has the degree 0, so that all of them coincide with the zero r-cycle, then the system is said to be *trivial*. However, in the important case where $\mathfrak{V}(R)$ is connected, as is the case with a regular local ring R according to Theorem 1, Corollary 2, it can be easily seen that all cycles in a continuous system $\Xi(R)$ must have the same degree, so that if for any one N the cycle $\Xi(N)$ has the degree 0, then the system $\Xi(R)$ must be trivial. To see this, one need only to observe that the set of all positive cycles of a given degree d in $\bar{S}^n$ is a closed subset in $\mathfrak{Z}(\bar{S}^n)$ and so is the set of all positive cycles in $\bar{S}^n$ of degrees different from d; the inverse images of these two subsets under the mapping $N \to \Xi(N)$ must then be two disjoint closed subsets in $\mathfrak{V}(R)$ whose union coincides with $\mathfrak{V}(R)$, and hence one of them must be empty. This means that if d is the degree of any one cycle in $\Xi(R)$, then all the other cycles in $\Xi(R)$ must have the same degree d.

If there exist two non-trivial continuous systems of positive r-cycles $\Xi'(R)$ and $\Xi''(R)$ such that we have the equation $\Xi(N) = \Xi'(N) + \Xi''(N)$ for every N in $\mathfrak{V}(R)$, then we shall say that the system $\Xi(R)$ *splits* into the two systems $\Xi'(R)$ and $\Xi''(R)$, or that the pair $(\Xi'(R), \Xi''(R))$ is a splitting of $\Xi(R)$. Such a splitting is said to be $\bar{R}$-disjoint at a valuation ring N if

$\Xi'(N)$ and $\Xi''(N)$ are $\bar{R}$-disjoint, i. e. the supports of $\Xi'(N)$ and $\Xi''(N)$ are disjoint sets in the Zariski $\bar{R}$-topology, and it is said to be *everywhere $\bar{R}$-disjoint* or simply *$\bar{R}$-disjoint* if it is $\bar{R}$-disjoint at every N in $\mathfrak{V}(R)$. In terms of the associated forms, a splitting of $\Xi(R)$ can be given by a splitting of the system of forms $\Phi_R(U)$ into two systems of forms $\Phi'_R(U)$ and $\Phi_R''(U)$, so that for every N in $\mathfrak{V}(R)$ we have the equation $\Phi_N(U) \doteq \Phi'_N(U)\Phi''_N(U)$, where the symbol $\doteq$ denotes equality up to proportionality factors; and we observe that in case $r = 0$, the cycles $\Xi'(N)$ and $\Xi''(N)$ are $\bar{R}$-disjoint if and only if their associated forms $\Phi'_N(U)$ and $\Phi''_N(U)$ are relatively $\bar{R}$-prime, i. e. there does not exist a splitting form in $\bar{R}[U]$ which has common factors with both $\Phi'_N(U)$ and $\Phi''_N(U)$. Finally, a continuous system of cycles is said to be *$\bar{R}$-connected* if it does not have any $\bar{R}$-disjoint splitting.

If Z is a positive r-cycle in S^n, rational over K, then the algebraic set of cycles $\bar{Z}(R)$ forms in a natural way a continuous system parametrized by R, which we shall also denote by the same symbol $\bar{Z}(R)$, there being not much danger of confusion on this score. In this case, we shall say that the cycle Z *generates* the continuous system $\bar{Z}(R)$. If the cycle Z is the sum of two positive r-cycles Z' and Z'', also rational over K, then it is clear that we have the equation $\bar{Z}(N) = \bar{Z}'(N) + \bar{Z}''(N)$ for every N in $\mathfrak{V}(R)$; hence the system $\bar{Z}(R)$ splits into the two systems $\bar{Z}'(R)$ and $\bar{Z}''(R)$. Thus we see that a splitting of the cycle Z induces a splitting of the system $\bar{Z}(R)$ generated by Z. The Extended Principle of Degeneration, which we shall now formulate, asserts that conversely every $\bar{R}$-disjoint splitting of the system $\bar{Z}(R)$ can be obtained in this way. We shall prove this principle here only for the case of a complete regular local ring, although it is very likely that it holds generally for an arbitrary complete local domain.

THEOREM 2 (Extended Principle of Degeneration). *Let $(R, \mathfrak{p})$ be a complete regular local ring, with the quotient field K and the residue field $\bar{R}$, and let Z be a positive cycle in S^n, rational over K; if Z is K-connected, then the continuous system of cycles $\bar{Z}(R)$ is $\bar{R}$-connected.*

The proof of this theorem will be given in the next section. Here we shall show how the General Connectedness Theorem can be obtained from it by a very simple argument.

THEOREM 3 (General Connectedness Theorem). *Let $(R, \mathfrak{p})$ be a complete local domain, with the quotient field K and the residue field $\bar{R}$, and let Z be a positive cycle in S^n, rational over K; if Z is K-connected, then the point set $|\bar{Z}(R)|$ is $\bar{R}$-connected.*

Proof. In case R is regular, this theorem follows almost immediately from Theorem 2. In fact, if there exist two disjoint $\bar{R}$-closed subsets Ω' and Ω'' in $\bar{S}^n$ such that $|\bar{Z}(R)|$ is contained in their union and has non-empty intersections with both of them, and if we denote by $\Xi'(N)$ and $\Xi''(N)$ the restriction of $\bar{Z}(N)$ in Ω' and Ω'' respectively, then we obtain two continuous systems $\Xi'(R) = \{\Xi'(N) \mid N \in \mathfrak{V}(R)\}$ and $\Xi''(R) = \{\Xi''(N) \mid N \in \mathfrak{V}(R)\}$ which are clearly non-trivial, and the equation $\bar{Z}(N) = \Xi'(N) + \Xi''(N)$ defines evidently an $\bar{R}$-disjoint splitting of the system $\bar{Z}(R)$ into the two systems $\Xi'(R)$ and $\Xi''(R)$; it follows then from Theorem 2 that Z cannot be K-connected. It remains to show that the general case of an arbitrary complete local domain can be reduced to this special case.

According to Cohen [6], Theorem 16, there exists a subring R_0 in R such that R_0 is a complete regular local ring having the same dimension and the same residue field as R, and such that R is a finite module over R_0; this implies that K is a finite algebraic extension of the quotient field K_0 of R_0. Let $Z_1 = Z, Z_2, \cdots, Z_s$ be the complete set of conjugates of Z over K_0, and let $K_1 = K, K_2, \cdots, K_s$ and $R_1 = R, R_2, \cdots, R_s$ be the corresponding sets of conjugates of K and R respectively over K_0; let K' be the compositum of all K_i and let R' be the ring generated by all the R_i. Since R' is a finite module over R_0 and since R_0 is complete, it follows from [3], III, Proposition 8, that R' is a complete local domain; since the local rings R_i all have the same residue field $\bar{R}$, we can attach R' to Λ by identifying its residue field also with $\bar{R}$, so that R' dominates all R_i as well as R_0. Furthermore, it is easily seen that for any valuation ring in K_0 dominating R_0, any extension of it in K' must dominate R'; and a similar statement holds also for valuation rings in K_i dominating R_i. We set $Z_0 = \sum_{i=1}^{s} Z_i$, so that Z_0 is rational over K_0; since Z is K-connected, it follows that Z_0 is K_0-connected. In fact, any K_0-irreducible component X_0 in Z_0 is the sum of the complete set of conjugates over K_0 of a K-irreducible component X in Z, and two such components X_0 and Y_0 in Z_0 will certainly meet if the corresponding components X and Y in Z meet. It is sufficient to show that $|\bar{Z}_0(R_0)| = |\bar{Z}(R)|$; for, since R_0 is a complete regular local domain, $|\bar{Z}_0(R_0)|$ must be $\bar{R}$-connected by what we have just shown above, and hence $|\bar{Z}(R)|$ must also be $\bar{R}$-connected.

Since every valuation ring in K_0 dominating R_0 can be extended to a valuation ring in K' dominating R', it follows that $\bar{Z}_0(R_0) = \bar{Z}_0(R')$, and since every valuation ring in K_i dominating R_i can also be extended to a valuation ring in K' dominating R', it follows that $\bar{Z}_i(R_i) = \bar{Z}_i(R')$. It is sufficient to show that $|\bar{Z}(R)| = |\bar{Z}_i(R_i)|$ for every i; for then $|\bar{Z}_0(R_0)|$

$= | \bar{Z}_0(R') | = \bigcup_i | \bar{Z}_i(R') | = \bigcup_i | \bar{Z}_i(R_i) | = | \bar{Z}(R) |$. To prove this we observe first that R_0 contains a coefficient ring k ([6], Theorem 9 and Theorem 11) which is also a coefficient ring for every R_i and is invariant under the isomorphism between R and R_i over R_0. Let a and a' be any two conjugate elements over K_0 in R and R_i respectively, and let b be an element in k such that $\bar{b} = \bar{a}$ (such an element b exists by the definition of a coefficient ring); then $a - b$ is in $\mathfrak{p}$, and hence by isomorphism $a' - b$ is in $\mathfrak{p}_i$ (the maximal prime ideal in R_i), which show that $\bar{b} = \bar{a}'$. Thus any two conjugate elements over K_0 in R and R_i must have the same residue in $\bar{R}$. Let y be the associated point of Z, and let y' be the associated point of Z_i; if $f(Y)$ and $f'(Y)$ are two conjugate forms over K_0 in $R[Y]$ and $R_i[Y]$ respectively, then by isomorphism we have the relation $f(y) = 0$ if and only if $f'(y') = 0$, and it follows from what we have just shown that the residue forms $\bar{f}(Y)$ and $\bar{f}'(Y)$ must coincide. Since $\bar{y}(R)$ and $\bar{y}'(R_i)$ are defined by such residue forms, it follows that the two sets $\bar{y}(R)$ and $\bar{y}'(R_i)$ must coincide; hence the two algebraic sets of cycles $\bar{Z}(R)$ and $\bar{Z}_i(R_i)$ must coincide, and hence $| \bar{Z}(R) | = | \bar{Z}_i(R_i) |$.

We observe here that if we are only concerned with the absolute geometry and assume the absolute connectednesss of Z, then it would be sufficient to invoke the simpler Principle of Degeneration in [4] instead of Theorem 2; only we must then invoke directly the connectedness of $\mathfrak{V}(R)$, which will also be used in the proof of Theorem 2. In fact, in this case Z would be N^*-connected for every N in $\mathfrak{V}(R)$, and therefore by that principle $\bar{Z}(N^*) = \bar{Z}(N)$ would be $\bar{N}$-connected and hence also $\bar{R}$-connected. Then, instead of using Theorem 2, we need only to observe in the first part of the above proof that since $\mathfrak{V}(R)$ is connected, all cycles in each of the systems $\Xi'(R)$ and $\Xi''(R)$ must have the same degree, so that every cycle in $\bar{Z}(R)$ must split into two $\bar{R}$-disjoint parts. Of course, since in this case the hypothesis is preserved under any extension of the local domain R to another complete local domain dominating it, one sees readily that the point set $| \bar{Z}(R) |$ must in fact be absolutely connected.

We shall conclude this section with a remark on the uniqueness of the type of splittings which arise from the Hensel's Lemma. Let R be a valuation ring in a field K, with the residue field $\bar{R}$, and let $F(X)$ be a form in $R[X]$; if $F(X) = G(X)H(X)$ is a splitting of $F(X)$ in $R[X]$ (or in $K[X]$) such that $\bar{G}_R(X)$ and $\bar{H}_R(X)$ are relatively prime forms in $\bar{R}[X]$, then $G(X)$ and $H(X)$ are uniquely determined by the forms $\bar{G}_R(X)$ and $\bar{H}_R(X)$ up to proportionality factors in K. Therefore, if a, b, c are the *leading coefficients* (leading coefficient = coefficient of the term with the lowest "weight") in

$F(X)$, $G(X)$, $H(X)$ respectively, then the additional stipulation of the splitting $a=bc$ in R or K determines the forms $G(X)$ and $H(X)$ uniquely. For proof, one need only observe that since $\bar{G}_R(X)$ and $\bar{H}_R(X)$ are relatively prime, $G(X)$ and $H(X)$ must also be relatively prime, and the uniqueness of the splitting $F(X)=G(X)H(X)$ then follows from the unique factorization property of the polynomial ring $K[X]$. It is convenient to express this uniqueness by saying that the splitting $F(X)=G(X)H(X)$ is determined uniquely by the *initial condition* $(\bar{G}_R(X), \bar{H}_R(X); b, c)$.

5. Proof of Theorem 2. We shall first prove a lemma on a relation between a complete regular local ring and the completion of its quadratic valuation ring.

Let $(R, \mathfrak{p})$ be a complete regular local ring of dimension t, with the quotient field K and the residue field $\bar{R}$, and let $x_1, \cdots, x_t$ be the elements in a minimal basis for $\mathfrak{p}$. We denote by Q a system of representatives of $\bar{R}$ in R; apart from the condition that every element in $\bar{R}$ is the $\mathfrak{p}$-residue of exactly one element in Q, the elements in Q can be chosen arbitrarily, but for convenience we shall assume that Q contains the elements 0 and 1. We denote by $Q[x]=Q[x_1, \cdots, x_t]$ the set of all polynomials in $x_1, \cdots, x_t$ with coefficients in Q, and we shall say that an element in $Q[x]$ is *monic* if the coefficient of the leading term (i. e. the term with the lowest weight, each x_i being assigned the weight i) is 1. We denote by $Q(x)$ the set of all quotients $f(x)/g(x)$ of two elements in $Q[x]$ such that $g(x)$ is monic and that the polynomials $\bar{f}(T)$ and $\bar{g}(T)$ are relatively prime (where $T=(T_1, \cdots, T_t)$ is a system of indeterminates), and we shall say that an element $f(x)/g(x)$ in $Q(x)$ is homogeneous of degree i if both $f(x)$ and $g(x)$ are forms in x and $i=$ degre of $f(x)$ — degree of $g(x)$. We shall now associate with the system of elements $x=(x_1, \cdots, x_t)$ a system of independent variables $\boldsymbol{x}=(\boldsymbol{x}_1, \cdots, \boldsymbol{x}_t)$ over $\bar{R}$ in Λ; then we can associate with each element $f(x)/g(x)$ in $Q(x)$ the element $\bar{f}(\boldsymbol{x}/\bar{g}(\boldsymbol{x})$ in the field $\bar{R}(\boldsymbol{x})$, which we shall call the *homogeneous residue* of $f(x)/g(x)$. It is easily seen that, by virtue of our choice of the elements in $Q(x)$ made above, this association of the elements in $Q(x)$ with their homogeneous residues in $\bar{R}(\boldsymbol{x})$ is a one-to-one mapping of $Q(x)$ onto $\bar{R}(\boldsymbol{x})$, which maps the subset $Q[x]$ onto the subring $\bar{R}[\boldsymbol{x}]$.

We denote by $Q[(x)]$ the set of all series of the type $\sum_{i=0}^{\infty} a^{(i)}$, where each term $a^{(i)}$ is a homogeneous element of degree i in $Q(x)$, and we denote by $Q[[x]]$ the subset in $Q[(x)]$ consisting of all such series in which each term

$a^{(i)}$ is a homogeneous degree i in $Q[x]$. Now, since R is a complete regular local ring, it can be easily shown that every element a in R can be represented *in a unique manner* by a series $a = \sum_{i=0}^{\infty} a^{(i)}$ in $Q[[x]]$, while conversely it is obvious that any series in $Q[[x]]$ is convergent in the topology of R and hence represent an element in R; this shows that we can set $R = Q[[x]]$, and we shall say that the series $\sum_{i=0}^{\infty} a^{(i)}$ is the *power series development* of the element a (with respect to the given choice of the basis elements $x_1, \cdots, x_t$ and the system Q), although it is in general not a power series in the usual sense. Similarly, if we denote by $(M, \mathfrak{q})$ the $\mathfrak{p}$-adic valuation ring of R, then every element in the completion M^* of M can be represented *in a unique manner* by a series in $Q[(x)]$, while conversely it is also obvious from the definition of the $\mathfrak{p}$-adic valuation that any series in $Q[(x)]$ is convergent in the topology of M and hence represent an element in M^*; therefore we can set $M^* = Q[(x)]$, and we shall say that the series in $Q[(x)]$ corresponding to an element in M^* is the *power series development* of this element.

If $a = \sum_{i=0}^{\infty} a^{(i)}$ is any element in M^*, then the first non-zero term $a^{(\mu)}$ in the series is called the *initial element* of a and the number μ is called the *order* of a; the homogeneous residue $\boldsymbol{a}^{(\mu)}$ of $a^{(\mu)}$ is also called the homogeneous residue of a and will be denoted also by $\boldsymbol{a}$. For elements of order zero in M^*, i. e. units in M^*, the mapping $a \to \boldsymbol{a}$ induces an isomorphism between the residue field $M^*/M^*\mathfrak{q}$ and the subfield in $\bar{R}(\boldsymbol{x})$ consisting of all homogeneous elements of order zero; we can therefore attach M^* to Λ by identifying $M^*/M^*\mathfrak{q}$ with this subfield, and we observe that with this attachment of M^* the homogeneous residue $\boldsymbol{a}$ of any element a of order zero coincides with its residue $\bar{a}$ modulo $M^*\mathfrak{q}$. Geometrically, we can consider the system $\boldsymbol{x} = (\boldsymbol{x}_1, \cdots, \boldsymbol{x}_t)$ as the system of homogeneous coordinates of a generic point ζ over $\bar{R}$ of the center $\bar{x}(R)$ of the quadratic transform $x(R)$ of R, so that we can set $\bar{R}(\zeta) = M/\mathfrak{q} = M^*/M^*\mathfrak{q}$. However, we stress here that the elements $\boldsymbol{x}_i$ are not in $\bar{R}(\zeta)$, and that the symbol $\boldsymbol{x}$ denotes the system of t elements $(\boldsymbol{x}_1, \cdots, \boldsymbol{x}_t)$, not the point ζ, whose homogeneous coordinates are after all only determined up to a proportionality factor.

If $F(X)$ is a polynomial (or form) in $M^*[X] = M^*[X_0, X_1, \cdots, X_n]$, then we can consider the *power series development* $F(X) = \sum_{i=0}^{\infty} F^{(i)}(X)$, where $F^{(i)}(X)$ is a polynomial (or form) in X with coefficients which are homogeneous elements of degree i in $Q(x)$; the first non-vanishing term $F^{(\lambda)}(X)$

in this series is then called the *initial polynomial* (or *form*) of $F(X)$, and the polynomial (or form) $\mathbf{F}^{(\lambda)}(X)$ obtained from $F^{(\lambda)}(X)$ by replacing every coefficient in it by its homogeneous residue will be called the *homogeneous residue polynomial* (or *form*) of $F(X)$ and will be denoted also by $\mathbf{F}(X)$. it is clear that in case $\mathbf{F}^{(0)}(X)$ does not vanish, we have the equation $\mathbf{F}(X) = \mathbf{F}^{(0)}(X) = \bar{F}(X)$.

Let η be a point of dimension $t-s$ over $\bar{R}$ in the center $\bar{x}(R)$ of the quadratic transform $x(R)$; without any loss of generality, we can assume that $\eta_1 = 1$ and that $\eta_{s+1}, \cdots, \eta_t$ are independent variables over $\bar{R}$. Since $\bar{x}(R)$ coincides with the projective space $\bar{S}^{t-1}$, there exist $s-1$ forms $\phi_2(X), \cdots, \phi_s(X)$ in $\bar{R}[X]$ of the same degree μ such that the elements $\phi_2(\zeta)/\zeta_1{}^{\mu}, \cdots, \phi_s(\zeta)/\zeta_1{}^{\mu}$ constitute a basis for the maximal prime ideal in the specialization ring $\mathfrak{Q}(\zeta/\eta, \bar{R})$ in $\bar{R}(\zeta)$ of $\bar{S}^{t-1}$ at η. If $f_2(X), \cdots, f_s(X)$ are forms in $R[X]$ such that $\bar{f}_i(X) = \phi_i(X)$, then the elements $x'_1 = x_1$, $x'_2 = f_2(x)/x_1{}^{\mu-1}, \cdots, x'_s = f_s(x)/x_1{}^{\mu-1}$ constitute a basis for the maximal prime ideal in $\mathfrak{Q}(x(R)/\eta)$. Since η is a specialization of ζ over $\bar{R}$, M is the quotient ring of $\mathfrak{Q}(x(R)/\eta)$ with respect to the prime ideal $\mathfrak{Q}(x(R)/\eta) \cap \mathfrak{q}$, and since the ideal $\mathfrak{Q}(x(R)/\eta) \cap \mathfrak{q}$ must then have the rank 1 and evidently contains the prime ideal generated by $x'_1 = x_1$ in $\mathfrak{Q}(x(R)/\eta)$, it must coincide with the latter. Furthermore, it is easily seen that $\mathfrak{Q}(x(R)/\eta)/\mathfrak{Q}(x(R)/\eta) \cap \mathfrak{q}$ is isomorphic to $\mathfrak{Q}(\zeta/\eta, \bar{R})$ and hence can be identified with the latter, so that $M/\mathfrak{q} = \bar{R}(\zeta)$ is the quotient field of $\mathfrak{Q}(x(R)/\eta)/\mathfrak{Q}(x(R)/\eta) \cap \mathfrak{q}$.

Consider now the completion $(R_1, \mathfrak{p}_1)$ of $\mathfrak{Q}(x(R)/\eta)$ and denote by $\mathfrak{q}_i$ the prime ideal generated by x'_1 in R_1 (note that x'_1 is also an element of a basis for $\mathfrak{p}_1$); since $\mathfrak{q}_1 = R_1(\mathfrak{Q}(x(R)/\eta) \cap \mathfrak{q})$, it follows that $R_1/\mathfrak{q}_1$ is the completion of $\mathfrak{Q}(x(R)/\eta)/\mathfrak{Q}(x(R)/\eta) \cap \mathfrak{q} = Q(\zeta/\eta, \bar{R})$. Let $(M', \mathfrak{q}')$ be the quotient ring of R_1 with respect to $\mathfrak{q}_1$, which is a real discrete valuation ring. Since M is the quotient ring of $\mathfrak{Q}(x(R)/\eta)$ with respect to $\mathfrak{Q}(x(R)/\eta) \cap \mathfrak{q} = \mathfrak{Q}(x(R)/\eta) \cap \mathfrak{q}'$, it can be embedded canonically as a subring in M', and if we attach M' to Λ compatibly with the attachement of M to Λ, then it is clear that M' dominates M. Consider now the completion M'^* of M'; it is important for our present purpose to observe that the completion M^* of M *can be embedded canonically in M'^* as a subspace.* In fact, according to [12], Lemma 7, we have the relation $R_1 \cap \mathfrak{q}'^i = R_1(\mathfrak{Q}(x(R)/\eta) \cap \mathfrak{q}^i)$ for every positive integer i; it follows that

$$\begin{aligned}\mathfrak{Q}(x(R)/\eta) \cap (M \cap \mathfrak{q}'^i) &= \mathfrak{Q}(x(R)/\eta) \cap (R_1 \cap \mathfrak{q}'^i) \\ &= \mathfrak{Q}(x(R)/\eta) \cap R_1(\mathfrak{Q}(x(R)/\eta) \cap \mathfrak{q}^i) = \mathfrak{Q}(x(R)/\eta) \cap \mathfrak{q}^i,\end{aligned}$$

and hence $M \cap \mathfrak{q}'^i = \mathfrak{q}^i$. This shows that M' contains M as a subspace, and hence M'^* contains M^* as a subring.

Let $a = \sum_{i=\lambda}^{\infty} a^{(i)}$ be an element in M^*, expressed as a power series with $a^{(\lambda)} \neq 0$. We want to find the condition that this element a be contained in R_1; this question makes sense because both M^* and R_1 are subrings in M'^*. It is evidently sufficient to assume that every element $a^{(i)}$ is contained in $\mathfrak{Q}(x(R)/\eta)$; we shall show that this condition is also necessary. If a is contained in R_1, then a/x_1^λ is also contained in R_1; hence the $M'^*\mathfrak{q}'$-residue of a/x_1^λ, being also the $\mathfrak{q}_1$-residue of a/x_1^λ, is contained in the completion of $\mathfrak{Q}(\zeta/\eta, \bar{R})$. On the other hand, since a/x_1^λ and $a^{(\lambda)}/x_1^\lambda$ have the same $M'^*\mathfrak{q}'$-residue, and since the $M'^*\mathfrak{q}'$-residue of $a^{(\lambda)}/x_1^\lambda$, being also the $\mathfrak{q}$-residue of $a^{(\lambda)}/x_1^\lambda$, is contained in $\bar{R}(\zeta)$, it follows that the $M'^*\mathfrak{q}'$-residue of $a^{(\lambda)}/x_1^\lambda$ must be contained in both $\bar{R}(\zeta)$ and the completion of $\mathfrak{Q}(\zeta/\eta, \bar{R})$. Since $\bar{R}(\zeta)$ is the quotient field of $\mathfrak{Q}(\zeta/\eta, \bar{R})$, it follows from a well-known property of local rings that the $M'^*\mathfrak{q}'$-residue of $a^{(\lambda)}/x_1^\lambda$ must be contained in $\mathfrak{Q}(\zeta/\eta, \bar{R})$. Since this means that the $\mathfrak{q}$-residue of $a^{(\lambda)}/x_1^\lambda$ is contained in $\mathfrak{Q}(\zeta/\eta. \bar{R})$, it follows from the uniqueness of the power series development of a that $a^{(\lambda)}/x_1^\lambda$ must be contained in $\mathfrak{Q}(x(R)/\eta)$, and hence $a^{(\lambda)}$ itself must be contained in $\mathfrak{Q}(x(R)/\eta)$. Now since the element $a - a^{(\lambda)}$ is also an element in R_1, we can apply the same argument to it as we did to a and conclude that $a^{(\lambda+1)}$ must also be contained in $\mathfrak{Q}(x(R)/\eta)$. Thus, continuing in this way, we conclude that every element $a^{(i)}$ is contained in $\mathfrak{Q}(x(R)/\eta)$.

Since R is the intersection of all the rings $\mathfrak{Q}(x(R)/\eta)$ as η runs through all points in $\bar{x}(R)$ (in fact, it is sufficient to consider only point of dimension $t-2$ over $\bar{R}$), it follows that an element in M^* is in R if and only if it is contained in the completion of $\mathfrak{Q}(x(R)/\eta)$ for every point η (of dimension $t-2$ over $\bar{R}$) in $\bar{x}(R)$. We shall state this result as a lemma for later reference.

LEMMA 5.1. *An element in M^* is in R if and only if it is contained in the completion of $\mathfrak{Q}(x(R)/\eta)$ for every point η (of dimension $t-2$ over $\bar{R}$) in $\bar{x}(R)$.*

Consider now the canonical mapping of M'^* onto $M'^*/\mathfrak{q}'^* = M'/\mathfrak{q}'$, and let M_1 be the subring in M'^* consisting of all elements whose images under this mapping are in $R_1/\mathfrak{q}_1$; it is clear that M_1 contains $\mathfrak{q}'^*$ as a minimal prime ideal, and that $M_1/\mathfrak{q}'^* = R_1/\mathfrak{q}_1$ and $M_{1,\mathfrak{q}'^*} = M'^*$. In case $t = s = 2$, the local ring $R_1/\mathfrak{q}_1$ has the dimension 1 and hence is a complete real discrete valuation ring; it follows then that in this case M_1 is a discrete valuation ring of rank 2, obtained from the composition of the valuations of M'^* and $R_1/\mathfrak{q}_1$, and that it is complete. This ring M_1 will be useful to us later; we observe here that M_1 dominates R_1.

We proceed now to prove Theorem 2.

Proof of Theorem 2. We observe first that, just as in the proof of the Principle of Degeneration in [4], we can without any loss of generality restrict ourselves to the case where Z is K-irreducible. For the condition for the supports of two positive cycles to have a common point can be expressed by a system of "universal" equations in the coefficients of their associated forms, and it is clear that such a condition is always preserved under a specialization. If Z is the sum of two positive cycles Z' and Z'', both rational over K, such that their supports meet, then the support of $\bar{Z}'(N)$ and $\bar{Z}''(N)$ must also meet for every N in $\mathfrak{V}(R)$, so that $\bar{Z}'(N)$ and $\bar{Z}''(N)$ cannot be $\bar{R}$-disjoint; it follows then that any $\bar{R}$-disjoint splitting of the system $\bar{Z}(R)$ will induce an $\bar{R}$-disjoint splitting of at least one of the two systems $\bar{Z}'(R)$ and $\bar{Z}''(R)$. Thus the validity of our theorem for every K-irreducible positive cycle will imply its validity in the general case.

Next, we show that we can reduce our theorem to the special case where Z is a positive 0-cycle. Let $u^{(i)}{}_j$, $i=1,\cdots,r$, $j=0,1,\cdots,n$, be $r(n+1)$ independent variables over K, and set $u^{(i)}=(u^{(i)}{}_0,u^{(i)}{}_1,\cdots,u^{(i)}{}_n)$, $i=1,\cdots,r$ (r being the dimension of the cycle Z). The ring $R[[u]]$ of all power series in $u^{(i)}{}_j$ with coefficients in R is easily seen to be a complete regular local ring (of dimension $t+r(n+1)$, t being the dimension of R), and we can attach it to Λ by identifying the residue field of $R[[u]]$ with $\bar{R}$, so that $R[[u]]$ dominates R. Let $F(U^{(0)},U^{(1)},\cdots,U^{(r)})$ be the associated form of Z, and consider the form $F(U^{(0)},u)=F(U^{(0)},u^{(1)},\cdots,u^{(r)})$ obtained from it by substituting the indeterminates $U^{(1)},\cdots,U^{(r)}$ by $u^{(1)},\cdots,u^{(r)}$; it is clear that since $F(U^{(0)},U^{(1)},\cdots,U^{(r)})$ is K-irreducible, $F(U^{(0)},u)$ must be $K(u)$-irreducible. Since the coefficients in $F(U^{(0)},u)$ are homogeneous of the same degree in the variables $u^{(i)}{}_j$, it is easily seen that $F(U^{(0)},u)$ must be irreducible also over the power series ring $K[[u]]$; and since $K[[u]]$ is a unique-factorization ring, this implies that $F(U^{(0)},u)$ is irreducible over the quotient field $K((u))$ of $K[[u]]$, and hence also irreducible over the quotient field $R((u))$ of $R[[u]]$. The form $F(U^{(0)},u)$ therefore determines a positive 0-cycle Z_u in S^n which is rational and irreducible over $R((u))$; and we observe that since Z_u is contained in the support of Z, the cycle $\bar{Z}_u(N_1)$ must be contained in the support of the cycle $\bar{Z}(N_1)$ for every valuation ring N_1 in $\mathfrak{V}(R[[u]])$. If $\bar{Z}(N)=\Omega'(N)+\Omega''(N)$ $(N\in\mathfrak{V}(R))$ is an $\bar{R}$-disjoint splitting of $\bar{Z}(R)$, then we denote by $\Omega'_u(N_1)$ and $\Omega''_u(N_1)$ the restrictions of

$\bar{Z}_u(N_1)$ in $|\Omega'(N_1)|$ and $|\Omega''(N_1)|$ respectively; the equation $Z_u(N_1) = \Omega'(N_1) + \Omega''_u(N_1)$ $(N_1 \in \mathfrak{V}(R[[u]]))$ then defines an $\bar{R}$-disjoint splitting of $\bar{Z}_u(N_1)$, provided it is not trivial for at least one N_1. But this is easily seen to be the case when we take for N_1 the quadratic valuation ring of $R[[u]]$; for in this case $\bar{u}^{(i)}(N_1)$, $i = 1, \cdots, r$, are independent generic points in S^n over a field of rationality for $\bar{Z}(N_1)$, and $\bar{Z}_u(N_1)$ is simply the intersection cycle of $\bar{Z}(N_1)$ with the linear subspace of dimension $n - r$ defined by the r dual hyperplanes of the points $\bar{u}^{(i)}(N_1)$. Thus the validity of our theorem for the 0-cycle Z_u implies its validity for the r-cycle Z.

Finally, we reduce our theorem to the case where the dimension t of R is equal to 2, whereby we note that the case $t = 1$ (Z being now a 0-cycle) reduces essentially to the Hensel's Lemma. We observe first that it is permissible to replace R by the complete local domain $R(v)^*$ obtained from R by the adjunction of $t - 2$ independent variables $v = (v_1, \cdots, v_{t-2})$ over K; for the cycle Z will remain irreducible over $K(v)$, while any $\bar{R}$-disjoint splitting of $\bar{Z}(R)$ will induce an $\bar{R}(\bar{v})$-disjoint splittings of $\bar{Z}(R(v)^*)$ through the mapping $N \rightarrow N \cap K$ of $\mathfrak{V}(R(v)^*)$ into $\mathfrak{V}(R)$. We can therefore apply Lemma 2.9 (taking $d = t - 2$) and consider the local domain R_v^* introduced there, which is a complete regular local ring of dimension 2; since K is algebraically closed in the quotient field of R_v^* and Z is irreducible over K, it follows that Z is also irreducible over the quotient field of R_v^*. On the other hand, since R_v^* dominates R, the system $\bar{Z}(R_v^*)$ is a subsystem of $\bar{Z}(R)$; it follows that an $\bar{R}$-disjoint splitting of $\bar{Z}(R)$ will induce also such a splitting of $\bar{Z}(R_v^*)$, which must be of course also $\bar{R}(\bar{v}, \bar{u})$-disjoint, using the notation in Lemma 2.9. Thus the validity of our theorem for the case $r = 0$, $t = 2$, will imply its validity in the general case.

We therefore have to prove our theorem only for the case where Z is a K-irreducible positive 0-cycle and R is a complete regular local ring of dimension 2; we assume that there exists an $\bar{R}$-disjoint splitting of the system $\bar{Z}(R)$ and we shall show that the cycle Z must then be reducible over K, in contradiction to our hypothesis. In terms of the associated form $F(U)$, the given $\bar{R}$-disjoint splitting of $\bar{Z}(R)$ can be expressed by a splitting $\bar{F}_N(U) \doteq \Phi_N(U)\Theta_N(U)$ of the system of forms $\bar{F}_R(U)$ into two systems $\Phi_R(U)$ and $\Theta_R(U)$ such that $\Phi_N(U)$ and $\Theta_N(U)$ are relatively prime over $\bar{R}$ for every valuation ring N in $\mathfrak{V}(R)$, and our theorem will clearly follow if we show that there is then a splitting $F(U) = G(U)H(U)$ in $R[U]$ such that $\bar{G}_N(U) \doteq \Phi_N(U)$ and $\bar{H}_N(U) \doteq \Theta_N(U)$ for every N in $\mathfrak{V}(R)$. Let y be the associated point of Z, and consider the rational transform $y(R)$; since R is a regular local ring of dimension 2, there exists according to Theorem 1, Corollary 1, a finite quadratic transform $z(R)$ of a certain order π

(along a suitable sequence of valuations) such that the correspondence from $z(R)$ to $y(R)$ is everywhere regular. In case $\pi=0$, i.e., $z(R)=R$, the center $\bar{y}(R)$ is a uniquely determined point and the corresponding form is just the residue form $\bar{F}(U)$; it follows that $\bar{F}_N(U)=\bar{F}(U)$ for every N in $\mathfrak{V}(R)$, so that all forms in the system $\bar{F}_R(U)$ are essentially the same. Now, there are only a finite number of ways in which the form $\bar{F}(U)$ can split into two factors and the set of all N such that the splitting $\bar{F}_N(U)=\Phi_N(U)\Theta_N(U)$ coincides with a given splitting of $\bar{F}(U)$ is easily seen to be a closed subset in $\mathfrak{V}(R)$. Since according to Theorem 1, Corollary 2, the space $\mathfrak{V}(R)$ is connected, it follows that all splittings in the given $\bar{R}$-disjoint splitting of the system $\bar{F}_R(U)$ must coincide with one single splitting $\bar{F}(U)\doteq\Phi(U)\Theta(U)$; furthermore, since $\bar{F}(U)$ is rational over $\bar{R}$ and the splitting is $\bar{R}$-disjoint, it follows that both $\Phi(U)$ and $\Theta(U)$ must be rational over $\bar{R}$. By Hensel's Lemma for the local ring R, there exist forms $G(U)$ and $H(U)$ in $R[U]$ such that $F(U)=G(U)H(U)$, $\bar{G}(U)\doteq\Phi(U)$, and $\bar{H}(U)\doteq\Theta(U)$. This proves our assertion for $\pi=0$. We shall prove our assertion in the general case by induction on the order π of $z(R)$; we therefore assume that our assertion has already been proved for the case where $y(R)$ is an everywhere regular transform of a finite quadratic transform of R of order less than π.

It is well-known that R is a unique -factorization ring; for convenience, we shall assume that the coefficients in $F(U)$ have been so chosen that they are all in R, but without any common factor. Consider the homogeneous residue form $\boldsymbol{F}(U)$ of $F(U)$; since $\boldsymbol{F}(U)$ differs from $\bar{F}_M(U)$ only by a proportionality factor in $\bar{R}(\boldsymbol{x})$ (we recall that $M=Q(x(\mathrm{R})/\zeta)$ and ζ is the point in $\bar{x}(R)$ with $(\boldsymbol{x}_1,\cdots,\boldsymbol{x}_t)$ as homogeneous coordinates), the given splitting $\bar{F}_M(U)\doteq\Phi_M(U)\Theta_M(U)$ induces a splitting $\boldsymbol{F}(U)\doteq\Phi(U)\Theta(U)$ in $\bar{R}[\boldsymbol{x}]$, with $\Phi(U)\doteq\Phi_M(U)$ and $\Theta(U)\doteq\Theta_M(U)$. We assume now that (A) *the leading coefficient* α *in* $\boldsymbol{F}(U)$ *does not vanish*; if we denote by δ the highest common factor (in $\bar{R}[\boldsymbol{x}]$) of all coefficients in $\boldsymbol{F}(U)$, and choose the forms $\Phi(U)$ and $\Theta(U)$ (which are each determined up to a proportionality factor in $\bar{R}(\boldsymbol{x})$) so that the coefficients of each form are all in $\bar{R}[\boldsymbol{x}]$ but without any common factor, then we have $\boldsymbol{F}(U)=\delta\Phi(U)\Theta(U)$ up to a proportionality factor in $\bar{R}$, which can be absorbed in δ. If β and γ are the leading coefficients in $\Phi(U)$ and $\Theta(U)$ respectively, we have then $\alpha=\delta\beta\gamma$; we assume that (B) *the elements* δ, β, γ *are relatively prime.* The properties (A) and (B) assumed here, if not already present, can be easily obtained as follows. Let $w=(w_0,\mathrm{w}_1,\cdots,w_n)$ be a system of independent variables over K, and consider the complete local ring $R(w)^*$, which is also regular and has the dimension 2, attached to Λ by identifying the elements $\bar{w}=(\bar{w}_0,\bar{w}_1,\cdots,w_n)$ with a system of independent variables over $\bar{R}$. We apply to $F(U)$ the linear transformation which leaves $U_1, \cdots, U_n$ invariant but carries U_0 into $\sum_{i=0}^{n} w_i U_i$, and apply to $\boldsymbol{F}(U)$, $\boldsymbol{\Phi}(U)$, $\boldsymbol{\Theta}(U)$ the corresponding transformation whereby $U_1, \cdots, U_n$ remain invariant but U_0 is carried into $\sum_{i=0}^{n} \bar{w}_i U_i$; then the elements $\boldsymbol{F}(\bar{w})$, $\boldsymbol{\Phi}(\bar{w})$, and $\boldsymbol{\Theta}(\bar{w})$ becomes the leading coefficients of the transformed forms of $\boldsymbol{F}(U)$, $\boldsymbol{\Phi}(U)$, and $\boldsymbol{\Theta}(U)$ respectively. It is clear that our problem is invariant under such a linear transformation. Now, it is obvious that (A) holds for the transformed forms; while as to (B), we need only to observe that any common factor (in $\bar{R}[\boldsymbol{x}]$) of all coefficients in $\boldsymbol{F}(U)$ appears with the minimum power in $\boldsymbol{F}(\bar{w})$ (so that we can assume δ to be relatively prime to β and γ), and that since $\boldsymbol{\Phi}(U)$ and $\boldsymbol{\Theta}(U)$ are relatively prime with their coefficients determined as indicated above, the elements $\boldsymbol{\Phi}(\bar{w})$ and $\boldsymbol{\Theta}(\bar{w})$ must be relatively prime (so that we can assume β and γ to be relatively prime). On the other hand, since the field K is algebraically closed in the quotient field of $R(w)^*$, any splitting $F(U) = G(U)H(U)$ of $F(U)$ in $R(w)^*[U]$ must be already a splitting in $R[U]$, so that the validity of our assertion for $R(w)^*$ will imply its validity for R. Thus we have shown that we can assume the properties (A) and (B) without any loss of generality.

Let λ be the order of $F(U)$ in its power series development as a form in $M^*[U]$, and let a be the leading coefficient $F(U)$; by (A), the order of a must also be λ and we have $\boldsymbol{a} = \alpha = \beta\gamma\delta$, so that λ is also the degree of α (as a form in $\bar{R}[\boldsymbol{x}]$). Let μ, ν, ϵ be the degrees of β, γ, δ respectively, so that $\lambda = \mu + \nu + \epsilon$.

We now invoke a property of the local ring R which is dependent on its dimension being 2. According to [8], Satz 9 (see item (b) in the proof), any splitting of $\boldsymbol{a}$ into relatively prime elements is induced by a corresponding splitting of a in R. Since β, γ, δ are relatively prime, there exist therefore elements b, c, d in R such that $a = bcd$, $\boldsymbol{b} = \beta$, $\boldsymbol{c} = \gamma$, $\boldsymbol{d} = \delta$. Let $\delta = \delta_1 \cdots \delta_s$ be the factorization of δ into powers of distinct prime elements in $\bar{R}[\boldsymbol{x}]$; there exist then elements $d_1, \cdots, d_s$ in R such that $d = d_1 \cdots d_s$, $\boldsymbol{d}_1 = \delta_1, \cdots, \boldsymbol{d}_s = \delta_s$. We denote by ϵ_i the order of d_i, which is the degree of δ_i, so that $\epsilon = \epsilon_1 + \cdots + \epsilon_s$. We observe that since each δ_i is a power of an irreducible form in $\bar{R}[\boldsymbol{x}]$, it determines a point $\xi^{(i)}$ in $\bar{x}(R)$, unique up to conjugates over $\bar{R}$, which is algebraic over $\bar{R}$, so that $Q(x(R)/\xi^{(i)})$ has the dimension 2.

Let η be a point in $\bar{x}(R)$; following the notations introduced in the proof of Lemma 5.1, we shall denote the completion of $Q(x(R)/\eta)$ by R_1, and we observe that R_1 is now a complete regular local ring of dimension $\leqq 2$ (the dimension is equal to 1 if and only if η is a generic point of $\bar{x}(R)$ over $\bar{R}$). Let $z'(R_1)$ be the finite quadratic transform of R_1 along the subsequence of those valuations in the definition of $z(R)$ which dominate $Q(x(R)/\eta)$, or rather the sequence of the extensions of these (real discrete) valuations to R_1; it is easily seen that the correspondence from $z'(R_1)$ to $z(R_1)$ is regular everywhere, from which it follows that the correspondence from $z'(R_1)$ to $y(R_1)$ is also regular everywhere. Since the order of the quadratic transform $z'(R_1)$ is evidently less than π, we can apply our induction hypothesis. We recall that if N is any valuation ring dominating R, then by definition $\Phi_N(U) = \Phi_{K \cap N}(U)$ and $\Theta_N(U) = \Theta_{K \cap N}(U)$, and that $\Phi_{R_1}(U)$ and $\Theta_{R_1}(U)$ are subsystems of $\Phi_R(U)$ and $\Theta_R(U)$ respectively obtained by restricting N to valution rings dominating R_1. It is clear that the equation $\bar{F}_N(U) = \Phi_N(U)\Theta_N(U)$, for every N dominating R_1, defines a splitting of $\bar{F}_{R_1}(U)$ into the two systems $\Phi_{R_1}(U)$ and $\Theta_{R_1}(U)$, which is clearly $\bar{R}$-disjoint and hence also $\bar{R}(\eta)$-disjoint. By induction hypothesis, there exists therefore a splitting $F(U) = G_\eta(U)H_\eta(U)$ in $R_1[U]$ such that $G_\eta(U)$ generates the system $\Phi_{R_1}(U)$ and $H_\eta(U)$ generates the system $\Theta_{R_1}(U)$. Now, this splitting $F(U) = G_\eta(U)H_\eta(U)$ in $R_1[U]$ implies a splitting of the leading coefficient a in $F(U)$ into two factors in R_1, which we shall now proceed to investigate. Without any loss of generality, we can assume for convenience that $\eta_1 \neq 0$, as we did in the proof of Lemma 5.1; then all coefficients in $F(U)$ are divisible by x_1^λ in R_1, and since R_1 is a unique factorization ring, there exist positive integers λ' and λ'' $(\lambda' + \lambda'' = \lambda)$ such that all coefficients in $G_\eta(U)$ are divisible by $x_1^{\lambda'}$ and all coefficients in $H_\eta(U)$ are divisible by $x_1^{\lambda''}$. If we set $F^0(U) = (1/x_1)^\lambda F(U)$, $G^0(U) = (1/x_1)^{\lambda'} G_\eta(U)$, and $H^0(U) = (1/x_1)^{\lambda''} H_\eta(U)$, then we have evidently the splitting $F^0(U) = G^0(U)H^0(U)$ in $R_1[U]$. We set $a^0 = (1/x_1)^\lambda a$, $b^0 = (1/x_1)^\mu b$, $c^0 = (1/x_1)^\nu c$, $d^0 = (1/x_1)^\epsilon d$, and $d_i^0 = (1/x_1)^{\epsilon_i} d_i$; then the elements a^0, b^0, c^0, d^0, d_i^0 are all in R_1, and a^0 is the leading coefficient in $F^0(U)$. The situation divides into two cases, which we shall treat separately.

(1) The point η coincides with one of the points $\xi^{(1)}, \cdots, \xi^{(s)}$, say $\xi^{(j)}$. In this case, it is easily seen that the elements b^0, c^0, d_i^0 $(i \neq j)$ are all units in R_1; for example, the p_1-residue of b^0 is the value of β for $\boldsymbol{x}_1 = 1$, $\boldsymbol{x}_2 = \eta_2/\eta_1$, which is not zero since β and δ_j are relatively prime in $\bar{R}[\boldsymbol{x}]$. The splitting $F^0(U) = G^0(U)H^0(U)$ induces therefore a splitting of d_j^0 into two factors in R_1. We maintain that this splitting of d_j^0 is induced by a splitting $d_j = d'_j d''_j$ in R; in other words, if ϵ' and ϵ'' are the orders of d'_j and d''_j respectively, then the given splitting of d_j^0 coincides with the splitting $d_j^0 = (d'_j/x_1^{\epsilon'})(d''_j/x_1^{\epsilon''})$ up to units in R_1. To prove this, it is sufficient to show that if e is a prime factor of d_j, of order ρ in M, then $e^0 = e/x_1^\rho$ must be a prime element in R_1; for we can then apply this argument to every

prime factor of d_j and obtain a complete factorization of d_j^0 in R_1, and our assertion then follows from the unique factorization property of R_1. First, we note that e^0 is a prime element in $R[x_2/x_1]$; for, if e^0 splits into two factors in $R[x_2/x_1]$, they must be of the form $e'/x_1^{\rho'}$ and $e''/x_1^{\rho''}$ with e' and e'' in R, and again the unique factorization property in R shows that we must have $\rho = \rho' + \rho''$ and hence $e = e'e''$. Next, since $\mathfrak{Q}(x(R)/\eta)$ is the quotient ring of $R[x_2/x_1]$ with respect to a prime ideal which contains the element e^0, it follows from a well-known property of quotient ring that e^0 must remain a prime element in $Q(x(R)/\eta)$. Now, consider the residue rings R/Re and $Q(x(R)/\eta)/Q(x(R)/\eta)e^0$; since $Q(x(R)/\eta)e^0 \cap R = R[x_2/x_1]e^0 \cap R = Re$, we can embed the former canonically in the latter. It is clear that both rings are local domains of dimension 1, and that their residue fields can be identified with $\bar{R}$ and $\bar{R}(\eta)$ respectively. Since R is complete, R/Re is also complete, and since $\bar{R}(\eta)$ is algebraic over $\bar{R}$, it follows ([6], Theorem 8) that $Q(x(R)/\eta)/Q(x(R)/\eta)e^0$ is a finite module over R/Re and hence ([3], § III, Proposition 8) must also be complete. This implies that

$$Q(x(R)/\eta)/Q(x(R)/\eta)e^0 = R_1/R_1e^0,$$

so that R_1/R_1e^0 is a local domain; this shows that e^0 is a prime element in R_1. It is clear that since $d'_j/x_1^{\epsilon'}$ and $d''_j/x_1^{\epsilon''}$ coincide up to units in R_1 with the leading coefficients in $G^0(U)$ and $H^0(U)$ respectively, we can take as the respective leading coefficients in the forms $G_\eta(U)$ and $H_\eta(U)$ any multiples of d'_j and d''_j in R which are factors in a splitting of a in R.

(2) The point η is distinct from the points $\xi^{(1)}, \cdots, \xi^{(s)}$. In this case, the form $\boldsymbol{F}(U)$ does not vanish identically at the point η, and hence at least one coefficient in $F^0(U)$ must be a unit in R_1; and this implies that also both $G^0(U)$ and $H^0(U)$ must have at least one coefficient which is a unit in R_1. If we denote by $\bar{F}^0(U)$, $\bar{G}^0(U)$, and $\bar{H}^0(U)$ the respective residue forms in $\bar{R}_1[U]$, then we must have $F^0{}_N(U) \doteq \bar{F}^0(U)$, $G^0{}_N(U) \doteq \bar{G}^0(U)$, and $H^0{}_N(U) \doteq \bar{H}^0(U)$ for every valuation ring N dominating R_1; it follows that $\Phi_N(U) \doteq \bar{G}^0(U)$ and $\Theta_N(U) \doteq \bar{H}^0(U)$ for every N dominating R_1, and hence in particular $\Phi_{M_1}(U) \doteq \bar{G}^0(U)$ and $\Theta_{M_1}(U) \doteq \bar{H}^0(U)$, where M_1 is the valuation ring of rank 2 introduced after Lemma 5.1. On the other hand, since M'^* is a quotient ring of M_1, $\Phi_{M_1}(U)$ and $\Theta_{M_1}(U)$ are specializations of $\Theta_{M'^*}(U) = \Phi_M(U)$ and $\Phi_{M'^*}(U) = \Theta_M(U)$ respectively over the specialization $\zeta \rightarrow \eta$ over $\bar{R}$; it follows that $\bar{G}^0(U)$ and $\bar{H}^0(U)$ are up to proportionality factors equal to the forms obtained from $\Phi(U)$ and $\Theta(U)$ respectively by setting $(\boldsymbol{x}_1, \boldsymbol{x}_2) = (\eta_1, \eta_2)$. Now, since β and γ have no common factor in $\bar{R}[\boldsymbol{x}]$, at least one of them does not vanish at the point η. If β does not vanish at η, then the leading coefficient in $\bar{G}^0(U)$ does not vanish and hence

the leading coefficient in $G^0(U)$ must be a unit in R_1; since in this case the elements b and d^0 are units in R_1, the leading coefficient in $H^0(U)$ must be equal to c up to a unit in R_1. On the other hand, if γ does not vanish at η, then the same argument with $G^0(U)$ and $H^0(U)$ interchanged shows that c^0 and d^0 are units in R_1, and that the leading coefficient in $G^0(U)$ must be equal to b up to a unit in R_1. Going over to the forms $G_\eta(U)$ and $H_\eta(U)$, we conclude that wec an take as their respective leading coefficients any multiples of b and c in R which are factors in a splitting of a in R.

We set $d' = d'_1 \cdots d'_s$ and $d'' = d''_1 \cdots d''_s$, so that we have $d = d'd''$, and consider the splitting $a = bd' \cdot cd''$ in R. We shall now normalize the forms $G_\eta(U)$ and $H_\eta(U)$, for every point η in $\bar{x}(R)$, by taking bd' as the leading coefficient in $G_\eta(U)$ and cd'' as the leading coefficient in $H_\eta(U)$. This choice of the leading coefficients is permissible in view of what we have just shown above. For $\eta = \zeta$, we obtain a splitting $F(U) = G_\zeta(U)H_\zeta(U)$ in $M^*[U]$; incidentally, we observe that this is the splitting according to Hensel's Lemma with the intial conditions $(\Phi_M(U), \Theta_M(U); bd', cd'')$, in the terminology introduced at the end of the section 4. We shall now show that $G_\zeta(U) = G_\eta(U)$ and $H_\zeta(U) = H_\eta(U)$ for every point η in $\bar{x}(R)$, in the sense that they are the same forms in $M'^*[U]$, so that all the splittings $F(U) = G_\eta(U)H_\eta(U)$ are in reality one and the same. This will prove our theorem. For, then the forms $G_\zeta(U)$ and $H_\zeta(U)$ must be both in $R[U]$ according to Lemma 5.1 (applied to every coefficient in them); if we set $G(U) = G_\zeta(U)$ and $H(U) = H_\zeta(U)$, then $F(U) = G(U)H(U)$ is a splitting in $R[U]$ and it is clear that $\bar{G}_N(U) \doteq \Phi_N(U)$ and $\bar{H}_N(U) \doteq \Phi_N(U)$ for every valuation ring N in K dominating R.

It remains therefore to show that $G_\zeta(U) = G_\eta(U)$ and $H_\zeta(U) = H_\eta(U)$, for every point η in $\bar{x}(R)$, whereby we shall use again the notations introduced in the proof of Lemma 5.1. Consider the valuation ring M_1 of rank 2 introduced after Lemma 5.1. It is clear that

$$\bar{G}_{\zeta,M'^*}(U) \doteq \bar{G}_{\zeta,M}(U) \doteq \Phi_M(U) \doteq \Phi_{M'^*}(U)$$

and

$$\bar{H}_{\zeta,M'^*}(U) \doteq \bar{H}_{\zeta,M}(U) \doteq \Theta_M(U) \doteq \Theta_{M'^*}(U);$$

since M'^* is a quotient ring of M_1, it follows that both $\bar{G}_{\zeta,M_1}(U)$ and $\Phi_{M_1}(U)$ are specializations of $\Phi_M(U)$ over the specialization $\zeta \to \eta$ over $\bar{R}$, and that both $\bar{H}_{\zeta,M_1}(U)$ and $\Theta_{M_1}(U)$ are specializations of $\Theta_M(U)$ over the specialization $\zeta \to \eta$ over $\bar{R}$. Since $\bar{x}(R)$ is a non-singular curve, the specializations of both $\Phi_M(U)$ and $\Theta_M(U)$ over the specialization $\zeta \to \eta$ over $\bar{R}$ must be unique; it follows that $\bar{G}_{\zeta,M_1}(U) \doteq \Phi_{M_1}(U)$ and $\bar{H}_{\zeta,M_1}(U) \doteq \Theta_{M_1}(U)$. On the other hand, by the inductive hypothesis, we have $\bar{G}_{\eta,M_1}(U) \doteq \Phi_{M_1}(U)$ and $\bar{H}_{\eta,M_1}(U)$

$\doteq \mathfrak{G}_{M_1}(U)$. This shows that $\bar{G}_{\zeta,M_1}(U) \doteq \bar{G}_{\eta,M_1}(U)$ and $\bar{H}_{\zeta,M_1}(U) \doteq \bar{H}_{\eta,M_1}(U)$. Recalling that the forms $G_\zeta(U)$ and $G_\eta(U)$ have the same leading coefficient bd', and that the forms $H_\zeta(U)$ and $H_\eta(U)$ have the same leading coefficient cd'', we see that the two splittings $F(U) = G_\zeta(U)H_\zeta(U)$ and $F(U) = G_\eta(U)H_\eta(U)$ in $M_1[U]$ have the same initial condition

$$(\bar{G}_{\zeta,M_1}(U), \bar{H}_{\zeta,M_1}(U); bd', cd'') = (\bar{G}_{\eta,M_1}(U), \bar{H}_{\eta,M_1}(U); bd', cd'');$$

it follows from the remark at the end of section 4 that $G_\zeta(U) = G_\eta(U)$ and $H_\zeta(U) = H_\eta(U)$. This concludes the proof of Theorem 2.

THE JOHNS HOPKINS UNIVERSITY.

REFERENCES.

[1] Abhyankar, S., and Zariski, O., "Splitting of valuations in extensions of local domains. I," *Proceedings of the National Academy of Sciences*, vol. 41 (1955), pp. 84-90.

[2] Abhyankar, S., "On the valuations centered in a local domain," *American Journal of Mathematics*, vol. 78 (1956), pp. 321-48.

[3] Chevalley, C., "On the theory of local rings," *Annals of Mathematics*, vol. 44 (1943), pp. 690-708.

[4] Chow, W. L., "On the principle of degeneration in algebraic geometry," *Annals of Mathematics*, vol. 66 (1957), pp. 70-79.

[5] ———, "On the theorem of Bertini for local domains," *Proceedings of the National Academy of Sciences*, vol. 44 (1958), pp. 580-584.

[6] Cohen, I. S., "On the structure and ideal theory of complete local rings," *Transactions of the American Mathematical Society*, vol. 59 (1946), pp. 54-106.

[7] Krull, W., "Ein Satz ueber primaere Integritaetsbereiche," *Mathematische Annalen*, vol. 103 (1930), pp. 450-465.

[8] ———, "Zur Theorie der kommutativen Integritaetsbereiche," *Journal fuer die reine und angewandte Mathematik*, vol. 192 (1954), pp. 230-252.

[9] Murre, J. P., "On a connectedness theorem for a birational transformation at a simple point," *American Journal of Mathematics*, vol. 80 (1958), pp. 3-15.

[10] Nagata, M., "Note on a chain condition for prime ideals," *Memoirs of the College of Science, University of Kyoto*, Series A, vol. 32 (1959), pp. 85-90.

[11] Zariski, O., "Foundations of a general theory of birational correspondences," *Transactions of the American Mathematical Society*, vol. 53 (1943), pp. 490-542.

[12] ———, "Analytical irreducibility of normal varieties," *Annals of Mathematics*, vol. 49 (1948), pp. 352-361.

[13] ———, "A simple analytical proof of the fundamental property of birational transformations," *Proceedings of the National Academy of Sciences*, vol. 35 (1949), pp. 62-66.

[14] ———, "Theory and applications of holomorphic functions on algebraic varieties over arbitrary ground fields," *Memoirs of the American Mathematical Society*, No. 5, 1951.

[15] ———, "The connectedness theorem for birational transformations," *Algebraic Geometry and Topology*, A symposium in honor of S. Lefschetz, Princeton, 1957, pp. 182-188.

ON THE REAL TRACES OF ANALYTIC VARIETIES.*[1]

By Wei-Liang Chow.

To B. L. van der Waerden on his 60th birthday

1. Let V be a complex-analytic variety of complex dimension n which is the complexification of its real trace V^0, so that V^0 is a real-analytic variety of real dimension n, and let X be a complex-analytic subvariety of complex dimension r in V such that $X^0 = X \cap V^0$ has the real dimension r (and hence $\bar{X} = X$). If we denote by $\phi(X)$ and $\phi(V^0)$ the homology classes mod 2 determined by X and V^0 respectively as topological cycles mod 2 in V, then their intersection class $\phi(X) \cdot \phi(V^0)$, with due regard for their supports, can be considered as an element in $H_{2r-n}(V^0, \mathbf{Z}_2)$. On the other hand, X^0 is also a topological cycle mod 2 in V^0 and hence determines an element $\psi(X^0)$ in $H_r(V^0, \mathbf{Z}_2)$. It is natural to ask about the relation between these two elements $\phi(X) \cdot \phi(V^0)$ and $\psi(X^0)$ in $H_*(V^0, \mathbf{Z}_2)$, since roughly speaking they represent respectively the algebraic intersection and the geometric intersection of X and V^0. We shall give here an answer to this question for the case where V and X are algebraic. Our result is surprisingly simple: We have the relation

$$\phi(X) \cdot \phi(V^0) = \psi(X^0)^2 \tag{1}$$

in the sense of numerical equivalence mod 2 for algebraic classes in $H_*(V^0, \mathbf{Z}_2)$. In case V^0 has the property that every homology class mod 2 is representable by an algebraic cycle, as for example in the case of a real Grassmannian, then the relation (1) holds in the sense of homology mod 2.

We can probably best illustrate our result by an interesting application, which was in fact the original motivation of our present inquiry. Let V and V^0 be respectively the complex and real Grassmannians of all s-spaces in the complex and real spaces of dimension m; if we consider V and V^0 as the respective classifying spaces of the complex and real s-vector bundles, then the universal dual Chern classes $c_1, \cdots, c_s$, and the universal dual Whitney classes $w_1, \cdots, w_s$, are represented by certain Schubert varieties $C_1, \cdots, C_s$ and their real traces $C_1{}^0, \cdots, C_s{}^0$ respectively. The precise sense in which the Whitney classes can be considered as the "real traces" of the Chern classes

* Received March 20, 1963.

[1] This work was partially supported by a research grant of the National Science Foundation.

has been made clear by Borel and Haefliger [1] by means of a mapping ρ of the complex-analytic cycles in V into $H_*(V^0, \mathbf{Z}_2)$, so that one has the relation $\psi(C_j^0) = \rho(C_j)$. Now, if we choose the dimension m so that V^0 is orientable and hence the Poincaré duality (over $\mathbf{Z}$) holds in V^0, then the universal dual Pontrjagin classes p_j are given by $\theta(C_j) \cdot \theta(V^0)$, $j = 1, \cdots, s$, where $\theta(C_j)$ and $\theta(V^0)$ denote the homology classes over $\mathbf{Z}$ determined by C_j and V^0 respectively as cycles over $\mathbf{Z}$ in V, and the intersection classes $\theta(C_j) \cdot \theta(V^0)$ are to be considered as elements in $H_*(V^0, \mathbf{Z})$. Since the elements $\phi(C_j) \cdot \phi(V^0)$ can be identified with the reduction mod 2 of the elements $\theta(C_j) \cdot \theta(V^0)$, our result then yields the following well-known relations between the Pontrjagin classes and the Whitney classes (or rather their duals):

$$p_j = \omega_j^2 \mod 2.$$

A topological proof of these relations using transgressive elements can be found in [2], Appendix II.

2. Our proof of (1) is based on a theorem (Theorem 1) which gives a relation between the intersection of a cycle X with its conjugate and its intersection with V^0. Although in this paper we shall need this theorem only for the case of algebraic cycles in an algebraic variety, we shall prove it here for the general case, both on account of its own interest as well as for the sake of possible generalization of our result.

We begin by making a few comments on the intersection theory for an analytic variety, either complex or real, and the relations between the two cases. It is possible to develop an intersection theory for complex-analytic cycles in a complex-analytic variety along lines roughly similar to the now classical intersection theory for algebraic cycles in an algebraic variety. Such an intersection theory, just as in the algebraic case, is essentially unique, and it is compatible with the topological homology theory in the sense that the mapping ϕ (or θ) which assign to each complex-analytic cycle in V the homology class determined by it in $H_*(V, \mathbf{Z}_2)$ (or $H_*(V, \mathbf{Z})$) is multiplicative, i.e., $\phi(X) \cdot \phi(Y) = \phi(X \cdot Y)$ whenever $X \cdot Y$ is defined. Recently, Borel and Haefliger [1], on the basis of a homology theory with supports previously developed by Borel and Moore, were able to obtain by topological methods an intersection theory for complex-analytic cycles in a complex-analytic variety. Since in this theory the intersection multiplicities are defined by means of homology classes with supports, there is no question here of compatibility and the mapping ϕ is so to speak "built-in." For our present purpose it is immaterial which theory is used; the only property we need here is the criterion for multiplicity by means of small deformations contained in the following lemma.

Before stating our lemma, we add a few words in explanation of a notation used in it. Let N be a neighborhood of the origin in the complex n-space and let $e=(e_1,\cdots,e_n)$ be a point in N; then we denote by T_e the translation in N defined by the mapping $z\to z+e$. As a transformation in N, T_e is of course only defined in $N\cap T_{-e}N$ and maps it onto $N\cap T_eN$; however, given any neighborhood N' of the origin such that $\bar{N}'\subset N$, we have the relation $N'\subset T_eN$ for all sufficiently small e. If X is any cycle in N, then we shall denote by T_eX the cycle in $N\cap T_eX$ which is the image under T_e of the restriction $X\cap T_{-e}N$ of X in $N\cap T_{-e}N$. Thus T_eX is strictly speaking not a cycle in N, but for any N' such that $\bar{N}'\subset N$, the restriction of T_eX in N' is a cycle for all sufficiently small e. If N is a neighborhood of a point p in a complex-analytic variety V such that there exists a system of local coordinates in V at p which is valid in N, and if X is any cycle in N, then we can extend the above definitions of T_e and T_eX to this more general case by means of the local coordinate system; and finally, if X is any cycle in V, we shall for simplicity write T_eX for $T_e(X\cap N)$.

LEMMA 1. *Let V be a complex-analytic variety of dimension n, and let X and Y be complex analytic prime cycles of dimension r and $n-r$ respectively in V which intersect properly at a point p; let N be a neighborhood of p in V such that X and Y intersect only at p in N and such that there exists a system of local coordinates in V at p which is valid in N. Then there is a neighborhood N' of p in N such that for almost all points e in a sufficiently small neighborhood of p, the number of intersections of T_eX with Y in N' is exactly equal to $i(X\cdot Y,p)$.*

Remark. The expression "almost all" means here "all except for a proper complex-analytic subset."

Proof. The property to be proved being local, we can without any loss of generality take V to be the coordinate neighborhood N, or equivalently take $V=N$ to be a neighborhood of the origin p in the complex n-space with the coordinates $z=(z_1,\cdots,z_n)$; furthermore, we can assume that both X and Y are locally irreducible at p, for otherwise we can treat each irreducible component separately, replacing N by a smaller neighborhood if necessary (so small that any two different local components of X or Y at p represent different subvarieties in N). Now, just as in the abstract algebraic intersection theory, we can introduce the product cycle $X\times Y$ in the product variety

$V \times V$, and show that the number of intersection points of X and T_eY in N' is the same as the number of intersection points of $X \times Y$ with $T_e\Delta$ in $N' \times N'$, where Δ is the diagonal of $V \times V$ and T_e as applied to Δ is the local translation induced in $N \times N$ by the application of the original T_e to the second factor alone. It follows then that our lemma can be reduced to the special case where Y is the linear variety in N defined by the equations $z_1 = \cdots = z_r = 0$. The condition that X and Y intersect properly at p implies that the defining ideal of X (in the power series ring of n variables $z_1, \cdots, z_n$) is regular with respect to the given coordinate system, with $z_1, \cdots, z_r$ as independent variables. We have therefore, after a suitable linear transformation of the coordinates $z_{r+1}, \cdots, z_n$ if necessary, the well-known "parametric" representation of X by equations of the form $P(z_{r+1}) = 0$, $z_i = f_i/D$, $i = r+2, \cdots, n$, where $P(z_{r+1})$ is an irreducible "distinguished polynomial" in z_{r+1} with coefficients which are power series in $z_1, \cdots, z_r$, the function D is the discriminant of the polynomial P, and the functions $f_{r+2}, \cdots, f_n$ are polynomials in z_{r+1} with coefficients which are power series in $z_1, \cdots, z_r$. This representation is valid in a sufficiently small polycylindrical neighborhood $N' = \{z \mid |z_i| < d_i\}$ of p, and is to be understood in the sense that the points in $X \cap N'$ consist precisely of all solutions of these equations in N' for which the systems of values of $z_1, \cdots, z_r$ are not zeroes of D, as well as all limits of these solutions as the set of values of $z_1, \cdots, z_r$ approaches a zero of D in N'. (We observe that this implies in particular that $X \cap N'$ has no limiting points on the boundary of N' outside of the subset satisfying the conditions $z_i = d_i$ for $i = 1, \cdots, r$.) Therefore, if t is the degree of the polynomial $P(z_{r+1})$, for any set of values of $z_1, \cdots, z_r$ with $|z_i| < d_i$ which is not a zero of D, the t distinct zeroes $P(z_{r+1})$ will give rise in this representation to t distinct points in $X \cap N'$, and these are the only points in $X \cap N'$ having the given set of values for the coordinates $z_1, \cdots, z_r$. Now, let $N'' \subset N'$ be a neighborhood of p such that for every point e in N'' we have the relation $N' \subset T_eN$ (we can evidently assume that $\bar{N}' \subset N$), then for every e in N'' the restriction of T_eY to N' is the variety defined by the equations $z_1 = e_1, \cdots, z_r = e_r$. It follows that for any such point e for which the subsystem $e_1, \cdots, e_r$ is not a zero of D, the intersection points of X with T_eY in N' are precisely the t points obtained from the above representation of X by setting $z_1 = e_1, \cdots, z_r = e_r$. Our lemma $i(X \cdot Y, p) = t$ follows then from [1], 4.10, by observing that T_eY intersects X transversally in N' and that the intersection of the set $\{z \mid |z_i| \leqq e_i, i = 1, \cdots, r\}$ with $X \cap N'$ is compact.

3. In case of a real-analytic variety V^0 the situation is somewhat similar if one takes everything over $\mathbf{Z}_2$ instead of $\mathbf{Z}$. Here we have a mapping ψ which assigns to each real-analytic cycle in V^0 the homology class determined by it in $H_*(V^0, \mathbf{Z}_2)$, the reason for taking $\mathbf{Z}_2$ being the fact that the real-analytic varieties are not always orientable. However, there is a difficulty due to the fact that in the real-analytic case the sheaf of ideals of a subvariety is not necessarily coherent, in contrast to what is known in the complex-analytic case. This leads to the unfortunate situation, first pointed out by H. Cartan, that it is not always possible to resolve globally a real-analytic subvariety into irreducible components in a reasonable manner; and since the analytic cycles are defined in terms of irreducible subvarieties, this makes it difficult to give a meaningful definition of the analytic cycles. One could remedy this situation by restricting oneself to the "coherent" subvarieties, i. e. subvarieties whose sheaves of ideals are coherent, but this would have the drawback that even some real-algebraic subvarieties defined by single real-algebraic equations would be excluded. A satisfactory compromise seems to be the notion of a C-analytic subvariety, first introduced by H. Cartan [4], which roughly speaking is a real-analytic subvariety which has a global complexification. As has been shown by Whitney and Bruhat [5], the C-analytic subvarieties have many of the "good" properties of the complex-analytic subvarieties; in particular, such a subvariety admits a unique decomposition into C-irreducible components, whereby a C-analytic subvariety is said to be C-irreducible if it is not the union of two distinct C-analytic subvarieties.

Although the way seems to be thus opened for an intersection theory for real-analytic cycles in a real-analytic variety, there does not seem to be at this moment such a theory which runs more or less parallel to the complex-analytic case and is independent of it. However, as Borel and Haefliger have observed [1], it is possible to derive an intersection theory over $\mathbf{Z}_2$ for real-analytic cycles from the complex-analytic case by using the complexifications, with the result that the mapping ψ mentioned above is also multiplicative. To explain this briefly, let V be the complexification of V^0 (the existence of this complexification is assured if we assume that V^0 is paracompact); then a mapping α can be defined which assigns to every complex-analytic cycle X (over $\mathbf{Z}$ or $\mathbf{Z}_2$) of complex dimension r in V a real-analytic cycle X^0 over $\mathbf{Z}_2$ of real dimension r in V^0, and the intersection theory in V^0 is precisely so defined as to make the mapping α multiplicative for the subset of all complex-analytic cycles in V which are invariant under conjugation (this set includes the complexifications of all real-analytic cycles in V^0). This mapping α is defined for irreducible subvariety X by setting $\alpha(X) = X^0 = X \cap V^0$ in case

X^0 has the real dimension r, and setting $\alpha(X) = 0$ otherwise; this definition is then extended to complex-analytic cycles in V by linearity and reduction mod 2. The composite mapping $\psi \circ \alpha$ of complex-analytic cycles in V into $H_*(V^0, \mathbf{Z}_2)$ is then the mapping ρ mentioned in Section 1. For our present purpose it is significant to observe that a complex-analytic variety V can also be considered as a real-analytic variety, and the complex-analytic cycles in V, taken over $\mathbf{Z}_2$, are also real analytic cycles; and that the intersection theory for real-analytic cycles over $\mathbf{Z}_2$ in V as a real-analytic variety, when restricted to the complex-analytic cycles in V over $\mathbf{Z}_2$, coincides with the intersection theory for complex-analytic cycles in V as a complex-analytic variety, taken mod 2.

THEOREM 1. *Let V be a complex-analytic variety of complex dimension n which is the complexification of a real-analytic variety V^0, and let X be a complex-analytic cycle in V such that X and $\bar{X}$ intersect properly in V at every component containing a real point; then X and V^0 intersect properly in V as real-analytic cycles over $\mathbf{Z}_2$, and we have the relation $\alpha(X \cdot \bar{X}) = X \cdot V^0$.*

Proof. Since the properties asserted here are linear over $\mathbf{Z}_2$, it is sufficient to restrict ourselves to the case where X is a complex-analytic subvariety in V. Let n and r denote the complex dimensions of V and X respectively. Let p be a real point in $X \cap \bar{X}$, and let $z_j = x_j + iy_j$, $j = 1, \cdots, n$, be a system of local coordinates of V at p such that $x_1, \cdots, x_n$ form a system of local coordinates of V^0 at p. Let X be defined in the neighborhood of p by the equations $f_k(z) = g_k(z) + ih_k(z) = 0$, $k = 1, \cdots, s$, where f_k are convergent power series with complex coefficients and g_k and h_k are convergent power series with real coefficients; then $\bar{X}$ is defined by the equations $\bar{f}_k(z) = g_k(z) - ih_k(z) = 0$. Then the analytic set $X \cap \bar{X}$ is determined in the neighborhood of p by the equations $g_k(z) = h_k(z) = 0$, and our assumption asserts that every component of the solutions of these equations has the complex dimension $2r - n$. Now, let V^* be a complexification of V considered as a real-analytic variety of real dimension $2n$, and let $x_1^*, y_1^*, \cdots, x_n^*, y_n^*$ be a system of local coordinates of V^* at p such that $x_1, y_1, \cdots, x_n, y_n$ constitute its real part. If we set $g_k(z) = g_k'(x, y) + ig_k''(x, y)$ and $h_k(z) = h_k'(x, y) + ih_k''(x, y)$, then the complexification X^* of X in V^* is determined locally at p by the equations $g_k'(x^*, y^*) - h_k''(x^*, y^*) = g_k''(x^*, y^*) + h_k'(x^*, y^*) = 0, k = 1, \cdots, s$; and it is clear that the complexification V^{0*} of V^0 in V^* is determined locally at p by the equations $y_1^* = \cdots, y_n^* = 0$.

If we observe that $g_k''(x, 0) = h''(x, 0) = 0$ and hence also $g_k''(x^*, 0) = h_k''(x^*, 0) = 0$, then the analytic set $X^* \cap V^{0*}$ is determined locally at p

by the equations $g_k'(x^*,0)=h_k'(x^*,0)=0$, $k=1,\cdots,s$. On the other hand, since $g_k(z)$ and $h_k(z)$ have real coefficients, they remain unchanged if we first restrict the variables z to their real parts x and then replace these real variables x again by the complex variables z, i.e. $g_k(z)=g_k'(z,0)$ and $h_k(z)=h_k'(z,0)$. Replacing the variables z by x^* it follows that every component of the solutions of the equations $g_k'(x^*,0)=h_k'(x^*,0)=0$, $k=1,\cdots,s$, in the neighborhood of p has the complex dimension $2r-n$. This shows that X^* and V^{0*} intersect properly in V^* at every component containing a real point, and by definition, this means that X and V^0 intersect properly in V.

To prove the relation $\alpha(X\cdot\bar{X})=X\cdot V^0$, we shall show that both sides have the same multiplicity at each component. We need only consider those components of $X\cap V^0$ having a real dimension $2r-n$, for any component of a lower dimension will have complexifications in V and in V^* which are not proper components in $X\cap\bar{X}$ and $X^*\cap V^{0*}$ respectively and hence will have by definition the multiplicity zero in both $\alpha(X,\bar{X})$ and $X\cdot V^0$. Let C^0 then be such a component, and let C and C^{0*} be its complexifications in V and V^* respectively, so that they are proper components in $X\cap\bar{X}$ and $X^*\cap V^{0*}$ respectively; let p be a point in C^0 which is simple in C (hence also in C^0) and does not lie in any other component of $X\cap V^0$. If L is a local linear subvariety at p of complex dimension $2n-2r$, defined over $\boldsymbol{R}$ (i.e. by setting to zero $2r-n$ linear forms of the local coordinates z with real coefficients), which is transversal to C at p, then the intersection cycle $L\cdot X$ is defined locally in a neighborhood of p, and we have the relations $i((L\cdot X)\cdot(\overline{L\cdot X}),p)_L=i(X\cdot\bar{X},C)$ and $i((L\cdot X)\cdot L^0,p)_L=i(X\cdot V^0,C^0)$. Since our problem is essentially local (i.e. we only need to consider X as defined in a neighborhood of p), this shows that we can restrict ourselves to the case where $2r-n=0$ and hence $C=C^0=p$. According to Lemma 1, the multiplicity $i(X\cdot\bar{X},p)$ of $X\cdot\bar{X}$ at p is equal to the number of solutions in a suitable neighborhood N of the origin of the equations $f_k(z+2ie)=\bar{f}_k(z)=0$, $k=1,\cdots,s$, for almost all sufficiently small real n-vectors e. If we perform a local coordinate translation by replacing z by $z+ie$, then the equations will have the form $f_k(z+ie)=\bar{f}_k(z-ie)=0$, $k=1,\cdots,s$, but the number of solutions in N will remain the same provided e is taken sufficiently small. If we set $F_k(z)=f_k(z+ie)$ then our equations will have the form $F_k(z)=\bar{F}_k(z)=0$, $k=1,\cdots,s$; hence, if we set $F_k(z)=G_k(z)+iH_k(z)$, where G_k and H_k are power series with real coefficients, then our equations can be written in the form $G_k(z)=H_k(z)=0$, $k=1,\cdots,s$. On the other hand, again by Lemma 1, the multiplicity $i(X^*\cdot V^{0*},p)$ is equal to the number of solutions in N (replacing N by a smaller neighborhood if necessary) of the equations $g_k'(x^*, e) - h_k''(x^*, e) = g_k''(x^*, e) + h_k'(x^*, e) = 0$, $k = 1, \cdots, s$, for almost all sufficiently small real n-vectors e. If we set

$$G_k(z) = G_k'(x, y) + iG_k''(x, y) \text{ and } H_k(z) = H_k'(x, y) + iH_k''(x, y),$$

then we have the relations $G_k'(x, y) - H_k''(x, y) = g_k'(x, y + e) - h_k''(x, y + e)$, and $G_k''(x, y) + H_k'(x, y) = g_k''(x, y + e) + h_k'(x, y + e)$, since both sides are the real and imaginary parts respectively of the function $F_k(z) = f_k(z + ie)$. It follows that $G_k'(x, 0) = g_k'(x, e) - h_k''(x, e)$ and $H_k'(x, 0) = g_k''(x, e) + h_k'(x, e)$, and hence $G_k'(x^*, 0) = g_k'(x^*, e) - h_k''(x^*, e)$ and $H_k'(x^*, 0) = g_k''(x^*, e) + h_k'(x^*, e)$, so that $i(X^* \cdot V^{0*}, p)$ is equal to the number of solutions in N of the equations $G_k'(x^*, 0) = H_k'(x^*, 0) = 0$, $k = 1, \cdots, s$. Now, since both G_k and H_k are real, a similar argument as before shows that $G_k(z) = G_k'(z, 0)$ and $H_k(z) = H_k'(z, 0)$, so that $i(X^* \cdot V^{0*}, p)$ is equal to the number of solutions in N of the equations $G_k(z) = H_k(z) = 0$, $k = 1, \cdots, s$. This shows that $i(X \cdot \bar{X}, p) = i(X^* \cdot V^{0*}, p)$; taking both sides mod 2, we conclude that $i(X \cdot V^0, p)$ is equal to the multiplicity of p in $\alpha(X \cdot \bar{X})$. This concludes the proof of Theorem 1.

Remark. The equation $i(X \cdot \bar{X}, p) = i(X^*, V^{0*}, p)$ shows that X and $\bar{X}$ are transversal at p in V if and only if X^* and V^{0*} are transversal at p in V^*, which is so if and only if X and V^0 are transversal at p in V. This can be easily verified by direct calculation of the equations of the tangent spaces of X, $\bar{X}$, and V^0 at p.

4. From now on we shall assume that V is algebraic and quasi-projective, and that the cycles involved are also algebraic. The reason for making this restriction is that the algebraic cycles in a quasi-projective algebraic variety can be "moved around" up to equivalence so that any two or finite number of them can be made to intersect properly (in fact even transversally, but we shall not need this). We shall state the properties we need here in a lemma as follows.

LEMMA 2. *Let V be a quasi-projective complex-algebraic variety, defined over **R**, which is a complexification of its real trace V^0, and let X be a complex-algebraic cycle in V which is defined over **R**; then*

(a) *there exists a complex-algebraic cycle X', defined over **R** and rationally equivalent to X over **R**, such that X' intersect properly in V any given cycle Y which is defined over **R** (or over **C**), and*

(b) *there exists a complex-algebraic cycle X'', rationally equivalent to X, such that X'' and $\overline{X''}$ intersect properly in V.*

Proof. We shall apply a result in [3]; for this purpose we introduce a universal domain $\boldsymbol{K}$ which contains $\boldsymbol{C}$ as an "admissible" field, and denote by $V_{\boldsymbol{K}}$, $X_{\boldsymbol{K}}$, $Y_{\boldsymbol{K}}$, the extensions of V, X, Y, respectively over $\boldsymbol{K}$, so that $V_{\boldsymbol{K}}$, $X_{\boldsymbol{K}}$, $Y_{\boldsymbol{K}}$, are algebraic variety and cycles respectively in the sense of [3] and are defined over $\boldsymbol{R}$, and V, X, Y are obtained from $V_{\boldsymbol{K}}$, $X_{\boldsymbol{K}}$, $Y_{\boldsymbol{K}}$, respectively by restriction to points which are rational over $\boldsymbol{C}$. We now apply Lemma 2, Corollary 4, in [3] with $\boldsymbol{R}$ as the ground field, and obtain a cycle $X(u)$ in V, defined over a purely transcendental extension $\boldsymbol{R}(u) = \boldsymbol{R}(u_1, \cdots, u_m)$ of $\boldsymbol{R}$ and rationally equivalent to $X_{\boldsymbol{K}}$ over $\boldsymbol{R}(u)$, such that $X(u)$ intersects $Y_{\boldsymbol{K}}$ properly in $V_{\boldsymbol{K}}$ and specializes to $X_{\boldsymbol{K}}$ over the specialization $u \rightarrow 0$ over $\boldsymbol{R}$. Let $f_i(u)$ be a system of elements in $\boldsymbol{R}[u]$ which represent the associated coordinates of the cycle $X(u)$, and let μ be a real m-vector which is not a common zero of $f_i(u)$; then there is a unique specialization $X(\mu)$ of $X(u)$ over the specialization $u \rightarrow \mu$ over $\boldsymbol{R}$, and $X(\mu)$ is defined over $\boldsymbol{R}$ and is rationally equivalent to $X_{\boldsymbol{K}}$ over $\boldsymbol{R}$. Now, the condition for two algebraic cycles in $V_{\boldsymbol{K}}$ to intersect improperly can be expressed by the vanishing of a finite number of polynomials in the associated coordinates of the cycles with coefficients in $\boldsymbol{R}$. There exists therefore a set of polynomials $g_j(u)$ in $\boldsymbol{R}[u]$ such that $X(\mu)$ intersects $Y_{\boldsymbol{K}}$ properly in $V_{\boldsymbol{K}}$ if and only if μ is not a common zero of $g_j(u)$. If we choose the real m-vector μ so that it is not a common zero of the set $f_i(u) g_j(u)$, and denote by X' the restriction of $X(\mu)$ to rational points over $\boldsymbol{C}$, then X' is a cycle in V satisfying (a). Next, let μ be a complex m-vector which is not a common zero of $f_i(u)$; then there exist unique specializations $X(\mu)$ and $X(\bar{\mu})$ of $X(u)$ over the specializations $u \rightarrow \mu$ and $u \rightarrow \bar{\mu}$ respectively over $\boldsymbol{R}$, and both $X(\mu)$ and $X(\bar{\mu})$ are defined over $\boldsymbol{C}$ and are rationally equivalent to $X_{\boldsymbol{K}}$ over $\boldsymbol{C}$; and it is clear that $X(\bar{\mu})$ is the conjugate of $X(\mu)$. If we now take a system of variables $v = (v_1, \cdots, v_m)$ over $\boldsymbol{R}$ in $\boldsymbol{K}$ which are independent with respect to u over $\boldsymbol{R}$, then the cycle $X(v)$ has the same properties as $X(u)$, except that the variables u are now replaced by v; furthermore, again by [3], Lemma 2, Corollary 4, the cycles $X(u)$ and $X(v)$ intersect properly in $V_{\boldsymbol{K}}$. It follows then, just as before, that there exists a set of polynomials $h_k(u, v)$ in $\boldsymbol{R}[u, v]$ such that $X(\mu)$ and $X(\nu)$ intersect properly in $V_{\boldsymbol{K}}$ if and only if the pair μ, ν is not a common zero of $h_k(u, v)$. If we choose the complex m-vector μ so that the pair μ, $\bar{\mu}$ is not a common zero of the set $f_i(u) h_k(u, v)$ (or equivalently of the set $f_i(u) f_j(v) h_k(u, v)$), and denote by X'' the restriction of $X(\mu)$ to rational points over $\boldsymbol{C}$, then X'' is a cycle in V satisfying (b). This concludes the proof of Lemma 2.

We are now in a position to prove the result announced at the beginning of this paper.

THEOREM 2. *Let V be a quasi-projective complex-algebraic variety, defined over $\mathbf{R}$, which is a complexification of its real trace V^0, and let X be a complex-algebraic cycle in V which is defined over $\mathbf{R}$; then $\phi(X)\cdot\phi(V^0)$ and $\rho(X)^2$, both as elements in $H_*(V^0,\mathbf{Z}_2)$, are numerically equivalent with respect to real-algebraic homology classes in $H_*(V^0,\mathbf{Z}_2)$.*

Proof. By Lemma 2(a), there exists a complex-algebraic cycle X' in V, defined over $\mathbf{R}$ and rationally equivalent to X over $\mathbf{R}$, such that X and X' intersect properly. It follows then from [1], Theorem 5.3, that

$$\rho(X\cdot X')=\rho(X)\cdot\rho(X')=\rho(X)^2.$$

On the other hand, by Lemma 2(b), there exists a complex-algebraic cycle X'' in V, rationally equivalent to X, such that X'' and $\overline{X''}$ intersect properly in V. It follows then from Theorem 1 that $X''\cdot V^0=\alpha(X''\cdot\overline{X''})$ and hence $\phi(X)\cdot\phi(V^0)=\phi(X'')\cdot\phi(V^0)=\rho(X''\cdot\overline{X''})$. It remains only to show that $\rho(X\cdot X')$ and $\rho(X''\cdot\overline{X''})$ are numerically equivalent with respect to real-algebraic homology classes in $H_*(V^0,\mathbf{Z}_2)$. Let Y^0 be a real-algebraic cycle of the complementary real dimension to $\alpha(X\cdot X')$ in V^0 (i.e., if n and r are the real dimension of V^0 and $\alpha(X)$ respectively, then the real dimension of Y^0 should be $n-(2r-n)=2n-2r$), and let Y be its complexification in V; in view of Lemma 2(a), we can assume that Y intersect both $X\cdot X'$ and $X''\cdot\overline{X''}$ properly in V. Since $X\cdot X'$ and $X''\cdot\overline{X''}$ are rationally equivalent, they must have the same intersection number with Y; it follows that $\alpha(X\cdot X')$ and $\alpha(X''\cdot\overline{X''})$ must have the same intersection number with $Y^0=\alpha(Y)$, and hence $\rho(X\cdot X')$ and $\rho(X''\cdot\overline{X''})$ must have the same intersection number with $\psi(Y^0)$. This proves our theorem.

We conclude this paper by mentioning one further application of our result. Assume that $H_n(V,\mathbf{Z}_2)$ has a numerical equivalence base $B_1,\cdots,B_t$ consisting of complex-algebraic cycles defined over $\mathbf{R}$, and let $M=(m_{ij})$ be its intersection matrix; then the cycle V^0 is numerically equivalent over $\mathbf{Z}_2$ to a cycle of the form $\sum_{i=1}^{t} n_iB_i$. Now, Theorem 2 provides us with the numerical relations $\phi(V^0)\cdot\phi(B_j)=\rho(B_j)^2$, or $\sum_{i=1}^{t} n_i m_{ij}=m_{jj}$, $j=1,\cdots,t$, from which we can easily calculate the n_i and hence the self intersection number $\phi(V^0)^2$ as a cycle over $\mathbf{Z}_2$ in V. For example, in case V is the complex Grassmannian, it is well-known that $H_n(V,\mathbf{Z}_2)$ has a base consisting of the Schubert varieties of complex dimension $n/2$ (none in case n is odd), each of which is the dual of some other or itself, and a simple calculation shows that the number $\phi(V^0)^2$ is equal mod 2 to the total number of Schubert varieties of dimension n, or equivalently the number of those which are self-dual.

THE JOHNS HOPKINS UNIVERSITY.

REFERENCES.

[1] A. Borel and A. Haefliger, "La classe d'homologie fondamentale d'un espace analytique," *Bulletin de la Société Mathématique de France*, vol. 89 (1961), pp. 461-513.

[2] A. Borel and F. Hirzebruch, "Characteristic classes and homogeneous spaces, II," *American Journal of Mathematics*, vol. 81 (1959), pp. 315-382, see Appendix II.

[3] W. L. Chow, "On equivalence classes of cycles in an algebraic variety," *Annals of Mathematics*, vol. 64 (1956), pp. 450-479.

[4] H. Cartan, "Variétés analytiques-réelles et variétés analytiques complexes," *Bulletin de la Société Mathématique de France*, vol. 85 (1957), pp. 77-100.

[5] H. Whitney and F. Bruhat, "Quelques propriétés fondamentales des ensembles analytiques-réels," *Commentarii Mathematici Helvetici*, vol. 33 (1959), pp. 132-160.

ON UNMIXEDNESS THEOREM.*

By Wei-Liang Chow.

B. Dwork and K. Ireland, in their recent work on the zeta function of an algebraic variety, have raised the question whether the unmixedness theorem holds for the doubly or multiply projective space, or rather for the (generalized) Segre representation of such a space as a projective variety. In this paper we shall give an affirmative answer to this question. However, in course of our investigation, it turns out that our method is susceptible of a bit of generalization; in fact, we shall now express our result as a theorem on the validity of the unmixedness theorem for the "Segre" product of two Noetherian graded rings, assuming that unmixedness theorem holds for the two rings themselves. This generalization is not nominal, for, apart from giving a clearer picture of the nature of our reasoning, it yields also some additional results of quite a bit of interest, as we shall see presently.

1. Let V be a projective variety, defined over a field K, and let R be the homogeneous coordinate ring of V. We shall say that the unmixedness theorem holds for V, if the unmixedness theorem holds for homogeneous ideals in R. We recall that the unmixedness theorem is said to hold for homogeneous ideals if the following assertion is true: If a homogeneous ideal $\mathfrak{a}$ in R generated by r elements has the height r, then it is unmixed, i. e., has no embedded prime divisor. Let $\mathfrak{M}$ be the (maximal) ideal in R generated by all elements of positive degrees, and let $R_{\mathfrak{M}}$ be the quotient ring of R with respect to $\mathfrak{M}$; since every prime divisor of $\mathfrak{a}$ is homogeneous and hence must be contained in $\mathfrak{M}$, it follows that $\mathfrak{a}$ is unmixed if and only if $R_{\mathfrak{M}}\mathfrak{a}$ is unmixed. Now, it is easily seen that $R_{\mathfrak{M}}$ has a system of parameters consisting of homogeneous elements in R; hence the validity of the unmixedness theorem for homogeenous ideals implies that any such system of parameters must be distinct. Conversely, it is well-known that if $R_{\mathfrak{M}}$ possesses one distinct system of parameters (in which case $R_{\mathfrak{M}}$ is said to be Macaulay), then every system of parameters in $R_{\mathfrak{M}}$ must be distinct and the unmixedness theorem holds in $R_{\mathfrak{M}}$. Thus we see that the unmixedness theorem holds for homogeneous ideals in R if and only if $R_{\mathfrak{M}}$ is Macaulay; in other words, the unmixedness theorem holds for V if and only if $R_{\mathfrak{M}}$ is Macaulay.

* This work was partially supported by the National Science Foundation under Grant NSF-G7030.

Now, let U be another projective variety, defined over the same field K; let S be the homogeneous coordinate ring of U, and let $\mathfrak{N}$ be the ideal in S generated by all elements of positive degrees. Let T be the tensor product of R and S over K; we note that both R and S can be canonically embedded as subrings in T, and that their gradings induce in T a bi-grading, so that T is a bigraded ring. If we set $R=\sum_{i=0}^{\infty} R_i$ and $S=\sum_{i=0}^{\infty} S_i$, where R_i and S_i are the K-modules of homogeneous elements of degree i in R and S respectively, then we have $T=\sum_{i,j=0}^{\infty} T_{i,j}$ with $T_{i,j}=R_iS_j$, and the ideal $\mathfrak{D}=T(\mathfrak{M},\mathfrak{N})$ is the maximal ideal in T generated by all homogeneous elements of positive bi-degrees. As we shall see later, it can be easily shown that if both $R_{\mathfrak{M}}$ and $S_{\mathfrak{N}}$ are Macaulay, then $T_{\mathfrak{D}}$ is also Macaulay. Now, T can be considered as the homogeneous coordinate ring of a projective variety $J(V,U)$ which is sometimes called the join of V and U; briefly, $J(V,U)$ may be described as the variety generated by all the lines joining a point of V with a point of U, assuming that the ambient projective spaces of V and U are embedded as two skew linear subspaces in a third (bigger) projective space. Thus, if the unmixedness theorem holds for both V and U, then it holds also for $J(V,U)$. However, we are interested here not so much in the join $J(V,U)$ as in the product variety $V\times U$ represented projectively as a subvariety in the Segre variety representing the product of their ambient spaces; we shall call the so represented $V\times U$ the Segre product of V and U. More generally, we shall consider the Segre product $V^{(d)}\times U^{(e)}$, where $V^{(d)}$ and $U^{(e)}$ are the images of V and U respectively under birational transformations determined by the linear systems of hypersurfaces of degrees d and e respectively. The homogeneous coordinate ring of $V^{(d)}\times U^{(e)}$ is the subring $A=\sum_{i=0}^{\infty} T_{id,ie}$ in T generated by all homogeneous elements of bi-degrees (id,ie) for all $i=0, 1,\cdots$; and the ideal $\mathfrak{L}=A\cap\mathfrak{D}$ is the maximal ideal in A generated by all homogeneous elements of positive degrees. Our main problem is to determine the condition under which the ring $A_{\mathfrak{L}}$ is Macaulay, assuming that both $R_{\mathfrak{M}}$ and $S_{\mathfrak{N}}$ are Macaulay. The conditions we shall give will be easily seen to be satisfied in case both R and S are polynomial rings over K, endowed with their usual grading; and since polynomial rings are homogeneous coordinate rings of projective spaces, this answers the question posed at the beginning of this paper.

However, as mentioned above, our investigations can be extended readily to more general situations. Instead of homogeneous K-algebras arising from algebraic varieties, we shall only assume that R and S are Noetherian graded algebras over a common field K (with the restrictions $R_0 = S_0 = K$); it is easily seen that everything we have said above, except for the geometrical interpretations, holds without any change in this more general situation, provided we assume that $R_0 = S_0 = K$. The main point in this generalization is that the rings involved are no longer assumed to be generated by the homogeneous elements of the first degree. This last point enables us to apply our result to the following case, which I understand from Dwork is of interest in his study of the zeta-function of a complete intersection: We take for R and S as before the polynomial rings over K, but for one (or both) of these two rings we shall assign a different grading from the usual one; for example, if $y_1, \cdots, y_n$ are the variable generating ring S, we shall assign to each y_i an arbitrary positive integer g_i as its degree. It is clear that this case is a generalization of the case of two projective spaces, and we shall see later that our main result remains valid here.

We shall follow generally the terminology of Nagata in his recent book on *Local Rings* (referred to as LR) since this terminology represents what is now more or less standard in the field of commutative algebra. The only exception is that we shall speak of a homogeneous ideal in a graded ring instead of a graded ideal, which is the term used in LR. For the sake of convenience of the reader, we shall give occasional reference to LR for some of the well-known results used in our argument. In this connection we should like to point out that, in course of our investigation, we have found it convenient to introduce the concepts of an H-local ring and of an H-Macaulay ring. These are natural generalizations to the case of a graded ring of the usual concepts of a local ring and of a Macaulay ring, and they are more convenient to use in questions concerning homogeneous elements and ideals in a graded ring. We shall develop these notions here only to the extent needed for our present purpose, but it is likely that they may prove also useful in other connections.

2. In this section we shall develop some results about a certain type of ideals in a Noetherian ring, which we shall need later. Let R be a Notherian ring; we recall that a sequence of elements $u_1, \cdots, u_r$ in R is said to be an R-sequence if $R(u_1, \cdots, u_r) \neq R$ and if u_i is not a zero-divisor for the R-module $R/R(u_1, \cdots, u_{i-1})$ for every $i = 1, \cdots, r$, and that this property is independent of the arrangement of the elements of the system into a sequence, so that we can speak of a (finite) set of elements forming an R-sequence without any specification about their arrangement into a sequence.

We begin with the simple observation that if two elements u and v in R form an R-sequence, then $R(uv) = R(u) \cap R(v)$. In fact, if an element au in $R(u)$ is contained in $R(v)$, then au corresponds to the zero element in the R-module $R/R(v)$; since u is not a zero-divisor for $R/R(v)$, a itself must correspond to the zero element and hence is contained in $R(v)$; this shows that au is contained in $R(uv)$. We are interested in generalizing this simple result to a situation which involves two systems of edements as well as more intricate conditions. Let $u_1, \cdots, u_t$ and $v_1, \cdots, v_t$ be two systems of elements in R satisfying the following two conditions, where m and n are two fixed positive integers $\leqq t$: (A) Every set containing at most m elements among the $u_1, \cdots, u_t$ and at most n elements among the $v_1, \cdots, v_t$ forms an R-sequence, and (B) the ideal $R(u_1, \cdots, u_t)$ is generated by any set of m elements among the $u_1, \cdots, u_t$, and the ideal $R(v_1, \cdots, v_t)$ is generated by any set of n elements among the $v_1, \cdots, v_t$. We set $w_i = u_i v_i$, $i = 1, \cdots, t$, and consider the ideal $R(w_1, \cdots, w_t)$; we shall show (Lemma 3) that, in case $t < m + n - 1$, this ideal can be expressed as the intersection of all ideals of the form $R(u_{k_1}, \cdots, u_{k_a}, v_{k_{a+1}}, \cdots, v_{k_t})$, where the indices $k_1, \cdots, k_t$ are a rearrangement of the integers from 1 to t. Before we do this, we shall first prove a result (Lemma 1) which will be useful to us later. In what follows in this section, we shall set for convenience

$$R(u) = R(u_1, \cdots, u_m), \qquad R(v) = R(v_1, \cdots, v_n),$$
$$R(u, v) = R(u_1, \cdots, u_m, v_1, \cdots, v_n),$$

and we observe that by virtue of (B), we have the relatons

$$R(u) = R(u_1, \cdots, u_t), \qquad R(v) = R(v_1, \cdots, v_t),$$
$$R(u, v) = R(u_1, \cdots, u_t, v_1, \cdots, v_t).$$

LEMMA 1. *Asume that $t \leqq m + n - 2$, and let z be an element in R which is contained every one of the ideals $R(u_{k_1}, \cdots, u_{k_a}, v_{k_{a+1}}, \cdots, v_{k_t})$, for all partitions of the integers from 1 to t into two sets $(k_1, \cdots, k_\alpha)$ and $(k_{\alpha+1}, \cdots, k_t)$ and for all values of α from $t - n + 1$ to $m - 1$. Then there exist elements $c_{i_1 \cdots i_m}$ in R such that*

$$z = \sum_{i_1 < \cdots < i_m} c_{i_1 \cdots i_m} v_{i_1} \cdots v_{i_m} \bmod R(u)$$

Proof. Since z is evidently contained in $R(u, v)$, we can set $z = z' + z''$, where z' is contained in $R(u)$ and z'' is contained in $R(v)$; we set $z'' = \sum_{i=1}^{n} c_i v_i$.

Consider a typical term $c_i v_i$, in z'' and take an ideal

$$I_i = R(u_{k_1}, \cdots, u_{k_{m-1}}, v_{k_m}, \cdots, v_{k_t}),$$

where the $t-m+1$ ($\leqq n-1$) indices $k_m, \cdots, k_t$ are to be taken among the integers from 1 to n but distinct from the index i; we denote by J_i the ideal generated in R by all the elements $v_1, \cdots, v_n$ except v_i. Since z is by assumption contained in I_i, the element $c_iv_i + z' = z - \sum_{j \neq i} c_jv_j$ must be contained in $I_i + J_i$, and hence the element c_iv_i must be contained in $R(u) + J_i$. Since $(R(u) + J_i) : R(v_i) = R(u) + J_i$, it follows that c_i is contained in $R(u) + J_i$; hence, after transferring to z' a term belonging to $R(u)$ if necessary, we can set $z'' = \sum_{i<j} c_{ij}v_iv_j$, where the indices i and j run from 1 to n. If $m = 2$, we can stop here for we will have already arrived at the desired expression for z''; if $m > 2$, we can proceed as follows. Consider a typical term $c_{ij}v_iv_j$ in z'', and take an ideal $I_{ij} = R(u_{k_1}, \cdots, u_{k_{m-1}}, v_{k_m}, \cdots, v_{k_t})$, where the $t-m+1$ indices $k_m, \cdots, k_t$ are now to be taken among the integers from 1 to n or $n+1$ in case $t-m+1 \leqq n-2$ or $t-m+1 > n-2$ respectively, but distinct from both i and j; we observe that since $m > 2$, the condition $t-m+1 > n-2$ implies $t > n$, so that element v_{n+1} exists. We denote by J_{ij} the ideal generated in R by all the elements $v_1, \cdots, v_n$ or $v_1, \cdots, v_{n+1}$ in the respective case except v_i and v_j. Since z is by assumption contained in I_{ij}, the element $c_{ij}v_iv_j + z'$ must be contained in $I_{ij} + J_{ij}$, and hence the element $c_{ij}v_iv_j$ must be contained in $R(u) + J_{ij}$. Since

$$(R(u) + J_{ij}) : R(v_i) = R(u) + J_{ij} \text{ and } (R(u) + J_{ij}) : R(v_j) = R(u) + J_{ij},$$

it follows that c_{ij} is contained in $R(u) + J_{ij}$; hence, after transferring to z' a term belonging to $R(u)$ if necessary, we can set $z'' = \sum_{i<j<k} c_{ijk}v_iv_jv_k$, where the indices run from 1 to n or to $n+1$ in case $t-m+1 \leqq n-2$ or $t-m+1 > n-2$ respectively.

Thus we can proceed step by step. In general, the situation can be described as follows. If $t-m+1 = n-s+1$ ($2 \leqq s \leqq n$), no new v's need be introduced in addition to $v_1, \cdots, v_n$ during the first s steps; for, as long as the degree of the expression for z'' is less than s, the $t-m+1$ elements $v_{k_m}, \cdots, v_{k_t}$ can be chosen among the $v_1, \cdots, v_n$ and still be distinct from those involved in any one given term in the expression for z''. Thus, after the first s steps (note that $t-m+1 = n-s+1$ implies $s \leqq m$), we obtain for z'' an expression of the form $z'' = \sum_{i_1 < \cdots < i_s} c_{i_1 \cdots i_s} v_{i_1} \cdots v_{i_s}$, where the indices all run from 1 to n. After that, if $s < m$, we can proceed further, but we need to introduce one new element among the $v_{n+1}, \cdots, v$ at each step. In fact, at he beginning of the $(s+1)$-th step there are only $n-s$ elements among the $v_1, \cdots, v_n$ which are distinct from those involved in any one given term in the expression for z''; therefore we need to bring in v_{n+1} to make up the set of $t-m+1=n-s+1$ elements $v_{k_m},\cdots,v_{k_t}$, and we observe that this is possible since $s<m$ implies $t-s+1>t-m+1=n-s+1$ and hence $t>n$. Generally, for any integer r such that $s<r<m$, we have after the r-th step an expression for z'' of the form $z''=\sum_{i_1<\cdots<i_r} c_{i_1\cdots i_r}v_{i_1}\cdots,v_{i_r}$, where the indices run from 1 to $n+r-s$. For the $(r+1)$-th step, since there are only $(n+r-s)-r=n-s$- elements among the $v_1,\cdots,v_{n+r-s}$ which are distinct from those involved in any one given term in the expression for z'', we need to bring in $v_{n+r-s+1}$ to make up the set $v_{k_m},\cdots,v_{k_t}$; and again we observe that this is always possible since $r<m$ implies $t-r+1>n-s+1$ and hence $t>n+r-s$. Thus we can always proceed until after m steps we obtain for z'' an expression of the form $z''=\sum_{i_1<\cdots<i_m} c_{i_1\cdots i_m}v_{i_1}\cdots v_{i_m}$; this proves our lemma.

LEMMA 2. *We have the relation* $R(u)R(v)=R(u)\cap R(v)$; *in particular, in case either* $m=1$ *or* $n=1$ *we have the relation* $R(w)=R(u)\cap R(v)$.

Proof. To prove the relation $R(u)R(v)=R(u)\cap R(v)$, we have to show that $R(u)\cap R(v)$ is the ideal generated by all elements of the form u_iv_j, for $i=1,\cdots,m$, and $j=1,\cdots,n$. It is clearly sufficient to show that if $z=\sum_{j=1}^{n} c_jv_j$ is an element in $R(v)$ which is also contained in $R(u)$, then we can choose the elements c_j so that they all are contained $R(u)$. Assume that for any given integer k $(0<k<n)$ the elements c_j with $j<k$ are all contained in $R(u)$, we shall that c_k can be so chosen (without affecting the c_j with $j<k$) that it is also contained in $R(u)$; this would of course prove our assertion. In fact, since z as well as all c_j with $j<k$ are contained in $R(u)$, the element $c_kv_k=z-\sum_{j\neq k} c_jv_j$ is contained in $R(u)+R(v_{k+1},\cdots,v_n)$; and since $(R(u)+R(v_{k+1},\cdots,v_n)):R(v_k)=R(u)+R(v_{k+1},\cdots,v_n)$, it follows that c_k is contained in $R(u)+R(v_{k+1},\cdots,v_n)$. If we set $c_k=c_k'+\sum_{j>k} c_j'v_j$, where c_k' is contained in $R(u)$, then we have for z the expression $z=\sum_{j<k} c_jv_j+c_k'v_k+\sum_{j>k}(c_j+c_j'v_k)v_j$, in which the coefficients of all v_j with $j<k+1$ are contained in $R(u)$; our assertion is therefore proved. The last statement of the lemma follows by observing that, in case $m=1$, the element u_iv_j is contained in $R(u_jv_j)=R(w_j)$ for every $i=1,\cdots,t$, and hence $R(w)=R(u)R(v)$.

LEMMA 3. *In case* $t<m+n-1$, *then*

(1) $R(w_1,\cdots,w_l)=\bigcap R(u_{i_1},\cdots,u_{i_a},\ v_{i_{a+1}},\cdots,v_{i_t})\cap R(u)\cap R(v)$,

where the intersection extends over all partitions of the integers from 1 *to* t *into two sets* $(i_1,\cdots,i_\alpha)$ *and* $(i_{\alpha+1},\cdots,i_t)$, *for all values of* α *from* $t-n+1$ *to* $m-1$.

Proof. We use double induction on the pair of integers (m,n); as initial step of our induction, we take the case where either $m=1$ or $n=1$, and observe that in this case the formula (1) reduces to the formula $R(w)=R(u)\cap R(v)$ proved already in Lemma 2. We therefore assume that both m and n are greater than 1, and that our lemma holds true in case either m or n is replaced by a smaller integer. It is sufficient to show that

$$(2)\qquad R(w_1,\cdots,w_t)=R(w_1.\cdots,w_{t-1},u_t)\cap R(w_1,\cdots,w_{t-1},v_t).$$

To see this, consider the residue ring $\bar{R}=R/R(v_t)$ and denote generally by a bar the image in $\bar{R}$ of an element in R; it is easily seen that the systems $\bar{u}_1,\cdots,\bar{u}_{t-1}$ and $\bar{v}_1,\cdots,\bar{v}_{t-1}$ satisfy the conditions (A) and (B) with the integers t, m, n replaced either by $t-1$, m, $n-1$ respectively in case $t>m$, or by $t-1$, $m-1$, $n-1$ respectively in case $t=m$. In case $t>m$, we have the inequality $t-1<m+(n-1)-1$, while in case $t=m$ the inequality $t-1<(m-1)+(n-1)-1$ will also be satisfied except when $n=2$. Thus, except when $t=m$, $n=2$, the induction hypothesis is always satisfied; on the other hand, in case $t=m$, $n=2$, we have the relation

$$\bar{R}(\bar{w}_1,\cdots,\bar{w}_{t-1})=\bar{R}(\bar{u})\cap\bar{R}(\bar{v})=\bar{R}(\bar{u}_1,\cdots,\bar{u}_{m-1})\cap\bar{R}(\bar{v}_1)$$

by the last assertion of Lemma 2, which is a special case of formula (1), as we have just noted above. Therefore, either by induction hypothesis or by Lemma 2, we have the relation

$$(3)\qquad \begin{aligned}\bar{R}(\bar{w}_1,\cdots,\bar{w}_{t-1}) &=\cap\,\bar{R}(\bar{u}_{i_1},\cdots,\bar{u}_{i_\alpha},\bar{v}_{i_{\alpha+1}},\cdots,\bar{v}_{i_{t-1}})\\ &\cap\,\bar{R}(\bar{u}_1,\cdots,\bar{u}_{t-1})\cap\bar{R}(\bar{v}_1,\cdots,\bar{v}_{t-1})\end{aligned}$$

where the intersection extends over all partitions of the integers from 1 to $t-1$ into two sets $(i_1,\cdots,i_\alpha)$ and $(i_{\alpha+1},\cdots,i_{t-1})$, for all values of α from $t-n+1$ to $\inf(m-1,t-2)$. Going over to the ring R, we obtain the relation

$$(4)\qquad \begin{aligned}&R(w_1,\cdots,w_{t-1},v_t)=\\ &\bigcap R(u_{i_1},\cdots,u_{i_\alpha},v_{i_{\alpha+1}},\cdots,v_{i_{t-1}},v_t)\cap R(u_1,\cdots,u_{t-1},v_t)\\ &\qquad\cap R(v_1,\cdots,v_t).\end{aligned}$$

Similarly, applying the same argument to the rseidue ring $R/R(u_t)$, we obtain the relation

$$R(w_1,\cdots,w_{t-1},u_t) = \tag{5}$$
$$\bigcap R(u_{j_1},\cdots,u_{j_\beta},\ v_{j_{\beta+1}},\cdots,v_{j_{t-1}},u_t) \cap R(u_1,\cdots,u_t) \cap R(v_1,\cdots,v_{t-1},u_t),$$

where β runs from $\sup(t-n,1)$ to $m-2$. If we substitute (4) and (5) into (2) and eliminate some redundant terms, we obtain (1).

To prove (2), consider an element z in R such that zu_t is contained in $R(w_1,\cdots,w_{t-1},v_t)$; we have to show that there exists an element z_1 in $R(w_1,\cdots,w_{t-1},v_t)$ such that $zu_t-z_1u_t$ is contained in $R(w_1,\cdots,w_{t-1})$. Going over to the residue ring $\bar{R}$ introduced above, our assumption implies that $\bar{z}\bar{u}_t$ is contained in $\bar{R}(\bar{w}_1,\cdots,\bar{w}_{t-1})$; in view of (3), this implies that $\bar{z}\bar{u}_t$ is contained in each $\bar{R}(\bar{u}_{i_1},\cdots,\bar{u}_{i_\alpha},\bar{v}_{i_{\alpha+1}},\cdots,\bar{v}_{i_{t-1}})$ as well as in $\bar{R}(\bar{v}_1,\cdots,\bar{v}_{t-1})$. Since $\bar{u}_{i_1},\cdots,\bar{u}_{i_\alpha},\bar{v}_{i_{\alpha+1}},\cdots,\bar{v}_{i_{t-1}},\bar{u}_t$ forms an $\bar{R}$-sequence, it follows that $\bar{z}$ must be contained in $\bar{R}(\bar{u}_{i_1},\cdots,\bar{u}_{i_\alpha},\bar{v}_{i_{\alpha+1}},\cdots,\bar{v}_{i_{t-1}})$; similarly, we conclude that $\bar{z}$ is contained in $\bar{R}(\bar{v}_1,\cdots,\bar{v}_{t-1})$. If $t=m$, then $\bar{u}_1,\cdots,\bar{u}_{t-1}$, $\bar{u}_t$ also form an $\bar{R}$-sequence, and hence $\bar{z}$ is contained also in $\bar{R}(\bar{u}_1,\cdots,\bar{u}_{t-1})$; it follows then from (3) that $\bar{z}$ is contained in $\bar{R}(\bar{w}_1,\cdots,\bar{w}_{t-1})$, and hence z is contained in $R(w_1,\cdots,w_{t-1},v_t)$, so that we can take z_1 to be z itself. We can therefore assume that $t>m$. According to Lemma 1, applied to the ring $\bar{R}$, the element $\bar{z}$ can be expressed in the form $\bar{z}'+\bar{z}''$, where $\bar{z}'$ is contained in $\bar{R}(u_1,\cdots,\bar{u}_{t-1})$ and $\bar{z}''$ has the form

$$\sum_{i_1<\cdots<i_m} \bar{c}_{i_1\cdots i_m}\bar{v}_{i_1}\cdots\bar{v}_{m}.$$

Going over to the ring R, we conclude that $z=z'+z''$, where z' is contained in $R(u_1,\cdots,u_{t-1},v_t)$ and z'' has the form $\sum_{i_1<\cdots<i_m} c_{i_1\cdots i_m}v_{i_1}\cdots v_{i_m}$, where the indices run from 1 to $t-1$. Now, since u_t is contained in $R(u_{i_1},\cdots,u_{i_m})$, for every such set of indices $i_1,\cdots,i_m$, it follows that $v_{i_1}\cdots v_{i_m}u_t$ is contained in $R(w_{i_1},\cdots,w_{i_m})$, and hence $z''u_t$ is contained in $R(w_1,\cdots,w_{t-1})$; this shows that $zu_t-z'u_t$ is contained in $R(w_1,\cdots,w_{t-1})$. Going back again to the residue ring $\bar{R}$, we see that $\bar{z}'\bar{u}_t$ is also contained in $\bar{R}(w_1,\cdots,w_{t-1})$; hence, by the same argument as we have applied to $\bar{z}\bar{u}_t$ before, we conclude that $\bar{z}'$ is contained in every $\bar{R}(\bar{u}_{i_1},\cdots,\bar{u}_{i_\alpha},\ \bar{v}_{i_{\alpha+1}},\cdots,\bar{v}_{i_{t-1}})$ and in $\bar{R}(\bar{v}_1,\cdots,\bar{v}_{t-1})$. Since $\bar{z}'$ is also contained in $\bar{R}(\bar{u}_1,\cdots,\bar{u}_{t-1})$, it follows from (3) that $\bar{z}'$ is contained in $\bar{R}(\bar{w}_1,\cdots,\bar{w}_{t-1})$. Going over to R, this means that z' is contained in $R(w_1,\cdots,w_{t-1},v_t)$; this concludes the proof of Lemma 3.

Lemma 3 ceases to be generally valid in case $t\geqq m+n-1$; in fact, in this case the right side of (1) reduces to $R(u)\cap R(v)$, which is by Lemma 2 the ideal generated by all the elements of the form $u_i v_j$, while the left side can be a smaller ideal, as can be seen from the following simple example. Let $m=n=2$, and $t=3$; take for u_1, u_2, v_1, v_2 any set of four elements forming an R-sequence, and set $u_3=u_1+u_2$, $v_3=v_1+v_2$. Then $R(w_1, w_2, w_3)$ is the ideal generated by the elements u_1v_1, u_2v_2, $u_1v_2+u_2v_1$, which is smaller than the ideal $R(u) \cap R(v) = R(u_1, u_2) \cap R(v_1, v_2) = R(u_1v_1, u_2v_2, u_1v_2, u_2v_1)$. On the other hand, it can be easily verified that every element in $R(u) \cap R(v)$ which can be expressed as a form of degree 3 in the u_1, u_2, v_1, v_2 is contained in $R(w_1, w_2, w_3)$; for example, the element

$$u_1{}^2 v_2 = u_1(u_3-u_2)v_2 = u_1u_3v_2 - u_1w_2$$
$$= u_1u_3(v_3-v_1) - u_1w_2 = u_1w_3 - u_3w_1 - u_1w_2$$

is contained in $R(w_1, w_2, w_3)$. Therefore, if we denote by $R(w, (u,v)^3)$ the sum of the ideal $R(w_1, w_2, w_3)$ with the ideal $R(u_1, u_2, v_1, v_2)^3$, we obtained the equation $R(w_1, w_2, w_t) = R(u) \cap R(v) \cap R(w, (u,v)^3)$. In case R is the polynomial ring $K[u_1, u_2, v_1, v_2]$ over a field K generated by the indeterminates u_1, u_2, v_1, v_2, this equation gives a primary decomposition of $R(w_1, w_2, w_3)$, in which $R(w, (u,v)^3)$ is an embedded component belonging to the maximal prime ideal $R(u_1, u_2, v_1, v_2)$. The situation we have just described holds in general for $t=m\,j\,n-1$.

LEMMA 4. *For $t=m+n-1$, we have the relation*

$$(6) \qquad R(w_1, \cdots, w_t) = R(u) \cap R(v) \cap R(w, (u,v)^t),$$

where $R(w, (u,v)^t) = R(w_1, \cdots, w_t) + R(u,v)^t$.

Proof. It is sufficient to show that every element z in $R(u) \cap R(v)$ which is a "monomial" of degree t is contained in $R(w_1, \cdots, w_t)$. Since $t=m+n-1$, the monomial z must have either a degree $\geqq n$ in the u's or a degree $\geqq m$ in the v's; assume the first case and set $z=au_{i_1}\cdots u_{i_n}v_j$. We maintain that every such monomial can be expressed as a sum of such monomials in which all the indices $i_1, \cdots, i_n$ are distinct. In fact, assume that $i_1, \cdots, i_\alpha$ $(\alpha<n)$ are distinct and let $k_1, \cdots, k_m$ be m distinct positive integers $\leqq t$ which are distinct from $i_1, \cdots, i_\alpha$ (observe that $t-\alpha \geqq m$), then we can express every one of the factors $u_{i_{\alpha+1}}, \cdots, u_{i_n}$ in z by a linear combination of $u_{k_1}, \cdots, u_{k_m}$, and obtain an expression of z as a sum of monomials each of which contains at least $\alpha+1$ distinct u's; after a finite number of steps, we arrive at an expression of z as a sum of monomials each of which contains n distinct u's. Thus we only need to prove our assertion for the case where the indices $i_1, \cdots, i_n$ in the expression for z are distinct;

but in this case we can express v_j as a linear combination of $v_{i_1}, \cdots, v_{i_n}$, say $v_j = b_1 v_{i_1} + \cdots + b_n v_{i_n}$, and obtain the equation

$$z = a u_{i_1} \cdots u_{i_n}(b_1 v_{i_1} + \cdots + b_n v_{i_n})$$
$$= a b_1 u_{i_2} \cdots u_{i_n} w_{i_1} + \cdots + a b_n u_{i_1} \cdots u_{i_{n-1}} w_{i_n}$$

which shows that z is contained in $R(w_1, \cdots, w_t)$.

Remark. Since the above proof uses only the condition (B) but not the condition (A), Lemma 4 therefore holds also without the condition (A).

3. Let R be a Noetherian graded ring, and let $R = \sum_{i=0}^{\infty} R_i$ be the expression of R as a direct sum of modules of homogeneous elements of different degrees; it is clear that R_0 is a Notherian subring in R and that each R_i is also an R_0-module. We shall say that R is *H-local* if it has only one maximal homogeneous ideal. It is easily seen that R is H-local if and only if the subring R_0 is local, and that if such is the case and if $\mathfrak{M}$ is the maximal homogeneous ideal of R, then $\mathfrak{M}_0 = R_0 \cap \mathfrak{M}$ is the maximal ideal of the local ring R_0. The concept of an H-local ring is a natural generalization of that of a local ring (and reduces to it in case $R = R_0$), and as such it has many of the properties of a local ring provided we consider only homogeneous ideals. We shall introduce here some notions and a few results about an H-local ring which we shall need for our present purpose; these results can be proved easily by generalizing the proofs of the corresponding results for a local ring, but it is even easier to derive the former from the latter by using the basic relation $\mathfrak{a} = R \cap R_{\mathfrak{M}}\mathfrak{a}$ for any homogeneous ideal $\mathfrak{a}$ in R. This relation $\mathfrak{a} = R \cap R_{\mathfrak{M}}\mathfrak{a}$ for a homogeneous ideal $\mathfrak{a}$ in R follows from (and is in fact equivalent to) the well-known fact (LR, 8.7) that every homogeneous ideal in R has a primary decomposition consisting of only homogeneous primary ideals, which together with their associated prime ideals must then all be contained in $\mathfrak{M}$. If m is the height of the maximal homogeneous ideal $\mathfrak{M}$ of an H-local ring R, then a system of elements $x_1, \cdots, x_m$ in R is said to be a *system of parameters* for $\mathfrak{M}$ if the ideal $R(x_1, \cdots, x_m)$ is primary for $\mathfrak{M}$; and a system of parameters for $\mathfrak{M}$ is said to be *distinct* if it forms an R-sequence. We shall consider here only system of *homogeneous* parameters for $\mathfrak{M}$, i.e. a system of parameters consisting of homogeneous elements in R; then the relations

$$R(x_1, \cdots, x_i) = R \cap R_{\mathfrak{M}}(x_1, \cdots, x_i)$$

for $i = 1, \cdots, m$, show that a system of homogeneous elements $x_1, \cdots, x_m$ in R is a system of parameters for $\mathfrak{M}$ if and only if it is such for $R_{\mathfrak{M}}$, and that such a system of parameters for $\mathfrak{M}$ is distinct if and only if it is distinct as a system of parameters for $R_{\mathfrak{M}}$. It follows then from a well-known result for local ring (LR, 25.5) that if one system of homogeneous parameters for $\mathfrak{M}$ is distinct, then every system of homogeneous parameters for $\mathfrak{M}$ is distinct. We note incidentally that the definitions given here are compatible with the terminology in LR provided the elements $x_1, \cdots, x_m$ are homogeneous; without the homogeneity condition, our definitions would be somewhat stricter than those given in LR.

We shall be concerned in this paper only with the special case of an H-local ring R where $R_0 = K$ is a field; equivalently we can characterize this case by the condition that all homogeneous elements of positive degrees in R generate a maximal ideal $\mathfrak{M}$ in R, which must then be the only maximal homogeneous ideal in R. A simple but important consequence of this assumption is that, if m is the height of $\mathfrak{M}$, then there exist homogeneous elements $x_1, \cdots, x_m$ in $\mathfrak{M}$ such that the ideal $R(x_1, \cdots, x_r)$ has the height r; this implies in particular that $x_1, \cdots, x_m$ form a system of homogeneous parameters for $\mathfrak{M}$, since $\mathfrak{M}$ is the only homogeneous prime ideal in R having a height $\geqq m$. We shall prove this assertion by induction, assuming that the homogeneous elements $x_1, \cdots, x_r$ $(r < m)$ have already been chosen in $\mathfrak{M}$ so that $R(x_1, \cdots, x_r)$ has the height r. Let $\mathfrak{p}_1, \cdots, \mathfrak{p}_s$ be the minimal prime divisor of $R(x_1, \cdots, x_r)$ having the height r; we recall that they are all homogeneous ideals and hence are all contained in $\mathfrak{M}$. Consider any one ideal $\mathfrak{p}_i$; since $\mathfrak{p}_i$ does not contain $\bigcap_{j \neq i} \mathfrak{p}_j$ and since both ideals are homogeneous, there exists a homogeneous element z_i in $\bigcap_{j \neq i} \mathfrak{p}_j$ which is not contained in $\mathfrak{p}_i$. Let d_i be the degree of the homogeneous element z_i (observe that z_i as an element in $\mathfrak{M}$ has a positive degree), and set $c_i = d_1 \cdots d_s / d_i$; then the element $z_1^{c_1} + \cdots + z_s^{c_s}$ is a homogeneous element (of degree $d_1 \cdots d_s$) in $\mathfrak{M}$ and is not contained in any one of the ideal $\mathfrak{p}_1, \cdots, \mathfrak{p}_s$. If we take this element to be x_{r+1}, then clearly $R(x_1, \cdots, x_{r+1})$ must have the height $r+1$; this proves our assertion.

An H-local ring R is said to be H-Macaulay if its maximal homogeneous ideal $\mathfrak{M}$ has a distinct system of homogeneous parameters. The existence of a homogeneous system of parameters for $\mathfrak{M}$ in the special case $R_0 = K$ enables us to prove the following lemma; without the assumption $R_0 = K$, the "if" part still holds, but the "only if" part would be no longer generally true, as will be shown by a simple example below.

LEMMA 5. *Let* $R = \sum_{i=0}^{\infty} R_i$ *be a Notherian graded ring such that* R_0 *is* *a field; then the unmixedness theorem holds for homogeneous ideals in R if and only if R is H-Macaulay.*

Proof. If R is H-Macaulay, then $R_{\mathfrak{M}}$ is Macaulay and hence the unmixedness theorem hold in $R_{\mathfrak{M}}$; the relation $\mathfrak{a} = R \cap R_{\mathfrak{M}}\mathfrak{a}$ for any homogeneous ideal in R shows that then the unmixedness theorem holds for homogeneous ideals in R. Conversely, assume that the unmixedness theorem holds for homogeneous ideals in R, and let $x_1, \cdots, x_m$ be a system of homogeneous parameters for the maximal homogeneous ideal $\mathfrak{M}$ in R, with the property that $R(x_1, \cdots, x_i)$ has the height i for $i = 1, \cdots, m$; we have just noted above that such a system of homogeneous parameter exists. Since $R(x_1, \cdots, x_{i-1})$ has the height $i - 1$ and does not have any embedded prime divisors, all its prime divisors must have the height $i - 1$; it follows then that x_i is not contained in any of the prime divisors of $R(x_1, \cdots, x_{i-1})$ and hence is not a zero-divisor for the R-module $R/R(x_1, \cdots, x_{i-1})$. This shows that the system of homogeneous parameters $x_1, \cdots, x_m$ is distinct, so that R is H-Macaulay.

To show that an H-local ring does not necessarily have a system of homogeneous parameters for its maximal homogeneous ideal, even when the unmixedness theorem holds in it, consider the following example. Let $P[x] = P[x_1, \cdots, x_m]$ be the polynomial ring of m variables over a real discrete valuation ring P, graded as usual by assigning to each variable x_i the degree 1; it is clear that $P[x]$ is H-local, and that if t is a generator of the maximal ideal in P, then the maximal homogeneous ideal in $P[x]$ is generated by the $m + 1$ elements $x_1, \cdots, x_m, t$. Furthermore, since the elements $x_1, \cdots, x_m, t$ clearly form *a* $P]x[$-sequence, they form a distinct system of homogeneous parameters for the maximal homogeneous ideal in $P[x]$, so that $P[x]$ is H-Macaulay. (In fact, it is well-known that $P[x]$ is locally Macaulay.) Now, let $\mathfrak{n}$ be the homogeneous ideal in $P[x]$ generated by the single element tx_m and consider the residue ring $R = P[x]/\mathfrak{n}$; it is clear that R as a graded ring is also H-local and that the maximal homogeneous ideal $\mathfrak{M}$ in R is generated by the images $\bar{x}_1, \cdots, \bar{t}_m, \bar{x}$ in R of $x_1, \cdots, x_m, t$ respectively. The relations $\bar{x}_m{}^2 = \bar{x}_m(\bar{x}_m + \bar{t})$ and $\bar{t}^2 = \bar{t}(\bar{x}_m + \bar{t})$ show that the ideal $R(\bar{x}_1, \cdots, \bar{x}_{m-1}, \bar{x}_m + \bar{t})$ is primary for $\mathfrak{M}$; on the other hand, since the ideal $R(\bar{x}_1, \cdots, \bar{x}_{m-1})$ has the height $m - 1$ and since $\bar{x}_m + \bar{t}$ is not contained in either of the two prime divisors $R(\bar{x}_1, \cdots, \bar{x}_{m-1}, \bar{x}_m)$ and $R(\bar{x}_1, \cdots, \bar{x}_{m-1}, \bar{t})$ of $R(\bar{x}_1, \cdots, \bar{x}_{m-1})$, it follows that $\mathfrak{M}$ has the height m and hence the elements $\bar{x}_1, \cdots, \bar{x}_{m-1}. \bar{x}_m + \bar{t}$ form a system of parameters for M. In fact, our argument shows that $\bar{x}_m + \bar{t}$ is not a zero-divisor for the

R-module $R/R(\bar{x}_1, \cdots, \bar{x}_{m-1})$, and since $\bar{x}_i$ is evidently not a zero-divisor for the R-module $R/R(\bar{x}_1, \cdots, \bar{x}_{i-1})$ for $i = 1, \cdots, m-1$, it follows that $\bar{x}_i, \cdots, \bar{x}_{m-1}, \bar{x}_m + \bar{t}$ form a distinct system of parameters for $\mathfrak{M}$. This implies that $R_{\mathfrak{M}}$ is Macaulay and hence the unmixedness theorem holds in $R_{\mathfrak{M}}$; and the basic relation $\mathfrak{a} = R \cap R_{\mathfrak{M}}\mathfrak{a}$ for any homogeneous ideal in R shows then that the unmixedness theorem holds for homogeneous ideals in R. However, since the element $\bar{x}_m + \bar{t}$ is not homogeneous, the system $\bar{x}_1, \cdots, \bar{x}_{m-1}, \bar{x}_m + \bar{t}$ is not a system of homogeneous parameters for $\mathfrak{M}$; in fact, we maintain that such a system of parameters cannot exist. For any system of homogeneous parameters for $\mathfrak{M}$ cannot consist of positive degrees only, for then all elements in the system would be contained in $R(\bar{x}_1, \cdots, \bar{x}_m)$, which has the height $m-1$. On the other hand, if such a system contains a homogeneous element of degree 0 in $\mathfrak{M}$, the image system in the residue ring $R/R(\bar{t})$ contains less than m non-zero elements and yet must generate an ideal which is primary for the ideal $\mathfrak{M}/R(t)$; but this is impossible, for $R/R(t)$ is isomorphic to the polynomial ring of m variables over the field $P/P(t)$ and in this isomorphism the ideal $\mathfrak{M}/R(t)$ corresponds to the ideal generated by the m independent variables.

Let $R = \sum_{i=0}^{\infty} R_i$ and $S = \sum_{i=0}^{\infty} S_i$ be two Noetherian graded rings which both contain the same field K and satisfy the condition $R_0 = S_0 = K$; in other words, R and S are Noetherin graded K-algebras which both contain K as the subring of all homogeneous elements of degree 0. Let $\mathfrak{M}$ and $\mathfrak{N}$ be the maximal homogeneous ideals in R and S respectively, which we recall are the ideals generated by all homogeneous elements of positive degrees in R and S respectively. Consider the tensor product $T = R \otimes S$ of R and S over K; as usual, we shall consider both R and S embedded canonically as subalgebras in T, so that the tensor product between elements in R and S can be expressed as ordinary product in T. The gradings of R and S induce in a natural way a bi-grading of $T = \sum_{i,j=0}^{\infty} T_{i,j}$, where $T_{i,j}$ is the K-module generated by all products of homogeneous elements of degree i in R with homogeneous elements of degree j in S. An element in $T_{i,j}$ will be said to be homogeneous of bi-degree (i, j), and the bi-degree (i, j) is said to be positive if either i or j is positive; we note that $T_{0,0} = K$ and that $\mathfrak{O} = (\mathfrak{M}, \mathfrak{N})$ is the maximal prime ideal in T generated by all homogeneous elements of positive bi-degrees. Furthermore, we can make T into a graded ring by assigning to each element in $T_{i,j}$ the degree $i + j$, so that we can set $T = \sum_{i=0}^{\infty} T_i$ with $T_i = \sum_{j+k=i} T_{j,k}$; and we note again that $T_0 = K$ and that $\mathfrak{D}$ is the ideal generated in T by all homogeneous elements of positive degrees.

LEMMA 6. *If $a_1, \cdots, a_r$ is an R-sequence in R and $b_1, \cdots, b_s$ is an S-sequence in S, then $a_1, \cdots, a_r, b_1, \cdots, b_s$ is a T-sequence in T.*

Proof. We use induction on the total number $r+s$, the assertion being trivially true in case $r = s = 0$. Since by induction hypothesis the system $a_1, \cdots, a_r, b_1, \cdots, b_{s-1}$ is already a T-sequence in T, we have only to show that b_s is not a zero-divisor for the T-module $T/T(a_1, \cdots, a_r, b_1, \cdots, b_{s-1})$. Now, there is a ring isomorphism between $T/T(a_1, \cdots, a_r, b_1, \cdots, b_{s-1})$ and the tensor product $(R/R(a_1, \cdots, a_r)) \otimes (S/S(b_1, \cdots, b_{s-1}))$, and the image of b_s in $S/S(b_1, \cdots, b_{s-1})$ under the canonical mapping is by assumption not a zero-divisor; our assertion follows then from the fact that in the tensor product of two algebras over a field any regular element in one of the algebras cannot become a zero-divisor.

An almost immediate consequence of Lemma 6 is that if both R and S are H-Macaulay, then the tensor product T is also H-Macaulay. In fact, if the system $x_1, \cdots, x_m$ is a distinct system of homogeneous parameters for $\mathfrak{M}$ and the system $y_1, \cdots, y_n$ is a distinct system of homogeneous parameters for $\mathfrak{N}$, then the combined system $x_1, \cdots, x_m, y_1, \cdots, y_n$ is by Lemma 6 a T-sequence in T; and since this combined system evidently generate in T an ideal which is primary for the ideal $\mathfrak{D} = T(\mathfrak{M}, \mathfrak{N})$, it is a distinct system of homogeneous parameters for $\mathfrak{D}$. Geometrically, this implies that if the unmixedness theorem holds for two varieties U and V, then it holds also for their join $J(U, V)$.

As mentioned before, we are interested here not so much in the tensor product as in what we shall call the Segre product of R and S over K. For any pair of positive integers d and e, we define the Segre product of order (d, e) of R and S over K to be the Noetherian graded subalgebra $A = \sum_{i=0}^{\infty} A_i = \sum_{i=0}^{\infty} T_{id,ie}$ in the tensor product T; for brevity we shall simply speak of the Segre product of R and S in case the base field K as well as the order (d, e) are clear from the context. We observe that the grading of A is induced not only by the grading of T as a graded ring $T = \sum_{i=0}^{\infty} T_i$ but also by the bi-grading of T as a bi-graded ring $T = \sum_{i,j=0}^{\infty} T_{i,j}$, only in the latter case the induced bi-grading of A becomes a simple grading on account of the absence of any elements (except 0) of bi-degrees different from the form (id, ie). It is clear that $A_0 = K$, and that the ideal $\mathfrak{L} = A \cap \mathfrak{D}$ is the ideal in A generated by all homogeneous elements of positive degrees and hence is the maximal homogeneous ideal in A.

We wish to construct a system of parameters for the ideal $\mathfrak{L}$ consisting of homogeneous elements of the same degree in A. As we have shown before, there exists a system of homogeneous parameters for the ideal $\mathfrak{M}$ in R; moreover, by replacing each element in such a system by a suitable power of itself, we can obtain a system of parameters for $\mathfrak{M}$ in which every element is homogeneous of the degree fd, where f can be any multiple of a suitable positive integer. Similarly, we can obtain a system of parameters for $\mathfrak{N}$ consisting of homogeneous elements of degree fe in S, where f again can be any multiple of a suitable positive integer. We now take for f an arbitrary but fixed integer which satisfies both these conditions; let $x_1, \cdots, x_m$ be a system of parameters for $\mathfrak{M}$ consisting of homogeneous elements of degree fd in R, and let $y_1, \cdots, y_n$ be a system of parameters for $\mathfrak{N}$ consisting of homogeneous elements of degree fe in S. Next, let $u_1, \cdots, u_{m+n-1}$ be linear combinations in the $x_1, \cdots, x_m$ with coefficients in K such that every m elements among them are linearly independent (as linear forms over K); it is easily seen that then every m elements among the $u_1, \cdots, u_{m+n-1}$ form also a system of parameters for $\mathfrak{M}$. Similarly, let $v_1, \cdots, v_{m+n-1}$ be linear combinations in the $y_1, \cdots, y_n$ with coefficients in K such that every n elements among them are linearly independent (as linear forms over K); then every n elements among the $v_1, \cdots, v_{m+n-1}$ form also a system of parameters for $\mathfrak{N}$. In order to insure the existence of these linear combinations, we shall assume that the field K contains sufficiently large number of elements (for fixed integers m and n), e. g., K is infinite; however, this restriction is not serious, for not only can it often be met in applications by a change of the ground field, but also it can be removed in the general case by the usual trick of adjunction of a variable to the field K. Now, we set $w_i = u_i v_i$, $i = 1, \cdots, m+n-1$; it is clear that $w_1, \cdots, w_{m+n-1}$ are all bi-homogeneous elements of bi-degree (fd, fe) in T and hence are homogeneous elements of degree f in A. We shall show presently that $w_1, \cdots, w_{m+n-1}$ form a system of parameters for $\mathfrak{L}$. Before doing this, we shall prove a few lemmas concerning the relations between the ideals in T and the ideals in A.

LEMMA 7. *For any ideal $\mathfrak{a}$ in A, we have the relation $A \cap T\mathfrak{a} = \mathfrak{a}$.*

Proof. Let B be the K-submodule in T generated by all homogeneous elements of bi-degrees not of the type (ld, le) for any l; then we have the direct sum $T = A + B$ as K-modules, and it is easily seen that B is also an A-module. Now, let $a_1, \cdots, a_r$ be a base of the ideal $\mathfrak{a}$, and let $z = \sum_{i=1}^{r} c_i a_i$ be an element in $A \cap T\mathfrak{a}$, where the c_i are elements in T. If we set $c_i = c_i' + c_i''$, with c_i' in A and c_i'' in B, then we have the relation $z = \sum_{i=1}^{r} c_i' a_i + \sum_{i=1}^{r} c_i'' a_i$; since $\sum_{i=1}^{r} c_i'' a_i$ is an element in both A and B, it must be equal to 0. Therefore $z = \sum_{i=1}^{r} c_i' a_i$, which shows that z is in $\mathfrak{a}$.

LEMMA 8. *If $\mathfrak{P}$ is a bi-homogeneous prime ideal in T which does not contain either $T\mathfrak{M}$ or $T\mathfrak{N}$, then $\mathfrak{P}$ is the radical of $T(A \cap \mathfrak{P})$.*

Proof. Since $\mathfrak{P}$ does not contain either $T\mathfrak{M}$ or $T\mathfrak{N}$, it cannot contain all the x_i or all the y_i; we can therefore assume that, say, x_j and y_k are not in P. Let E be the multiplicatively closed set generated by x_j and y_k, i.e., E is the set of all monomials in x_j, y_k; since E does not intersect $\mathfrak{P}$, we have the usual relations of a quotient ring, $T \cap T_E\mathfrak{P} = \mathfrak{P}$ and $T \cap (T_E(A \cap \mathfrak{P})) = T(A \cap \mathfrak{P})$. Let $p_1, \cdots, p_s$ be a base for $\mathfrak{P}$ consisting of bi-homogeneous elements, and let $(g_1, h_1), \cdots, (g_s, h_s)$ be their bi-degrees respectively; then, using the notations introduced above, the ideal $A \cap \mathfrak{P}$ contains the elements $p_i^{def} x_j^{dh_i} y_k^{eg_i}$, $i = 1, \cdots, s$, for they have the bi-degrees $(df(eg_i + dh_i), ef(dh_i + eg_i))$, $i = 1, \cdots, s$, respectively. Then it is easily seen that $T_E(A \cap \mathfrak{P})$ must contain the elements p_i^{def}, $i = 1, \cdots, s$, and hence $T(A \cap \mathfrak{P}) = T \cap (T_E(A \cap \mathfrak{P}))$ must also contain these s elements. This shows that $\mathfrak{P}$ is contained in the radical of $T(A \cap \mathfrak{P})$ and hence must coincide with it.

LEMMA 9. *If $\mathfrak{P}$ is a bi-homogeneous prime ideal in T which does not contain either $T\mathfrak{M}$ or $T\mathfrak{N}$, then $\mathfrak{P}$ and $A \cap \mathfrak{P}$ have the same height.*

Proof. Any chain of distinct prime ideals $\mathfrak{P} \supset \mathfrak{P}_1 \supset \cdots \supset \mathfrak{P}_t$ will contract in A to a chain of prime ideals $A \cap \mathfrak{P} \supset A \supset \mathfrak{P}_1 \supset \cdots \supset A \cap \mathfrak{P}_t$, and Lemma 8 shows that the ideals in the latter chain must be distinct; hence the height of $\mathfrak{P}$ is not greater than that of $A \cap \mathfrak{P}$. To prove the converse, consider first a prime ideal $\mathfrak{p}$ in A; it follows easily from Lemma 7 that there must be at least one minimal prime divisor $\mathfrak{P}'$ of $T\mathfrak{p}$ which contracts in A to $\mathfrak{p}$. By Lemma 8, $\mathfrak{P}'$ coincides with the radical of $T\mathfrak{p}$ and hence must be the only minimal prime divisor of $T\mathfrak{p}$. Now, if $A \cap \mathfrak{P} \supset \mathfrak{p}_1 \supset \cdots \supset \mathfrak{p}_t$ is a chain of distinct prime ideals in A, then we can obtain a chain of prime ideals $\mathfrak{P} \supset \mathfrak{P}_1 \supset \cdots \supset \mathfrak{P}_t$ in T such that each $\mathfrak{P}_i$ is a minimal prime divisor of $T\mathfrak{p}_i$; it follows from what we have just shown that each $\mathfrak{P}_i$ is the unique minimal prime divisor of $T\mathfrak{p}_i$ which contracts in A to $\mathfrak{p}_i$, and hence the ideals in the latter chain must be distinct. This shows that the height of $A \cap \mathfrak{P}$ is not greater than that of $\mathfrak{P}$.

LEMMA 10. *The elements* $w_1, \cdots, w_{m+n-1}$ *form a system of homogeneous parameters for maximal homogeneous ideal* $\mathfrak{L}$ *in* A.

Proof. We can choose the system $x_1, \cdots, x_m$ so that $R(x_1, \cdots, x_r)$ has the height r for every $r \leqq m$, and choose the system $y_1, \cdots, y_n$ so that $S(y_1, \cdots, y_s)$ has the height s for every $s \leqq n$; then it can be shown that $T(x_1, \cdots, x_r, y_1, \cdots, y_s)$ has the height $r + s$ (the proof is simple, but we shall not go into details here). In particular, the ideal $T(x_1, \cdots, x_{m-1}, y_1, \cdots, y_{n-1})$ has the height $m + n - 2$, and hence (LR, 9.3) every one of its minimal prime divisors must have the height $m + n - 2$. If $\mathfrak{P}$ is any one such prime divisor, then $\mathfrak{P}$ cannot contain the element $x_m y_n$; for, otherwise $\mathfrak{P}$ must contain either x_m or y_n, which would imply that either $T(x_1, \cdots, x_m, y_1, \cdots, y_{n-1})$ or $T(x_1, \cdots, x_{m-1}, y_1, \cdots, y_n)$ must have the height $m + n - 2$. Since we can choose the u's and the v's so that $w_1 = u_1 v_1$ (or any other w_i) is equal to $x_m y_n \bmod T(x_1, \cdots, x_{m-1}, y_1, \cdots, y_{n-1})$, it follows that $\mathfrak{P}$ cannot contain w_1 and hence the ideal $A \cap \mathfrak{P}$ is a proper subset in $\mathfrak{L}$. Now, since $\mathfrak{P}$ does not contain either $T\mathfrak{M}$ or $T\mathfrak{N}$, Lemma 9 shows that $A \cap \mathfrak{P}$ must have the same height $m + n - 2$ as $\mathfrak{P}$ itself; this shows that the height of $\mathfrak{L}$ must be at least $m + n - 1$. On the other hand, it is clear that the system $u_1, \cdots, u_{m+n-1}$ and $v_1, \cdots, v_{m+n-1}$ introduced in this section satisfy the condition (B) in Section 2 (with $t = m + n - 1$); hence we can apply Lemma 4, by the Remark after that lemma. The application of Formula (6) in Lemma 4 to our present situation shows that

$$T(w_1, \cdots, w_{m+n-1}) = T(x) \cap T(y) \cap T(w, (x, y)^{m+n-1});$$

hence, by Lemma 7, we have the decomposition

$$\begin{aligned} A(w_1, \cdots, w_{m+n-1}) &= A \cap T(w_1, \cdots, w_{m+n-1}) \\ &= (A \cap T(x)) \cap (A \cap T(y)) \cap (A \cap T(w, (x, y)^{m+n-1})). \end{aligned}$$

Since it can be easily seen that a sufficiently high power of every element in $\mathfrak{L}$ is contained in every one of the three ideals appearing in the expression on the right hand side of this equation, it follows that $A(w_1, \cdots, w_{m+n-1})$ is primary for $\mathfrak{L}$. This shows that the height of $\mathfrak{L}$ is at most $m + n - 1$ and hence must be equal to $m + n - 1$; therefore the elements $w_1, \cdots, w_{m+n-1}$ form a system of homogeneous parameters for $\mathfrak{L}$.

4. Our aim is to investigate the validity of the unmixedness theorem for homogeneous ideals in the Segre product A of R and S, assuming that the unmixedness theorem holds for homogeneous ideals in both R and S. It turns out that a certain additional condition on both R and S has to be introduced, which proves to be both necessary and sufficient (except in some very special cases). In order to be able to formulate our result properly, we shall first make a few remarks about this condition.

As before, let $R = \sum_{i=0}^{\infty} R_i$ be a Noetherian graded ring with $R_0 = K$, and let $\mathfrak{M}$ be the ideal generated in R by all elements of positive degrees. For simplicity, we shall call a system of parameters for $\mathfrak{M}$ *homogeneous* if it consists of homogeneous elements of the same degree in R, and this common degree is then called the *degree* of the system. Let $x_1, \cdots, x_m$ be a homogeneous system of parameters of degree fd for$M$; we shall say that this system has the d-exponent ϵ (or simply the exponent ϵ in case $d=1$) if ϵ is the smallest integer with the property that every homogeneous element in R of degree ld with $l \geqq \epsilon f$ is contained in $R(x_1, \cdots, x_m)$. In case R is a homogeneous ring (i.e., $\mathfrak{M}$ is generated by homogeneous elements of dsgree 1) and $fd = 1$, then the exponent of the homogeneous system of parameters $x_1, \cdots, x_m$ is simply the exponent of the primary ideal $R(x_1, \cdots, x_m)$. We shall show that, under the assumption that R is H-Macaulay, if any one homogeneous system of parameters of degree fd for $\mathfrak{M}$ has the d-exponent ϵ, then every homogeneous system of parameters of degree fd will have the same d-exponent ϵ. It is sufficient to show that if the system $x_1, \cdots, x_m$ has d-exponent $\leqq \epsilon$, then the same is true of the homogeneous system of parameters $x_1, \cdots, x_{m-1}, z$ obtained from it by replacing x_m by another homogeneous element of degree fd in R. Let a be a homogeneous of degree ld in R with $l \geqq \epsilon f$; since the ideal $R(x_1, \cdots, x_{m-1}, z)$ is primary for $\mathfrak{M}$, there is a positive integer β such that $ax_m{}^\beta$ is contained in $R(x_1, \cdots, x_{m-1}, z)$. We have then a relation of the form $ax_m{}^\beta = a_1x_1 + \cdots + a_{m-1}x_{m-1} + a_mz$, where $a_1, \cdots, a_m$ are homogeneous elements of degree $(l + \beta f - f)d$ in R. Since $(l + \beta f - f) \geqq \epsilon f$, we have the relation $a_m = b_1x_1 + \cdots + b_mx_m$, by our assumption about the d-exponent of the system $x_1, \cdots, x_m$. Combining these two relations, we obtain the relation $(a_1 + b_1z)x_1 + \cdots + (a_{m-1} + b_{m-1}z)x_{m-1} + (b_mz - ax_m{}^{\beta-1})x_m = 0$, from which we conclude that $b_mz - az_m{}^{\beta-1}$ is contained in $R(x_1, \cdots, x_{m-1})$ and hence $ax_m{}^{\beta-1}$ is contained in $R(x_1, \cdots, x_{m-1}, z)$. We can then repeat the same argument for $ax_m{}^{\beta-1}$, and conclude after β steps that a itself is contained in $R(x_1, \cdots, x_{m-1}, z)$.

It is easily seen that the d-exponent does not in general remain unchanged if we replace a homogeneous system of parameters for $\mathfrak{M}$ of degree fd by another one of degree $f'd$ with a different f'. On the other hand, we shall show that, always under the assumption that R is H-Macaulay, if any one (and hence every) homogeneous system of parameters of degree fd for $\mathfrak{M}$ has a d-exponent $\leqq m$, then the same is true for every homogeneous system of parameters of degree $f'd$ for any f'. It is clearly sufficient to show this for the case where either f is a multiple of f' or f' is a multiple of f; futhermore, in view of what we have just shown above, it is sufficient to show this for any two particular homogeneous systems of parameters of degrees fd and $f'd$. It is therefore sufficient to show that the system $x_1, \cdots, x_m$ has a d-exponent $\leqq m$ if and only if the system $x_1^s, \cdots, x_m^s$ (for any positive integer s) has a d-exponent $\leqq m$.

Assume first that the system $x_1, \cdots, x_m$ has a d-exponent $\leqq m$, and let a be a homogeneous element of degree ld with $l \geqq msf$. Since evidently $l \geqq mf$ we have the relation $a = a_1x_1 + \cdots + a_mx_m$, where $a_1, \cdots, a_m$ are homogeneous elements of degree $(l-f)d$; since $l-f \geqq (s-1)mf \geqq mf$, the elements $a_1, \cdots, a_m$ are themselves contained in $R(x_1, \cdots, x_m)$. Thus proceeding in this way, we conclude after $\delta = m(s-1)+1$ steps that $a = \sum a_{i_1 \cdots i_\delta} x_{i_1} \cdots x_{i_\delta}$. Since every term in this expression must contain the s-th power of at least one of the x_i, it follows that a is contained in $R(x_1^s, \cdots, x_m^s)$. Conversely, assume that the system $x_1^s, \cdots, x_m^s$ has a d-exponent $\leqq m$, and let a be a homogeneous element of degree ld with $l \geqq mf$. Consider the product $a(x_1 \cdots x_m)^{s-1}$, which is a homogeneous element of degree $(l+(s-1)mf)d$; since $l+(s-1)mf \geqq msf$, this element $a(x_1 \cdots x_m)^{s-1}$ must be contained in $R(x_1^s, \cdots, x_m^s)$, so that we have the $a(x_1 \cdots x_m)^{s-1} = \sum_{i=1}^{m} a_i x_i^s$. If a were not contained in $R(x_1, \cdots, x_m)$, then this relation could be written as a non-trivial relation in the $x_1, \cdots, x_m$ with coefficients which are not in $R(x_1, \cdots, x_m)$; in fact this relation would not be trivial (i.e., both sides being identical as forms in the $x_1, \cdots, x_m$) because the left side could contain no s-th power of any one of the $x_1, \cdots, x_m$ while each term in the right side would contain the s-th power of at least one of them. Since such a relation cannot exist (LR, 25.2), we conclude that a must be contained in $R(x_1, \cdots, x_m)$.

Thus we have shown that the property of having a d-exponent $\leqq m$ is independent of the choice of the homogeneous system of parameters for $\mathfrak{M}$, (provided of course it has a degree which is a multiple of d) and hence can be considered as a property of the H-Macaulay graded ring R itself. We shall express this property by saying that the H-Macaulay ring R is *d-proper*, or simply *proper* in case $d=1$. It is easily seen from our above argument that if R is d-proper, then it is also d'-proper for any d' which is a multiple of d; in particular, if R is proper, then it is d-proper for every d.

We are now in a position to prove our main theorem.

THEOREM. *Let R and S be Noetherian graded K-algebras with $R_0 = S_0 = K$, and let m and n be the heights of the prime ideals generated by all the homogeneous elements of positive degrees in R and S respectively; we assume that both R and S are H-Macaulay. Then the Segre Product A of order (d, e) of R and S over K is H-Macaulay if R is d-proper (unless $n = 1$) and S is e-proper (unless $m = 1$); and this condition is also necessary in case R and S contain non-zero homogeneous elements of all degrees which are multiples of d and e respectively. Furthermore, when R and S are respectively d-proper and e-proper, then the H-Macaulay ring A is proper.*

Proof. Using the results and notations developed in Section 3, our theorem, except for the last assertion, is equivalent to the statement that the system of parameters $w_1, \cdots, w_{m+n-1}$ for $\mathfrak{L}$ is distinct if and only if the stated condition is satisfied. Now, as we have observed before, the system of parameters $w_1, \cdots, w_{m+n-1}$ is distinct if and only if the ideal $A(w_1, \cdots, w_t)$ is unmixed for every $t = 1, \cdots, m+n-2$. By Lemma 7, we have the relation $A(w_1, \cdots, w_t) = A \cap T(w_1, \cdots, w_t)$; therefore, if $T(w_1, \cdots, w_t) = \bigcap_i \mathfrak{Q}_i$ is a (not necessarily shortest) primary decomposition of $T(w_1, \cdots, w_t)$, i.e., a representation of $T(w_1, \cdots, w_t)$ as an intersection of primary ideals, then $A(w_1, \cdots, w_t) = \bigcap_i (A \cap \mathfrak{Q}_i)$ is a primary decomposition of $A(w_1, \cdots, w_t)$. In order that $A(w_1, \cdots, w_t)$ be unmixed, it is therefore necessary and sufficient that every ideal $A \cap \mathfrak{Q}_i$ which is not redundant in the primary decomposition has the height t.

We now apply the results of Section 2 to the ring T, and observe that the systems $u_1, \cdots, u_t$ and $v_1, \cdots, v_t$, as constructed in Section 3, satisfy the conditions stated in Section 2. We note here a small discrepancy in convention between Section 2 and our present situation, namely in Section 2 the integer t is always not less than m or n, while now t may be any integer from 1 to $m + n - 2$. However, this discrepancy in convention can be easily taken care of; for, in case t is less than m or n, we can replace m or n respectively by t, and one verifies easily that this change of m or n will not invalidate the inequality $t \leqq m + n - 2$ except when $n = 1$ or $m = 1$ respectively (after the change in case both m and n are changed). Therefore we can apply formula (1) of Lemma 3 to our present situation for all values of t from 1 to $m + n - 2$:

$$T(w_1,\cdots,w_t) = \bigcap_{(i)} T(u_{i_1},\cdots,u_{i_\alpha}, v_{i_{\alpha+1}},\cdots,v_{i_t}) \cap T(u_1,\cdots,u_t) \cap T(v_1,\cdots,v_t), \tag{7}$$

where the intersection extends over all partitions (i) of the integers from 1 to t into two sets $(i_1,\cdots,i_\alpha)$ and $(i_{\alpha+1},\cdots,i_t)$, for all values of α from $\sup(t-n+1,0)$ to $\inf(m-1,t)$. We observe that in case $t<m$ the term $T(u_1,\cdots,u_t)$ no longer coincides with $T(u_1,\cdots,u_m)=T(x_1,\cdots,x_m)$ and is in fact redundant in (7); a similar statement holds of course for the term $T(v_1,\cdots,v_t)$ in case $t<n$.

Now, since each ideal $T(u_{i_1},\cdots,u_{i_\alpha}, v_{i_{\alpha+1}},\cdots,v_{i_t})$ is generated by t elements in a distinct system of parameters for $\mathfrak{O}$, it has the height t. It follows then from the validity of the unmixedness theorem for homogeneous ideals in T that this ideal has a (shortest) primary decomposition

$$T(u_{i_1},\cdots,u_{i_\alpha}, v_{i_{\alpha+1}},\cdots,v_{i_t}) = \bigcap_{j=1}^{\alpha(i)} \mathfrak{Q}_{(i),j},$$

in which every ideal $\mathfrak{Q}_{(i),j}$ has the height t. Furthermore, since this is a bi-homogeneous ideal in T, every ideal $\mathfrak{Q}_{(i),j}$ is also bi-homogeneous. If we assume for convenience $m\leqq n$ and set $T(x)=T(x_1,\cdots,x_m)$ and $T(y)=T(y_1,\cdots,y_n)$, then

$$\begin{aligned} T(w_1,\cdots,w_t) &= \bigcap_{(i)} \bigcap_{j=1}^{\alpha(i)} \mathfrak{Q}_{(i),j} && (t<m),\\ &= \bigcap_{(i)} \bigcap^{\alpha(i)} \mathfrak{Q}_{(i),j} \cap T(x) && (m\leqq t<n),\\ &= \bigcap_{(i)} \bigcap_{j=1}^{\alpha(i)} \mathfrak{Q}_{(i),j} \cap T(x) \cap T(y) && (n\leqq t). \end{aligned}$$

If we set $\mathfrak{q}_{(i),j}=A\cap\mathfrak{Q}_{(i),j}$, then it follows from the relation $A(w_1,\cdots,w_t)=A\cap T(w_1,\cdots,w_t)$ that

$$\begin{aligned} A(w_1,\cdots,w_t) &= \bigcap_{(i)} \bigcap_{j=1}^{\alpha(i)} \mathfrak{q}_{(i),j} && (t<m),\\ &= \bigcap_{(i)} \bigcap_{j=1}^{\alpha(i)} \mathfrak{q}_{(i),j} \cap (A\cap T(x)) && (m\leqq t<n),\\ &= \bigcap_{(i)} \bigcap_{j=1}^{\alpha(i)} \mathfrak{q}_{(i),j} \cap (A\cap T(x)) \cap (A\cap T(y)) && (n\leqq t). \end{aligned} \tag{8}$$

Since all the ideals $\mathfrak{Q}_{(i),j}$ are bi-homogeneous and none of them contains either $T\mathfrak{M}$ or $T\mathfrak{N}$, it follows from Lemma 9 that the contraction $\mathfrak{q}_{(i),j}$ in A

of every $\mathfrak{Q}_{(i),j}$ has the height t. On the other hand, it is clear that both $A \cap T(x)$ and $A \cap T(y)$ are primary ideals for $\mathfrak{L}$, for a sufficiently high power of every element in $\mathfrak{L}$ is easily seen to be contained in both $A \cap T(x)$ and $A \cap T(y)$. It follows then that $A(w_1, \cdots, w_t)$ is unmixed if and only if the ideal $A \cap T(x)$ and $A \cap T(y)$, whenever either or both occur in the above primary decomposition of $A(w_1, \cdots, w_t)$, are in effect redundant. We have therefore to prove that this is so if and only if the stated condition is satisfied.

Consider the ideal $A \cap T(x)$. If $n = 1$, then $t < m$ and hence $A \cap T(x)$ does not appear in the right side of (8), so that there is nothing to prove. Assume that $n > 1$, and take $t \geqq m$ (which is possible in case $n > 1$); let z be any element in A which is contained in every one of the ideals $\mathfrak{q}_{(i),j}$. Since z is evidently contained in every one of the ideals $T(u_{i_1}, \cdots, u_{i_a}, v_{i_{a+1}}, \cdots, v_{i_t})$ appearing in the expression on the right side of (7), it follows from Lemma 1 that z is mod $T(x)$ equal to a sum of elements of the form $cv_{i_1} \cdots v_{i_m}$. Furthermore, since z is contained in A, each such element $cv_{i_1} \cdots v_{i_m}$ must itself be a sum of terms of the form $abv_{i_1} \cdots v_{i_m}$, where a is a homogeneous element of degree ld in R with $l \geqq mf$, and b is a homogeneous element of degree $(l - mf)e$ in S. Conversely, if a is any homogeneous element of degree ld in R with $l \geqq mf$, and if we take for b any homogeneous element of degree $(l - mf)e$ in S, then the element $abv_{i_1} \cdots v_{i_m}$ is contained in every ideal $T(u_{i_1}, \cdots, u_{i_a}, v_{i_{a+1}}, \cdots, v_{i_t})$ appearing in the the right side of (7); and since this element has the bi-degree (ld, le), it is contained in A and hence is contained in every one of the ideals $\mathfrak{q}_{(i),j}$. Therefore, the ideal $A \cap T(x)$ is redundant in (8) if and only if for every homogeneous element a of degree ld in R with $l \geqq mf$, the element $abv_{i_1} \cdots v_{i_m}$ is contained in $T(x)$, for all choices of b and $v_{i_1}, \cdots, v_{i_m}$. We recall now that, for any ideal $\mathfrak{a}$ in R, the product of an element a in R with a non-zero element in S is contained in $T\mathfrak{a}$ if and only if a itself is contained in $\mathfrak{a}$; in fact, in view of the canonical isomorphism $T/T\mathfrak{a} = (R/\mathfrak{a}) \otimes S$, this follows from the well-known fact that the product in T of a non-zero element in R with a non-zero-element in S cannot be zero. If we apply this to the ideal $R(x)$ in R and observe that $bv_{i_1} \cdots v_{i_m}$ is a non-zero element in S, we conclude that $A \cap T(x)$ is redundant in (8) if and (in case S contains non-zero homogeneous elements of all degrees which are multiples of e) only if every homogeneous element of degre ld in R with $l \geqq mf$ is containd in $R(x)$, i.e., R is d-proper. Similarly, we can show that, unless $m = 1$ (in which case there is no condition), the ideal $A \cap T(y)$ is redundant in (8) if and (in case R contains non-zero homogeneous elements of all degrees which are multiples of d) only if S is e-proper. This jroves our theorem except for the last assertion.

To prove the last assertion, consider a homogeneous element z of degree l in A with $l \geqq (m+n-1)f$; since z as an element in T has the bi-degree (ld, le) and hence is a sum of elements of the type ab, where a and b are homogeneous elements of degrees ld and le in R and S respectively, it is sufficient to show every such element ab is contained in $T(w_1, \cdots, w_{m+n-1})$. Since R is d-proper and we have the inequality $l \geqq mf$, the element a is contained in $R(x)$ and hence can be exprssed in the form $a = \sum_i a_i x_i$ where the a_i are homogenous elements of degree $(l-f)d$ in R. If $n=1$, then we stop here; if $n>1$, then $l-f \geqq mf$ and hence each a_i can be expressed in the form $a_i = \sum_j a_{ij} x_j$, so that we can set $a = \sum_{i,j} a_{ij} x_j x_i$, where the a_{ij} are homogeneous elements of degree $(l-2f)d$ in R. Thus proceeding in this manner we obtain after n steps an expression for a of the form

$$a = \sum_{i_1, \cdots, i_n} a_{i_1 \cdots i_n} x_{i_1} \cdots x_{i_n}.$$

Similarly, we can obtain for the element b an expression of the form

$$b = \sum_{j_1, \cdots, j_m} b_{j_1 \cdots j_m} y_{j_1} \cdots y_{j_m}.$$

Then the element ab can be expressed as the sum of elements each of which is a monomial of degre $m+n$ in the x's and y's; hence, ab is contained in the ideal $T(w, (u, v)^{m+n})$, in the notation of Lemma 4. Since ab is evidently contained in both $T(x) = T(u)$ and $T(y) = T(v)$, it follows from Lemma 4 that ab is contained in $T(w_1, \cdots, w_{m+n-1})$.

We conclude this paper by giving a few samples of the application of our result. Let $R = K[x_1, \cdots, x_m]$ be a polynomial ring generated by m variables $x_1, \cdots, x_m$ over a field K. The usual way of making R into a graded ring is to assign each variable x_i the degree 1; we shall generalize the the situation here by assigning each variable x_i an arbitrary positive integer d_i as its degree, so that a "monomial" of the form $x_1^{r_1} \cdots x_m^{r_m}$ is a homogeneous element of degree $r_1 d_1 + \cdots + r_m d_m$ in R. It is clear that $R_0 = K$ in the so graded ring R and that m is the height of the maximal prime ideal $\mathfrak{M}$ generated by all homogeneous elements of positive degrees in R. Furthermore, it is well-known (LR, 25.10) that R is locally Macaulay and hence is of course also H-Macaulay, Finally, R is d-proper for any positive integer d. To see this, let $f = d_1 \cdots d_m$, and $f_i = \prod_{j \neq i} d_j$, and set $u_i = x_i^{df_i}$, so that $u_1, \cdots, u_m$ form a homogeneous system of parameters of degree fd for $\mathfrak{M}$; it is sufficient to show that an element of the form $z = x_1^{r_1} \cdots x_m^{r_m}$ of degree $ld = r_1 d_1 + \cdots + r_m d_m$ with $l \geqq mf$ is contained in the ideal $R(u_1, \cdots, u_m)$. Now, it is clear that the inequality $r_1 d_1 + \cdots + r_m d_m \geqq mfd$ implies the inequality $r_i d_i \geqq fd$ for at least one i, and hence the inequality $r_i \geqq f_i d$ for at least one i; and this implies that $x_i^{r_i}$ is contained in $R(u_i)$ and hence z is contained in $R(u_i)$. Thus we have shown that the graded ring R satisfies the hypothesis of our main theorem. If we take for S a polynomial ring $S = K[y_1, \cdots, y_n]$ generated by n variables $y_1, \cdots, y_n$ over K, graded by assigning each variable y_j an arbitrary positive integer e_j as its degree, and let e be any positive integer, then our main theorem shows that the Segre product A of order (d, e) of R and S is a proper H-Macaulay ring.

Finally, we give a couple of simple examples where the condition in our main theorem is not satisfied. Let $R = K[x_1, x_2, x_3]$ be the homogeneous coordinate ring of the plane curve defined by the equatoon $x_3^3 - x_1^2 x_2 = 0$. It is clear that x_1, x_2 form a distinct system of homogeneous parameters for the maximal homogeneous ideal $\mathfrak{M}$ in R. Since any relation between x_1, x_2, x_3 must be a multiple of the basic relation $x_3^3 - x_1^2 x_2 = 0$, the element x_3^2 cannot be contained in $R(x_1, x_2)$; this shows that R cannot be proper. By our main theorem, the Segre product A of R with any H-Macaulay ring of altitude > 1, say the polynomial ring $S = K[y_1, y_2]$ of two variables, cannot be H-Macaulay. In fact, the elements

$$w_1 = x_1 y_1, \quad w_2 = x_2 y_2, \quad w_3 = (x_1 + x_2)(y_1 + y_2)$$

form a system of homogeneous parameters for the maximal homogeneous ideal $\mathfrak{L}$ of A; but this system is not distinct, for the element $x_3^2 y_1 y_2$ in A is not in $R(w_1, w_2)$ while the product $x_3^2 y_1 y_2 \cdot w_3 = x_3^2 (y_1 + y_2)(y_2 w_1 + y_1 w_2)$ is contained in $R(w_1, w_2)$. The next example shows that even in case $m = 1$ the condition in our main theorem need not be satisfied. Let $R = K[x_1, x_2]$ be the homogeneous coordinate ring of the irreducible 0-dimensional variety defined by the equation $x_2^2 - 2x_1^2 = 0$ over a prime field K of characteristic $\neq 2$. It is clear that R is H-Macaulay with x_1 as a parameter for $\mathfrak{M}$, but R is not proper since x_2 is not contained in $R(x_1)$. Therefore, by our main theorem, the Segre product A of R with $S = K[y_1, y_2]$ is not H-Macaulay. We note that A is here the homogeneous coordinate ring $K[z_1, z_2, z_3, z_4]$ of the variety defined by the equation $z_3^2 - 2z_1^2 = 0$, $z_4^2 - 2z_2^2 = 0$, $z_1 z_4 - z_2 z_3 = 0$; the first two equations define four lines in the projective 3-space, while the last equation excludes two of these four lines.

THE JOHNS HOPKINS UNIVERSITY.

On meromorphic maps of algebraic varieties

By WEI-LIANG CHOW

Let X and Y be algebraic varieties, and let F be a meromorphic map of X into Y; then F is algebraic and is in fact a rational map of X into Y (see [2]). Under certain circumstances the hypothesis here can be reduced, in the sense that the result still holds even when the meromorphic map F is defined only in some open subset N in X. Situations of this nature have been studied recently by H. Hironaka and H. Matsumura [3], W. Barth [1], and H. Rossi [4]. Their main results deal with meromorphic functions on X (i.e., $Y = P^1$), and assert that for certain varieties X any meromorphic function defined in a connected neighborhood N of an algebraic subvariety Z of positive dimension is (the restriction to N of) an algebraic or even rational function on X. This theorem has been proved for the projective space, and for abelian varieties under the assumption that Z generates X. The methods of Barth and Rossi are function-theoretic, with Barth relying heavily on the theory of pseudo-concave spaces of Andreotti, while Hironaka and Matsumura obtain their results as corollaries of their study of formal rational functions.

In this paper we shall recapture and substantially generalize these results by proving a rather general theorem of this nature for a certain class of algebraic homogeneous spaces. Of particular interest seems to be the fact that this class includes those algebraic varieties which represent irreducible symmetric homogeneous spaces, such as the grassmannian, as well as the subvarieties in it commonly associated with the symplectic and orthogonal groups. The basic idea of our method stems from the observation that the property of an analytic subvariety or function being algebraic should be reflected already by its behavior in the neighborhood of a single point and hence can be studied from the viewpoint of local analytic geometry. After a few preliminaries in § 1, we shall give such a local criterion of algebraicity in § 2, where we have formulated our result in abstract setting, using the terminology of algebroid geometry in the sense of Chevalley. Although this terminology is now subsumed in the language of formal geometry of Grothendieck, we prefer to use it here because we shall deal only with strictly local concepts and hence have no use for the sophisticated global approach of the latter.

* This research was partially supported by National Science Foundation under Grant GP-8048.

Using this criterion, we then give in § 3 a proof of the main theorem, which we feel is extremely simple and direct, and we complement it by a theorem giving some sufficient conditions for the map F to be not only algebraic but also rational. In the final section, we turn to the special case of the projective space. Here we can prove a more general theorem about a certain type of analytic subvarieties in $N \times Y$, which has some rather interesting consequences.

1. Algebroid geometry

Just as in algebraic geometry (after Weil), we introduce here first a *universal domain* Ω, which is an algebraically closed field of infinite degree of transcendency over the prime field in it; only here, for reasons which will soon be apparent, we stipulate that this degree of transcendency be more than countably infinite. An example of such a universal domain is of course the complex field, whose degree of transcendency over the rational field has the cardinality of the continuum. Correspondingly, when we speak of a field in Ω, or simply speak of a field, we shall (unless explicitly mentioned otherwise) understand it to be a subfield K in Ω such that Ω has an uncountably infinite degree of transcendency over K. We shall denote by S^n the affine space of dimension n, or simply the n-space, over the universal domain Ω, each point in S^n being given by an ordered n-tuple of elements in Ω, which are then called the coordinates of the point. Let $x = (x_1, \cdots, x_n)$ be a system of n indeterminates (we recall that they are not elements in Ω); then, for any point $p = (p_1, \cdots, p_n)$ in S^n, we associate with it a system of n indeterminates to be denoted by $x - p = (x_1 - p_1, \cdots, x_n - p_n)$, and consider the ring $\Omega[[x - p]]$ of formal power series in the indeterminates $x_1 - p_1, \cdots, x_n - p_n$ with coefficients in Ω. An *algebroid variety* V at p in S^n is then given by a prime ideal in $\Omega[[x - p]]$, which we shall denote by $I(V)$; and V is said to have the *dimension* r if r is the altitude of the local ring $\Omega[[x - p]]/I(V)$. A field K in Ω (in the sense stipulated above) is said to be a *field of definition* for V, if p is a rational point over K and if $I(V)$ has a base consisting of elements in $K[[x - p]]$; if we set $I_K(V) = K[[x - p]] \cap I(V)$, then the second condition can be expressed by the equation $I(V) = \Omega[[x - p]]I_K(V)$. Given any algebroid variety V at p, there is always a field of definition K for V which is a countably extension of the prime field in Ω; in fact, if $f_1, \cdots, f_m$ is a base of the ideal $I(V)$ in $\Omega[[x - p]]$, then we can obtain such a field K by the adjunction to the prime field of all the coefficients of the power series $f_1, \cdots, f_m$ as well as the coordinates of the point p.

More generally, we consider an *algebroid set* W at p in S^n, which is defined by a radical ideal $I(W)$ in $\Omega[[x - p]]$. Corresponding to the unique representation of the ideal $I(W)$ as the intersection of its minimal prime divisors $I(W) = \bigcap_i P_i$, we shall consider W as the union of the algebroid varieties V_i defined by P_i, which are called the *components* of W. A field K in Ω is said to be a *field of definition* for W if p is rational over K and if $I(W)$ is the radical of the ideal $\Omega[[x - p]]I_K(W)$, where we set $I_K(W) = K[[x - p] \cap I(W)$; then we shall say that W is defined over K by the ideal $I_K(W)$, which evidently determines $I(W)$. Just as in the case of an algebroid variety, we can construct a field of definition for any algebroid set W from a base of the ideal $I(W)$; however, we note that while the existence of a base for $I(W)$ in $K[[x - p]]$ is sufficient for W to be defined over K, this condition is here not necessary. In fact, it is clear that for any field K in Ω and any ideal Q in $K[[x - p]]$, the algebroid set defined by the radical of the ideal $\Omega[[x - p]]Q$ is defined over K by the radical of Q. In case an algebroid set W at p is defined over a field K by a prime ideal $I_K(W)$, we shall say that W is *K-irreducible*. In general, if an algebroid set W is defined over K, then the unique representation of $I_K(W)$ as the intersection of its minimal prime divisors in $K[[x - p]]$ shows that W is the union of K-irreducible algebroid sets, which are then called the *K-components* of W. Finally, we note that in case an algebroid set W is an algebroid variety (i.e., has only one component) then a field of definition for W as an algebroid variety is also one for W as an algebroid set, but not conversely in general. This calls for a certain caution in the use of this expression for such a situation, but we shall see to it that the context will make the meaning clear in every case.

Since every element in the polynominal ring $\Omega[x]$ can be expressed in a unique way as a power series in $x_1 - p_1, \cdots, x_n - p_n$ with coefficients in Ω, for any point p in S^n, we can embed $\Omega[x]$ canonically as a subring in $\Omega[[x - p]]$. We recall that algebraic varieties and algebraic sets are defined by prime ideals and radical ideals respectively in $\Omega[x]$, with similar usage concerning their fields of definition. Let U be an algebraic set in S^n, defined by a radical ideal $I(U)$ in $\Omega[x]$, and let p be a point in U; then the algebroid set U_p at p in S^n, defined by the ideal $\Omega[[x - p]]I(U)$, is said to be *induced* by U at p, and a component of U_p is said to be a *branch* of U at p. If U is an algebraic variety of dimension r, then it is well-known that the ideal $\Omega[[x - p]]I(U)$ is the intersection of prime ideals of depth r, so that every branch of U at p has also the dimension r. On the other hand, given any algebroid set W at p in S^n, we define its *algebraic closure* $\tilde{W}$ to be the algebraic set defined by the (radical) ideal $I(\tilde{W}) = \Omega[x] \cap I(W)$ in $\Omega[x]$; it is clear that the algebraic closure of an algebroid variety is an algebraic variety, and that the algebraic closure of an algebraic set is the union of the algebraic closures of its components. If V is an algebroid variety at p of dimension r, then its algebraic closure $\tilde{V}$ must have a dimension at least equal to r. This follows from the fact that the canonical injection of $\Omega[x]/I(\tilde{V})$ into $\Omega[[x-p]]/I(V)$ induces a Ω-homomorphism of the completion of the local ring of $\Omega[x]/I(\tilde{V})$ at p onto the local ring $\Omega[[x-p]]/I(V)$, so that the depth of $I(\tilde{V})$ is at least equal to the depth of $I(V)$. As to the fields of definition, we observe here the following. If K is a field of definition for an algebraic set U and if p is a rational point over K in U, then the algebroid set U_p is also defined over K, being in fact defined by the ideal $\Omega[[x-p]]I_K(U)$, with $I_K(U) = K[x] \cap I(U)$. Similarly, if K is a field of definition for an algebroid set W at p, then its algebraic closure $\tilde{W}$ is also defined over K; this follows from the fact that a power of every element in $I(W)$ can be expressed in a unique manner as a linear combination of elements in $I_K(W)$ with coefficients which are elements of a linear base of Ω over K, and hence a power of every element in $I(\tilde{W})$ is also such a linear combination of elements in $I_K(\tilde{W}) = K[x] \cap I_K(W)$. The same argument shows that if an algebroid variety V at p is defined over K as a variety, then its algebraic closure $\tilde{V}$ is also defined over K as a variety.

An algebroid variety V at p in S^n is said to be *algebraic* if it is a branch of an algebraic variety at p, aud an algebroid set W at p in S^n is said to be *algebraic* if every component of W is algebraic. It is clear that if V is algebraic, then its algebraic closure $\tilde{V}$ is the uniquely determined algebraic variety of which V is a branch at p; thus V is algebraic if and only if $\tilde{V}$ has has the same dimension as that of V. Similarly, if an algebraic set W at p is algebraic, then every component of it is a branch of its algebraic closure $\tilde{W}$ at p; and in case every component of W has the same dimension r, then W is algebraic if and only if $\tilde{W}$ has also the dimension r. The last statement holds in particular when W is K-irreducible for some field K; and we note also that in this case W is algebraic if any one of its components is algebraic, in view of the fact that all components of W are conjugates of each other over K.

Up to now we have restricted ourselves to varieties and sets in S^n, mainly for the sake of simplicity. However, since every algebraic variety has at any point a local representative which is an affine variety, it is clear that all our definitions and results apply just as well to the case where the ambient variety P is the projective space or any algebraic variety, with a few obvious modifications. In particular, if V is an algebroid variety at a point p in P, then its algebraic closure with respect to a local representative of P at p can be extended in a unique manner to an algebraic subvariety in P, which we shall again call the algebraic closure of V (in P) and denote by $\tilde{V}$. It is easily seen that the property of such an algebroid variety V being algebraic is independent of the choice of the local representative of P at p. All this is to be understood tacitly when we apply our results to global situations in § 3.

2. A criterion for algebraicity

Let K be a field in Ω and let p be a rational point over K in S^n. If W is an algebroid set at p in S^n, then the (radical) ideal $K[[x-p]] \cap I(W)$ defines an algebroid set V at p. We shall say that the algebroid set V is the *carrier* of W over K, and that W *spans* V over K. It is clear that $I_K(V) = K[[x-p]] \cap I(W)$, and that V is K-irreducible if W is an algebroid variety at p. This terminology is suggested by the analogous situation in algebraic geometry where an algebraic set U, defined by a (radical) ideal $I(U)$ in $\Omega[x]$, would generated over K an algebraic system of algebraic sets, and the carrier of this system is then the algebraic set defined by the ideal $K[x] \cap I(U)$. There is in fact a simple relation between the two cases. If $\tilde{V}$ and $\tilde{W}$ are the algebraic closures of V and W respectively, then the equations

$$I_K(\tilde{V}) = K[x] \cap I(V) = K[x] \cap I(W) = K[x] \cap I(\tilde{W})$$

show that $\tilde{V}$ is the carrier of $\tilde{W}$ over K. This leads us naturally to expect that if W is algebraic, then V would also be algebraic. We shall show that this is in fact true, and thus obtain a criterion for algebraicity which will play an essential role in our investigation.

THEOREM 1. *Let V be an algebroid set at p in S^n, defined over a field K, and let W be an algebroid set at p which is contained in V and spans V over K. If W is algebraic, then V is also algebraic.*

PROOF. Without any loss of generality, we can assume that V is K-irreducible, say of dimension r, and that W is an algebroid variety, say of dimension $s \leqq r$. By assumption, the algebraic closure $\tilde{W}$ of W has the same dimension s and W is a branch of $\tilde{W}$ at p; and we have noted above that the algebraic closure $\tilde{V}$ of V is the carrier of $\tilde{W}$ over K. Since $\tilde{W}$ is an algebraic variety, it is defined over a finitely generated extension L of K, say of dimension d over K; then W is defined over an algebraic extension L_1 of L, which has of course also the dimension d over K. We shall carry our proof by double induction on the dimensions r and d, for a given s; and we note that for $r = s$ our theorem follows from the observation that then W is a component of V, and that $d = 0$ is possible only if $r = s$. Let u be an element in L which is a variable over K (note that we can assume $d > 0$), and set $K' = K(u)$. Then L (and hence also L_1) has the dimension $d - 1$ over K'. Let V' be the carrier of W over K'; since

$$I_K(V) = K[[x - p]] \cap I(W) \subset K'[[x - p]] \cap I(W) = I_{K'}(V')\ ,$$

it follows that V' is contained in V. As before, with K' replacing K as the ground field, we conclude that the algebraic closure $\tilde{V}'$ of V' is the carrier of $\tilde{W}$ over K'. Now, if V' coincides with V, so that W spans V over K', then V must be algebraic by induction hypothesis, since L_1 has a dimension $< d$ over K'. On the other hand, if V' does not coincide with V and hence is a K'-irreducible algebroid set of dimension $t < r$, then again by induction hypothesis, V' must be algebraic and hence $\tilde{V}'$ has also the dimension t. Since

$$I_K(\tilde{V}) = K[x] \cap I(\tilde{W}) = K[x] \cap I_{K'}(\tilde{V}')\ ,$$

one sees that $\tilde{V}'$ spans $\tilde{V}$ over K, and since K' has the dimension 1 over K, it follows that $\tilde{V}$ must have a dimension at most equal to $t + 1 \leqq r$. This implies again that V is algebraic, and hence concludes the proof of our theorem.

3. The main theorem

In this section the universal domain Ω is the field of all complex numbers, which we shall briefly call the complex field. In the complex domain, the algebraic sets are subsumed under the more general category of analytic sets, which are defined locally as the common zeroes of analytic functions. Just as in the case of an algebraic set, an analytic set A in S^n will induce at a point p in it an algebroid set V, each component of which is then called a *branch* of A at p. In the present case, the so induced algebroid set V at p will have the special property that its ideal $I(V)$ in $\Omega[[x - p]]$ has a base consisting of convergent power series. We can then associate with V a local analytic set at p, defined as the common zeroes of these convergent power series, which coincides with A in a sufficiently small neighborhood N of p. We recall the well-known fact that if this neighborhood N is taken sufficiently small, then every irreducible component of the analytic set $N \cap A$ contains the point p; when such is the case, we shall say that V *represents* A in N. We note that if V represents A in N and if V' is the algebroid set induced by A at a point p' in $N \cap A$, then every component of V' is an analytic continuation of a component of V, both components being considered as local analytic varieties.

Our main result concerns a homogeneous algebraic variety. We recall that an algebraic variety X is said to be *homogeneous* if there is an algebraic group variety G which acts transitively on X, i.e., there is a rational map $(g, x) \to g(x)$ of $G \times X$ into X such that $g(g'(x)) = g \cdot g'(x)$ and $e(x) = x$, where $g, g' \in G$, $x \in X$, and e is the unit element in G, and such that for every pair $x, x' \in X$ there exists an element $g \in G$ with $g(x) = x'$. Let Z be an algebraic subvariety in X, and let p be a point in Z. We denote by $G_{p,Z}$ the set of all elements in G which carry p to a point in Z. It is easily seen that $G_{p,Z}$ is an algebraic set and is irreducible in case the isotropic subgroup of G at p is connected. Let G_Z be the subgroup in G generated by the set $G_{p,Z}$; we note that G_Z is an algebraic subgroup and is independent of the choice of the reference point p. We shall say that Z *generates* X if $G_Z = G$, and it is easily seen that Z generates X if and only if $X = G_Z(p)$, where p can be any point in Z and $G_Z(p)$ is the orbit of p under G_Z (this justifies the term "generate"). We note that in case $X = G$ with G acting on it by left translation, then Z generates X in our sense if and only if it generates G in the usual sense. Finally, we observe that if $G^0_{p,Z}$ is the component in $G_{p,Z}$ containing the unit element e, then a simple dimensionality argument shows that $G^0_{p,Z}$ generates the entire group G if and only if $G_{p,Z}$ generates G.

It is clear that if the isotropic subgroup of G on X contains a maximal connected closed subgroup in G, then every subvariety of positive dimension in X must generate X. This is the case not only for simple abelian varieties and projective spaces, but also for all those algebraic varieties which represent irreducible homogeneous spaces, such as the grassmannian, as well as those subvarieties in it commonly associated with the symplectic and orthogonal groups (usually denoted by $\mathbf{Sp}(n)/\mathrm{U}(n)$ and $\mathbf{SO}(2n)/\mathrm{U}(n)$). For all these varieties our results will apply to every subvariety of positive dimension.

THEOREM 2. *Let X be a homogeneous algebraic variety, let Z be a complete algebraic subvariety of positive dimension in X which generates X, and let N be a connected neighborhood of Z in X. If Y is an algebraic variety and if F is a meromorphic map of N into Y, then F is algebraic.*

PROOF. Let G be the group acting on X, let p be a point in Z, and let $G^0_{p,Z}$ be the component in $G_{p,Z}$ containing the unit element e, in the notation introduced before. Our assumption that Z generates X implies that there is an integer t such that every point in X can be expressed in the form $g_t^{(-1)^t} \cdots g_1^{-1} \cdot g_0(p)$, where $g_0, g_1, \cdots, g_t$ are suitable elements in $G^0_{p,Z}$. Without any loss of generality we can assume that F is holomorphic at p; for, if Z does not contain such a point p, we can replace Z by a transform $g(Z)$ which does, X being homogeneous, whereby we take the element g in G so near to e

that $g(Z)$ still lies in N. For convenience, we identify F with its graph in $N \times Y$; we set $p^* = (p \times Y) \cap F$, and denote by V the algebroid variety at p^* induced by F. We have only to prove that V is algebraic.

Let K be a field such that $G, X, Z, G^0_{p,Z}, Y, p, p^*$, and V are all defined over K. We recall that the universal domain here is the complex field, and that we can obtain such a field K by a countable extension of the rationals. Let $u_0, u_1, \cdots, u_t$ be independent generic points in $G^0_{p,Z}$. Then the point $u_t^{(-1)^t} \cdots u_1^{-1} \cdot u_0(p)$ is a generic point of X over K, in view of our choice of the integer t. For $i = 0, 1, \cdots, t$, we set $K_i = K(u_{i+1}, \cdots, u_t)$ with $K_t = K$ and set $p_i = u_t^{(-1)^t} \cdots u_{i+1}^{(-1)^{i+1}}(p)$ with $p_t = p$; and we denote by Z_i the locus of the point $u_t^{(-1)^t} \cdots u_1^{-1} \cdot u_0(p)$ over K_i. We observe that $Z_0 \subset \cdots \subset Z_t = X$, and that Z_i spans Z_{i+1} over K_{i+1} and contains both p_i and p_{i+1}. It is clear that if we take the points $u_1, \cdots, u_t$ sufficiently near to e, then all points p_i will lie in N and the map F will be holomorphic at every one of them. We shall assume this, although we shall presently make a more specific choice of these points. We then set $p_i^* = (p_i \times Y) \cap F$, and denote by V_i and V_i' the algebroid sets at p_i^* and p_{i+1}^* respectively induced by the analytic set $(Z_i \times Y) \cap F$ in $N \times Y$. It is clear that $V_t = V$ and $V_i' = (Z_i \times Y) \cap V_{i+1}$. We shall prove our theorem by showing that every V_i is algebraic, and we shall do this by induction on the index i. For this purpose we have to make a more specific choice of the points $u_t, \cdots, u_1$, successively in that order.

We denote by Z_i' the locus of the point $u_i^{(-1)^i} \cdots u_1^{-1} \cdot u_0(p)$ over K, and take a neighborhood N' of p such that $\bar{N}' \subset N$ and every component of the analytic set $N' \cap Z_i'$ contains the point p, for every $i = t, \cdots, 1, 0$. We have then the equation

$$Z_i = u_t^{(-1)^t} \cdots u_{i+1}^{(-1)^{i+1}}(Z_i') ,$$

and if we set

$$N_i = u_t^{(-1)^t} \cdots u_{i+1}^{(-1)^{i+1}}(N') ,$$

then every component in $N_i \cap Z_i$ will contain the point p_i. We set $K_t^* = K_t = K$, and take for u_t a generic point of $G^0_{p,Z}$ over K_t^* so near to the unit element e in G that $\bar{N}_{t-1} \subset N$ and $p_t = p \in N_{t-1}$, which we can evidently do. Then we take a field of definition K_{t-1}^* for V_{t-1} which is a countable extension of $K_t^*(u_t) = K_{t-1}$, and take for u_{t-1} a generic point of $G^0_{p,Z}$ over K_{t-1}^* so near to e that $\bar{N}_{t-2} \subset N$ and $p_{t-1} \in N_{t-2}$; and so on. In general, having chosen K_{i+1}^* and u_{i+1}, we take a field of definition K_i^* for V_i which is a countable extension of $K_{i+1}^*(u_{i+1})$, and take for u_i a generic point of $G^0_{p,Z}$ over K_i^* so near to e that $\bar{N}_{i-1} \subset N$ and $p_i \in N_{i-1}$. In this way, we construct the sequence of fields $K_t^*, \cdots, K_1^*$ and the sequence of points $u_t, \cdots, u_1$.

We now make two observations. In the first place, it is clear that every component in the analytic set $(N_i \times Y) \cap (Z_i \times Y) \cap F$ contains the point p_i^*, and that

$$p_{i+1}^* \in (N_i \times Y) \cap (Z_i \times Y) \cap F .$$

This implies that V_i as a local analytic set represents $(Z_i \times Y) \cap F$ in $N_i \times Y$, and since V_i' is the algebroid set induced by $(Z_i \times Y) \cap F$ at the point p_{i+1}^* in $N_i \times Y$, it follows that every component of V_i' is an analytic continuation of a component of V_i. Secondly, since $V_i' = (Z_i \times Y) \cap V_{i+1}$ and since $Z_i \times Y$ is defined over $K_i = K_{i+1}(u_{i+1})$, it follows that V_i' is defined over $K_{i+1}^*(u_{i+1})$; and since Z_i spans Z_{i+1} over K_{i+1}, we conclude that V_i' spans V_{i+1} over K_{i+1}^*.

We can now finish our proof in a few lines. Assume that V_i is algebraic. Since every component of V_i' is an analytic continuation of a component of V_i, the algebroid set V_i' is also algebraic, and since V_i' spans V_{i+1} over K_{i+1}^*, it follows from Theorem 1 that V_{i+1} is also algebraic. Thus, in order to prove that $V = V_t$ is algebraic, we have only to show that V_0 is algebraic. Now, V_0 is the algebroid set at p_0 induced by the analytic set $(Z_0 \times Y) \cap F$ in $N \times Y$. If we take the points $u_1, \cdots, u_t$ all sufficiently near to e, then

$$Z_0 = u_t^{(-1)^t} \cdots u_1^{-1}(Z)$$

can be made arbitrarily near to Z and hence lies entirely in N; then $(Z_0 \times Y) \cap F$ is a compact analytic subvariety in the algebraic variety $X \times Y$ and hence must be algebraic. This concludes the proof of Theorem 2.

As we have mentioned at the beginning of this paper, Theorem 2 not only includes as special cases all known results in this direction (so far as we know), but also substantially generalizes them. In order to see this, we shall investigate the conditions for our meromorphic map F to be not only algebraic, but also rational.

The situation being as in Theorem 2, we start with its conclusion that the given map F is algebraic. Let $\tilde{V}$ be the algebraic closure of V in $X \times Y$, and let π be the projection of $\tilde{V}$ onto X; then F is a component variety in the open subset $\pi^{-1}(N) = (N \times Y) \cap \tilde{V}$ in $\tilde{V}$. We assume that $\tilde{V}$ does not define a rational transformation of X into Y, so that π is not birational. Let L be the function field of $\tilde{V}$ over K (we recall that $\tilde{V}$ is defined over K), and let the function field of X over K be embedded as a subfield in L by means of the rational map π. Let W be the normalization of X in L (i.e., the derived normal model of L over X); let φ be the rational map of W onto X associated with the normalization, and let T be the birational correspondence of W onto $\tilde{V}$ so that $\varphi = \pi \cdot T$. Since the component varieties in $\varphi^{-1}(N)$ and $\pi^{-1}(N)$ correspond to each other under T, and since W is analytically irreducible everywhere, there is an isolated component N^* in $\varphi^{-1}(N)$ which corresponds under T to F in $\tilde{V}$. Hence, the map $\varphi = \pi \cdot T$ is biholomorphic everywhere in N^* except possibly for a proper analytic subset in N^*. Moreover, since neither φ nor φ^{-1} has fundamental points, and since both N and N^* are normal, it follows that φ is biholomorphic everywhere in N^*. If we set $Z^* = \varphi^{-1}(Z) \cap N^*$, then φ is unramified at every point in Z^*, and $\varphi^{-1}(Z)$ contains Z^* as an isolated component. Let u be a generic element of G over K, and consider the transform $u(Z)$ of Z; since $u(Z)$ spans X over K, the algebraic set $\varphi^{-1}(u(Z))$ consists of a complete set of conjugate varieties over $K(u)$, and if we denote one of them by C, then for every element g in G every component of $\varphi^{-1}(g(Z))$ is a specialization of C over K. Since this implies in particular that Z^* is a specialization of C over K, it follows that φ is unramified at every point of C. Now, if W is not an unramified covering of X, the set of all points in W at which φ is ramified is an algebraic set Δ of codimension 1, by the Theorem of Purity of branch locus. It follows that no specialization of C over K can intersect Δ unless it is entirely contained in Δ; and therefore for every g in G no component of $\varphi^{-1}(g(Z))$ can intersect Δ unless it is entirely contained in Δ. Going down to X, this implies that for every g in G the transform $g(Z)$ cannot intersect $\varphi(\Delta)$ unless it is entirely contained in $\varphi(\Delta)$. As this would contradict our assumption that Z generates X, we conclude that W must be an unramified covering of X. This of course cannot occur if X has a trivial fundamental group, e.g., if X is a (complete) rational variety, so that in this case we can conclude that F must be a rational map. In the general case, we observe that $\varphi^{-1}(Z)$ consists of at least two isolated components such that all of them are specializations of C over K and one of them is Z^*. This implies that the intersection $Z^* \cdot \varphi^{-1}(Z)$, in the sense of algebraic equivalence classes, must be zero, and hence by the projection formula the self-intersection $Z \cdot Z$ of Z in X must also be zero. In case Z has a dimension $< r/2$, where r is the dimension of X, this of course says nothing, but in case Z has a dimension $\geqq r/2$, this imposes a condition for F to be not a rational map. For example, if X is a simple abelian variety of dimension r, and if Z has a dimension $\geqq r/2$, then $Z \cdot Z$ is never zero, so that in this case F must be a rational map. We summarize our results in the following theorem, which can be considered as a complement to Theorem 2.

THEOREM 3. *If either X is a rational variety or Z has a non-zero self-*

intersection, then the conclusion of Theorem 2 can be strengthened to the effect that F is the restriction to N of a rational transformation of X into Y.

4. The projective spaces

The situation is particularly simple in case the variety X is the projective space. In fact, in this case we can take $t = 1$ in the proof of Theorem 2 and the entire argument shrinks into an almost trivial application of Theorem 1. The basic reason for this is the fact that the projective space is doubly homogeneous, i.e., every pair of points can be carried into every other pair by a projective transformation; for then the transform $u(Z)$ of Z by a generic element u of the isotropic subgroup of G at p (or rather the connected component of it) spans the entire X, and hence the transform $u_1^{-1}(Z) = u_1^{-1} \cdot u_0(p)$, of which $u(Z)$ is a specialization, must also span X. We shall now show that, by a slight modification of the same technique, we can prove in this case a more general theorem about analytic subvarieties in $N \times Y$, which has some rather interesting consequences. We denote by pr_N the projection of $N \times Y$ onto N, and we shall say that the map pr_N is proper over an analytic subvariety F in $N \times Y$ if the restriction of pr_N to F is a proper map. We recall that when pr_N is proper over F, then the image set $\mathrm{pr}_N F$ is an analytic subvariety in N.

THEOREM 4. *Let Z be an algebraic subset of dimension r in the projective space X of dimension n, and let N be a connected neighborhood of Z in X. Let Y be an algebraic variety, and let F be an analytic subvariety in $N \times Y$ such that pr_N is proper over F, and that the image $\mathrm{pr}_N F$ intersects Z and has a dimension $> n - r$. Then F is algebraic.*

PROOF. Let p be a point in $Z \cap \mathrm{pr}_N F$, and consider the isotropic subgroup G_p of X at p, i.e., the group of all projective transformations of X leaving p fixed. Then, for every element g in G_p which is sufficiently near to the identity, the transform $g(Z)$ lies in N. Since $\mathrm{pr}_N F$ intersects $g(Z)$ and has a dimension $> n - r$, it follows that $g(Z) \cap \mathrm{pr}_N F$ is a compact analytic subset of positive dimension in N and hence is algebraic. Moreover, since pr_N is proper over F, this implies that $(g(Z) \times Y) \cap F$ is also a compact analytic subset in $N \times Y$ and hence is also algebraic. Let p^* be a point in F such that $\mathrm{pr}_N p^* = p$, and let V and V_g be the algebroid subsets at p^* induced by F and $(g(Z) \times Y) \cap F$ respectively. We note that V_g is algebraic as long as g is sufficiently near to the identity. Let K be a field such that X, Z, Y, p. G_p, p^*, and V are all defined over K, and let u be a generic point of G_p over K, chosen so near the identity that V_u is algebraic. Since, for any point q

in F, there is an element g in G_p such that $g(Z)$ contains the point $\mathrm{pr}_N q$, and hence $(g(Z) \times Y) \cap F$ contains the point q, and since $(g(Z) \times Y) \cap F$ is a specialization of $(u(Z) \times Y) \cap F$ over K, it follows that $(u(Z) \times Y) \cap F$ spans F over K and hence V_u spans V over K. Since V_u is algebraic, our theorem follows then from Theorem 1.

In case Y reduces to a point, we can identify $X \times Y$ with X and identify F with $\mathrm{pr}_N F$. We then obtain the following result of Rossi [4], with a few minor improvements.

COROLLARY. *Let Z be an algebraic set of dimension r in the projective space X of dimension n, and let N be a connected neighborhood of Z in X. Then any analytic subvariety of dimension $> n - r$ in N which intersects Z is algebraic.*

Next, we observe that Theorem 4 by its very nature extends immediately to the case where X is any algebraic subvariety in a projective space P, say of dimension m, and Z is the intersection of X with an algebraic subset Z' of dimension $r + m - n$ in P. In fact, we can take a neighborhood N' of Z' in P such that $N' \cap X \subset N$, and replace F by its restriction F' in $(N' \cap X) \times Y$; then F' is an analytic subvariety in $N' \times Y$ such that $\mathrm{pr}_{N'} F' = N' \cap \mathrm{pr}_N F$ intersects Z' and has a dimension $> n - r = m - (r + m - n)$, and we can apply Theorem 4 with P, Z', N', F' replacing X, Z, N, F, respectively. We shall say that an algebraic subset Z in a projective variety X is a *proper intersection* if it can be represented in the form $Z = Z' \cap X$ just described for some embedding of X in a projective space P. We note that this notion of a proper intersection is a generalization of the usual notion of a complete intersection, which requires in addition that Z' be the intersection of $n - r$ hypersurfaces in P. We have then the following generalization of Theorem 4.

THEOREM 5. *Let X be a projective algebraic variety of dimension n, let Z be an algebraic subset of dimension r in X which is a proper intersection, and let N be a connected neighborhood of Z in X. Let Y be an algebraic variety, and let F be an analytic subvariety in $N \times Y$ such that pr_N is proper over F and that the image $\mathrm{pr}_N F$ intersects Z and has a dimension $> n - r$. Then F is algebraic.*

Finally, one naturally asks whether Theorem 4 can be extended to an arbitrary homogeneous algebraic variety X with a suitable "generating" condition for Z, so that it becomes a full generalization of Theorem 2. This can be done in case $t = 2$ in the proof of Theorem 2, by a slight modification of the proof of Theorem 4; however, for the general case, the application of our method presents some complications and the situation seems to call for further investigation. We hope to be able to come back to this and other related questions on some future occasion.

THE JOHNS HOPKINS UNIVERSITY

REFERENCES

[1] W. BARTH, *Fortsetzung meromorpher Funktionen in Tori und komplex-projektiven Raeumen*, Invent. Math. **5** (1968), 42-62.
[2] W-L. CHOW, *On compact complex analytic varieties*, Amer. J. Math. **71** (1949), 893-914.
[3] H. HIRONAKA and H. MATSUMURA, Formal functions and formal imbeddings. Preprint.
[4] H. ROSSI, *Continuation of subvarieties of projective varieties*, to appear in Amer. J. Math.

(Received October 24, 1968)

ON THE ALGEBRAICITY OF CERTAIN RINGED SPACES*

By Wei-Liang Chow

Many years ago we proved a theorem which says that a closed analytic subspace in the projective space is algebraic. Since then the theorem has received a number of other proofs. Of particular interest is the proof given by Serre in his well-known paper entitled "Géométrie algébrique et géométrie analytique" (often quoted as GAGA), in which this theorem appears as a corollary to a systematic treatment of the relations between analytic and algebraic sheaves. Under the circumstances, it would seem slightly absurd to offer here yet another proof of this old theorem. A few words of explanation are therefore in order.

We shall offer here a proof which depends essentially on a simple estimate of the "postulation" type. While this method is in spirit closely related to the geometric proof used in our original paper, it is more elementary in that very little use is made of the topological properties of the space. However, what makes this new proof worthy of attention is the fact that, with a few appropriate modifications, it applies also to the case of a non-archimedean valued field. In fact, we shall derive the result for both the complex-analytic case and the non-archimedean case from a general theorem, which states that if a ringed space of a certain type is immersed (see below) in a projective space, then it is essentially algebraic. In view of the radically different nature of the topology of the ground field in the two cases, it will be a basic feature of our treatment that direct use of this topology be avoided as much as possible. For this reason, we have replaced the usual concept of embedding by a definition of "immersion" which makes no use of the topology of the ground field. Such an immersion is a weaker concept than the usual embedding, but it is sufficient for our purpose.

Our approach can be used to prove other results of a similar nature. These will be presented in a subsequent paper.

We should like to mention that the possibility of extending the GAGA-type results to the non-archimedean case was indicated by Kiehl

* Research partially supported by a grant from National Science Foundation.

(Invent. Math., vol. 2, p. 191). This has been carried out in all generality by Ursula Koepf in her 1973 Muenster thesis.

I.

The object of our study is a G-ringed space $(X, \mathfrak{T}, \mathcal{O})$. We recall that such a space consists of the following objects: (1) A topological space X, which serves as the underlying space for the G-ringed space. (2) A Grothendieck topology $\mathfrak{T}$ on the category of certain open sets in X, called the admissible sets, and the "coverings" in $\mathfrak{T}$ are certain open coverings of admissible sets in X, called admissible coverings. (3) A sheaf of rings $\mathcal{O}(\mathfrak{T})$ on $\mathfrak{T}$, called the structural sheaf. More specifically, the structural sheaf $\mathcal{O}(\mathfrak{T})$ of X is defined by a contravariant functor $U \to \mathcal{O}(U)$ from admissible sets in $\mathfrak{T}$ to rings, which satisfies the usual sheaf condition F1 and F2 (in the notation of R. Godement: Théorie des Faisceaux) with respect to the admissible coverings in $\mathfrak{T}$. We note that the conditions F1 and F2 are not required to be satisfied with respect to coverings which are not admissible.

From the structural sheaf $\mathcal{O}(\mathfrak{T})$ we construct a sheaf space $\mathcal{O}_X$ over X as follows, whereby we observe that the construction is independent of the particular choice of the admissible coverings. By the definition of a sheaf, there is a canonical homomorphism $\rho_U{}^V$ of $\mathcal{O}(V)$ into $\mathcal{O}(U)$, for any pair of admissible sets U, V such that $U \subset V$. For every point x in X, we define the "stalk" $\mathcal{O}_x$ over x to be the inductive limit $\mathcal{O}_x = \lim_{x \in U} \mathcal{O}(U)$ of the family of rings $\mathcal{O}(U)$ with respect to the ordered system of all admissible sets containing the point x; then, for every admissible set U containing x, there is a canonical homomorphism $\rho_x{}^U$ of $\mathcal{O}(U)$ into $\mathcal{O}_x$. The sheaf space $\mathcal{O}_X$ is defined to be the union $\mathcal{O}_X = \bigcup_{x \in X} \mathcal{O}_x$ of the stalks over all points in X, endowed with the topology which has as a base for open sets all sets of the form $\{\rho_x{}^U(f) | x \in U\}$, for every admissible set U and every element f in $\mathcal{O}(U)$. To be sure, in order that the projection map of $\mathcal{O}_X$ onto X be continuous, we have to replace the original topology of X by the possibly coarser topology which has the admissible sets in $\mathfrak{T}$ as a base for open sets. However, to avoid unnecessary complications which are irrelevant for our present purpose, we shall assume that the admissible sets in $\mathfrak{T}$ already form a base for the original topology of X. Following the usual notations, we shall denote by $\Gamma(U, \mathcal{O}_X)$ the ring of all sections of $\mathcal{O}_X$ over the admissible set U. If f is any element in $\mathcal{O}(U)$, the correspondence $x \to f_x =$

$\rho_x{}^U(f)$ defines a map of U into $\mathcal{O}_X$, which is easily seen to be a section of $\mathcal{O}_X$ over U. There is therefore a canonical homomorphism of $\mathcal{O}(U)$ into $\Gamma(U, \mathcal{O}_X)$ for every admissible set U in T. We shall assume that this canonical homomorphism is injective, so that $\mathcal{O}(U)$ can be considered as a subring in $\Gamma(U, \mathcal{O}_X)$; and we shall express this by saying that the structural sheaf $\mathcal{O}(U)$ is Γ-*injective.*

From now on we shall make the following assumptions (without any further mention): (1) There is a fixed ground field k such that every ring $\mathcal{O}(U)$ is an algebra over k and every homomorphism between two such rings is a k-homomorphism. It follows then that the rings $\mathcal{O}_x$ are also k-algebras, and the homomorphisms $\rho_x{}^U$ are also k-homomorphisms. In fact, unless mentioned explicitly otherwise, all homomorphisms between k-algebras are understood to be k-homomorphisms. Furthermore, for simplicity, we shall also assume that the ground field k is algebraically closed, although much of our results are valid without this assumption or with appropriate modifications. On the other hand, we shall avoid any direct use of any topology that might exist in the field k, even though such a topology may have played a role in the original definition of the G-ringed space X. (2) Every stalk $\mathcal{O}_x$ is a local ring. More specifically, the situation is as follows. For any point x in an admissible set U, let $\mathcal{O}(U)_x$ be the quotient ring of $\mathcal{O}(U)$ with respect to the multiplicative set of all elements f such that $f_x = \rho_x{}^U(f)$ is a unit in $\mathcal{O}_x$. Then, for any admissible set V containing U, the canonical homomorphism $\rho_U{}^V$ of $\mathcal{O}(V)$ to $\mathcal{O}(U)$ can be extended to a homomorphism of $\mathcal{O}(V)_x$ to $\mathcal{O}(U)_x$, which we shall denote by the samc symbol. If $\mathcal{O}_x$ is a local ring, then $\mathcal{O}(U)_x$ is also a local ring for every admissible set U containing x, and the canonical homomorphism of $\mathcal{O}(V)_x$ to $\mathcal{O}(U)_x$ is also local. The local ring $\mathcal{O}_x$ itself can be identified with the inductive limit of the family of rings $\mathcal{O}(U)_x$ with respect to the ordered system of all admissible sets containing x.

Let $\mathfrak{J}$ be a sheaf of ideals on a G-ringed space X. We shall associate with $\mathfrak{J}$ a G-ringed space $X(\mathfrak{J})$, which is called the *subspace* in X defined by the sheaf $\mathfrak{J}$. The underlying topological space of $X(\mathfrak{J})$ is the subspace of X (as an ordinary topological space) consisting of all points in X such that $\mathfrak{J}_x \neq \mathcal{O}_x$. Next, an open set in $X(\mathfrak{J})$ is said to be admissible if and only if it is the restriction $U_{X(\mathfrak{J})}$ to $X(\mathfrak{J})$ of an admissible set U in X. Similarly, an open covering of $X(\mathfrak{J})$ is said to be admissible if and only if it is the restriction to $X(\mathfrak{J})$ of an admissible covering in X. This defines then a Grothendieck topology $\mathfrak{T}_{\mathfrak{J}}$ on $X(\mathfrak{J})$. Finally, in order to define the structural sheaf $\mathcal{O}(\mathfrak{T}_{\mathfrak{J}})$ of the G-ringed space $X(\mathfrak{J})$, we first consider the presheaf $\mathcal{O}'(\mathfrak{T}_{\mathfrak{J}})$ defined by the functor $U_{X(\mathfrak{J})} \to \mathcal{O}(U)/\mathfrak{J}(U)$ on admissible sets in $\mathfrak{T}_{\mathfrak{J}}$, the canonical homomorphisms being the restrictions of the maps $\rho_U{}^V$ to the admissible sets $V_{X(\mathfrak{J})}$ and $U_{X(\mathfrak{J})}$ in $\mathfrak{T}_{\mathfrak{J}}$. The structural sheaf $\mathcal{O}(\mathfrak{T}_{\mathfrak{J}})$ of $X(\mathfrak{J})$ is then the sheaf on $\mathfrak{T}_{\mathfrak{J}}$ generated by the presheaf $\mathcal{O}'(\mathfrak{T}_{\mathfrak{J}})$. It is easily seen that for any point x in $X(\mathfrak{J})$, the stalk at x of the sheaf space $\mathcal{O}_{X(\mathfrak{J})}$ is isomorphic to the residue ring $\mathcal{O}_{X,x}/\mathfrak{J}_x$. The canonical homomorphism of $\mathcal{O}'(\mathfrak{T}_{\mathfrak{J}})$ into $\mathcal{O}(\mathfrak{T}_{\mathfrak{J}})$ is in general not injective. In order to give a criterion for the injectivity of this homomorphism, we introduce the following definition. Let $\mathfrak{A}$ be an ideal in $\mathcal{O}(U)$, and consider the ideal $\mathfrak{A}_x = \mathfrak{A} \otimes \mathcal{O}_x$ in $\mathcal{O}_x$ for any point x in U. The ideal $\mathfrak{A}$ is said to be *complete* if an element f in $\mathcal{O}(U)$ is in $\mathfrak{A}$ if (and only if) f_x is in $\mathfrak{A}_x$ for every point x in U. A sheaf $\mathfrak{J}$ of ideals is said to be *complete* if the ideal $\mathfrak{J}(U)$ in $\mathcal{O}(U)$ is complete for every admissible set U in $\mathfrak{T}$. One verifies readily that if the sheaf of ideal $\mathfrak{J}$ is complete, then the homomorphism of $\mathcal{O}'(\mathfrak{T}_{\mathfrak{J}})$ into $\mathcal{O}(\mathfrak{T}_{\mathfrak{J}})$ is injective.

In accordance with our aim of minimizing the use of the topology of our space, we introduce here a slightly different definition of the dimension of a G-ringed space, called the r-dimension (short for ring-dimension). The *r-dimension* of a (local) G-ringed space X is defined to be $\max_{x \in X}$ (dim $\mathcal{O}_{X,x}$), where dim $\mathcal{O}_{X,x}$ is the Krull dimension (i.e. altitude) of the local ring $\mathcal{O}_{X,x}$. We shall assume that our space X has a finite r-dimension. It is clear that the r-dimension of a subspace $X(\mathfrak{J})$ is at most equal to the r-dimension of X; furthermore, if the sheaf $\mathfrak{J}$ has the property that $\mathfrak{J}_x$ is a non-zero-divisor (i.e. $\mathfrak{J}_x$ does not consist entirely of zero-divisors) in $\mathcal{O}_{X,x}$ for every x, then the r-dimension of $X(\mathfrak{J})$ is less than the r-dimension of X.

A complex-analytic space in the usual sense is of course an example of our G-ringed space. Another example is a holomorphic space over a complete non-archimedean valued field, and it is for the purpose of including this case in our treatment that we have introduced a Grothendieck topology on our space. The theory of holomorphic spaces over a non-archimedean valued field has been studied with great success by Grauert and Remmert (and their disciples) in recent years, following the initial fundamental work of Tate. It will be seen that the properties we shall impose on our space all hold for a holomorphic space, both in the complex-analytic case and in the non-archimedean case.

Following the usual practice, we shall from now on use the notation

$\mathcal{O}_X$ to denote both the structural sheaf $\mathcal{O}(\mathfrak{T})$ and the sheaf space $\mathcal{O}_X$ as defined above, and we shall refer to the whole thing as the structural sheaf of the G-ringed space X. The subscript will sometimes be dropped in case there is no ambiguity about the space under consideration.

A *morphism* φ of a G-ringed space $(X, \mathfrak{T}, \mathcal{O}_X)$ into a G-ringed space $(X', \mathfrak{T}', \mathcal{O}_{X'})$ consists of a pair of maps (φ_0, φ_1), where (1) φ_0 is a continuous map of X into X' such that the Grothendieck topology $\varphi_0^{-1}(\mathfrak{T}')$ is contained in $\mathfrak{T}$, and (2) φ_1 is a homomorphism (= local k-homomorphism) of the sheaf space $\mathcal{O}_{X'} \circ \varphi_0$ into the sheaf space $\mathcal{O}_X$ which carries $\mathcal{O}_{X'}(U')$ into $\mathcal{O}_X(\varphi_0^{-1}(U'))$ for every admissible set U' in $\mathfrak{T}'$. This is of course the usual definition of a morphism of a G-ringed space into another. On the other hand, for our present purpose, we shall introduce the following definition of "immersion," which is a slight modification of the usual definition of "injective immersion," in that the condition that φ_1 be a surjection is replaced by a weaker one.

Definition 1. A morphism $\varphi = (\varphi_0, \varphi_1)$ of $(X, \mathfrak{T}, \mathcal{O}_X)$ into $(X', \mathfrak{T}', \mathcal{O}_{X'})$ is said to be an *immersion* of X in X' if (1) φ_0 is injective and (2) for every point x in X there exists a system of parameters in $\mathcal{O}_{X,x}$ consisting of elements in $\varphi_1(\mathcal{O}_{X'} \circ \varphi_0)_x$.

As we shall see later, this definition enables us to speak of an immersion of a holomorphic space in a projective space without having to make the latter itself into a holomorphic space. An immersion φ of X into X' is said to be an *embedding* if $\varphi_0^{-1}(\mathfrak{T}') = \mathfrak{T}$ and the homomorphism of $\mathcal{O}_{X'}(U')$ into $\mathcal{O}_X(\varphi_0^{-1}(U'))$ is surjective for every U' in $\mathfrak{T}'$. It will turn out later that although we are primarily interested in embedded spaces (in a projective space), it is nevertheless sufficient to consider the embedding as a mere immersion.

Let φ be a morphism of X into X'. If $\mathfrak{F}$ is a sheaf of $\mathcal{O}_{X'}$-modules on X', we can define a sheaf $\mathfrak{F}_X$ of $\mathcal{O}_X$-modules on X which is the inverse image of $\mathfrak{F}$ under φ (or the lift of $\mathfrak{F}$ to X). We define first a presheaf $\hat{\mathfrak{F}}$ as follows. For any admissible set V in the Grothendieck topology $\varphi_0^{-1}(\mathfrak{T}')$, we set $\hat{\mathfrak{F}}(V) = \mathcal{O}_X(V) \otimes \mathfrak{F}(\varphi_0(V))$, where the tensor product is taken over $\mathcal{O}_{X'}(\varphi_0(V))$. We have then a presheaf $V \to \hat{\mathfrak{F}}(V)$ on the Grothendieck topology $\varphi_0^{-1}(\mathfrak{J}')$, and the canonical map $\mathfrak{F}(\varphi_0(V)) \to \hat{\mathfrak{F}}(V)$ is a sheaf homomorphism, which we shall denote also by φ_1. Next, we extend this presheaf $\hat{\mathfrak{F}}$ to a presheaf over the Grothendieck topology $\mathfrak{T}$, to be denoted by the same symbol $\hat{\mathfrak{F}}$. For any admissible set U in $\mathfrak{T}$, we define $\hat{\mathfrak{F}}(U)$ to be the inductive limit $\hat{\mathfrak{F}}(U) = \lim_{U \subset V}$

$\hat{\mathfrak{F}}(V)$ with respect to the ordered system of all admissible sets V in $\varphi_0^{-1}(\mathfrak{X}')$ containing U. The presheaf $U \to \hat{\mathfrak{F}}(U)$ on $\mathfrak{X}$ is then the desired extension. Finally, the sheaf $\mathfrak{F}_X$ is defined to be the sheaf associated with the presheaf $\hat{\mathfrak{F}}$. The homomorphism φ_I of $\mathfrak{F}$ into $\hat{\mathfrak{F}}$ can be considered in a natural way as a homomorphism of $\mathfrak{F}$ into $\mathfrak{F}_X$. In case $\mathfrak{F}$ is an invertible sheaf, it is readily seen that $\mathfrak{F}_X$ is then also an invertible sheaf. In this case, the homomorphism φ_1 can be described locally as follows. Let V be an admissible set in $\varphi_0^{-1}(\mathfrak{X}')$ such that $\mathfrak{F}(\varphi_0(V))$ is a free $\mathcal{O}(\varphi_0(V))$-module of rank 1 and hence isomorphic to the ring $\mathcal{O}_{X'}(\varphi_0(V))$ as a module. The map $\varphi_1\colon \mathfrak{F}(\varphi_0(V)) \to \hat{\mathfrak{F}}(V)$ is then essentially the same as the map $\varphi_1\colon \mathcal{O}_{X'}(\varphi_0(V)) \to \mathcal{O}_X(V)$. We shall find this remark useful later.

II.

Let $S = S^N$ be the projective space of dimension N over the ground field k. We shall endow S with its Zariski topology, which also serves as its Grothendieck topology, and the structural sheaf of S is the usual one associated with the Zariski topology. Thus, the stalk $\mathcal{O}_{S,p}$ at a point p in S is the ring consisting of all quotients of two homogeneous elements of the same degree in the polynomial ring $k[u] = k[u_0, u_1, \ldots, u_N]$, with denominators not vanishing at p. In other words, the G-ringed structure of S is its usual Zariski structure. More generally, we can consider any algebraic scheme over k as a G-ringed space in the same way; this applies in particular to the subschemes in S.

Let φ be a morphism of a G-ringed space X into S. For every point x in X, the restriction φ_{1x} to x of φ_1 is a homomorphism of $(\mathcal{O}_S \circ \varphi_0)_x$ into $\mathcal{O}_{X,x}$, which can also be considered as a homomorphism of $\mathcal{O}_{S,\varphi_0(x)}$ into $\mathcal{O}_{X,x}$. Consider a homogeneous polynomial $f(u)$ in $k[u]$ with the following property: For every x in X there exists a homogeneous polynomial $g(u)$ of the same degree as $f(u)$ such that the image of $f(u)/g(u)$ under $\varphi_{1,x}$ is zero. The set of all polynomials having this property generates a homogeneous ideal $A(\varphi)$ in $k[u]$, and the subscheme in S defined by this ideal $A(\varphi)$ is called the *algebraic hull* of X in S (relative to the morphism φ) and will be denoted by $A(X)$, or $A_\varphi(X)$ in case it is advisable to specify the morphism involved. If we denote by $\iota = (\iota_0, \iota_1)$ the injection morphism of $A(X)$ into S, then we can factorize $\varphi = \iota \circ \varphi'$, where φ' is a morphism of X into $A(X)$. If φ is an immersion (embedding), then φ' is also an immersion (embedding), and we shall then say that X is an immersed (embedded) space in S. The following definition of algebraicity seems to be appropriate.

Definition 2. An immersed G-ringed space X in S is said to be ***algebraic*** if the dimension of (the scheme) $A(X)$ is equal to the r-dimension of X.

It is easily seen that the dimension of $A(X)$ is at least equal to the r-dimension of X. Hence, for an immersed G-ringed space X to be algebraic, it is sufficient to show that the dimension of $A(X)$ is at most equal to the r-dimension of X. To see this, we observe first that the homomorphism $\varphi_{1,x}$ of $\mathcal{O}_{S,\varphi_0(x)}$ into $\mathcal{O}_{X,x}$ induces a homomorphism of $\mathcal{O}_{A(X),\varphi_0(x)}$ into $\mathcal{O}_{X,x}$, which for simplicity we shall also denote by $\varphi_{1,x}$. Let n be the altitude of $\mathcal{O}_{A(X),\varphi_0(x)}$, and let $h_1, \ldots, h_n$ be a system of parameters of $\mathcal{O}_{A(X),\varphi_0(x)}$; then the ideal generated in $\mathcal{O}_{X,x}$ by the elements $\varphi_{1,x}(h_1), \ldots, \varphi_{1,x}(h_n)$ will have a radical which contains a system of parameters in $\mathcal{O}_{X,x}$ and hence must coincide with the maximal prime ideal in $\mathcal{O}_{X,x}$. It follows (from the Altitude Theorem of Krull) that the altitude of $\mathcal{O}_{X,x}$ is at most equal to n, which proves our assertion.

When we refer to a holomorphic subspace X in a projective space S, whether in the complex case or in the non-archimedean case, we must of course first endow S with a holomorphic structure. Then X as a subspace will be defined by a holomorphic sheaf of ideals $\mathfrak{J}$, which then defines the holomorphic structure on X, and the injection map of X into S is then an embedding. The subspace X is usually said to be algebraic if the sheaf $\mathfrak{J}$ defining X is the extension of an algebraic sheaf of ideals, and it is easily seen that this is the case (assuming that X is irreducible or equi-dimensional) if and only if the algebraic hull of X has the same dimension as X itself. On the other hand, the algebraic hull of X is defined even if we consider the embedding of X as a mere immersion into S; and we observe that any immersion of X into the holomorphic space S is also an immersion of X into the algebraic space S (i.e. endowed with its Zariski structure), and the algebraic hull of X is the same in either case.

Let $\mathcal{L}$ be the canonical invertible sheaf in S, and denote by $\mathcal{L}_X{}^m$ the lift to X of the sheaf $\mathcal{L}^m$. The map φ_1 of $\mathcal{L}^m$ into $\mathcal{L}_X{}^m$ induces on S a homomorphism of $\mathcal{L}^m(S)$ into $\mathcal{L}_X{}^m(X)$ as k-modules. We shall determine the kernel of this homomorphism. We recall first that $\mathcal{L}^m(S)$ can be identified with the module $k[u]_m$ of all homogeneous polynomials of degree m in $k[u]$. More generally, for any (Zariski) open set W in S, the $\mathcal{O}_S(W)$-module $\mathcal{L}^m(W)$ can be represented by the set of all quotients g/h of two homogeneous polynomials in $k[u]$, with $\deg g - \deg h = m$ and h not vanishing anywhere in W. Let x be a point in X, and let $h^{(x)}$ be an element in $k[u]_m$ which does not vanish at $\varphi_0(x)$; such a polynomial $h^{(x)}$ certainly exists. If we choose W so that $\varphi_0(x)$ is contained in W and $h^{(x)}$ does not vanish anywhere in W, then $h^{(x)}$ can serve as a generator of $\mathcal{L}^m(W)$, and the map $g/h \to g/hh^{(x)}$ defines an isomorphism between $\mathcal{L}^m(W)$ and $\mathcal{O}_S(W)$ as $\mathcal{O}_S(W)$-modules. As we have shown before, the map φ_1: $\mathcal{L}^m(W) \to \mathcal{L}_X{}^m(\varphi_0{}^{-1}(W))$ can then be represented through this isomorphism by the map φ_1: $\mathcal{O}_S(W) \to \mathcal{O}_X(\varphi_0{}^{-1}(W))$. Let f be an element in $\mathcal{L}^m(S)$; then its restriction to W can be represented by the element $f/h^{(x)}$ in $\mathcal{O}_S(W)$. If $\varphi_1(f) = 0$ in $\mathcal{L}_X{}^m(X)$, then $\varphi_1(f/h^{(x)}) = 0$ in $\mathcal{O}_X(\varphi_0{}^{-1}(W))$; this implies that in the map $\varphi_{1,x}$: $\mathcal{O}_{S,\varphi_0(x)} \to \mathcal{O}_{X,x}$ the element $(f/h^{(x)})_x$ is mapped into zero. Since this is true for any point x in X, this implies that f is in the kernel $A(\varphi)$ of φ. On the other hand, since the structural sheaf of X is assumed to be Γ-injective, every element in $A(\varphi)$ is mapped by φ_1 into zero. Thus we have shown that the kernel of the homomorphism φ_1: $\mathcal{L}^m(S) \to \mathcal{L}_X{}^m(X)$ is the submodule $k[u]_m \cap A(\varphi)$, and hence the image $\varphi_1(\mathcal{L}^m(S))$ of $\mathcal{L}^m(S)$ in $\mathcal{L}_X{}^m(X)$ is isomorphic to the quotient module $k[u]_m/(k[u]_m \cap A(\varphi))$.

Now, the residue ring $k[u]/A(\varphi)$ is simply the graded ring associated with the projective scheme $A(X)$, and $k[u]_m/(k[u]_m \cap A(\varphi))$ is isomorphic to the module $(k[u]/A(\varphi))_m$ of elements of degree m in $k[u]/A(\varphi)$. It is well-known in algebraic geometry that if t is the dimension of $A(X)$, then $\dim_k(k[u]/A(\varphi))_m$ is given, for sufficiently high m, by a polynomial $\chi(m)$ of degree t in m, called the Hilbert polynomial. Therefore, if we can show that $\dim_k L_X{}^m(X) \leq Cm^s$ for an infinite number of positive integers m, for some constant C and an integer $s \leq r$-dim X, then dim $A(X) \leq r$-dim X and hence X must be algebraic. Thus we have proved the following lemma.

LEMMA 1. *Let X be an immersed G-ringed space. If there is a constant C and an integer $s \leq r$-dim X such that $\dim_k \mathcal{L}_X{}^m(X) \leq Cm^s$ for an infinite sequence of integers m, then X is algebraic.*

The next question is naturally how one can get an estimate of the dimension of $\mathcal{L}_X{}^m(X)$ as described in Lemma 1. An answer will be given by what we shall call a distinct system of sections. For the purpose of later generalization, we shall treat the case of an arbitrary invertible sheaf on a G-ringed space X, not necessarily the lift of an invertible from an immersion into S^N. We begin with an observation on a possible extension of the structural sheaf, which is often desirable in our context.

We shall say that an element f in a ring R is *rationally dependent* on a subring R_1 in R, if there are elements g and h in R_1 such that $gf = h$. A subring R_1 in a ring R is said to be *rationally closed* in R. if every element in R which is rationally dependent on R_1 is already contained in R_1. In general, for any subring R_1 in R, there is a smallest subring R_2 in R which contains R_1 and is rationally closed in R. We shall call this ring R_2 the *rational closure* of R_1 in R. Applying these concepts to the subring $\mathcal{O}(U)$ in $\Gamma(U, \mathcal{O}_X)$, we obtain the rational closure of $\mathcal{O}(U)$ in $\Gamma(U, \mathcal{O}_X)$, which we shall call the *Γ-closure* of $\mathcal{O}(U)$ and shall denote by $\tilde{\mathcal{O}}(U)$. It is easily verified that the functor $U \to \tilde{\mathcal{O}}(U)$ defines a sheaf on $\mathfrak{X}$, which has the same sheaf space $\mathcal{O}_X$ as the structural sheaf $\mathcal{O}(\mathfrak{X})$; we shall call this sheaf the Γ-closure of $\mathcal{O}(\mathfrak{X})$ and shall denote it by $\tilde{\mathcal{O}}(\mathfrak{X})$. The sheaf $\mathcal{O}(\mathfrak{X})$ is said to be *Γ-closed* if it coincides with its Γ-closure. Since $\mathcal{O}(\mathfrak{X})$ have the same sheaf space $\mathcal{O}_X$, the r-dimension of a space X is not changed if we replace $\mathcal{O}(\mathfrak{X})$ by $\tilde{\mathcal{O}}(\mathfrak{X})$ as the structural sheaf of X. The G-ringed space X is said to be *Γ-closed* if its structural sheaf is Γ-closed.

Let $\mathcal{L}$ be an invertible sheaf on X, and let f be an element in $\mathcal{L}(X)$. We shall associate with f a sheaf of ideals $\mathcal{J}_f$ as follows. Let U be an admissible set in $\mathfrak{X}$. If $\mathcal{L}(U)$ is a free $\mathcal{O}_X(U)$-module of rank 1 and if we fix a basis element h for $\mathcal{L}(U)$, then the restriction f_U of f to U can be represented by an element $[f_U]$ in $\mathcal{O}_X(U)$, defined by the equation $f_U = [f_U]h$. We denote by $(f)_U$ the ideal in $\mathcal{O}_X(U)$ generated by this element $[f_U]$, and we note that this ideal $(f)_U$ is independent of the choice of the basis element h. If U is such that $\mathcal{L}(U)$ is not a free $\mathcal{O}_X(U)$-module of rank 1, then we let $(f)_U$ be the zero ideal in $\mathcal{O}_X(U)$. The map $U \to (f)_U$ then defines a presheaf of ideals on $\mathfrak{X}$, and the sheaf $\mathcal{J}_f$ is defined to be the sheaf generated by this presheaf. We observe that $\mathcal{J}_f(U) = (f)_U$ whenever $\mathcal{L}(U)$ is a free $\mathcal{O}_X(U)$-module of rank 1. If x is any point in such an admissible set U and if we take h_x to be the basis for $\mathcal{L}_x$, then $f_x = [f_U]_x h_x$, so that f_x can be represented by the element $[f_U]_x$ in $\mathcal{O}_{X,x}$. It is easily seen that the ideal $(f)_x$ generated by the element $[f_U]_x$ in $\mathcal{O}_{X,x}$ is independent of the choice of the element h as well as the choice of the admissible set U; in fact, the map $x \to (f)_x$ represents the sheaf space of $\mathcal{J}_f$.

Consider the subspace $X_1 = X(\mathcal{J}_f)$ in X defined by the sheaf $\mathcal{J}_f$, and let $\psi = (\psi_0, \psi_1)$ be the injection morphism of X_1 into X. We recall that the homomorphism ψ_1 first carries $\mathcal{O}_X(U)$ into $\mathcal{O}_X(U)/\mathcal{J}_f(U)$ and then follows with the canonical homomorphism of $\mathcal{O}_X(U)/\mathcal{J}_f(U)$ into $\mathcal{O}_{X_1}(U)$, the structural sheaf $\mathcal{O}_{X_1}$ of X_1 being the sheaf generated by the presheaf $U \to \mathcal{O}_X(U)/\mathcal{J}_f(U)$. Let $\mathcal{L}_{X_1}$ be the lift to X_1 of $\mathcal{L}$ through the morphism ψ; the homomorphism ψ_1: $\mathcal{L} \to \mathcal{L}_{X_1}$ then induces a homomorphism of $\mathcal{L}^m(X)$ into $\mathcal{L}_{X_1}{}^m(X_1)$, which we shall also denote by ψ_1 for simplicity. On the other hand, the operation of multiplication by the element f induces a homomorphism of $\mathcal{L}^{m-1}(X)$ into $\mathcal{L}^m(X)$. We shall need the following lemma.

Lemma 2. *Let $\mathcal{L}$ be an invertible sheaf on a G-ringed space X, and let f be an element in $\mathcal{L}(X)$. If $\mathcal{O}_X$ is Γ-closed and if f is nowhere a zero-divisor (i.e. f_x is not a zero-divisor for any point x in X), then the sheaf $\mathcal{J}_f$ is complete and the sequence*

$$\mathcal{L}^{m-1}(X) \xrightarrow{f} \mathcal{L}^m(X) \xrightarrow{\psi_1} \mathcal{L}_{X_1}{}^m(X_1)$$

is exact, where $X_1 = X(\mathcal{J}_f)$.

Proof. We first show that $\mathcal{J}_f$ is complete. Let g be an element in $\mathcal{O}_X(U)$, U being an admissible set, and assume that g_x is in $(f)_x$ for every point x in U. Then, for any point x in U, there is an element $h^{(x)}$ in $\mathcal{O}_{X,x}$ such that $g_x = h^{(x)}[f_U]_x$; this element $h^{(x)}$ is uniquely determined, since $[f_U]_x$ is not a zero-divisor. Further, there exist an admissible set V with $x \in V \subset U$ and an element h in $\Gamma(V, \mathcal{O}_X)$ with $h_x = h_x = h^{(x)}$ such that $g_y = h_y[f_U]_y$ for every point y in V. Since $[f_U]_y$ is not a zero-divisor in $\mathcal{O}_{X,y}$, it follows that $h_y = h^{(y)}$ for every point y in V. This shows that the map $x \to h^{(x)}$ is an element h in $\Gamma(U, \mathcal{O}_X)$, and we have the equation $g = h[f_U]$.

To prove the exactness of the sequence, let $U = \{U_i\}$ be an admissible covering in $\mathfrak{T}$ such that $\mathcal{L}(U_i)$ (and hence also $\mathcal{L}^m(U_i)$) is a free $\mathcal{O}_X(U_i)$-module of rank 1. For every i, we take a basis element b_i for $\mathcal{L}(U_i)$, and take $b_i{}^m$ to be the basis element for $\mathcal{L}^m(U_i)$. If g is an element in $\mathcal{L}^m(X)$, then g_{U_i} is represented by the element $[g_{U_i}]$ in $\mathcal{O}_X(U_i)$ defined by the equation $g_{U_i} = [g_{U_i}]b_i{}^m$. The homomorphism ψ_1: $\mathcal{L}^m(U_i) \to \mathcal{L}_{X_1}{}^m(U_{i,X_1})$ can then be represented by the homomorphism ψ_1: $\mathcal{O}_X(U_i) \to \mathcal{O}_X(U_i)/\mathcal{J}_f(U_i) = \mathcal{O}_X(U_i)/(f)_{U_i} \to \mathcal{O}_{X_1}(U_{i,X_1})$. Since $\mathcal{J}_f$ is complete, the second homomorphism here is injective; hence $\psi_1(g) = 0$ implies that $[g_{U_i}]$ is contained in the ideal $(f)_{U_i}$ for every i. There exists then an element h_i in $\mathcal{O}_X(U_i)$ such that $[g_{U_i}] = h_i[f_{U_i}]$. If U_j is another set in the covering U, then there are elements c_{ij} and c_{ji} in $\mathcal{O}_X(U_i \cap U_j)$ such that $b_i = c_{ij}b_j$, $b_j = c_{ji}b_i$, $c_{ij}c_{ji} = 1$ in $U_i \cap U_j$. Then we have the equations $[g_{U_i}] = c_{ij}{}^m[g_{U_j}]$ and $[f_{U_i}] = c_{ij}[f_{U_j}]$, and hence $h_i[f_{U_j}] = h_i c_{ji}[f_{U_i}] = c_{ji}[g_{U_i}] = c_{ij}{}^{m-1}[g_{U_j}] = c_{ij}{}^{m-1}h_j[f_{U_j}]$; since $[f_{U_j}]$ is not a zero-divisor, we have the equation $h_i = c_{ij}{}^{m-1}h_j$ in $U_i \cap U_j$. If we now assign to every U_i the element $h_i b_i{}^{m-1}$ in $\mathcal{L}^{m-1}(U_i)$, we obtain an element h in $\mathcal{L}^{m-1}(X)$ such that $g = hf$. This concludes the proof of our lemma.

For any invertible sheaf $\mathcal{L}$ on X, we denote by $\mathcal{L}^+(X) = \bigoplus_{m>0} \mathcal{L}^m(X)$ the graded ring of the sections of $\mathcal{L}^m$ over X, for all positive integers m. If $f_1, \ldots, f_s$ are homogeneous elements in $\mathcal{L}^+(X)$, we denote by $(f_1, \ldots, f_s)_U$ the ideal in $\mathcal{O}_X(U)$ generated by the ideals $(f_1)_U, \ldots, (f_s)_U$, U being an admissible set. As before, we denote by $\mathcal{J}(f_1, \ldots, f_s)$ the sheaf of ideals generated by the presheaf defined by the map $U \to (f_1, \ldots, f_s)_U$, and we denote by $(f_1, \ldots, f_s)_x$ the stalk of $\mathcal{J}(f_1, \ldots, f_s)$ at the point x. The subspace of X defined by this sheaf of ideals $\mathcal{J}(f_1, \ldots, f_s)$ will be denoted by $X(f_1, \ldots, f_s)$, or in short X_s (in case there is no ambiguity about the system $f_1, \ldots, f_s$ involved), and we note that at each point x in X_s its structural ring $\mathcal{O}_{X_s,x}$ is the residual ring $\mathcal{O}_{X,x}/(f_1, \ldots, f_s)_x$.

Definition 3. A system of homogeneous elements $f_1, \ldots, f_t$ in $\mathcal{L}^+(X)$ is said to be a *distinct system* (of length t) if for every i $(1 \leq i \leq t)$ and for every point x in $X_{i-1} = X(f_1, \ldots, f_{i-1})$, the ideal $(f_i)_x$ is not a zero-divisor modulo the ideal $(f_1, \ldots, f_{i-1})_x$. A distinct system $f_1, \ldots, f_t$ is said to be *maximal* if X_t is empty.

It is easily seen that if $f_1, \ldots, f_t$ is a distinct system, then the r-dimension of $X(f_1, \ldots, f_t)$ is at most equal to r-dim $X - t$. It follows that if a distinct system has a length greater than the r-dimension of X, then it must be maximal. We note that if $f_1, \ldots, f_t$ is a maximal distinct system, then $(f_t)_x$ is the unit ideal for every point x in X_{t-1}. If the structural sheaf of X_{t-1} is Γ-closed, this implies that the restriction $f_t(X_{t-1})$ of f_t to X_{t-1} has an "inverse" element in $\mathcal{L}_{X_{t-1}}{}^{-m_0}(X_{t-1})$, if m_0 is the degree of f_t. Then the multiplication by the element $f_t(X_{t-1})^m$ induces an isomorphism of $\mathcal{O}_{X_{t-1}}(X_{t-1})$ onto $\mathcal{L}_{X_{t-1}}{}^{mm_0}(X_{t-1})$ for every positive integer m.

The following lemma gives an answer to the question posed after Lemma 1. In the proof we shall find it convenient to replace the structural sheaf $\mathcal{O}_{X_i}$ of each X_i by its Γ-closure $\tilde{\mathcal{O}}_{X_i}$. Although this extension of the structural sheaf will result in a corresponding extension of the sheaf $\mathcal{L}_{X_i}$, we shall for the sake of simplicity retain the same symbol for the so extended sheaf. We note that since the sheaf space of every X_i remains unchanged, a distinct system in the original $\mathcal{L}^+(X)$ remains such after the extension of $\mathcal{L}$ and its properties will not be essentially changed by the subsequent extensions of the other $\mathcal{L}_{X_i}$.

LEMMA 3. *Let X be a G-ringed space and let $\mathcal{L}$ be an invertible sheaf on X. If there exists a maximal distinct system $f_1, \ldots, f_{t+1}$ of length $t+1$ in $\mathcal{L}^+(X)$, and if $\dim_k \tilde{\mathcal{O}}_{X_i}(X_i)$, $i = 1, \ldots, t$, are all finite, then there exist positive integers m_0 and C such that $\dim_k \mathcal{L}^{mm_0}(X) \leq Cm^t$ for all positive integers m.*

Proof. Let m_i be the degree of f_i, and set $m_0 = m_1 \cdots m_{t+1}$. We shall prove our lemma by showing that for every $i = 0, 1, \ldots, t$, we have a relation $\dim_k \mathcal{L}_{X_i}{}^{mm_0}(X_i) \leq C_i m^{t-i}$ for some positive integer C_i. For $i = 0$ this is the desired result with $C = C_0$, while for $i = t$ this follows from the above mentioned isomorphism between $\tilde{\mathcal{O}}_{X_t}(X_t)$ and $\mathcal{L}_{X_t}{}^{mm_0}(X_t)$, the isomorphism being induced by multiplication by $f_{t+1}(X_t)^{n_{t+1}}$, with $n_{t+1} = m_0/m_{t+1}$. We shall prove our relation by induction on the integer i in the reversed order from t to 0; that is, we assume its validity for $i+1$ and shall show its validity for i. We set $n_i = m_0/m_i$, and denote by f_{i+1} the homomorphism of $\mathcal{L}_{X_i}{}^{(m-1)m_0}(X_i)$ into $\mathcal{L}_{X_i}{}^{mm_0}(X_i)$ obtained by multiplication with the element $f_{i+1}(X_i)^{n_i}$. We denote by $\psi^{(i+1)}$ the injection morphism of X_{i+1} into X_i; the homomorphism $\psi_1{}^{(i+1)}\colon \mathcal{L}_{X_i}{}^{mm_0} \to \mathcal{L}_{X_{i+1}}{}^{mm_0}$ then induces a homomorphism of $\mathcal{L}_{X_i}{}^{mm_0}(X_i)$ into $\mathcal{L}_{X_{i+1}}{}^{mm_0}(X_{i+1})$, which we shall denote also by $\psi_1{}^{(i+1)}$. By lemma 2, we have the exact sequence of k-modules

$$\mathcal{L}_{X_i}{}^{(m-1)m_0}(X_i) \xrightarrow{f_{i+1}} \mathcal{L}_{X_i}{}^{mm_0}(X_i) \xrightarrow{\psi_1{}^{(i+1)}} \mathcal{L}_{X_{i+1}}{}^{mm_0}(X_{i+1}),$$

and hence the inequality

$$\dim_k \mathcal{L}_{X_i}{}^{mm_0}(X_i) \leq \dim_k \mathcal{L}^{(m-1)m_0}(X_i) + \dim_k \mathcal{L}_{X_{i+1}}{}^{mm_0}(X_{i+1}).$$

Summing over all m down to 0, we obtain the inequality

$$\dim_k \mathcal{L}_{X_i}{}^{mm_0}(X_i) \leq \dim_k \tilde{\mathcal{O}}_{X_i}(X_i) + \sum_{j=1}^{m} \dim_k \mathcal{L}_{X_{i+1}}{}^{jm_0}(X_{i+1}),$$

and hence by induction hypothesis

$$\dim_k \mathfrak{L}_{X_i}{}^{mm_0}(X_i) \leq \dim_k \tilde{\mathfrak{O}}_{X_i}(X_i) + C_{i+1}m^{t-1} \leq C_i m^{t-i},$$

where we set $C_i = \dim_k \tilde{\mathfrak{O}}_{X_i}(X_i) + C_i$. This proves our assertion and hence the lemma.

III.

The next question is of course the existence of a distinct system. For this purpose we shall introduce a few finiteness properties about our space X. In fact, we have up to now assumed very little about our space X, and it is clear that we cannot expect to proceed much further without any additional assumptions. We therefore introduce the following definitions.

(1) An admissible set U in a G-ringed space is said to be *flat* if $\mathfrak{O}_x$ is a flat $\mathfrak{O}(U)_x$-module for every point x in U. An admissible covering of an admissible set in X is said to be *flat* if every set in the covering is flat. A G-ringed space X is said to be *flat* if every admissible covering of X admits a flat admissible refinement.

(2) If U is an admissible set in a G-ringed space X, then a subring R in $\mathfrak{O}(U)$ is said to be *quasi-Noetherian* if every complete ideal in R is the intersection of a finite number of primary ideals. If V is an admissible set in X which contains U, we shall denote by $\mathfrak{O}(V, U)$ the image ring of $\mathfrak{O}(V)$ in $\mathfrak{O}(U)$ under the homomorphism $\rho_U{}^V$. Then V is said to be *quasi-Noetherian relative to U* if the ring $\mathfrak{O}(V, U)$ is quasi-Noetherian. Let $\mathfrak{V} = \{V_j\}$ ($j \in J$, an indexing set) be an admissible covering of X, and let $\mathfrak{U} = \{\mathfrak{U}_i\}$ ($i \in I$) be an admissible refinement of $\mathfrak{V}$; we have then a map $\tau: I \to J$ such that $U_i \subset V_{\tau(i)}$ for every $i \in I$. We shall say that the admissible covering $\mathfrak{U}$ is a *quasi-Noetherian* refinement of $\mathfrak{V}$ if every $V_{\tau(i)}$ is quasi-Noetherian relative to U_i. The space X itself is said to be locally *quasi-Noetherian* if every admissible covering of X admits a quasi-Noetherian refinement.

We observe that if an admissible covering $\mathfrak{U}$ is a quasi-Noetherian refinement of an admissible covering $\mathfrak{V}$, then every admissible refinement of $\mathfrak{U}$ is also a quasi-Noetherian refinement of $\mathfrak{V}$. To prove this statement, we introduce another way to express the quasi-Noetherian property, whereby U and V are as before two admissible sets such that $U \subset V$. An ideal $\mathfrak{A}$ in $\mathfrak{O}(V)$ is said to be *U-complete* if it contains every element f in $\mathcal{O}(V)$ with the property that $f_x \in \mathfrak{A} \otimes \mathcal{O}_x$ for every point x in U. Then $\mathcal{O}(V, U)$ is quasi-Noetherian if and only if every U-complete ideal in $\mathcal{O}(V)$ is the intersection of a finite number of primary ideals. If W is an admissible set contained in U, then a W-complete ideal in $\mathcal{O}(V)$ is clearly also U-complete. It follows then that if V is quasi-Noetherian relative to U, then it is also such relative to W; this proves our assertion.

(3) A G-ringed space X is said to be *c-finite* (short for covering-finite) if every admissible covering of X admits a finite admissible refinement. This is of course the usual compactness definition for a topological space. However, in our present context where we have to deal with non-archimedean holomorphic spaces, it is advisable to give this property a different name. Similarly, the space X is said to be *c-countable,* if every admissible covering of X admits a countable admissible refinement.

LEMMA 4. *Let X be a G-ringed space which is flat, quasi-Noetherian, and c-finite (c-countable). Let $\mathcal{L}$ be an invertible sheaf on X, and let $\mathcal{J}$ be a quasi-coherent sheaf of ideals on X. Then there is a finite (countable) set of points in X such that an element in $\mathcal{L}^+(X)$ is a non-zero-divisor modulo $\mathcal{J}$ everywhere in X if and only if it is such at every point in this set.*

Proof. Let $\mathfrak{B} = \{V_j\}$ $(j \in J)$ be an admissible covering of X such that each $\mathcal{L}(V_j)$ is a free $\mathcal{O}(V_j)$-module of rank 1. Then, for every integer $m > 0$, each $\mathcal{L}^m(V_j)$ is also a free $\mathcal{O}(V_j)$-module of rank 1; hence, if f is an element in $\mathcal{L}^m(X)$, it can be represented in each V_j by an element $[f_{V_j}]$ in $\mathcal{O}(V_j)$, with respect to a module basis for $\mathcal{L}^m(V_j)$. If x is any point in V_j, then $[f_{V_j}]_x$ is a zero-divisor modulo $\mathcal{J}_x$ as an element in $\mathcal{O}_x$ if and only if f_x is a zero-divisor modulo $\mathcal{J}_x$ as an element in the $\mathcal{O}_x$-module $\mathcal{L}_x{}^m$. Since the space X is flat, we can assume that the covering $\mathfrak{B}$ is flat, replacing it by a refinement if necessary. Since X is quasi-Noetherian, there is a quasi-Noetherian refinement $\mathfrak{U} = \{U_i\}$ $(i \in I)$ of $\mathfrak{B}$, and since X is c-finite (c-countable), we can assume that $\mathfrak{U}$ is finite (countable). In view of the quasi-coherence of the sheaf $\mathcal{J}$, our lemma therefore will follow from the following assertion: If U and V are admissible sets in X such that $U \subset V$ and if V is flat and quasi-Noetherian relative to U, and if $\mathfrak{A}$ is an ideal in $\mathcal{O}(V)$, then there is a finite set of points x_i such that an element f in $\mathcal{O}(V)$ is a non-zero-divisor modulo $\mathfrak{A}$ everywhere in U if and only if f_{x_i} is such in $\mathcal{O}_{x_i}$ for every i.

To prove this assertion, consider the U-complete ideal $\tilde{\mathfrak{A}}$ generated by $\mathfrak{A}$ in $\mathfrak{O}(V)$. Since $\mathfrak{O}(V)$ is quasi-Noetherian relative to U, we have the representation $\tilde{\mathfrak{A}} = \bigcap^s_{i=1} \mathfrak{q}_i$, where the $\mathfrak{q}_i$ are primary ideals, and we can assume that the intersection on the right hand side is irredundant. Let $\mathfrak{p}_1, \ldots, \mathfrak{p}_s$ be the primary ideals belonging to $\mathfrak{q}_1, \ldots, \mathfrak{q}_s$ respectively. We maintain that, for every $i = 1, \ldots, s$, there is a point x_i in U such that every element in $\mathfrak{p}_i$ vanishes at x_i. In fact, assume that one of the $\mathfrak{p}_i$, say $\mathfrak{p}_1$, does not have this property, so that there is no point in U at which every element in $\mathfrak{p}_1$ vanishes. Then, for every point x in U, there is an element $h^{(x)}$ in $\mathfrak{p}_1$ which is a unit at x. If g is any element in the intersection $\bigcap_{i\neq 1} \mathfrak{q}_i$, then $g_x h_x{}^{(x)}$ is contained in $\tilde{\mathfrak{A}} \otimes \mathfrak{O}_x$, and hence g_x must be contained in $\tilde{\mathfrak{A}} \otimes \mathfrak{O}_x$, for every point x in U. Since $\tilde{\mathfrak{A}}$ is U-complete, this implies that g is contained in $\tilde{\mathfrak{A}}$; this would imply that $\mathfrak{q}_1$ is actually redundant in the intersection $\tilde{\mathfrak{A}} = \bigcap^s_{i=1} \mathfrak{q}_i$, in contradiction to our assumption. Therefore, there exists a set of points $x_1, \ldots, x_s$ such that every element in $\mathfrak{p}_i$ vanishes at x_i, and we shall show that this set has the desired property.

Let f be an element in $\mathfrak{O}(V)$, and let $\mathfrak{n}$ be the ideal in $\mathfrak{O}(V)$ consisting of all elements g such that gf is contained in $\tilde{\mathfrak{A}}$. Then we have the exact sequence of $\mathfrak{O}(V)$-modules: $\mathfrak{n} \to \mathfrak{O}(V) \xrightarrow{f} \mathfrak{O}(V)/\tilde{\mathfrak{A}}$, where the first map is injection and the second map is induced by the multiplication by f followed by going over to the residue ring. Assume that f_{x_i} is not a zero-divisor modulo $\tilde{\mathfrak{A}} \otimes \mathfrak{O}_{x_i}$ for any i, so that f is not contained in any $\mathfrak{p}_i$. Then every element in $\mathfrak{n}$ must be contained in every $\mathfrak{q}_i$, and hence $\mathfrak{n}$ is contained in $\tilde{\mathfrak{A}}$. Now, for every point x in U, $\mathfrak{O}(V)_x$ is a flat $\mathfrak{O}(V)$-module, by a well-known property of the quotient ring; and since V is flat, $\mathfrak{O}_x$ is a flat $\mathfrak{O}(V)_x$-module. It follows that $\mathfrak{O}_x$ is a flat $\mathfrak{O}(V)$-module, and hence the above exact sequence remains exact upon tensorisation by $\mathfrak{O}_x$. Noting that $\mathfrak{O}_x/\tilde{\mathfrak{A}} \otimes \mathfrak{O}_x$ is canonically isomorphic to the tensorisation of $\mathfrak{O}(V)/\tilde{\mathfrak{A}}$ by $\mathfrak{O}_x$, we have then the exact sequence $\mathfrak{n} \otimes \mathfrak{O}_x \to \mathfrak{O}_x \xrightarrow{f_x} \mathfrak{O}_x/\tilde{\mathfrak{A}} \otimes \mathfrak{O}_x$, where the second homomorphism is induced by multiplication by f_x followed by going over to the residue ring. Since $\mathfrak{n} \otimes \mathfrak{O}_x \in \tilde{\mathfrak{A}} \otimes \mathfrak{O}_x$, it follows that f_x is not a zero-divisor modulo $\tilde{\mathfrak{A}} \otimes \mathfrak{O}_x$. This concludes the proof of our lemma.

We are in a position to prove our main result.

THEOREM. *Let X be a flat, quasi-Noetherian, c-finite (c-countable if k is non-countable) G-ringed space which is immersed in a projective space, and assume that $\dim_k \tilde{\mathfrak{O}}_{X(\mathfrak{J})}(X(\mathfrak{J}))$ is finite for every subspace defined by a coherent sheaf of ideals $\mathfrak{J}$ on X. Then X is algebraic.*

Proof. Let $\mathfrak{L}$ be the canonical invertible sheaf of the projective space S^N in which X is immersed by an immersion φ, and let $\mathfrak{L}_X$ be the lift of $\mathfrak{L}$ to X. In view of Lemma 1 and Lemma 3, we have only to show the existence of a maximal distinct system in $\mathfrak{L}_X{}^+(X)$ of length at most equal to r-dim $X + 1$. We shall construct such a maximal distinct system by induction, assuming that a distinct system $f_1, \ldots, f_i$ has already been constructed. If this system is already maximal, then we have nothing more to prove; we therefore assume that it is not maximal and proceed to extend it by another element f_{i+1}. Let x_j be the finite (countable) set of points given by Lemma 4, for the sheaf $\mathfrak{L}_X$ with reference to the ideal sheaf $\mathfrak{J}(f_1, \ldots, f_i)$, and consider the points $\psi_0(x_j)$ in S^N. We recall that the k-module $\mathfrak{L}(S)$ can be represented through a coordinate system in S^N by the system of all linear forms in N indeterminates. Since the ground field k is infinite (non-countable), there is at least one linear form g which does not vanish at any one of the finite (countable) set of points $\varphi_0(x_j)$. Let f_{i+1} be the element in $\mathfrak{L}_X(X)$ which is the image of g under φ_1; then f_{i+1} is a non-zero-divisor modulo $\mathfrak{J}(f_1, \ldots, f_i)$ at every point x_j and hence by Lemma 4 is a non-zero-divisor modulo $\mathfrak{J}(f_1, \ldots, f_i)$ everywhere in $X(f_1, \ldots, f_i)$. Then the system $f_1, \ldots, f_{i+1}$ is a distinct system in $\mathfrak{L}_X{}^+(X)$. Thus we can proceed until we reach a maximal system, but this must happen at the latest when the length becomes r-dim $X + 1$, for every distinct system of this length must be maximal.

COROLLARY. *If X is a closed holomorphic subspace in a projective space, either over the complex field or over a complete non-archimedean valued field, then it is algebraic.*

THE JOHNS HOPKINS UNIVERSITY

CORRECTION TO "ON THE ALGEBRAICITY OF CERTAIN RINGED SPACES"*

By WEI-LIANG CHOW

There are several mistakes in the section III of our paper which make the proof of Lemma 4 given there almost unintelligible. The basic ideas of the proof are sound, but in the process of formulating them in abstract general terms we unfortunately made some mistakes. We take this opportunity not only to correct these mistakes but also to add a few remarks which we hope will improve the exposition.

(A) Page 376. Replace the lines 14–19 (i.e. the first six lines under the heading (2)) by the following:

(2) Let U, V be admissible sets in a G-ringed space X with $U \subset V$. Then V is said to be quasi-Noetherian relative to U if for every ideal $\mathfrak{A}$ in $\mathcal{O}(V)$ there is a finite admissible covering $\{U_i\}$ of U such that the ideal generated by $\mathfrak{A}$ in each $\mathcal{O}(U_i)$ is the intersection of a finite number of primary ideals, whose prime ideals each has at least one zero point in U_i.

(B) Omit the last six lines on page 376 and the first seven lines on page 377, and replace them by the following paragraph:

We remark here that this definition is formulated with a view of applying it to the case of a complex-analytic variety. In this case we can take U to be a relatively compact set whose closure $\bar{U}$ lies in V. Then, given any ideal $\mathfrak{A}$ in $\mathcal{O}(V)$, for any point x in $\bar{U}$ there is an open subset U_x in V such that the ideal generated by $\mathfrak{A}$ in $\mathcal{O}(U_x)$ has a finite base. By repeated application of this fact, as well as the fact that the ring of the germs of analytic functions at x is Noetherian, we can show by the usual argument that the ideal generated by $\mathfrak{A}$ in $\mathcal{O}(U_x)$ is the intersection of a finite number of primary ideals, provided U_x is taken sufficiently small. Since $\bar{U}$ is compact, it follows that there is a covering of U by a finite number of such open sets.

Manuscript received January 31, 1985.

American Journal of Mathematics, **101** (1979), 364–379.

The last condition about the zero points of the prime ideals follows from the validity of the Nullstellensatz in the complex analytic case. In case of a holomorphic space over a non-archimedean valued field, the condition in this definition is trivially fulfilled by taking $V = U$. For in this case we have a covering by affinoid sets, and the affinoid algebra of an affinoid set is Noetherian and the Nullstellensatz holds here.

(C) Page 377, line 21 from top, replace "and only if it is such" by "it does not vanish."

Page 377, lines 1 and 2 from bottom, replace "and only if f_x is such in $\mathfrak{O}_{x_i}$" by "f does not vanish at x_i."

(D) Page 378, replace the entire page except the last five lines by the following:

Let $\mathfrak{A}'$ be the ideal generated by $\mathfrak{A}$ in $\mathfrak{O}(U)$, and let $\mathfrak{A}' = \bigcap_{i=1}^{s} \mathfrak{q}_i$ be the representation of $\mathfrak{A}'$ as an intersection of primary ideals. Let $\mathfrak{p}_1, \ldots, \mathfrak{p}_s$ be the prime ideals belonging to the primary ideals $\mathfrak{q}_1, \ldots, \mathfrak{q}_s$ respectively. In view of the validity of the Nullstellensatz in $\mathfrak{O}(U)$, there exists a set of points $x_1, \ldots, x_s$ in $\mathfrak{O}(U)$ such that every element in $\mathfrak{p}_i$ vanishes at x_i. We shall show that this set has the desired property. Let f be an element in $\mathfrak{O}(U)$ which does not vanish at any of the s points $x_1, \ldots, x_s$. If g is an element in $\mathfrak{O}(U)$ such that gf is in $\mathfrak{A}'$, then g must be in $\mathfrak{q}_i$ for every $i = 1, \ldots, s$, and hence must be contained in $\mathfrak{A}'$. This implies that the map $\mathfrak{O}(U)/\mathfrak{A}' \xrightarrow{f} \mathfrak{O}(U)/\mathfrak{A}'$ induced by the multiplication with f is injective, so that we have the exact sequence $0 \to \mathfrak{O}(U)/\mathfrak{A}' \xrightarrow{f} \mathfrak{O}(U)/\mathfrak{A}'$. Now, for any point x in U, $\mathfrak{O}(U)_x$ is a flat $\mathfrak{O}(U)$-module, by a well-known property of the quotient ring, and since U is flat, $\mathfrak{O}_x$ is a flat $\mathfrak{O}(U)_x$-module. It follows that $\mathfrak{O}_x$ is a flat $\mathfrak{O}(U)$-module, and hence the above exact sequence (considered as a sequence of $\mathfrak{O}(U)$-modules) remains exact upon tensorization by $\mathfrak{O}_x$. This implies that f_x is not a zero-divisor modulo $\mathfrak{A}' \otimes \mathfrak{O}_x$. This concludes the proof of our lemma.

THE JOHNS HOPKINS UNIVERSITY

Invent. math. 86, 115–130 (1986)

Inventiones mathematicae
© Springer-Verlag 1986

Formal functions on homogeneous spaces

Wei-Liang Chow

Johns Hopkins University, Department of Mathematics, Baltimore, MD 218, USA

Let X be a homogeneous algebraic variety on which a group G acts, and let Z be a subvariety of positive dimension in X. Assume that Z generates X in the sense that $G_Z^0 = G$, where G_Z^0 is the subgroup in G generated by the component $G_{p,Z}^0$ containing the unit element in the set $G_{p,Z}$ of all elements in G which carry a given point p in Z each to another point in Z. One asks whether a formal rational function on X along Z is the restriction along Z of an algebraic function (or even a rational function) on X. In a paper [1] some years ago, the author gave an affirmative answer to this question, under the assumption that the subvariety Z is complete, but only for the complex-analytic case with the formal function replaced by the usual analytic function defined in a neighborhood of Z. The question remains whether the result holds also for the formal functions in the abstract case over any ground field. We had then some thoughts on this question, but we did not pursue them any further as we did not see a way to reach the desired conclusion at the time. In a recent paper [3], Faltings raised this same question and gave a partial answer to it in a slightly different formulation. This result of Faltings led us to reconsider this question again, and this time we are more fortunate. In fact, we have been able not only to solve the problem, but also to do it by using essentially the same method we used in our original paper.

Let us begin by taking a look at our original proof of the theorem in question, which is Theorem 2 in our original paper [1], except that Theorem 2 deals with the (apparently) more general case of a meromorphic map F of X into an algebraic variety Y instead of a meromorphic function. This proof is based on a criterion for algebraicity given in Theorem 1 in that paper, which says roughly that a local formal variety at a point p (called an algebroid variety there) is algebraic if it is spanned by a system of algebraic subvarieties through p. This theorem is of course purely algebraic. Next, in order to apply this criterion, we introduce a sequence of subvarieties Z_i, $i=0, 1, \ldots, t$, in X with each Z_i spanning the next one Z_{i+1}, in the following way. The fact that $G_Z^0 = G$ implies that there exist independent generic points $u_0, u_1, \ldots, u_t$ in $G_{p,Z}^0$

over the ground field K such that $u_t^{(-1)^t}\dots u_1^{-1}u_0(p)$ is a generic point of X over K. For $i=0, 1, \dots, t$, we set $K_i=K(u_{i+1}, \dots, u_t)$ with $K_t=K$, and set $p_i=u_t^{(-1)^t}\dots u_{i+1}^{(-1)^{i+1}}(p)$ with $p_t=p$; then the variety Z_i is defined as the locus of the point $u_t^{(-1)^t}\dots u_1^{-1}u_0(p)$ over K_i. We observe that $Z_0\subset\dots\subset Z_t=X$, and that Z_i spans Z_{i+1} over K_{i+1} and contains both p_i and p_{i+1}. Assume now that the ground field K is the complex number field, and let F be a meromorphic map of a neighborhood N of Z into an algebraic variety Y. If the points $u_0, u_1, \dots, u_t$ are chosen sufficiently near to the unit point e of G, then all the points p_i will lie in N and (with a suitable choice of p) the map F will be holomorphic at every one of them. We can then go over to the graph of F, to be again denoted by the same symbol F, in the product space $N\times Y$; and we set $p_i^*=(p_i\times Y)\cap F$, and denote by V_i and V_i' the analytic sets at p_i^* and p_{i+1}^* respectively induced by the analytic set $(Z_i\times Y)\cap F$ in $N\times Y$. Also, to preserve the independence of the generic points u_i in the new context, we replace the fields K_i by the similarly constructed fields K_i^*, as described in our paper. It follows then that V_i' as an analytic set at p_{i+1}^* spans V_{i+1} over K_{i+1}^*. If we now apply the criterion in Theorem 1 and observe that the algebraicity of V_i implies that of V_i', we conclude that if V_i is algebraic, then V_{i+1} is also algebraic. Thus, in order to prove that $V=V_t$ is algebraic, we have only to show that V_0 is algebraic. Now, V_0 is the analytic set at p_0 induced by the analytic set $(Z_0\times Y)\cap F$ in $X\times Y$. If we take the points $u_1, \dots, u_t$ all sufficiently near to e, then $Z_0=u_t^{(-1)^t}\dots u_1^{-1}(Z)$ can be made arbitrarily near to Z and hence lies entirely in N; then $(Z_0\times Y)\cap F$ is a compact analytic subvariety in the algebraic variety $X\times Y$ and hence must be algebraic. Finally, since $V=V_t$ is the analytic set at $p^*=p_t^*$ induced by the analytic set $F=(X\times Y)\cap F$, it follows that F is also algebraic.

We see from the above resumé of our proof that in order to generalize it to the abstract case of a formal variety, we must replace the last part of our proof by something of an analogous nature. Let us consider for the sake of simplicity the case of a single analytic function f defined in a neighborhood of Z, instead of the map F; in fact, as we shall see later, the general case can be reduced to the consideration of the ideals generated by such functions. In the abstract case this function f is a formal function on X around Z. Let U be an affine open subset in X containing a part of Z, and let $(x_1, \dots, x_n)$ be a system of affine coordinates for U; let $b_1(x), \dots, b_s(x)$ be a basis for the ideal J in $K[x]$ defining Z in U. Then f can be represented in U as a power series in the functions $b_1(x), \dots, b_s(x)$ with coefficients which are polynomials in the x's modulo J. In the complex-analytic case, this power series is convergent in a neighborhood N of Z; hence, if the points $u_1, \dots, u_t$ are taken sufficiently near to e, then a part of the subvariety Z_0 will be contained in U and f will then define a function on this part of Z_0. However, also in the abstract case we can have convergence here, so that f can also define a function on Z_0; only this time the convergence is understood in another sense. Let the field K_0 be endowed with a non-archimedean valuation over K which is centered at the unit point e in G (this description is not entirely accurate, but will suffice in this preliminary discussion); let the algebraic closure of K_0 be also endowed with the uniquely determined extension of this valuation, and let K_0^* be its completion with respect to this valuation. If q is any algebraic point over K_0 in U which is contained in Z_0, then the formal function f as a power series in the elements $b_1(q), \ldots, b_s(q)$ is convergent with respect to this valuation and hence has a uniquely determined value in K_0^* at each such point q. It follows that the formal function f induces a function on Z_0 in U, defined at every algebraic point over K_0 in U. In fact, it turns out that the set of all points on Z_0 in U which are defined over K_0^* forms an affinoid set, if we consider Z_0 as a holomorphic space over the non-archimedean valued field K_0^*, and the function induced on this affinoid set by the formal function f is an affinoid function. Since also over a non-archimedean valued field we have the theorem (see [2]) that a closed holomorphic subspace in a complete algebraic variety is also algebraic, we have now all the means we need to generalize our original proof of Theorem 2 to the abstract case.

II.

Let X be an algebraic variety, and let Z be a proper subvariety in X. We consider the formal completion $\hat{X}$ of X along Z. We recall that $\hat{X}$ is a ringed space with Z as its base space and with $\mathcal{O}_{\hat{X}} = \varprojlim \mathcal{O}_X/\mathcal{J}^j$ as its structural sheaf, where $\mathcal{J}$ is the defining sheaf of Z on X. An element f in the ring of sections $\Gamma(Z, \mathcal{O}_{\hat{X}})$ is called a formal regular function on X along Z. Locally such a formal function f can be described as follows. Let K be a field of definition for both X and Z; let U be an affine open subset in an affine covering of X, and let $K[x] = K[x_1, \ldots, x_n]$ be its coordinate ring (n being the dimension of the embedding affine space for U, which could be an affine part of the ambient projective space in case X is a projective variety). If J is the ideal in $K[x]$ defining Z locally in U, then f can be represented by an element in the J-adic completion $K[x]^\wedge$ of $K[x]$, and we can hence set $f = \sum_{j=0}^{\infty} f_j(x)$, where $f_j(x)$ is in J^j for every j. We note that each $f_j(x)$ can be expressed as a homogeneous polynomial of degree j in the basis elements $b_1(x), \ldots, b_s(x)$ of J with coefficients which are elements in $K[x]$ modulo J.

In analogy to the case of an algebraic subscheme, we shall consider any coherent sheaf $\mathcal{Q}$ of ideals on $\hat{X}$ as defining a formal subspace V in $\hat{X}$. This V is in a sense the formal analogy of an analytic subvariety in a neighborhood of Z in X in the complex-analytic case. The base space of V consists of those points p in Z such that $\mathcal{Q}_p \neq \mathcal{O}_{\hat{X},p}$, and the structural sheaf of V is the quotient sheaf $\mathcal{O}_{\hat{X}}/\mathcal{Q}$, considered as a sheaf on the base space of V. In case the sheaf $\mathcal{Q}$ is contained in the defining ideal sheaf $\mathcal{J}$ of Z in X, then V has the entire space Z as its base space, but its structural rings are the rings in the quotient sheaf $\mathcal{O}_{\hat{X}}/\mathcal{Q}$. In the affine open subset U in X as described above, V will be represented by an ideal Q in $K[x]^\wedge$, and the residue ring $K[x]^\wedge/Q$ is the coordinate ring of V in U. As usual, the dimension of V is defined as the maximum of the Krull dimensions of the structural rings of V; in case V is locally irreducible at a point p, this dimension is also the dimension of the coordinate ring of V on a suitably chosen affine open subset U of X containing p. If we set $\tilde{Q}=K[x]\cap Q$, then the system of ideals $\tilde{Q}$ as U ranges through the various affine subsets in an affine covering of X along Z defines a coherent sheaf $\tilde{\mathcal{Q}}$ of ideals on X. The algebraic subscheme $\tilde{V}$ defined by the sheaf $\tilde{\mathcal{Q}}$ in X is called the *algebraic closure* of V. It is clear that the dimension of $\tilde{V}$ is at least equal to the dimension of V. In case the ideals in the sheaf $\mathcal{Q}$ are prime, in which case V can be considered as a formal subvariety, then the ideals in $\tilde{\mathcal{Q}}$ are also prime, so that $\tilde{V}$ is essentially an algebraic variety. We shall say that the formal subvariety V is *algebraic* if $\tilde{V}$ has the same dimension as V. It is easily seen that when such is the case, then Q is a component of the extension ideal $K[x]^{\wedge}\tilde{Q}$ of $\tilde{Q}$ in $K[x]^{\wedge}$.

Before proceeding any further, we shall insert here a few remarks about the algebroid varieties and sets. Locally at a given point p in Z, a formal function f on X around p can be expressed as a formal power series in $K[\![x-p]\!]$ or, in the terminology we used in [1], an algebroid function around p, assuming that p is contained in the affine set U. Similarly, the formal ideal Q in $K[x]^{\wedge}$ can be expressed as an ideal in $K[\![x-p]\!]$, so that a formal subspace (subvariety) V will define at p an algebroid set (subvariety) around p. Assume now that the field K is replaced by (or contained in) a complete non-archimedean valued field K^*, whereby here as well as below we shall always assume that the valuation involved is real valued and discrete, so that in particular the valuation ring R in K^* is Noetherian. As usual, we shall use the usual symbol $|\ |$ to denote the absolute value of an element in K^* or the norm of a point or a polynomial defined over K^*, the latter being defined as the maximum of the absolute values of the coordinates of the point or the coefficients of the polynomial respectively. Also, if an element is in the valuation ring R of K^* or if a point (polynomial) is defined over R, then we shall use a bar over it to denote its residue value or images modulo the maximal prime ideal in R. Now, over the complete valuation ring R it is possible for a power series $f(x-p)$ in $R[\![x-p]\!]$ to have a value at a point p' other than the origin p; in fact, the power series $f(p'-p)$ is convergent for every point p' within the unit radius of the point p, i.e. $|p'-p|<1$, assuming that the point p' is also defined over R. Actually, when such is the case, we can redevelop the power series $f(x-p)$ as a power series $f'(x-p')=f(x-p'+p'-p)$ in the coordinates of the point $x-p'$, i.e. as an element in the power series ring $R[\![x-p']\!]$, it being easily seen that the coefficients in such a redevelopment are all convergent in R. In fact, this process of redeveloping around the point p' induces an endomorphism $T_{p,p'}$ of the local ring $R[\![x-p]\!]$ into the local ring $R[\![x-p']\!]$, which in this case is actually an isomorphism, since conversely every element in $R[\![x-p']\!]$ can also be redeveloped around p into a power series in $R[\![x-p]\!]$.

Consider an ideal Q in the local ring $R[\![x-p]\!]$, and let V be the algebroid variety around p defined by the ideal $K^*[\![x-p]\!]Q$ in $K^*[\![x-p]\!]$. If Q has the property

$$R[\![x-p]\!]\cap K^*[\![x-p]\!]Q=Q, \tag{A}$$

then we shall say that V is defined over the ring R by the ideal Q. When such is the case, then the ideal Q is essentially determined by the algebroid variety

V; in fact, we can clearly replace Q by its radical without changing the variety V, and if we do that (i.e. if we require Q to be semi-prime), then Q will be uniquely determined by V. The condition (A) is equivalent to the condition that the canonical map of the residue ring $R[\![x-p]\!]/Q$ into the residue ring $K^*[\![x-p]\!]/K[\![x-p]\!]Q$ be injective. This in turn is equivalent to the condition that any element in $R[\![x-p]\!]$ which does not vanish at p cannot be a contained in any prime divisor of Q, i.e. any such element is not a zero-divisor modulo Q. If fact, if we denote by $(x-p)$ the prime ideal in $R[\![x-p]\!]$ consisting of all elements which vanish at p, and denote by $R[\![x-p]\!]_{(x-p)}$ the ring of quotients of $R[\![x-p]\!]$ with respect to the multiplicatively closed set of all elements not contained in $(x-p)$, then we have the equation

$$R[\![x-p]\!] \cap R[\![x-p]\!]_{(x-p)} Q = Q \tag{B}$$

if and only if every prime divisor of Q is contained in $(x-p)$. The completion of the local ring $R[\![x-p]\!]_{(x-p)}$ is isomorphic to the local ring $K^*[\![x-p]\!]$ and hence can be identified with it, and we have always the equation

$$R[\![x-p]\!]_{(x-p)} \cap K^*[\![x-p]\!] Q = R[\![x-p]\!]_{(x-p)} Q. \tag{C}$$

In view of (B) and (C), it follows that (A) holds if and only if every prime divisor of Q is contained in $(x-p)$. Geometrically, the condition (A) can be described by saying that in the analytic set defined by Q in the open unit ball around p every irreducible component passes through the point p. Lastly, we observe that the condition (A) can also be expressed as an intrinsic property of the residue ring $R[\![x-p]\!]/Q$, irrespective of its origination. In fact, if we denote by $(R[\![x-p]\!]/Q)_{(x-p)/Q}$ the ring of quotients of the residue ring $R[\![x-p]\!]/Q$ with respect to the multiplicatively closed set consisting of all elements not contained in the ideal $(x-p)/Q$, then it is canonically isomorphic to the residue ring of $R[\![x-p]\!]_{(x-p)}$ modulo the ideal $R[\![x-p]\!]_{(x-p)}Q$, and its completion can be canonically identified with the residue ring $K^*[\![x-p]\!]/K^*[\![x-p]\!]Q$. It follows that the condition (A) is equivalent to the condition that every element in $R[\![x-p]\!]/Q$ which is not contained in $(x-p)/Q$ is not a zero-divisor. We shall find this remark useful later.

Let p' be a point in the open unit ball around p, i.e. $|p-p'|<1$, and assume that every element in the ideal Q also vanishes at p'. If Q' is the image of Q in $R[\![x-p']\!]$ under the isomorphism $T_{p,p'}$, then the ideal $K^*[\![x-p']\!]Q'$ will define an algebroid set V' around p', which we shall call the redevelopment around p' of the algebroid set V around p. We want to show that if V is algebraic, then V' is also algebraic. It is clearly sufficient to show this for the case where the ideal Q is prime, and we note that in this case the condition (A) is automatically satisfied. Since $R[\![x-p]\!]$ is a Macaulay local ring, the ring of quotients $R[\![x-p]\!]_{(x-p)}$ is also a Macaulay local ring, and hence its completion $K^*[\![x-p]\!]$ is also a Macaulay local ring; it follows (see e.g. [5], p. 125) that the ideal $K^*[\![x-p]\!]Q$ is unmixed in the sense that all its prime divisors have the same depth, so that the algebroid variety V is also unmixed in the sense that all irreducible components of V have the same dimension. Consider now the algebraic closure $\tilde{V}$ of the algebroid set V around p; by definition, $\tilde{V}$ is defined by the ideal $K^*[x] \cap K^*[\![x-p]\!]Q$, and since every ideal in $K^*[x]$ is generated by an ideal in $R[x]$, it follows that $\tilde{V}$ is also defined by the ideal $R[x] \cap K^*[\![x-p]\!]Q = R[x] \cap Q$ in $R[x]$. Since the contraction $R[x] \cap Q'$ of the ideal Q' in $R[x]$ clearly contains the ideal $R[x] \cap Q$, and hence the contraction $K^*[x] \cap K^*[\![x-p']\!]Q'$ must also contain the ideal $R[x] \cap Q$, it follows that the algebroid set V' around p' is contained in the algebraic closure $\tilde{V}$ of V. Now, if V is algebraic, then $\tilde{V}$ has the same dimension as that of V, and since V' is contained in $\tilde{V}$, it must be also algebraic if it is unmixed and has the same dimension as that of V. That V' is unmixed follows from the fact that the ideal Q' is prime, by the same argument as in the case of V. It therefore remains to show that V' has the same dimension as V. Let $(x-p')$ be the prime ideal in $R[\![x-p]\!]$ consisting of all elements which vanish at the point p', and let $R[\![x-p]\!]_{(x-p')}$ be the ring of quotients of $R[\![x-p]\!]$ with respect to the multiplicatively closed set in $R[\![x-p]\!]$ consisting of all elements not contained in $(x-p')$. Then $R[\![x-p]\!]_{(x-p')}$ is a local ring with the ideal $R[\![x-p]\!]_{(x-p')}(x-p')$ as its maximum prime ideal, and it is in fact isomorphic to the local ring $R[\![x-p']\!]_{(x-p')}$, by an extension of the isomorphism $T_{p,p'}$. Consider the residue ring $R[\![x-p]\!]/Q$; let $(R[\![x-p]\!]/Q)_{(x-p')/Q}$ be the ring of quotients of $R[\![x-p]\!]/Q$ with respect to the multiplicatively closed set in $R[\![x-p]\!]/Q$ consisting of all elements not contained in the ideal $(x-p')/Q$. It is easily seen that the ring $(R[\![x-p]\!]/Q)_{(x-p')/Q}$ is canonically isomorphic to the residue ring of $R[\![x-p]\!]_{(x-p')}$ modulo the ideal $R[\![x-p]\!]_{(x-p')}Q$ and hence also isomorphic to the residue ring of $R[\![x-p']\!]_{(x-p')}$ modulo the ideal $R[\![x-p']\!]_{(x-p')}Q'$. Since $K^*[\![x-p']\!]$ is the completion of the local ring $R[\![x-p']\!]_{(x-p')}$, it follows that the altitude of $(R[\![x-p]\!]/Q)_{(x-p')/Q}$ is equal to the altitude of the residue ring $K^*[\![x-p']\!]/K^*[\![x-p']\!]Q'$, which is by definition the dimension of the algebroid set V' around p. Now, since Q is prime, $R[\![x-p]\!]/Q$ is a complete local domain; and since it is a homomorphic image of the Macaulay local ring $R[\![x-p]\!]$, it is unmixed, and we have for the prime ideal $(x-p')/Q$ the "dimension" formula (see [5], p. 125, (34.5) and (34.9)): altitude $R[\![x-p]\!]/Q =$ altitude $(R[\![x-p]\!]/Q)_{(x-p')/Q} +$ depth $(x-p')/Q$. On the other hand, it is clear that depth $(x-p')/Q =$ depth $(x-p')$ $=$ altitude R, irrespective of the choice of the point p', and we have just seen that altitude $(R[\![x-p]\!]/Q)_{(x-p')/Q}$ is equal to the dimension of V'. Thus we have shown that the algebroid set V' defined by Q at any point p' in the open unit ball around p has a dimension equal to a number $=$ altitude $R[\![x-p]\!]$ $-$ altitude R, independent of the choice of the point p'. This completes the proof of our assertion. We state our result here as a lemma for later reference.

Lemma 1. *Let V be an algebroid set around a point p, which is defined over R, and let V' be its redevelopment around a point p' in the open unit ball around p. If V is algebraic, then V' is also algebraic.*

Returning to the formal subspaces, we begin with a few remarks about the analogy between the formal subspaces and the classical analytic subvarieties in the complex-analytic case. In the usual definition of the formal completion $\hat{X}$ of a variety X along a subvariety Z, one considers as the base topological space only the subvariety Z, with the result that the various formal or algebraic subspaces of $\hat{X}$ can be obtained only from the structural sheaf $\mathcal{O}_{\hat{X}}$ of $\hat{X}$, and one cannot obtain any point in such a subspace beyond those in Z itself. This is in contrast to the complex-analytic case, where a coherent system of germs of analytic subvarieties in a neighborhood of Z can be realized actually as a closed subset of the appropriate dimension in a certain open subset containing Z in X. We shall now redress this situation, at least to a certain extent, by considering a special case where the given subvariety Z is contained in an algebraic system of algebraic subvarieties. By this we mean the following situation. Let W be an affine variety defined over K, and let w be a generic point of W over K, so that $K(w)$ is a regular extension of K. Let $Z(w)$ be a subvariety in X of the same dimension r as the dimension of Z, defined over $K(w)$; than $Z(w)$ generates over K an algebraic system of subvarieties (or rather r-cycles) in X. More precisely, there is a subvariety E in the product space $W \times X$, defined over K, such that $Z(w) = pr_X(E.(w \times X))$; and for every point w_1 in W such that the intersection $E.(w_1 \times X)$ is defined (and proper), the r-cycle $Z(w_1) = pr_X(E.(w_1 \times X))$ is a member of this algebraic system. We assume that there is a rational point w_0 over K in W such that $Z(w_0) = Z$, so that Z is contained in this algebraic system. Locally, in the affine open subset U introduced above, we first extend the ground field from K to $K(w)$, so that $K(w)[x]$ is now the coordinate ring of U, and then choose a basis $b_1(w, x), \ldots, b_s(w, x)$ for the defining ideal $J(w)$ for $Z(w)$ in U so that $b_i(w_0, x) = b_i(x)$ for $i = 1, \ldots, s$. We note that each $b_i(w, x)$ is a polynomial in $x_1, \ldots, x_n$ with coefficients in $K(w)$.

Now, let the field $K(w)$ be endowed with a non-archimedean valuation over K which is centered on the point w_0 in W; this implies that the valuation ring R in $K(w)$ dominates the local ring of the point w_0 in W, and it is well known that one can choose such a valuation to be real (and discrete) if one so wishes. In the affine open subset U introduced above, one can assume without any loss of generality that the coefficients in the basis elements $b_i(w, x)$ are all elements in R. We recall that the algebraic closure of the completion of $K(w)$ can also be endowed with the uniquely determined extension of this valuation, and we shall denote its completion with respect to this valuation by $K(w)^*$. Now, the algebraic variety $Z(w)$ over the valued field $K(w)^*$ can be considered as a holomorphic space over $K(w)^*$ in the sense of the theory of holomorphic spaces over a non-archimedean valued field, as developed by Tate, Grauert and Remmert, and it is a closed holomorphic space as defined in this theory if it is a complete algebraic variety. To describe the holomorphic space $Z(w)$ locally, one naturally replace the affine open subset U by an affinoid subset D in U consisting of all rational points q over $K(w)^*$ in U satisfying the condition $|q| \leqq 1$, assuming that the origin of the coordinate system in U is taken at a point on Z. We note that the affinoid set D contains all points in U which are algebraic over K, so that if we replace every affine set in an affine covering of X by the corresponding affinoid subset (i.e. "closed" unit sphere) in it, the resulting affinoid covering will contain all algebraic points over K in X, in particular all algebraic points over K in Z. We assume from now on that Z is a complete variety. We note that in this case the subvariety $Z(w)$ must also be complete, since a specialization of a non-complete variety cannot be complete. We hasten to add here a remark in order to avoid any misunderstanding. Here and in the following we shall treat the $x_1, \dots, x_n$ as if they were independent variables, while in fact they are elements determined only modulo the polynomials in the defining ideal for the variety X or rather for the embedding of U in the ambient affine space. However, since these polynomials have their coefficients in K, which is contained in the valuation ring R of $K(w)$, one sees readily that the properties in question are really independent of the choice of the elements x_i's modulo these polynomials, so that we may in fact treat the x_i's as if they were independent variables.

Let f be a formal regular function on X along Z. Such a function of course induces a rational function on Z. In the affine open subset U introduced above, f is represented by an element in $K(x)^\wedge$, so that we can set $f = \sum_{j=0}^{\infty} f_j(x)$ with $f_j(x)$ in J^j, and the rational function induced by f on Z is of course represented by the function $f_0(x)$. Now, if Z is embedded in an algebraic system defined by $Z(w)$ and if we consider $Z(w)$ as a holomorphic space over the valued field $K(w)^*$, then f will induce a function on $Z(w)$ which we maintain is holomorphic. In fact, if q is any point of $Z(w)$ in the affinoid subset D in U, then we have $b_i(w, q) = 0$ for every $i = 1, \dots, s$. Since $b_i(w, x)$ is a polynomial in x with coefficients in R, it follows that $\overline{b_i(w, q)} = b_i(\bar{q}) = 0$ for every i, so that $\bar{q}$ is a point in Z. This implies that $|b_i(q)| < 1$ for every i, and hence $f_j(q) \to 0$ as $j \to \infty$, so that the function $f(x)$ is defined at the point q. Thus the function f has a well defined value at every point q in $Z(w)$ which is contained in an affinoid set similar to D, and since Z is complete, the function f is defined at every point in $Z(w)$. As we have mentioned before, since the function $f(x)$ can be expressed as a power series at the point $\bar{q}$ in D and since $|\bar{q} - q| < 1$, $f(x)$ can be redeveloped around the point q as a power series $f'(x)$ in $K(w)^*[\![x - q]\!]$, which then defines an algebroid function around q. Although this power series is convergent in an "open" sphere of radius 1 around q, this is nevertheless not enough to tie all the local algebroid functions together into a holomorphic function. For this and also for later use we shall need a more precise determination of the function along $Z(w)$ induced by $f(x)$.

To determine more precisely the function on X along $Z(w)$ induced by the formal function $f(x)$ on X along Z, we set $b_i(x) = b_i(w, x) + (b_i(x) - b_i(w, x))$. Since $|b_i(x) - b_i(w, x)| < 1$ for every i, there is a real number $e < 1$ which is greater than the absolute values of all the coefficients in all the polynomials $b_i(x) - b_i(w, x)$, so that $|b_i(x) - b_i(w, x)| < e$ for every i. For each j we denote by $B_{jk}(x)$, $k = 1, \dots, N(j)$, the various monomials of degree j in the elements $b_1(x), \dots, b_s(x)$, and similarly we denote by $B_{jk}(w, x)$ the corresponding monomials of degree j in the elements $b_1(w, x), \dots, b_s(w, x)$. It follows then from the equation $b_i(x) = b_i(w, x) + (b_i(x) - b_i(w, x))$ that

$$B_{jk}(x) = \sum_{l=0}^{j} \sum_{m=1}^{N(l)} a_{jk,lm}(w, x) B_{lm}(w, x),$$

where for notational convenience we set $N(0) = 1$ and $B_{01}(w, x) = 1$, and we also set $a_{jk,jk}(w, x) = 1$ and $a_{jk,jm}(w, x) = 0$ for $m \neq k$. We note that $|a_{jk,lm}(w, x)| < e^{j-l}$. Since f_j is a polynomial of degree j in the elements $b_i(x)$

with coefficients which are polynomials in $K[x]$, we can set $f_j(x) = \sum_{k=1}^{N(j)} c_{jk}(x)\, B_{jk}(x)$. Substituting in this the above formula for $B_{jk}(x)$, we obtain the following equation

$$\begin{aligned} f_j(x) &= \sum_{k=1}^{N(j)} c_{jk}(x) \left(\sum_{l=0}^{j} \sum_{m=1}^{N(l)} a_{jk,lm}(w,x)\, B_{lm}(w,x) \right) \\ &= \sum_{l=0}^{j} \sum_{m=1}^{N(l)} \left(\sum_{k=1}^{N(j)} c_{jk}(x)\, a_{jk,lm}(w,x) \right) B_{lm}(w,x). \end{aligned}$$

It follows then that

$$\begin{aligned} f(x) &= \sum_{j=0}^{\infty} f_j(x) \\ &= \sum_{j=0}^{\infty} \sum_{l=0}^{j} \sum_{m=1}^{N(l)} \left(\sum_{k=1}^{N(j)} c_{jk}(x)\, a_{jk,lm}(w,x)\, B_{lm}(w,x) \right) \\ &= \sum_{l=0}^{\infty} \sum_{j=l}^{\infty} \sum_{m=1}^{N(l)} \left(\sum_{k=1}^{N(j)} c_{jk}(x)\, a_{jk,lm}(w,x)\, B_{lm}(w,x) \right) \\ &= \sum_{l=0}^{\infty} \sum_{m=1}^{N(l)} \left(\sum_{j=l}^{\infty} \sum_{k=1}^{N(j)} c_{jk}(x)\, a_{jk,lm}(w,x) \right) B_{lm}(w,x) \\ &= \sum_{l=0}^{\infty} \sum_{m=1}^{N(l)} C_{lm}(w,x)\, B_{lm}(w,x), \end{aligned}$$

where we set $C_{lm}(w,x) = \sum_{j=1}^{\infty} \sum_{k=1}^{N(j)} c_{jk}(x)\, a_{jk,lm}(w,x)$. We observe now that each $C_{lm}(w,x)$ is a strictly convergent series in the affinoid set D and hence defines there an affinoid function. Thus we have shown that the function induced by $f(x)$ on X along the variety $Z(w)$ is almost like a formal function on X along $Z(w)$. Just like a formal function on X along $Z(w)$, this function can be expressed locally as a power series in the basis elements $b_1(w,x), \ldots, b_s(w,x)$ of the ideal $J(w)$, but with the difference that the coefficients of the $B_{lm}(w,x)$ are strictly convergent series instead of polynomials. We can call such a function a quasiformal regular function on X around $Z(w)$.

In particular, for $l=0$ and hence $m=1$, the power series

$$f''(x) = C_{01} = \sum_{j=0}^{\infty} \sum_{k=1}^{N(j)} c_{jk}(x)\, a_{jk,01}(w,x)$$

is strictly convergent in the affinoid set D and defines there the same function on $Z(w)$ as that induced by the function $f(x)$. As this is true for every affinoid subset in X covering a part of $Z(w)$, it follows that f induces a holomorphic function on the entire subvariety $Z(w)$. We observe also that if $f'(x)$ is the redevelopment of $f(x)$ around the point q, then $f'(x) - f''(x)$ is an element in the ideal $K(w)^*[\![x-q]\!]\, J(w)$ generated in $K(w)^*[\![x-q]\!]$ by the ideal $J(w)$. More generally, if instead of a formal function we have a coherent sheaf of formal ideals on X around Z, then it will induce a coherent sheaf of ideals on $Z(w)$ as a holomorphic space over the valued field $K(w)^*$ and hence define a holomorphic subspace in $Z(w)$. In other words, a formal subscheme on X around Z will define a holomorphic subspace on $Z(w)$. Finally, since Z is a complete and hence also $Z(w)$ is complete, the holomorphic subspace in $Z(w)$ defined by a formal subscheme on X around Z will be a closed holomorphic space as defined in the theory of holomorphic space over a non-archimedean valued field (see for example the definitions given in [4]). As we have shown in [2], such a closed holomorphic space in a complete (projective) algebraic variety must also be algebraic.

To express our results in terms of local geometry, we set $p=\bar{q}$ and recall that a formal function $f(x)$ in U can be expressed locally at p as an algebroid function $f(x)$ around p and that this algebroid function around p can be redeveloped around q as an algebroid function $f'(x)$ around q. Furthermore, the function induces an affinoid function $f''(x)$ on $Z(w)$, and that the difference $f'(x)-f''(x)$ is an element in the ideal $K(w)^*[\![x-q]\!]J(w)$ generated by $J(w)$ in $K(w)^*[\![x-q]\!]$. More generally, consider an ideal Q in $K(w)^*[\![x-p]\!]$, and let the algebroid ideal Q be its redevelopment around q and let the affinoid ideal Q'' be the ideal induced by Q on $Z(w)$, both Q' and Q'' being ideals in $K(w)^*[\![x-q]\!]$. Then it follows from the above that Q' and Q'' are equal modulo the ideal $K(w)^*[\![x-q]\!]J(w)$. Since $J(w)$ is an ideal in $K(w)^*[x]$, it follows that Q' and Q'' have the same restrictions in $K(w)^*[x]$. Geometrically this implies that for any algebroid variety V on X around p, its redevelopment V' around q is algebraic if and only if the affinoid subvariety V'' around q induced by V on $Z(w)$ is algebraic. We shall state our results as a lemma in a form convenient for later reference.

Lemma 2. *Let X be an algebraic variety and let Z be a complete subvariety in X; let V be a formal subvariety on X around Z, and let K be a field of definition for X, Z, and V. Let $Z(w)$ be a generic element over K of an algebraic system of subvarieties in X which contains Z as a member, and let $K(w)$ be endowed with a non-archimedean valuation which is centered at Z. If q is a point in $Z(w)$ which is defined over the field $K(w)^*$, then the algebroid variety around q defined by V at q is algebraic.*

If F is a formal map of X around Z into an algebraic space Y, we shall study it by considering what essentially is its graph in the product space $X \times Y$, so that we can apply the above results to it. This is the same basic approach as in our earlier paper [1]. In the following we shall for simplicity restrict ourselves to the case of a formal rational function, which is essentially a formal map of X around Z to the projective line P. The apparently more general case of a formal map can be handled easily by a simple extension of our method.

Consider a formal rational function F on X along Z. We shall associate with such a function F a formal subvariety V in the product space $X \times P$, which is in a sense the graph of a formal map on $\hat{X}$ to the projective line P. This formal subvariety V will be defined by an ideal sheaf on the formal completion $(X \times P)^\wedge$ of $X \times P$ along the subvariety $Z \times P$, which is usually denoted as the blow-up of a coherent sheaf on X around Z associated with the formal function F. Let U be as before an affine open subset in an affine covering of X. We assume for simplicity that the variety X is non-singular. There exist then two invertible sheaves $\mathscr{G}$ and $\mathscr{H}$ on X such that in each such subset U the function F can be expressed as a quotient g/h, where g and h are sections of $\mathscr{G}$ and $\mathscr{H}$ respectively in U, provided the covering is suitably chosen with respect to the function F. The pair of sheaves $(\mathscr{G}, \mathscr{H})$ generates a coherent sheaf of ideals on $\hat{X}$, and V is the blow-up of this ideal sheaf and can be defined by an ideal sheaf $\mathscr{Q}$ on $X \times P$ around $Z \times P$. Let (z_0, z_1) be a system of homogeneous coordinates for P, and let T_0 and T_1 be the two affine subsets covering P defined by the conditions $z_0 \neq 0$ and $z_1 \neq 0$ respectively. In the affine subset $U \times T_0$ the function $f^{(0)} = (z_1/z_0)h - g$ is a formal function on $U \times T_0$ along $Z \times P$; similarly, in the affine subset $U \times T_1$ the function $f^{(1)} = h - (z_0/z_1)g$ is a formal function on $U \times T_1$ along $Z \times P$. Since the function $f^{(0)}$ is equal to the function $f^{(1)}$ multiplied by the factor z_1/z_0 in the intersection set $U \times (T_0 \cap T_1)$, it follows that they define the same formal ideal in this intersection set. As the set U runs through the subsets in an open affine covering of X, the sets $U \times T_0$, $U \times T_1$ form an open affine covering of $X \times P$ and we obtain in this way a coherent sheaf of formal ideals on $X \times P$ around $Z \times P$. This ideal sheaf is contained in the ideal sheaf $\mathscr{Q}$, and it coincides with the ideal Q at any point which is regular for F. We shall find this remark useful later.

III.

We return now to the set up in Sect. I, in order to complete our generalization of our original proof of Theorem 2 in [1]. We first recall the definitions introduced in Sect. I. Let X be an algebraic variety on which a group G acts and let Z be a complete subvariety in X. We assume as before that Z generates X in the sense that $G_Z^0 = G$. Let F be a formal rational function on X around Z. Without any loss of generality, we can assume that there is a point p in Z which is a regular point for the function F, so that the value $F(p)$ of the function F is well defined. In fact, if there is no such point p in Z, we can replace Z by its transform $w(Z)$ by a generic element w in G over K, and replace the function F by the function F_w induced by F on $w(Z)$. Although the function F_w is only a quasi-formal rational function, in the sense that it is locally the quotient of two quasi-formal regular functions, it can be easily seen that all our arguments will go through without any difficulties, only in that case the ground field K should of course be replaced by the valued field $K(w)^*$. Thus we can assume that Z contains a point p which is regular for the function F and that if the ground field K is a complete valued field with the valuation ring R, then G, X, Z, $G_{p,Z}^0$, F, p and $F(p)$ are all defined over R. Let $u_0, u_1, \ldots, u_t$ be independent generic points in $G_{p,Z}^0$ over K. We then define the fields K_i, the subvarieties Z_i, and the points p_i, $i = 0, 1, \ldots, t$, as before, but we shall now endow the field K_0 with a valuation which is over K in the sense that it is an extension of whatever valuation the field K may have (the trivial valuation in case the field K does not have any). We endow the field $K_0 = K(u_1, \ldots, u_t)$ with a valuation over K which is centered at the point $(e, \ldots, e)$

in the t-fold product of $G^0_{p,Z}$ with itself. More precisely, if we choose the generic points $u_1, \ldots, u_t$ of $G^0_{p,Z}$ so that each represents an affine system of coordinates in an affine set on $G^0_{p,Z}$ with the origin at e, then the valuation of K_0 has as its center on $K[u_1, \ldots, u_t]$ the ideal consisting of all functions which vanish at the origin. We denote by K_0^* the completion of the algebraic closure of K_0 with respect to the extension of this valuation in the algebraic closure. More generally, for every $i=1, \ldots, t$ the valuation of K_0 induces a valuation in each K_i, and K_0^* contains in a natural way the completion K_i^* of the algebraic closure of K_i with respect to the extension of this valuation in this algebraic closure, and we have of course the relation $K_i^* \subset K_{i-1}^*$. Also, we shall denote by R_i the valuation ring in K_i^* for every i, and we have of course the relation $R_i \subset R_{i-1}$. From now on we shall consider the field K_0^* as our "universal domain", so that all varieties can have points with coordinates in this universal domain, even though they may be defined initially over a much smaller field.

Let U be an affine subset in X containing the point p, and let $(x) =(x_1, \ldots, x_n)$ be a system of affine coordinates in U. It is clear that all the points $p_t=p, p_{t-1}, \ldots, p_0$ are in the subset U and hence can be represented by their coordinates in U, in view of the choice of the "universal domain" we have just made. We maintain now that for every i the point p_i is defined over the valuation ring R_i and that $\bar{p}_i=p$. For this it is sufficient to show that the point p is the only specialization of the point p_i over the specialization of the point $u_t^{(-1)^t} \ldots u_{i+1}^{(-1)^{i+1}}$ to e over K; for, if p_i were not in R_i, then at least one of its specializations must be infinite with respect to R_i. This assertion, in turn, is a clear consequence of the fact that X is a homogeneous space on which the group variety G acts; for the definition of such a homogeneous space implies that the map $G \times X \to X$ is defined everywhere, and hence in particular at the point $e \times p$, and the situation remains the same if we replace G by the subvariety $G^0_{p,Z}$ in it. We note also that

$$|p_i - p_{i+1}| = |p_i - p + p - p_{i+1}| \leqq \max(|p_i - p|, |p_{i+1} - p|) < 1.$$

Next, we shall show that the function F is defined at every one of the points p_i in the following sense. Locally at the point p the function F is represented as a quotient g/h of two formal power series g and h in the power series ring $K[\![x-p]\!]$, with $h(p) \neq 0$, so that the value $F(p)$ of F at p is well defined. In fact, since the function h has an inverse in $K[\![x-p]\!]$, we can express F itself as an element in $K[\![x-p]\!]$. Consider the function F now as a function over the field K_0^*, that is we shall consider F as an element in $K_0^*[\![x-p]\!]$. Since F is an element in $R_i[\![x-p]\!]$ and since $|p-p_i|<1$, we can redevelop F around the point p_i as an element F_i in $R_i[\![x-p_i]\!]$. Furthermore, since $|p_i - p_{i+1}|<1$, we can redevelop F_i around the point p_{i+1} as an element in $R_i[\![x-p_{i+1}]\!]$, and it is easily seen that the result coincides with F_{i+1}, so that it is actually an element in the ring $R_{i+1}[\![x-p_{i+1}]\!]$.

Just as in the complex case in [1], also here we have to make sure that the point p_{i+1} lies in an analytic branch of Z_i at the point p_i, only we shall proceed here in a slightly different manner. As in [1], we denote by Z_i' the locus of the point $u_i^{(-1)^i}\dots u_1^{(-1)}u_0(p)$ over K, so that we have the equation

$$Z_i = u_i^{(-1)^i}\dots u_{i+1}^{(-1)^{i+1}}(Z_i').$$

Let J_i' be the prime ideal in $K[x]$ defining the variety Z_i' in U, and consider the ideal $K[\![x-p]\!]J_i'$ in $K[\![x-p]\!]$, which is a semi-prime ideal and whose prime divisors represent the analytic branches of Z_i' at the point p. (We recall again that here as well as in the following we shall treat the elements of x as independent variables, in which case the ideal J_i' will include also the elements in the ideal in $K[x]$ which defines the embedding of U as an affine variety.) Extend now the ground field to the complete valued field K_i^*, and consider the ideal $R_i[\![x-p]\!]J_i'$ in $R_i[\![x-p]\!]$; since the ideal $R_i[\![x-p]\!]J_i'$ is the extension in $R_i[\![x-p]\!]$ of the ideal J_i' in $K[x]$, it has a basis consisting of elements in $K[x]$. Now, every element in $R_i[\![x-p]\!]$ is defined everywhere in the open unit ball B around p, and the ideal $R_i[\![x-p]\!]J_i'$ defines in B an analytic set W, which is the part of the variety Z_i' contained in B. Let q be any point in W, and let W_q be the locus of q over K; the equation $p=\bar{q}$ shows that p is a specialization of q over K and hence must be contained in W_q. Since the ideal $R_i[\![x-p]\!]J_i'$ has a basis consisting of elements in $K[x]$, the variety W_q is also contained in W. Since q is a generic point of W_q over K, there is an irreducible analytic set W_q' in W_q containing both p and q. Now, among the irreducible analytic components of W containing q, at least one must contain the analytic set W_q' and hence also must contain the point p. Thus we have shown that every point q in W is contained in an irreducible analytic component of W containing the point p, and hence every irreducible component of W must pass through the point p. As we have remarked in the proof of Lemma 1 in the previous section, this implies that validity of what we referred to there as the condition (A), or the equation:

$$R_i[\![x-p]\!] \cap K_i^*[\![x-p]\!]J_i' = R_i[\![x-p]\!]J_i'.$$

Now, consider the point $p' = u_{i+1}^{(-1)^i}(p)$ in B (note that $\overline{p'}=p$); this point p' is in W, since it is a specialization of $u_i^{(-1)^i}\dots u_1^{(-1)}u_0(p)$ over the specialization $u_0 \to e, \dots, u_{i-1} \to e, u_i \to u_{i+1}$ over K_i. If we apply the action of the element $u_i^{(-1)^i}\dots u_{i+1}^{(-1)^{i+1}}$ to Z_i', we obtain the variety Z_i which is defined in U by the transformed ideal J_i of J_i' in $K_i[x]$. The transform of the point p is the point p_i, the transform of the open unit ball B around p is the open unit ball B_i around p_i, and the transform of W is the analytic set W_i in B_i defined by the ideal $R_i[\![x-p_i]\!]J_i$ in $R_i[\![x-p_i]\!]$. The transform of the point p' is the point p_{i+1}. The condition (A) is now expressed by the equation

$$R_i[\![x-p_i]\!] \cap K_i^*[\![x-p_i]\!]J_i = R_i[\![x-p_i]\!]J_i. \tag{1}$$

Now, we go over to the product space $X \times P$. We recall that the formal function F gives rise to a formal subvariety V on $X \times P$ around $Z \times P$, which is in a sense the graph of a formal map on $\hat{X}$ to the projective line P. As described in the previous section, this formal subvariety V is defined by an ideal sheaf $\mathcal{Q}$ on the formal completion $(X \times P)^\wedge$ of $X \times P$ along the subvariety $Z \times P$. Using the notations introduced there, we take an affine subset U in X

which contains the point p, and consider the affine subset $U^*=U\times T_0$ in $X\times P$; since F is assumed to be regular at the point p, we can take the affine subset U so small that the sheaf $\mathcal{Q}$ is represented in U^* by the formal ideal generated by the one element $Q=(z_1/z_0)-F$, and the point $p^*=p\times F(p)$ is a point in V. Our aim is to prove that the formal subvariety V is algebraic. We note first that with a suitable choice of the affine subset U the points p_i are all contained in U; then the points $p_i^*=p_i\times F_i(p_i)$ are all contained in the affine subset U^*, with $p_t^*=p^*$. For notational convenience, we set $x_{n+1}=z_1/z_0$, so that the affine subset U^* has the coordinate system $(x^*)=(x_1,\ldots,x_n,x_{n+1})$; then, if the point p_i has the coordinates $(p_{i,1},\ldots,p_{i,n})$ in U, the point p_i^* will have the coordinates $(p_{i,1},\ldots,p_{i,n},p_{i,n+1})$ in U^*, with $p_{i,n+1}=F_i(p_i)$, whereby we note in particular that the point p^* has the coordinates $(p_{t,1},\ldots,p_{t,n},p_{t,n+1})$. Since Q is an element in the ring $K[\![x^*-p^*]\!]$, this element Q (together as always with the ideal in $K[x^*]$ defining the embedding of U^* in the ambient affine space) will define an algebroid variety around p^*, which we shall for convenience denote by V_t, in conformity with the notation $p_t^*=p^*$. It is of course sufficient to prove that the algebroid variety V_t is algebraic.

Consider now the ideal J_i in $K_i[x]$ which defines the variety Z_i in U, and its extension $R_i[\![x-p_i]\!]J_i$ in $R_i[\![x-p_i]\!]$. The same ideal J_i generates also an ideal in $R_i[x^*]$, which we shall denote by J_i^*, and it is clear that J_i^* defines in U^* the subvariety $Z_i\times P$ in $X\times P$. Next, consider the element $Q_i=x_{n+1}-F_i$ $=(x_{n+1}-p_{i,n+1})-(F_i-F_i(p_i))$ in the ring $R_i[\![x^*-p_i^*]\!]$; since for $i=0,\ldots,t-1$, each F_{i+1} is a redevelopment of F_i around p_{i+1}, each Q_{i+1} is also a redevelopment of Q_i around p_{i+1}^*. Let (Q_i,J_i^*) be the ideal generated by the elements Q_i and J_i^* in $R_i[\![x^*-p_i^*]\!]$. We claim that we have here also the property (A), which is now expressed by the equation

$$R_i[\![x^*-p_i^*]\!]\cap K_i^*[\![x^*-p_i^*]\!](Q_i,J_i^*)=(Q_i,J_i^*). \tag{2}$$

In fact, the map $x_1^*\to x_1,\ldots,x_n^*\to x_n,x_{n+1}^*\to F_i$ defines an endomorphism of $R_i[\![x^*-p_i^*]\!]$ onto $R_i[\![x-p_i]\!]$ which induces clearly an isomorphism of the residue ring $R_i[\![x_i^*-p_i^*]\!]/R_i[\![x^*-p_i^*]\!]Q_i$ onto $R_i[\![x-p_i]\!]$, and hence also induces an isomorphism of the residue ring $R_i[\![x^*-p_i^*]\!]/R_i[\![x^*-p_i^*]\!](Q_i,J_i^*)$ onto the residue ring $R_i[\![x-p_i]\!]/R_i[\![x-p_i]\!]J_i$. Recalling the remark in the proof of Lemma 1 in the previous section about the property (A) being an intrinsic property of the residue ring $K^*[\![x-p]\!]/K^*[\![x-p]\!]Q$, this isomorphism shows that Eq. (2) follows from Eq. (1). Now, let V_i be the algebroid set around p_i^* defined by the ideal $K_i^*[\![x^*-p_i^*]\!](Q_i,J_i^*)$, and let V_i' be the algebroid set around p_{i+1}^* defined by the ideal $K_i^*[\![x^*-p_{i+1}^*]\!](Q_{i+1},J_i^*)$; we note that $Z_i\times P$ contains both points p_i^* and p_{i+1}^*, and that the elements in the ideal J_i^* being polynomials can be considered as elements in both $K_i^*[\![x^*-p_i^*]\!]$ and $K_i^*[\![x^*-p_{i+1}^*]\!]$. The validity of Eq. (2) shows that the variety V_i is defined over the ring R_i by the ideal (Q_i,J_i^*); and since the ideal (Q_{i+1},J_i^*) is the redevelopment around p_{i+1}^* of the ideal (Q_i,J_i^*), the algebroid set V_i' is the redevelopment around p_{i+1}^* of the algebroid set V_i around the point p_i^*. It follows that we can apply the Lemma 1 in the previous section and conclude that if V_i is algebraic, then V_i' is also algebraic. Finally, since Z_i spans Z_{i+1} over K_{i+1}, the algebroid set V_i' spans the algebroid set V_{i+1} over K^*_{i+1}, whereby we observe that both are algebroid sets around the point same p^*_{i+1}. It follows then from Theorem 1 in [1] that if V'_i is algebraic, then V_{i+1} is also algebraic. Thus we have shown that if V_i is algebraic, then V_{i+1} is also algebraic.

We can now finish our proof in a few lines, almost exactly as we did in our original paper [1]. We have already shown for every, $i=0, 1, \ldots, t-1$, that if V_i is algebraic, then V_{i+1} is also algebraic. It follows then by induction that if V_0 is algebraic, then V_t is also algebraic, so that it remains only to prove that the algebroid variety V_0 is algebraic. Now, V_0 is the algebroid variety at p^*_0 defined by the function $Q_0=x_{n+1}-F_0$ and the equations defining $Z_0 \times P$ in U^*. We recall that Z_0 is the subvariety $u_t^{(-1)^t} \ldots u_1^{-1}(Z)$ in X, which is defined over the valued field K^*_0, and it generates over K an algebraic system of subvarieties in X which contains Z as a member at the center of its valuation. It follows that in the product space $X \times P$ the subvariety $Z_0 \times P$, which is also defined over the valued field K^*_0, generates over K an algebraic system of subvarieties which contains $Z \times P$ as a member at the center of its valuation. Since V_0 is the algebroid variety around p^*_0 induced at the point p^*_0 by the formal variety V on $X \times P$ around $Z \times P$, it follows from Lemma 2 that V_0 must be algebraic. This concludes the proof of our theorem.

In conclusion, we should like to remark that the method used in our present proof may have some interest in itself, in that it seems to represent a rather novel way of applying of the non-archimedean function theory to a problem in algebraic geometry. The idea is to introduce a non-archimedean valuation in the field of definition of a generic element of an algebraic family and to consider such an element as a holomorphic space over the so valued field. In fact, just such an approach to study the algebraic systems was suggested by the author in a lecture given at the University of Maryland quite a few years ago. We find this aspect of our method interesting because of its analogy with the classical complex-analytic case. However, we observe on the other hand that the application of the non-archimedean function theory in our present case can be replaced by using the Grothendieck GAGA in the following way. Let the variety $G^0_{p,Z}$ be denoted by H for convenience and let z be a generic point of Z over K, then consider the subvariety Y_0 in $H^r \times X$ defined over K by the point $(u_1, \ldots, u_t) \times u_t^{(-1)^t} \ldots u_1^{(-1)}(z)$, and its projection into the factor X. If we denote by $\hat{Y}_0$ the completion of Y_0 along $(e, \ldots, e) \times Z$, then the projection induces a map of $\hat{Y}_0$ into $\hat{X}$, and a formal function on $\hat{X}$ induces a formal function on $\hat{Y}_0$. The formal variety $\hat{Y}_0$ can be considered as a kind of fibre space over the complete ring R_0, and our variety Z_0 is essentially its generic fibre. Since the full fibre of $\hat{Y}_0$ at the origin is the complete variety Z, it follows from the Grothendieck GAGA that any formal object in $\hat{Y}_0$ is algebraic over R_0. Applying this to the product space $\hat{Y}_0 \times P$, one concludes that the formal function induced on $\hat{Y}_0$ by a formal rational function F on $\hat{X}$ is algebraic over K_0. This is in fact a slightly different way of saying that the function induced by F on Z_0 is algebraic. This connection with the Grothendieck GAGA is called to our attention by the referee, to whom we should like to express our thanks for this and other helpful suggestions.

References

1. Chow, W.-L.: On meromorphic maps of algebraic varieties. Ann. Math. **89**, 391-403 (1969)
2. Chow, W.L.: On the algebraicity of certain ringed spaces. Am. J. Math. **101**, 364-379 (1979). See also Correction to this paper. Am. J. Math. **107**, 759-760 (1985)
3. Faltings, G.: Formale Geometrie und homogene Räume. Invent. Math. **64**, 123-165 (1981)
4. Gerritzen, L., Grauert, H.: Die Azyklizität der affinoiden Überdeckungen. Global analysis, Papers in Honor of K. Kodaira, 159-184. Tokyo: University of Tokyo Press 1969
5. Nagata, M.: Local rings. New York: Interscience Publishers 1962

Oblatum 4-IV-1985 & 26-VII-1985 & 20-I-1986

Shiing-Shen Chern as Friend and Mathematician, A Reminiscence on the Occasion of His 80th Birthday

Wei-Liang Chow

Department of Mathematics

Johns Hopkins University

I met Shiing-Shen Chern for the first time in Hamburg sometime in the fall of 1934. We were at that time both students at the Hamburg University; Chern was at that time studying with Blaschke, while I was attending the lectures of Artin. I purposely use the rather vague expression "attending the lectures of Artin" in order to indicate the tentative nature of my sojourn in Hamburg at that time; more correctly, I should say that I was at that time strictly speaking a student at the University of Leipzig, but I decided to stay in Hamburg for personal reasons, as I shall explain later, and I took the opportunity to learn something by attending the lectures of Artin. In order to explain this rather strange lack of definiteness in my mathematical aspirations and also in order to understand the importance of Chern's influence at a critical juncture later in my life, let me say a few words about my rather unusual educational background.

Except for a very brief period, I never attended schools or colleges in China. Beginning at the age of five (1916) I was taught the standard Chinese classics by an old Chinese tutor and at the age of eleven I was taught to read and write English. However, I discovered very soon that the ability to read English provided me with the opportunity to acquire the knowledge about almost any subject I wanted to learn. Since the curricula in most Chinese universities at that time were modeled after those in the American universities and many of them often used books written by American professors, it was not difficult for me to find out the most commonly used text books in America on most subjects. Thus in this way I taught myself all sort of subjects from mathematics and physics to history and economics. This situation lasted from 1924 to 1926 when I succeeded in persuading my father to send me to study in the United States. At that time my main interest was political economy, and economics was still my major subject of study when I entered the University of Chicago in October 1929. However, during the next two years I began to have some serious doubts about taking economics as my major.

Already in my childhood I always wanted to be an electric engineer, although I did not really know at that time what that implies. Now, as my doubts about the wisdom of majoring in economics increased, I decided to change my major to physics, hoping that it would eventually lead me to engineering. Thus, when I graduated at the University of Chicago in 1930, my major was in physics. At about the same time I happened to read the book called *Pure Mathematics* by the famous English mathematician Hardy. This book opened the door to mathematics for me, although I was at that time still studying applied mathematics, hoping eventually to study physics. In summer 1931 I discussed studying mathematics with a graduate Chinese mathematics student who got his Ph.D. at Chicago and then spent a year in Princeton. He was very enthusiastic about Princeton (he attended the lectures of John Von Neumann there) and he advised me to go to

Princeton or even better to go to Goettingen in Germany which he thought was then the world center for mathematics. Therefore, with only a vague idea of studying mathematics, I went to Goettingen in October 1932. Although I had previously taken a course in German at the University of Chicago, it took me about three months to learn the German language sufficiently to enable me to understand the lectures. However, by that time, at the beginning of 1933, something happened in the German politics which would soon change drastically not only the university at Goettingen, but the entire Germany and in fact eventually the entire world, namely Hitler and his Nazi party came to power. Not knowing anything about the German politics, I was surprised to hear some students murmuring that Hermann Weyl would very probably leave Goettingen, which Weyl did that summer. At about the same time first Richard Courant and then Emmy Noether also left Goettingen. Thus the world mathematics center I hoped to come to study was essentially depleted.

Thus in the summer of 1933 I had to find a German university to replace Goettingen. The summer before in Chicago I had taken a course in modern algebra, in which the then relatively new text book called *Moderne Algebra* by van der Waerden was used, and I was very much impressed by the elegance of the presentation of the subject in that book, and upon learning that van der Waerden was a professor at the university in Leipzig, I went to Leipzig, hoping to study with van der Waerden. It was a stroke of luck for me that I went to Leipzig at just about the time when van der Waerden was writing the beginning of his series of papers entitled *Zur algebraischen Geometrie*, and as a result I was introduced to a subject I never heard of before. Van der Waerden was very kind to me; he told me to study the books of Severi and also the older books of Bertini and Enriques. Also, van der Waerden has the unusual talent of explaining even the most complicated mathematical theory in rather simple terms and he made me feel that my lack of knowledge of some important subjects in mathematics was not important as long as I was willing to learn; in fact, for the first time in my life I began to feel that I made the right choice to study mathematics.

I went to Hamburg in the summer of 1934 for summer vacation and there I met a young lady, Margot Victor, who eventually became my wife. It was a case of love at first sight, and I actually proposed marriage to Margot within a week after we met. Realizing that my feeling of love could not be expected to be reciprocated within a short period of time, I decided to stay in Hamburg to woo Margot. At the same time I would take this opportunity to attend the lectures of Artin, hoping to learn something about algebraic number theory, while at the same time I continued my study of algebraic geometry. (The German university system was very flexible at that time and allowed me the freedom of "wandering" from one university to another.) In time Chern and I became good friends, although we did not have much mathematical contacts with each other at that time. At the beginning of 1936, I went back to Leipzig to finish my dissertation with van der Waerden. When on July 10th, 1936 I married Margot Victor in Hamburg, Chern was present at the reception given by Margot's parents. (See the picture taken at that reception.)

The next time I saw Chern was some time in 1939 or perhaps in 1940 (I do not remember exactly), when he came to Shanghai for a visit. I learned from him that he could not even land in Shanghai when he came back to China in 1937 because of the fighting around Shanghai. I did not know what was the reason for his visit to Shanghai, but we did discuss the possibility of issuing a volume of the journal of the Chinese Mathematical Society. By that time I was forced by circumstances to abandon my mathematical research; after all I had to support not only my wife and two children, but also my parents-in-law, who were forced almost penniless to leave Germany by the racial policy of the

Nazi regime. Although my father was normally a very well-to-do man (after all he supported my study in the United State and Germany for almost nine years), his business and investments suffered serious set backs from the war and I realized that I had to stand on my own and earn a living by doing whatever business I could find.

The next time I met Chern was in the spring of 1946. The war was over by that time, and Chern had just come back from a successful journey of two years at the Institute for Advanced Study in Princeton, and he was called upon to organize a new institute of mathematics for the Academia Sinica in Nanking. On my part I had been practically entirely out of any mathematical activities for almost a decade; the last paper I wrote (1938) was inspired by reading an old paper on thermodynamics by Caratheodory, and although van der Waerden accepted the paper with some words of praise, including some kind comments by Caratheodory, but it was so long ago that I almost forgot about it. The contrast between us in our mathematical positions was very clear. Chern was by that time an established mathematician of international reputation, while I was at best a post-doctoral student, with the additional burden of almost a decade of neglect. My plan for the future, in so far one could make future plans in those confusing and uncertain days in China, was to develop an import and export business through my business connection with an uncle of my wife, Hans Victor, who was a successful business man in Germany before the Nazi regime came to power and was forced almost penniless to leave Germany at the age of sixty, but who nevertheless later built up his business in America. Realizing that I could not rely on what would normally be my inheritance from my father and observing also the uncertainty of any normal academic condition in China, I had to find a way to support not only my wife and two children, but also my mother-in-law (my father-in-law had died in the mean time).

Chern realized that in the rather confused condition in China after the war what was mostly needed by mathematicians in China was recent literature in their own fields of research. For this reason Chern collected a large number of reprints from mathematicians not only in his own fields of research, but also in related fields. For my part I was very happy to find among his collection of reprints most of the recent papers of Oscar Zariski who apparently had carried the modern algebraic geometry far beyond the stage reached by van der Waerden before the war. Chern also told me about the important work of Andre Weil, whose work would probably be published in book form in the near future. It was clear to me at that time that if I ever wanted to get back to mathematics, my first task was to study the papers of Zariski, which Chern very kindly lent to me. Chern also suggested that I spend a year at the Institute for Advanced Study in Princeton in order to get myself into the stream of modern mathematics, and he expressed the opinion that in spite of the loss of my past ten years, it was not too late for me to get started again in mathematics. In fact Chern even wrote a letter to Lefshetz, suggesting that an invitation be extended to me to visit Princeton.

It was clear to me that I had to make a fundamental decision, probably the most basic one in my life, whether I should quit my business activities and try to get back into mathematics. After all, I was then thirty five years old and it was more than ten years after I got my doctoral degree and I had hardly done anything since then. It seemed that I probably had missed my chance to be a mathematician, and I was so discouraged by this situation that I was ready to give up mathematics. However, the conversation with Chern had changed my mind and renewed my hope to stay in mathematics. I discussed the problem with Margot, who showed great understanding for my wish to go back to mathematics; she felt that I should follow my own inclination and judgement, and that whatever I decided to do, she would adjust her activities accordingly. After all, she had worked in her uncle's office for several years before her marriage and she could go back to work again if it would become necessary. We had at that time saved enough money from my past business activities to pay for our passages to the United States and to live for a year or two without any additional income, if necessary. We realized that we were taking a great risk, since the possibility of my achieving even a moderate degree of success in this new endeavor at that stage in my life was by no means assured, but we also felt that sometimes in life one must take bold actions. (As to my mother-in-law, I would provide her with enough funds so that she could go to Capetown in South Africa, where a sister of Margot lived, so that she would not be entirely alone.) Once we had made this decision, the next step was for me to wind up my business and put all my assets in liquid form; and while I waited to get the necessary passports and visa in order, I devoted my entire time and effort to mathematics, in particular to study the papers of Zariski. Originally we intended to leave for the United States sometime in the fall of 1946, but all the arrangements I had to make (plus the fact that I had to attend to some of business matters of my father who was old and in very poor health) delayed our departure until March of 1947.

I arrived in Princeton at the beginning of April in 1947 and was promptly admitted as a temporary member in the Institute for Advanced Study, thanks to the letter of Chern to Lefschetz. It was too late for me to get a stipend even for the forthcoming year, which was not too important for me at that time, but we (i.e. I and my family) could live in the Institute housing, which was important, for there was then a great shortage of housing due to the war. In the spring of 1948 I learned that van der Waerden was visiting at the Johns Hopkins University and I went there to see him. It turned out that there was a vacancy at Hopkins, and van der Waerden, always very kind to me, told me that he would recommend me for this position if I would be interested in it. As a result I went to Hopkins in 1948, and I stayed there until my retirement in 1977. I met Chern again in 1949 at the Institute for Advanced Study, where he was visiting, before he went to Chicago. What happened after that is now public mathematics history; as we all know, Chern was a professor at the University of Chicago until 1960 when he went to Berkeley and he has remained in Berkeley ever since. After Chern had gone to Berkeley, my contact with him was less frequent than when he was in Chicago, owing to the great distance between the east and west coasts, but I did make two summer trips to the Bay area to visit him. Chern has received some of the highest honors which a mathematician can possibly have, and I shall leave this aspect of his career to some of his collaborators or his former students, who are probably in a better position to give an adequate account of the citations leading to these honors. Chronologically, Chern has long passed the normal retirement age (we were both born in 1911). But a mathematician like Chern in a sense never retires. Even after passing his formal retirement age, Chern has been for several years the director of a research institute set up by the National Science Foundation in Berkeley. Also Chern went back to China several times, and I understand there is a research institute at his Alma mater, the Nankai University, of which Chern is the director. I think at the rate Chern is working to promote mathematics in China, he will be not only a leading mathematician of our generation, but also the father of modern mathematics in China.

I think I can best conclude this article by quoting a part of my letter to Chern on the occasion his formal retirement in 1979. Apart from the publication of his Selected Papers, there was an International Symposium in Global Analysis and Global Geometry in Berkeley in his honor, organized by the mathematics department at Berkeley and sponsored by the National Science Foundation. I was invited, and I would certainly have attended this Symposium, if I were not in Germany at that time. Our European trip, our first such trip since we left

Germany more than forty years ago, was planned a long time ago, and our relatives and friends all made plans to fit our time table that it was not possible for us to change our plans at the last minute. Since I could not personally congratulate him on this occasion, I wrote Chern a letter instead. The letter was written on June 20th, 1979, at a place called Schloss Elmau, a beautiful place in the Bavarian Alps, not far from the Austrian border, and I ended my letter with the following statements: "It is befitting that the National Science Foundation and your colleagues at Berkeley have chosen to honor you in this way, as you are no doubt one of the leading mathematicians of our generation. On a more personal level, I shall always remember that it was mainly due to your advice that I returned to mathematics after the war. Without this encouragement on your part at that critical moment in my life, I would probably never have made whatever modest contributions I may have made to mathematics. For this I owe you an enduring personal debt of gratitude."

Reprinted from Notice of the American Mathematical Society **43**(10) (1996) 1117–1118
© 1996 American Mathematical Society

WEI-LIANG CHOW, 1911–1995

S. S. Chern

After a long illness Wei-Liang died on August 10, 1995. He and I first met in Hamburg, Germany, in October 1934, when I had just come from China as an entering student while he was on his way from Göttingen to Leipzig in order to work with van der Waerden.

Wei-Liang came from a Mandarin family, one whose leading members realized the importance of the westernization of China. With their resources and enlightened view, the family produced during the turn of the century several leaders of Chinese society in different areas. Perhaps as a result of the family situation Wei-Liang did not attend a Chinese school. However, through private tutoring he was quite familiar with the Chinese language and Chinese history. His family position allowed him to go through college in the U.S., receiving his B.A. from the University of Chicago in 1931. His Chicago years gradually focused him to mathematics, and in 1932 he went to Göttingen, then one of the greatest mathematical centers in the world. Unfortunately political events in Germany during this period made his stay in Göttingen undesirable, and he decided to go to Leipzig to work with van der Waerden.

It was at this juncture that we met in Hamburg. The decline of Göttingen had the result of elevating Hamburg to a leading mathematical center in Germany. Her leading attraction was Emil Artin, the young professor who gave excellent lectures and whose interest extended over all areas of mathematics. Although Wei-Liang was a Leipzig student, the German university system allowed him to live in Hamburg. Besides the contacts with Artin, he had a more important objective, which was to win the love of a young lady, Margot Victor. They were married in 1936, and I was fortunate to be present at the wedding.

After their marriage Wei-Liang returned to China and became a professor of mathematics at the Central University in Nanking, then the Chinese capital. The next years China was at war, with the coastal provinces occupied by the Japanese.

We next saw each other in 1946 in Shanghai after the war ended. In a decade of war years Wei-Liang had practically stopped his mathematical activities, and the question was whether it was advisable or even possible for him to come back to mathematics.

His return to mathematics was most successful; I would consider it a miracle. He began by spending the years 1947–49 at the Institute for Advanced Study, after which he accepted a position at Johns Hopkins University, from which he retired in 1977. At Johns Hopkins he served as chairman for more than ten years. He was also responsible for the *American Journal of Mathematics*, a Hopkins publication and the oldest American mathematical journal.

Wei-Liang was an original and versatile mathematician, although his major field was algebraic geometry. He made several fundamental contributions to mathematics:

1. A fundamental issue in algebraic geometry is intersection theory. The Chow ring has many advantages and is widely used.
2. The Chow associated forms give a description of the moduli space of the algebraic varieties in projective space. It gives a beautiful solution of an important problem.
3. His theorem that a compact analytic variety in a projective space is algebraic is justly famous. The theorem shows the close analogy between algebraic geometry and algebraic number theory.
4. Generalizing a result of Caratheodory on thermodynamics, he formulated a theorem on accessibility of differential systems. The theorem plays a fundamental role in control theory.
5. A lesser-known paper of his on homogeneous spaces gives a beautiful treatment of the geometry known as the projective geometry of matrices and treated by elaborate calculations. His discussions are valid in a more general context.

Chow led a simple and secluded life, with complete devotion to mathematics and some other intellectual activities including philately. He was an authority on Chinese stamps and published a book on them. Margot and Wei-Liang had three daughters and a happy family life.

Undoubtedly he was an algebraic geometer of the first caliber. I once nominated him, with the support of Zariski, for membership in the National Academy of Sciences. Unfortunately it did not meet with success, and I think it was a loss to the Academy.

Reprinted from Notice of the American Mathematical Society **43**(10) (1996) 1119–1123
© 1996 American Mathematical Society

COMMENTS ON CHOW'S WORK

Serge Lang
Department of Mathematics, Yale University, New Haven, CT 06520, USA

Van der Waerden's prewar series of articles began an algebraization of Italian algebraic geometry. I was born into algebraic geometry in the immediate postwar period. This period was mostly characterized by the work of Chevalley, Chow, Weil (starting with his Foundations and his books on correspondences and abelian varieties), and Zariski. In the fifties there was a constant exchange of manuscripts among the main contributors of that period. I shall describe briefly some of Chow's contributions. I'll comment here mostly on some of Chow's works in algebraic geometry, which I know best.

Chow Coordinates

One of Chow's most influential works was also his first, namely, the construction of the Chow form, in a paper written jointly with van der Waerden [1]. To each projective variety, Chow saw how to associate a homogeneous polynomial in such a way that the association extends to a homomorphism from the additive monoid of effective cycles in projective space to the multiplicative monoid of homogeneous polynomials, and the association is compatible with the Zariski topology. In other words, if one cycle is a specialization of another, then the associated Chow form is also a specialization. Thus varieties of given degree in a given projective space decompose into a finite number of algebraic families, called Chow families. The coefficients of the Chow form are called the Chow coordinates of the cycle or of the variety. Two decades later he noted that the Chow coordinates can be used to generate the smallest field of definition of a divisor [12]. He also applied the Chow form to a study of algebraic families when he gives a criterion for local analytic equivalence [13]. He was to use them all his life in various contexts dealing with algebraic families.

In Grothendieck's development of algebraic geometry, Chow coordinates were bypassed by Grothendieck's construction of Hilbert schemes whereby two schemes are in the same family whenever they have the same Hilbert polynomial. The Hilbert schemes can be used more advantageously than the Chow families in some cases. However, as frequently happens in mathematics, neither is a substitute for the other in all cases. In recent times, say during the last decade, Chow forms and coordinates have made a reappearance due to a renewed emphasis on explicit constructions needed to make theorems effective (rather than having noneffective existence proofs, say) and for computational aspects of algebraic geometry whereby one wants not only theoretical effectiveness but good bounds for solutions of algebraic geometric problems as functions of bounds on the data. Projective constructions such as Chow's are very well suited for such purposes. Thus Chow coordinates reappeared both in general algebraic geometry and also in Arakelov theory and in diophantine applications. The Chow coordinates can be used, for example, to define the height of a variety, and to compare it to other heights constructed by more intrinsic, nonprojective methods as in [42, 43, 44]. They were used further in Arakelov theory by Bin Wang [46].

Chow coordinates were also used to prove a conjecture of Lie on a converse to Abel's theorem. See the papers by Wirtinger [47] and Chern [36].

Abelian Varieties and Group Varieties

(a) **Projective construction of the Jacobian variety.** In the fifties Chow contributed in a major way to the general algebraic theory of abelian varieties due to Weil (who algebraicized the transcendental arguments of the Italian school, especially Castelnuovo). For one thing, Chow gave a construction of the Jacobian variety by projective methods, giving the projective embedding directly and also effectively [18]. The construction also shows that when a curve moves in an algebraic family, then the Jacobian also moves along in a corresponding family.

(b) **The Picard variety.** Chow complemented Igusa's transcendental construction of the Picard variety by showing how this variety behaves well in algebraic systems, using his "associated form" [15]. He announced an algebraic construction of the Picard variety in a "forthcoming paper". Indeed, such a paper circulated as an unpublished manuscript a few years later [Ch55c] but was never published as far as I know.

(c) **Fixed part of an algebraic system.** Chow also developed a theory of algebraic systems of abelian varieties, defining the fixed part of such systems, i.e., that part which does not depend genuinely on the parameters [19, 20]. His notion of fixed part was used by others in an essential way, e.g., by Lang–Néron, who proved that for an abelian variety A defined over a function field K, the group of rational points of A in K modulo the group of points of the fixed part is finitely generated [41]. This is a relative version of the Mordell–Weil theorem.

(d) **Field of definition.** Chow gave conditions under which an abelian variety defined over an extension of a field k can actually be defined over k itself [19, 20]. Chow's idea was extended by Lang [40] to give such a criterion for all varieties, not just abelian varieties, and Weil reformulated the criterion in terms of cohomology (splitting a cocycle) [46].

Homogeneous Spaces

(a) **Projective embedding of homogeneous spaces.** Chow extended the Lefschetz–Weil proof of the projective embedding of abelian varieties to the case of homogeneous spaces over arbitrary group varieties, which may not be complete [25]. Chow's proof has been overlooked in recent years, even though interest in projective constructions has been reawakened, but I expect Chow's proof to make it back to the front burner soon just like his other contributions.

(b) **Algebraic properties.** Chow's paper [9] dealt with the geometry of homogeneous spaces. The main aim of this paper is to characterize the group by geometric properties. The latter could refer to the lines in a space, as in projective geometry, or to certain kinds of matrices, such as symmetric matrices. For instance, a typical theorem says: Any bijective adjacence-preserving transformation of the space of a polar system with itself is due to a transformation of the basic group, provided that the order of the space is greater than 1. Birational geometry is considered in this context.

The Chow Ring

In topology, intersection theory holds for the homology ring. In 1956 Chow defined rational equivalence between cycles on an algebraic variety and defined the intersection product for such classes, thus obtaining the Chow ring [21], which proved to be just as fundamental in algebraic geometry as its topological counterpart. The latest surfacing of the Chow groups is in [38].

Algebraic Geometry over Rings

In the late fifties began the extension of algebraic geometry over fields to algebraic geometry over rings of various type, partly to deal with algebraic or analytic families, but partly because of the motivation from number theory, where one deals with local Dedekind rings, p-adic rings, and more generally complete Noetherian local rings. Chow contributed to this extension in several ways. Of course, in the sixties Grothendieck vastly and systematically went much further in this direction, but it is often forgotten that the process had begun earlier. I shall mention here some of Chow's contributions in this direction.

(a) **Connectedness theorem.** In 1951 Zariski had proved a general connectedness theorem for specializations of connected algebraic sets. Zariski based his proof on an algebraic theory of holomorphic functions which he developed for this purpose. In [24] and [29] Chow gave a proof of a generalization over arbitrary complete Noetherian local domains, based on much simpler techniques of algebraic geometry, especially the Chow form.

(b) **Uniqueness of the integral model of a curve.** The paper [25] proved the uniqueness of the model of a curve of genus ≥ 1 and an abelian variety over a discrete valuation ring in the case of nondegenerate reduction.

(c) **Cohomology.** Invoking the theory of deformations of complex analytic structures by Kodaira–Spencer, the connectedness theorem, and Igusa's work on moduli spaces of elliptic curves, Chow and Igusa proved the upper semicontinuity of the cohomology over a broad class of Noetherian local domains [28]. Semicontinuity was proved subsequently in the complex analytic case by Grauert and by Grothendieck in more general algebraic settings. However, Chow and Igusa's contribution did not get the credit they deserved (cf. [39], Chapter 3, §12, and the bibliographical references given there, referring to work in the sixties but not to Chow–Igusa).

(d) **Bertini's theorem.** During that same period in the late fifties Chow extended Bertini's theorem to local domains [27].

(e) **Unmixedness theorem.** A homogeneous ideal defining a projective variety is said to be unmixed if it has no embedded prime divisors. Chow proved that the Segre product of two unmixed ideals is also unmixed, under fairly general conditions, in a ring setting [31].

Algebraicity of Analytic Objects

Chow was concerned over many years with the algebraicity of certain complex analytic objects. We mention two important instances.

(a) **Meromorphic mappings and formal functions.** In 1949 Chow proved the fundamental fact, very frequently used from then on, that a complex analytic subvariety of projective space is actually algebraic [8]. Twenty years later he came back to similar questions and proved in the context of homogeneous varieties that a meromorphic map is algebraic [32]. Remarkably and wonderfully, almost twenty years after that he came back once more to the subject and completed it in an important point [34]. I quote from the introduction to this paper, which shows how Chow was still lively mathematically: "Let X be a homogeneous algebraic variety on which a group G acts, and let Z be a subvariety of positive dimension. Assume that Z generates X [*in a sense which Chow makes precise*].... One asks whether a formal rational function on X along Z is the restriction along Z of an algebraic function (or even a rational function) on X. In a paper [32] some years ago, the author gave an affirmative answer to this question, under the assumption that the subvariety Z is complete, but only for the complex-analytic case with the formal function replaced by the usual analytic function defined in a neighborhood of Z . The question remains whether the result holds also for the formal functions in the abstract case over any ground field. We had then some thoughts on this question, but we did not pursue them any further as we did not see a way to reach the desired conclusion at the time. In a recent paper [37], Faltings raised this same question and gave a partial answer to it in a slightly different formulation. This result of Faltings led us to reconsider this question again, and this time we are more fortunate. In fact, we have been able not only to solve the problem, but also to do it by using essentially the same method we used in our original paper."

(b) **Analytic surfaces.** In a paper with Kodaira it was proved that a Khler surface with two algebraically independent meromorphic functions is a nonsingular algebraic surface [16].

Other Works in Algebraic Geometry

Chow's papers in algebraic geometry include a number of others, which, as I already asserted, I am less well acquainted with and won't comment upon, such as his paper on the braid group [7], on thc fundamental group of a variety [17], on rational dissections [22], and on real traces of varieties [30].

PDE

Chow's very early paper on systems of linear partial differential equations of first order [5] gives a generalization of a theorem of Caratheodory on the foundations of thermodynamics. This paper had effects not well known to the present generation of mathematicians, including me. It was only just now brought to my attention. An anonymous colleague wrote to the editor of the present collection of articles on Chow's work: "This paper essentially asserts the identity of the integral submanifold of a set of vector fields and the integral submanifold of the Lie algebra generated by the set of vector fields. This is widely known as 'Chow's theorem' in nonlinear control theory and is the basis for the study of the controllability problem in nonlinear systems. Controllability refers to the existence of an input signal that drives the state of a system from a given initial state to a desired terminal state. A more detailed exposition of the role of Chow's theorem, with several references, is provided in the survey paper [35]."

Works by Wei-Liang Chow are referenced [1] through [34].

[1] (with van der Waerden) *Zur algebraische Geome-try IX*, Math. Ann. **113** (1937), 692–704.

[2] *Die geometrische Theorie der algebraischen Funktionen für beliebige vollkommene Körper, Math. Ann.* **114** (1937), 655–682.

[3] *Einfacher topologischer Beweis des Fundamental-satzes der Algebra*, Math. Ann. **116** (1939), 463.

[4] *Über die Multiplizität der Schnittpunkte von Hyperflächen*, Math. Ann. **116** (1939), 598–601.

[5] *Über systemen von linearen partiellen Differentialgleichungen erster Ordnung*, Math. Ann. **117** (1939), 98–108.

[6] *On electrical networks*, J. Chinese Math. Soc. **2** (1940), 3–160.

[7] *On the algebraical braid group*, Ann. Math. **49**, no. 3 (1948), 654–658.

[8] *On compact complex analytic varieties*, Amer. J. Math. **71**, no. 4 (1949), 893–914.

[9] *On the geometry of algebraic homogeneous spaces*, Ann. Math. **50**, no. 1 (1949), 32–67.

[10] *Über die Lösbarkeit gewisser algebraischer Gleichungssysteme*, Comment. Math. Helv. **23**, no. 1 (1949), 76–79.

[11] *On the genus of curves of an algebraic system*, Trans. Am. Math. Soc. **65** (1949), 137–140.

[12] *On the defining field of a divisor in an algebraic variety*, Proc. Amer. Math. Soc. **1**, no. 6 (1950), 797–799.

[13] *Algebraic systems of positive cycles in an algebraic variety*, Amer. J. Math. **72**, no. 2 (1950), 247–283.

[14] *On the quotient variety of an abelian variety*, Proc. Nat. Acad. Sci. **38** (1952), 1039–1044.

[15] *On Picard varieties*, Amer. J. Math. **74**, no. 4 (1952), 895–909.

[16] (with Kodaira) *On analytic surfaces with two independent meromorphic functions*, Proc. Nat. Acad. Sci. **38**, no. 4 (1952), 319–325.

[17] *On the fundamental group of an algebraic variety*, Amer. J. Math. **74** (1952), 726–736.

[18] *The Jacobian variety of an algebraic curve*, Amer. J. Math. **76**, no. 2 (1954), 453–476.

[18A] *Remarks on my paper "The Jacobian variety of an algebraic curve"*, Amer. J. Math. **80** (1958), 238–240.

[19] *On abelian varieties over function fields*, Proc. Nat. Acad. Sci. **41** (1955), 582–586.

[20] *Abelian varieties over function fields*, Trans. Amer. Math. Soc. **78** (1955), 253–275.

[Ch55c] *Abstract theory of the Picard and Albanese varieties*, unpublished manuscript.

[21] *On equivalence classes of cycles in an algebraic variety*, Ann. Math. **64**, no. 3 (1956), 450–479.

[22] *Algebraic varieties with rational dissections*, Proc. Nat. Acad. Sci. **42** (1956), 116–119.

[23] *On the projective embedding of homogeneous spaces*, Lefschetz conference volume, Algebraic Geometry and Topology, Princeton University Press, 1957.

[24] *On the principle of degeneration in algebraic geometry*, Ann. Math. **66** (1957), 70–79.

[25] (with S. Lang) *On the birational equivalence of curves under specialization*, Amer. J. Math. **79** (1952), 649–652.

[26] *The criterion for unit multiplicity and a generalization of Hensel's lemma*, Amer. J. Math. **80**, no. 2 (1958), 539–552.

[27] *On the theorem of Bertini for local domains*, Proc. Nat. Acad. Sci. **44**, no. 6 (1958), 580–584.

[28] (with Igusa) *Cohomology theory of varieties over rings*, Proc. Nat. Acad. Sci. **44**, no. 12 (1958), 1244–1248.

[29] *On the connectedness theorem in algebraic geometry*, Amer. J. Math. **81**, no. 4 (1959), 1033–1074.

[30] *On the real traces of analytic varieties*, Amer. J. Math. **85**, no. 4 (1963), 723–733.

[31] *On the unmixedness theorem*, Amer. J. Math. **86** (1964), 799–822.

[32] *On meromorphic maps of algebraic varieties*, Ann. Math. **89**, no. 2 (1969), 391–403.

[33] *On the algebraicity of certain ringed spaces*, Amer. J. Math. **101** (2) (1979) 364–379.

[33A] Correction to *"On the algebraicity of certain ringed spaces"*, Amer. J. Math. **107** (3) (1985) 759–760.

[34] *Formal functions on homogeneous spaces*, Invent. Math. **86** (1986), 115–130.

[35] R. W. Brockett, *Nonlinear systems and differential geometry*, Proc. IEEE **64** (1976), 61–71.

[36] S. S. Chern, *Web geometry*, AMS Proc. Sympos. Pure Math., vol. **39** (1983), 3–10.

[37] G. Faltings, *Formale Geometrie und homogene Raüme*, Invent. Math. **64** (1981), 123–165.

[38] B. B. Gordon and J. P. Murre, *Chow motives of elliptic modular surfaces and threefolds*, preprint, Mathematics Institute, University of Leiden, 1996.

[39] R. Hartshorne, *Algebraic geometry*, Springer-Verlag, 1977.

[40] S. Lang, *Abelian varieties over finite fields*, Proc. Nat. Acad. Sci. USA **41**, no. 3 (1955), 174–176.

[41] S. Lang and A. Néron, *Rational points of abelian varieties over function fields*, Amer. J. Math. **81**, no. 1 (1959), 95–118.

[42] P. Philippon, *Sur des hauteurs alternatives I*, Math. Ann. **289** (1991), 255–283.

[43] ——, *Sur des hauteurs alternatives II*, Ann. Institut Fourier **44**, no. 4 (1994), 1043–1065.

[44] ——, *Sur des hauteurs alternatives III*, J. Math. Pures Appl. **74** (1995), 345–365.

[45] B. Want, *A note on Archimedean height pairing and Chow forms*, preprint, Brown University, 1996.

[46] A. Weil, *The field of definition of a variety*, Amer. J. Math. **78**, no. 3 (1956), 509–524.

[47] W. Wirtinger, *Lies Translationsmannigfaltigkeiten une Abelsche Integrale*, Monat. Math. u. Physik **46** (1938), 384–443.

Wei-Liang Chow's Bibliography

1937

[1] (with van der Waerden) Zur algebraische Geometrie IX, *Math. Ann.* **113** (1937) 692–704.

[2] Die geometrische Theorie der algebraischen Funktionen für beliebige vollkommene Körper, *Math. Ann.* **114** (1937) 655–682.

1939

[3] Einfacher topologischer Beweis des Fundamentalsatzes der Algebra, *Math. Ann.* **116** (1939) 463.

[4] Über die Multiplizität der Schnittpunkte von Hyperflächen, *Math. Ann.* **116** (1939) 598–601.

[5] Über systeme von linearen partiellen Differentialgleichungen erster Ordnung, *Math. Ann.* **117** (1939) 98–105.

1940

[6] On Electric Networks, *J. Chinese Math. Soc.* **2** (2) (1940) 321–339.

1948

[7] On the Algebraical Braid Group, *Ann. Math.* **49** (3) (1948) 654–658.

1949

[8] On Compact Complex Analytic Varieties, *Amer. J. Math.* **71** (4) (1949) 893–914.

[9] On the Geometry of Algebraic Homogeneous Spaces, *Ann. Math.* **50** (1) (1949) 32–67.

[10] Über die Lösbarkeit gewisser algebraischer Gleichungssysteme, *Comm. Math. Helv.* **23** (1) (1949) 76–79.

[11] On the Genus of Curves of an Algebraic System, *Trans. Amer. Math. Soc.* **65** (1949) 137–140.

1950

[12] On the Defining Field of a Divisor in an Algebraic Variety, *Proc. Amer. Math. Soc.* **1** (6) (1950) 797–799.

[13] Algebraic Systems of Positive Cycles in an Algebraic Variety, *Amer. J. Math.* **72** (2) (1950) 247–283.

1952

[14] On the Quotient Variety of an Abelian Variety, *Proc. NAS* (USA) **38** (1952) 1039–1044.
[15] On Picard Varieties, *Amer. J. Math.* **74** (4) (1952) 895–909.
[16] (with Kodaira) On Analytic Surfaces with Two Independent Meromorphic Functions, *Proc. NAS* (USA) **38** (4) (1952) 319–325.
[17] On the Fundamental Group of an Algebraic Variety, *Amer. J. Math.* **74** (1952) 726–736.

1954

[18] The Jacobian Variety of an Algebraic Curve, *Amer. J. Math.* **76** (2) (1954) 453–476.

1955

[19] On Abelian Varieties over Function Fields, *Proc. NAS* (USA) **41** (1955) 582–586.
[20] Abelian Varieties over Function Fields, *Trans. Amer. Math. Soc.* **78** (1955) 253–275.

[Ch55c] Abstract Theory of the Picard and Albanese Varieties, unpublished manuscript [Ch 55c] that is posted at http://mathnt.mat.jhu.edu/mathnew/Chow/bibliography.htm

1956

[21] On Equivalence Classes of Cycles in an Algebraic Variety, *Ann. Math.* **64** (3) (1956) 450–479.
[22] Algebraic Varieties with Rational Dissections, *Proc. NAS* (USA) **42** (1956) 116–119.

1957

[23] On the Projective Embedding of Homogeneous Varieties, Lefschetz conference volume, *Algebraic Geometry and Topology* (Princeton University Press, Princeton, 1957) 122–128.

[24] On the Principle of Degeneration in Algebraic Geometry, *Ann. Math.* **66** (1) (1957) 70–79.

[25] (With S. Lang) On the Birational Equivalence of Curves under Specialization, *Amer. J. Math.* **79** (1957) 649–652.

1958

[18A] Remarks on my paper "The Jacobian Variety of an Algebraic Curve," *Amer. J. Math.* **80** (1958) 238–240.

[26] The Criterion for Unit Multiplicity and a Generalization of Hensel's Lemma, *Amer. J. Math.* **80** (1958) 539–552.

[27] On the Theorem of Bertini for Local Domains, *Proc. NAS* (USA) **44** (6) (1958) 580–584.

[28] (with Igusa) Cohomology Theory of Varieties over Rings, *Proc. NAS* (USA) **44** (12) (1958) 1244–1248.

1959

[29] On the Connectedness Theorem in Algebraic Geometry, *Amer. J. Math.* **81** (4) (1959) 1033–1074.

1963

[30] On the Real Traces of Analytic Varieties, *Amer. J. Math.* **85** (4) (1963) 723–733.

1964

[31] On Unmixedness Theorem, *Amer. J. Math.* **86** (1964) 799–822.

1969

[32] On Meromorphic Maps of Algebraic Varieties, *Ann. Math.* **89** (2) (1969) 391–403.

1979

[33] On the Algebraicity of Certain Ringed Spaces, *Amer. J. Math.* **101** (2) (1979) 364–379.

1985

[33A] Correction to "On the Algebraicity of Certain Ringed Spaces", *Amer. J. Math.* **107** (3) (1985) 759–760.

1986

[34] Formal Functions on Homogeneous Spaces, *Invent. Math.* **86** (1986) 115–130.

Permission

[1] (with van der Waerden) Zur algebraische Geometrie IX, reprinted from *Math. Ann.* **113** (1937) 692–704. © 1937 Springer-Verlag.

[2] Die geometrische Theorie der algebraischen Funktionen für beliebige vollkommene Körper, reprinted from *Math. Ann.* **114** (1937) 655–682. © 1937 Springer-Verlag.

[3] Einfacher topologischer Beweis des Fundamentalsatzers der Algebra, reprinted from *Math. Ann.* **116** (1939) 463. © 1939 Springer-Verlag.

[4] Über die Multiplizität der Schnittpunkte von Hyperflächen, reprinted from *Math. Ann.* **116** (1939) 598–601. © 1939 Springer-Verlag.

[5] Über systeme von linearen partiellen Differentialgleichungen erster Ordnung, reprinted from *Math. Ann.* **117** (1939) 98–105. © 1939 Springer-Verlag.

[6] On Electric Networks, reprinted from *J. Chinese Math. Soc.* **2** (2) (1940) 321–339.

[7] On the Algebraical Braid Group, reprinted from *Ann. Math.* **49** (3) (1948) 654–658. © 1948 Princeton University Press.

[8] On Compact Complex Analytic Varieties, reprinted from *Amer. J. Math.* **71** (4) (1949) 893–914. © 1949 Johns Hopkins University Press.

[9] On the Geometry of Algebraic Homogeneous Spaces, reprinted from *Ann. Math.* **50** (1) (1949) 32–67. © 1949 Princeton University Press.

[10] Über die Lösbarkeit gewisser algebraischer Gleichungssysteme, reprinted from *Comm. Math. Helv.* **23** (1949) 76–79. © 1949 Birkhaeuser Publishers Ltd.

[11] On the Genus of Curves of an Algebraic System, reprinted from *Trans. Amer. Math. Soc.* **65** (1949) 137–140. © 1949 American Mathematical Society.

[12] On the Defining Field of a Divisor in an Algebraic Variety, reprinted from *Proc. Amer. Math. Soc.* **1** (6) (1950) 797–799. © 1950 American Mathematical Society.

[13] Algebraic Systems of Positive Cycles in an Algebraic Variety, reprinted from *Amer. J. Math.* **72** (2) (1950) 247–283. © 1950 Johns Hopkins University Press.

[14] On the Quotient Variety of an Abelian Variety, reprinted from *Proc. NAS* (USA) **38** (1952) 1039–1044.

[15] On Picard Varieties, reprinted from *Amer. J. Math.* **74** (4) (1952) 895–909. © 1952 Johns Hopkins University Press.

[16] (with Kodaira) On Analytic Surfaces with Two Independent Meromorphic Functions, reprinted from *Proc. NAS* (USA) **38** (4) (1952) 319–325.

[17] On the Fundamental Group of an Algebraic Variety, reprinted from *Amer. J. Math.* **74** (1952) 726–736. © 1952 Johns Hopkins University Press.

[18] The Jacobian Variety of an Algebraic Curve, reprinted from *Amer. J. Math.* **76** (2) (1954) 453–476. © 1954 Johns Hopkins University Press.

[18A] Remarks on my paper "The Jacobian Variety of an Algebraic Curve", reprinted from *Amer. J. Math.* **80** (1958) 238–240. © 1958 Johns Hopkins University Press.

[19] On Abelian Varieties over Function Fields, reprinted from *Proc. NAS* (USA) **41** (1955) 582–586.

[20] Abelian Varieties over Function Fields, reprinted from *Trans. Amer. Math. Soc.* **78** (1955) 253–275. © 1955 American Mathematical Society.

[21] On Equivalence Classes of Cycles in an Algebraic Variety, reprinted from *Ann. Math.* **64** (3) (1956) 450–479. © 1956 Princeton University Press.

[22] Algebraic Varieties with Rational Dissections, reprinted from *Proc. NAS* (USA) **42** (1956) 116–119.

[23] On the Projective Embedding of Homogeneous Varieties, reprinted from Lefschetz Conference Volume, *Algebraic Geometry and Topology* (Princeton University Press, Princeton, 1957) 122–128.

[24] On the Principle of Degeneration in Algebraic Geometry, reprinted from *Ann. Math.* **66** (1), (1957) 70–79. © 1957 Princeton University Press.

[25] (with S. Lang) On the Birational Equivalence of Curves under Specialization, reprinted from *Amer. J. Math.* **79** (1957) 649–652. © 1957 Johns Hopkins University Press.

[26] The Criterion for Unit Multiplicity and a Generalization of Hensel's Lemma, reprinted from *Amer. J. Math.* **80** (1958) 539–552. © 1958 Johns Hopkins University Press.

[27] On the Theorem of Bertini for Local Domains, reprinted from *Proc. NAS* (USA) **44** (6) (1958) 580–584.

[28] (with Igusa) Cohomology Theory of Varieties over Rings, reprinted from *Proc. NAS* (USA) **44** (12) (1958) 1244–1248.

[29] On the Connectedness Theorem in Algebraic Geometry, reprinted from *Amer. J. Math.* **81** (4) (1959) 1033–1074. © 1959 Johns Hopkins University Press.

[30] On the Real Traces of Analytic Varieties, reprinted from *Amer. J. Math.* **85** (4) (1963) 723–733. © 1963 Johns Hopkins University Press.

[31] On Unmixedness Theorem, reprinted from *Amer. J. Math.* **86** (1964) 799–822. © 1964 Johns Hopkins University Press.

[32] On Meromorphic Maps of Algebraic Varieties, reprinted from *Ann. Math.* **89** (2) (1969) 391–403. © 1969 Princeton University Press.

[33] On the Algebraicity of Certain Ringed Spaces, reprinted from *Amer. J. Math.* **101** (2) (1979) 364–379. © 1979 Johns Hopkins University Press.

[33A] Correction to "On the Algebraicity of Certain Ringed Spaces", reprinted from *Amer. J. Math.* **107** (3) (1985) 759–760. © 1985 Johns Hopkins University Press.

[34] Formal Functions on Homogeneous Spaces, reprinted from *Invent. Math.* **86** (1986) 115–130. © 1986 Springer-Verlag.

[35] Shiing Shen Chern as Friend and Mathematician, A Reminiscence on the Occasion of His 80th Birthday, reprinted from *Chern — A Great Geometer of the Twentieth Century* (Internat. Press, Hong Kong, 1992) 79–87. © 1992 International Press.

